CARBON DIOXIDE REMOVAL

PROCEEDINGS OF THE THIRD INTERNATIONAL CONFERENCE ON CARBON DIOXIDE REMOVAL

U.K.	Elsevier Science Ltd, The Boulevard, Langford Lane, Kidlington, Oxford OX5 1GB, U.K.
U.S.A.	Elsevier Science Inc., 655 Avenue of the Americas, New York, NY 10010, U.S.A.
JAPAN	Elsevier Science Japan, 9-15 Higashi-Azabu 1-chome, Minato-ku, Tokyo, 106 Japan

© 1997 Elsevier Science Ltd

First edition 1997

Library of Congress Cataloguing in Publication Data

A catalogue record for this book is available from the Library of Congress

British Library Cataloguing in Publication Data

A catalogue record for this book is available from the British Library

ISBN 0 08 0428 401

Printed and bound in Great Britain by BPC Wheatons Ltd, Exeter

PROCEEDINGS OF THE THIRD INTERNATIONAL CONFERENCE ON CARBON DIOXIDE REMOVAL

Cambridge, MA, U.S.A., 9–11 September 1996

Guest Editor:

Howard J. Herzog
MIT, Cambridge, MA, U.S.A.

Pergamon

AIMS AND SCOPE

The Journal will provide an interdisciplinary forum through which energy conversion, fuel conservation, and energy management can be treated as a coherent, integrated subject. Papers of high technical merit addressing both significant advances in the field and state-of-the-art developments are sought. All sources, forms, and uses of energy are appropriate, as are all conversion processes. This includes, but is not limited to, solar, nuclear, fossil, geothermal, wind, hydro, and biomass sources, process heat, electrolysis, heating and cooling, and electric drive applications, as well as commercial, industrial, transportation and residential uses. The conversion processes include, but are not limited to, thermoelectricity, thermionic processes, photoelectric processes, electrochemistry associated with fuel cells and various forms of batteries, magnetohydrodynamic conversion processes, advances related to dynamic processes, and other new and unconventional conversion processes. As the Journal is international in scope, papers dealing with state-of-the-art developments in various nations, or comparative studies of energy conversion and management in several countries will be accepted.

Annual Institutional Subscription Rates 1997: Europe, The CIS and Japan 2977 Dutch Guilders. All other countries US$1838. Associated personal subscription rates are available on request for those whose institutions are library subscribers. Dutch Guilder prices exclude VAT. Non-VAT registered customers in the European Community will be charged the appropriate VAT in addition to the price listed. Prices listed include postage and insurance and are subject to change without notice.

For orders, claims, product enquiries (no manuscript enquiries) please contact the Customer Support Department at the Regional Sales Office nearest to you:

The Americas: Elsevier Science Customer Support Department, P.O. Box 945, New York, NY 10010, U.S.A. [Tel: (+1) 212-633-3730/1-888 4ES-INFO. Fax: (+1) 212-633-3680. E-mail: usinfo-f@elsevier.com].

Japan: Elsevier Science Customer Support Department, 9-15 Higashi-Azabu 1-chome, Minato-ku, Tokyo 106, Japan [Tel: (+3) 5561-5033. Fax: (+3) 5561-5047. E-mail: kyf04035@niftyserve.or.jp].

Asia Pacific (excluding Japan): Elsevier Science (Singapore) Pte Ltd, No. 1 Temasek Avenue, 17-01 Millenia Tower, Singapore 039192 [Tel: (+65) 434-3727. Fax: (+65) 337-2230. E-mail: asiainfo@elsevier.com.sg].

Rest of the World: Elsevier Science Customer Service Department, P.O. Box 211, 1001 AE Amsterdam, The Netherlands [Tel: (+31) 20-485-3757. Fax: (+31) 20-485-3432. E-mail: nlinfo-f@elsevier.nl].

ENERGY CONVERSION AND MANAGEMENT

VOLUME 38 Suppl. 1997

CONTENTS

PROCEEDINGS OF THE THIRD INTERNATIONAL CONFERENCE ON CARBON DIOXIDE REMOVAL

SECTION 5. CHEMICAL UTILIZATION

Energy Convers. Mgmt is indexed/abstracted in Res. Alert, Biosis Data., CAB Inter., Cam. Sci. Abstr., Chem. Abstr. Serv., Curr. Cont./Eng. Tech. & Applied Sci., Eng. Ind., Environ. Per. Bibl., INSPEC Data., Curr. Cont. Sci. Cit. Ind., Curr. Cont. SCISEARCH Data., SSSA/CISA/ECA/ISMEC, Applied Sci. & Tech. Ind., Applied Sci. & Tech. Abstr.

0196-8904(1997)38:S;1-S

Pergamon

ISSN 0196-8904

ECMADL 38(S) S1–S690 (1997)

Pergamon

Energy Convers. Mgmt Vol. 38, Suppl., pp. S1–S2, 1997
© 1997 Elsevier Science Ltd. All rights reserved
Printed in Great Britain
0196-8904/97 $17.00 + 0.00

PII: S0196-8904(96)00236-1

INTRODUCTION

In June of 1992, the UN Framework Convention on Climate Change was drafted. It entered into force on March 21, 1994, and today the treaty has been signed and ratified by 159 states. According to Article 2 of the convention, *the ultimate objective of this convention ... is to achieve ... stabilization of greenhouse gas concentrations in the atmosphere*.

The major greenhouse gas is carbon dioxide (CO_2) and the major source of anthropogenic CO_2 is fossil fuel combustion. While mitigation measures such as improved energy efficiency and fuel switching to less carbon intensive fuels may suffice in the short-term, there is a high probability that additional mitigation strategies will be needed for stabilization in the longer-term. These longer-term strategies may include the large-scale use of renewable energy, increasing the application of nuclear energy, or recovering CO_2 from large stationary sources with the subsequent use or storage of the CO_2. This last option is the focus of a series of conferences, the International Conferences on Carbon Dioxide Removal (ICCDR), which are a forum to exchange technical information on CO_2 removal, storage, and utilization technologies and to promote research and development in this field.

ICCDR-1 was held in March 1992 in Amsterdam, and ICCDR-2 was held in Kyoto in October 1994. The Third International Conference on Carbon Dioxide Removal (ICCDR-3) met September 9-11 on the campus of the Massachusetts Institute of Technology (MIT) in Cambridge, MA. There were 250 delegates from 26 countries. The conference was hosted by the MIT Energy Laboratory and sponsored by the U.S. Department of Energy (DOE) and the Electric Power Research Institute (EPRI). Co-sponsors were the Research Institute of Innovative Technology (RITE) and the New Energy and Industrial Technology Organization (NEDO) in Japan, ABB Asea Brown Boveri in Switzerland, and the IEA Greenhouse Gas R&D Programme in the UK.

At ICCDR-3, two important changes were made for future conferences. First, the IEA Greenhouse Gas R&D Programme will become the sanctioning organization. Secondly, the name will change to the International Conference on Greenhouse Gas Control Technologies (GHGT). The next conference, GHGT-4, will be organized by ABB Asea Brown Boveri and will be held August 31-September 2, 1998, in Interlaken, Switzerland.

These proceedings contain 111 papers divided into seven sections. Section 1 contains papers from the invited speakers, who were asked to provide a context for our proceedings. The next five sections address the core technical topics of the conference, namely CO_2 removal (Section 2), storage (geological storage in Section 3, ocean storage in Section 4), and utilization (chemical in Section 5 and biological in Section 6). Finally, Section 7 addresses additional topics such as economics, full fuel cycle analysis, policy and implementation issues, and comparisons to other mitigation options.

Time did not allow us to conduct a formal peer review process, but we thought some peer review was required. Therefore, we asked each contributing author to review two papers submitted to this proceedings. In this way, we obtained two reviews for most papers and at least one review for every paper. This review process made significant contributions to the quality of the papers in these proceedings. I want to acknowledge our thanks to the contributing authors who participated in this review.

During the conference, the ICCDR-3 Organizing Committee presented three "Greenman Awards" to recognize individuals who made significant contributions to the field of CO_2 removal, storage, and utilization. The recipients received a reproduction of a Greenman and a certificate that read as follows:

The Greenman is an ancient celtic archetype of a human face peering though a growing foliage which is often depicted on buildings, churches, and cathedrals. It symbolizes the mysteries of creativity, compassion, healing, new beginnings, and especially our connection with nature and the power of humankind working together with nature, the cycles of nature, and "man and the forest." To honor contributions toward harnessing technology so that the human race can better live in harmony with the environment, the organizers of the Third International Conference on Carbon Dioxide Removal proudly present _____ with a Greenman award. The Greenman presented with this certificate is "Jeep's Leafface," from St. John the Divine Cathedral in New York City.

While there were many deserving candidates, the organizing committee recognized **Prof. Yoichi Kaya**, Keio University, Japan, for his leadership in promoting CO_2 removal, storage, and utilization research; **Dr. Meyer Steinberg**, Brookhaven National Laboratory, USA for his pioneering work in CO_2 removal, storage, and utilization research; and **Prof. Wim Turkenburg**, Utrecht University, The Netherlands for founding the ICCDR series and organizing ICCDR-1. Congratulations to all three for their well deserved award.

Finally, I want to acknowledge the many people who helped make ICCDR-3 a success. In addition to the sponsors and the contributing authors mentioned above, I want to thank the participants; members of the organizing and program committees listed below; my colleagues at the MIT Energy Laboratory including Anne Carbone, Patricia Connell, Stephen Connors, and Elisabeth Drake; and Cathi DiIulio and her co-workers at MIT Conference Services.

On behalf of the organizers of the next conference, we look forward to seeing you in Switzerland in 1998.

Howard J. Herzog
Editor

ICCDR-3 Organizing Committee
> Howard Herzog, Chair, Massachusetts Institute of Technology, USA
> Perry Bergman, U.S. Department of Energy, USA
> Richard Rhudy, Electric Power Research Institute, USA

ICCDR-3 Program Committee
> Jefferson Tester, Chair, Massachusetts Institute of Technology, USA
> Eric Adams, Massachusetts Institute of Technology, USA
> Michele Aresta, Bari University, Italy
> John Benemann, University of .California, Berkeley, USA
> Amit Chakma, University of Regina, Canada
> Baldur Eliasson, ABB Corporate Research, Switzerland
> Paul Freund, IEA Greenhouse Gas R&D Programme, UK
> Tomoyuki Inui, Kyoto University, Japan
> Olav Kaarstad, Statoil R&D Centre, Norway
> Hiroshi Komiyama, University of Tokyo, Japan
> Gilbert Stegen, Science Applications International Corp., USA
> Meyer Steinberg, Brookhaven National Laboratory, USA
> Kelly Thambimuthu, Natural Resources Canada
> Wim Turkenburg, Utrecht University, The Netherlands
> Yukio Yanagisawa, RITE, Japan

SECTION 1

INVITED PAPERS

Pergamon

PII: S0196-8904(96)00237-3

Energy Convers. Mgmt Vol. 38, Suppl., pp. S3–S12, 1997
© 1997 Elsevier Science Ltd. All rights reserved
Printed in Great Britain
0196-8904/97 $17.00 + 0.00

SUSTAINABLE DEVELOPMENT, CLIMATE CHANGE, AND CARBON DIOXIDE REMOVAL (CDR)

WIM C. TURKENBURG

Department of Science Technology and Society, Utrecht University
Padualaan 14, 3584 CH Utrecht, The Netherlands

ABSTRACT

In this article the characteristics and requirements of a sustainable energy system are described. Special attention is given to the need to reduce greenhouse gas emissions, especially CO_2. It is indicated that we may have to reduce the annual CO_2 emissions due to our energy consumption from 6 GtC at present to less than 3 GtC in the year 2100, and the cumulative CO_2 emissions between the years 1990 and 2100 to 550-750 GtC.

To reduce the CO_2 emissions, one option we can develop and apply is improvement of the energy efficiency with 50-90%. In addition strong efforts are needed to develop a sustainable energy supply system with low or no CO_2 emissions. Option number one, the use of renewable energy sources, offers a huge potential but major technological breakthroughs are required to allow a massive application of these sources at affordable costs, in a reliable way and in an acceptable manner. This probably also holds for another option, nuclear energy. The impact of a fuel switch from coal to oil and natural gas will depend on the recoverable amount of fossil fuels. Therefore, due attention should be given to Carbon Dioxide Removal (CDR), especially as it is the only option that may allow continuing large scale use of fossil fuels. Application of this option could prevent the emission of 300 GtC or more between the years 1990 and 2100. Studies in the Netherlands suggest its application would be accepted socially. One of the main implementation barriers is its high cost. In the power sector CDR may increase the electricity production costs with 30-100%, although there is room to reduce these costs. Lower cost CDR options are available outside the power sector, like CDR from natural gas recovery processes, from specific industrial processes and from the production of hydrogen from natural gas, coal or biomass. Consequently, CDR may be applied at a cost ranging from US$ 5-10 to 250 per ton Carbon avoided, depending on the option and circumstances involved. Given the pursuit of sustainability, further attention is needed not only for costs reductions but also for the environmental and safety aspects of CO_2 disposal, especially in aquifers and in the deep ocean. © 1997 Elsevier Science Ltd

KEYWORDS

Energy and sustainable development, climate change, CO_2 mitigation, carbon dioxide removal.

INTRODUCTION

One of the results of the United Nations Conference on Environment and Development, held in 1992 in Rio de Janeiro, was the acceptance of a global action plan ('Agenda 21') to achieve a sustainable development, socially, economically as well as ecologically. In this article the consequences of the pursuit of sustainability for the energy sector are investigated. Special attention is given to the issue of climate change, based on studies of the Intergovernmental Panel on Climate Change (IPCC) and the call of 1996 Conference of Parties of the UN Framework Convention on Climate Change to reduce the emission of greenhouse gases significantly. The potential consequences are investigated, focused on potentially required CO_2 emission reduction levels. Options to reduce the CO_2 emission are analyzed, especially the possibility to improve the efficiency of our energy use substantially and the possibilities of producing fuels, heat and electricity with low or no CO_2 emissions. It will be argued that, as part of a CO_2 mitigation strategy, due attention should be given to the option of Carbon Dioxide Removal (CDR). The mitigation potential of CDR is reviewed as well as its costs. Options to reduce the energy and cost penalties of CDR, in- and outside the power sector, are discussed. Subsequently, the social acceptance of CDR as a strategy to mitigate CO_2 emissions is assessed. Also remarks are made about the environmental impact of CDR. It is concluded that CDR is a viable option to reduce CO_2 emissions although further research, development and demonstration is required to improve its performance and to allow its further integration in a sustainable development of the energy system.

It should be noted that some parts of the article, written by the author, have recently been published in reports of the IPCC (Ishitani *et al.*, 1996) and the UN energy committee (UNCNRSEED, 1996).

ENERGY AND SUSTAINABLE DEVELOPMENT

Access to - and therefore adequate availability of - energy services is a prerequisite to achieving the socio-economic development that is required to improve the quality of life and satisfy basic human needs, including access to jobs, food, running water, housing, health services, education and communication: energy as a source of prosperity. An adequate supply of energy is urgently needed in developing countries. It is also a prerequisite for a sustainable socio-economic development in the industrialized world and in the countries with economies in transition. In addition the supply of energy should be secure and reliable. Therefore, attention should be given to: (a) the dependence on imports of energy carriers from unevenly distributed energy resources; (b) the vulnerability of energy supplies to severe accidents or disruptions in the energy system or a major change in the socio-cultural environment in which such systems must operate; © the exhaustion of scarce energy resources for which alternative options must to be developed in time. To achieve the desired economic and social development, it is also essential that access to energy services and the supply of energy be realized at affordable costs; that the development and application of energy sources and technologies be realized in a socially acceptable manner; and that the development of energy sources creates opportunities for local employment and industrial activities.

Meanwhile, the further development of energy systems should in itself be coherent with the pursuit of sustainability: it should not endanger the quality of life of present and future generations and should not exceed the carrying capacity of supporting ecosystems. This means that the production and consumption of energy should be clean and safe. It also means that the use of scarce resources to fulfil present needs for energy services should not compromise the ability of future generations to meet their needs for the same services. This in turn requires an efficient use of resources and a timely development of alternative resources. An efficient use of resources is also required to reduce the production of wastes. Finally, it is important that the development of short term options to fulfil our energy needs in an environmentally sound way does not hinder the development options that in the longer term contribute better to sustainability.

As concluded in Agenda 21 "much of the world's energy is currently produced and consumed in ways that could not be sustained if technology were to remain constant and if overall quantities were to increase substantially". Therefore, new energy strategies towards sustainability must be developed and implemented. As indicated by the UN Energy Committee (UNCNRSEED, 1996) the main characteristics of a new energy path towards sustainability should be:
  a) A more efficient use of energy and energy-intensive materials;
  b) Increased use of renewable sources of energy;
  c) More efficient and clean production and use of fossil fuels;
  d) Fuel substitution, from high-carbon to low-carbon and non-carbon-based fuels.

CLIMATE CHANGE

One of the major problems that the development of a sustainable energy system faces is the emission of greenhouse gases, especially CO_2, and the associated risk of a severe climate change. Since the industrial revolution the atmospheric concentration of carbon dioxide (CO_2) and other greenhouse gases is increasing strongly. In the case of CO_2, the concentration was about 280 ppmv in the years 1000-1750. Nowadays it is about 360 ppmv with a rate of change of +1.5 ppmv per year (Houghton *et al.*, 1996). Almost certainly the increase is due to human activities, especially the deforestation of land and the consumption of fossil fuels. Due to our energy consumption, at present about 22 Gigatonne CO_2 (equivalent to 6 GtC) is emitted to the atmosphere annually. If we continue to fulfil our energy needs like we did in the past, according to the IPCC (Leggett *et al.*, 1992) the CO_2 emission might rise to about 15-20 GtC in the year 2050 and about 20-35 GtC in the year 2100, assuming "more realistic world population projections" (Jefferson, 1996). Consequently, the cumulative emission of CO_2 between 1990 and 2100 might range from about 1450 to 2200 GtC. According to the IPCC, this may lead to major changes in the Earth's climate which in turn could cause severe consequences for mankind and nature (Houghton *et al.*, 1996; Watson *et al.*, 1996).

Based on the assessments of the IPCC, and within the context of the United Nations, in 1992 the UN Framework Convention on Climate Change was set up which entered into force on March 21, 1994. In September 1996 the treaty had been signed and ratified by 159 states. According to article 2 of the convention, "the ultimate objective of this convention (..) is to achieve (..) stabilization of greenhouse gas concentrations in the atmosphere". On the level at which this stabilization should be achieved, the convention says that it should be a level "that would prevent dangerous anthropogenic interference with the climate system". On the time-frame within which this level should be achieved, the convention says that this time-frame should be "sufficient to allow ecosystems to adapt naturally to climate change, to ensure that food production is not threatened and to enable economic development to proceed in a sustainable manner". Due to our limited knowledge about the climate system and the intrinsic impossibility to forecast the climate behavior in relation to human activities, it will not be easy to achieve international agreement about a translation of these preconditions in quantitative goals. However, based on the assessments of the IPCC, the Government of the Netherlands has proposed the following operationalization (VROM, 1996):
1. The global mean temperature increase should remain below 2°C above the pre-industrial level.
2. The rate of increase of the global mean temperature should not exceed the value of 0.1 °C per decade.
3. When establishing a global rate of greenhouse gas emission reduction, account must be given to its technical and economic feasibility.
4. As a consequence the concentration of greenhouse gases in the atmosphere should be stabilized before the end of the next century at a level widely below twice the pre-industrial concentrations.

In line with this approach, it can be argued that before the year 2100 the concentration of CO_2 in the atmosphere should be stabilized at or below a level of about 450 ppmv. To achieve this, the anthropogenic global emission of CO_2 should probably be reduced to a level below 3 GtC per year before the end of the next century. Moreover, taking into account our uncertain knowledge about the carbon cycle, the cumulative emission of CO_2 between the years 1990 and 2100 should be limited to about 550 - 750 GtC (Houghton *et al.*, 1996). Earlier we have seen that business-as-usual might result in cumulative emissions ranging from 1450-2200 GtC. Therefore, options and policies have to be developed to prevent the atmospheric emission of 700-1600 GtC in the next century.

According to article 3.1 of the UN-FCCC "the Parties should protect the climate system (...) in accordance with their common but differentiated responsibilities and respective capabilities. Accordingly, the developed country parties should take the lead in combating climate change and the adverse effects thereof". Consequently, a global reduction of the annual CO_2 emissions with more than 50% probably implies a reduction in the industrialized countries with at least 80%, an average of approximately 1.5% per year between 1990 and 2100. Assuming that the GDP of these countries will increase annually with 2.0-3.0% per year, as a result the CO_2 emission per unit of GDP should be reduced with 3.5-4.5% per year on average.

OPTIONS TO REDUCE THE CO_2 EMISSIONS

Assuming a business-as-usual development of the demand and supply of energy similar to scenario SA90 of the IPCC (Houghton, 1990), the emission of CO_2 might increase 1% annually and the cumulative CQ emission between the years 1990 and 2100 could be about 1600 GtC. Based on this scenario, in Table 1 a number of options are presented to reduce the CO_2 emission to sustainable levels as well as their potential contribution.

Table 1. Potential contribution of options to reduce the cumulative CO_2 emissions in the period 1990-2100 compared to a business-as-usual scenario (cumulative emissions 1600 GtC), based on a variety of recently published assessment studies (Turkenburg, 1995). Note that the contributions are mutually dependent.

- energy and material efficiency improvement	200-600 GtC
- renewable energy sources	200-600 GtC
- nuclear fission	100-300 GtC
- nuclear fusion	0- 25 GtC
- switching from coal to oil or natural gas	0- 300 GtC
- CO_2 recovery and storage	100-300 GtC
- afforestation	50-100 GtC

An important option to prevent the emission of CO_2 is a strong reduction of the energy intensity of our economy, largely by improving the efficiency of energy (and material) consumption. The potential of efficiency improvement is large. Recently it has been assessed that in a country like the Netherlands, depending on the sector involved, the improvement could be 50-90% within a time-frame of 60-70 years (VCE, 1996). On average this improvement could be about 75%, equivalent to a factor 4. Similar results have been found in other studies. Following these figures, and taking into account the potential effects of structural changes in the economic system on energy demand, the energy intensity can in principle be decreased with 2.0-2.5% per year, at least till the middle of the next century. Under business-as-usual conditions it is expected that the energy intensity will decrease with about 1% per year. Therefore, major efforts and policy measures are needed to improve this figure to 2.0-2.5%. In practice, however, it will already be a major challenge to reduce the energy intensity with 1.5-2.0% per year. As a consequence, the indicated reduction of CO_2 emissions can be achieved only if energy sources and technologies are used with low or no CO_2 and which - given the pursuit of sustainable development - are clean, safe, reliable, efficient, affordable, and socially acceptable.

Option number one, of course, is the use of renewable energy sources. The potential of these sources is huge and their future looks promising. According to World Bank staff estimates new renewable energy sources can fulfil more than half our energy needs in the middle of the next century although it may take twenty years from now before massive application can be achieved in a competitive way (World Bank, 1992). A similar view has been presented recently by Shell (Kassler, 1994). These assessments are in line with a number of other analyses including a study of the World Energy Council (WEC, 1993) and of Johansson et al. (1993). In these studies, apart from hydro power, biomass and wind energy sources are expected to play an important role within a few decades later on followed by solar energy conversion, especially photovoltaics. However, major technological developments are required to allow wide application of these sources at affordable costs, in a reliable way, and in an acceptable manner.

Similar remarks can be made concerning the use of nuclear energy. The globally installed capacity of nuclear power plants is at present 340 GW_e. In 1995 the contribution of nuclear to the total energy supply was almost 6% and to the total electricity supply about 17%. To allow nuclear energy to play a substantial role in the reduction of CO_2 emissions, the generating capacity should increase by a factor 10. As a result at the end of the next century at least half the electricity demand could be supplied by nuclear (Ishitani *et al.*, 1996). It could reduce the cumulative emissions of CO_2 between the years 1990 and 2100 with about 300 GtC. The present development, however, indicates another trend. The nuclear expansion, which peaked in the eighties at a level of about 30 GW$_e$/yr, has decreased strongly and is at present about 3 GW$_e$/yr. It illustrates the problems nuclear energy is facing today. Some of these problems are: the lack of public support, the safety of nuclear power plants and other nuclear facilities, the lifetime and disposal of nuclear waste, the diversion of fissile materials, and the costs of nuclear power. It might well be that the solution of these problems requires the development of a new nuclear technology which can be applied in an inherently safe and clean way, reduces the lifetime of longlived nuclear waste, is far less proliferation risky, and allows an economically competitive utilization (Turkenburg, 1996).

Because of the difference in carbon intensity of the fuels, switching from coal to oil or natural gas could reduce the CO_2 emission by 40%. This figure could increase to more than 50% if the higher conversion efficiency that can be achieved with natural gas instead of coal power generation is taken into account. Whether such a shift in fuel consumption could substantially contribute to a decrease of the cumulative CO_2 emissions till the year 2100 depends on the development of the energy demand and especially on the ultimate availability of oil and natural gas. A cumulative emission reduction by fuel switching of about 300 GtC might be feasible if, instead of coal, an additional amount of e.g., 10.000 EJ of oil and 20.000 EJ of natural gas would be recovered and utilized. Whether this can be achieved is unknown yet (Nakicenovic, 1996).

From this overview we may conclude that in principle a number of options can be developed to reduce the emission of CO_2 in a competitive way. Also we must recognize that it is uncertain whether these options can be developed in time and whether they can be applied in a cost-effective way at the required level. Therefore, as part of a CO_2 mitigation strategy, due attention should also be given to Carbon Dioxide Removal (CDR).

CARBON DIOXIDE REMOVAL

There are several arguments why the option of CDR should be developed:
- Development of CDR fits well in a precautionary approach as agreed upon in the UN Framework Convention on Climate Change;
- CDR can already be applied on short and intermediate terms;
- CDR could be a cost-effective answer to a high carbon-tax;
- Application of CDR could be necessary on longer terms if other mitigation options fail;
- Combined with the production of hydrogen from fossil fuels, CDR could be a low cost mitigation option;
- CDR is the only greenhouse gas mitigation option that may allow continuing large scale use of fossil fuels.

The development and application of CDR should of course be in agreement with the pursuit of sustainability. Consequently CDR should fulfil the following criteria:
- enough potential to deliver a substantial contribution to the reduction of CO_2 emissions during longer periods of time;
- efficient use of energy and material efficient;
- economically affordable;
- socially acceptable;
- environmentally sound.

Following these criteria, we will take a closer look at the potential of CDR.

CO_2 emission reduction potential of CDR

In Table 2 an overview is presented of the potential of CO_2 utilization and storage options as discussed in the literature. Low and high estimates are given.

Table 2.   Low and high estimates of the potential of CO_2 utilization and storage options.

- Utilization (incl. EOR)	0.2	-	1	GtC/yr
- Exhausted gas wells	90	-	400	GtC
- Exhausted oil wells	40	-	100	GtC
- Saline aquifers	90	-	>1000	GtC
- Ocean disposal	400	-	>1200	GtC

The potential to utilize recovered CO_2 for e.g., enhanced oil recovery is interesting but small, annually maybe 0.2-1 GtC. The potential to store CO_2 in depleted oil and natural gas fields is much larger. Estimates range from 130 GtC to 500 GtC, depending on the recoverable amount of oil and gas. Another option is disposal in saline aquifers with an estimated potential ranging from about 90 GtC to more than 1000 GtC, due to different assumptions about the necessity of having a structural trap to assure safe and sustained disposal and about aspects like the volume of aquifers, the percentage of the aquifer to be filled and the density of CO_2 under reservoir conditions (Hendriks, 1994). The ocean is the largest potential repository for CO_2. It contains already nearly 40.000 GtC. Eventually it will absorb perhaps 85% of the CO_2 that is released to the atmosphere from human activities (Houghton *et al.*, 1990). However, not much is known about the environmental effects of storing CO_2 in the ocean, either directly (via CDR) or indirectly (via the atmosphere). In this situation, following a very prudent approach, the environmental space to dispose of CO_2 in the ocean might be restricted to 1% of the natural background. However, if we would accept a maximum deviation of the pH of ocean waters with 0.2 units, the buffer capacity can be estimated at 1200 GtC (Spencer, 1993). These figures suggest that the environmental space to dispose of CO_2 from CDR in the ocean might be limited to several hundred GtC or even zero, depending on the total amount of CO_2 that is emitted to the atmosphere due to human behavior and absorbed eventually by the ocean.

Based on these options, recent analyses suggest that the application of CDR could prevent the emission of 300 GtC or more between the years 1990 and 2100 (Ishitani *et al.*, 1996).

Energy and cost penalties of CDR

There are several technologies to capture CO_2 from flue gases. Relatively much attention is given to scrubbing of CO_2 from power plant exhaust gases with a regenerable chemical solvent. Starting with a conventional coal-fired power plant of 600 MW_e and a coal-to-busbar efficiency of 41% (LHV), this approach might decrease the conversion efficiency to 30% in the modified plant if the CO_2 emission is reduced from 840 g/kWh tot 120 g/kWh. The cost of electricity production would then increase by about 80%, which is equivalent to US$ 150/tC avoided. Starting with a natural gas fired, combined cycle plant of 600 MW_e and a conversion efficiency of 52% (LHV), the efficiency might decrease to 45% in the modified plant if the CO_2 emission is reduced from 410 g/kWh to 70 g/kWh. As a result the costs of electricity production might increase by about 50%, which is equivalent to US$ 210/t C avoided.

R&D efforts are focused to minimizing the energy and cost penalties of CO_2 capturing. An important option to reduce these penalties is to start with higher conversion efficiency of the base case plant. Other options are the development of new amines and chemical solvents, the development of a new packing of the absorber, the use of gas absorption membranes (to reduce the size of the plant) and an improved integration of the recovery process into the generation of power (to reduce heat losses). One interesting option under investigation is the combustion of carbon fuels in an O_2/CO_2 atmosphere which would result in a flue gas that essentially is CO_2. When applied to a coal fired power station (ICGCC), the decrease of conversion efficiency might be limited to about 7% and the mitigation cost to less than US$ 80/t C avoided (Hendriks, 1994).

To reduce the energy and cost penalties, an attractive approach could be to decarbonize the fuels before they are utilized. A frequently suggested scheme is the development of an IGCC power plant integrated with CDR. In this scheme, the gasifier off-gas is converted with steam to CO_2 and hydrogen and then separated to make a fuel gas that is essential hydrogen and can be burned in the combined cycle to generate electricity. In this process, because of its high partial pressure, the waste gas CO_2 can be recovered by a physical solvent for which regeneration hardly requires energy. Starting from an original ICGCC plant with a conversion efficiency of 44% (LHV), this scheme might decrease the efficiency to about 37% if the CO_2 emission is reduced from 800 g/kWh to 80 g/kWh. Due to the recovery, the costs of electricity production might increase with about 35%, which is equivalent to US\$ 75/t C avoided (Hendriks, 1994). It should be noted that also in this case there are options to minimize the energy and cost penalties; one possibility is integration of CDR with fuel cells technologies to generate electricity (Jansen et al., 1992).

After recovery the CO_2 has to be transported and stored or disposed of, which adds to the removal costs. Pipeline transportation costs of CO_2 are US\$ 3-12/t C per 100 km, depending on e.g., the size and capacity of the pipeline. At longer distances transportation with tankers could reduce these costs. The costs of underground storage may vary from US\$ 5 to 30 per ton of Carbon, depending on local circumstances (Hendriks, 1994). If the CO_2 is disposed of in the ocean, the costs are probably marginally higher (US DOE, 1993).

In the field of CDR, the interest has been focused mainly on the removal of CO_2 from power plants. From an economic point of view, however, several other options are more attractive and deserve more attention. One of theses options is CDR from fossil fuel recovery processes. An example is the Statoil project in which CO_2 that is removed from the gas of the Sleipner-West field (CO_2 content 9%), is not vented to the atmosphere but captured and stored in an aquifer 1 km below the sea bottom. The injection has started the second half of 1996. The injection rate will be about 1 Mt CO_2 per year. The associated costs are estimated to add maybe 1% to the total production costs of the fuel gas (Kaarstad, private communication, 1996). On a small scale similar projects are executed in Canada (Chakma, 1997). In Indonesia it is planned to recover and store huge amounts of CO_2 from the Natuna gas field (CO_2 content 71%) that will come available by cryogenic separation of the gas into sales gas (mostly methane) and eventually LNG and waste gas (mostly carbon dioxide). The waste gas will be compressed to a supercritical state, transported by pipeline and injected into nearby aquifers. The LNG deliveries and the disposal of CO_2 could commence within less than 8 years (Pertamina & Exxon, 1996).

Another option is CDR from large scale industrial processes, like the production of ammonia from natural gas in the fertilizer industry. In this production process large quantities of CO_2 are produced and separated, but at present most of it is vented to the atmosphere instead of captured and stored. Also in refineries gas streams with a high CO_2 concentration can be identified, like in hydrogen manufacturing and in residue gasification, of which the CO_2 can be recovered and stored at relatively low costs. Another process that might be interesting is the reduction of iron ore to pig iron in the steel industry using blast furnaces. The gas of the blast furnaces contains most of the carbon introduced in the process. From this gas CO_2 can be removed using chemical absorption techniques. It has been analyzed that in these and other cases the costs to mitigate the CO_2 emissions might range between 30 and 130 US\$/t C avoided, depending on the industrial process involved (Farla et al., 1995).

As soon as it is attractive to produce hydrogen as an energy carrier to be used in e.g., fuel cell car vehicles, CDR combined with the production of hydrogen from carbon based fuels could become one of the most attractive mitigation option. Starting from natural gas, the hydrogen can be produced by steam reforming or partial oxidation and separation of the CO_2. Sequestration of the CO in depleted gas wells might result in an incremental hydrogen production cost of 10-25% relative to the case where the separated CO_2 is vented (Blok et al., 1996; Kaarstad et al., 1997). This percentage is substantially lower than the percentages calculated for the power sector. If, as a result of reservoir repressurization, the injection of CO_2 could be combined with enhanced natural gas recovery, the penalty for sequestration might even be reduced to an incremental cost of several percent (Blok et al., 1996).

Social acceptance of CDR

In the period 1993-1995, in the Netherlands the opinions, attitudes and preferences of the people concerning electricity supply and the greenhouse problem has been investigated (Daamen *et al.*, 1996). One of the questions people had to answer was: how should we bridge the gap between demand and supply of electricity in the year 2010. Six options were presented: building coal plants, building coal plants integrated with CDR; building natural gas plants; building nuclear power plants; reducing the electricity demand by a number of governmental incentives (package I); reducing the electricity demand even further by very strong governmental measures and incentives (package II). Each respondent had to select two options. The results are shown in Table 3. From Table 3 it is clear that the people of the Netherlands prefer a combination of three options: saving (package I), power production from natural gas, and power production from coal combined with carbon dioxide removal. In 1993 also the impact of providing information about the disadvantages of each option was investigated. As shown in the table, this didn't shift the preferences fundamentally.

Table 3. Selection percentages per option as obtained from preference investigations in the Netherlands about bridging the gap between demand and supply of electricity in the year 2100, with and without the provision of information about the disadvantages of each option (Daamen *et al.*, 1996).

| | Nov./Dec. 1993 | | Nov. 1994 | Nov. 1995 |
	with info	without info	without info	without info
- coal	7 %	3 %	10 %	9 %
- coal + CDR	36 %	38 %	41 %	41 %
- natural gas	40 %	44 %	62 %	58 %
- nuclear	22 %	16 %	17 %	16 %
- savings (package I)	73 %	73 %	56 %	59 %
- savings (package II)	22 %	26 %	14 %	17 %

In another study the preferences of about 300 decision makers in the business sector of the Netherlands was investigated concerning options to mitigate the greenhouse problem. The results, shown in Table 4, indicate that after 'renewables' and 'energy savings', the options 'sequestration of CO_2 by reforestation' and 'CO_2 removal' are preferred above 'nuclear energy' and 'fuel switch'.

Table 4. Ranking of options to mitigate the greenhouse problem mentioned by decision makers in the business sector of the Netherlands (R&M, 1995).

- application of renewable energy sources	41 %
- more intense energy savings	35 %
- reforestation / growth of peat	27 %
- recovery and storage of CO_2	26 %
- application of nuclear energy	13 %
- switch from coal to oil and natural gas	13 %

These studies indicate that in the Netherlands CDR is perceived as an acceptable option. However, as we know from other examples, this doesn't guarantee that the support will remain when CDR really is implemented. Much will depend on (the perception of) especially the environmental and safety aspects of CO_2 storage and disposal.

Environmental soundness of CDR

To allow CDR to play a major role, due attention should be given to the environmental soundness of the technologies involved. Special attention is required for the reliability, safety and environmental consequences of CO_2 storage and disposal.

In 1986 volcanic CO_2 escaped from Lake Nyos in Africa, killing more than 1700 people. Could this also happen when large amounts of CO_2 are stored in the underground? Given the experience with underground storage of gases most probably not, provided that advanced planning, intensive control of the injection, adequate maintenance of the equipment and appropriate materials are applied. The risk of a CO_2 escape could however be an argument to dispose of CO_2 in aquifers located off-shore instead of on-shore, especially if these aquifers do not have a structural trap. Such an approach might also have a positive influence on the so-called NIMBY ('not in my back yard') effect. Other implications of CO_2 storage in the underground to be investigated are the dissolution of host rock, the sterilization of mineral resources and the effects on ground water.

In the case of CO_2 disposal in the ocean, apart from safety risks, we should know more about questions like the retention period of the CO_2, the impact on marine life of CO_2 disposal especially in the release area, and the tolerable change of the pH of ocean waters.

CONCLUSIONS

We conclude that CDR can reduce the CO_2 emission at a cost ranging from US$ 5-10 to 250 per ton Carbon avoided, depending on specific circumstances and conditions. Recent scenario and system analyses indicate that CDR could reduce the cumulative CO_2 emissions between the years 1990 and 2100 with more than 300 GtC. When applied in the power sector, CDR could increase the electricity production costs with 30-100%, but there is room for cost reductions. From an economic point of view, more attention should be given to low cost CDR options outside the power sector. On the longer term, CDR could be very attractive if hydrogen production from fossil fuels is going to play a major role. CDR seems also a socially acceptable option, although much will depend on the environmental performance of CO_2 storage. It is one of the reasons why more research should be done to the environmental impact of CDR. From this overview we conclude that CDR is a viable option to reduce CO_2 emissions. However, further research, development and demonstration is required to improve the performance of CDR and to allow its further integration in a sustainable development of the energy system.

ACKNOWLEDGMENT

The comments and suggestions of Kornelis Blok, Jacco Farla, Chris Hendriks and Howard Herzog are gratefully acknowledged, as well as the stimulating discussions with colleagues of the UN energy committee and the IPCC Working Group on Energy Supply Mitigation Options.

REFERENCES

Blok, K. *et al.* (1996). Hydrogen production from natural gas, sequestration of recovered CO_2 in depleted gas wells and enhanced natural gas recovery. *Energy*, to be published.

Chakma, A. (1997). Acid gas re-injection, a practical way to eliminate CO_2 emissions from gas processing plants, *Energy Convers. Mgmt*, this volume.

Daamen, D.D.L. *et al.* (1996). *Veronderstellingen, houdingen, en beleidsvoorkeuren van de Nederlander betreffende elektriciteitsvoorziening en broeikaseffect (Suppositions, attitudes and policy preferences of the Dutchman concerning electricity supply and greenhouse effect).* Werkgroep Energie- en Milieuonderzoek, Leiden University, NL.

Farla, J.C.M. *et al.* (1995). Carbon dioxide recovery from industrial processes. *Climatic Change*, 29, 439-461.

Hendriks, C. (1994). *Carbon Dioxide Removal from Coal-Fired Power Plants*, Kluwer Academic Press, Dordrecht, NL.

Houghton, J.T. *et al.*, eds. (1990). *Climate Change - The IPCC Scientific Assessment.* Cambridge University Press, Cambridge, UK.

Houghton, J.T. *et al.*, eds. (1996). *Climate Change 1995 - The Science of Climate Change.* Contribution of Working Group I to the Second Assessment Report of the IPCC. Cambridge University Press, Cambridge, UK.

Ishitani, H. *et al.* (1996). Energy supply mitigation options. In: *Climate Change 1995 - Impacts, Adaptions and Mitigation of Climate Change: Scientific-Technical Analyses* (R.T. Watson *et al.*, eds.), pp. 587-647. Cambridge University Press, Cambridge, UK.

Jansen, D. *et al.* (1992). CO_2 reduction potential of future coal gasification based power generation technologies. *Energy Convers. Mgmt,* 35, 365-372.

Jefferson, M. (1996). *The Intergovernmental Panel on Climate Change: the Second Assessment Report Reviewed.* World Energy Council, Wellington, New Zealand.

Johansson, T.B. *et al.* (1993). Renewable fuels and electricity for a growing world economy: defining and achieving the potential. In: *Renewable Energy; Sources for Fuels and Electricity* (T.B. Johansson *et al.*, eds.), pp. 1-71. Island Press, Washington, D.C.

Kaarstad, O. *et al.* (1997). Hydrogen and electricity from decarbonized fossil fuels, *Energy Convers. Mgmt,* this volume.

Kassler, P. (1994). *Energy for Development.* Shell International Petroleum Company, London, UK.

Leggett, J. *et al.* (1992). Emissions scenarios for IPCC: an update. In: *Climate Change 1992 - The Supplementary Report to The IPCC Scientific Assessment* (J. Houghton *et al.*, eds.), pp. 69-95. Cambridge University Press, Cambridge, UK.

Nakicenovic, N. (1996). Energy primer. In: *Climate Change 1995 - Impacts, Adaptions and Mitigation of Climate Change: Scientific-Technical Analyses* (R.T. Watson *et al.*, eds.), pp. 75-92. Cambridge University Press, Cambridge, UK.

Pertamina & Exxon (1996). LNG Project. Pertamina / Exxon Corporate, Natuna, Indonesia.

R&M (1995). *Kennis en houding van beslissers in het bedrijfsleven ten aanzien van de klimaatproblematiek (Knowledge and attitude of decision makers in the business sector concerning the climate change problem).* Research and Marketing, Heerlen, NL.

Spencer, D.F. (1993). Use of hydrate for sequestering CO_2 in the deep ocean. National Summer Conference of the American Institute of Chemical Engineers. Seattle, Washington.

Turkenburg, W.C. (1995). Energy demand and supply options to mitigate greenhouse gas emissions. In: *Climate Change Research: Evaluation and Policy Implications.* (S. Zwerver *et al.*, eds.), pp. 1013-1054. Elsevier Science, B.V., Amsterdam.

Turkenburg, W.C. (1996). *Kernenergie en Duurzame Ontwikkeling (Nuclear Energy and Sustainable Development).* Netherlands Energy Research Foundation, Petten, NL.

UNCNRSEED (1996). Report on the second session of the UN committee on new and renewable sources of energy and on energy for development. Economic and Social Council, Official Records 1996, Supplement No. 4. United Nations, New York.

US D.O.E. (1993). *The Capture, Utilization and Disposal of Carbon Dioxide from Fossil Fuel-Fired Power Plants,* Vols. I and II. US Department of Energy, Washington, DC.

VROM (1996). *Vervolgnota Klimaatverandering (Second Memorandum on Climate Change).* Sdu Uitgeverij, The Hague.

VCE (1996). *Verkenning Energieonderzoek (Energy Research Exploration).* Overlegcommissie Verkenningen, Amsterdam.

Watson, R.T. *et al.*, eds. (1996). *Climate Change 1995 - Impacts, Adaptions and Mitigation of Climate Change: Scientific-Technical Analyses.* Contribution of Working Group II to the Second Assessment Report of the IPCC. Cambridge University Press, Cambridge, UK.

WEC (1993). *Energy for Tomorrow's World.* St. Martin's Press, New York.

World Bank (1992). *Development and the Environment - World Development Report 1992.* Oxford University Press, UK.

Energy Convers. Mgmt Vol. 38, Suppl., pp. S13–S18, 1997
Published by Elsevier Science Ltd
Printed in Great Britain
0196-8904/97 $17.00 + 0.00

PII: S0196-8904(96)00238-5

United States Strategy for Mitigating Global Climate Change

Robert L. Kane
Program Manager, Global Climate Change
Office of Fossil Energy
Office of Planning and Environment
U.S. Department of Energy
Washington, DC 20585

Daniel E. Klein
Senior Consultant
ICF Resources Incorporated
9300 Lee Highway
Fairfax, VA 22031

ABSTRACT

Beginning with the adoption of the Framework Convention on Climate Change (FCCC) in 1992, which called for industrialized countries to reduce their greenhouse gas (GHG) emissions to 1990 levels in 2000, political pressures have been increasing to reduce GHG emissions. The Berlin Conference of the Parties (COP1) in March, 1995, increased this pressure for countries to commit to post-2000 emission reduction goals. Published by Elsevier Science Ltd

Until July, 1996, the primary focus of U.S. climate change policy was moderate, near-term reductions for the year 2000. The U.S. Climate Change Action Plan (CCAP), announced in October, 1993, began a comprehensive set of voluntary programs aimed at reducing these emissions. The most successful of these voluntary programs is the Climate Challenge program, a cooperative program between the U.S. Department of Energy (DOE) and the electric utilities, covering over 60 percent of the industry.

In July, 1996, the U.S. position changed and urged that future negotiations on emission reductions focus on "a realistic, verifiable, and binding medium term (2010-2020) emissions target". At the same time, the U.S. endorsed the "Geneva Declaration", which endorses the science findings of the Intergovernmental Panel on Climate Change (IPCC) and permits flexibility for individual countries to design the most cost-effective approaches to emission reductions.

Issues of availability, cost, and timing dictate that fossil fuel use will continue to drive domestic and international societies for many years. It follows that efforts to reduce GHG emissions will have to consider the worldwide energy requirements, the efficiency and means of meeting those requirements, and/or associated GHG emissions. Within DOE, the Office of the Assistant Secretary for Fossil Energy has important responsibilities for meeting these twin challenges of reducing greenhouse gas emissions while ensuring cost-effective and abundant supplies of fossil energy. This paper explores some of DOE's fossil-related technology and climate-related activities.

KEYWORDS

climate change, greenhouse gases, electric utilities, voluntary programs, fossil energy, energy efficiency, energy research, CO_2 recovery

DEVELOPMENT OF THE U.S. CLIMATE CHANGE ACTION PLAN (CCAP)

In April 1993, President Clinton announced the U.S. commitment to return GHG emissions in 2000 to their 1990 levels. President Clinton also instructed his Administration to prepare an action plan to achieve this goal and continue the trend of reduced emissions. *The Climate Change Action Plan* (CCAP), published in October, 1993, consists of about 50 distinct but interrelated federal initiatives.

A majority of these initiatives seek to reduce or avoid GHG emissions via influencing patterns of energy demand and supply. In addition, special programs are also employed for methane emission reduction and recovery, reduction of minor GHGs (HFC, PFC, and N_2O), and enhancement of carbon sequestration via forestry actions.

While the CCAP is based almost entirely on existing legislation as of 1992, it breaks new ground in fostering cooperative approaches between the government and private sector. Also notable is the particular emphasis on innovative approaches to policy development and implementation. President Clinton stated in his April, 1993 address, "I am instructing my Administration to produce a cost-effective plan. . . that can continue the trend of reduced emissions. This must be a clarion call, not for more bureaucracy or regulation or unnecessary costs, but instead for American ingenuity and creativity to produce the best and most cost-efficient technology."

The CCAP has several defining characteristics. The plan is comprehensive, targeting all sectors of the economy and all greenhouse gases. The plan develops partnerships with American businesses to foster cooperative approaches and cost-effective results. It is designed for rapid implementation and adaptability to changing circumstances. And while the CCAP was designed to achieve the President's goal with domestic actions alone, it recognizes the significant potential for cost-effective emissions reductions in other countries by including a U.S. Initiative on Joint Implementation as a pilot program intended to gain experience in evaluating investments in other countries for emission reduction benefits.

GREENHOUSE GAS EMISSION TRENDS

While the CCAP's primary goal was to return U.S. GHG emissions in the year 2000 to 1990 levels, progress to date is mixed. Based upon EIA emissions data, the relatively flat emissions trend of the 1988-1992 period looks to be the exception more than the rule. Currently, carbon emissions are significantly above both 1990 levels and the levels estimated for the CCAP budget.

These difficulties were outlined in the U.S.'s *Climate Action Report*, submitted in September 1994 as required under the Framework Convention on Climate Change. That report noted that since the original CCAP projections were prepared, economic growth was more robust, oil prices were lower, and Congressional funding was less than requested. It now seems likely that the U.S. will fall significantly short of the CCAP goal for 1990 emission levels in the year 2000. The Climate Action Report suggested that because of these developments, the U.S. may need to implement additional measures to meet its commitment of returning emissions to their 1990 levels by the year 2000.

The U.S., however, is not alone in missing this target for GHG emissions. Most industrialized countries will not achieve the year 2000 aim. In the *Annual Energy Outlook 1995*, EIA projects global carbon emissions to grow from about 6 billion metric tons in 1990 to over 8 billion metric tons by 2010, representing an annual growth rate of 1.5 percent. In contrast, the average growth in U.S. emissions is only about one percent annually.

POLITICAL PRESSURES FOR FURTHER ACTIONS

These rising trends in GHG emissions, both in the U.S. and worldwide, are creating political pressures to reduce GHG emissions. The CCAP goal to reduce emissions in the year 2000 to 1990 levels was set to be consistent with the 1992 Framework Convention on Climate Change (FCCC). However, this goal was seen as an *initial* step for developed countries under the FCCC. Under the Framework Convention, a broad *ultimate* objective was stated that GHG concentrations in the atmosphere should be stabilized at a level that would prevent dangerous anthropogenic interference with the climate system.

Last year, at the first Conference of Parties (COP1), the "Berlin Mandate" initiated the process of examining actions that might be needed in the post-2000 period.   Whereas the Framework Convention's initial step and the U.S. CCAP near-term goals were specific to the year 2000 and not necessarily beyond, the Berlin Mandate more strongly suggested that tougher actions post-2000 might be needed, and that the industrialized countries were expected to take the lead in these efforts.

These pressures for further actions took another step forward at the second Conference of Parties (COP2), held in Geneva in July, 1996. Undersecretary of State Timothy Wirth recommended that future negotiations focus on an agreement that sets a realistic, verifiable, and binding emissions target for the medium term, around 2010-2020. Without recommending specific targets and timetables, the U.S. position stressed the need for flexible and cost-effective measures, including international trading. At COP2, Ministers of most countries endorsed the "Geneva Declaration," which calls for a legally binding approach to next steps, and that this legal framework be completed in time for adoption at COP3, to be held in Kyoto, Japan in December, 1997.

The Geneva Declaration endorses the Second Assessment Report of the Intergovernmental Panel on Climate Change (IPCC) as currently the most comprehensive and authoritative assessment of the science of climate change. According to the IPCC report, unless substantial reductions in emissions are made, GHG concentrations will continue to climb, and that these higher concentrations will lead to projected changes in the climate system that will result in significant and often adverse impacts on the human health, welfare, and the environment. Thus, the Geneva Declaration provides a linkage between the underlying science and the policy objective of the Framework Convention.

ELECTRIC UTILITIES AND THE CLIMATE CHALLENGE PROGRAM

Like the voluntary actions of the CCAP, future targets and timetables that might emerge from future international negotiations are likely to include all gases and all sectors of the economy. While there are several GHGs, CO_2 emissions dominate with about 85 percent of the total, with methane accounting for much of the rest. Importantly from a DOE perspective, most of the CO_2 emissions result from combustion of coal, oil, and gas. In addition, a significant part of U.S. methane emissions result from energy production and transportation.

The electric utility sector may be relatively more vulnerable to controls on GHG emissions. In generating power for their residential, commercial, and industrial customers, electric utilities account for over one-third of the carbon emissions in the U.S. (and a slightly lower percentage of U.S. total GHG emissions net of sequestration). Hence, their size alone would suggest that utilities would be a significant part of any plan to reduce GHGs. And with a relatively small number of emitters and the utilities' history as a regulated industry, many have come to see them as a potential target for GHG emission reductions.

The centerpiece of the utilities' response to the climate change issue is the "Climate Challenge" program. Climate Challenge is a joint initiative between the U.S. Department of Energy (DOE) and the electric utility industry to voluntarily reduce greenhouse gas emissions. The initiative, announced as a foundation action under the Climate Change Action Plan, consists of voluntary commitments by electric utilities to undertake actions to reduce, avoid, offset or sequester GHG emissions. As a partnership between DOE and the electric utilities, Climate Challenge utilities are moving to reduce their GHG emissions using a wide range of emission reduction options and innovative approaches.

The origins of Climate Challenge lie in the energy industry's pledge of cooperation with the U.S. Congress to pursue voluntary GHG emission reduction efforts as part of the 1992 Energy Policy Act (EPAct). EPAct had already set the stage for voluntary reporting of GHG emissions and emission reductions under its Section 1605(b). Secretary O'Leary's vision of transforming the DOE's role and functions provided additional momentum for a voluntary program. As the Secretary stated in the *National Journal* (November 5, 1994, p. 2566), "Voluntary is not a dirty word." The central concept behind Climate Challenge is that flexibility significantly lowers costs compared to mandatory programs because utilities face diverse circumstances and place different values on specific technologies and emission reduction options.

Widespread participation in Climate Challenge and the vigorous support by its members exemplifies a new era of cooperation between DOE and the electric utility industry. In April 1994, DOE signed a Climate Challenge Memorandum of Understanding (MOU) with the Edison Electric Institute (EEI), American Public Power Association (APPA), National Rural Electric Cooperative Association (NRECA), Large Public Power Council (LPPC), and the Tennessee Valley Authority. This MOU established Climate Challenge's guiding principles and actions to be undertaken by both DOE and the electric utility industry.

Climate Challenge commitments are formalized in individual Participation Accords with the utilities. These Participation Accords contain specific commitments describing the actions that the utility and DOE have each committed to undertake under the Climate Challenge Program. The types of commitments are broad enough that any utility can participate, whether large or small, with or without generation facilities, and having all kinds of resource mixes and load growth. "Flexibility" is a key word in all of these efforts. The participants agree to periodically report their individual progress and the obstacles that they have encountered, and they can modify the accords as needed.

As of October, 1996, about 600 electric utilities had signed 114 Participation Accords with DOE, specifying the actions they would be taking. These utilities represent over 60 percent of 1990 U.S. electric utility generation and utility carbon emissions. As additional utilities enter into Participation Accords with DOE, the share of the industry's generation and carbon emissions covered by Climate Challenge utilities will continue to rise.

In the Participation Accords so far signed, the Climate Challenges utilities are pledging a wide range of GHG reduction activities, in aggregate about 44 million metric tons of carbon equivalent. About half of the pledged GHG reductions stem from supply-side activities, coming as the result of improvements in nuclear plant availability, fossil heat rates, renewable energy sources, transmission and distribution modifications, fuel switches to natural gas from coal and oil, and others. Substantial GHG reductions are also pledged from demand-side management programs, landfill and coal mine methane capture, forest carbon programs, international programs, and others.

Yet the "challenge" of Climate Challenge is not merely in terms of tons of carbon reduced, but also in terms of forging partnerships to reassert technological leadership in the areas of energy and environment, to advance economic growth, and to show that the government and private industry do much better working together voluntarily when compared to "command and control" approaches.

Climate Challenge is still in its infancy, about three years old. Yet in that brief time, it has garnered the support of most of the electric utility industry, demonstrated the value of voluntary and flexible approaches, and is making a substantial contribution to the Administration's Climate Change Action Plan. The utilities' proactive behavior has made them a leader on this important global issue. DOE is proud to call them "partners."

This type of voluntary approach has attracted widespread, bipartisan, and international support. The U.S. government strongly supports voluntary programs and believes that those who take early action should receive credit toward future emission reduction requirements.

RESEARCH ON GLOBAL CLIMATE CHANGE

Climate change research remains a high priority for the United States. The U.S. Global Change Research Program (USGCRP) budget for FY 1996 invests about $1.6 billion in global change research. The USGCRP focuses on the scientific study of the Earth system and its components. Global change research provides short- and long-term benefits to the nation by:

- contributing to fundamental scientific knowledge regarding the interactions of the Earth's climate system;

- documenting and assessing potential changes in the Earth system and the implications of these changes on climate, ecosystem health, and future resource availability; and

- developing improved predictions of extreme weather events, thereby allowing actions to reduce the vulnerability of people and property to natural disasters.

USGCRP research is organized around a framework of observing and documenting change, understanding processes and consequences, predicting future changes, and assessing options for dealing with change.

FOSSIL ENERGY RESEARCH PROGRAM

While this research is undertaken, our economy must continue to move forward and grow. Fossil fuels – coal, oil, and gas – have long been a prime "engine" of our industrialized society, offering abundant supplies of energy at low costs. At the same time, most anthropogenic GHG emissions are related to the use of these fuels. Coal is the most widely used fuel for the generation of electricity worldwide because it tends to be readily available, easily transportable, and relatively inexpensive. Oil-derived products dominate transportation fuels.

There are, at present, few practical alternatives to most uses of fossil fuels. Non-GHG concerns may be limiting factors to the increased use of nuclear or hydroelectric generation technologies. Further, non-fossil fuels have also proven illusive for most transportation and industrial technologies.

These limitations make it clear that fossil fuels will continue to be a vital component of the energy mix for the foreseeable future. Even though advances are being made in technological alternatives to fossil fuels, issues of availability, cost, and/or timing dictate that fossil fuel use will continue to drive domestic and international societies for many years.

The United States and the world community are challenged to find new and better ways to meet the world's increasing needs for energy while reducing carbon dioxide, methane, and other greenhouse gases that can result from fossil energy production and use. New technologies can improve the efficiency of fossil fuel use, reduce fossil fuel demand, and enable fossil fuels to be burned more cleanly, thus reducing greenhouse gas emissions.

Today's concerns about climate change, environmental performance, and energy efficiency have created incentives for us to revisit the technology of power plant design. CO_2 emissions are, generally, a function of the amount and type of a given fuel burned rather than the amount of power produced. Higher efficiency means less fuel consumption and lower CO_2 emissions. The replacement of aging, power systems with high efficiency systems of equal capacity could result in significantly lower emissions of CO_2.

Much of the responsibility for the United States' response to these challenges rests with DOE. DOE's role in the research, development, demonstration and commercialization of high efficiency coal power systems will create the potential for sizable domestic and export markets for these technologies. High efficiency systems offer a means to minimize the environmental impact of economic growth or previous environmental neglect. In addition, by accommodating low-grade domestic coal worldwide, these technologies enable nations to increase their fuel flexibility and self-reliance.

By achieving greater generating efficiency, the fossil fuel option can be preserved in an economically and environmentally acceptable manner. DOE's research efforts have resulted in considerable improvement in the performance of advanced coal-fired systems. Over the next ten years or so, improvements in low emission boiler systems and pressurized fluidized bed combustion systems will boost efficiencies closer to 50 percent, with reduced levels of CO_2 and conventional pollutants such as SO_2 and particulates. Longer-term technologies include improvements in integrated gasification combined cycle and integrated gasification fuel cell systems. Introduction of these technologies would reduce CO_2 emissions by over 40 percent relative to conventional coal-fired systems, with conventional pollutant levels perhaps a tenth of current New Source Performance Standards.

The growing demand for energy, coupled with increasing environmental concerns, will create a sizable global market for clean coal technologies, consisting largely of high efficiency power systems or components of these systems. The size of this world market amounts to hundreds of billions of dollars. As the current world leader in clean coal technology development, demonstration, and production, the U.S. is positioned to serve a large segment of the export market of these technologies.

Central to the theme of this conference are DOE's efforts related to advanced technology for the recovery, reuse, and disposal of CO_2. This program focuses on the development of technically, economically, and ecologically sound methods to capture, reuse, and dispose of CO_2 produced by coal-fired combustion systems.

Current approaches to the recovery, reuse, and disposal of CO_2 are too costly and/or require too much energy. Once CO_2 is recovered, there are currently no practical means available to ensure that the recovered CO_2 is kept from reentering the atmosphere for sufficiently long periods of time (on the order of hundreds of years) to reduce climate change concerns.

At present, these research efforts are meant to serve as an *alternate* approach in case other strategies for carbon dioxide mitigation prove to be inadequate or will not achieve a prescribed level of CO_2 emissions reduction in a timely manner. For CO_2 recovery, reuse, and disposal to become a practical option for helping to meet future greenhouse gas emission reduction goals, research is required to develop novel and more cost-effective solutions to the problem.

Until recently, a primary focus of the U.S. climate policy was moderate near-term reductions, with a specific aim for the year 2000. Actions taken at COP2 — the "Geneva Declaration" in particular — have shifted this focus more to the medium term (2010-2020), and perhaps requiring greater GHG reductions. These policy actions provide a context for stepping up research efforts on advanced technologies. In this light, research on CO_2 recovery, reuse, and disposal may provide another policy option for government and industry.

SUMMARY

Under the President's Climate Change Action Plan, DOE is participating in the development of strategies and partnerships to meet the challenges of sustaining worldwide economic growth while reducing GHG emissions. Given DOE's responsibility for issues related to the production and use of oil, natural gas, and coal, DOE plays a major strategic role in balancing the needs for economic sources of energy with the need to protect the domestic and global environment.

Central to DOE's strategy for reducing GHG emissions is the use of existing knowledge of current power systems to develop economically palatable and advanced solutions. By encouraging the widespread availability of these advanced systems, both developed and developing countries may share in the benefits. Furthermore, by promoting the advancement of these methods within the United States, the Nation may be better able to incorporate them into current operations while creating economically exportable technologies for the next generation.

Pergamon

Energy Convers. Mgmt Vol. 38, Suppl., pp. S19–S23, 1997
© 1997 Elsevier Science Ltd. All rights reserved
Printed in Great Britain
0196-8904/97 $17.00 + 0.00

PII: S0196-8904(96)00239-7

Japanese Strategy for Mitigating Global Warming

Yoichi Kaya

Keio University Shounan Fujisawa Campus,
Endou 5322, Fujisawa-shi, Japan, 150

Abstract

This paper illustrates first what the government of Japan intends to do with a few official pledges on mitigating global warming, such as the action plan and the famous idea of New Earth 21 program, toward long term stabilization of CO_2 concentration in the air. In practice, however, there are a number of factors which may disturb realization of the targets envisaged in these plans. Limitation to advancement of further energy conservation, not only in industrial sectors but also in other sectors where individual consumers play key roles, is certainly serious. Another crucial factor is the nuclear power. Recent accident at the experimental FBR, Monju, has influenced the public acceptance of nuclear power very seriously, and the government is now at the stage of re-designing the future scenario of nuclear power, taking into account both the opinions in favor of nuclear and those of anti-nuclear. Nevertheless most of energy experts believe that nuclear power is safe and also economical enough to be maintained for the intermediate future at least in Japan so that strong policy measures are to be introduced to improve public acceptance of nuclear power.

The second part of this paper describes the future of so called "new energy" which the government of Japan is eager to develop and also the limits to development of these "new energy". The third part is contributed to description of R&D of long term technological options and also difficulties in realizing these options. It is also worth noting that Japan, both government and industries, has much interest in CO_2 removal from flue gas of power plants and its disposal. © 1997 Elsevier Science Ltd

Keyword

CO_2 emisshion; energy ddemand; energy conservation; nuclear power plant; new energy; R&D

1. CO2 Stabilization Target and Present Situation

The Framework Convention on Climate Change (FCCC) asks developed countries defined in ANEX I to stabilize their CO_2 emission in 2000 at the level of 1990. Well prior to ratification of FCCC the government of Japan announced its action plan in 1989 to stabilize CO_2 emission per capita by 2000, and reported to the Secretariat of FCCC in 1994 that Japan would be by and large able to attain the target. This expectation is based upon the long term energy demand-supply outlook made in 1990 by the Advisory Committee on Energy, MITI, which expects decarbonization and energy conservation will advance in 1990's as in the same pace as in 1980's. Actually, however, this expectation seems to have been too optimistic.

Shown in Table 1.1 are the rates of change in three important factors of CO_2 emission in 1980's, in the scenario B(strong policy case) of MITI outlook for 1990-2000 and in 1991-1994. We notice that despite lower economic growth than expected during this period the growth of CO_2 emission is remarkable in 1990's due to little energy conservation: actually the energy intensity was worsened! The sectoral analysis of energy demands shows that the worsening of energy intensity in 1990's is seen in all sectors; i.e. industrial, transportation, commercial and residential ones. There may be several causes for the increase in energy demand and CO_2 emission: hotter summer weather than before, less rainfall (which gives rise to increase in summer cooling demand and decrease in output of hydro-power) , and collapse in prices of some commodities (particularly in those of products of electronic and machine manufacturers) .

Those described above are however still minor. The major cause in worsening the situation was in authors opinion that little concern on energy conservation has prevailed nationwide due to stagnant or even lowering energy prices. As an example look at the situation of automobile sector. The oil price went down in the former half of 1980's and since then it has been stagnant. Table 1.2 shows relative changes in the levels of variables connected to fuel consumption for automobiles. The variable D can be interpreted as an indicator of the relative burden to each driver of automobile. It is less than half now when compared to the level in 1970: in other words consumers have much less economic incentives for conserving energy now than even well before the oil shock. As one of the results the average size of automobiles has become larger and larger monotonously and now average engine capacity of an automobile reached the level of 1,900cc as is seen in Table 1.2. The government is now doing efforts for promoting energy conservation by introducing energy conservation targets for the improvement of average fuel economy of automobiles (8.5 % by 2000) and for those of energy per unit production in industries (more than 1 % per year) , and reinforcing the energy conservation law, but the author wonder how effective they are in practice.

One but probably only good aspect in energy conservation is improvement of electric power generation efficiency. The average efficiency at the generator outpout is now 38.5 %, already higher than those in most countries in the world.

Now most Japanese power companies are enthusiastic in introducing combined cycle power plants with LNG as fuels. The total capacity of combined cycle power plants will reach at the level of 24 GW by 2004, about 10 % of the tatal power generation capacity of Japan, and the thermo-electric conversion efficiency of the combined cycle plants coming into operation around 2004 will be 50% (HHV base) , about 30 % higher than those of present power plants(1). Considering that combined cycle plants use natural gas of which CO_2 emission per unit energy is the lowest among fossil fuels, the impact of increase in combined cycle plants on reduction in total CO_2 emission will be considerably large.

Another serious problem we should notice is how decarbonization of fuels will advance in future. As is seen in Table 1.1 decarbonization advanced as expected even after 1990, ,but we see the situation will be worsened in the coming years.

The crucial factor in talking about decarbonization of fuels is nuclear power. Japan has been promoting nuclear power since 1960's and now the total capacity of operating nuclear power plants is 40 GWe. As nuclear power plants have high operation rates the share of nuclear power in total electric power generation (i.e. kwh)is now close to 30 %, despite that the capacity of nuclear power plants is only 20 % of the total capacity of power plants in Japan. MITI's long term energy outlook forsees 72 GW nuclear power plants will be in operation by 2010 which will provide about 20 % of the primary energy of this country.

However the trouble occured in experimental fast breeder reactor called Monju last December (leakage of sodium from heat exchanger , despite no leakage of radioactivities) gave rise to nationwide expansion of anti-nuclear movements. On August 4, 1996 the votes by the residents on the construction of nuclear power plants were made for the first time in the history of this country at Maki Town, Niigata Prefecture and the result was negative. (60 % were anti-nuclear). Based upon this result the governor of the town confirmed his will of not sellig the land owned by the town, located in the center of the candidate site.

Legally speaking the votes by the residents do not prohibit construction of nuclear power plants at the site, but it is in practice almost prohibitive for the power company to do so. More serious is that this result will have considerable negative effects on the public opinions in other areas where construction of nuclear power plants are being planned. Power companies and the government do a lot of efforts for improving the situation, but it is not an easy task. It seems therefore very improbable that the plan of constructing 72 GWe nuclear power plants in total be realaized by 2010. Also taking into account that decarbonization of energy in 1980's in Japan was realized mainly due to increase in number of nuclear power plants, the author is pessimistic about the future advancement of decarbonization in Japan, at least in the near term future.

2. Long Term Efforts

Limits to "new energy" development

The government of Japan is eager to develop so called "new energy" which includes renewable energy, energy recovered from wastes, and energy from highly efficient new energy conversion plants (actually cogeneration and fuel cells). The outlook by MITI envisages their future scenario around 2010, as shown in Table 2.1. After adding some small resources to this table the total of "new energy" in 2010 will be around 5.8 % of the total primary energy supply of Japan.

The question is whether this scenario can be realized and to what extent they will be expanded after 2010. My frank opinion is pessimistic for both questions. The first reason is that most of "new energy" in Table 2.1 comes from the conventional sources being utilized for years in the past (solar heaters, energy recovered from pulp wastes) which are already limited in use. The second reason is that photovoltaics and cogeneration will be expanded but again limited in use due to the following reasons.

In case of photovoltaics the time changeability of its output is one of the serious limiting factors. The peak of night demand occurs in the early evening of summer, and it is around 90 % of the daypeak throughout the year. Since the total capacity of the power system is determined by the daypeak and photovoltaics cannot provide any output at night, conventional power plants should have the capacity of at least around 90 % of th systm capacity. Then the potential of photovoltaics connected to the power system is limited to only 10 % in capacity, unless its unit cost is lower than average fuel cost of conventional power plants (it is highly improbable in the near future). Taking the low operation rate of photovoltaics compared to those of conventional power plants contribution of photovoltaics to the total power may be very limited.

Cogeneration seems more promising than photovoltaics, but we should notice that Japan is located rather south and therefore its heat demand in residential and commercial sectors is much lower than other developed countries. It means average heat / power ratio of the demand in these sectors is so low (around 2 or lower) that the advantage of cogeneration of providing both heat and power may not be fully utilized in Japan. (For more detailed discussions, see (1)) Cogeneration in industrial sector may be in better situation, but we should notice that Japanese industries made a lot of efforts after the oil shocks for energy conservation; in other words if there are rooms for cogeneration to be economically profitable, most of them may have already been exploited by the concerned industries.

Efforts for long term R&D of innovative technologies

The above discussions indicate that the possibility of reducing CO_2 emission in near future is very limited. The ultimate target being to attain stabilization of CO_2 concentration in the air as mentioned in FCCC, we should find ways for substantial reduction in CO_2 emission: this will require R&D of innovative energy and related technologies.

The government of Japan announced New Earth 21 plan in 1989, which aims at developing innovative technologies to enable stabilization of CO_2 concentration in the air. Since then the government has been doing efforts for long term technologies under various R&D programs. The most important among them is New Sunshine project in MITI in which two programs are remarkable as a part of NW21. One is the program called WENET(World Energy NETwork) and the other is Eco Energy City program.

The basic concept of WENET is to develop technologies of collecting, converting and transporting solar energy (direct solar, hydro and/or wind). One of the important factors is hydrogen technology which will be used widely in future. Eco Energy City program intends to develop technologies for utilizing ambient energy sources in urban and suburban areas such as waste water, river, underground, etc. Development of superheat pumps is a part of this program.

 In addition to these programs the government of Japan is trying to breed technology seeds by providing larger scale grants to univerisities and research institutes than before. From 1995 fiscal year the government launched new grant programs for science and technology R&D in 5 ministries, namely Ministry of Education and Research, Science and Technology Agency, MITI, Ministry of Health and Welfare and Ministry of Agriculture and Fishery. The total amount is 30-40 billion Yen per year. The unit size of the grants differs from ministry to ministry, but between 50 and 300 million Yen (between 0.5 and 3 million dollars) per year. The grants in MITI and Science and Technology Agency includes those for energy and environment areas, and the author expects innovative ideas for effectively reducing CO_2 emission will emerge and be developed in these grants in the near future.

3. Conclusion: position of CO_2 removal/disposal technologies

The issue of global warming is of so long term character that efforts which can sustain for years are required. The author is convinced that development of innovative technologies for energy efficiency improvement and expansion of non-fossil fuel use is indispensable for substantial reduction of CO_2 emission and Japan is enthusiastic in promoting R&D of such innovative technologies. It is however not easy to develop these technologies in time so that we should also develop reliever type technologies in parallel with them as tools in the contingency plan. CO_2 removal from flue gas and its disposal is, the author believes, the most effective one among such technologies and we should do efforts for R&D of the related technologies as soon as possible. In this sense the author firmly believes importance of ICCDR.

Table 1.1 Rates of Change in three factors related to CO_2 emission in Japan (%)

	carbon intensity*1	energy intensity*2	GDP	CO_2 emission*3
1980-1989 data	- 0.7	- 2.0	+ 3.7	+ 1.0
scenario B of MITI, 1990-2000	- 0.5	- 2.1	+ 3.0	+ 0.4
1991-1994 data	- 0.6	+ 1.3	+ 1.1	+ 1.8

*1: CO_2 emission / primary energy
*2: primary energy / GDP
*3 rate of change in CO_2 emission = sum of those of carbon intensity, energy intensity and GDP.

Table 1.2 Changes in important variables, 1970 - 1993

	the level in 1993
A. gasoline retail price 1970 = 100	250
B. consumer price index 1970 = 100	304
C. per capita GDP 1970 = 100	171
D = A / B / C	0.48
E. average engine capacity of automobiles, 1974 = 100	127 (1900 cc)

References

1. Kaya, Y.: Natural gas for electricity and cogeneration, 3rd IEA gas technology conference, Berlin, (September 2-4, 1996)
2. Kaya, Y.: Role of CO_2 disposal, ICCDR-2, Kyoto, (September, 1994)

Pergamon

Energy Convers. Mgmt Vol. 38, Suppl., pp. S25–S30, 1997
© 1997 Elsevier Science Ltd. All rights reserved
Printed in Great Britain
0196-8904/97 $17.00 + 0.00

PII: S0196-8904(96)00240-3

Lower CO$_2$ Emissions Through Better Technology

Gernot Gessinger
Corporate R&D Programs
ABB Asea Brown Boveri Ltd., Zurich, Switzerland

Abstract

The natural emissions of CO$_2$ are approximately 200 Gigatons carbon per year. The same amount is taken out of the atmosphere every year. The man-made emissions are approximately 8 Gigatons C/year, 2 Gigatons from deforestation and 6 from the use of fossil fuels. While this amount seems low, it is the only part which is not offset by carbon removal, and it is growing rapidly as the world economy expends.

1.8 Gt C/year are due to electricity generation from combustion of fossil fuels; 4.2 Gt C/year are caused by use of fossil fuels in transportation, industry and private homes.

Three strategies are available to reduce the amount of CO$_2$

- Prevention: avoid formation of CO$_2$ by higher efficiencies in electricity generation, transmission and use

- Recycling of CO$_2$ by capture, utilization and disposal/storage

- Use of renewable sources of energy like hydro, solar, wind, biomass, etc., and use of nuclear power.

The potential to prevent an increase from today's 1.8 Gt C/year in 2010 is 1.5 Gt C/year, if best available technologies will be utilized in the power industry.

The potential to recycle CO$_2$ is much higher, but its utilization hinges on costs of fossil fuel generated electricity, which would increase by 50 - 100 %.

Man-made CO$_2$ Emissions

In the global carbon dioxide cycle of nature, generation and absorption of large amounts of carbon dioxide in perfect equilibrium: 200 Gt C are generated each year by plant and soil respiration and decomposition and the ocean, and they are matched by an equal amount of CO$_2$ absorbed by plant photosynthesis and by the oceans. Man is disturbing this equilibrium by generating yearly 8 Gt of CO$_2$, of which only 4.5 Gt are reabsorbed by nature.

6 of the 8 Gt are caused by electricity generation (1.8 Gt C/yr) and by transportation and industry and domestic use (4.2 Gt C /yr).

ABB is in the business of electric power generation, its transmission and distribution to industrial and private users; but ABB is also involved in providing technologies to industrial users, and finally it is one of the main suppliers of systems and services in the transportation business. It therefore works with vital technologies which can be considered the causes of pollution by CO_2, but it also does develop technologies which help to reduce the amount of CO_2 generation.

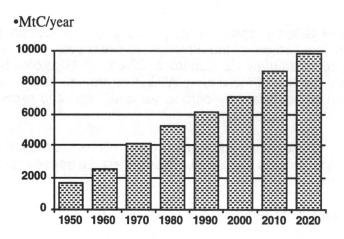

•MtC/year

Fig. 1: Man-made CO2 emissions - combustion of fossil fuels

Figure 1 shows the evolution of man-made CO_2 emissions caused by the combustion of fossil fuels - coal, oil and gas. It can be fairly safely predicted that more electricity needs to be generated as the needs in less developed countries are growing rapidly. Only nuclear power and hydro-power do not generate CO_2, although there are other environmental concerns that have to be considered for these technologies. Fossil fuels are often the only available resource in many countries, and the choice of fuel very often depends on the availability and the economics of exploitation and transportation.

This conference deals with the technologies which will be required to recycle CO_2 (that is the capture of CO_2 for utilization and the disposal of CO_2). The current costs for capturing CO_2 range from 35 to 114 $/t CO_2 (absorption and adsorption, respectively). The global potential for utilization of CO_2 in chemical manufacturing and for enhanced oil recovery is estimated with 0.2-0.5 GtC/year. ABB Lummus Crest has developed CO_2 recovery technology from Kerr-McGee and applied in a chemical plant which concurrently operates as a power generation plant at Shady Point in Oklahoma.

There is enough potential world-wide to dispose of or store CO_2 in gas wells or in the ocean, although the costs are very high at present: the incremental electricity cost increase is 48 to 58%.

CO_2 can be reduced by fuel switching: a pulverized coal-fired power plant produces 0.83 kg CO_2 per kWh of electricity generated, whereas this number drops to 0.41 in a natural gas fired combined cycle plant.

In this paper we will argue that much more can be done now by <u>using best available technologies</u>, and more by using technologies in the future to prevent the formation of CO_2, by more efficient utilization of fossil fuels, be it in the combustion process itself, or be it in the transmission and end-use of electric energy.

CO_2 Prevention in Electricity Generation

Coal has been the most widely used fuel since the beginning of the industrial revolution. Even today 40 % of all electricity is generated from coal, however very often with a very low thermal efficiency around 30%. It might appear logical to apply the easy solution and to simply switch all coal plants to natural gas. For some countries this is a valid option. But others will have to rely on coal simply because
- coal is so available, for such a long time yet, and it is so evenly distributed all over the world,
- coal is so important for the national economies and foreign exchange balance of such giant countries such as China and India, but also Russia and the USA.

Coal, therefore, will remain the most important fuel for electricity generation. Several coal-fired technologies are available today (Pulverized Coal-fired Power Plants, Pressurized Fluidized-bed Combustion), which reach thermal efficiencies of 42 - 47%, and these values will climb to 50 - 52% by the year 2010. Best available technologies by then will be steam temperatures of 700° C , and requiring advanced high temperature materials (superalloys) currently only used in gas turbines. The cycle efficiency can be further increased with a working medium other than water (water - ammonia mixture as in the KALINA cycle).

The overall fuel-to-busbar efficiency of power plants can be increased by further increasing the top cycle temperature and by raising the cycle efficiency with reheat and/or regenerative features or by using heat exchangers with small temperature differentials and/or high specific heat transfer coefficients. The largest potential and the largest achievements towards high efficiencies are offered by **natural gas fired combined cycle plants**. Since the first introduction of industrial gas turbines in 1939 the inlet temperature has steadily climbed from 550° C to around 1250° C today. There is no reason why this trend should not continue. If the gas turbine gets coupled with a heat recovery steam generator and a steam turbine, then today thermal efficiencies of 55 to 58.5% are possible. This number could well reach values from 62-65% by 2010° C. ABB, some years ago, made a technological quantum leap with its GT 24/26 gas turbine family. This turbine makes use of the reheat concept, which means that in a 2-stage combustor the hot gas can work consecutively (staged) on two turbine stages connected to one rotor. By using the reheat concept higher efficiencies can be reached without going to higher turbine inlet temperatures. To reach the goal of 65% efficiency by 2010° C, a combination of the following technologies will be necessary

- reheat gas turbine
- higher gas temperatures
- further advanced high temperature materials and advanced cooling concepts
- reheat steam turbine.

To sum it up, best available technology will be available to raise the plant efficiency from 30% to 65%.

CO₂ Prevention in Transmission & Distribution

High voltage and medium voltage systems transmit electric energy with losses of about 15% today. Best available technologies in 2010 will have reduced these losses to 10% by
- making more use of transformerless HVDC transmission systems which reduce AC losses by 40-50%
- power systems control avoiding losses from local overloads.

CO₂ Prevention in Electricity Consumption

This presents by far the biggest savings potential. Three quarters of electricity load in industry is typically motor-driven. Two thirds of this load is connected with pumps, with ventilation fans, and with compressors for cooling and heating, ventilation and air conditioning. Much of the distributed electrical energy is today simply *lost in the electromechanical conversion, in mechanical throttling valves and in gear boxes*. Power electronics and variable speed drives, VSDs, are key technologies to avoid such energy-consuming throttling devices. ABB sells a growing number of the VSDs around the world. Many industry categories, e.g. pulp&paper, have already converted from constant-speed 50 or 60 Hz motor systems to variable speed systems based on lowest operating costs. Often the available power can thus be doubled compared with throttled systems, provided the whole system is optimized and not the motor alone.

CO₂ Prevention Potential in Optimized System

Let us assume we combine optimized end use systems making more intelligent use of electric energy, with highest efficient transmission of power, which in itself is generated in the most efficient fossil-fired power plant. If the inefficiently utilized application using inefficiently generated electric power is associated with an amount of 100% CO₂, then this amount can be brought down to a mere 35% by optimized end use of electricity and use of highest efficient power generation and distribution technology (Fig.2).

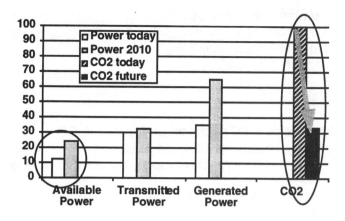

Fig. 2: CO2 Prevention potential in optimized system

Figure 2 also shows that we have to move towards a new paradigm which considers whole systems of energy generation and end use. Traditional electric utilities have been optimized to generate large amounts of electric energy in the most efficient way. Large industrial users of electric power and independent power producers are more likely to apply this systems approach.

Where could all these new technologies bring us to if they were not only available but also applied?

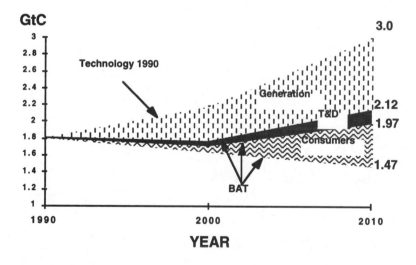

Fig. 3: Emissions from electricity generation and use

Figure 3 shows the CO_2 emissions due to electricity generation and use from 1990 to 2010. Assuming the same energy mix as in 1990 the CO2 emissions would increase from 1.8 to 3 GtC/year. Applying best available technologies this number can be reduced to 1.47 GtC/year.

Conclusions

- Only small market potential exists for captured and recycled CO_2.
- There is a large world-wide capacity for CO2 disposal.
- Disposal of CO2 would lead to an increase of electricity prices by 50 to 60%.
- Best available new technologies can reduce CO2 emissions from electricity generation and use by 30 to 50%.

Pergamon

Energy Convers. Mgmt Vol. 38, Suppl., pp. S31–S35, 1997
© 1997 Elsevier Science Ltd. All rights reserved
Printed in Great Britain
0196-8904/97 $17.00 + 0.00

PII: S0196-8904(96)00241-5

TECHNOLOGY RESPONSES TO GLOBAL CLIMATE CHANGE CONCERNS: THE BENEFITS FROM INTERNATIONAL COLLABORATION

JOHN TILLEY

International Energy Agency
75775 Paris Cedex 16, France

Mr Chairman, Ladies and Gentlemen,

It is a pleasure to speak on behalf of the International Energy Agency, at this 3rd conference in the ICCDR series. The conference is very timely. Governments from around the world have just met for the second time at Ministerial level to discuss progress in implementing the UN Framework Convention on Climate Change. They also discussed possible next steps, post-2000, in addressing the climate change challenges.

This conference provides a very valuable opportunity to take stock of activities in a potentially important area of climate change responses - CO_2 capture, disposal and utilisation. I am pleased to see that the program also sets out to examine the question "where to next?" In the IEA's view this must include an assessment of how to strengthen basic and applied research in this field, and an examination of what are the priority areas.

In this presentation, I will address some of the key factors which the IEA believes will set the framework for future action in the energy sector to respond to the climate change challenge. I will concentrate on the technology dimension, the factors influencing technology choice and the role of international collaboration in enhancing technology progress. © 1997 Elsevier Science Ltd

Energy Sector Emissions

The IEA's World Energy Outlook projections illustrate how energy use will grow in the absence of effective policies to alter established patterns of energy production and use. CO_2 emissions from energy in OECD countries could rise between 20 and 30 percent above 1990 levels in 2010: those from countries outside the Annex I group of countries could rise between 90 and 150 percent.

In fact, whatever assumptions we made about economic growth, energy prices and energy efficiency, emissions projections in our Outlook rose substantially. In the simplest terms, this message confirms that the world's economy is highly geared to the use of fossil fuels.

The work underlying the IEA model does suggest that there is room for policies that could result in faster than expected efficiency improvements, which would reduce the rate of growth in emissions.

But, even theoretical "no regrets" policies, when viewed against the projections of the IEA's World Energy Outlook, will not be adequate to stabilise, much less reduce, energy-related CO_2 emissions in the OECD by 2010.

If more is to be done to reduce greenhouse gas emissions, costs will clearly have to be incurred.

At the same time there must be realistic expectations which do not disregard the inherent rigidities in the energy system. Realistic opportunities which are properly defined will vary widely from country to country, even within the OECD.

In the short to medium-term this means faster deployment of existing energy technologies which emit fewer greenhouse gases, as well as those which use energy more efficiently. Additional policy measures will, however,be needed to enhance the market opportunities for many of these options under current energy prices.

However, there are limits to potential energy efficiency improvements and many fuel switching options; more innovative solutions will be needed in the longer term if the current goals of Climate Change Convention are to be met.

Factors Influencing Technology Choice

Apart from cost and reliability, technology choice including longer term options, will be influenced by a number of factors. The features of the energy scene which determine the level and pattern of energy-related greenhouse gas emissions include the structure and pattern of energy demand; the fuel mix; the technology mix; the age and turnover rate of capital stock; import and export balances. These will all impact on the attractiveness of technology options. Since these elements also differ substantially between countries, we can also expect to see different technology choices and priorities at the national level.

The nature of the current infrastructures for delivery of services will also be critical. Infrastructures such as buildings, road networks and energy grids embody the technology choices and the patterns of energy use of an immense web of economic agents. For the most part, such choices did not systematically take into account their energy implications, still less their environmental implications. But, they will shape our way of life and patterns of energy use for decades. They will also influence our selection of longer term technologies.

The lengthy lifetimes of plant and equipment and the long lead times for introducing new technologies will also have a significant impact on technology choice.

Carbon Dioxide Capture, Disposal and Use - The Challenges

It is in this context that CO2 capture, disposal and use remains an attractive option for the medium to longer term, particularly if the current pattern of energy supplies continues, based on the existing fossil fuel infrastructure and the reliability of associated technologies.

However, in the case of CO2 capture, disposal and use there are still significant challenges to be overcome in order to bring the technology into the market place as a commercial option. Most of the technology options for capture, storage and use result in considerable cost increases. As energy producers and consumers seek to reduce greenhouse gas emissions from their activities, market focus will favour the most cost-competitive solutions. While CO_2 capture, disposal and use has the potential to play a significant role, costs will need to come down.

Also, technology reliability and maintainability will need to be established - as with other emerging technologies. This means the full range of technology risks must be clearly understood as early as possible. The more significant risks must be reduced and the information communicated to potential customers for the technology. This will entail technology demonstrations.

The environment impacts will also need to be fully understood. This will involve assessment of localised impacts as well as the regional and global impacts, particularly in the case of the innovative, ocean disposal options. The results of these studies must also be communicated widely, so that decisions can be taken on a sound basis.

Benefits of International Collaboration

Further action will be required to mobilise and stimulate the substantial efforts needed in these areas.

While much of the core work on CO2 capture, disposal and use is undertaken at the national level, international mechanisms are playing an important role in enhancing technology progress and in improving awareness of the CO2 capture, disposal and use options. Continuing international collaboration will be needed to bring these technologies successfully to the market.

The ICCDR conference series is one such vehicle to link R&D efforts and disseminate R&D results. So too, is the IEA Implementing Agreement on the GHG R&D Programme.

In the IEA's view we can expect considerable benefits from such international collaboration.

Collaboration enable participants to share the costs of projects. Countries and groups of researchers can pool their R&D resources to achieve a better outcome than could be achieved alone. In many cases sharing costs and pooling resources has accelerated the development and deployment of new technologies whilst reducing the R&D costs of each participant.

International collaboration reduces the duplication of activities, reducing costs and speeding progress, for example by clarifying methodologies and facilitating the checking of research results. Independent reproduction of results is often not economically feasible for expensive demonstration plants and even for many bench-top experiments. Furthermore, ongoing international collaboration means that research results can be better disseminated because, for example, important tacit knowledge is better transferred or because methods and reporting procedures are standardised.

International collaboration also provides a mechanism for hedging bets on technically risky research projects, which is especially important for expensive longer term research. Collaboration provides an alternative to attempting to go it alone on all fronts, or to choosing between uncertain alternatives. This means that a number of paths can be collectively pursued in depth.

The cost of such collaboration is, of course, that an organisation or a country may forgo an uncertain chance of gaining an advantage over competitors. The further the technology is from the market place the less concern there is likely to be about this potential cost of collaboration. On the contrary, it is often seen as mutually beneficial to collaborate to speed up the diffusion of information about scientific advances and innovations.

If, in the course of your discussions, you identify areas of R&D which would benefit from international collaboration, the IEA through its programme of collaborative research, offers a flexible mechanism for interested players to pool their scarce R&D resources for their mutual benefit, and for the long-term benefit of industry and consumers alike. We are at your service.

Two such collaboration programmes are already in place in the field of CO2 capture, disposal and utilisation. I would like to mention them briefly.

IEA Technology Collaboration

The IEA Greenhouse Gas R&D Programme has the objective to evaluate (on a full fuel cycle basis) the technical and economic feasibility and environmental impacts of technologies for the abatement, control, utilisation and disposal of carbon dioxide and other greenhouse gases derived from fossil fuel use. Leading from this, another of the goals of the programme is to identify targets for R&D in this field and to facilitate practical activities. The programme is also encouraging a more broadly based use of a systems approach to assessment of greenhouse mitigation options, and is examining ways to help identify technologies suitable for application under AIJ projects. You will hear more about the work of this group during the Conference.

Climate Technology Initiative

Another technology collaboration vehicle which has been established by the IEA is the Climate Technology Initiative. This programme is helping to support the technology related aspects of the Climate Change Convention. You can find out more about the CTI on the IEA's Homepage (Http://www.iea.org).

One of the seven elements of the CTI focuses on greenhouse gas capture and disposal, including the role of capture and disposal options as part of a hydrogen fuel chain based on fossil fuels. This task includes an assessment of the feasibility of developing longer-term technologies in these fields and ways to strengthen relevant basic and applied research.

The CTI is designed to promote national and international policies that can improve the framework for science and technology delivery systems. It will contribute to meeting climate change objectives through the demonstration of new and improved technologies, and through national and international efforts of governments and industries to strengthen technological progress.

The Initiative has a short and a long term focus. The shorter term focus is on enhancing markets for currently available climate friendly technologies. The longer term focus is on stimulating the research, development and diffusion of new and improved technologies that can contribute to meeting the FCCC goal of stabilising concentrations of greenhouse gases in the atmosphere.

In particular, the CTI aims to:

- promote awareness of technology related activities already underway to assist with responses to climate change concerns;

- identify and share expertise and experiences between countries already working on particular topics, as well as with countries having limited expertise in particular areas;

- identify gaps in national and multilateral technology programmes which could be addressed in order to strengthen climate response strategies; and

- strengthen and undertake practical collaboration activities between countries to make technology responses to climate change concerns more effective.

Concluding Remarks

Mr Chairman, in summing up, although the climate change issue has been actively discussed in government and industry circles now for almost a decade, discussion is only now turning actively to market and sectoral realities. We are likely to remain highly dependent on fossil fuels for power, heat and transport; existing infrastructures will shape climate change responses for a considerable time. Innovative solutions will eventually be needed.

Technology will play an important role in achieving longer term greenhouse gas emission reductions. Capture, disposal and utilisation of greenhouse gases provides one potentially significant way to reduce greenhouse gas emissions.

Intensified efforts are needed to speed up the otherwise lengthy technology development and deployment process so as to realise those potential technology contributions. Enhanced international co-operation involving all of the players offers the opportunity to speed up that process by :

- sharing the costs of research, development and demonstration; and

- sharing the lessons learned and so avoiding costly replication of often unproductive R&D.

Increased awareness by decision makers in government and industry about the potential of the CO_2 capture, disposal and utilisation option, and about its reliability and acceptability, will be critical.

ICCDR plays an important role in that process.

Mr. Chairman, I wish you a successful conference.

SECTION 2

CO_2 SEPARATION AND RECOVERY

Energy Convers. Mgmt Vol. 38, Suppl., pp. S37–S42, 1997
© 1997 Elsevier Science Ltd. All rights reserved
Printed in Great Britain
0196-8904/97 $17.00 + 0.00

Pergamon

PII: S0196-8904(96)00242-7

RESEARCH AND DEVELOPMENT ISSUES IN CO$_2$ CAPTURE

AXEL MEISEN and XIAOSHAN SHUAI

Department of Chemical and Bio-Resource Engineering
2324 Main Mall
The University of British Columbia
Vancouver, BC Canada V6T 1Z4

ABSTRACT

The principal technologies for CO$_2$ capture (including absorption, adsorption, membrane separation, cryogenic separation) and for producing rich CO$_2$ process streams by CO$_2$/O$_2$ combustion cycles are reviewed in summary form and their present limitations are identified. The latter include uncertainties regarding their technical design, short- and long-term operational problems, and economic feasibility. In particular, the current lack of knowledge of physical, chemical and rate data, contact mechanisms, solvent, adsorbent and membrane stability, corrosion, fouling, and by-product generation are considered. These problems together with the need to improve capture rates, raise capture efficiencies and lower capture costs are then used for suggesting directions of future research and development.

© 1997 Elsevier Science Ltd

KEYWORDS

CO$_2$ capture, CO$_2$ separation, research and development.

INTRODUCTION

The role of carbon dioxide (CO$_2$) in global warming is one of the most important contemporary environmental issues and it is therefore necessary to have available technology which minimizes the discharge of CO$_2$ into the atmosphere. Amongst the anthropogenic sources of CO$_2$, electric power stations utilizing fossil fuels (especially coal and heavy hydrocarbons), petroleum refineries, natural gas plants and certain chemical plants are the largest single-point sources of CO$_2$ and therefore deserve particular attention (Riemer and Ormerod, 1995). In the aforementioned cases, the CO$_2$ is discharged into the atmosphere in the form of mixtures with other constituents (principally N$_2$, H$_2$O, O$_2$, CO, SO$_x$, NO$_x$ and/or particualtes) near atmospheric pressure and at elevated temperatures (typically well above 100 °C). The present paper therefore commences with a review of the principal CO$_2$ capture technologies, which may be suitable for such mixtures from the aforementioned industrial sources, and then describes combustion process changes yielding highly concentrated CO$_2$ exit streams which may not require further separation. Important gaps in knowledge are then identified to stimulate further research and development so that effective and economical technologies will be available when they are needed. Ultimate use or disposal of CO$_2$ are not subjects of this paper.

CO$_2$ CAPTURE TECHNOLOGIES

Absorption Processes

Chemical and/or physical absorption processes are widely used in the petroleum, natural gas and chemical industries for the separation (or capture) of CO_2. Chemical absorption is based on reactions between CO_2 and one or more basic absorbents such as aqueous solutions of mono-, di- or tri-ethanol amines, diisopropanol amine, sodium hydroxide, sodium carbonate and potassium carbonate.   Typical physical solvents are methanol, N-methyl-2-pyrrolidone, polyethylene glycol dimethylether, propylene carbonate and sulfolane.   Since absorption increases with pressure, operating pressures are dictated by CO_2 capture specifications and economics; in practice, operating pressures generally exceed several atmospheres. The CO_2-bearing gas is typically contacted counter-currently with the descending absorbent in a tower equipped with packings or plates.

A favorable characteristic of absorption is that, with the choice of proper absorbents, the absorption can be reversed by sending the CO_2-rich absorbent to a desorber (or stripper) where the pressure is reduced and/or the temperature is raised. The regenerated absorbent is then returned (after pressurization and/or cooling) to the absorption tower thereby creating a fully continuous process. The CO_2 is usually obtained in relatively concentrated form near-atmospheric pressure and it is saturated with water vapor. When acid gases other than CO_2 are present in the feed to the absorption tower, they leave the desorber together with the CO_2 unless they form heat-stable salts with the absorbent or special measures are taken to effect differential absorption of the acidic constituents.

Inherent drawbacks of chemical absorption processes are their limited CO_2 loadings and significant energy requirements resulting from the reaction stoichiometry and the heats of reaction, respectively.   Furthermore, they require extensive equipment for circulating large volumes of liquid absorbents and for heat exchange. Nevertheless, absorption is generally competitive for large-scale applications, especially when CO_2 occurs in high pressure mixtures having constituents which react reversibly with the absorbents. As stated before, flue gases are typically near atmospheric pressure and contain O_2, CO, SO_x and/or NO_x many of which are detrimental to absorbents.

There are presently only two major industrial installations in which CO_2 is captured from industrial flue gases by means of chemical solvents; they are the ABB Lummus Crest (1995) designs based on aqueous monoethanolamine solutions at Trona (California, USA) and Shady Point (Oklahoma, USA).   The plants are claimed to be working well but given their limited number and the purity of the flue gases, absorption is not yet well proven technology for the capture of CO_2 from flue gases.

Adsorption Processes

Adsorption processes are based on significant intermolecular forces between gases (including CO_2) and the surfaces of certain solid materials (such as molecular sieves and activated carbon). Depending on the temperature, partial pressures, surface forces and adsorbent pore sizes, single or multiple layers of gases may be adsorbed and the adsorption may be selective. The adsorbents are normally arranged as packed beds of spherical particles. In pressure swing adsorption (PSA), the gas mixture flows through the beds at elevated pressures and low temperatures until the adsorption of the desired constituent approaches equilibrium conditions at the bed exit. The beds are then regenerated by stopping the flow of the feed mixture, reducing the pressure and elutriating the adsorbed constituents with a gas having low adsorptivity. Once regenerated, the beds are ready for another adsorption cycle. By contrast, in temperature swing adsorption (TSA), the adsorbents are regenerated by raising their temperature utilizing a hot inert gas or external heating.   Riemer and Webster (1994) found that PSA was superior to TSA in all cases due to the high energy requirements and low speed of regeneration in TSA.

The inherent advantage of adsorption processes is their relatively simple albeit unsteady state operation.   Nevertheless, adsorption is not yet a highly attractive approach for CO_2 removal in the large-scale industrial treatment of flue gases because the capacity and CO_2-selectivity of available adsorbents is low.   There is, however, the possibility that adsorption may become attractive when combined with another capture technology.

Membrane Processes

Gas separation membranes.   Gas separation membranes are solids and operate on the principle that their porous structure permits the preferential permeation of mixture constituents.   The main design and operational parameters of membranes are their selectivity and permeability.   Membrane separation of CO_2 from light hydrocarbons has met with considerable success in the petroleum, natural gas and chemical industries because of the inherent simplicity resulting from steady state operation, absence of moving parts and modular construction (see Kesting and Fritzsche, 1993).   The CO_2-bearing gas mixture is introduced at elevated pressure into the membrane separator consisting typically of a large number of hollow cylindrical membranes arranged in parallel.   The CO_2 passes preferentially through the membranes and is recovered at reduced pressure on the shell side of the separator.

Gas separation membranes have thus far not been widely explored for CO_2 capture from flue gases due to the comparatively high mixture flows and the need for flue gas pressurization. Feron et al. (1992) found in their exploratory study that a two-stage system was needed to achieve good separation and that the costs were double those of conventional amine separation processes.   The principal cost component was compression energy.

Gas absorption membranes.   Gas absorption membranes consist of microporous solid membranes in contact with a liquid absorbent.   The gas component to be separated diffuses through the solid membrane and is then absorbed into and removed by the liquid absorbent. This arrangement results in independent control of gas and liquid flows and minimization of entrainment, flooding, channeling and foaming.   The equipment also tends to be more compact than conventional membrane separators.   Some investigators (Feron et al. 1995 and 1992, Nishikawa et al. 1995) considered gas absorption membranes for CO_2 capture from flue gases and found them to be promising but still requiring considerably more research.

Cryogenic Processes

Cryogenic separation of gas mixtures involves compressing and cooling the gas mixtures in several stages to induce phase changes in CO_2 and, in the case of flue gases, invariably other mixture components (US Department of Energy, 1993). Depending on the operating conditions (the CO_2 critical temperature and triple point are 31.1 °C and - 56.6 °C, respectively), the CO_2 may arise as a solid or liquid together with other components from which is may be distilled.   Water vapor in the CO_2 feed mixture leads to the potential formation of solid CO_2 clathrates and ice which, together with solid CO_2 particles, can result in major plugging problems.

The basic advantage of cryogenic processes is that, provided the CO_2 feed is properly conditioned, high recovery of CO_2 and other feed constituents is possible.   This may also facilitate the final use or sequestering of CO_2.   However, cryogenic processes are inherently energy intensive.

CO$_2$ / O$_2$ Combustion Cycles

The use of oxygen (or oxygen enriched air) significantly improves the combustion of fossil fuels (particularly coal) because it leads to higher combustion rates, higher combustion temperatures and therefore thermal efficiencies, lower thermal NO$_x$ formation and smaller combustion equipment since atmospheric nitrogen is absent or reduced. To moderate the combustion temperatures, part of the product gases which are rich in CO$_2$ can be recycled. Depending on the composition of the fuel and the ultimate disposition of the CO$_2$, the product gases may need to undergo further treatment including particulate, SO$_x$ and NO$_x$ removal (Nakayama *et al.*, 1992). In the case of pulverized coal combustion, drying of the recycle gas was also found to be a major consideration.

An important inherent disadvantage of the CO$_2$/O$_2$ combustion process is the need for oxygen which is still relatively expensive to produce from air. It has been estimated (US Department of Energy, 1993), that over 20% of the net electric power output of a conventional fossil fuel power plant would be needed for the production of oxygen.

R&D DIRECTIONS FOR CO$_2$ CAPTURE TECHNOLOGIES

The previous sections have identified the principal CO$_2$ capture technologies which appear to be promising for industrial applications. Although these technologies are basically feasible from a technical perspective, their efficiency, reliability, long-term performance and economics are still uncertain. Most technologies are expensive and their costs can only be reduced if the technologies are better understood. Important research and development issues will therefore be identified. All processes suffer from the fact that off gases from the principal industrial CO$_2$ sources (and particularly electric power plants) are near atmospheric pressure; in many cases they also contain particulates (such as fly ash). It is therefore highly desirable to investigate process modifications which result in high pressure CO$_2$ mixtures. Furthermore, particulate removal (especially at elevated temperatures) is essential for many situations. Both of these measures are important but fall outside the scope of this paper.

Absorption Processes

The stability of most chemical and physical absorbents for CO$_2$ mixtures containing O$_2$, CO, SO$_x$ and NO$_x$ is poorly understood (especially under conditions of elevated temperatures and long-term use). This is an important research problem which is amenable to solution by small-scale laboratory or pilot plant studies. If solvent instability is excessive, the principal reactants (likely O$_2$, SO$_x$ and NO$_x$) need to be removed prior to CO$_2$ separation. There are currently no mature technologies for doing so efficiently and economically. Since the formation of some heat stable salts and absorbent degradation products is inevitable and since they can lead to corrosion, fouling and/or foaming, processes are also needed for the continuous or intermittent purification of partially contaminated absorbents. Such processes are lacking. The safe disposal of the contaminants is another largely unsolved problem.

Since CO$_2$ capture at low pressure is more difficult than at high pressure and therefore inherently more expensive, it is essential to have available more precise data on the physical properties of solvents, reaction rates (to determine enhancement factors) and packing characteristics (to avoid flooding and minimizing pressure drops). In particular, more needs to be known about CO$_2$ desorption (stripping) from loaded absorbents since it strongly influences the process economics and absorbent degradation. Most emphasis has, thus far, been placed on absorbers rather than desorbers.

Basic and applied research on novel absorbents, absorbent reactions and packings is therefore needed.

Adsorption Processes

Low adsorption capacity and selectivity of currently available adsorbents are the principal impediments for making adsorption competitive with absorption. Since these are inherent deficiencies of conventional adsorbents, targeted research on developing new adsorbents for CO_2 is warranted but difficult to realize. There are no materials which seem to have high selective adsorptivity for CO_2 and which have not yet been investigated. However, continuous exploration of new high-temperature polymers may provide a breakthrough. In addition to adsorptivity, adsorbent stability to chemicals associated with the CO_2, temperature cycling and pressure swings need to be established. Optimization of operating conditions (i.e., pressure ratios, cycle times, step durations, etc.) and adsorbent structures (i.e., packing geometry, ordered laminated structures, etc.) deserve attention but are unlikely to improve adsorption processes sufficiently to make them highly competitive for large-scale CO_2 capture.

Membrane Processes

Conventional organic membranes suffer from low CO_2 permeability and some lack of selectivity. Furthermore, they are not well suited for high temperature operations (as required for the direct CO_2 capture from flue gases in electric power plants). There is consequently a strong need for developing metal, alumina and ceramic membranes. High temperature polymeric membranes, possibly supported by inorganic substrates, are another approach. The considerable pressure drops in the direction of main gas flow (as opposed to pressure drops across the membranes) are another serious impediment in the adoption of conventional membrane technology and may be overcome by more advanced membrane configurations.

Gas absorption membranes hold considerable promise provided their physical and chemical stability can be assured since they have higher selectivities and permeabilities than conventional membranes. Their chemical stability problems are similar to those already mentioned for absorption processes and concerted research on absorbent stability will therefore also benefit gas absorption membranes.

Cryogenic Processes

Cryogenic processes are inherently difficult to apply to CO_2 capture because CO_2 occurs in conjunction with other gases (e.g., H_2O, SO_x and NO_x) which severely interfere with cooling and cause corrosion, fouling and plugging. Furthermore, the phase behavior of CO_2 is complex and easily leads to the formation of solids which plug equipment and severely reduce heat transfer rates. In light of these limitations and the high cost of refrigeration, cryogenic processes can probably only be used in special circumstances and as an adjunct to other processes.

CO_2 / O_2 Combustion Cycles

Carbon dioxide capture from the combustion of fossil fuels using O_2 or O_2-enriched air will only be viable technically if flue gas recycle is not necessary or the particulate content of the recycle stream is low. Due to the higher combustion temperatures, combustion with O_2 or O_2-enriched air requires major changes in combustion equipment and materials; it is therefore particularly well suited for new installations. Alternatively, if flue gas recycle is necessary (as is likely in retrofits), the particulate loading of the recycle stream must be low either because the fuel has a low ash content or efficient high-temperature gas cleaning equipment is used. Such equipment is already under development to reduce particulate discharges into the atmosphere and CO_2 capture therefore provides an additional incentive for this work. If economic and ultimate CO_2-disposal considerations indicate that water vapor needs to be removed from the combustion gases (as suggested by Nakayama *et al.*, 1992), then CO_2/O_2 combustion cycles become more difficult since moisture gives rise to operational problems such as corrosion.

Before undertaking detailed technical research and development on CO_2/O_2 combustion cycles, it seems preferable to concentrate on economic studies of the type conducted by the US Department of Energy (1993) and Riemer *et al.* (1993).

CONCLUSIONS

Promising technologies for the large-scale capture of CO_2 from industrial point sources (such as the petroleum, natural gas, chemical and electricity generating industries) were assessed and absorption processes were found to be most practicable given the state of current knowledge and the existence of two industrial plants. However, major research and development are needed to determine particularly absorbent stability. Other technologies (adsorption, membrane separation, cryogenic separation and CO_2/O_2 combustion cycles) require major additional research which is identified. All CO_2 capture processes would be enhanced by making modifications in upstream processes so that the CO_2 mixtures are produced at elevated pressures.

REFERENCES

ABB Lummus Crest (1995), CO_2 Recovery From Flue Gases, ABB Lummus Crest Inc., 12141 Wickchester, Houston, TX 77079-9570, USA.

Feron, P.H.M. and Jansen, A.E. (1995). Capture of carbon dioxide using membrane gas absorption and reuse in the horticultural industry. *Energy Convers. Mgmt.*, **36**, No. 6-9, pp. 411-414.

Feron, P.H.M., Jansen, A.E. and Klaassen, R. (1992). Membrane technology in carbon dioxide removal. *Energy Convers. Mgmt.*, **33**, No. 5-8, 421-428.

Kesting, R.E. and Fritzsche, A.K. (1993). *Polymeric Gas Separation Membranes,* John Wiley & Sons, Inc. New York.

Nakayama, S., Noguchi, Y., Kiga, T., Miyamae, S., Maeda, U., Kawai, M., Tanaka, T., Koyata, K. and Makino, H. (1992). Pulverized coal combustion in O_2/CO_2 mixtures on a power plant for CO_2 recovery. *Energy Convers. Mgmt.*, **33**, No. 5-8, pp. 379-386.

Nishikawa. N., Ishibashi, M., Ohta, H., Akutsu, N., Matsumoto, H., Kamata, T. and Kitamura, H. (1995). CO_2 Removal by hollow-fiber gas-liquid contactor. *Energy Convers. Mgmt.*, **36**, No. 6-9, pp. 415-418.

Riemer, P.W.F., Webster, I.C., Ormerod, W.G. and Audus, H. (1994). Results and full fuel cycle study plans from the IEA greenhouse gas research and development programme. *Fuel* **73**, No. 7, 1151-1158.

Riemer, P.W.F and Ormerod, W.G. (1995). International perspectives and the results of carbon dioxide disposal and utilization studies. *Energy Convers. Mgmt.*, **36**, No. 6-9, 813-818.

US Department of Energy (July 1993). The capture, utilization and disposal of carbon dioxide from fossil fuel-fired power plants. DOE/ER30194.

Ref: AMe/ICCX1028.DOC (1996)

Energy Convers. Mgmt Vol. 38, Suppl., pp. S43–S44, 1997
Pergamon

PII: S0196-8904(96)00243-9

0196-8904/97 $17.00 + 0.00

CO$_2$ abatement investigations under the Framework Programme of the RTD policy of the EU, in particular under the Joule II - Programme (1993 to 1995)

R. Pruschek
Universität Essen
Universitätsstr. 2-17
D-45141 Essen, Germany

OUTLINE OF THE JOULE II PROGRAMME

The JOULE II Programme is one of the 15 specific programmes of the third Framework Programme of the RTD (Research and Technological Development) policy of the European Union. Area II of the JOULE II Programme, "Minimum emission power production from fossil fuel sources", deals with the "Minimisation of CO$_2$ Emissions" and with "Advanced Clean Coal Technology R&D".

In this Area, two integrated projects, the "COMBINED CYCLE PROJECT" and the "POWDER COAL COMBUSTION PROJECT", have been set up to run for three years (1993 to 1995) in order to promote more efficient coal to electricity conversion. The COMBINED CYCLE project deals with coal gasification and coal combustion with air, oxygen enriched air and pure oxygen with flue gas recirculation, as applied to combined cycles. The POWDER COAL COMBUSTION project deals with advanced combustion techniques, both atmospheric and pressurised. Scientists from eleven European countries were involved in the Combined Cycle Project and from ten countries in the Powder Coal Combustion Project. Leading European coal research institutes, industry and academic institutions have participated in the research programme.

The work carried out under the JOULE II Programme mainly addressed advanced, clean and efficient processes for environmentally sustainable fossil fuel conversion, specifically in relation to greenhouse gas emissions. The following objectives were pursued:

- Development of combined cycles based on advanced coal gasification and coal combustion technologies.
- Development of hot gas cleaning systems for fuel gases and for flue gases.
- Assessment of simultaneous emission profiles for flue gases (such as N$_2$O, unburnt hydrocarbons, etc.) and development of methods for decreasing the emission of ozone precursors from fossil fired plant.
- Increase of the overall efficiency of conventional and combined cycle technologies.
- Development of fossil fuelled power plants with minimum emissions of all pollutants and safe and stable disposal of CO$_2$.

The last item includes the projects covering CO$_2$ abatement or CO$_2$ removal, respectively.

PROJECT ORGANISATION

A total of fourteen contracts have been awarded and these have been organised into two related Projects. Seven of the contracts have been grouped together to be coordinated as the POWDER COAL COMBUSTION PROJECT and another seven contracts form the COMBINED CYCLE PROJECT. These projects are funded jointly by the participants and the European Commission. The Powder Coal Combustion Project has a total budget of 7.6 MECU, of which the EC funds 4.5 MECU, and for the Combined Cycle Project the figures are 12.4 and 7.5 MECU, respectively. The total commitment is therefore 20 MECU.

Eight contracts directly deal with CO$_2$ removal concepts (see Fig. 1).

Three contracts of the Powder Coal Combustion Project related to CO$_2$ abatement were coordinated by the International Flame Research Foundation (IFRF) in the Netherlands: JOU2-0062, entitled "Pulverised coal combustion systems for CO$_2$ capture"; JOU2-0093, "IFRF carbon dioxide concentration research programme"; and JOU2-0220 "Coal combustion in advanced burners for minimal emissions and carbon dioxide reduction technologies".

Another contract, JOU2-0153, coordinated by Deutsche Montan Technologie (DMT) in Germany, dealt with "Evaluation of pressurised coal combustion with CO$_2$ recirculation".

European Commission JOULE II - Programme: " Clean Coal Technology R&D"

POWDER COAL COMBUSTION PROJECT

JOU2-0062	**Atmospheric Pulverised Coal Combustion - O₂/CO₂ Firing**
JOU2-0093	11 Partners: Babcock Energy Ltd.(UK), Air Products PLC (UK/DE), University of Ulster (UK), University of Naples (I), International Flame Research Center (NL), Deutsche Vereinigung für Verbrennungsforschung (DE), British Flame Research
JOU2-0220	Committee (UK), Comité Francais de la FRIF (FR), Comitato Nazionale Italiano per la Ricerca Sulle Fiamme (IT), Nederlands Vereninging voor Vlamonderzoek (NL), Riso National Laboratories (DK)

JOU2-0153   **Pressurised Pulverised Coal Combustion - O₂/CO₂ Firing**
4 Partners: Ruhruniversität Bochum (DE), Technische Universität Braunschweig (DE), National Technical University of Athens (GR), Institut Francais du Petrole (FR)

COMBINED CYCLE PROJECT

JOU2-0185   **IGCC + CO Shift, Phys. Scrubbing, Membranes** ★
4 Partners: Siemens (DE), Universität-GH Essen (DE), ECN (NL), IEN (PL)
JOU2-0158   **IGCC + CO Shift, Membrane Reactor (WIHYS)** ★
5 Partners: ECN (NL), Siemens (DE), Universität-GH Essen (DE), University of Limerick (IR), NTU Athens (GR)

JOU2-0128 **IGCC, REVAP - O₂/CO₂ Firing**	(REVAP=Recuperative Evaporative Cycle)
7 Partners: Vrije Universiteit Brussel (BE), RWTH Aachen (DE), ALCE SC. (BE), University of Liege (BE), Chem. Eng. Inst. (GR), Cranfield Inst. Tech. (UK), Royal Inst. Tech. (S)	

JOU2-0154   **Hybrid Combined Cycles with PFBC - O₂/CO₂ Firing**
6 Partners: DMT (DE); RWTH Aachen (DE), TU Delft (NL), ★ British Coal CRE (UK), University of trento (IT), University of Ulster (UK)

★Presentations at ICCDR-3

Fig. 1: Topics of the contracts dealing with CO_2 removal within the JOULE II - Programme "Clean Coal Technology R&D".

There are four more contracts dealing with CO_2 removal within the scope of the Combined Cycle Project. Two of these in the Project Area "IGCC With CO_2 Removal" have been coordinated by Universität Essen in Germany: JOU2-0185, entitled "Coal-fired multi-cycle power generation systems for minimum noxious gas emission, CO_2 control and CO_2 disposal" and JOU2-0158 "An attractive option for CO_2 control in IGCC systems: Water gas shift with integrated hydrogen / carbon dioxide separation (WIHYS) process". A predesign study of an IGCC with CO_2 control by CO_2 Rectisol wash and combined cycles incorporating syngas/oxygen firing with CO_2 recycling was carried out. In addition, an integrated hydrogen/carbon dioxide separation process including development of ceramic (catalytic) membrane materials and catalysts for the water gas shift reaction was under development (WIHYS).

A single contract JOU2-0128, entitled: "Efficient CO_2 capture through a combined steam and CO_2 gas turbine cycle" forms the Cycle Development Project Area, coordinated by University of Brussels, Belgium. The project covers investigations on CO_2/steam gasification in a fluidised bed gasifier, basic research work on CO_2/steam gasification in a thermobalance, studies on CO_2/steam gas turbines as well as CO_2/steam argon gas turbines; CO_2 capture by separation from the exhaust stack; fuel preparation; direct recovery of CO_2; operation of a test rig; burner design and burner diagnostics; as well as cycle diagnostics.

A further single contract JOU2-0154, entitled: "Hybrid combined cycles with pressurised fluidised bed combustor" forms the Combustion Project Area of the Combined Cycle Project, which is coordinated by DMT, Germany. The main objectives of this Project Area were development of concepts, comprising cycle schemes with a coal fired pressurised bed combustor and featuring an afterburner for maximum efficiency and minimum carbon dioxide emission. Two approaches have been followed by using existing gas turbines only as well as longer term concepts using recycled flue gas/oxygen mixtures as oxidant in order to facilitate removal of CO_2 from the flue gas.

The Clean Coal Technology Area was extended in the course of the JOULE II Programme, and three more Projects dealing with:" Atmospheric combustion of pulverised coal based blends for power generation"; "Integrated hot fuel gas cleaning for advanced gasification combined cycle processes"; and "Novel approaches in advanced combustion (Pressurised Systems)", respectively, were set up. In these projects also the cocombustion of biomass with coal as a further means of CO_2 emission reduction was considered.

The results of the work carried out under the quoted contracts will be published in the final report on the JOULE II - Programme.

Energy Convers. Mgmt Vol. 38, Suppl., pp. S45–S50, 1997
© 1997 Elsevier Science Ltd. All rights reserved
Printed in Great Britain
0196-8904/97 $17.00 + 0.00

Pergamon

PII: S0196-8904(96)00244-0

DEVELOPMENT REQUIREMENTS FOR ABSORPTION PROCESSES FOR EFFECTIVE CO₂ CAPTURE FROM POWER PLANTS

Eur. Ing Dr Colin L Leci

ICEL Developments Ltd, 15 Fairview Way, Edgware HA8 8JE Middx UK

ABSTRACT

Effective commercial liquid absorption technology currently applicable for CO_2 capture from power plant flue gases uses a conventional chemical solvent, MEA(monoethanolamine), which was developed over 60 years ago as general non selective solvent to remove acid gases from natural gas streams. The application to CO_2 capture from almost atmospheric pressure flue gases required modification of the technology on account of the oxygen content to incorporate inhibitors to resist solvent degradation and equipment corrosion, as well as upstream treatment of the flue gases to remove both SO_x and NO_x present. The solvent strength used by such processes is relatively low in comparison with high strength modern chemical solvents. The resultant effect is large equipment and high regeneration energy requirements which is a parasitic load on the power plant. Using current technology for CO_2 capture from flue gas derived from conventional pulverized coal fired power stations increases the nameplate capacity with a consequential increase of 50% in electricity prices. The development requirements to replace the current available technology were identified by sensitivity analysis as higher solution concentration which results in smaller equipment size and a lower thermal regeneration energy demand and consequential reduction in the overall power plant nameplate capacity. © 1997 Elsevier Science Ltd

KEYWORDS

CO₂ Removal, Flue Gas, Solvent Development, MEA, Sensitivity Analysis, Solution Concentration, Parasitic Load

INTRODUCTION

A sensitivity analysis based on the criteria adopted by the IEA Greenhouse Gas R & D Programme (IEAGHG., 1992) was undertaken to identify the key parameters for the major development requirements of current commercial absorption technology for CO₂ capture from flue gases based on the MEA solvent model.

Numerous studies (Riemer., 1993) have been performed to determine the effect of CO₂ capture on the operating economics of conventional pulverised fuel power stations; these have identified substantial increases in levelised generating cost and specific investment cost. It was shown that the levelised generating cost, incorporating CO₂ removal, result in an increase of 40-80% over conventional plant, investment cost rising by 74%, Tables 1 and 2.

The capital cost increase relative to a conventional plant of 74.5%, is the sum of the CO₂ capture plant (46.5%) the SO₂ removal plant (11%) (to protect the CO₂ system) and the increase in nameplate capacity (17%), the latter resulting from the parasitic load on the power plant by the CO₂ capture plant.

Operational and maintenance power plant levelised cost together with the additional fuel cost increased by approximately 27%. The contribution of the CO_2 capture plant fixed and variable cost, represent 6% and 2.6% respectively of the total levelised cost, making a total contribution of 9%. It is abundantly clear that there are significant cost and energy savings to be made by improving both solvent morphology and equipment. Additional benefits will also be gained by such an improvement resulting in an overall reduction of the parasitic load on the overall power plant nameplate capacity.

MAJOR VARIABLES INFLUENCING OPERATING COST OF CO_2 CAPTURE PLANT.

The breakdown of CO_2 capture plant operating costs (Leci., 1992) attribute 67% to steam requirements for solvent regeneration and reclamation. In order of magnitude this is followed by makeup chemicals 15% (a quarter of the steam cost), cooling water 8%, power required to overcome the additional pressure drop caused by treating the flue gas 5% and solvent circulation power 3%.

BASIS OF SENSITIVITY ANALYSIS

This analysis utilized the identical economic models derived for the earlier reported studies (Riemer., 1993, Leci., 1992) for the overall plant export capacity of 500 MW$_{so}$. It focuses on the impact of reducing the MEA solvent circulation rate by increasing the solvent concentration. The MEA plant configuration is completely identical, including the use of an upstream cold water scrubber which not only removes the NO_x and SO_x but also a substantial quantity of water contained in the flue gas. The operational temperature of the plant ensured a maintained water balance.

Substantial savings appeared possible by focusing attention on two major areas. The first relates to a reduction in steam, cooling water and power consumption which are a function of the solvent circulation rate which is dependent on the CO_2 flue gas partial pressure. The second area relates to makeup chemicals where both higher solvent degradation resistance is necessary and carryover losses must be minimised. This can be achieved using less sensitive solvents which also exhibit low vapour pressures coupled with improvements in equipment design. It is not part of the analysis to deal with this latter issue.

SENSITIVITY ANALYSIS

Effect of Increased Solution Concentration

Increasing solvent strength from the conventional 30% wt MEA in aqueous solution to 100 %wt reduces solvent circulation rate by 70% Fig 1. Above 70% wt the gain becomes less significant with increasing circulation. Current commercial restrictions on using higher MEA concentrations are due to both excessive corrosion and solvent chemistry.

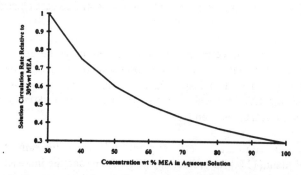

Fig .1. Decrease in circulation rate with increasing concentration

Total CO_2 capture plant operating costs decrease substantially with increasing solution strength. Overall, the operating cost decreases from US$ 94m at the 30% base case to US$ 46m at 70% wt. The low pressure (LP) steam contribution decreases from 66% for the base case to 58% at 70% wt, Fig.2, as the overall operating costs halve. Less significant in monetary terms, but following the same pattern as expected, is exhibited by cooling water. As the percentage contribution to operating cost by LP steam and cooling water decrease, that of makeup increases from 15.7% for the base case to 19.6% at 70% wt, whilst the percentage contribution of the blower power, also increases, from 5.3% to 10.7%, Fig 3 . In terms of relative variation in operating costs with increasing solvent concentration, it is clear that substantial savings are brought about in steam, cooling water and pump power, by increased solution strength. No significant saving occurs relative to makeup.

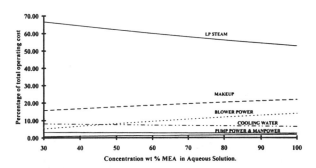

Fig.2. Variation in percentage operating cost for CO_2 removal from flue gas with increasing solution concentration

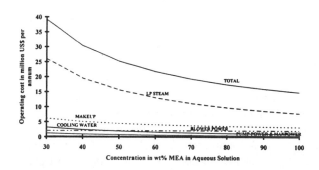

Fig.3. Variation in operating cost with increasing solution concentration

The total decrease in operating cost is unfortunately marginalised due to less significant reductions in make up costs Fig 4. With increasing solution concentration, for the same power plant export capacity, a decrease in both the overall equipment size and number of parallel streams occurs. However, the decrease in total installed cost (TIC) as a relative percentage of the base case CO_2 plant , although significant, does not match that of the operating costs. At 70% wt. the TIC decreases by 40% over the base case Fig 5, whilst operating costs decrease to 49% Fig 4.

Clearly, substantial benefits are to be gained by developing a solvent which can operate without degradation, corrosion, etc., at high concentrations. As alluded to earlier, development must be focused on investigation into suitable alternative organic solvents which can be used at very high concentrations, similar to physical solvents, but reacting in a chemical manner which have a very low heat of absorbtion and regeneration. This analysis has shown, that the regeneration energy is a prime factor in the operating costs of such plants. Not only will gains be made in reduced equipment cost as the size of the CO_2 capture plant decreases, but also a substantial reduction of the parasitic load on the power plant to supply both power and steam to the CO_2 capture plant occur with the associated reduction in both the overall power plant nameplate capacity and its operating costs.

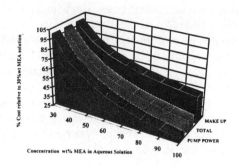

Fig.4. Relative variation in operating cost with increasing solution concentration

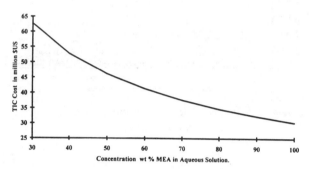

Fig.5. Decrease in CO_2 capture plant capital cost with increasing solution concentration

Overall Effect On Power Costs

The specific investment costs for a 500 MW_{so} coal fired pulverized fuel power station, were determined for both 50% and 70%wt MEA cases and compared with conventional 30% MEA base case as previously reported Table 1. The data takes into consideration appropriate fees, land purchase, site preparation and contingencies, to arrive at a total capital required for the total plant as in the original study. The total capital requirement has then been divided by the export plate capacity to arrive at a specific investment cost. At 50% wt MEA, the power plant name plate capacity is 615 MW, a substantial reduction over the base case requirement of 700 MW. The parasitic load equivalent on the power plant decreases from 200 MW to 115 MW, a reduction of over 40%. The specific investment cost at 50% wt is $US 1462 per KW_{so}, a 38% increase over the conventional zero CO_2 capture condition, and only 79% of that required for 30% wt MEA. In terms of capital cost, the overall capital investment was 38% more than the conventional zero CO_2 capture case and only 79% of the 30% MEA base case.

Table 1. Specific Investment Cost variation with MEA solvent concentration

	No CO₂ Capture(Riemer., 1993, Leci., 1992)	CO₂ Capture with 30% wt MEA	CO₂ Capture with 50% wt MEA	CO₂ Capture with 70% wt MEA
Power Plant Nameplate MW	500	700	615	584
Specific Investment Cost ($/kW$_{es}$)	1058	1842	1462	1348
Overall Capital Investment Cost $US million	528.9	921.2	731	674
Percentage increase in overall capital cost	0	74.2	38.2	27.4
Power Plant $US million	464.2	617.5	552.9	529.4
Percentage increase in Power Plant cost	0	33	19	14
CO₂ Capture Plant $US million	-	246.3	134.3	102.3
Percentage decrease in CO₂ Capture Plant cost relative to 30%wt MEA	-	0	45.5	58.5
SO₂ Capture plant $US million	64.7	56.5	43.9	42.3
Percentage decrease in SO₂ Capture Plant cost relative to 30% MEA	-	0	22.3	25.1

Table 2 highlights the overall significance on power production costs. To avoid substantial penalties in achieving global CO₂ reduction from conventional power plant using absorption processes higher solvent concentrations are essential.

Table 2. Levelised Generating Cost in mills/kWh$_{so}$

	No CO₂ Capture (Riemer., 1993, Leci., 1992)	CO₂ Capture with 30% MEA	CO₂ Capture with 50% MEA	CO₂ Capture with 70% MEA
Total Discounted Cost	53.4	80.4	66.8	61.9
Percentage increase relative to zero CO₂ capture	0	50	25	16
Decrease relative to CO₂ capture with 30% MEA	-	0	17	23

The most significant factor emerges that by maintaining higher solvent concentrations substantial power cost reduction is achieved as the equipment for the CO₂ capture plant becomes smaller in size and the parasitic load of the CO₂ capture plant for steam and power on the overall power plant capacity reduces, which in turn reduces fuel demand and hence the generated quantity of CO₂ to be removed.

MEA was used for this analysis primarily because it is currently the only commercial solvent capable of being used even though it is known as having a very high regeneration energy compared with higher series amines, such as MDEA. Regeneration energy requirements for 50% wt MDEA are less than half of those of a 30% wt MEA (Brown., 1982). The requirements of 50% wt MEA solution are only 40% lower than a 30% wt MEA, hence an additional 10% saving on regeneration energy could be achieved, if possible, by switching solvents with the associated knock on effects as indicated earlier. Indeed, apart from physical solvents, no alkanolamines solvents currently operate at concentrations greater than 50%wt.

OVERVIEW

It has clearly been illustrated that the most significant area for future R & D for solvent absorption processes for effective CO₂ capture from conventional pulverised fuel power stations, is that of the development of a solvent which can be used at flue gas low operating pressures at high solution concentration. The significance being twofold in that not only are the utility requirements for such a stand alone plant considerably reduced, but more significant, is that the parasitic demands made on the power plant in terms of both power and steam by the former is drastically reduced. The inter-government conference in August 1996 on global warming stated that the targets previously set to come into force within the next twenty years for CO_2 reduction, will not be met. The sensitivity analysis has clearly indicated that to reduce the greenhouse effect of CO_2 extensive efforts are urgently required to develop new improved solvents processes which operate at high concentrations with low thermal energy requirements that could be rapidly retrofitted to existing conventional power generation plants to meet the desired inter governmental targets.

REFERENCES

Brown F.C. and Leci C.L.(1982). *Criteria for Selecting CO₂ Removal Processes* - Proceedings No 210 The Fertiliser Society, London

IEAGHG (1992) *Document 9/13*

Leci.C.L and Goldthorpe.S.A. (1992). Assessment of CO_2 removal from power station flue gas.   *Energy Convers. Mgmt.* 33 477-485

Riemer.P. (1993). *The capture of carbon dioxide from fossil fuel fired power stations.* Restricted Circulation IEAGHG/SR2

Energy Convers. Mgmt Vol. 38, Suppl., pp. S51–S56, 1997
© 1997 Elsevier Science Ltd. All rights reserved
Printed in Great Britain
0196-8904/97 $17.00 + 0.00

Pergamon

PII: S0196-8904(96)00245-2

CO$_2$ CAPTURE PROCESSES -
OPPORTUNITIES FOR IMPROVED ENERGY EFFICIENCIES

Amit Chakma

Environmental Systems Engineering, Faculty of Engineering

The University of Regina, Regina, Saskatchewan, Canada

ABSTRACT

CO$_2$ capture from flue gas streams and disposal into geological formations has been considered as a technically feasible but a costly option for the reduction of CO$_2$ emission into the atmosphere. CO$_2$ capture is the major cost component. Therefore, there is considerable incentive in finding energy efficient and thus less costly processes for the capture of CO$_2$ as compared to the conventional monoethanolamine (MEA) based processes. In this paper, some strategies for reduced energy consumption in a chemical solvent based separation process are identified and their impacts on the overall process are discussed. © 1997 Elsevier Science Ltd

KEYWORDS

CO$_2$ Capture, Energy Efficiency, Mixed Solvent, Enhanced Stripping

INTRODUCTION

Among the several options avaialable for the removal of CO$_2$ from flue gas streams absorption by chemical solvents is by far the most popular one. However, other options may also provide some attractive alternatives to the chemical absorption process. For example, a study by MIT (1989) found that O$_2$ combustion combined with flue gas recycling may be a cheaper option. So far nearly all the feasibility studies have considered "off-the-shelf" MEA based technologies for CO$_2$ removal from flue gases and concluded that CO$_2$ removal and disposal would be very expensive (Smelser et. al, 1991; Padamsy and Railton, 1993). The implication of these studies are that CO$_2$ removal is not a cost effective option. However, conclusions based on the adaptation of "off-the-shelf" technology suitable for natural gas processing for flue gas separation with little modification should not be accepted as the "final word" in this matter. A number of process modifications can be made which could reduce the CO$_2$ separation cost. Some of these options are described in the following sections of this paper.

OPTIMIZED PROCESS DESIGN

A simplified flowsheet of a chemical solvent based CO_2 separation unit is shown in Figure 1. The flowsheet is a slightly modified version of a typical MEA unit for the treatment of flue gas streams. However, there are a few exceptions. First, is the presence of a pre-contactor between the blower and the absorber, and second, there are two steam injection lines in the stripper.

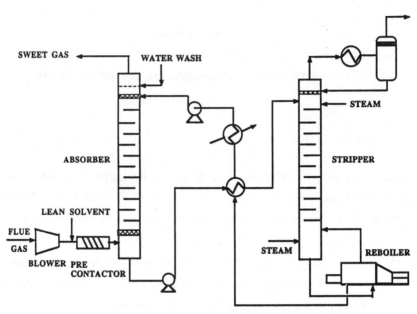

Figure 1:   A simplified flowsheet of an optimized CO_2 separation unit.

Pre-Contactor

The pre-contactor is an on line mixer where the flue gas stream is mixed with lean solvent solution prior to its entrance into the bottom section of the absorber. Its main role is to reduce the height of the absorber by absorbing the bulk of the CO_2 from the flue gas stream. Lean solvent is injected into the flowing gas stream upstream of the pre-contactor through a high efficiency nozzle in the form of fine droplets. Both the flue gas stream and the solvent droplets flow co-currently. Intimate mixing of the amine droplets and the flowing flue gas stream occur during their passage through the pre-contactor. A properly designed pre-contacting system can easily provide separation equivalent to half an equilibrium stage to a full equilibrium stage. This implies that the number of actual stages required can be reduced by 5 to 10 stages.

Since the fine amine droplets are now loaded with CO_2, their carry-over into the main contacting section of the absorber column can have serious negative effects on the separation process. There is a need to place a mist elimination device at the bottom of the absorber. The mist eliminator adds to the overall pressure drop and thus increases the workload of the blower.

Live Steam Injection

The flowsheet shown in Figure 1 also indicates external steam injection at two points, one at the bottom, and one at the top of the regenerator tower. The purpose of live steam injection is two fold. First, it enhances the stripping process by providing additional heat and stripping vapor further to those generated in the reboiler, and secondly, it reduces the solvent degradation rate by reducing the acid gas loading of the solvent before it enters the reboiler. The net effect is that the cyclic capacity of the solvent, expressed as the difference in the acid gas loading of the lean and the rich solvent increases, thus reducing the solvent circulation rate for a given separation duty. Typical cyclic capacity of MEA for a standard design is about 0.2 mole of CO_2/mole of MEA. With live steam injection this can be increased to over 0.3 mole of CO_2/mole of MEA, which represents a reduction of solvent circulation rate in excess of 50%.

MIXED SOLVENTS

The use of mixed solvents in the natural gas processing industry is now very common. Many of the older plants using conventional solvents have been converted to mixed solvents either to increase the throughput of the plant or to solve operations problems such as corrosion. However, in many cases the switch over to mixed solvents has also led to considerable energy savings. Most of the proprietary solvents marketed by the major solvent manufacturers are based on mixed amines. By judicious choice of an amine mixture, process efficiency of the gas separation plants can be enhanced significantly.

Regeneration Energy Requirements:

The main source of energy consumption in a chemical solvent based CO_2 separation process is the regeneration process. As much as 80% of the total energy is consumed during solvent regeneration.
The total energy required to regenerate a CO_2 loaded solvent can be expressed as follows:

Total Energy = Heat of Reaction + Sensible Heat + Latent Heat of Vaporization of Water
                + Latent Heat of Vaporization of the Solvent (Partial).

In the regeneration step, first the rich solvent temperature must be raised to the stripper temperature by sensible heat transfer. The amount of heat required for this process is dictated by the specific heat capacity of the solvent which does not vary much among the various solvents. In addition, the water component of the solvent must also be vaporized to generate the stripping vapor. While the specific heat capacity and the latent heat of vaporization of water remains the same for all solvents, the energy required for this step depends on the proportion of water present in a given solvent. The higher the water content, the greater the energy requirement for this step. For example, the energy required for the vaporization of water would be greater for a 30 wt% MEA solution than that of a 50 wt% TEA solution.

Finally, part of the solvent itself would also be vaporized. The amount of energy required by various solvents for this part can be compared to the latent heats of vaporization. Finally, sufficient heat must be provided to break up the CO_2-solvent complex formed during the absorption process. This can be accounted for by the heat of reaction.

Heats of reaction or enthalpies of solution for various solvents are provided in Table 1.

Table 1:   Heats of reaction (enthalpies of solution) for various solvents

Solvent	Concentration (M)	CO_2 Loading	Enthalpy of Solution (kJ/mol of CO_2)
MEA	5 (30 wt%)	0.4	72
DEA	3.5 (36 wt%)	0.4	65
TEA	3.35 (50 wt%)	0.5	62
MDEA	4.28 (50 wt%)	0.5	53.2

As can be seen from Table 1, MEA's heat of reaction (enthalpy of solution) is higher than that of all the other solvents listed. As a result more energy must be provided to regenerate MEA than the other less reactive solvents.

In addition to the enthalpy of solution, MEA's latent heat of vaporization is also higher than that of other less active solvents as shown in the Table 2 below.

Table 2:   Latent Heats of Vaporization for various solvents.

Solvent	Heat of Vaporization (kJ/kg)
MEA	826
DEA	670
TEA	535
MDEA	550

Based on the heats of reaction and the latent heats of vaporization, it is clear that MDEA would offer better energy savings compared to MEA. However, the problem is that the mass transfer rate of MDEA is lower than that of MEA.

Table 3 shows typical values of the first order rate constant for the overall reaction between various amines and CO_2. Among the conventional amines, MDEA has the lowest k_1'' of all. Therefore, its absorption rate for CO_2 is also lower than for other amines.

Table 3:   Typical values of the overall forward rate constant for CO_2-amine reactions.

AMINE TYPE	k_1'' (mols/L.s)
MEA	7600
DEA	1500
DIPA	400
TEA	16.8
MDEA	9.2

In general, the higher the rate of reaction, the greater the mass transfer rates. As can be seen from Table 3, one can choose various amines and their blends to achieve a desired mass transfer rate. For example, assuming the diffusional resistances to be the same, MEA will provide the highest mass transfer rate while MDEA will result in the lowest rate among the amines listed in Table 3. This is why MEA is considered to be the best solvent for CO_2 removal from flue gases. This higher reactivity also implies that more heat is required to regenerate the MEA solvent.

At a first glance, one might expect that the overall reaction rate of CO_2 with a blend of different amines will depend primarily on the proportion of each amine present in the mixture. However, this is not always the case. Due to the interactions between various amines and their reacted species, the overall mass transfer rate can be enhanced significantly by the addition of rate promoters or catalysts. Regeneration energy requirement for such a mixed solvent can be reduced by as much as 30% compared to that of MEA (Chakma, 1995).

Mass Transfer Enhacement with Mixed Amines

The mass transfer enhancement with mixed amines may be explained by the so called "shuttle mechanism". The shuttle mechanism for the case of CO_2 absorption in aqueous mixtures of primary and tertiary amines is illustrated in Figure 2. At the gas-liquid interface both the amines react with CO_2. However, due to its higher reactivity, more of the primary amine is depleted at the interface than the tertiary amine. Figure 2 neglects the amount of tertiary amine that reacts with CO_2 at the interface for the sake of simplicity. Thus it depicts only the reaction between CO_2 and the primary amine RNH_2 at the interface. The reacted primary amine in its protonated (RNH_3^+) and carbamate forms ($RNCOO^-$) travels from the interface to the bulk of the solution and then transfers the CO_2 to the un-reacted tertiary amine R_3N. The primary amine is therefore regenerated and return to the interface to pick up more CO_2. This enhances the overall absorption rate.

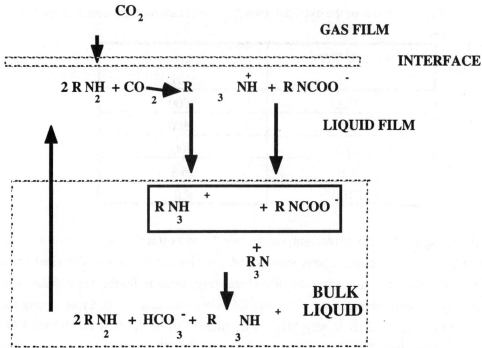

Figure 2.   A schematic representation of the shuttle mechanism for CO_2 absorption
               by an aqueous solution containing a primary and secondary amine.

MEMBRANE - SOLVENT HYBRID PROCESS

The pre-contactor shown in Figure 1 can be replaced by a conventional membrane as has been suggested for the separation of CO_2 from high pressure natural gas streams. Since off-the-shelf membrane based processes are pressure driven, for a membrane-solvent hybrid process to be used in the separation of CO_2 from flue gas streams, pressure of the flue gas needs to be increased significantly. The cost of such an operation outweighs any benefit to be gained from the reduced solvent circulation rate. As such, membrane-solvent hybrid processes are not suitable for CO_2 separatiuon from flue gas streams.

CONCLUSION

A number of options for the reduction of regeneration energy required in a chemical solvent based CO_2 separation process have been identified. They include the use of a pre-contactor, injection of external steam and the utilization of mixed solvents. The combined effect of all these options can reduce the regeneration energy requirement by as much as 30% compared to the conventional MEA process.

REFERENCES

1.   A. Chakma, Energy Convers. Mgmt., 36 (6-9), 427-430 (1995).

2.   MIT Report MIT-EL-89-003, MIT, Cambridge, MA (1989).

3.   R. Padamsey and J. Railton, Energy Convers. Mgmt., 34 (9-11), 1165-1175 (1993).

4.   C. Smelser, S.C. Stock and G.J. McCleary, EPRI Report IE-7365, Vol. 1 (1991).

Pergamon

Energy Convers. Mgmt Vol. 38, Suppl., pp. S57–S62, 1997
© 1997 Elsevier Science Ltd. All rights reserved
Printed in Great Britain
0196-8904/97 $17.00 + 0.00

PII: S0196-8904(96)00246-4

Development of Energy Saving Technology for Flue Gas Carbon Dioxide Recovery in Power Plant by Chemical Absorption Method and Steam System

Tomio Mimura [1], Hidenobu Simayoshi [1], Taiichiro Suda [2], Masaki Iijima [3], and Sigeaki Mituoka [4]

1) Environmental Research Center, Kansai Electric Power Co.
   11-20 Nakoji 3-chome, Amagasaki, Hyogo 661, Japan
2) General Office of Nuclear and Fossil Power Production, Kansai Electric Power Co.
   3-22 Nakanoshima 3-chome, Kita-ku, Osaka 530, Japan
3) Mitsubishi Heavy Industries, Ltd.
   3-1 Minatomirai 3-chome, Nishi-ku, Yokohama, Kanagawa 220-84, Japan
4) Hiroshima Research & Development Center, Mitsubishi Heavy Industries, Ltd.
   6-22, Kan-non-shinmachi 4-chome, Nishiku, Hiroshima 733, Japan

The most important issue involved with the chemical absorption method for recovering carbon dioxide from a power plant's flue gas is to develop energy-efficient absorbents. KS-1 absorbent was presented at ICCDR-2 in Kyoto. After the conference, efforts to develop energy-efficient absorbents have been made and a new absorbent KS-2 was developed as a result and its performance confirmed by the pilot plant. KS-2 has similar energy efficiency as KS-1 both of which require 20% less energy than MEA. KS-2 is however more stable than KS-1 and a more efficient absorbent for low CO_2 content flue gas. Detailed steam system was also discussed in this conference and accurate calculation results were presented.

© 1997 Elsevier Science Ltd

KEYWORDS

Flue gas ; carbon dioxide ; recovery ; power plant ; chemical absorption

PREFACE

The main theme of the chemical absorption method is reduction of its high regeneration energy requirement. The regeneration energy required for CO_2 recovery by MEA is about 900kcal/kg-CO_2, which is equivalent to about 20% of the boiler combustion energy of the thermal power plant.

With the aim of developing energy efficient absorbent which is essential for the power plant CO_2 recovery, laboratory screening tests were focusing on the sterically hindered amines (SHA).

Already 80 alkanolamine specimens have been tested at the laboratory and the promising absorbents have been selected among the sterically hindered amine

(SHA) and subjected to bench-scale tests. Subsequent performance tests which have been carried out by all utility companies in Japan from April, 1995 in a pilot plant showed excellent characteristics of KS-1. Compared with MEA, KS-1 requires approximately 20% less energy to recover CO$_2$ and this result was presented at ICCDR-2 in Kyoto. [1]

THE BASIC CHARACTERISTICS OF KS-2

After the development of KS-1, KS-2 was developed and had its performance confirmed.

With simulated gas containing 10 vol% CO$_2$ at atmospheric condition, high CO$_2$ loading (0.63 mol CO$_2$/mol amine) is obtained by using KS-2.

Molecular structure of KS-2 has higher degree of steric hindrance, hence it is presumed that the regeneration energy of KS-2 may be small.

Heat of Absorption

<u>Test Procedure</u>   CO$_2$ gas (100 vol%) was made to react with alkanolamine 30 wt% solution in a vacuum flask.

Time-temperature curves were obtained, and compared with the heat curve obtained by heating absorbent with a pipe heater.

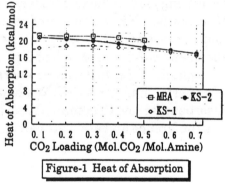

Figure-1 Heat of Absorption

<u>Test Results</u>   The reaction heat of KS-2 is smaller than that of MEA, but higher than that of KS-1 as shown in Figure-1.

Equilibrium Data

Comparison of the equilibrium data are shown in Figure-2.

Equilibrium data shows that KS-2 has high CO$_2$ absorption capacity under flue gas condition and very low CO$_2$ loading under regeneration condition compared to monoethanol amine (MEA) and KS-1.

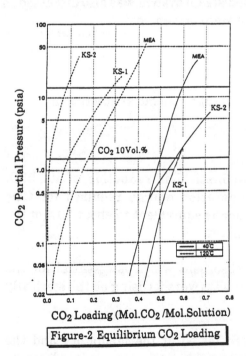

Figure-2 Equilibrium CO$_2$ Loading

BENCH SCALE TESTS AND PILOT PLANT TESTS

Bench scale tests were carried out to confirm the stability as well as corrosion rate and also to compare regeneration energy.

KS-2 showed to be very stable with no sign of corrosion and regeneration energy smaller than other absorbents.

Based on these data, pilot plant test was carried out. Figure-3 shows the regeneration energy required for CO_2 recovery by KS-2 absorbent during 1000 hours continuous running tests. The regeneration energy of KS-2 is about 700 kcal/kg-CO_2 recovered which is about 20% less than monoethanol amine (MEA) solution, and this figure is almost same as that of KS-1 absorbent.

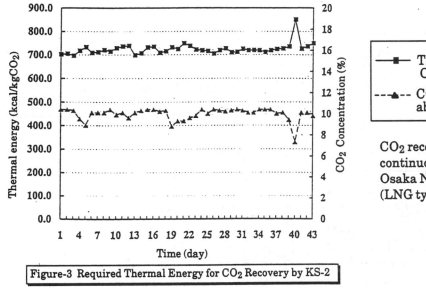

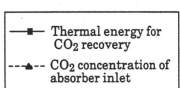

CO₂ recovery pilot plant continuous operation results in Osaka Nanko Power station (LNG type)

Figure-3 Required Thermal Energy for CO₂ Recovery by KS-2

Heat stable salt contents were measured during 1000hour test period and compared in Figure-4.

Based on the heat stable salt analysis, KS-1 and KS-2 are both very stable and may not degradate by itself.

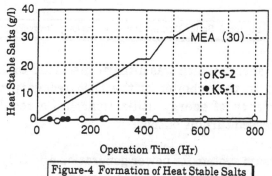

Figure-4 Formation of Heat Stable Salts

OPTIMUM STEAM SYSTEM FOR POWER PLANT FLUE GAS CO₂ RECOVERY

A process of recovering CO_2 from boiler flue gas requires a large volume of low pressure steam for regeneration of the solution. It is necessary to secure the means of obtaining such large volume of low pressure steam out of steam system in a power plant.

Such steam system in a power plant is expressed by the enthalpy-entropy curve as shown in Fig. 5. [1]

As a system is assumed in which $3{\sim}4$ kg/cm² abs. of steam is extracted from the low pressure turbine in the steam system and is used in a reboiler for the CO_2 recovery process, the energy of the steam at Portion Ⓖ in the enthalpy-entropy curve of Fig. 5, can be given to the reboiler. [1]

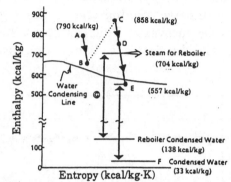

Figure-5 Enthalpy-Entropy Curve Power Plant Steam System

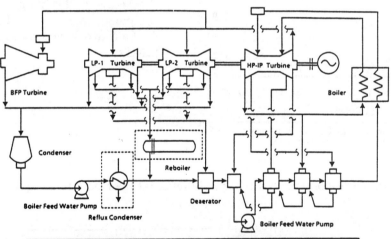

Figure-6 Boiler Turbine Power Plant CO₂ Recovery Steam System

The energy which was hitherto discarded to the outside by the condenser can now be used in the reboiler of the CO_2 recovery process to enhance the consolidated efficiency of energy utilization. As a result, more steam will be available for increased turbine output.

A steam system for power generation and CO_2 recovery which is built taking these points into view, is shown in Figure-6, and the exhaust heat of reboiler and reflux condenser is used for the power plant feed water heating.

CALCULATION OF TURBINE OUTPUT AND POWER PLANT POWER LOSS BY CO_2 RECOVERY

Based on flow scheme of Fig.7 precise calculation of turbine output was made. Turbine pressure is lowered by extraction of reboiler steam flow. And recovery of waste heat of reboiler is added to power plant feed water at two points. Power plant heat balance is calculated in this simulation. In addition to the heat balance, turbine exhaust loss is calculated according to change of turbine speed by each turbine's exhaust loss curve.

KS-2 absorbent not only uses smaller amounts of steam for regeneration but also lowers the regeneration temperature. Thus steam with a lower pressure level can be used. It can be realized from both of the results of the reduction of regeneration energy and lowering of the regeneration temperature that the lowering of a power plant output can be prevented to a great extent.

Further, by a combined use of the KP-1 packing described in ICCDR-2, a reduction in the power consumption within the plant was made possible as shown in Table-1. Power plant power reduction is about 5.4~5.8% when KS-2 absorbent and KP-1 packing are used with optimum steam system in the natural gas fired power station, and about 6.8% when KS-2 solvent and KP-1 packing are used with optimum steam system in the coal fired power station.

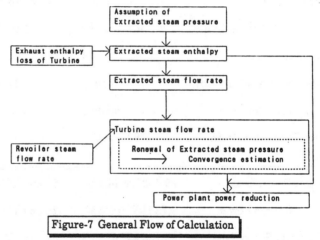

Figure-7 General Flow of Calculation

CONCLUSIONS

(1) KS-2 absorbent which is suitable to recover carbon dioxide from a fossil power plant's flue gas was developed. The thermal energy required for CO_2 recovery by KS-2 is as about 700kcal/kg-CO_2, which is same as that for KS-1 but drastically lower than that for the conventional absorbent MEA.

(2) Power plant power reduction is about 5~6 percentage, if KS-2 absorbent and KP-1 packing are used, with optimum steam in natural gas-fired power plant.

REFERENCE

[1]   T. Mimura et al, Energy Convers. Mgmt 36 (1995) p397

Table-1   Power Plant Flue Gas CO_2 Recovery Details

Description	Natural Gas-Fired Plant-1	Natural Gas-Fired Plant-2	Coal-Fired Plant
Generator Output	600 MW	600 MW	900 MW
Flue Gas Flow Rate	1,622,000 m³N/H	1,622,000 m³N/H	2,600,000 m³N/H
CO_2 Content	8.55 Mol.%	8.55 Mol.%	13.3 Mol.%
CO_2 Recovery	90 %	90 %	90 %
Recovered CO_2	245 Tons/H	245 Tons/H	611 Tons/H
CO_2 Recovery Energy	700 kcal/kg CO_2	700 kcal/kg CO_2	660 kcal/kg CO_2
Steam Enthalpy for Reboiler	696 kcal/kg	707 kcal/kg	671 kcal/kg
Reboiler Steam Flow Rate	308 Tons/H	308 Tons/H	737 Tons/H
Reflux Condenser Energy Utilization	32 %	31 %	14 %
Low Press. Turbine Outlet Steam Enthalpy	556 kcal/kg	557 kcal/kg	557 kcal/kg
No.3 Heater Steam	37 Tons/H	16 Tons/H	17 Tons/H
No.3 Heater Steam Enthalpy	661 kcal/kg	646 kcal/kg	648 kcal/kg
No.4 Heater Steam	42 Tons/H	74 Tons/H	28 Tons/H
No.4 Heater Steam Enthalpy	696 kcal/kg	704 kcal/kg	685 kcal/kg
Reboiler Drain Enthalpy	138 kcal/kg	138 kcal/kg	138 kcal/kg
Reboiler Condenser Drain Enthalpy	92 kcal/kg	92 kcal/kg	92 kcal/kg
Generator Power Loss	29,400 kW (4.9%)	31,800 kW (5.3%)	75,028 kW (8.3%)
Blower and Pump Power	2,725 kW (0.45%)	2,725 kW (0.45%)	6,545 kW (0.73%)
CO_2 Recovery Power Loss Total	32,125 kW (5.35%)	34,525 kW (5.75%)	81,573 kW (9.03%)
CO_2 Liquefaction Power	29,250 kW (4.88%)	29,250 kW (4.88%)	72,946 kW (8.1%)
CO_2 Recovery and Liquefaction Power Plant Power Loss	61,375 kW (10.2%)	63,775 kW (10.6%)	154,519 kW (17.1%)

Note : Power Plant loss is generator output power basis.

Energy Convers. Mgmt Vol. 38, Suppl., pp. S63–S68, 1997
© 1997 Elsevier Science Ltd. All rights reserved
Printed in Great Britain
0196-8904/97 $17.00 + 0.00

Pergamon

PII: S0196-8904(96)00247-6

EVALUATION OF TEST RESULTS OF 1000 m³N/h PILOT PLANT FOR CO₂ ABSORPTION USING AN AMINE-BASED SOLUTION

Norio Arashi
Hitachi Research Laborarory, Hitachi Ltd.
7-1-1 Omika-cho, Hitachi-shi, Ibaraki-ken, 319-1 Japan

Naoki Oda
Kure Works, Babkock Hitachi K.K.
6-9 Takara-machi, Kure-shi, Hiroshima-ken, 737 Japan

Mutsuo Yamada
Kure Research Laboratory, Babkock Hitachi K.K.
6-9 Takara-machi, Kure-shi, Hiroshima-ken, 737 Japan

Hiromitsu Ota,   Satoshi Umeda   Motoaki Tajika
Energy and Environment R&D Center, Tokyo Electric Power Company
4-1 Egasaki-cho, Tsurumi-ku, Yokohama-shi, Kanagawa-ken, 230 Japan

ABSTRACT

Pilot plant tests of the CO₂ chemical absorption process (capacity:1000m3N/h) were carried out under various operation conditions.   In the tests, a newly developed absorption solution was used.   Test results were evaluated by using the overall mass transfer coefficient KGa of the absorption process. The collected data were applied to the evaluation program which calculates plant equipment specifications and design configurations.   After the computer aided study the CO₂ absorption plant for 1000MW thermal power plant was designed in which CO₂ removal is 90%.   The construction area for of the CO₂ absorption plant for the 1000MW power plant will be about 170m×110m .

© 1997 Elsevier Science Ltd

KEYWORDS

CO₂; chemical absorption; amine-based solution; flue gas; thermal power plant; mass transfer coeficient;

INTRODUCTION

In recent years, global warming due to CO₂ release by human activities has attracted much attention. About half of the CO₂ is delivered from thermal power stations and industries through fossil fuel utilization process.   One option to reduce emissions is recovering CO₂ from the stack gas of the emission sources.   CO₂ recovery plants in which alcanole amine is used as an absorption solution are already being used at some thermal power stations (Clair *et al.*, 1983, Kaplan, 1982).   A problem has been, however, the deterioration of absorbent by SO₂ in the flue gas (Barchas *et al.*, 1992). Subsequently, Tokyo Electric Power Company and Hitachi Ltd. developed an amine based absorption solution which is not affected as much by SO₂   (Yamada *et al.*, 1996).

The 1000m³N/h pilot plant tests in which this new absorption solution was used were carried out at the

Yokosuka Thermal Power Station of Tokyo Electric Power Company. The objectives of these pilot plant tests were to prove the applicabilities of the new absorption solution to SO_2 containing flue gas and to obtain the engineering data needed for design work.   The test data showed the solution had sufficient characteristics of SO_2 durability and of energy saving (Ota *et al.*, 1996).   This paper describes analytical studies done using the test results, from the viewpoint of mass transfer in the absorption and regeneration processes.   The computer aided study for a 1000MW coal fired thermal power plant is also presented.

PILOT PLANT

The pilot plant was installed at the Yokosuka Power Station which burns a COM (coal oil mixture).   A flow sheet of the plant is illustrated in Fig.1.   1000m³N/h of flue gas are extacted from the main flue gas stream of a 265MW power generating boiler.

The gas is previously cooled to about 40°C to decrease its moisture content before being introduced into the absorber.   The absorber contains a ring-shaped stainless steel packing (1 inch in diameter) to get effective contact of the solution with the gas.   In the absorber the CO_2 in the flue gas is absorbed by the solution at temperatures between 40°C and 50°C. CO_2 loaded solution is then extracted from the bottom of the absorber and transfered to the regenerator through a heat exchanger in which the solution temperature is raised to between 100°C and 110°C.   In the regenerator, the solution contacts with steam which comes from the reboiler and CO_2 is stripped from the solution.   The mixture of steam and CO_2 exits from the top of the regenerator and is cooled in the condenser to separate the CO_2.   The purity of the recovered CO_2 is up to 99%.

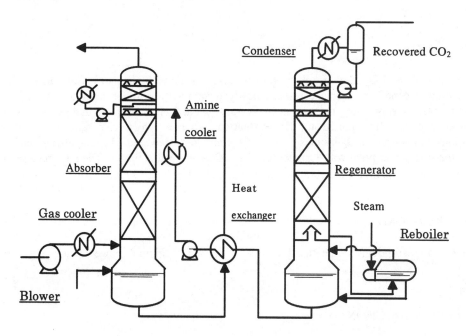

Fig. 1   Flow sheet of pilot plant

RESULTS AND DISCUSSION

CO_2 Removal and Solution CO_2 Loading

CO_2 removal was measured for various operation conditions such as flow rate of circulating absorption solution, flow rate of steam for regeneration of CO_2 loaded solution, and flow rate of flue gas. Fig.2 (a) shows the effects of solution flow rates on CO_2 removal. CO_2 removal increases with increase of solution flow rate under 1800L/h and gradually decreases over 1800L/h. In ordinary absorption operation, the higher the solution flow rate is, the higher the CO_2 removal is, because the wetted surface of the packing material for gas-liquid contact increases with solution flow rate. In the tests, as the steam supply was kept constant, at 250 kg/h, and the heat required for regeneration of solution was insufficient when the flow rate was increased. Fig.2 (b) shows the CO_2 loading on the solution after absorbing CO_2 (rich solution) and after stripping CO_2 by steam (lean solution). The CO_2 loading on the lean solution increases when the solution flow rate exceeds 2000 l/h. When CO_2 loading in the lean solution is high, the driving force of mass transfer from gas to solution becomes small, so the CO_2 removal can not be compensated by the increase of solution flow rate.

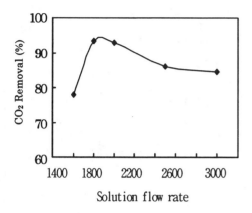

Fig.2 (a)   Relation between CO_2 removal and solution flow rate. Steam flow rate, 250kg/h

Fig.2 (b)   Relation between CO_2 loading and solution flow rate. Steam flow rate, 250kg/h

Fig.3 (a) indicates the effects of the steam flow rate for regeneration of rich solution on CO_2 removal. CO_2 removal increases with increase of steam flow rate. This can be explained as follows. The more the amount of steam which is supplied to the regenerator, the lower the CO_2 concentration of the gas phase stream in the regenerator becomes. So, CO_2 can be more easily stripped from the solution when the steam flow increases. As a consequence, the CO_2 loading on the lean solution decreases with the increase of steam flow rate as can be seen from Fig.3 (b).

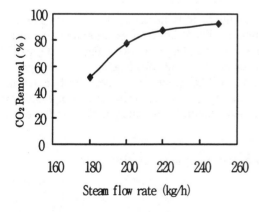
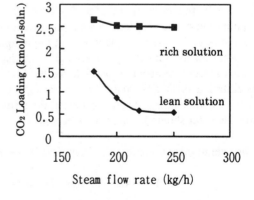

Fig. 3 (a)   Relation between CO_2 removal and steam flow rate. Solution flow rate, 2000 l/h

Fig.3 (b)   Relation between CO_2 loading and steam flow rate. Solution flow rate, 2000 l/h

Mass Transfer Coefficient

Experimental results were next analyzed from the viewpoint of mass transfer to obtain information for plant design. Mass transfer of the CO_2 from gas to solution is represented by the overall mass transfer coefficient $K_{OG}a$.

In this analysis temperature distribution in the absorber should be investigated first.   When the solution absorbs CO_2, the temperature of the solution rises due to absorption reaction heat.   On the other hand, as evaporation of water from solution occurs, the solution temperature is lowered by loss of latent heat from the solution.   Consequently temperature distribution is formed in the absorber.   Fig.4 is an example calculated temperature profile in the absorber.

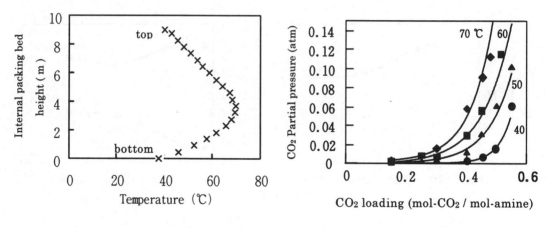

Fig.4 Temperature distribution in absorber.
Gas flow rate, 1000m3/h; CO_2 concentration
in gas, 11%; Solution flow rate, 2000 l/h;
Steam flow rate, 250 kg/h.

Fig.5 CO_2 equilibrium on solution.

As shown in Fig.4, the temperature when solution enters the absorber from the top is initially 40℃, and gradually rises with the depth of the packing bed height.   At about one third of the packing bed height from the bottom, the temperature reaches its maximum value of 70℃, then gradually drops towards the bottom of the absorber.

To clarify the mass transfer at this temperature distribution, it is necessary to measure the CO_2 equilibrium of the CO_2 loaded solution at each temperature.   Fig.5 shows the equilibrium curve of the solution used in the pilot plant tests at the temperatures corresponding to those of Fig.4. CO_2 partial pressure on the solution becomes high with rising of temperature and CO_2 loading .   At the part in which solution temperature is high, as the differences of the CO_2 concentration between flue gas and equilibrium surface become less, the transport of CO_2 from gas to solution decreases.   For evaluation of an accurate mass transfer coefficient, the influence of temperature destribution in the absorber on the mass transfer should be taken into account.

The overall mass transfer coefficient $K_{OG}a$ is generally expressed by next equation.

$$1 / K_{OG}a = 1 / k_{G}a + H / k_{L}a \qquad (1)$$

where $k_{G}a$ and $k_{L}a$ are the mass transfer coefficient in gas and liquid respectively, and H is equivalent to the Henrry's constant.   This value is varied with the CO_2 loading on solution, because the partial pressure equilibrium to the solution is not proportional to the CO_2 as can be seen from Fig.5.   In this study the oerall mass transfer coefficient $K_{OG}a$ is measured as follows.   Another expresion of $K_{OG}a$ is :

$$K_{OG}a = N_{OG} \cdot G_m / z \qquad\qquad (2)$$

where N_{OG} is the overall mass transfer unit, G_m is the molar gas velocity in the absorber, z is the height of the internal packing bed in the absorber. N_{OG} is calculated as follows:

$$N_{OG} = \int dy / (y - y^*) \qquad\qquad (3)$$

where y is the CO_2 concentration in flue gas, y^* is the CO_2 concentration equilibrium with CO_2 loaded solution, dy is the difference of CO_2 concentration at inlet and outlet of elemental bed height of packing dz. As the temperature of solution varies with z, N_{OG} of Eq.(3) was calculated using the equilibrium data shown in Fig.5.

Fig.6 (a) shows the relation between $K_{OG}a$ and solution flow rate at constant steam flow rate 250 kg/h. The change of $K_{OG}a$ with solution flow rate resembles that of CO_2 removal. $K_{OG}a$ varies from about 50 to 110 kmol/m³·h·atm with the solution flow rate.   Fig.6 (b) shows the relation between $K_{OG}a$ and steam flow rate. $K_{OG}a$ increases as the steam flow rate is raised.   The reason why $K_{OG}a$ changes with the steam flow rate is considered as follows.   The H in Eq.(1) decreases with decrease of CO_2 loading on the lean solution by increase of steam flow rate, and then $K_{OG}a$ increases as clarified from Eq.(1).

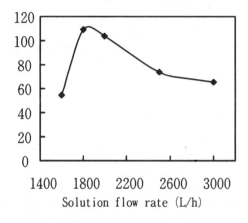

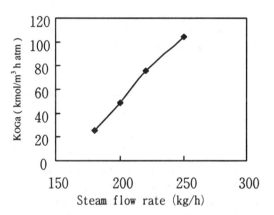

Fig.6 (a)   Variation of $K_{OG}a$ with solution flow rate.
Steam flow rate, 250kg/h; Gas flow rate, 1000m³N/h

Fig.6 (b)   Variation of $K_{OG}a$ with steam flow rate.
Solution flow rate, 2000 l/h; Gas flow rate, 1000m³N/h

Computer Aided Study of 1000MW Plant

The pilot plant test data were applied to a computer aided study of the CO_2 removal plant for the thermal power plant with a capacity of 1000MW.   The study provides the design aid soft ware.

In this study the number of absorbers was selected as an essential parameter which dominates plant characteristics such as absorber diameter, plant arrangement, plant cost and so on.   From the total volume of flue gas and number of absorbers the diameter of the absorber was decided.   The packing bed height was calculated by the mass transfer coefficent $K_{OG}a$ which was obtained in the pilot plant tests.   The configuration of the absorber was then determined from these dimensional values.   For the regeneration process, pairing of absorbers was determined in the same way as for the absorption process. Other equipment specifications such as heat for exchangers, booster blowers and pumps were determined by the design aid program.

The plant layout was depicted from the viewpoint of engineering features.   Fig.7 shows the a view of the CO_2 removal plant for a 1000MW coal fired thermal power station.

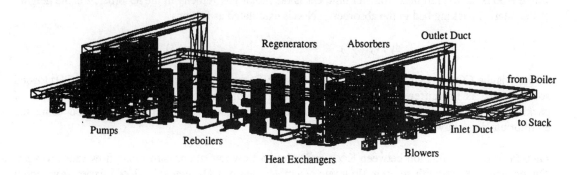

Regenerators   Absorbers   Outlet Duct

from Boiler

Pumps   Reboilers   Heat Exchangers   Blowers   Inlet Duct   to Stack

111.79m

171.65m

Fig.7   Plant layout of CO_2 removal plant for 1000MW thermal power station

Eight pairs of absorbers and regenerators were installed.   Each absorber was equipped with a booster blower, and each regenerator with a pump.   Two pairs of absorbers and regenerators had one heat exchanger.

The construction area required for this plant is about 170m $\times$ 110m square.

CONCLUSION

As a result of analysis of the data from the CO_2 absorption pilot plant, the overall mass transfer coefficient K_OGa was obtained.   It was also clarified that to evaluate K_OGa the temperature distribution which was formed in the absorber had to be taken into account.   The K_OGa was in the range of 50-110 kmol / m^3 h atm.   Application of K_OGa to a computer aided study of the CO_2 removal plant for the 1000MW thermal power plant, clarified that the consturuction area of the CO_2 removal plant was about 170 m $\times$ 110 m.

REFERENCES

Barchas, R. and R. Davis (1982). The Kerr-McGee / ABB Lumus Crest Technology for the recovery of CO_2 from stack gases. Proceedings of the first international conference on cabon dioxide removal. <u>33</u>, 133.

Clair, J. H. and W. F. Simister (1983). Process to recover CO_2 from flue gas gets first large-scale tryout in Texas. Oil & Gas J. Feb. 14, 109-113.

Kaplan, L. J. (1982). Cost-saving process recovers CO_2 from power-plant flue gas. Chemical Engineering. Nov. 29, 30-31.

Ota, H., S. Umeda, M. Tajika, M. Nishimura, M. Yamada, A. Yasutake and J. Izumi (1996). CO_2 removal technology on chemical absorption and physical adsorption method. Proceedings of the Joint IEW/JSER Internationl Conference on energy, economy and environment. June 25-27, 235-240.

Yamada, M. , K. Murakami, N. Oda, A. Mori, M. Nishimura, M. Ishibashi and H. Ota (1996). CO_2 removal technology from flue gases containing SO_2 at thermal power plants. J. of the Japan Institute of Energy. <u>75</u>, in press.

Pergamon

Energy Convers. Mgmt Vol. 38, Suppl., pp. S69–S74, 1997
© 1997 Elsevier Science Ltd. All rights reserved
Printed in Great Britain
0196-8904/97 $17.00 + 0.00

PII: S0196-8904(96)00248-8

THE ENHANCEMENT OF THE RATE OF ABSORPTION OF CO_2 IN AMINE SOLUTIONS DUE TO THE MARANGONI EFFECT

J. BUZEK, J. PODKAŃSKI, K. WARMUZIŃSKI

Institute of Chemical Engineering, Polish Academy of Sciences, 44-100 Gliwice,
ul. Bałtycka 5, Poland

ABSTRACT

Results of the experimental and theoretical investigations of the occurrence of cellular convection during absorption of CO_2 in amine solutions (Marangoni effect) are presented. Experiments in CO_2-aqueous solutions of MEA and CO_2-aqueous solutions of DEA systems show that cellular convection indeed occurs during chemisorption and that it can increase mass transfer rate significantly enough to lower the absorber height. The influence of certain parameters characterizing mass transfer (e.g. mass transfer coefficient in the liquid phase, initial concentrations of reactants, presence of surface-active agents) on the intensity of cellular convection is analyzed. Theoretical considerations confirm the possibility of the occurrence of cellular convection in the systems investigated. Experimental correlations are given which describe the influence of the effect in terms of the mass transfer enhancement factor. The results can be directly employed in the design of absorbers for CO_2 capture. © 1997 Elsevier Science Ltd

KEYWORDS

Absorption; amines; cellular convection; CO_2; Marangoni effect; Marangoni instability

CELLULAR CONVECTION

The experimental and theoretical results described in the paper concern the liquid surface instabilities (cellular convection) during gas absorption in liquids accompanied by a chemical reaction. The surface instabilities are of interest to chemical engineers - both scientifically and practically oriented (Sawistowski, 1971; Golovin, 1992). This is mainly due to the potential influence they can have on a transfer rate between phases. Surface instability, if it does occur, can increase mass transfer rates during gas chemisorption in liquids by a factor of two and more (for liquid-liquid systems, a factor as high as 75 is reported).

Cellular convection is a spontaneous phenomenon occurring in liquids near the interface during heat or mass transfer (mass transfer can be purely physical or can be accompanied by a chemical reaction). The term "cellular convection" is frequently used to describe two types of phenomena: well-ordered, stable, regular convective structures or, alternatively, violent, turbulent eruptions (or irruptions) of liquid interface resulting in the interface deformation. Our attention will be focused on the former pattern, as in gas-liquid systems the latter is not likely to occur.

The primary reason for the occurrence of instability can be local disturbances of temperature and/or concentration near the interface. These disturbances can lead to fluctuations of density or fluctuations of the surface tension and result in the convective movements. Fig. 1 presents possible mechanisms of the initiation of cellular convection.

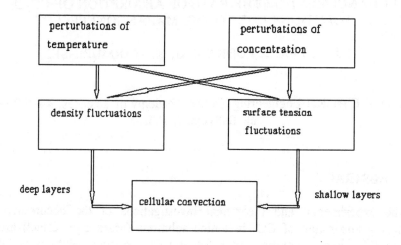

Fig. 1:   Mechanisms of the occurrence of cellular convection.

Our investigations were focused on the practically important mechanism inducing the surface instability during gas chemisorption in liquids - the so called Marangoni instability (MI): the concentration disturbances on the interface affect the local values of surface tension, which may, in turn, produce instability.

The absorption system CO_2-amines (MEA and DEA) was selected for experimental and theoretical examination as it reveals properties which can lead to the occurrence of MI. The chemisorption of CO_2 leads to the formation of products which increase the surface tension (Fig. 2). The absorption is never uniform and areas of lower surface tension (lower concentration of products) expand. A fresh solution which has even lower product concentration comes up to the surface, resulting in the formation of "cells" of convective nature (Fig. 2).

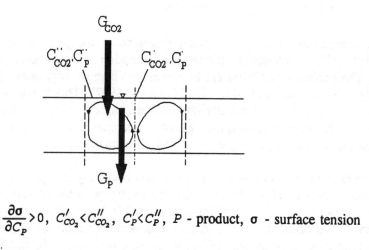

$$\frac{\partial \sigma}{\partial C_P} > 0, \quad C'_{CO_2} < C''_{CO_2}, \quad C'_P < C''_P, \quad P \text{ - product}, \quad \sigma \text{ - surface tension}$$

Fig. 2:   Convective cells during gas chemisorption in liquid.

The investigations of MI can be divided into several groups depending on the problem to be solved and the investigation technique used in a particular study. The occurrence of MI during gas chemisorption can be detected: 1. directly, using visualization methods or 2. indirectly, making use of the fact that MI usually enhances the transfer rate. Mathematical models (3rd approach) can also give the answer to the question whether or not the surface instability should be taken into consideration. Similarly, there are three ways of studying the influence of MI on the rate of mass transfer: 1. direct measurement of the rate of mass transfer which is then compared with the experiment in which the MI is damped, 2. indirect method where the additional, trace and inert component is introduced into the system and its transfer rate is measured and then compared with the experiment in which MI is damped, 3. mathematical modelling of cellular convection.

The research on MI has been carried out in the Institute of Chemical Engineering of the Polish Academy of Sciences since the early seventies (Warmuziński *et al.*, 1995). Since then many experimental and theoretical work has been done by a number of researchers: comprehensive experimental investigation of MI during chemisorption of gases in liquids, experiments in a wetted-wall column (Buzek, 1979, 1983); - investigation of the influence of MI on the mass transfer rate during chemisorption of gases in liquids in the industrial scale equipment (packed columns) (Podkański, 1990); - theoretical modelling of cellular convection during chemisorption (Warmuziński, 1990, 1991a, 1991b).

EXPERIMENTAL AND THEORETICAL INVESTIGATIONS

Wetted-wall Columns

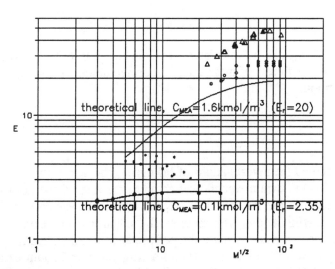

Fig. 3:   Absorption of CO_2 in MEA solutions, selected experimental results in the wetted-wall column, pure CO_2; $\triangle$ - C_{MEA}=1.6 kmol/m³, $\circ$ - C_{MEA}=1.6 kmol/m³ + surface active agent; * - C_{MEA} = 0.1 kmol/m³, $\bullet$ - C_{MEA} = 0.1 kmol/m³ + surface active agent.

First studies concerned the occurrence of MI during chemisorption of CO_2 in MEA solutions in a laboratory-scale wetted-wall column. The experiments were carried out under conditions corresponding to the operating parameters of industrial packed columns used for CO_2 removal.

Abnormally large mass-transfer coefficients in the liquid phase were obtained (Fig. 3). It was shown that this is the consequence of cellular convection. The results were compared with the experiments in which the MI was damped by an addition of a surface active agent-Teepol (direct method of studying the influence of MI on the rate of mass transfer was employed) and with theoretical predictions. The dominant role in the initiation of the phenomenon is played by the changes of surface tension caused by fluctuations of concentration of ionized products of chemical reaction at the gas-liquid interface (the solution is not ionized before the absorption).

The occurrence of MI is strongly influenced by the chemical reaction rate. For infinite reaction rates the cellular convection disappears. This is due to the fact that the infinite reaction rate at the surface diminishes the fluctuations of concentration of ionized products of the reaction thus diminishing the driving force of cell formation. In order to support this conclusion the system H_2S-MEA-water was studied. The experimental results agreed well with the theoretical predictions. There was also no influence of the addition of surface active agent on mass transfer rate (unlike in the experiments with the CO_2-MEA-water system).

The system CO_2-NaOH-water was used to investigate the influence of preliminary ionization of the solution on MI. It turned out that in case of ionized, dissolved substrates of reaction already present in the system cellular convection does not influence the mass transfer rate or is eliminated altogether.

The influence of MI on the mass transfer rate in the case studied depends mainly on the hydrodynamics of the system (which can be described by the physical mass transfer coefficient $(k_l)_{theor}$ in the absence of MI), the initial amine concentration C^o_{MEA} and on the partial pressure of a gas absorbed p_{CO2} (Eq. 1; $(k_l)_{theor}$ in m/s, C^o_{MEA} in kmol/m^3 and p_{CO2} in Pa). This influence was described by a coefficient γ which is a ratio of $(k_l)_{exp}$ (physical mass transfer coefficient in the presence of the MI) to $(k_l)_{theor}$, and thus describes the enhancement of physical absorption.

$$\gamma - 1 = 6.59 x 10^6 (k_l)_{theor}^{0.75} (C^o_{MEA})^{1.08} (p_{CO_2})^{-0.62} \tag{1}$$

Thus the coefficient $(k_l)_{exp}$ can be calculated from the equation $(k_l)_{exp} = \gamma (k_l)_{theor}$ and used for the calculation of the coefficient of mass transfer during chemisorption $(k_l^*)_{exp}$ according to the theory of chemisorption.

Packed Columns

Further studies concerned the investigations of MI during chemisorption in packed columns. As in the case of experiments with a wetted-wall column, the rate of chemisorption accompanied by MI was compared with the rate of chemisorption when MI was damped by the addition of a surface active agent (Fig. 4) and with theoretical predictions. The experiments were carried out in the apparatus designed to simulate the performance of an industrial packed column (D = 0.1 m, height up to 0.3 m, 15 mm ceramic Raschig rings or 10 mm ceramic spheres, gas phase - pure CO_2 and CO_2 diluted in the air, liquid phase - aqueous solutions of MEA and DEA, initial amine concentration up to 0.4 kmol/m^3, liquid flow rate up to 10 kg/m^{2}s, temp. 25°C, atmospheric pressure). The main conclusions are as follows:

MI does occur in packed columns during chemisorption and can considerably increase mass transfer rate in certain parts of the column. MI can easily by damped by the addition of a surface active agent; this leads to the conclusion that the phenomenon is caused by surface tension forces. The increase in mass transfer rate due to cellular convection for the entire

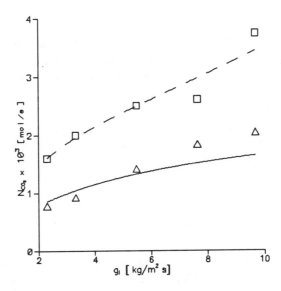

Fig. 4:   The influence of MI on the rate of absorption of CO_2 in MEA solutions in the packed column; N_{CO2} - rate of absorption of CO_2, g_l - liquid flow rate; □ - absorption with accompanying MI, △ - absorption with MI damped by the addition of surface active agent, solid line - theoretical mass transfer rates (no enhancement due to MI included in calculations), dashed line - mass transfer rates calculated using Eq. 2; column height = 0.105 m, column diameter = 0.1 m, 15 mm Raschig rings, air flow rate = 2.32×10^{-2} mol/s, initial MEA concentration = 0.4 $kmol/m^3$, initial CO_2 concentration = 0.331 kmol CO_2/kmol air

experimental column can exceed 100 %. The increase in the inlet amine concentration increases the influence of cellular convection on mass transfer rate. This is probably due to the effect of the amine concentration on the intensity of mass transfer rate, which in turn influences the convection intensity. The increase in mass transfer rate due to cellular convection is limited by the depletion of the solution or by the increase in concentration of the product of the reaction along the column. This was clearly seen for higher columns. The effect of the product concentration on cellular convection has been difficult to isolate. However, this influence seems to be small according to earlier studies. The influence of cellular convection on mass transfer rate is less noticeable for ceramic spheres than for Raschig rings. The key parameter in this case is probably the mass transfer coefficient $(k_l)_{theor}$ which is higher for Raschig rings in comparison with ceramic spheres. The increase in $(k_l)_{theor}$ enhances the effect of cellular convection on mass transfer rate. It is probably linked directly, as in the case of the amine concentration, to the effect of mass transfer rate on the intensity of cellular convection. The influence of cellular convection on mass transfer rate is similar in the case of MEA and DEA solutions.

Experimental results for the CO_2-MEA-water system were correlated according to the Eq. 2:

$$\gamma - 1 = 4.60 \times 10^6 (k_l)_{theor}^{0.87} (C_{MEA}^o)^{0.56} (p_{CO_2})^{-0.53} \qquad (2)$$

Theoretical Predictions

A mathematical model was developed to study MI during chemisorption. The model furnished a lot of useful information on the necessary conditions for MI to manifest itself, on the morphology of the cells and on the influence of MI on the mass transfer rate.

According to the model, the time of transient absorption necessary for the cellular convection to occur is very short - below 0.01 s for MEA and below 0.1 s for DEA. The time of the inception of cellular convection decreases with increasing initial amine concentration and partial CO_2 pressure. The initial depletion of the solution does not have any noticeable influence on the moment when cellular convection begins to manifest itself.

CONCLUSIONS

The possibility of the occurrence of MI during the absorption of CO_2 in MEA and DEA solutions was confirmed for the operating conditions similar to those prevailing in industrial equipment. MI can significantly enhance mass transfer rate (by 100 % in certain parts of the column), but it can easily be damped by a surface active agent. Therefore, amine solutions used for absorption should not be contaminated by even traces of surfactants.

Theoretical calculations confirmed the occurrence of MI in the experiments with both wetted-wall and packed columns. They indicate that MI can occur along the entire packed column except for the parts where the depletion of the solution or of the gas phase is significant.

REFERENCES

Buzek J. (1979). Cellular Convection During Absorption with Chemical Reaction. *Zesz. Nauk. Politech. Śląsk.*, 90 (in Polish) 1-109.

Buzek J. (1983). Some Aspects of the Mechanism of Cellular Convection. *Chem. Eng. Sci.*, 38, 155-160.

Golovin A. A. (1992). Mass Transfer Under Interfacial Turbulence: Kinetic Regularities. *Chem. Eng. Sci.*, 47, 2069-1080.

Podkański J. (1990). Effect of Cellular Convection on the Process of Chemical Absorption in Packed Columns. *Ph.D. Thesis*, Polish Academy of Sciences, Institute of Chemical Engineering (in Polish).

Sawistowski H. (1971), Interfacial Phenomena. In: *Recent Advances in Liquid-Liquid Extraction*, (C. Hanson, Ed.), Pergamon Press.

Warmuziński K., Buzek J. (1990). A Model of Cellular Convection During Absorption Accompanied by Chemical Reaction. *Chem. Eng. Sci.*, 45, 243-254.

Warmuziński K., Tańczyk M. (1991a). Oscillatory Marangoni Instability During Absorption Accompanied by Chemical Reaction. *Chem. Eng. Sci.*, 46, 2031-2039.

Warmuziński K., Tańczyk M. (1991b). Marangoni Instability During the Absorption of Carbon Dioxide into Aqueous Solutions of Monoethanolamine. *Chem. Eng. Process.*, 30, 113-121.

Warmuziński K., Buzek J., Podkański J. (1995). Marangoni Instability During Absorption Accompanied by Chemical Reaction. *Chem. Eng. J.*, 58, 151-160.

Pergamon

Energy Convers. Mgmt Vol. 38, Suppl., pp. S75–S80, 1997
© 1997 Elsevier Science Ltd. All rights reserved
Printed in Great Britain
0196-8904/97 $17.00 + 0.00

PII: S0196-8904(96)00249-X

MASS TRANSFER STUDIES OF HIGH PERFORMANCE STRUCTURED PACKING FOR CO$_2$ SEPARATION PROCESSES

Adisorn Aroonwilas and Paitoon Tontiwachwuthikul[*]
Industrial Systems Engineering
Faculty of Engineering, University of Regina
Regina, Saskatchewan, Canada S4S 0A2
Tel: (306) 585-4726; Fax: (306) 585-4855

ABSTRACT

The feasibility of using high efficiency structured packing in CO$_2$ separation processes with a newly proposed solvent, 2-amino-2-methyl-1-propanol (AMP), was investigated. The packing performance was evaluated by CO$_2$ absorption into sodium hydroxide and aqueous AMP solutions over ranges of main operating variables, i.e. up to 10 kPa partial pressure of CO$_2$, 0.29 - 0.61 m^3/m^2 s gas rate, 4.87 - 14.18 m^3/m^2 h liquid rate, and 1.1 - 2.0 kmol/m^3 liquid concentration. Compared with random packings, the structured packing provides 10 to 33 times higher overall mass transfer coefficient. This indicates possibility of using the structured packing for CO$_2$-AMP absorption system.

© 1997 Elsevier Science Ltd

KEYWORDS

CO$_2$ absorption, structured packing, mass transfer, AMP

INTRODUCTION

The typical CO$_2$ absorption process consists, basically, of absorbing CO$_2$ from a gas phase into a liquid solvent in the absorber and liberating the absorbed CO$_2$ from the solvent at the regeneration unit. The employed absorbents are simply classified in two categories; physical and chemical solvents. However, chemical solvents are preferred for absorption processes. Alkanolamines are considered the most extensively used chemical solvents (Maddox, 1985). Among all alkanolamines, MEA is the most popular solvent (DuPart *et al.*, 1993).

Recently, a new class of acid gas treating solvents called sterically hindered amines has been disclosed (Kohl and Riesenfeld, 1985). Of these hindered amines, 2-amino-2-methyl-1-propanol (AMP) is the most promising solvent since it has some excellent characteristics compared with the primary amine MEA. On the basis of stoichiometry, AMP can react with CO$_2$ at a theoretical ratio of one mole CO$_2$ per mole of amine. This becomes a superior characteristic of the hindered amine to the conventional MEA whose each mole can react with only one-half mole of CO$_2$. Besides its outstanding absorption capacity, AMP also induces less corrosion which is considered the major operational problem in the conventional CO$_2$ absorption plants (Veawab, 1995).

The replacement of the conventional alkanolamine MEA with the hindered amine AMP is very attractive. However, one limitation for use of AMP is that its absorption rate is lower than that of MEA (Alper, 1990). To make the replacement feasible, the rate of CO$_2$ absorption into the new AMP

---

[*] Author to whom correspondence should be addressed.

solution must be improved. An approach that might lead to the improved efficiency of a CO_2 absorption process using AMP solution is modifications to the interior of the absorber in order to increase the contact area between gas and liquid phases. With successive experiences in using structured packing for distillation and gas dehydration applications (Hausch *et al.*, 1992 and Kean *et al.*, 1991), it is expected that use of structured packing would improve mass transfer performance in the new CO_2-AMP absorption system.

Therefore, the primary objective of this study was to investigate the role of a high-efficiency structured packing in improving the efficiency of CO_2 absorption process. The performance of the structured packing was mainly evaluated by performing CO_2 absorption into sodium hydroxide (NaOH) solutions. The absorption of CO_2 into an aqueous solution of AMP was also conducted in order to ensure the capability of the structured packing for CO_2 absorption.

OVERALL MASS TRANSFER COEFFICIENT ($K_G a_v$)

To evaluate the performance of packings, the overall mass transfer coefficient per unit volume ($K_G a_v$) has to be determined. This term can be defined as:

$$K_G a_v = \{G_I / [P(y_A - y_A^*)]\} \{dY_A/dZ\} \tag{1}$$

In this study, CO_2 absorption was conducted in a tested column. The CO_2 concentration in the gas phase along the column was measured, interpreted in term of mole ratio (Y_A) and subsequently plotted as functions of column height (Z) called CO_2 concentration profile. The slope of the profile expressing concentration gradient (dY_A/dZ) is used for $K_G a_v$ determination.

EXPERIMENT

The absorption experiments were conducted in an 1.77 m high and 0.019 m ID absorption column. The column was packed with *EX type* laboratory structured packing provided by Sulzer Brothers Limited, Winterthur, Switzerland. The total height of the packing section was approximately 1.10 m. Besides the absorption column, the auxiliary equipment including liquid feed and storage tanks, digital gas flowmeters, and a liquid rotameter was required.

The CO_2 absorption experiments commenced by preparing feed solutions at desired concentration and warming up measurements such as gas flowmeters and CO_2 analyzer. A mixture of air and CO_2 were then introduced through the mass flowmeters into the bottom of the column. At the same time, the prepared solution from the feed tank was pumped to the top of the column. The absorption process was operated until steady state conditions were reached. Then, the CO_2 concentration profile along the column was measured and recorded. The CO_2 rich solution at the column bottom was simultaneously sampled in order to determine the amount of absorbed CO_2 in the liquid phase and subsequently verify the CO_2 absorption rates calculated from the gas phase data. The compositions of the liquid samples were determined by using the methods described in Aroonwilas's Master Thesis (1996).

RESULTS AND DISCUSSION

Twenty-seven absorption runs was conducted. The results were plotted as profiles of CO_2 concentration and liquid composition along the packed column and were subsequently interpreted in term of $K_G a_v$ values. These values are reported as functions of the main operating variables. The $K_G a_v$ values for structured packing and random packing are also compared here.

Effects of CO_2 Operating Variables

Figure 1 shows the overall K_Ga_v values for CO_2-NaOH and CO_2-AMP systems as functions of CO_2 partial pressure. In both systems, the K_Ga_v values tend to move upwards when the partial pressure decreases from 10 to 1 kPa. The CO_2-NaOH system gives higher mass transfer coefficients approximately 3 to 4 times more than the AMP system. This is basically caused by the faster rate of reaction between CO_2 and hydroxide ions.

Liquid flow rate has a significant effect on the overall K_Ga_v value in CO_2-NaOH absorption. The mass transfer coefficient is directly proportional to the liquid irrigation rate as shown in Fig. 2. Doubling the liquid flow rate from 4.87 to 9.73 m^3/m^2 h increases the K_Ga_v value by 31 %. This trend is probably due to the increasing wetted area of packing, which was reported to be a function of liquid flow rate by Nawrocki et al, 1991.

The effect of absorbent (NaOH) concentration on the overall K_Ga_v value at various CO_2 partial pressures is shown in Fig. 3. It is obvious that the coefficient increases with NaOH concentration. This is basically due to an increasing chance for CO_2 to contact and subsequently react with hydroxide ions.

Comparison With Random Packing

In comparison with the K_Ga_v values for random packings summarized by Strigle (1987), structured packing used for CO_2-NaOH system provides significantly higher mass transfer coefficient. Relative K_Ga_v values, representing ratio of the coefficient for tested structured packing to those for random packings, range from 10 to 33 depending upon size and type of packings.

For CO_2-AMP system, the mass transfer coefficients for tested structured packing were compared with those for random packing reported by Tontiwachwuthikul et al. (1989). It was found that structured packing gives the higher overall K_Ga_v value, approximately 6 times. These results show it is possible to use structured packing to make effective CO_2-AMP absorption system.

CONCLUSIONS

The overall mass transfer coefficient for structured packing is a function of three main operating variables, i.e. it decreases with CO_2 partial pressure; it increases with liquid flow rate and increases with absorbent concentration. However, it should be noted that CO_2 partial pressure shows small effect on the mass transfer coefficient in CO_2-AMP absorption system.

In the CO_2-NaOH system, the EX type structured packing provides an excellent overall mass transfer coefficient, which is higher than the coefficient for random packings by a factor of 10 to 33. In the CO_2-AMP system, the structured packing shows at least 6 fold superior performance to random packing (1/2" Berl Saddles).

ACKNOWLEDGEMENTS

The financial supports of Canada Centre for Mineral and Energy Technology (CANMET), the Natural Sciences and Engineering Research Council of Canada (NSERC), Sulzer Brothers Ltd. (Switzerland), Saskatchewan Power Corporation, Prairie Coal Ltd., and Wascana Energy Inc. are gratefully acknowledged.

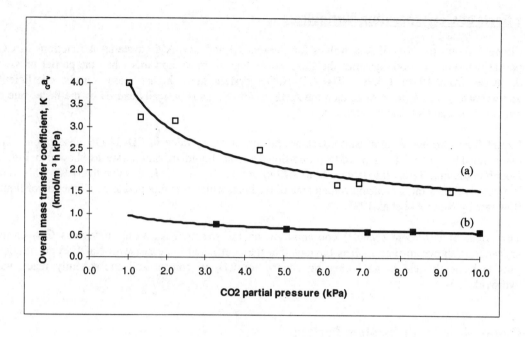

Fig. 1. Effect of CO_2 partial pressure on the overall mass transfer coefficient.
(a) CO_2 - NaOH(b) CO_2 - AMP

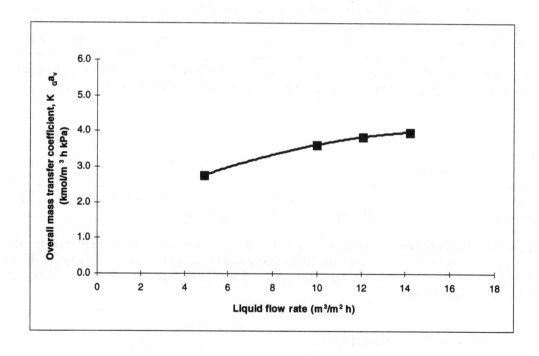

Fig. 2. Effect of liquid flow rate on the overall mass transfer coefficient.
(CO_2 partial pressure = 1.0 - 1.2 kPa and [OH⁻] = 1kmol/m³)

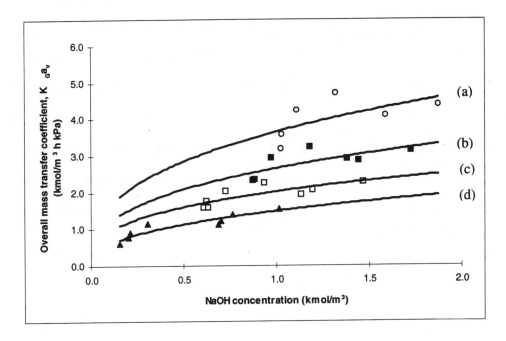

Fig. 3. Effect of absorbent concentration on the overall mass transfer coefficient at different CO_2 partial pressure.

(a) CO_2 partial pressure = 1.0 kPa (b) CO_2 partial pressure = 2.5 kPa
(c) CO_2 partial pressure = 5.0 kPa (d) CO_2 partial pressure = 9.1 kPa

NOMENCLATURE

a_v = effective interfacial area per unit volume of packing (m^2/m^3)
G_I = inert gas molar flow rate ($kmol/m^2$ h)
K_G = overall mass transfer coefficient ($kmol/m^2$ h kPa)
P = total pressure (kPa)
Y_A = mole ratio of component A in gas bulk (mol/mol)
y_A = mole fraction of component A in the gas bulk (mol/mol)
y_A^* = gas phase mole fraction of component A in equilibrium with the concentration of component A in the liquid phase (mol/mol)
Z = packing height (m)

REFERENCES

Alper, E. (1990). Reaction Mechanism and Kinetics of Aqueous Solutions of 2-Amino-2-methyl-1-propanol and Carbon Dioxide. *Ind. Eng. Chem. Res.*, 29, 1725-1728.

Aroonwilas, A. (1996). *High Efficiency Structured Packing for CO₂ Absorption Using 2-Amino-2-methyl-1-propanol (AMP)*. M.A.Sc. Thesis, University of Regina, Regina, Saskatchewan, Canada.

DuPart, M. S., Bacon, T. R. and Edwards, D. J. (1993). Part2-Understanding Corrosion in Alkanolamine Gas Treating Plants. *Hydrocarbon Processing* , 72(5), 89-94.

Hausch, G. W., Quotson, P. K. and Seeger, K. D. (1992). Structured Packing at High Pressure. *Hydrocarbon Processing* , 71(4), 67-70.

Kean, J. A., Turner, H. M. and Price, B. C. (1991). Structured Packing Proven Superior for TEG Gas Drying. *Oil & Gas Journal*, 89(38), 41-46.

Kohl, A. L. and Riesenfeld, F. C. (1985). *Gas Purification*. 4th Edition, Gulf Publishing Company, Houston.

Maddox, R. N. (1985). *Gas Conditioning and Processing*. Vol. 4, 3rd Edition, Campbell Petroleum Series, Norman, Oklahoma.

Nawrocki, P. A., Xu, Z. P. and Chuang, K. T. (1991). Mass Transfer in Structured Corrugated Packing. *Can. J. Chem. Eng.*, 69, 1336-1343.

Strigle, R. F., Jr., (1987). *Random Packing and Packed Towers; Design and Applications*. Gulf Publishing Company, Houston.

Tontiwachwuthikul, P., Meisen, A. and Lim, C.J. (1989). Separation of CO_2 Using Sterically Hindered Amine Solutions. *Proc. Int. Conf. on Recent Developments in Petrochemical and Polymer Technologies*, Bangkok, Thailand, 1-38.

Veawab, A. (1995). *Studies of Material Corrosion by Gas Treating Solvents*. M.A.Sc. Thesis, University of Regina, Regina, Saskatchewan, Canada.

Energy Convers. Mgmt Vol. 38, Suppl., pp. S81–S86, 1997
© 1997 Elsevier Science Ltd. All rights reserved
Printed in Great Britain
0196-8904/97 $17.00 + 0.00

Pergamon

PII: S0196-8904(96)00250-6

SEPARATION OF CARBON DIOXIDE FROM OFFSHORE GAS TURBINE EXHAUST

O. FALK-PEDERSEN and H. DANNSTRÖM

Kværner Water Systems a.s
P.O.Box 1017, 3204 Sandefjord, Norway

ABSTRACT

The introduction of a CO_2-tax in 1991, on offshore combustion of natural gas has led to an increased interest in both energy conservation and the possibility of separating CO_2 from gas turbine exhaust. In this paper a possible process will be presented. The result of the assessment is that an amine absorption process using membrane gas/liquid contactors in both the absorber and the desorber, in combination with a combined cycle power generation unit with 40% recycling of the exhaust gas and a CO_2 compression unit, is best suited for CO_2 removal among the options studied. © 1997 Elsevier Science Ltd

KEYWORDS

Carbon dioxide; amine; membranes; exhaust gas; offshore.

INTRODUCTION

In 1989 the Norwegian Government initiated the objective that the total CO_2 emission in Norway will be stabilised at the 1989 level in the year 2000. This led to the introduction of the CO_2-tax in 1991. The total CO_2-tax paid by the Norwegian oil companies is assumed to be MNOK 2700-2800 (US$ 369-382 million) in 1996, which is equal to 53 US$/tonne generated CO_2. This motivated the Norwegian oil companies to study new methods and technologies for reduction of the total CO_2 emissions.

In 1992, Kværner Engineering a.s. Environmental initiated a joint effort program for CO_2 separation between major North Sea operators and the Norwegian authorities. This study is at present in phase four and will be finished in November 1996. This paper will present the development activities in the study and the preliminary results.

The most crucial constraint when developing possible solutions for a CO_2 removal process for offshore installations is that occupying volume and weight is expensive.

The project started with a feasibility study where different possible solutions for removal of CO_2 were evaluated. The basis for the study has been an LM2500 PE gas turbine. The conclusions were that an amine absorption process in combination with 40% recycling of the exhaust was the best process, as shown in fig. 1.

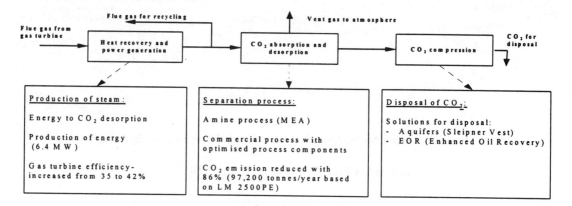

Fig. 1. Overview of the selected process

THE CO₂ REMOVAL PROCESS

The main target of the study is to optimise the process with respect to size, weight and cost. The development work focused on optimisation of the following topics: the absorption unit, the desorption unit and the total process.

<u>Power Generation Concepts</u>

The main condition that the power generation unit must fulfil, is to make the net power output from an LM 2500 simple cycle gas turbine (approximately 21 MW) available regardless of the power and heat requirements of the CO_2 removal process. This necessitated a waste heat recovery unit (WHRU).

Alternative power generation processes were examined with respect to efficiency and CO_2 emission, as shown in table 1.

Table 1. Comparison of power generation processes.

	LM2500 PE	LM2500 PE with combined cycle	LM2500 PE with 86% removal of CO_2
Energy to gas turbine (GT) [MW]	61.0	61.0	61.0
El. power from GT [MW]	21.3	21.3	21.3
El. from steam turbine [MW]	0	8.0	6.4
El. utilized in the CO_2 process[MW]			2.3
El. available for platform use [MW]	21.3	29.3	25.4
Efficiency (El.) [%]	35	48	42
CO_2 emission [1000 tonnes/year]	113	113	15,8
CO_2 -tax [MNOK/year] (1996)	40	40	5,6

Combined cycle with 40% recycling of exhaust is the best power generation system of the options studied in connection with removal of CO_2 on offshore installations. Both General Electric and Rolls Royce have in general agreed that the recycle ratio (40%) will not significantly influence gas turbine performance, though detailed testing is necessary to verify this.

Detailed results and information from this part of the study can be found in the paper "Assessment of Power Generation Concepts on Oil Platforms in Conjunction with CO_2 Removal" (Bjerve et al.,1994).

Optimisation of the absorption process

Membrane gas/liquid contactor. The separation is caused by the presence of an absorption liquid on one side of the membrane which selectively removes certain components from the gas stream on the other side of the membrane, as shown in fig. 2 a

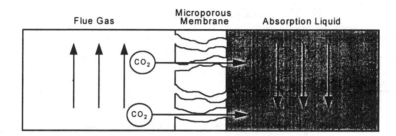

Fig. 2 a Principle of membrane gas/liquid contactors

The use of a membrane gas/liquid contactor has several advantages over conventional contacting equipment (e.g. packed columns): (1) the membrane gas/liquid contactor will be compact due to a high packing density (m^2/m^3), (2) the operation of the contactor is independent of gas- and liquid flow rate, (3) there are no foaming, channeling, entrainment or flooding.

Membrane gas-/liquid contactor used as absorber. The development started with a verification of the principle and an implementation study. The scope of work was to verify the three most critical parameters: possible wetting of the membrane, the total mass transfer coefficient and the project economy. The study concludes that: (1) GORE-TEX® ePTFE membranes are the only membranes that can be used, as shown in table 2, (2) the mass transfer is high and very close to the value calculated in our in-house simulation program, (3) conventional membrane modules cannot be used due to poor mechanical stability, low active membrane surface and high production cost, (4) the use of a membrane gas/liquid contactor as absorber has a significant effect on the internal rate of return.

Table 2.   Test results, absorber testing

Polymer	Wetting	Mass Transfer Coefficient
PP[1]	After some hours	Not measured
PES[2] coated with PDMS[3]	After 6 days	Very low
PP coated with PDMS	After 7 days	Very low
GORE-TEX® ePTFE [4]	No	High, in agreement with estimated mass transfer

---

[1] PP	=	polypropylene
[2] PES	=	polyether sulphone
[3] PDMS	=	polydimethylsiloxane (silicon rubber)
[4] PTFE	=	polytetrafluoroethylene (teflon)

The positive results from the testing initiated a cooperation between W.L. Gore & Associates, GmbH and Kværner. Gore is mainly responsible for the membrane development and Kværner is responsible for the total process.

A new phase of the development was initiated, which included theoretical studies and test work. The theoretical studies and the test work were performed by two external organisations and Gore and/or Kværner. To gain a neutral evaluation of the technology , there was no exchange of information between the external organisations. The conclusions from the absorber development are: (1) the membrane and the membrane module are chemically stabile and not wetted by the absorption liquid, (2) a new type of membrane module with high packing density, good mechanical stability and relatively low production cost has been developed, (3) 72% reduction in size and 66% reduction in weight compared with a conventional column, as shown in table 3 and fig. 2b

Table 3. Comparison of conventional absorber and membrane gas/liquid contactor
(97,200 tonnes/year based on LM 2500PE and 86% removal).

	Column	Membrane
Size (incl. wash water) (m)	ø4.5 x 21.8	4.0 x 4.5 x 5.0
Weight (tonnes)	70	24

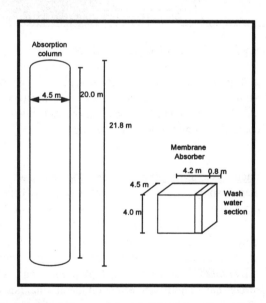

Fig. 2b Comparison of conventional absorber and membrane gas/liquid contactor
(97,200 tonnes/year based on LM 2500PE and 86% removal).

Membrane gas/liquid contactor used as desorber. In order to further minimize space and weight requirements of the CO_2 separation process, the project included the testing of several different desorber concepts. Of these, the membrane gas/liquid contactor was found to have the best potential and the project group agreed to continue the development.

Tests were carried out using GORE-TEX® ePTFE membranes provided by W.L. Gore & Associates GmbH.

The main conclusions from the test work are: (1) the membrane is chemically stable and not wetted by the hot absorption liquid, (2) the mass transfer is relatively high and close to the estimated value, (3) compared with a conventional column the size is reduced by 78% and the weight by 66%, as shown in table 4.

Table 4. Comparison of conventional desorber and membrane gas/liquid contactor
(97.200 tonnes/year based on LM 2500PE and 86% removal).

	Column	Membrane
Size (m)	ø2.2 x 22.0	1.6 x 1.6 x 3.0
Weight (tonnes)	20.5	7.0

Process optimisation. The possibility of using the membrane contactor in both the absorber and the desorber opens up a variety of options with respect to process optimisation compared to conventional absorber/desorber columns. Among others:

⇒ Reduced energy consumption due to optimised water/amine ratio and possible use of other chemicals.

⇒ Reduced corrosion, no steel in the absorber or stripper and elimination of the reboiler.

⇒ Reduced degradation of the solvent due to the elimination of the reboiler, the solvent is not heated by the reboiler, it is heated by the steam in the membrane gas/liquid desorber.

Optimisation of the Layout and Project Economy

Because space is very expensive offshore, the layout study has been a very important part of the project. Fig. 3 shows the CO_2 removal process.

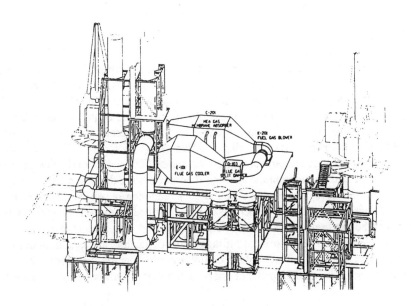

Fig.3   Plant Layout

The economy of the process depends on a number of factors which vary from platform to platform. The most important factors are: (1) the CO_2 tax, (2) the number of gas turbines on the platform, (3) the value of the CO_2 used for EOR, (4) the value of the electricity. The economic feasibility seems promising but dependant upon the number of gas turbines on the platform.

EFFECT ON THE TOTAL NORWEGIAN EMISSION OF CO_2

Fig. 4 shows the prognosis for the emission of CO_2 from the petroleum activity in the North Sea and the environmental effect of installing CO_2 removal processes on 5 and 10 gas turbines (LM 2500PE). The total Norwegian CO_2 emission is approximately 35,500 tonnes/year (1993). The contribution from the offshore gas turbines is approximately 6,800 tonnes/year, which is 83% of the total emission from the offshore activity.

Total CO₂ emission from the oil and gas production in the North Sea

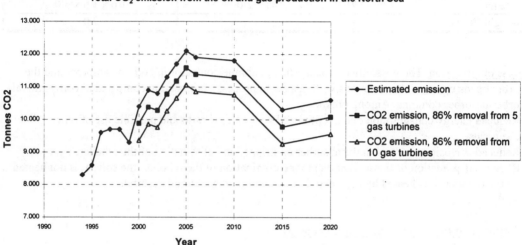

Fig. 4   Prognosis for the emission of CO_2 from the petroleum activity in the North Sea and the environmental effect of installing CO_2 removal processes on 5 and 10 gas turbines, Melhus [2].

CONCLUSIONS

The object of the project is to investigate which CO_2 removal process will be optimal for offshore use. The preliminary conclusions from the study are: (1) an amine absorption process is the best process, (2) 40% of the exhaust gas should be recycled to the gas turbine inlet, to reduce the exhaust gas flow and to increase the CO_2 concentration entering the absorber, (3) membrane gas/liquid contactors are the best alternative to conventional absorption columns, (4) the potential for the membrane gas/liquid contactor used as absorber has to be confirmed by pilot testing using real flue gas, (5) compared with conventional absorber columns the use of membrane gas/liquid contactors will give a 72% reduction in size and 66% reduction in weight, (6) it is indicated that membrane gas/liquid contactors are the best alternative to conventional desober columns with 78% reduction in size and 66% reduction in weight, (7) the potential for the membrane gas/liquid contactor used as desorber has to be confirmed by more lab. tests and pilot testing, (8) it is indicated that the total separation process can be optimised with respect to energy concumption, corrosion and degradation of the solvent , if membrane contactors are used in both the absorber and the desorber, (9) if the suggested technology could be made to work, the economic feasibility of the project seems promising but dependant upon the number of turbines on the platform.

REFERENCES

Y. Bjerve and O. Bolland (1994). Presented at the International Gas Turbine and Aeroengine Congress and Exposition, The Hague, Netherlands

M. S. L. Melhus (1996). The Norwegian Petroleum Directorate, Norway

Energy Convers. Mgmt Vol. 38, Suppl., pp. S87–S92, 1997
© 1997 Elsevier Science Ltd. All rights reserved
Printed in Great Britain
0196-8904/97 $17.00 + 0.00

Pergamon

PII: S0196-8904(96)00251-8

Implementations of Advisory System for the Solvent Selection of Carbon Dioxide Removal Processes

Christine Chan and Patrick Lau

Dept. of Computer Science/Energy Informatics Laboratory, University of Regina, Regina, SASK, S4S 0A2, Canada

ABSTRACT

The Solvent Selection Advisory System (SSAS) is a decision support system for aiding users in the preliminary selection of optimal solvents for carbon dioxide removal processes given different user specification and plant conditions. This paper describes an inference-network representation of the Solvent Selection Advisory System which has been previously implemented as a rule-based system. Two expert system shells, G2 and GDA from Gensym Corp. are used, and the implementations on the two shells are compared. Some advantages and disadvantages in the two representation approaches are discussed. © 1997 Elsevier Science Ltd

KEYWORDS

selection of solvents, implementations, rule-based system, block-based system

INTRODUCTION

Our work focuses on a specific design problem in the chemical engineering discipline of solvent selection for carbon dioxide removal processes. We have built a prototype system for solvent selection called the Solvent Selection Advisory System. The system can function as both an advisory and a tutorial system. The selection of solvents is based on heuristics and visual illustration of the problem help users learn about the domain. We used the real-time expert systems shells, G2 and GDA from Gensym Corporation, U.S.A., for developing prototypes of the system. G2 is an object-oriented tool that facilitates development of rule-based systems, whereas GDA is an object-oriented graphical software developed on the G2 platform. Since GDA is built on the G2 platform, conceptual model expressible in G2 should be expressible in GDA. A major difference between them is that in G2, inference knowledge is stored in rules while in GDA it is stored in inference networks.

PROBLEM DOMAIN AND CONCEPTUAL MODELLING

The sources of knowledge which formed the basis of the Solvent Selection Advisory System (SSAS) include both the authority literature in the field [Astarita et al. 1983 , Kohl and Riesenfeld 1985 , Tennyson and Schaaf 1977] and the expert's understanding of the domain. In the process of CO_2 absorption, the gas comes into contact with the liquid solvent. A state is reached in which absorption (from the gas to the liquid) and desorption (from the liquid to the gas) are taking place at equal rates so that the concentrations of both phases remain constant. That state is referred to as equilibrium. Based on the equilibrium property of the feed and product gas, the domain has been divided into subregions in

[Astarita et al. 1983, Tennyson and Schaaf 1977]. The expert reinterpreted the divisions suggested and constructed a new diagram consisting of 13 regions which served as basis of discussion between the knowledge engineer and the expert [Chan et al. 1995]. Assuming normal temperatures, the expert divided the domain into thirteen regions using the equilibrium property of the gases as the criteria.

Since equilibrium is a function of partial pressures of the CO_2 concentration in the raw and product gas, the two parameters are used as primary indices to get an approximate decision on the choice of solvents. In other words, for each region, there are several candidate solvents; the final choice depends on considerations of other parameters.

A conceptual model of the system was developed using the Inferential Modelling Technique. Details about the formulation of the conceptual model are found in [Chan et al. 1995]. The conceptual model for the SSAS domain consists of inference structures which represent selection heuristics elicited from the expert. An inference structure consists of conceptual and physical objects and parameters invoked in a partial order for accomplishing a specific task or objective.

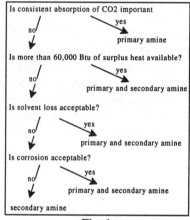

Fig. 1

In a real life application, solvent loss and corrosion need to be considered. For example, a sample task structure for region 9 shown in Fig. 1 invokes a number of conceptual objects including the need for consistent absorption of CO_2, availability of surplus heat, and acceptability of solvent loss and corrosion. The objectives of the task structure include gathering information and making a recommendation. The invocation illustrates the expert's reasoning for a sample region in the domain and is guided by knowledge at the strategy level. From Fig. 1, we can see that the two tasks of gathering information and making recommendation are inextricably intertwined.

The resultant conceptual model is an aggregation of the inference structure representing knowledge of all thirteen regions in the domain [Chan 1992]. Two prototypes have been developed based on the model, one on G2 and the other on GDA, the latter is primarily done for exploration and comparison purposes.

IMPLEMENTATION IN G2

In the G2 implementation of SSAS, hereafter called SSAS$_{G2}$, we used if-then rules and initial rules to implement both the heuristics and the program control rules. The initial rules are basically if-then rules which are invoked when the knowledge base is being started or restarted. These are used for checking input validity so that unacceptable partial pressure values are screened before other rules are activated. In G_2, a premise in a rule expresses conditions in the form of logical expressions, and a consequent consists of a set of actions such as invoke a workspace, conclude that a variable is of a certain value, inform the operator, manager, or supervisor, change the color of an icon, set the value of a variable, etc.

Take for example region 9, which is shown in Fig. 1 and identified by the partial pressure values in the product and raw gases of:

- 0.1 ≤ partial pressure of product gas <<0.22
  2 ≤ partial pressure of raw gas <<80
- 0.22 ≤ partial pressure of product gas << 2
  2 ≤ partial pressure of raw gas << 9
- 2 ≤ partial pressure of product gas << 6
  2 ≤ partial pressure of raw gas << 6

When the partial pressure values are found to be within these ranges, the logical variable *reg9* is set to true. Then the workspace which displays a question on the secondary parameter of "need for consistent absorption" is shown and user input elicited. If the user answered negatively, the fact that *reg9* is true and the value of n in the variable of *con-ab* (for "need for consistent absorption") would satisfy the following rule:

If reg9 and con-ab is n then
(a screen which poses the three questions on surplus heat,
solvent loss, and corrosiveness of a solvent is displayed).

If the user again responds negatively to all three queries, the following rules will be successfully evaluated:

If reg9 and avail-sur-heat is n
and solvent-loss is n and corrosion is n then inform the operator that
"Attention: the recommended solvent is secondary amine (c.f. 100 %)..."

where, *con-ab, avail-sur-heat, solvent-loss*, and *corrosion* are the symbolic variables for the parameters of "need for consistent absorption", "availability of surplus heat", "whether solvent loss is acceptable", and "if corrosion is tolerated" respectively; and the *inform* action displays the recommendation on the *Message Board Workspace*. Based on heuristics, a certainty factor (c.f.) of 100 % is included in the recommendation, which means the expert is 100 % certain this is the choice given the user specified inputs. Details about the G2 implementation can be found in [Chan and Tontiwachwuthikul, 1995].

IMPLEMENTATION IN GDA

The inference structure in the conceptual model presented in Fig. 1 is represented as inference networks in GDA which are made up of symbols and connection lines. The symbols of an inference network include 1) square for assertion, 2) circle for conclusion, 3) circle in a square for intermediate conclusion, 4) logical connectives like AND and OR, 5) query block for users and system communication, and 6) display block for displaying conclusions. There are more than a hundred predefined blocks in GDA (see for example Fig. 2).

Entry-block   Query-block   Logical AND   Display-block

Fig. 2

DEVELOPMENT OF GDA VERSION OF SSAS

The development process involves the three steps of (1) clone, (2) configure, and (3) connect. All the predefined graphical blocks are stored on the GDA palette shown in Fig. 2. The first step is to clone a block from the palette and then place it on the workspace. Then, the block is configured to fit user need by customizing the parameters in the table associated with the block and connecting it with other blocks. For example, to create a rule "IF A and B then C", the user needs to clone an entry-block from the palette and configure the block by naming it as A. Then, an entry block B is cloned and configured (i.e. named) and an AND gate and a conclusion block are similarly cloned and configured. The three blocks A, B, C, and AND gate are then connected together. Hence the development of "If A and B then C" results in the structure shown in Fig. 3.

This implemention process was used and the conceptual model shown in Fig. 1 can be developed (see for example Fig. 4.) In SSAS$_{GDA}$ there are altogether thirteen subworkspaces, eleven of which constitute the knowledge base of the system.   Each region was implemented as an inference network and then connected together.   The GDA implementation of region 9 is shown in Fig. 5.   In addition to the workspace for the inference network of the thirteen regions, the GDA version of SSAS also includes a subworkspace that provides information on the problem domain to the user and a subworkspace that allows the user to input values to activate the inference process.

COMPARISON OF SSAS$_{G2}$ AND SSAS$_{GDA}$

The rule-based representation in G2 and the block based representation in GDA each has its strengths and weaknesses. In contrast to the network representation in GDA, rules in G2 are not connected to other rules statically like the blocks in GDA. The rules in G2 allows for flexible inferencing. The block based representation in GDA facilitates implementation of systems that contain static knowledge that involves minimal inferencing and derivation of new information.

An advantage of the GDA implementation is that it allows explicit representation of the results of knowledge acquisition. Take for example region 9, the knowledge acquisition results show that the choice involving the three parameters can be represented as a decision table shown in Table 1. In SSAS$_{G2}$, the table was converted into rules.

Table 1.

Decisions	1	2	3	4	5	6	7	8
More than 60,000 Btu of surplus heat available	Yes	Yes	Yes	Yes	No	No	No	No
Can tolerate solvent loss	Yes	Yes	No	No	Yes	Yes	No	No
Corrosion is acceptable	Yes	No	Yes	No	Yes	No	Yes	No
Recommendation Screens	S1	S2	S3	S4	S5	S6	S7	S8

In SSAS$_{GDA}$, a decision table (such as Table 1) can be represented explicitly by blocks as shown in Fig. 5 which includes two parts, that for signal generation and that for recommendation presentation. The signal generation unit is controlled by a three-output-query-action-block which generates appropriate signals to the recommendation presentation unit according to the user's inputs. The user's inputs on the choice of parameters are elicited with an input screen as shown in Fig. 6. And the recommendation presentation unit invokes the appropriate recommendation screen to the user according to the control signal provided by the signal generation unit. For example, if the user checks off the buttons on the input screen so that the system is informed that "more than 60,000 Btu of surplus heat available," "solvent loss is acceptable", and "corrosion is acceptable" are the input values, then the signal generation unit generates a binary control signal like "111" on the inference paths to notify the recommendation presentation unit that the recommendation for "True True True" or "Yes Yes Yes" applies. The recommendation presentation unit then invokes the corresponding recommendation of "S1" to the user (see Fig. 5) and the output screen shown in Fig. 7 is displayed. This representation more accurately reflects the knowledge acquisition results.

CONCLUSION

We have used GDA primarily as an exploration exercise because of the visual similarity between the inference networks and the inference structures developed during conceptual modeling. While explicit representation is a strength in representing knowledge in some regions, the inference network precludes flexible inferencing implicit in a production system architecture. In other words, all the possible inference paths through a set of if-then statements need to be explicitly specified in GDA. This can be cumbersome when the number of if-then statements is huge in a knowledge base. By contrast, the rule-based representation in G2 does not have this limitation. We have found both G2 and GDA to be useful and user-friendly platforms for creating expert systems. Each of them has its pros and cons and the choice depends on the requirements of the problem domain.

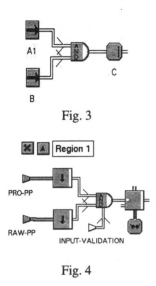

Fig. 3

Fig. 4

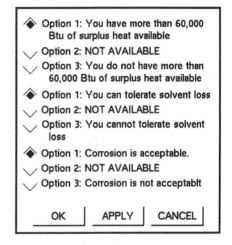

Fig. 6. Input Screen

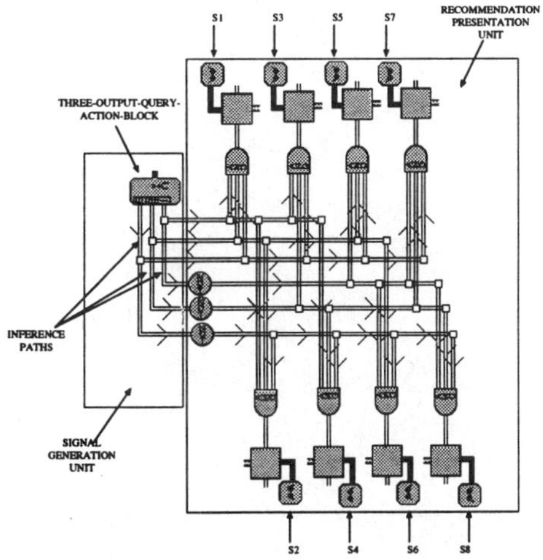

Fig. 5. Table 1 in block-based representation

Fig. 7. Recommendation Screen invoked by S1

REFERENCES

1. *A Guided Introduction to GDA (For version 3.0 of G2, GDA Beta release 1.0)* (1992), Gensym Corporation.
2. Astarita, G., Savage, D.W., and Bisio, A. (1983) *Gas Treating with Chemical Solvents.* New York: Wiley.
3. Chan, C.W. (1992) *Inferential Modelling Technique for Acquisition and Analysis of Expertise,* unpublished Ph.D. dissertation, Simon Fraser University, B.C. Canada.

4. Chan, C.W. and Tontiwachwuthikul, P. (1995), "Expert System for Solvent Selection of CO_2 Separation Processes", *Expert Systems with Applications: An International Journal*, 8 (1), p.33-46.

5. Chan, C.W., Tontiwachwuthikul, P., Cercone, N. (1995), "Knowledge Engineering for a Process Design Domain", *International Journal of Expert Systems: Research and Applications*, 8 (1), p.47-76.

6. Kohl, R., and Riesenfeld, R.C. (1985) *Gas Purification. Houston*: Gulf Publishing.

7. Tennyson, R.N. and Schaaf, R.P. (1977) Guidelines can help choose proper processes. *The Oil and Gas Journal*, 10, 78-86.

 **Pergamon**

PII: S0196-8904(96)00252-X

Energy Convers. Mgmt Vol. 38, Suppl., pp. S93–S98, 1997
© 1997 Elsevier Science Ltd. All rights reserved
Printed in Great Britain
0196-8904/97 $17.00 + 0.00

THE PRODUCTION OF CARBON DIOXIDE FROM FLUE GAS BY MEMBRANE GAS ABSORPTION

P.H.M. FERON and A.E. JANSEN

TNO-Institute of Environmental Sciences, Energy Research and Process Innovation
P.O. Box 342, 7300 AH, Apeldoorn, THE NETHERLANDS

ABSTRACT

The use of membrane gas absorption for carbon dioxide production from flue gases is discussed with special reference to the combined supply of heat and carbon dioxide to greenhouses. Novel absorption liquids are introduced which show an improved performance in terms of system stability and mass transfer compared to monoethanolamine when used in combination with commercially available, inexpensive polyolefin membranes. It is shown that the combined supply of heat and carbon dioxide from cogeneration plants to greenhouses will lead to significant primary energy savings. Carbon dioxide can be produced by the membrane gas absorption process and delivered to greenhouses at lower cost than current supply methods. © 1997 Elsevier Science Ltd

KEYWORDS

Membrane gas absorption; carbon dioxide; capture; greenhouses; cogeneration.

INTRODUCTION

Membrane gas absorption has been identified as a promising new technology for the recovery of carbon dioxide from flue gas streams (Feron et al., 1992; Feron, 1994; Matsumoto et al, 1994, Nii and Takeuchi, 1994). It is based on the combination of two processes, membrane separation and gas absorption. An optimal membrane gas absorption system makes use of the benefit of equipment compactness resulting from hollow fibre membranes and the benefit of process selectivity resulting from the chemical absorption process. The principle is shown in Fig. 1.

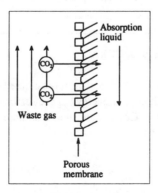

Fig. 1   CO$_2$-membrane gas absorption principle

A three-fold to ten-fold equipment size reduction should be achievable through the use of hollow fibre porous membranes. The reduced size of the equipment compared to packed columns should also lead to reduced equipment costs. Operational benefits include the avoidance of channeling, foaming, flooding and entrainment. This means that e.g. in case of a membrane gas absorption system there will be no need for a wash section to recover droplets being carried over from the absorber. This will give to an additional reduction in size and costs.

NOVEL LIQUIDS FOR CO₂ MEMBRANE GAS ABSORPTION

In the current membrane gas absorption application, carbon dioxide is brought into contact with a suitable absorption liquid via a porous, hydrophobic membrane. The essential element in a membrane gas absorption system is a stable combination of a membrane and an absorption liquid. It has been shown by various researchers (Matsumoto et al, 1995; Falk-Pedersen et al, 1995; Nishikawa et al, 1995) that a stable membrane gas absorption system cannot be achieved with the combination of commercially available polyolefin membranes and the usual absorption liquid, monoethanolamine (MEA). This has hampered the development of membrane gas absorption for the recovery of carbon dioxide from flue gas. TNO has discovered a range of liquids which show stable operation with commercially available polyolefin membranes (Jansen and Feron, 1995). These liquids, named CORAL (**CO₂ R**emoval **A**bsorption **L**iquid), are suitable for carbon dioxide recovery from flue gas but also from other gas streams, such as biogas and natural gas. From this range of liquids a number of liquids have been selected and assessed for CO₂-removal from flue gases. This selection has been based on the measurement of the mass transfer coefficient as this determines the size of the equipment and hence the investment costs. Three type of liquids have been chosen for further evaluation. This evaluation entails the determination of the equilibrium curves and measurement of mass transfer at several conditions.

CO₂-solubility

Figure 2 presents an example of an equilibrium curve for a CORAL solution, which contains 2 M of CO₂ absorbing components (CORAL20). In Fig. 2 also typical conditions for the absorber and stripper outlet are shown. Typical cyclic loadings for CO₂-removal from flue gases will vary between 0.2 and 0.3 mole CO₂ per mole of CO₂-binding component at the absorber outlet depending on the carbon dioxide concentration.

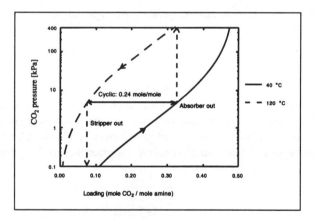

Fig. 2   Equilibrium curve for CORAL20.

Table 1 gives an estimate for the energy consumption of the process based on a temperature difference of 10 K over the stripper and a reflux ratio of 2. The desorption energy has been estimated from the equilibrium curve.
Energy consumption and cyclic loading of CORAL are similar to those of MEA which makes CORAL a good alternative to MEA.

Table 1. Estimated energy consumption for new CORAL aborption liquids

Desorption	60 kJ/mole CO_2
Evaporation	80 kJ/mole CO_2
Reheating	40 kJ/mole CO_2
Total	180 kJ/mole CO_2

Measurement of mass transfer

The performance of the membrane gas absorption system using CORAL liquids can be characterised by a mass transfer coefficient. The mass transfer coefficient for the CORAL liquids has been determined in a small scale pilot plant under the relevant conditions. The pilot plant allows for investigation of the absorption process and the desorption process. The design specifications of the pilot plant are shown in Table 2.

Table 2. Specification of CO_2-removal pilot plant

Inlet CO_2-concentration	3 - 40 %
Gas flow rate	0.4 - 4 m^3/hr
CO_2-removal	90%
Liquid flow rate	1-20 l/hr

The pilot plant flow sheet is shown in Fig. 3. CO_2 and air are mixed via mass flow controllers and fed to a prototype cross-flow membrane absorber (DAM-module). In this absorber, the gas stream CO_2 is transferred to a liquid stream in a countercurrent fashion. The rich liquid is fed to the top of a stripper via a plate heat exchanger. The stripper temperature is maintained at a temperature of around 105 °C by an electric heating element. As a result of the temperature increase of the absorption liquid, CO_2 is liberated from the solution. Water is recovered in the condensor and trickles back into the stripper. The lean liquid exits the stripper at the bottom and is fed back to the membrane absorber via the heat exchanger and a storage vessel.

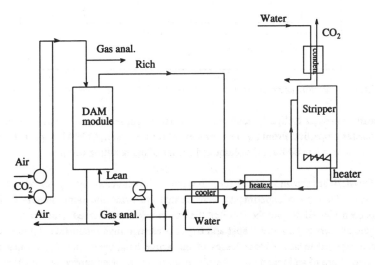

Fig. 3   Pilot plant flow sheet

The specifications of the prototype cross-flow membrane module designed according to a TNO patent are shown in Table 3.

Table 3. Specifications prototype cross-flow membrane module

Membrane area	0.275 m^2
Fibre material	polypropene
Pore size	0.2 μm
Fibre diameter (inside/outside)	0.6/1.0 mm

Results

In general, the mass transfer coefficient is dependent on a large number of parameters:
— Temperature, which not only influences the reaction kinetics but also the transport properties of the aborption liquid,
— CO_2 concentration on the gas side,
— Concentration of CO_2-binding components,
— Hydrodynamic conditions in the absorber.

The performance of a membrane gas absorption system based on monoethanolamine can be estimated from first principles as this system is well documented (see e.g. [4]). For the CORAL liquids this exercise cannot yet be carried out. It is however possible to compare the actual performance of CORAL liquids with theoretical estimates for an MEA-solution of identical composition. The results of this comparison are shown in Fig. 4 for solutions containing 2 M of CO_2 binding components. The mass transfer coefficient for 2 M MEA has been estimated at the mean conditions in the membrane absorber. Results of a number of experiments have been expressed as a mass transfer coefficient as a function of the liquid loading.

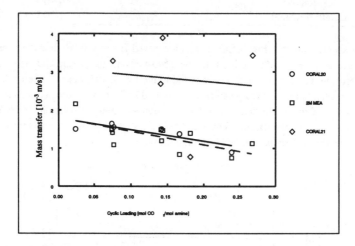

Fig. 4   Performance comparison between CORAL liquids and MEA at 2 M

CORAL20 appears to possess a slightly better mass transfer properties than 2 M MEA, whereas CORAL21 has superior mass transfer properties. From the point of view of mass transfer, CORAL appears to be good alternative for MEA. This should result in additional volume reductions of the resulting equipment.

Furthermore, CORAL20 has shown stable performance over a discontinuous testing period of four months. During this period absorption/desorption experiments were carried out using the same amount of liquid. No loss in performance has been noticed despite the fact that the experiment were carried with a mixture of air and carbon dioxide. The liquid appears to be more robust and oxygen-resistant than monoethanolamine which suffers from degradation when exposed to air. Vapour losses of the active component can also be neglected as the CO_2-absorbing components are of an ionic nature. Results from corrosion tests carried out at stripper conditions have revealed that the use of carbon steel is quite adequate for all of the novel absorption liquids. This will lead to a cost advantage over MEA systems.

CO₂-PRODUCTION FOR GREENHOUSES

Current research has focused on the production of carbon dioxide for horticultural applications (Feron and Jansen, 1995). In the Netherlands carbon dioxide is used to promote plant growth in greenhouses. Depending on the crop, the production can be increased by 25% by increasing the CO_2-concentration to 500 ppm. In temperate climates, it is necessary to heat these greenhouses which makes the horticultural industry in the Netherlands a large consumer of natural gas. Future heat supply systems will be largely based on cogeneration plants (gas engines and gas turbines) as significant energy savings can be achieved. Energy savings can be increased if the heat supply is coupled to carbon dioxide supply. This will incite growers to buy heat from the cogeneration plant rather than to use their own boilers. These boilers are needed to cover peak demand and emergencies. Also the amount of available carbon dioxide per unit of heat will be larger in case of combined production of heat and electricity, thus allowing for improved production. The heat demand and CO_2 demand are generally anti-cyclic. During periods of high heat demand (during winter and at night) little or no CO_2 is needed to maintain the desired CO_2 levels in the greenhouse. During periods of high CO_2 demand (during summer and during the day) little or no heat is needed. The thermal energy of the cogeneration plant can either be used for greenhouse heating or for the production of carbon dioxide. This will also extend the operating time of the cogeneration plant to daytime hours which is an additional benefit in terms of capacity credits. A schematic drawing of a cogeneration plant supplying heat and carbon dioxide to greenhouses in shown in Fig. 5.

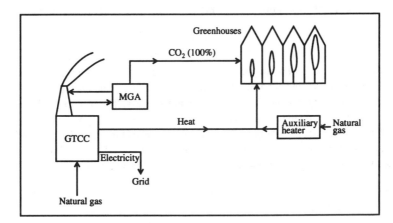

Fig. 5   Cogeneration plant supplying heat and CO_2 to greenhouses

Energy savings

Table 4 gives some estimates for the energy savings for a greenhouse area of 100 ha (Annual heat demand: 1.9 PJ). The energy savings are relative to the situation in which heat and CO_2 are supplied by a gas-fired heater. The predicted savings are based on the comparison between separate generation of electricity ($\eta = 0.4$) and heat ($\eta = 0.95$) and cogeneration by a gas turbine combined cycle (GTCC, $\eta_{th} = 0.35$, $\eta_e = 0.45$). The energy required for CO_2-production is taken to be 5 GJ/tonne.

Table 4.   Energy savings due to heat supply from a GTCC with and without CO_2-supply for a greenhouse area of 100 ha

Option	Heat delivered by cogeneration plant	Primary energy savings	CO_2 avoided
Heat supply from gas turbine combined cycle	54%	1.4 PJ/yr (29.5 %)	81 kton/yr
Heat and CO_2 supply from gas turbine combined cycle	80%	2.2 PJ/yr (31.4 %)	123 kton/yr

Table 4 shows that the primary energy savings are about 50% higher when the power plant supplies heat as well as carbon dioxide.

Economic perspective

Although the energy saving potential shown in Table 4 is large, the CO_2 produced should not be at a higher cost than the currently available sources. These sources are either production of carbon dioxide by burning natural gas or supply of carbon dioxide by tanker. Table 4 shows cost estimates for the available sources and the production of carbon dioxide by a membrane gas absorption process. The costs are based on an annual CO_2-production of 32.5 ktonne from the exhaust of a gas turbine combined cycle (CO_2-concentration = 3.5%).

— Natural gas: NLG 0.22 per m^3 equivalent with NLG 7/GJ
— Reboiler duty 5 GJ/tonne CO_2
— Steam: NLG 7/GJ

Table 5.   Costs of CO_2-delivery methods (1 NLG = 0.6 US$)

CO_2-delivery method	Costs (NLG/tonne)
Burning of natural gas	120
Delivery by tanker	200
CO_2-production from flue gas with MGA	105

Table 5 indicates that the production of CO_2 from flue gases using membrane gas absorption and delivery to the greenhouses is economical compared to the alternatives. The costs do not yet take into account profits which arise from the increased capacity credit resulting from the increased operating time of the power plant.

Development project

A development project is currently underway in which a scaled up version of the existing pilot plant will be built (2 kg/hr CO_2). The goal of this development is an assessment of the economic feasibility of membrane gas absorption for carbon dioxide production for the horticultural industry. The project is primarily funded by the Netherlands agency for energy and the environment (NOVEM) but financial support has also been received from representatives of the energy industry, a plant manufacturer, a membrane manufacturer and XTO-membrane technology, a joint venture between TNO and X-Flow, a membrane manufacturer. The latter company aims to commercialise the TNO cross-flow membrane module design for contacting duties.

CONCLUSIONS

Membrane gas absorption is a cost-effective technique for the production of carbon dioxide from flue gases. Novel absorption liquids are introduced which lead to a stable system in combination with porous polyolefin membranes. These liquids also show improved mass transfer compared to monoethanolamine.

The combined supply of heat and carbon dioxide from cogeneration plants to greenhouses leads to significant energy savings (±30%) compared to the simultaneous supply by a natural gas-fired heater. Carbon dioxide produced by the membrane gas absorption process can be delivered to greenhouses at lower cost than conventional methods.

REFERENCES

Falk-Pedersen, O., Y. Bjerve, G. Glittum and S. Rønning (1995). Separation of carbon dioxide from offshore gas turbine exhaust. Energy Convers. Mgmt. 36, 393-396

Feron, P.H.M., A.E. Jansen and R. Klaassen (1992). Membrane technology in carbon dioxide removal. Energy Convers. Mgmt. 33, 421-428

Feron, P.H.M. (1994). Membranes for carbon dioxide recovery from power plants. Ch. 26 in "Carbon Dioxide Chemistry: Environmental Issues" (eds. J. Paul, C.M. Pradier), The Royal Society of Chemistry, Cambridge

Feron, P.H.M. and A.E. Jansen (1995). Capture of carbon dioxide using membrane gas absorption and reuse in the horticultural industry. Energy Convers. Mgmt. 36, 411-414

Jansen, A.E. and P.H.M. Feron (1995). Method for gas absorption across a membrane. International Patent Application Number PCT/NL95/00116

Matsumoto H., T. Kamata, H. Kitamura, M. Ishibashi, H. Ohta, and N. Nishikawa (1994). Fundamental study of CO_2 removal from the flue gas of thermal power plant by hollow-fiber gas-liquid contactor. Ch. 27 in Carbon Dioxide Chemistry: Enviromental Issues (eds. J. Paul, C.M. Pradier), The Royal Society of Chemistry, Cambridge

Nishikawa, N., M. Ishibashi, H. Ohta, N. Akutsu, H. Matsumoto, T. Kamata and H. Kitamura (1995). 'CO_2 removal by hollow-giber gas-liquid contactor', Energy Convers. Mgmt., 36, 415-418

Nii, S. and H. Takeuchi (1994). Removal of CO_2 and/or SO_2 from gas streams by a membrane absorption method. Gas Separation & Purification, 8, 107-114

Energy Convers. Mgmt Vol. 38, Suppl., pp. S99–S104, 1997
Published by Elsevier Science Ltd
Printed in Great Britain
0196-8904/97 $17.00 + 0.00

Pergamon

PII: S0196-8904(96)00253-1

A NOVEL CARBON FIBER BASED MATERIAL AND SEPARATION TECHNOLOGY

TIMOTHY D. BURCHELL AND RODDIE R. JUDKINS

Oak Ridge National Laboratory
Oak Ridge, Tennessee 37831-6088, USA

ABSTRACT

Our novel carbon fiber based adsorbent material shows preferential uptake of CO_2 over other gases. The material has a unique combination of properties, which include a large micropore volume, a large BET surface area, and electrical conductivity. These properties have been exploited to effect the separation of CO_2 from a model gas (CH_4). Enhanced desorption is achieved using an electrical current passed through the material at low voltage. The manufacture, characterization, and CO_2 adsorption behavior of the materials is reported here, along with our novel electrical swing separation technology.   Published by Elsevier Science Ltd

KEYWORDS

Activated carbon fibers, porous composites, gas separation, electrical swing adsorption.

INTRODUCTION

The removal of CO_2 is of significance to the combustion of fossil fuels which releases large volumes of CO_2 to the environment. Several options exist to reduce CO_2 emissions, including substitution of nuclear power for fossil fuels, increasing the efficiency of fossil plants, and capturing the CO_2 prior to emission to the environment. All of these techniques have the attractive feature of limiting the amount of CO_2 emitted to the atmosphere, but each has economic, technical, or societal limitations. Gas separation is, therefore, a relevant technology in the field of energy production. A novel separation system based on a parametric swing process has been developed that utilizes the unique combination of properties exhibited by our carbon fiber composite molecular sieve (CFCMS).

The CFCMS is a monolithic activated carbon composed of petroleum pitch-derived carbon fiber and a phenolic resin-derived binder (Burchell et al., 1994). The binder phase content is quite low and provides a monolithic structure by bonding the fibers at their contact points only, thus rendering the CFCMS macroporous. The CFCMS is a highly adsorbent material with very little resistance to bulk gas flow. Experiments conducted at ORNL have shown the CFCMS to have a high affinity for carbon dioxide compared to conventional granular activated carbons (Burchell and Judkins, 1996). Moreover, granular carbon systems are subject to attrition due to abrasive wear in service, and channeling of the gas being processed as a result of inhomogeneous packing in the beds. The use of the CFCMS would permit the employment of, for example, horizontally-oriented vessels with controlled flow of the gas with the contaminants/diluents to be adsorbed through the adsorbent without risk of channeling and bypass flows. Here, we present the results of several experiments which demonstrate the performance of CFCMS and the electrical swing adsorption process for CO_2 separation.

EXPERIMENTAL

The carbon fiber composite molecular sieve (CFCMS) synthesis route is illustrated in Fig.1. Isotropic pitch-derived carbon fibers are mixed with powdered phenolic resin and water to form a slurry. The slurry is transferred to a molding tank and the water drawn through a porous screen under vacuum. The resultant green artifact is dried, cured at 60°C in air, and stripped from the mold screen. The composite is cured at ~150°C in air prior to carbonization at 650°C in an inert gas. The final synthesis stage involves activation of the composite in moisture saturated He in the temperature range 800-950°C.

Porosity characterization was performed using nitrogen adsorption and mercury intrusion. Nitrogen adsorption isotherms were measured at 77 K using our Autosorb-1 instrument. Micropore size analysis used a variety of methods, including the Brunauer, Emmett, and Teller (BET) (Brunauer et al., 1938) method for surface area, the Dubinin-Astakhov (DA) (Bansal et al., 1988) method for micropore radius, and the t-method for micro-pore volume. CO_2 adsorption isotherms for CFCMS were obtained using the Autosorb-1 apparatus over the pressure range 0.1 to 760-mm Hg and at temperatures of 30, 60, and 100 °C. High pressure CO_2 adsorption data (850 psi max.) were obtained using a gravimetric apparatus, courtesy of Westvaco Corporation.

A schematic diagram of our experimental breakthrough apparatus is shown in Fig. 2. The feed gas flows through a CFCMS sample (25.4-mm diameter and 76.2-mm long) to a vent. A sample of the downstream gas is fed to a mass spectrometer allowing on-line monitoring of the exit gas composition. Flowmeters (F1 & F2) are positioned on the inlet and outlet sides of the CFCMS sample. Electrical leads are connected to each end of the sample creating an electric circuit through the sample. On saturation of the CFCMS with, for example, carbon dioxide, immediate desorption can be accomplished by application of very low voltages (in our experiments we have used 0.5-1 volt) across the CFCMS adsorbent.

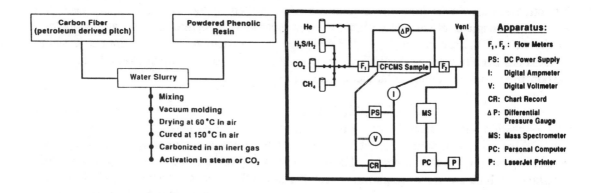

Fig. 1. The CFCMS synthesis route.   Fig. 2. Schematic drawing of our break-through apparatus.

RESULTS AND DISCUSSION

A series of CFCMS cylinders were prepared and activated to burn-offs ranging from 9 to 36% and the BET surface area and micropore size/volume determined from the N_2 adsorption isotherms. Table 1 reports the mass and average burn-off for each of the four cylinders (25-mm diameter x 75-mm length), and their BET surface area, micropore volume (t-method), and mean micropore radius (DA method). Samples were taken from the top (T) and bottom (B) of each cylinder for analysis. Where the measured BET surface areas were widely different between the top and bottom of the cylinders a repeat measurement was performed.

Table 1. Micropore analysis data for activated CFCMS samples

Specimen	Mass (g)	Burn-off (%)	BET Area (m^2/g)	Pore Volume [t-method] (cm^3/g)	DA Pore Radius (nm)
21-11 T	11.96	9	485	-	0.70
21-11 B	11.96	9	540	0.212	0.65
21-2B T	11.01	18	770	0.282	0.72
21-2B B	11.01	18	1725	0.603	0.70
21-2B B (repeat)	11.01	18	961	0.328	0.71
21-2D T	9.86	27	939	0.305	0.75
21-2D B	9.86	27	2470	0.866	0.75
21-2D B (repeat)	9.86	27	2477	0.791	0.75
21-2C T	8.86	36	923	0.235	0.87
21-2C B	8.86	36	2323	0.723	0.75
21-2C B (repeat)	8.86	36	856	0.270	0.75

At high burn-offs there was a tendency for one end of the cylinder to exhibit higher BET surface areas than the other end. The activation fixture used in this work was designed to distribute the saturated He along the length of the CFCMS cylinder. Despite this, activation to high burn-off does result in a non-uniform activation as indicated by the BET data. The BET surface area increases with burn-off, approaching 2500 m^2/g at >25% burn-off. The micropore size (DA pore radius) is apparently less sensitive to burn-off (Table 1) and increases only slightly over the weight loss range reported here. The pore volume (t-method) varies with the BET surface area, increasing with burn-off. The observed variations in BET surface area and micropore volume and radius are in agreement with our previous data (Burchell and Judkins, 1996).

Adsorption isotherms were obtained at temperatures of 30, 60, and 100°C at pressures up to one atmosphere for CO_2 and typical data are shown in Fig. 3. All of the CFCMS samples analyzed adsorbed less CO_2 at 60 and 100°C than at 30°C. At 100°C the amount of CO_2 adsorbed was approximately one third that adsorbed at 30°C. Figure 4 shows CO_2 adsorption isotherms at 25°C for CFCMS specimens 21-11 and 21-2B over the pressure range 0.5-58 bar (8-850 psi).

The measured volummetric (Fig. 3) and gravimetric (Fig. 4) adsorption capacities at one atmosphere for CO_2 are in good agreement for the CFCMS specimens. At one atmosphere, approximately 100 mg of CO_2 per g of CFCMS was adsorbed, rising to >490 mg/g (Fig. 4 specimen 21-2B).

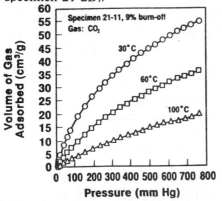

Fig. 3. Carbon dioxide adsorption isotherms at 30, 60 and 100 °C on CFCMS (9% burn-off).

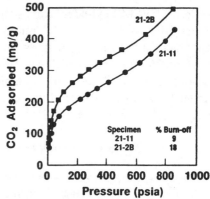

Fig. 4. Carbon dioxide isotherms at 298K on CFCMS activated to different burn-offs (9 and 18%).

The CO_2 adsorption data discussed above suggest that CFCMS might provide for the effective separation of CO_2 from gases such as CH_4 or flue gas mixtures. To determine the efficacy of CFCMS for this purpose, several steam-activated samples were tested in our breakthrough apparatus (Fig. 2). Figure 5 shows a breakthrough plot for CO_2 on CFCMS specimen 21-11. Initially the CFCMS sample is purged with helium to drive out any entrained air. The input gas is then switched to CO_2 at a flow of 0.1 standard liters per minute (slpm) and the He concentration falls to zero. The CO_2 concentration stays constant at a low level because the CO_2 is adsorbed. After approximately nine minutes, the CO_2 concentration begins to rise as the CFCMS becomes saturated with CO_2 (i.e., breakthrough occurs). The CO_2 adsorption capacity can thus be calculated from the gas flow rate and the breakthrough time. In this instance, specimen 21-11 adsorbed 0.9 liters of CO_2. A typical breakthrough plot for a CH_4/CO_2 mixture is shown in Fig. 6. Any entrained air is initially driven out with a He purge. The input gas is then switched to a 2:1 mixture of CH_4/CO_2 at a flow rate of 0.33 slpm. The outlet stream He concentration decreases and the CH_4 concentration increases rapidly (i.e., CH_4 breaks through). Adsorption of CO_2 occurs and, therefore, the CO_2 concentration remains constant at a low level for approximately six minutes before the CO_2 concentration begins to increase (i.e., "breakthrough" occurs). Breakthrough experiments were performed on four CFCMS samples with burn-offs ranging from 9 to 36%. In the case of the pure CO_2 gas experiments, the amount of CO_2 adsorbed exceeded 0.8 liters on 0.037 liters of CFCMS. The maximum CO_2 uptake during mixed gas breakthrough studies was observed for the lowest burn-off specimen, and was 0.73 liters of CO_2 adsorbed on a 0.037 liter CFCMS sample.

CFCMS has a continuous carbon skeleton which imparts electrical conductivity to the material. The carbon fibers used in the synthesis of CFCMS have, according to their manufacture, an electrical resistivity of 5 milliohm·cm. Figure 7 is a plot of the current-voltage and power-temperature characteristics of a 2.5-cm diameter, 7.5-cm long CFCMS cylinder. At an applied d.c. potential of one volt, approximately 5 amps flows through the CFCMS causing a temperature increase to 50-60°C. The CFCMS samples electrical resistance is thus 0.2 ohm and the resistivity is 131 milliohm·cm. This resistivity is considerably greater than that of the fibers, and is attributed to contact resistance and the lower electrical conductivity of the phenolic resin-derived carbon binder. We have utilized the electrical properties of CFCMS to effect a rapid desorption of adsorbed gases in our breakthrough apparatus. The process has been named electrically enhanced desorption, and the benefit of this technique is demonstrated in Fig. 8 where the CO_2 and CH_4 gas concentrations in the outlet gas stream of our breakthrough apparatus (Fig. 2) are shown. Three adsorption/desorption cycles are shown in Fig. 8. In the first and second cycles (A and B in Fig. 8) desorption is caused by the combined effect of an applied voltage (1 volt) and a He purge gas. In the third cycle (C in Fig. 8) desorption is caused only by the He purge gas. A comparison of cycles B and C indicated that the applied voltage reduces the desorption time to less than one-third of that for the He purge gas alone (cycle C). Clearly, the desorption of adsorbed CO_2 can be rapidly induced by the application of an electric potential. A separation may thus be effected on the basis of an "electrical swing."

An "electrical swing" separation system would appear to have inherent advantages over both pressure and temperature swing adsorption systems because separation can be achieved without altering system pressure or applying external heat.

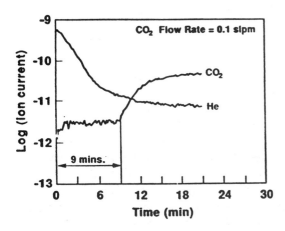

Fig. 5.   Typical CO_2 breakthrough plot CFCMS sample 21-11 (9% burn-off).

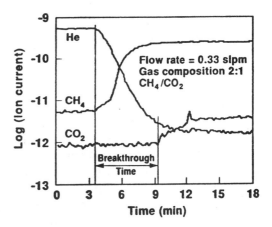

Fig. 6.   Typical CO_2/CH_4 breakthrough on plots on CFCMS sample 21-11 (9% burn-off).

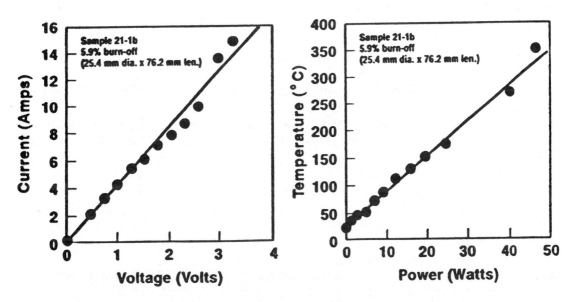

Fig. 7.   The current-voltage and power-temperature relationship for CFCMS (sample 21-2B, 18% burn-off, 25-mm diameter x 76-mm length).

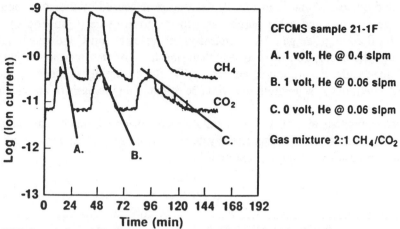

Fig. 8. CO_2/CH_4 breakthrough plots on CFCMS showing the benefit of electrically enhanced desorption.

CONCLUSIONS

A porous monolithic activated carbon material (CFCMS) has been developed that is strong, rigid, and which overcomes problems associated with operation using granular adsorbents. The open structure of CFCMS results in a permeable material which offers little resistance to the free-flow of fluids. The material has a unique combination of properties, including reasonable compressive strength, electrical conductivity, and a large micropore volume. CO_2 isotherms have been obtained for samples of CFCMS both volumetrically and gravimetrically. At 30°C and atmospheric pressure the CFCMS material has a CO_2 uptake of >50 cm^2/g (>100 mg/g). The CO_2 uptake is reduced at elevated temperature, and falls to approximately 20 cm^2/g (40 mg/g) at 100°C. However, the adsorption of CO_2 increases with increasing pressure such that at 58 bar and 25°C the mass of CO_2 adsorbed increases to >490 mg/g. A series of breakthrough experiments were performed on CFCMS specimens and their ability to selectively remove CO_2 demonstrated. The unique combination of properties of CFCMS has been exploited to effect the rapid desorption of CO_2 from the material using electrically enhanced desorption under a low applied voltage.

REFERENCES

Bansal, R.C., J-B Donnet and F. Stoeckli (1988). *Active Carbon*, Pub. Marcell Dekker, Inc., New York.

Brunauer, S., P.H. Emmett and E. Teller (1938). *J. Am. Chem. Soc.*, Vol. 60, p. 309.

Burchell, T.D. and R.R. Judkins, (1996). *Energy Convers. Mgmt*, Vol. 37, Nos 6-8, pp. 947-954.

Burchell, T.D., C.E. Weaver, F. Derbyshire, Y.Q. Fei and M. Jagtoyen (1994). "Carbon Fiber Composite Molecular Sieves: Synthesis and Characterization." In: *Poc. Carbon '94*, Granada, Spain, July 3-8, 1994. Pub. Spanish Carbon Group, July 1994.

ACKNOWLEDGEMENTS

Research sponsored by the U.S. Department of Energy, Office of Fossil Energy, Advanced Research and Technology Development Materials Program [DOE/FE AA 15 10 10 0, Work Breakdown Structure Element ORNL-1(E)] under contract DE-AC05-96OR22464 with Lockheed Martin Energy Research Corp.

Pergamon

PII: S0196-8904(96)00254-3

Energy Convers. Mgmt Vol. 38, Suppl., pp. S105–S110, 1997
© 1997 Elsevier Science Ltd. All rights reserved
Printed in Great Britain
0196-8904/97 $17.00 + 0.00

ROLE OF ACTIVATED CARBON PELLETS IN CARBON DIOXIDE REMOVAL

S.C. Sarkar, A. Bose

Advanced Centre of Cryogenic Research-Calcutta
P.O. Jadavpur Universirty, P.B. No. 17005
Calcutta-700 032 (India)

ABSTRACT

The removal of carbon dioxoide from gas/air streams is more often becoming necessary in many industries for different purposes. In cryogenic air separation plant, air has to be free from carbon dioxide before its liquefaction otherwise blockage due to freezing of heat exchange equipment would result. Enrichment of methane in biogas to have fuel of higher calorific value can be achieved by removing carbon dioxide. Elimination of carbon dioxide from the flue gas helps to increase its calorific value as well as to eliminate the greenhouse gas. The carbon dioxide thus generated can be utilised as an effective refrigerant.

Carbon dioxide could be removed in several ways such as by chemical separation, membrane separation, cryogenic separation as well as by adsorption, the latter playing a vital role if suitable adsorbent material is available.

We have developed activated carbon pellets from coconut shell and also designed a PSA adsorption system for studying the effectiveness of the activated carbon pellets in separating carbon dioxide.

This paper describes the physical properties of the carbon pellets, preliminary findings on carbon dioxide adsorption and the development of an adsorption system for gas separation/purification.   © 1997 Elsevier Science Ltd

KEYWORDS
Activated carbon, porosity, break through, pressure swing adsorption, gas separation.

INTRODUCTION

Carbon dioxide removal may be achieved by cryogenic separation, chemical separation, membrane separation or adsorption methods. Of all these methods, the adsorption-desorption method becomes a promising one when low cost adsorbent materials are available.

Activated Carbon can be produced from any carbonaceous material such as wood, lignite, anthracite, bone etc. A high quality gas adsorption grade activated carbon can be produced from coconut shell a waste product (sarkar et al 1996). In India, a huge quantity of activated carbon is imported. Activated carbon may be of different form such as powdered, granular or pelletised material . For gas separation and purification, regular shaped pellets are preferable to irregular granular material due to a lower fluid pressure drop. In this context the development of activated carbon pellets from coconut shell becomes a necessity.

PREPARATION OF ACTIVATED CARBON PELLETS

i)   Coconut shell particles (0.5-1.0 Cm) were carbonised by destructive distillation in a furnace at 550 C for one hour for completion of the carbonisatioin process.

ii)   Powdered char was then mixed with phenol-formaldehyde resin as the binder. The proportionate weight of binder and char was adjusted in such as a way so as to give reasonable strength in the finished pellets. The pellets were produced from the char resin mixture in a pelletizer machine by application of pressure. The pellets were of cylindrical shape having a diameter of 0.5 Cm and length of 0.4 Cm.

iii) These carbon pellets were then treated in an activating furnance and steam at a pressure of 1.5 bar was passed through it at a temperature of 750 C for 2.0 hour to produce the ACP-750-2.0, grade acticated carbon pellets.

Time and temperature of carbonisation and activation have been fixed after conducting several experiments so as to produce a high surface area at a good yield of finished product.

CHARACTERIZATION OF ACTIVATED CARBON DEVELOPED

Active carbons are characterised by their strong adsorption capacity which occurs mostly in the micropores. The corrersponding adsorption process is described by Dubinin theory for the filling of micropores. Active carbons also contain mesopores and macropores. Therefore, surface area measurements and pore volume determinations are important parameters in assessing the quality of gas adsorption grade activated carbon.

Macroporosity and mesoporosity can be assessed by mercury porosimetry, a simple technique based on the penetration of mercury into pores at a pressure of about 4000 bar. This work was undertaken usingthe QUANTA CHROME instrument. The total pore volume measured by this technique is $0.2852 cm^3/g$ of the activated carbon sample.

To assess pores below 18A, a nitrogen adsorption technique was utilised. A QUANTA CHROME AUTOSORB instrument was used to measure multipoint surface area, pore volume and the adsorption isotherm. The results are given below in Tables 1 and 2. The Fig.-1 represents the adsorption desorption isotherm.

Table - 1. Surface area and micropore volume of activated carbon pellet.

Sample	Surface area	Micropore volume
ACP-750-2.0	$1018.0 m^2/g$	$0.4008 cm^3/g$

Table - 2 : Adsorption desorption isotherm.

P/P$_o$	Volume Cm^3/g STP
0.0957	301.680
0.1958	329.201
0.3030	347.041
0.3957	358.443
0.5005	368.723
0.6043	377.489
0.6788	385.241
0.7976	394.671
0.8997	408.998
0.9930	426.134
0.8959	417.382
0.8008	410.821
0.6956	403.105
0.6040	396.607
0.5020	385.312
0.3981	359.377
0.2983	346.983
0.1983	330.371
0.0988	304.025

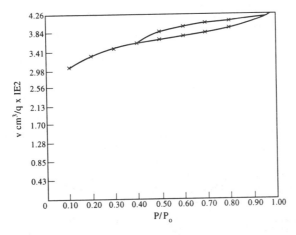

Fig. 1. Isotherm.

Pore volume determination and adsorption isotherm data are essential for understanding the adsorption process. To determine the void fraction in the adsorption column, it is necessary to have a knowledge of true density and bulk or apparent density of the pellets. Bulk density and the true density of the sample were found to be 0.53g/cm^3 and 1.2802g/cm^3 respectively.

GAS ADSORPTION SYSTEM

For measuring the effectiveness of carbon dioxide removal by activated carbon, a preliminary experiment was conducted in a Cryocalorimeter developed under the project. Fig.-2 shows a schematic of the calorimeter construction. For adsorption experiments the adsorption cell was utilized. The cylindrical adsorption cell is of 14.7cm length, 2.25cm internal diameter and 2.3cm outer diameter and it can withstand pressure of upto 10 bar.

The cell is closed at both ends by bolted flanges using teflon gaskets. A
nichrome wire heater coil (F) shown in Fig.-2 was not necessary for the
present experiment. The gas can enter the adsorption chamber through a

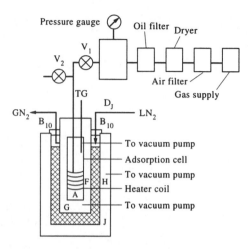

Fig. 2. Experimental set up of cryocalorimeter.

thin stainless pipe of 0.2 Cm outer diameter when valve V_1 is open and
valve V_2 is closed. The copper constantan thermocouple sensor (TC) is
used to measure the temperature of the adsorption chamber. All joints
are made vacuum tight by bakelite plugs and adhesive.

The adsorption cell was first filled with 21g of activated carbon pellets
and carbon dioxide gas from the cylinder was passed to the cell at
different pressure in the range of 3.5 bar. After adsorption was
completed as indicated by steady temperature, pressure was released and
the desorbed gas passed through valve V_2 (with valve V_1 remaining
closed). Rise in temperature of the cell was noted in each case. It was
observed that with the increase of pressure, adsorption was increased.
Beyond a pressure of 3.5 bar, adsorption did not increase substantially.
Amount of adsorption is indicated by the heat of adsorption and hence
by a rise in temperature. The adsorption of carbon dioxide at 3.5 bar
pressure and its desorption was repeated several times with the same
sample without further activation. It was found that the rise in
temperature in each case was about 12 C. Therefore, it can be concluded
that the amount of adsorption is sufficiently high at moderate pressure
and that the process is reversible. Hence it can be operated
continuously if two adsorbers can be made to operate consequtively. The
temperature rises by 12 C within 1.5 to 2.0 minutes, indicating that the
break through takes places in this period. Using the same adsorption
cell and the activated carbon pellets temperature rise due to air at 3.5
bar was noted to be 2.0 C. Therefore, from a mixture of air and carbon
dioxide, the carbondioxide will be more readily adsorbed.

Various designs of pressure swing adsorption systems are available in
the literature (Doong et al 1986, Hossen et al 1986, Mitchall 1971;
Mutusim et al 1992, Shendalamen et al 1972, Skarstrom et al 1959,
Turnock et al 1971). Following the above experimental findings, a gas
adsorption system based on Pressure Swing Adsorption. (PSA) has been
designed and fabricated using stainless steel as the construction
material. Fig.-3 shows a schematic of this system.

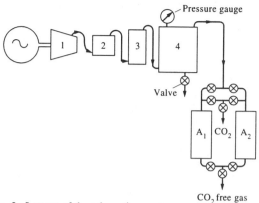

Fig. 3. Layout of the adsorption system.

Compressed air/gas containing carbon dioxide is passed through an air filter (1), oil filter (2) and finally through a silicagel chamber (3) to remove any dust, oil and moisture in it. The dry air/gas is then stored in the buffer vessel (4) from where it enters the actual adsorber zone. There are two adsorbers A_1 and A_2. when A_1 is in operation, A_2 is in regeneration and vice versa, thereby ensuring continuous removal of carbon dioxide (CO_2). Both adsorbers are cylindrical in shape having a diameter of 5 cm and length of 35 cm. Switch over time is   to   be fixed after determining the break through point i.e. the incoming and outgoing gases are to be analysed by Gas Chromatograph for finding the saturation time. The entire process can then be made automatic by using Solenoid valves controlled by a programmable sequence timer to be designed for this purpose.

RESULTS AND DISCUSSION

The carbon dioxide adsorption capacity of the carbon pellets at moderate pressure is quite high. It can be seen from the Table - 3 that easily liquefiable gases having large cohesive forces, are more readily adsorbed (Rakshit, 1970).

Table -3 **Volume of gas adsorbed per 1g of coconut charcoal at 15 C under 1 atm.**

Gases	Vol. adsorbed (cm^3)	B.P. (K)	Critical Temp.
O_2	8.2	90.2	154
N_2	8.0	77.3	126
H_2	4.7	20.3	133
CH_4	16.2	111.7	190
CO_2	48.0	-	304

Hence carbon dioxide can be removed from its mixture with methane, oxygen, nitrogen, hydrogen etc. by the pressure swing adsorption system discussed previously. Moreover, an alternative refrigeratioin cycle could be developed based on the use of a carbon dioxide-carbon adsorption unit.

CONCLUSION

Preliminary experimental findings are presented on an activated carbon based gas adsorption system for CO_2 separation. Using this system, it is planned to separate CO_2 from gas mixtures containing air and methane and to develop an alternate refrigeration cycle.

ACKNOWLEDGEMENT

Financial assistance from the Department of Science and Technology, Govt. of India is greatefully acknowledged. Authors are also thankful to the technical staff members of the Advanced Centre of Cryogenic Research - Calcutta namely, S. Jana, N.R. Chakraborty, A. Mukherjee, A. Sarkar, H.P. Dey, S. Das, B. Ghosh, S. Sarkar and Dr. D. Kundu of CGCRI-Calcutta for their help during fabrication work and in conducting the experimental study.

REFERENCES

Doong, S.J. and Young, R.T. (1986). Bulk separation of multi-component gas mixture by pressure swing adsorption pore/surface diffusion and equilibrium models. A.I.Ch.E.J. 42 397.

Hossen, M.M. Ruthven, D.M. and Raghavan, N.S. (1986). Air separation by pressure swing adsorption on a carbon molecular sieve. Chem. Engineering Science, 41, 1393.

Mitchall, J.E. and Schendalam, C.H. (1971). a study of heatless adsorption in the model system carbon dioxide in helium A.I.Ch.E. Symp. Ser. 69, 25.

Mutuzin, Z.Z. and Brown, J.J. .(1992). Multicomponent Pressure Swing Adsorption for non isothermal, non equilibrium conditions, Trans. I Chem. E. 70, 346.

Rakshit P.C. (1970) Physical Chemistry.

Shendalaman, L.H. and Mitchall, J.E. (1972). A study of heatless adsorption system in the model system carbon dioxide in helium. Chem. Engineering Science, 27, 1449.

Skarstrom, C.w. 1959. Use of adsorption phenomena in automated plant type gas analyser. Ann. N.Y. Academic Science, 72, 751.

Sarkar, S.C. and Bose, A. 1996. Activated carbon from cononut shell for cryogenic application-a review. J. Min. met and Fuels, 44, 117.

Turrock, P.H. and kadlec R.H. (1971). Separation of nitrogen and methance via periodic adsorption. A.I.Ch.E.J., 17, 225.

 **Pergamon**

Energy Convers. Mgmt Vol. 38, Suppl., pp. S111–S116, 1997
© 1997 Elsevier Science Ltd. All rights reserved
Printed in Great Britain
0196-8904/97 $17.00 + 0.00

PII: S0196-8904(96)00255-5

Development of Hollow Fiber Membranes
for CO₂ Separation

Y.Tokuda, E.Fujisawa, N.Okabayashi, N.Matsumiya, K.Takagi, H.Mano,

K.Haraya* and M.Sato*

Membrane laboratory, Research Institute of Innovative Technology for the Earth (RITE)

9-2, Kizugawadai, Kizu-cho, Soraku-gun, Kyoto 619-02 Japan

*National Institute of Materials and Chemical Research

Higashi 1-1, Tsukuba, Ibaraki 305 Japan

ABSTRACT

RITE is carrying out a research project to develop membranes for carbon dioxide separation. In the project, we study Cardo type polyimides. Cardo polyimides show good CO_2/N_2 separation factor. We developed new Cardo polyimide which has higher permselectivity, by introduction of a CO_2 affinitive functional group. Furthermore we developed wet-spinning method of the new Cardo polyimide hollow fiber, which has sponge-type support structure and inner defectless skin layer. The fiber has both good permselectivity and practical mechanical strength. Since this spring, bench scale test of CO_2 separations by this hollow fiber was started. © 1997 Elsevier Science Ltd

KEYWORDS

carbon dioxide separation; Polyimides; Cardo; membrane; asymmetric hollow fiber; wet-spinning; bench scale test; simulation

INTRODUCTION

Progress in the development of membranes with a high gas permselectivity for CO_2/N_2 pair is required in order to recover CO_2 from the combustion gases exhausted from stationary CO_2 sources such as the thermal power plants. In the Chemical CO_2 Fixation Project of RITE, we are developing "CO_2 global recycling system". By the system, CO_2 which concentrated by high gas permselectivity hollow fiber membranes reacts with hydrogen to synthesize methanol and other useful chemical substances. As a element technology of this project, we have been developing asymmetric hollow fiber membrane of polyimide with "Cardo" [1] constitutions. In this study, we will report on the improvement of materials, spinning of asymmetric hollow fibers, and bench scale test of CO_2 separations.

IMPROVEMENT OF MATERIALS

Cardo polyimide contains a bulky bis-phenylfuluorene moiety as the loop like moiety. So that Cardo polyimide have characteristics such as a high gas permeability, a high solubility , and a high heat-resistivity. In this study, at first step, we developed main chemical structure of polyimide. At second step, we improved permselectivity by functional group introduction on Cardo constitution.
Fig. 1 shows chemical structure and permselectivity for CO_2/N_2 pair of Cardo polyimide. Permselectivity was measured by high vacuum time-lag method. Cardo polyimides, particularly the ones which have 3,3',4,4'-Benzophenonetetracarboxylic dianhydride as a monomer (PI-BT), show outstanding behavior in the gas permselectivity.
Fig. 2 shows chemical structure and CO_2 separation properties of PI-BT. The selectivity of PI-BT on Fig.2 is higher than Fig.1. The results of Fig.1 was measured by pure CO_2 and N_2, but Fig.2 was by CO_2/N_2 mixed gas. Cardo polyimides usually show higher selectivities at mixed gas measurement, and it's called "Mixed gas effect". We guess that competitive sorption of CO_2 and N_2 cause to the effect.
Fig. 2 also shows chemical structure and CO_2 separation properties of PI-BT-X which introduced several functional groups. These functional groups have CO_2 affinity. As a result of functional

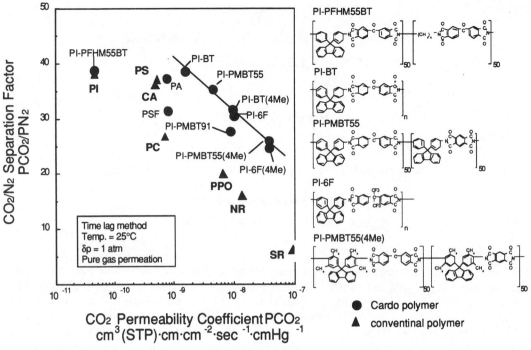

Fig. 1 Performance of Cardo Type Polymeric Membranes

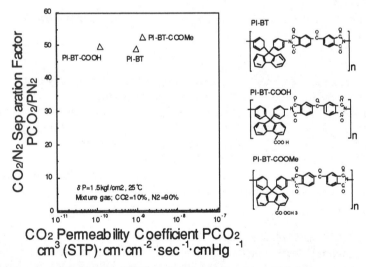

Fig.2 Permselectivity of PI-BT with functional group

group introduction, a PI-BT-X with corboxyl radical on Cardo constitutions (PI-BT-COOH) showed low permeability for CO₂. But PI-BT-X with methylcorboxyl radical (PI-BT-COOMe) showed especially high permselectivity for CO₂/N₂ pair.

ASYMMETRIC HOLLOW FIBER SPINNING

We have investigated the wet-spinning conditions to produce asymmetric hollow fibers with inner skin layers from PI-BT-COOMe as materials. On this investigation, for practical use of hollow fibers, its mechanical strength was considered. Main purpose of the investigation is to make hollow fibers with defectless thin skin layer to improve gas permeabilities without losing selectivities.

SPINNING: Fiber was spun using well-known wet spinning apparatus. [2] Solution (dope) of $16.7 \sim$ 25wt% PI-BT-COOMe in N-Methyl-2-pyrrolidon (NMP), N,N-Dimethylacetamide (DMAc) or mixtures of these solvents were prepared in a bath at 160℃. The solution were pumped through a outer slit of spinneret by 2ml/min, and bore liquid (water) were pumped through a inner tube of the spinneret by 1.5ml/min. The diameter of spinneret is 0.4mm (outer diameter of slit)-0.26mm (inner diameter of slit)-0.13mm (diameter of inner tube) or 1.0-0.6-0.3. After $50 \sim 700$mm distance in the controlled air (Air gap), the fiber wss immersed in a $25 \sim 60$℃ water bath where coagulation occurs. The fiber was continuously rolled on turn-table by $21 \sim 27$m/min. The fiber was drenched in water for complete solvent exchange. Finally dried in air or continuously dried in $100 \sim 230$℃ for $1 \sim 3$min.

GAS PERMEATION TEST: Fiber was fabricated to module, which was $20 \sim 30$ fibers and $150 \sim$ 170mm effective length. 10% CO_2 and 90% N_2 mixture gas was fed to inside of the fiber at 5kgf/cm²G. Permeation side of the fiber 0kgf/cm²G. Permeated gas composition was measured by G.C. and the flow rate was measured by soap film flowmeter.

EXTENSION TEST: A SHIMADZU AGS-100D was used to measure extension properties as practical mechanical strength. Samples were dried in a desiccator for 1 day. Then fixed between sample holders, which has 50mm distance. The extension rate is 10mm/min. The testing was done 10times/sample and testing temperature is 25℃.

RESULTS: Fig. 3 shows effects of polymer concentration of dope to fiber permselectivity and extension modulus. Practically, extension modulus is required over 10% to fabricate to a module. For the requirement, to raise polymer concentration of dope to 20wt% cause extension modulus 10%. By fiber sectional observation, the cause of extension modulus progress is spongenized structure of support layer. Sponge-type structure shows higher extensional strength than finger structure. However polymer high concentration lowers CO_2 permeability, because of thick skin layer.

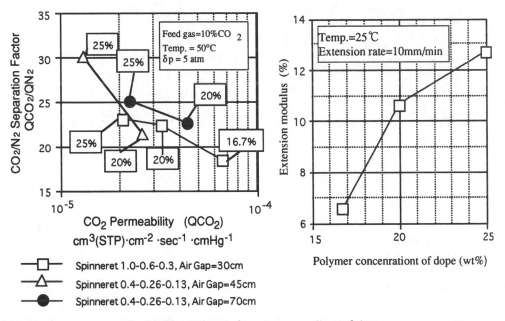

Fig.3 Effect of polymer concenrationt of dope

Fig. 4 shows effects of temperature of coagulation bath to fiber permselectivity and extension modulus. Coagulation at 60 ℃ causes higher CO_2 permeability, but lower separation factor. Temperature of coagulation bath is temperature where outer skin layer formation. So the temperature of coagulation control outer skin layer structure. Fig. 5 shows inner skin structure

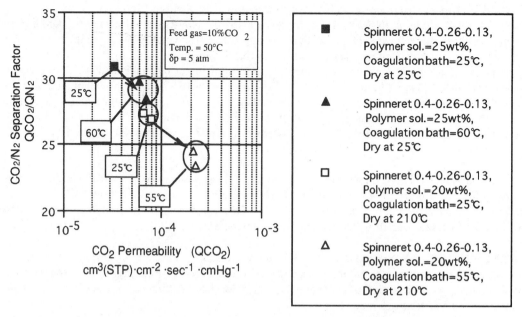

Fig.4 Effect of temperature of coagulation bath

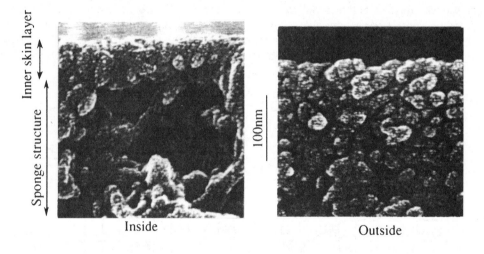

Fig.5 Inner skin layer and Outer surface structure

coagulated by $10\,°C$ bore water and outer surface structure coagulated by $60\,°C$ coagulation bath. Inner skin has continuous polymer skin layer, but outer surface is formed by polymer nodules. And gas may pass through gaps between nodules. The variation of permselectivity by temperature of coagulation bath is caused by outer surface porosity.

Variety of solvents has effect on permselectivity. [3] PI-BT-COOMe is soluble by NMP and DMAc. Fig. 6 shows effects of solvent type of polymer solutions. DMAc has lower solubility than NMP. So each polymer solutions contact to coagulant under same condition, DMAc solution coagulate quickly. As a result, DMAc solution form thinner skin layer, and CO_2 permeability of the fiber raise. On the other hand, defects on the thinner skin layer increase and lead to lower separation factor.

As a result, the relation between the skin layer thickness and formation of defect through wet-spinning conditions was clarified. We were also able to produce hollow fiber with practical mechanical strength, which has good balance between CO_2 permeability and CO_2/N_2 selectivity. The

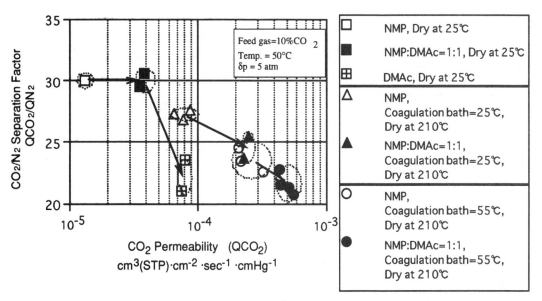

Fig.6 Effect of solvent type

characteristics of the improved spinning conditions were as follows:
(1) High polymer concentration of dope solution for high mechanical strength
(2) Outer surface coagulation by 60℃ water to prevent from forming outer skin layer
(3) NMP-DMAc solvents system of the dope solution for ultrathin inner skin layer

BENCH PLANT TEST OF THE MEMBRANES

We have started scale up study this spring. As a first step, we fabricated bench scale test plant for CO_2/N_2 separation. And modules with membrane area of $8m^2$ (inner diameter 0.2mm, effective length 0.5m, 25600 fibers) were also fabricated. Using this module, we will test the long-term stability, the effects of gas flow patterns in the module and the effects of impurities such as water.

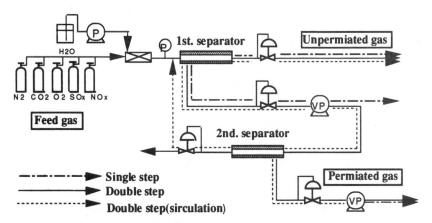

Fig.7 Schematic diagram of Bench test plant

SIMULATIONS

We have performed computer simulations of the gas separation (CO_2/N_2) using the membrane properties measured to predict a possibility (Table 1). In this simulation, temperatures were varied.

Table 1.Simulations of the gas separation

Temp.		Feed gas	1st. module Permeated gas	2nd. module Permeated gas
25℃	CO_2 concentration	10%	46.4%	89.1%
	Gas flow rate	1.6Nm³/hr	0.23Nm³/hr	0.11Nm³/hr
	Reuqired membrane area		13.4m²	1.52m²
50℃	CO_2 concentration	10%	41.5%	81.5%
	Gas flow rate	1.6Nm³/hr	0.26Nm³/hr	0.12Nm³/hr
	Reuqired membrane area		7.73m²	1.30m²

Simulatin parameters; $QCO_2=2.0 \times 10^{-4}$ and $QCO_2/QN_2=25$ at 50 ℃, $QCO_2=1.5 \times 10^{-4}$ and $QCO_2/QN_2=40$ at 25℃, 1st. membrane CO_2 recovery rate= 66.7%, 2nd. membrane CO_2 recovery rate= 90%, Pressure (feed/permiate)=1.05/0.105 (kgf/cm² abs.), Feed gas composition $CO_2/N_2=10\%$ / 90%, Feed gas rate =1.6Nm³/hr. Cascade; Double step (No sirculation). Flow in modules; Cross flow

CONCLUSIONS

Cardo polyimides have high CO_2/N_2 permselectivity. We developed new Cardo polyimide (PI-BT-COOMe) which has CO_2 affinitive methylcorboxyl functional group. It shows highest CO_2/N_2 selectivity. ($PCO_2/PN_2=52$, $PCO_2=1.5 \times 10^{-9}$cm³(STP)cm/cm² sec cmHg, at 25℃)

And we also developed wet spinning method of PI-BT-COOMe. By the method, we could spin hollow fiber membrane which has good balance of CO_2/N_2 permselectivity ($QCO_2/QN_2=40$, $QCO_2=1.5 \times 10^{-4}$cm³(STP)/cm² sec cmHg, at 25 ℃) and good mechanical strength. The characteristics of our method are to form skin layer by wet phase inversion at inside surface of hollow fiber, and to form sponge-like support layer by high polymer concentration of dope solution.

REFERENCES

1)V.V.Korshak et al.(1974), *J. Macromol.Sci.-Rev.Macromal.Chem.*, **c11(1)**, 45
2)XIANGQUN MIAO et al.(1996), *Separation Science and technology*, **31(2)**, 141-172
3)Ingo Pinnau(1993), *Polymer for Advanced Technologies*, **5**, 733-744

This work was supported by New Energy and Industrial Technology Development Organization (NEDO).

Pergamon

Energy Convers. Mgmt Vol. 38, Suppl., pp. S117–S122, 1997
© 1997 Elsevier Science Ltd. All rights reserved
Printed in Great Britain
PII: S0196-8904(96)00256-7
0196-8904/97 $17.00 + 0.00

PRESSURIZED FLUIDIZED BED COMBUSTION AND GASIFICATION OF COAL USING FLUE GAS RECIRCULATION AND OXYGEN INJECTION

J.Andries , J.G.M. Becht and P.D.J. Hoppesteyn

Laboratory for Thermal Power Engineering,
Department of Mechanical Engineering and Marine Technology,
Delft University of Technology, Mekelweg 2,
2628 CD Delft, The Netherlands

ABSTRACT

A combined cycle incorporating pressurized fluidized bed combustion of coal with pure oxygen and recycled flue gas followed by pressurized combustion of the fuel gas using pure oxygen, is a process which produces electricity and heat with a high efficiency and a flue gas with a high CO_2 concentration. CO_2 recovery from this process can be done more easily than from conventional coal conversion processes. In this way a high efficiency coal conversion process can be realized with low CO_2 emissions. The specific aim of the project is to assess the technical feasibility of this process. For this purpose a 1.6 MW_{th} pressurized fluidized bed combustion test rig has been modified to study experimentally the gasification process and the pressurized combustion of the resulting low calorific value fuel gas. The objectives of the project, the modified test rig and the experimental results obtained from combustion experiments with coal, recirculated flue gas and pure oxygen are described. The conversion efficiencies and the emission of harmfull components measured during these experiments are analyzed and compared with values obtained during combustion and gasification using air.

© 1997 Elsevier Science Ltd

KEYWORDS

gasification; combustion; gasturbine; low calorific value gas; pressurized fluidized bed;coal.

INTRODUCTION

Pressurized fluidized bed combustion is limited in its capabilities to be used in advanced combined cycle schemes because of the maximum possible temperature of the flue gas which is to be expanded in a gas turbine. By operating the pressurized fluidized bed reactor as a gasifier, the product gas can be burned in a gas turbine combustion chamber enabling the use of advanced gas turbines with a high turbine inlet temperature. Using recirculated flue gas and pure oxygen in the fluidized bed reactor and additional pure oxygen for the pressurized combustion of the low calorific value fuel gas in the gas turbine combustion chamber, a combined cycle process can be realized which produces a flue gas with a high CO_2 content, which makes efficient CO_2 recovery from the flue gas possible. Assuming that the recovered CO_2 can be disposed of in an environmentally acceptable manner, a scheme can be realized in which coal is converted into electricity and heat with a relatively high efficiency and minimal CO_2 emissions into the atmosphere. The specific aim of the work described here is to assess the technical feasibility of the process.

The results described in this paper were obtained during the first phase of the commissioning of the modified installation. During these tests the fluidized bed reactor was operated as a combustor and the experiments were aimed at commissioning and testing of the oxygen injection-, the flue gas recirculation- and the heat removal system.

INSTALLATION

Since a number of years the Laboratory for Thermal Power Engineering has been doing research on pressurized fluidized bed combustion using the test rig shown in Fig. 1. The main design data of the fluidized bed reactor are given in Table 1.

Table 1 Main design data

diameter bed	0.4 m
max pressure	10 bar
max bed height	2 m
freeboard height	4 m
fluidization velocity	0.8 m/s
max thermal capacity	1.6 MW

Table 2 Composition of the coal

on dry	wt %
ash	14.1
volatiles	33.4
C	69.2
H	4.9
N	1.3
O	8.3
S	1.7
moisture	6.2

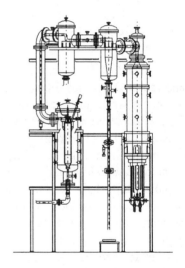

Fig. 1 The Delft PFBC/G test rig.

The existing test rig is being modified in phases to enable experiments with pressurized fluidized bed gasification of coal using flue gas recirculation and oxygen injection. The modifications consist of changes to the reactor vessel and the addition of an oxygen supply system, a high temperature ceramic filter, a high temperature ammonia removal system, a pressurized topping combustor, a scrubber, a booster compressor and an atmospheric combustor.

A schematic of the final configuration of the modified installation is shown in Fig. 2. The components to be used during phase 2 (pressurized topping combustor and downstream heat exchanger) were operational at the end of March 1996, the components to be used during phase 3 (ceramic filter, upstream heat exchanger and ammonia removal system) will be added in September 1996.

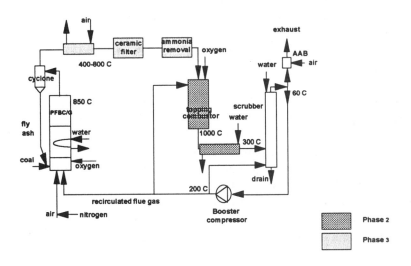

Fig. 2 The Delft PFBC/G test facility.

Oxygen is supplied to the laboratory in liquid form by road tankers and kept in a cryogenic storage tank equipped with an evaporation system. If the test rig is operating at full load, 320 kg/h of oxygen is needed. During continuous operation at these condition the storage tank has to be filled every two days. The use of pure oxygen demands special care for the system lay-out and the choice of material. The detailed design and construction have been done in close cooperation with the oxygen supplier (Air Products UK). The oxygen supplied to the bed is mixed with air (when enriched air is used as fluidization gas) or recirculated flue gas (when a CO_2/O_2 mixture is used as fluidization gas) inside a nozzle with small, radially outward directed, orifices located in the center of the gas distributor plate. The recirculated flue gas must be free of combustible gases and oil to ensure a safe mixing process. This implies the use of a booster compressor which delivers an oil free gas.

Recirculation of the flue gas implies recirculation of water vapour formed during the combustion of the coal and/or the fuel gas. The water concentration can become as high as 30 vol % at 8 bar. Because of the high dew point of the resulting flue gas this could result in severe corrosion in the system. Furthermore the water concentration in the flue gas strongly influences the composition of the fuel gas produced by the gasifier. To control the water vapour concentration a scrubber has been installed. By injecting water and cooling the resulting gas below the dew point, the water vapour concentration can be controlled and be kept below a few percent. The scrubber will also remove SO_2 and remaining dust particles from the gas stream thus protecting the booster compressor from erosion and corrosion.

The booster compressor is able to deliver a 2 bar pressure increase at system pressures ranging from 3 to 8 bar. The amount of recirculated flue gas, which is used for fluidizing the gasifier and cooling the pressurized combustor depends on the operating conditions of the test rig. To simplify the control of the operation of the compressor, the excess capacity of the compressor is recirculated via the scrubber. The compressor chosen, is a centrifugal one with a vaneless diffusor which can be used to compress CO_2 - rich flue gas as well as air.

The produced low calorific value fuel gas is presently cleaned by a cyclone dust separator and combusted in a pressurized topping combustor (PTC) using pure oxygen as oxidant. In September of this year a ceramic high temperature filter will be installed upstream of the PTC.

The PTC, shown in Fig. 3, consists of a specially designed burner mounted inside a slightly modified GE Frame 5 combustor can (with a diameter of 0.28 m).

The walls of the can are cooled by recirculated flue gas. The flue gases leave the pressure vessel containing the PTC at a temperature of 1000 °C. This temperature is limited because of constraints of the materials used in the test rig and is controlled by the amount of recirculated flue gas. An annular, water cooled heat exchanger downstream of the PTC cools the flue gases to about 300 °C.

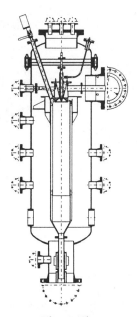

The PTC burner creates an axi-symmetric diffusion flame, where pure oxygen is led to the centre of the burner and low calorific value fuel gas is led around the oxygen. The burner has a conical flame-holder. Pure oxygen flows at an angle of 45 ° with the axis of the combustor along the flame-holder. The main part of the LCV fuel gas is led into the combustor through an annular space at the end of the flame-holder, directed along the axis of the combustor. The LCV fuel gas is ignited by a pilot torch, which is located at the centre of the burner. The pilot torch is fed with premixed air and methane. The pilot torch pre-ignites a small part of the LCV fuel gas which enters the combustor through 24 holes in the flame-holder. The pre-ignited LCV fuel gas ignites the main LCV fuel gas at the end of the flame-holder. A very stable flame is created by this staged ignition. Due to the conical shape of the flame-holder and the directions of the LCV fuel gas and oxygen flows an internal recirculation zone is created which gives a short and stable flame.

Fig. 3 The pressurized topping combustor.

EXPERIMENTS

During the commissioning of the modified test rig a number of combustion experiments have been done to assess the operation of the oxygen injection- the flue gas recirculation- and the heat removal system.. All experiments have been done at a bed temperature of 850 °C (measured at 0.4 m above the air distributor plate), a fluidization velocity of 0.8 m/s, with air (21 % oxygen), oxygen-enriched air (24 % oxygen) or oxygen-enriched recirculated flue gas (24 and 27 % oxygen) as fluidization gas and Kiveton Park coal premixed with Middleton limestone (97 % $CaCO_3$) at a Ca/S ratio of 2.3 as fuel. The composition of the Kiveton Park coal is given in table 2.

The main operating parameters and measured concentrations are averaged values obtained during at least 2 hours of steady state operation. The emission values are measured in dried flue gas and are given here as they are measured.

Fig. 4 shows the oxygen and carbon dioxide concentrations as a function of time during the change from 'once through' operation to recirculation

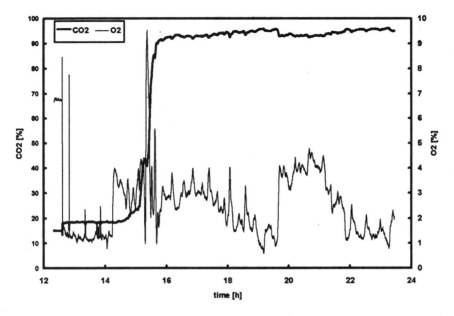

Fig. 4 CO_2 and O_2 concentrations as a function of time (as measured, dry).

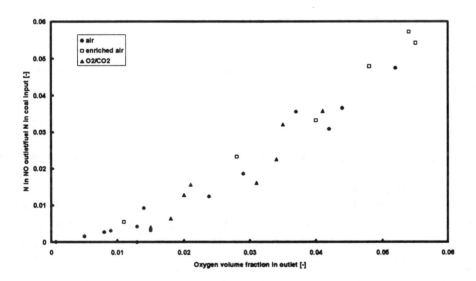

Fig. 5 Net conversion of fuel nitrogen into NO-nitrogen as a fuction of the oxygen concentration in the outlet.

Fig. 5 shows the net conversion of fuel nitrogen into NO nitrogen as a function of the outlet oxygen concentration.

CONCLUSIONS

Preliminary conclusions based on the results from the combustion experiments are:
* there seems to be no influence of the amount of recirculated NO on the emission of NO.
* the input of fuel-N and the oxygen level in the outlet determine together the emission of NO.
* the oxygen level at the inlet of the reactor seems to have no influence on the NO emission.
* the processes which determine the net emission of NO are not influenced by the replacement of nitrogen by carbon dioxide.

FUTURE WORK

During the second half of 1996 gasification experiments will be done using air, oxygen-enriched air and oxygen-enriched recirculated flue gas. During these experiments the carbon conversion in the fluidized bed and the combustion characteristics of the low calorific value fuel gas in the PTC will be determined. The results will be used to assess the technical feasibility of the proposed process.
In 1997 and following years the test rig will be used to assess the gasification characteristics of biomass and biomass-coal mixtures with air and steam in the pressurized fluidized bed gasifier and the combustion characteristics of the LCV gas in a gasturbine combustion chamber (1,2).

ACKNOWLEDGEMENT

This research is funded in part by the European Commission in the framework of the non-nuclear energy programme JOULE II (Project JOU2-CT92-0154).

REFERENCES

1.   Andries,J. and Hein, K.R.G., 'Co-gasification of biomass and coal in a pressurized fluidised bed gasifier', paper presented at *Developments in Thermochemical biomass conversion*, Banff, Canada, May 20 - 24, 1996
2.   Andries,J., Hoppesteyn,P.D.J., and Hein, K.R.G., 'Pressurised combustion of biomass-derived, low calorific value, fuel gas', paper presented at *Developments in Thermochemical biomass conversion*, Banff, Canada, May 20 - 24, 1996

Energy Convers. Mgmt Vol. 38, Suppl., pp. S123–S127, 1997
© 1997 Elsevier Science Ltd. All rights reserved
Printed in Great Britain
0196-8904/97 $17.00 + 0.00

Pergamon

PII: S0196-8904(96)00257-9

TRIAL DESIGN FOR A CO_2 RECOVERY POWER PLANT BY BURNING PULVERIZED COAL IN O_2/CO_2

M. OKAWA and N. KIMURA[1]   Electric Power Development Co., Ltd.
T. KIGA and S.TAKANO[2]   Ishikawajima-Harima Heavy Industry Co., Ltd.
K. ARAI[3]   Nippon Sanso Corp.
M. KATO[4]   Center for Coal Utilization, Japan

1, 15-1, Ginza 6-chome, Chuo-ku, Tokyo, 104 JAPAN
2, 2-16, Toyosu 3-chome, Koto-ku, Tokyo, 135 JAPAN
3, 4-320, Tsukagoshi, Saiwai-ku, Kawasaki, 210 JAPAN
4, 2-3-4, Ohkubo, Shinjuku-ku, Tokyo 169, JAPAN

ABSTRACT

A CO_2 recovery power plant based on pulverized coal O_2/CO_2 combustion uses pure oxygen and flue gas recycled through flue gas ducts as an oxidant to burn pulverized coal in an O_2/CO_2 mixture gas. Through this burning process, the system increases the CO_2 concentration in the flue gas to more than 90% and recovers the CO_2 gas directly. In this study, we designed a test for a power generation plant under the assumption that flue gas with a CO_2 concentration of 90% or more would be directly disposed of underground. We also investigated the thermal efficiency and economy of the CO_2 recovery power plant . © 1997 Elsevier Science Ltd

KEYWORDS

O_2/CO_2 combustion; recycled exhaust gas; underground disposal; optimum oxygen purity; thermal efficiency; economy.

RESEARCH OF GENERAL SYSTEM

First, we researched the optimum combustion process. The O_2/CO_2 pulverized coal-fired power plant consists of four main processes; oxygen generation, O_2/CO_2 combustion, flue gas treatment, and CO_2 recovery/disposal. Equipment capacity and power consumption vary depending on the flue gas extraction position for recycling and the configuration of the components. The most efficient and technical possible processes are shown in Figure 1. We adopted the wet recycling method rather than the dry recycling method (which requires gas coolers and dehumidifiers with larger capacities) for the combustion exhaust gas and the pulverized coal conveyance gas. A non-leak heat exchanger was used because oxygen leaking into the flue gas line would increase the power required for oxygen generation. However, due to its economic disadvantages when used at high temperatures, we adopted a conventional regenerative heat exchanger for the flue gas.

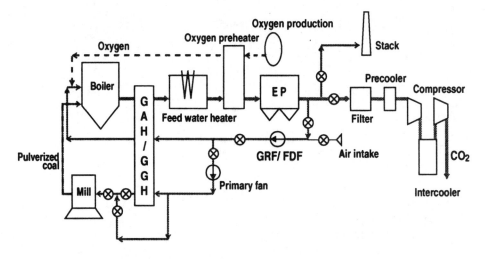

Fig. 1.   Pulverized coal-fired power plant using O_2/CO_2 combustion

Next came investigation of the best oxygen producing equipment. There are four methods of producing oxygen for the oxygen generation process: the cryogenic separation method, the adsorption method, the membrane separation method, and the chemical separation method. From these four methods we selected the cryogenic air separation method as the oxygen generation technology because it has given satisfactory results in large-scale plants.

Since the oxygen purity affect the economy at this system, optimum values must be selected.   As the purity becomes higher, the oxygen generation power increases, but the power required to compress the recovered gas can be reduced due to the increased CO_2 concentration. Fig-2 shows the relationship between the purity of generated oxygen and sum of power consumption for generating oxygen and CO_2 treatment. Therefore, we selected 97.5% as the optimum oxygen purity, at which total power consumption can be reduced to the lowest level.(Takano et al.,1995)

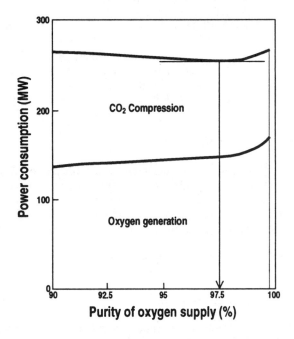

Fig. 2. Relationship between oxygen purity and sum of power consumption

Last, we examined the required area for this power plant. We assumed that CO_2 would be disposed of underground by gas compression, so that this system would not need DeSOx and DeNOx plants. This means that the required area for this power plant would not be any larger, even though additional area is required for the oxygen generator and the CO_2 recovery compressor. Consequently, this power plant can be constructed in an area equivalent to that used for a conventional pulverized coal-fired power plant.

EVALUATION OF ECONOMY

The following figures show a comparison of thermal efficiency and economy when a power plant is constructed using this technology. The assumption for evaluation of thermal economy is that recovered CO_2 is to be compressed and disposed of underground. The power plant condition is the supercritical steam condition (246 kg/cm^{2}g, 538/566°C) in a 1000MW class pulverized coal-fired plant. Figure 3 shows the results of comparison. Evaluation was based on conventional air combustion methods, including the amine absorption method, which has a good possibility of being put into practical use.

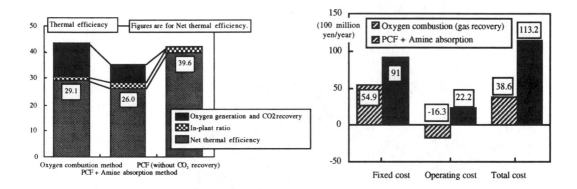

Fig. 3. Evaluation of economy

In the oxygen combustion method, the amount of flue gas decreases because no nitrogen mixes into the combustion air. It therefore reduces thermal heat loss in the flue gas while improving boiler efficiency, and also reduces power consumption in the plant. This means that gross efficiency is higher than with conventional air combustion. However, it requires a great deal of power consumption for oxygen generation and CO_2 compression within the plant, so the net efficiency falls from the 39.6% of a conventional power plant to 29.1%. On the other hand, the amine absorption method requires less power consumption within the plant compared to the oxygen combustion method. However, it reduces the gross efficiency, because of a large amount of turbine bleed steam for regenerating absorbent. As a result, the net efficiency is about 3% lower than that of the oxygen combustion method. In a comparison of annual costs, the oxygen combustion method was found to have reduced operating and equipment costs due to non-use of DeSOx and DeNox plants as well reduced fuel costs due to the rise in the net thermal efficiency. When the cost for the oxygen generator and CO_2 compressor are added, however, annual costs are estimated to rise about 3.8 billion yen compared to a conventional power plant. In the amine absorption method, costs can be reduced. Because steam turbine and generator are down-sized due to large amount of turbine bleed steam is used for regeneration of absorbent. However, it requires a high-performance DeSOx plant. When the amine absorber and CO_2 compressor added to this, annual costs would increase by some 11.3 billion yen over those of a conventional power plant.

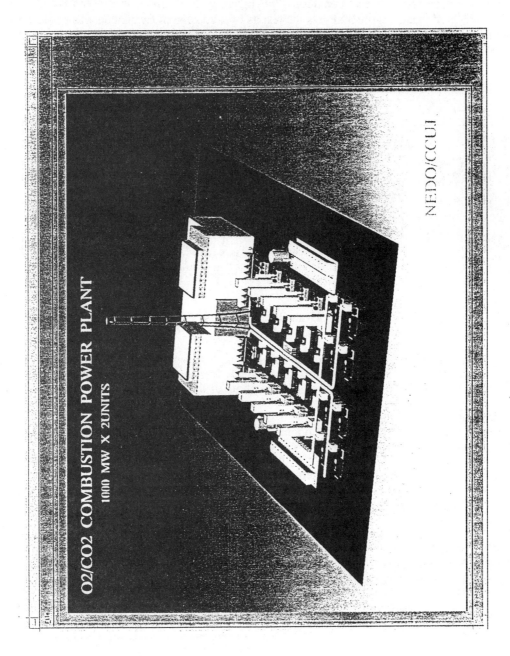

Fig.4 O$_2$/CO$_2$ combustion power plant

CONCLUSION

The results of this study are as follows:

1. Research of equipment's configurations showed that a CO_2 recovery power plant using pulverized coal O_2/CO_2 combustion should be feasible by combining current technologies.
2. The optimum oxygen purity should be 97.5% for CO_2 recovery by O_2/CO_2 combustion.
3. A CO_2 recovery power plant using pulverized coal O_2/CO_2 combustion can be constructed in an area equivalent to that of a conventional pulverized coal-fired power plant.
4. A CO_2 recovery pulverized coal-fired power plant using the O_2/CO_2 combustion method should exhibit excellent thermal efficiency and economy as a CO_2 recovery method.

As the results mentioned above indicate that the CO_2 recovery method using O_2/CO_2 combustion should prove to be very effective in a pulverized coal-fired power plant. However, many issues still remain that must be examined, so further study will be continued.

ACKNOWLEDGMENT

This project was supported by New Energy and Industrial Technology Development Organization,

REFERENCES

S. Takano, T. Kiga, Y. Endo, S. Miyamae and K. Suzuki (1995). CO_2 Recovery from PCF Power Plant with O_2/CO_2 combustion Process, IHI Engineering Review;
K. Omata, N. Kimura, T. Kiga, S. Takano and S. Shikisima (1994). The characteristics of Pulverized Coal Combustion in O_2/CO_2 mixture for CO_2 Recovery, Second International Conference on Carbon-Dioxide Removal

Pergamon

Energy Convers. Mgmt Vol. 38, Suppl., pp. S129–S134, 1997
© 1997 Elsevier Science Ltd. All rights reserved
Printed in Great Britain
0196-8904/97 $17.00 + 0.00

PII: S0196-8904(96)00258-0

CHARACTERISTICS OF PULVERIZED-COAL COMBUSTION IN THE SYSTEM OF OXYGEN/RECYCLED FLUE GAS COMBUSTION

T. KIGA and S. TAKANO
Combustion Engineering Dept., Ishikawajima-Harima Heavy Industries Co., Ltd.
3-2-16, Toyosu, Koto-ku, Tokyo, Japan

N. KIMURA, K. OMATA and M. OKAWA
R&D Dept., Electric Power Development Co., Ltd.
6-15-1, Ginza, Chuo-ku, Tokyo, Japan

T. MORI
Investigation and Research Dept., Institute of Research and Innovation
1-6-8, Yushima, Bunkyo-ku, Tokyo, Japan

M. KATO
Technical Development Dept., Center for Coal Utilization, Japan
2-3-4, Ohkubo, Shinjuku-ku, Tokyo, Japan

ABSTRACT

Concerning to the system of oxygen/recycled flue gas combustion for the removal of carbon dioxide from pulverized-coal firing power plants, combustion characteristics of pulverized coal and the heat absorption performance of boiler furnace were studied. To determine the ignition characteristics in the CO_2-rich atmosphere, the flame propagation speed of pulverized-coal cloud was measured in a microgravity combustion chamber. The results revealed that the flame propagation speed in an O_2/CO_2 atmosphere was markedly low compared with that in O_2/N_2 and O_2/Ar, and that it was improved by increasing the O_2 concentration. NO_X and SO_2 emissions from the system were investigated in the industrial-scale combustion test facilities. NO_X emission was not so largely decreased by staging as air-blown combustion, while a considerable part of sulfur was absorbed in ash in the system. As for the heat absorption performance of the boiler furnace, a 3-dimensional numerical analysis was applied to a large utility boiler furnace. It was found that a furnace designed for normal air-blown combustion was just large enough for the oxygen/recycled flue gas combustion with a realizable ratio of flue gas recycling. © 1997 Elsevier Science Ltd

KEYWORDS

CO_2 recovery; O_2/CO_2 combustion; pulverized coal; combustion improvement; microgravity; oxygen-enriched combustion; burner; furnace; heat absorption.

INTRODUCTION

Oxygen/recycled flue gas combustion is expected to be one of promising systems on CO_2 recovery from pulverized-coal firing power plants (Nakayama *et al.*, 1992). In the system, pulverized coal is fired in a CO_2-based atmosphere while it is normally burnt in air which is a N_2-based oxidant. Since the CO_2 is a well-known radiative gas having a higher specific heat than N_2, it is supposed that the combustion characteristics in the system will be considerably different from that in a normal combustion system. Also the oxidant contains the recycled NO_X and SO_2 in it, and the concentration of O_2 can be changed by varying the amount of recycled flue gas. Studies were conducted therefore to clarify the characteristics of pulverized-coal combustion in the system.

Our previous studies on the area were the laboratory-scale combustion test using a vertical electrically heated flow reactor (Koyata *et al.*, 1991), the numerical computational analysis on its test results (Nozaki *et al.*, 1995) and the industrial-scale combustion test using a 1.2 MW[thermal] horizontal cylindrical furnace (Kimura K. *et al.*, 1993, Kimura N. *et al.*, 1994). These studies offered us a lot of information on the combustion characteristics in oxygen/recycled flue gas. Among them, our great concern was flames with a vague ignition point and less luminosity resulting in higher unburnt carbon. Why is the ignition not so good?

How much NO_X and SO_2 does the system emit? How do such flames with less luminosity effect the heat absorption performance of boiler furnace? This paper introduces our activities to answer the above-mentioned questions.

MEASUREMENT OF FLAME PROPAGATION SPEED

Experimental studies on flame propagation speed in a coal-dust cloud were performed to fundamentally investigate the ignition characteristics of pulverized coal in CO_2-rich atmosphere. Testing was carried out in the microgravity environment using drop shaft facilities of Japan Microgravity Center (JAMIC) to obtain a spatially homogenous distribution of coal particles, to make them quiescent and to avoid the unexpected mixing by the natural convection which would reduce the accuracy of measurements.

Experimental

A capsule loading an experimental apparatus falls down in the drop shaft. The total length of the drop shaft is 710 m, and its primary zone of 490 m is used as a free fall zone while the last 220 m is used as a braking zone. The capsule consists of three parts that are a bus module, a payload module and a thruster module. The experimental apparatus is settled in an inner capsule contained in the payload module. The inner capsule is disconnected from the payload module during dropping. This provides an about 10 seconds' high-quality microgravity environment of less than 1.0×10^{-4} g.

Fig. 1 shows the schematics of the experiment apparatus. Pulverized-coal particles under 200 mesh (74 μm) were loaded on a cathode in a cylindrical combustion chamber with an inside diameter of 230 mm and a height of 245 mm. A part of particles is levitated by an electrostatic force of 1 - 5 kV and passed through the earthed grid by their inertia to be dispersed in the combustion chamber. The amount of levitated particles was measured by the attenuation of a He-Ne laser beam through coal dust cloud. After several seconds for coal particles to be almost stopped, a hot wire ignitor located at the center of the combustion chamber was activated for ignition. The flame propagation behavior was recorded by a 16 mm high-speed cinecamera at a speed of 400 fps (frames per second).

Experiments were carried out in O_2/CO_2, O_2/N_2 and O_2/Ar atmospheres at different oxygen concentration. Coal used in the experiment was Japanese high-volatile bituminous, named Taiheiyo, whose properties are listed in Table 1. So as to obtain a suitable dispersion, pulverized activated carbon was added by 5%.

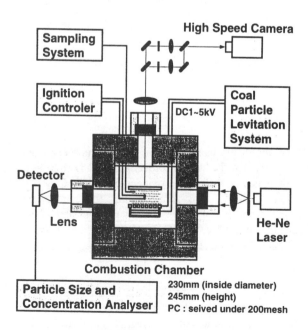

Fig. 1.   The schematics of the experiment apparatus for flame propagation testing.

Table 1.   Properties of Taiheiyo coal.

Heating value (ad)	28.4 MJ/kg
Proximate analysis (ad)	
Moisture	3.3%
Ash	13.3%
Volatile Matter	45.6%
Fixed carbon	37.8%
Ultimate analysis (dry)	
C	68.0%
H	5.5%
O	11.6%
N	1.1%
S	0.1%

Results and Discussion

Fig. 2 shows the flame propagation behavior observed in O_2/N_2 and O_2/CO_2 atmosphere with an oxygen concentration of 40%. Since, as shown here, the movement of the flame front could be easily determined and the flame front propagated nearly spherically and almost linearly from ignition point, the flame propagation speed could be successfully measured.

Fig. 3 summarizes the experimentally determined flame propagation speed obtained in O_2/CO_2, O_2/N_2 and O_2/Ar atmospheres at different oxygen concentrations. The figure shows that the flame propagation speed increases as the oxygen concentration increases in all cases as expected. The figure also indicates that the flame propagation speed is much smaller in O_2/CO_2 atmosphere than in O_2/N_2 and O_2/Ar and that in O_2/Ar was the highest. The results support the fact that flames with a vague ignition point were obtained in the oxygen/recycled flue gas combustion at industrial-scale combustion test facilities. The mixture of oxygen and carbon dioxide has the highest specific heat. For example, the 40% oxygen mixture with carbon dioxide has 46.5 J/kmol as average specific heat at 1,000 K and 1 atm. This is 38% higher than that of the mixture of oxygen and nitrogen and 76% higher than that of the mixture of oxygen and argon. Because the difference of thermal conductivity of each gas mixture is not large, the difference in specific heat may be the major reason for slower flame propagation speed in the O_2/CO_2 atmosphere.

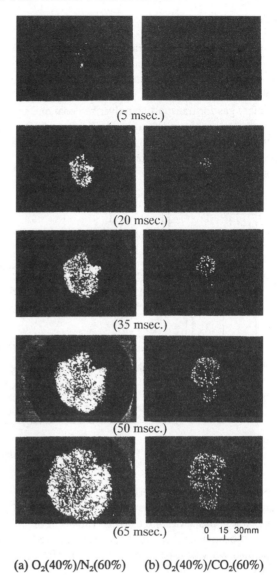

(5 msec.)

(20 msec.)

(35 msec.)

(50 msec.)

(65 msec.)   0 15 30mm

(a) $O_2(40\%)/N_2(60\%)$ (b) $O_2(40\%)/CO_2(60\%)$

Fig. 2.   Flame propagation behavior.

NO_X AND SO_2 EMISSIONS

Our previous studies on the industrial-scale combustion test facilities showed that NO_X emissions from the oxygen/recycled flue gas combustion system were reduced to about 25% when compared with conventional air-blown combustion. On the other hand, the staged combustion is normally applied to coal-fired boilers to reduce NO_X emissions, and they are reduced also to about 25% when compared to non-staged combustion. Accordingly, staged combustion tests for the system were carried out in the industrial-scale testing. The conversion of sulfur was also investigated in the test since previous industrial-scale tests indicated that the system produces remarkably lower SO_2 emission.

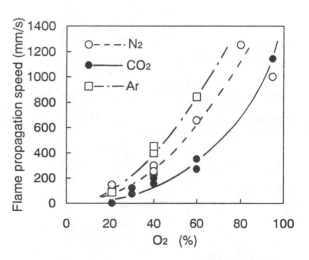

Fig. 3.   Effect of O_2 concentration and atmospheric gases components on flame propagation.

Experimental

Fig. 4 shows the flow diagram of the industrial-scale combustion test facilities. The furnace is a horizontal cylindrical type with a water wall lined with refractory, and its size is 1.3 m in inside diameter and 7.5 m in length. The exhaust gas enters a stack through a gas cooler, a tubular air preheater, a multi-cyclone dust collector and a bag filter. A part of it is taken out downstream of the bag filter and recycled to be used as the primary gas for transporting pulverized coal and secondary gas for combustion. Oxygen is supplied   using an evaporator system and a liquid oxygen tank. Part of it is mixed with the secondary gas and another part can be directly injected into the burning area through the center of pulverized-coal burner (Fig. 5). A part of the secondary gas was taken out and injected into the furnace through staging air ports to obtain staged combustion.

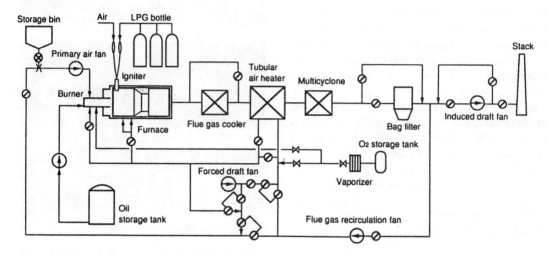

Fig. 4.   The flow diagram of industrial-scale combustion test facilities.

Combustion tests were performed at the firing rate of 100 kg/h using three kinds of bituminous coal, whose properties are listed in Table 2. For NO_x measurement, a non-dispersive infrared type analyzer was used so as to avoid the effect of CO_2.

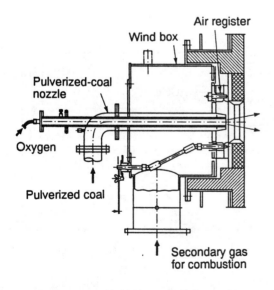

Fig. 5.   Pulverized-coal burner.

Table 2.   Properties of coals for the industrial-scale.

Coal	-	A	B	C
Heating value (ad)	MJ/kg	28.3	29.9	29.5
Proximate analysis (ad)				
Moisture	%	3.0	2.0	2.2
Ash	%	13.8	13.3	12.0
Volatile Matter	%	26.1	41.2	36.2
Fixed carbon	%	57.1	43.5	49.6
Ultimate analysis (dry)				
C	%	72.2	72.0	72.7
H	%	4.3	5.5	4.9
O	%	7.5	7.1	8.5
N	%	1.4	1.2	1.2
S	%	0.45	0.96	0.38
Ash properties				
Al_2O_3	%	37.4	19.8	18.5
SiO_2	%	46.3	45.4	73.2
TiO_2	%	1.6	2.6	0.8
P_2O_5	%	1.5	0.2	0.3
Fe_2O_3	%	5.5	4.9	2.4
CaO	%	4.3	12.0	1.2
MgO	%	1.1	1.6	0.6
Na_2O	%	0.6	3.8	0.2
K_2O	%	0.4	0.5	0.9
SO_3	%	1.3	7.3	0.7

Results and Discussion

Fig. 6 shows the relationship between the excess oxygen ratio λ_1 at the burner and the conversion ratio η_N of fuel nitrogen into NO_X when coal A was fired with or without staging. A very high value of η_N over 30% was recognized in air-blown combustion without staging, while it was remarkably small in oxygen/recycled flue gas combustion. This is mainly due to the decomposition of recycled NO_X in the flame. On the other hand, a decreasing λ_1 by staging reduces η_N in air-blown combustion rapidly down to 25% of the value without staging, while the reduction of NO_X in oxygen/recycled flue gas combustion was a half at most. The reason is thought to be that being different from NO_X recycled via the burner, the NO_X recycled in the staging gas was not decomposed because there is a lack of reducing components like hydrocarbons.

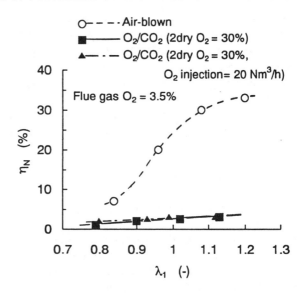

Fig. 6. The relationship between the excess oxygen ratio λ_1 at the burner and the conversion ratio η_N of fuel nitrogen into NO_X.

Table 3 shows the conversion of sulfur based on the measurements of SO_2 concentration at several positions and sulfur content in ash collected from ash hoppers for three kinds of coal. Although the mass balance of sulfur could not be closed completely, it is supposed from the table that the condensation of sulfate or sulfite in the duct where the gas temperature was low enough and the absorption of sulfur in ash resulted in the reduction of SO_2 emissions from a stack in oxygen/recycled flue gas combustion.

HEAT ABSORPTION IN FURNACE

In order to evaluate the effect of flames with less luminosity and the increase in the concentration of radiative gas CO_2 and H_2O in the oxygen/recycled flue gas combustion on the heat absorption performance in furnace, the 3-dimensional numerical analysis was applied to a 1,000 MW, super-critical pressure, pulverized-coal firing utility boiler furnace. A computational fluid dynamics code used here was VEGA-3, Viscous Flow Elliptic Solver with Grid Generation Algorithm 3-D version (Endo *et al.*, 1993).

Table 3.   Conversion of sulfur.

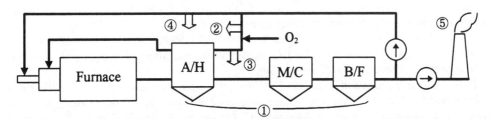

Coal	Coal A		Coal B		Coal C	
Combustion type	Air-blown	O_2/CO_2	Air-blown	O_2/CO_2	Air-blown	O_2/CO_2
Sulfur (kg/h) ①	0.11	0.13	0.16	0.23	0.04	0.04
②	-	0.03	-	0.16	-	0.04
③	-	0.04	-	Not analyzed	-	Not analyzed
④	-	0.01	-	0.03	-	0.03
⑤	0.30	0.16	0.71	0.35	0.29	0.15
Total	0.41	0.37	0.87	0.77	0.33	0.26
Sulfur In (kg/h)	0.43		0.92		0.37	

The furnace has 32 burners in 4 stages on a front wall and 24 burners in 3 stages on a rear wall. The size of the furnace is 14.5 m in depth, 32.9 m in width and 67.5 m in height. A furnace of 57.5 m in height with the same cross section was also analyzed to ascertain the possibility of reducing furnace size. The analysis was conducted under conditions of staging ratio of 30% and 15% of total oxygen to be injected directly to the burner, changing O_2 concentration in wind box. As shown in Fig. 7, the heat absorption in furnace increases as wind box O_2 concentration increases, and under the operation condition of wind box O_2 concentration 30%, heat absorption is almost same as that in air-blown combustion. The result has also revealed the possibility of reducing furnace size, if O_2 concentration is increased.

CONCLUSIONS

Combustion characteristics of pulverized coal in the system of oxygen/recycled flue gas combustion were generally made clear on ignition, NO_X and SO_2 emissions and heat absorption in furnace. From the results, combined with our trial designing of 1,000 MW pulverized coal plant (Okawa *et al.*, 1996), the system was proved to be basically feasible to remove CO_2 from coal-firing power plant. However, on the detail of combustion mechanism and condensation of corrosive materials, many issues need further examination. We will pursue research into these subjects.

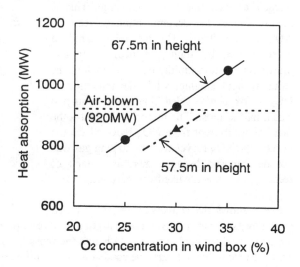

Fig. 7.   Heat absorption in furnace.

ACKNOWLEDGMENT

This work was performed in the frame of the project "Environmentally-friendly coal combustion technology - O_2/CO_2 combustion technology" of "Next-generation coal utilization technology", sponsored by New Energy and Industrial Technology Development Organization (NEDO).

REFERENCES

Endo, Y., S. Miyamae and Y. Ando (1993).   3-D numerical analysis for a p.c. firing boiler. *JSME-ASME International Conference on Power Engineering-93, September, 1993, Tokyo, Japan.*

Kimura, K., T. Kiga, S. Miyamae, Y. Noguchi and T. Tanaka (1993).   Experimental studies on pulverized coal combustion with oxygen/flue gas recycle for CO_2 recovery. *JSME-ASME International Conference on Power Engineering-93, September, 1993, Tokyo, Japan.*

Kimura, N., K. Omata, T. Kiga, S. Takano and S. Shikishima (1994).   The characteristics of pulverized coal combustion in O_2/CO_2 mixtures for CO_2 recovery. *Second International Conference on Carbon Dioxide Removal, October, 1994, Kyoto, Japan.*

Koyata, K., T. Tanaka, T. Kiga and K. Suzuki (1991).   Pulverized coal combustion in O_2/CO_2 atmosphere. *Eighth Annual International Pittsburgh Coal Conference, October, 1991, Pittsburgh, PA.*

Nakayama, S., Y. Noguchi, T. Kiga, S. Miyamae, U. Maeda, M. Kawai, T. Tanaka, K. Koyata and H. Makino (1992).   Pulverized coal combustion in O_2/CO_2 mixtures on a power plant for CO_2 recovery. *First International Conference on Carbon Dioxide Removal, March, 1992, Amsterdam, Netherlands.*

Nozaki, T., S. Takano, T. Kiga, K. Omata and N. Kimura (1995).   The analysis of the flame in O_2/CO_2 pulverized coal combustion. *Second International Symposium on CO_2 Fixation and Efficient Utilization of Energy, October, 1995, Tokyo, Japan.*

Okawa, M, K. Omata, N. Kimura, T. Kiga, S. Takano, K. Arai and M. Kato (1996).   Trial design for a CO_2-recovery power plant by burning pulverized coal in O_2/CO_2. *Third International Conference on Carbon Dioxide Removal, September, 1996, Boston, MA.*

Energy Convers. Mgmt Vol. 38, Suppl., pp. S135–S140, 1997
© 1997 Elsevier Science Ltd. All rights reserved
Printed in Great Britain
0196-8904/97 $17.00 + 0.00

Pergamon

PII: S0196-8904(96)00259-2

INTEGRATING O$_2$ PRODUCTION WITH POWER SYSTEMS TO CAPTURE CO$_2$

B.J. Jody, E.J. Daniels, and A.M. Wolsky

Argonne National Laboratory
9700 South Cass Avenue
Argonne, Illinois 60439

ABSTRACT

Chemical cycles for separating oxygen (O$_2$) from air were developed many years ago. These cycles involve a chemical reaction to capture O$_2$ from the air and a change in the operating conditions to effect a controlled breakdown of the newly formed product to release the O$_2$ and regenerate the original species. These cycles are generally more expensive than cryogenic separation of air, partly because they consume high-temperature thermal energy (500-850°C). The chemical cycles can be integrated with high-temperature power cycles to provide efficient heat cascading and recovery because the different temperature levels at which they require thermal energy are compatible with the levels encountered in high-temperature power cycles. The O$_2$ can be used in the combustion process to generate a CO$_2$-rich stream that is more readily separable for production of commercial-grade CO$_2$. This paper presents a preliminary discussion of such integrated systems to facilitate the capture of CO$_2$, aimed at reviving interest in these cycles.

© 1997 Elsevier Science Ltd

KEYWORDS

power generation, O$_2$ combustion, CO$_2$ capture, air separation, water splitting.

INTRODUCTION

As major emitters of CO$_2$, fossil-fuel-fired power generation systems are excellent targets for CO$_2$ capture. Technologies that may be employed to capture the CO$_2$ from flue gas include chemical solvents, cryogenic methods, membranes, physical absorption, and physical adsorption methods. In general, capturing CO$_2$ from flue gas is expensive and energy-intensive, and it results in a substantial increase in the cost of power generation (Wolsky et al., 1993). Reasons for the high energy requirement and high cost include:

(1) CO$_2$ is present in the flue gas at a low partial pressure (Table 1).
(2) Both N$_2$ and CO$_2$ require low temperatures and high pressures to condense.
(3) The large number of species present in the flue gas and the corrosive nature of some of these species complicate all applicable separation processes.
(4) Condensing the water vapor in the flue gas increases the cooling load on the separation process.
(5) CO$_2$ and N$_2$ are relatively inert and do not participate in reversible reactions that can form readily separable products under moderate conditions.

The major challenge to recovering CO$_2$ from flue gas is its separation from N$_2$. An alternative approach is to separate the N$_2$ from the combustion air. Beyond the CO$_2$ capture advantage, combustion with O$_2$ reduces the energy loss associated with heating the N$_2$ from ambient temperature to the flue-gas exhaust temperature, which represents about 6% of the coal's heating

Table 1. Composition (Mol %) of flue gases generated when coal ($CH_{0.8}$), oil ($CH_{1.6}$), and natural gas (CH_4) are burned with 110% theoretical air

Fuel	CO_2	H_2O	O_2	N_2
Coal	15.4	6.2	1.8	76.6
Oil	12.9	10.3	1.8	74.9
Gas	8.7	17.4	1.7	72.1

value (over 2 percentage points lost in the overall thermal efficiency of the power plant). Burning a fuel in the absence of N_2 should also reduce the formation of NO_x. However, O_2 production is expensive and energy-intensive. About 25-30% of the electric power output of a fossil-fueled power plant is consumed by the O_2 production process, and the capital cost of the O_2 generation system is about 25-30% of a conventional plant's capital cost (Wolsky et al., 1993). In advanced plants, such as gasification/combined cycle, it is generally more cost-effective to use oxygen rather than air in the gasification of coal. However, the cost of oxygen is justified only in the gasification process. After gasification, air rather than O_2 is used for combustion.

OXYGEN PRODUCTION

Almost all of the O_2 produced on a commercial scale is separated from air — although separation from water is also possible. The most widely practiced technology for large-scale production of O_2 from air is cryogenic air separation. The typical equipment for O_2 generation consumes about 220-275 kWh of electricity per ton of gaseous O_2. Although there have been modest but persistent improvements in the efficiency of cryogenic air separation, no revolutionary improvement is expected. Other technologies for producing O_2 from air include thermochemical, membrane, and physical adsorption methods. Thermochemical technologies are also applicable for separating O_2 from water. Thermochemical cycles for the production of H_2 from water were actively investigated in the 1970s and 1980s; Table 2 lists some of these cycles. Because they require thermal energy at several temperature levels, they are good candidates for integration with various power generation systems where thermal energy is present at different temperatures. None of these cycles has reached the commercial stage, so reliable cost data are not available. Exotic construction materials may be required because of corrosion problems at the operating temperatures required. Little effort, if any, is now directed toward improving water-splitting technologies because the process is considerably more expensive than methane reforming for H_2 production. However, water splitting may be an efficient way to produce O_2 when integrated with a power generation plant, as shown in Fig. 1. Theoretical analysis of a generic system like that shown in Fig. 1 can illustrate the technical feasibility of such systems. For example, when one mole of natural gas (methane) is used as a fuel (high heating value — 890,156 J/Mol), it can split as much as 3.11 moles of water (heat of reaction 285,838 J/Mol) to generate 1.56 Mol of O_2. Stoichiometrically, it takes 2 kMol of O_2 to burn 1 kMol of CH_4. Therefore, as much as 77% of the combustion air required to burn the fossil fuel can be replaced with O_2 generated from splitting water. In the case of coal (23,850 J/g), the amount of fuel replaced is about 60%. Of course, inefficiencies encountered in real processes would reduce the amount of air replaced. The hydrogen produced from splitting the water can be burned with air to recover most of the energy that went into splitting the water, with some loss in exergy. The resulting combustion products are essentially H_2O and N_2, with small amounts of O_2. This stream can be utilized to recover commercial-grade N_2. Therefore, the overall reaction is equivalent to burning the fuel in air, and the integrated system achieves internal separation of the air. The total amount of air used in both combustion chambers is essentially the same as that used in the conventional combustion process.

Chemical cycles for separating O_2 from air have also been investigated (Ebbing, 1984; Dunbobbin, 1987), and others are under development (*Chemical Engineering* 1996). These cycles involve initiating a reaction to capture O_2 from air and then altering the operating conditions to achieve a breakdown of the newly formed product to release the O_2 so that it can be captured in a pure form. One of these cycles, the Moltox™ process, employs a mixture of Na/K

Table 2. Water Splitting Thermochemical Cycles

CYCLE	MAIN REACTIONS
Prime Cycle	$2HI_x \longrightarrow xI_2 + H_2$ (425 °C) $2H_2O + SO_2 + xI_2 \longrightarrow H_2SO_4 + 2HI_x$ (90 °C) $H_2SO_4 \longrightarrow SO_2 + H_yO + 0.5 O_y$ (850 °C)
LASL Hybrid Cycle	$SO_2 + 2H_2O \longrightarrow H_2SO_4 + H_2$ $H_2SO_4 \longrightarrow SO_3 + H_2O$ (850 °C) $SO_3 \longrightarrow SO_2 + 0.5 O_2$
IGT Hybrid Cycle	$6H_2O + SO_2 + CuO \longrightarrow CuSO_4.5H_2O + H_2$ $CuSO_4.5H_2O \longrightarrow CuSO_4 + 5H_2O$ (200 °C) $CuSO_4 \longrightarrow CuO + 0.5 O_2 + SO_2$ (850 °C)
MARK 9	$6FeCl_2 + 8H_2O \longrightarrow 2Fe_3O_4 + 12 HCl + 2H_2$ (650 °C) $2Fe_3O_4 + 3Cl_2 + 12 HCl \longrightarrow 6FeCl_3$ $+ 6H_2O + O_2$ (>200 °C) $6FeCl_3 \longrightarrow 6FeCl_2 + 3 Cl_2$ (350 °C)

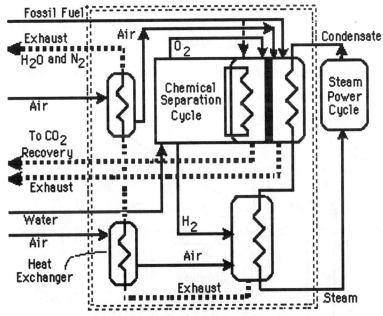

Fig. 1. Schematic diagram of an integrated chemical
water splitting cycle with a steam power plant

nitrites and nitrates to react with O_2 at about 450-650 °C and convert the nitrites to nitrates. The reaction is then reversed by increasing the temperature and/or decreasing the pressure — causing the nitrates to break down into nitrites and O_2 (Dunbobbin, 1987). The barium oxide/peroxide cycle, used commercially in the late 19th century for making O_2, also consists of two basic steps. First, barium oxide (BaO) reacts with O_2 upon exposure to air at about 500 °C to form barium peroxide (BaO₂); the BaO₂ is then heated to about 800 °C to reverse the reaction.

THERMOCHEMICAL AIR SEPARATION CYCLES

Both of these cycles are energy-intensive, stand-alone O_2 production processes. However, when they are integrated with high-temperature power systems, most of the thermal energy consumed in the air separation process can be returned to the power generation system, with some loss of exergy. These processes can also generate steam from their high-temperature waste heat. With steam recovery, the temperature-swing Moltox™ process reportedly consumes the equivalent of about 200-250 kWh of electricity per ton of O_2 recovered (Dunbobbin, 1987). Figure 2 is a schematic diagram of an integrated chemical cycle/conventional steam power cycle. Figure 3 is a schematic diagram of an integrated chemical cycle/conventional gas turbine power cycle.

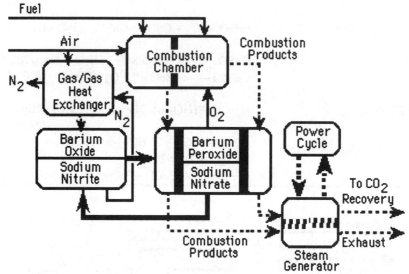

Fig. 2. Schematic diagram of a chemical air separation cycle with a steam power plant

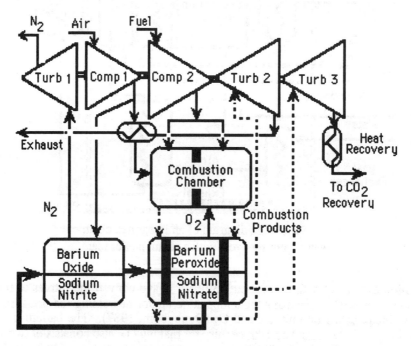

Fig. 3. Schematic diagram of a chemical air separation cycle
with an advanced Gas Turbine Power Plant

Integration of these high-temperature chemical cycles with high-temperature fuel cell systems (such as the molten carbonate and the solid oxide fuel cells) is another approach that could potentially be energy-efficient.

Figure 2 shows the difference between this design and that of its counterpart (without air separation): here, the N_2 exits the system at about 18°C above the ambient temperature (about 38°C), using state-of-the art heat exchanger technology; in the conventional case, N_2 exits at the burner exhaust temperature (about 150°C), as part of the flue gas mixture.   This change represents considerable energy savings.   For example, when the fuel is methane and 110% theoretical air is used in the combustion process, the energy savings will be about 3% of the methane's high heating value. If the efficiency of the power cycle is 33%, this represents a 1% increase in power production. Figure 2 also shows that the heat used by the separation process is returned to the system.

The separation process, however, places a second-law constraint on the system. The availability will be reduced primarily because of the irreversibilities associated with the combustion process (as well as other irreversibilities associated with the O_2 separation process) and because of the heat-transfer processes involved.   Assuming that the combustion chamber is adiabatic, the combined loss in availability caused by separating the air and by the combustion process can be represented by the following equation, where A_i is the availability of stream i and ΔA is the loss in availability or work production caused by these two processes (see Fig. 2):

$$\Delta A = (A_3 + A_2) - (A_1 + A_{10}) \tag{1}$$

In Fig. 1, and in the conventional combustion process, air is introduced at ambient temperature and pressure. Therefore, its availability is zero. So, in the conventional case:

$$\Delta A_{conv} = A(fuel) - A(combustion\ products)_{conv} \tag{2}$$

The equation for the integrated system is:

$$\Delta A_{integrated} = A(fuel) - A(combustion\ products)_{integrated} \\ - A(N_2\ by\text{-}product) \tag{3}$$

Therefore, the difference in the availability *loss* can be determined from:

$$\Delta A_{conv} - \Delta A_{integrated} = \Delta A_{difference}\ ,\ and$$
$$\Delta A_{difference}\ (per\ mole\ of\ fuel) = A(combustion\ products)_{integrated} \\ - A(combustion\ products)_{conv} + A(N_2\ by\text{-}product) \tag{4}$$

These equations can be used to identify the operating conditions (including degree of air separation required and level of recycling of depleted combustion products) for an integrated system so that it results in a net increase in the availability of the combustion products that are driving the power cycle. These conditions should result in:

$$A(CP)_{integrated} > A(CP)_{conv} - A(N_2\ by\text{-}product) \tag{5}$$

where $A(CP)$ is the availability of the combustion products per mole of fuel used.   The combustion products will be at a higher average temperature in the case of the integrated system because of the absence of N_2. In addition, the partial pressure of the CO_2, H_2O, and O_2 will be higher in the integrated case, assuming that the integrated and conventional combustion processes take place at the same total pressure. On the other hand, the partial pressure of N_2 in the combustion products will be much lower. All these terms affect the availability. Similar analysis can be developed for high-temperature Brayton-cycle-based systems (Fig. 3).

CONCLUSION

Thermochemical cycles for O$_2$ production from water and from air can be integrated with high-temperature power cycles to facilitate the production of commercial-grade CO$_2$. Such integrated systems can achieve internal separation of the air or water while minimizing the overall energy consumption of the integrated system. This process might (under the right operating conditions) be more efficient than the use of conventional O$_2$ separation technology to produce O$_2$ for use in conventional fossil-fuel-fired power generation systems. The lack of technical data at this time does not permit the evaluation of the economic competitiveness of such systems.

ACKNOWLEDGMENT

This work was supported by the U.S. Department of Energy, Assistant Secretary for Energy Efficiency and Renewable Energy, under Contract No. W-31-109-Eng-38.

REFERENCES

Chemical Engineering (1996). This sorbent cares only for oxygen, January.

Dunbobbin, B.R., and W.R. Brown (1987). Air separation by a high temperature molten salt process. *Journal of Gas Separation and Purification*, 1, 23-39.

Ebbing, D.D. (1984). *General Chemistry*, Houghton Mifflin Co., Boston, Mass.

Wolsky, A.M., E.J. Daniels, and B.J. Jody (1993).   CO$_2$ capture from oxygen-based combustion/gasification-based power plants. In: *A Research Needs Assessment for the Capture, Utilization, and Disposal of Carbon Dioxide from Fossil Fuel-Fired Power Plants*, Massachusetts Institute of Technology, Report # DOE/ER-30194, Vol. 2 of 2, July.

Energy Convers. Mgmt Vol. 38, Suppl., pp. S141–S146, 1997
© 1997 Elsevier Science Ltd. All rights reserved
Printed in Great Britain
0196-8904/97 $17.00 + 0.00

Pergamon

PII: S0196-8904(96)00260-9

HIGHLY EFFICIENT ZERO EMISSION CO$_2$-BASED POWER PLANT

E. IANTOVSKI and Ph. MATHIEU

University of Liège
Dept. of Nuclear Engineering and Power Plants
21, Rue Ernest Solvay - B-4000 Liège - Belgium

ABSTRACT

An attempt to make a zero emission power plant more profitable is presented. Thanks to complete elimination of waste gases, the use of special fuels with high heating value (for example, used lubricants), otherwise forbidden to be burnt, is now possible. These fuels have actually a negative price because local authorities pay for their incineration.
The performance of an original CO$_2$-based quasicombined cycle is calculated. Efficiencies are at the level of the best state-of-the-art combined cycles. The payback time is estimated at around 3 years, which seems to be acceptable even for a private owner. A first simple design of a gas turbine using CO$_2$ as working fluid is also presented. © 1997 Elsevier Science Ltd

KEY-WORDS: zero-emission power plants; CO$_2$-based turbomachinery; CO$_2$ removal; waste incineration

INTRODUCTION

The payback time of new power plants with reduced CO$_2$ emissions is currently in the range of 10 to 15 years [Riemer P. et al., 1993],[Iantovski E., 1994]. This is for sure too long. In order to increase the benefit of using CO$_2$-based plants some combined applications are proposed, like the use of CO$_2$ to enhance oil recovery [Iantovski et al, 1995].
Among many approaches to the CO$_2$ mitigation extensively investigated in [Blok et al., 1992][Riemer, 1993][Tamaura et al., 1995] and [Riemer et al., 1995], we propose here an original scheme, promising zero-emissions, and not a cut of CO$_2$ releases by 80 or 90%, as when using monoethanolamine scrubbing or membrane schemes. The removed CO$_2$ is then disposed of in a proper site, like in deep aquifers, in oceans or in empty gas fields. The exhaust gases been totaly recovered allows to burn not only coal and biomass but also dirty fuels, which otherwise would be considered as wastes. This is the case of the used lubricants and tyres or of the wastes and by-products of oil refinereries. Considering that around 2% of the processed crude oil is rejected as waste in oil refineries, the resulting low grade fuels would be produced at an annual rate of 100 Mtons. In addition large amounts of municipal solid wastes, which are currently disposed of in landfills, could be available for incineration (about 35 Mtons for the UK alone). Large sources of such bad fuels, which are to some extent noxious even before combustion and are polluting land and water, can now be used for power production. The noxious emissions abatement becoming a critical issue, the main goal of this paper is to present an outline and cycle of a zero-emission plant combining power generation and waste incineration.

LAYOUT OF THE PLANT

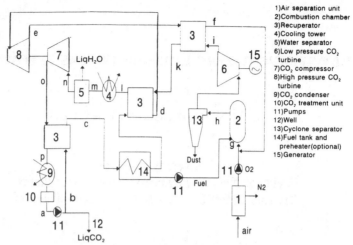

Fig.1: Zero-emission quasicombined cycle

1)Air separation unit
2)Combustion chamber
3)Recuperator
4)Cooling tower
5)Water separator
6)Low pressure CO_2 turbine
7)CO_2 compressor
8)High pressure CO_2 turbine
9)CO_2 condenser
10)CO_2 treatment unit
11)Pumps
12)Well
13)Cyclone separator
14)Fuel tank and preheater(optional)
15)Generator

The flowsheet of the plant is shown in Fig. 1 (Iantovski et al., 1996; Riemer et al., 1995). The proposed CO_2-based plant is comprised of a recuperative CO_2 gas turbine associated with a topping CO_2-like Rankine cycle. The CO_2 gas turbine is made of a quasi-isothermal compressor (3 stages with intercooling), a O_2/CO_2 combustion chamber and an expander which is not on the same shaft as the compressor. The heat content of the exhaust gas, mainly a mixture of CO_2 and H_2O, is transferred partly to the topping cycle and partly to the compressed CO_2 in a recuperator.

Another option is to use an oxygen blown gasifier of coal, biomass or low grade fuels in order to produce a syngas fuel. Here, instead of using water or steam in the gasification process, which generate large losses due to a very large heat of evaporation impossible to recuperate, CO_2 is used as a gasification agent.

EFFICIENCY CALCULATION

The cycle thermodynamics only slightly depend on the gaseous fuel composition. Therefore to illustrate some numerical results we will use the simplest fuel, methane CH_4, which is by the way created in landfills. In addition, the heating value of used lubricants is of the same order of magnitude of that of methane. Some other products of incineration, like municipal solid wastes, have much lower heating value. In this case more complicated calculations are needed. Also technical issues for gas clean up have to be addressed.

Figure 2 shows the cycle T-S diagram with the symbols of Fig. 1. The calculations have been based on computer simulation of the COOPERATE cycle [Iantovski et al., 1994].

The compressor and high pressure expander isentropic efficiencies are taken as 0.80 whilst the low pressure turbine and the CO_2 pump efficiencies are 0.85 and 0.60 respectively.

The electricity consumption of the air separation unit is taken equal to 0.2 kWh/kgO₂. Pressure losses in the heat exchangers are taken as 3% of inlet pressures. The resulting efficiency is quite high: 54.3%; that is at the level of the best modern combined cycles. After optimization with respect to the free parameters it might even increase. In the topping Rankine cycle the work of high pressure turbine is quite high, whereas the pumping work on liquid or supercritical CO_2 is very small. In the future, when the upper pressure is increased above 240 bar, the efficiency gain could even be higher. Such a high efficiency (above 50%) can be reached thanks to the low compression work on CO_2 near to the saturation line. In our case this total compression work from 4 to 240 bar is 181 kJ/kg (164 + 17). For an ideal gas

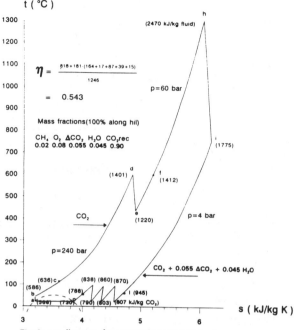

Fig.2: t-s diagram for a quasicombined cycle

isothermal compression from the same 4 bar up to 240 bar with a gas constant of 0.194 kJ/kgK, with a constant temperature of 320 K and a compressor efficiency of 0.80, a work of 317 kJ/kg is needed. The gain of 136 kJ/kg almost counterbalances the work consumed for oxygen production and compression (87 + 39). Therefore the low compression work may offset the oxygen penalty. The closer the compression process to the saturation line, the lower the work to increase the CO_2 density. When the saturation line is crossed, condensation takes place and CO_2 is then compressed at liquid state. **That is one of the great benefits of using CO_2.** It is reminded that the first proposal to use CO_2 as a better working fluid than steam for a Rankine cycle has been made by D.P. Hochstein in 1940, then confirmed by G. Angelino in 1967, N. Gasparovitch in 1969 and others.

THE CO₂-BASED TURBINE

Compressors operating on CO_2 have already been manufactured in the past. Also turbofans have been designed for the circulation of CO_2 used as coolant in military nuclear facilities in the UK and in France. However a one-shaft gas turbine operating on CO_2 as working fluid does not exist yet. Therefore, as CO_2-based expanders are used in the proposed quasicombined cycle, we wonder in which conditions an existing expander designed for operation on air can be used when CO_2 instead of air is the working fluid. This is looked at in the 2 following ways:

1. what are the performance and geometry changes of an air-based expander when CO_2 replaces air as the working fluid, all the other parameters being held unchanged;
2. what is the range of rotational speeds where the best flow similarity is obtained for CO_2 and air for a given geometry.

1. If CO_2 replaces air in a given geometry, only the working fluid properties are modified, the other quantities remaining the same as in an air-based expander. The results mentioned in table 1 are obtained for the first stage with a one-dimensional flow and show the following trends:

-   a reduction of the size (the mean diameter decreases from 1.85 m to 1.42 m, whilst the blade height remains almost the same);
-   a decrease of both the stage temperature drop (from 133.5 to 76°C) and pressure ratio (from 1.65 to 1.56) so that the number of stages increases by 1 (from 4 to 5);

$$M = U/\sqrt{\gamma r T}$$

$$\phi = C_a / U$$

$$\psi = \Delta h / U^2$$

$$R = \Delta h_{rotor} / \Delta h_{stage}$$

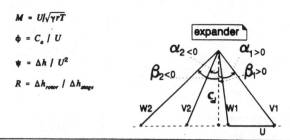

Expander	Fluid: Combustion products	Expander	Fluid: CO₂ + H₂O
data	results	data	results
cp = 1270 J/kgK	diameter = 1.85 m	cp = 1305 J/kgK	diameter = 1.42 m
m = 28 kg/kmol	blade height = 0.17 m	m = 44 kg/kmol	blade height = 0.19 m
$r_{p\theta}$ = 10.9	number of stages = 4	$r_{p\theta}$ = 10.9	number of stages = 5
ṁ = 500 kg/s	ΔT stage = 133.5 °C	ṁ = 500 kg/s	ΔT stage = 76 °C
ϵ_{m} = 0.9	U = 291.4 m/s	ϵ_{m} = 0.9	U = 222.7 m/s
M_{in} = 0.33	rp first stage = 1.65	M_{in} = 0.33	rp first stage = 1.56
φ = 0.75	α_1 = 66.2°	φ = 0.75	α_1 = 66.2°
ψ = 2	α_2 = -21.8°	ψ = 2	α_2 = -21.8°
R = 0.3	β_1 = 43°	R = 0.3	β_1 = 43°
-	β_2 = -60°		β_2 = -60°
N = 3000 rpm	v_1 = 541.4 m/s	N = 3000 rpm	v_1 = 413.8 m/s
TET = 1100 °C	v_2 = 235.4 m/s	TET = 1100 °C	v_2 = 180 m/s
	w_1 = 298.9 m/s		w_1 = 228.5 m/s
	w_2 = 437.3 m/s		w_2 = 334.3 m/s

Table 1: Comparison of CO₂ and air-based expanders

-     a decrease of all the velocities at the entrance of the first rotor wheel. On the other hand, the angles of the velocity triangles do not change at all and this is mainly due to identical inlet Mach numbers and to identical flow and work coefficients.

Similar conclusions are derived for a stand-alone compressor operating on superheated CO₂ [Mathieu et al., 1994], but they are not necessarily valid when the compression takes place near the saturation line as it is the case in the considered system.

2. If CO₂ replaces air in an existing expander, the best flow similarity is obtained when the inlet Mach numbers are the same (here 0.33). This is obtained when the rotational speed is reduced from 3000 to 2650 rpm, by some 12%.

A similar study is carried out for a CO₂ gas turbine in [Mathieu, 1995] and shows that, on the basis of a one-dimensional flow, a good similarity of CO₂ and air flows throughout the compressor and the expander is obtained with a reduction of the rotational speed of the air turbomachine by some 20 % of its design value. In practice, that means that a gas turbine designed for a rotational speed of 3600 rpm or 60 Hz like in the USA could be used almost as such with CO₂ instead of air in Europe (3000 rpm, 50 Hz). In the present scheme (see Fig. 1), the compressor and the high pressure expander are associated in a one-shaft gas generator, without power generation. Hence, there is no danger of compressor surge related to the fuel low heating value like in an industrial gas turbine.

As a conclusion, it cannot be claimed that a CO_2 expander as well as a CO_2 gas turbine is an existing machine available off-the-shelf on the market. However, with the proper tools and funding, the design and the fabrication of such machines is quite feasible within 2 or 3 years by a gas turbine manufacturer.

INCINERATION FOR THE USE OF NEGATIVE PRICE FUELS

The combustion of used lubricants in gas turbines is currently under study. Some of them could be burnt without previous gasification just as ordinary liquid fuels, according to the scheme of Fig. 1 mentioned in (Iantovski, 1996) and not reproduced here for a sake of conciseness.

The combination of power plant with incineration provides sources of money income: the payment by local Municipal Authorities for harmless incineration of wastes and incomes for sold electricity.

In every power plant (see Fig. 3), the cost of electricity (c/kWh) depends linearly upon the fuel price, which for fossil fuels covers a range from 2 $/GJ (coal) to 4 $/GJ (oil, gas). The two straight lines E and F represent the highly efficient zero-emissions combined cycles, described here and in [Mathieu, 1995], whilst the line C represents the less efficient Integrated Gasification Combined Cycle.

The difference between the price and cost of electricity in both cases E and F is rather small, therefore the payback time will be rather long, unacceptable for a private owner or utility. When the specific investment is assumed as 1850 $/kW (this is typical for IGCC) for the considered quasicombined cycle, for a targeted profit of 3 c/kWh, the payback time is 62000 h (1850/0.03), that is around 10 years (for a utilization rate of 6000 hours/year).

Fig.3: Power plants economics

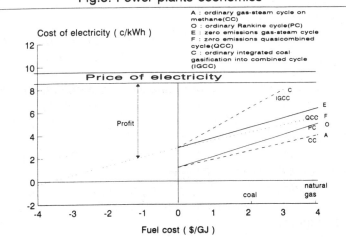

When the fuel is not be paid but instead brings money for its use (negative fuel price), the combination of electricity generation with waste incineration drastically changes the situation. Taking the example of Massachusetts in USA, we know that at the gas filling station everyone, changing oil, has to pay 20 cents per gallon of oil or about 1.8 $/GJ. It represents the mentioned negative price. In this case, the profit is 3 times higher than with the normal fuel price of 4 $/GJ. It means that the payback time might be 2 to 3 years, which looks acceptable.

In USA, with at least 1 Mton of used lubricants per year, it is possible to replace external power supply of filling stations and road systems by self-produced power, cleanly generated with the proposed zero-emission power plants. The amount of used lubricants is enough for the feed of 100 local power plants of 10 MW each.

CONCLUSION

We have presented an original zero-emission CO_2-based combined cycle plant combining a regenerative gas cycle with a topping Rankine-like cycle. Its efficiency is above 50%, at the level of a state-of-the-art combined cycle, when high grade fuels are used. In addition of no pollutant emissions, the proposed plant is able to burn or gasify low grade fuels which might be otherwise considered as wastes. Therefore, the combination of waste incineration and electricity production is a dubble source of profit: the income from the electricity sale and the negative price of the fuel.

The approach is worth of further investigation in order to build a demonstration plant in a near future.

ACKNOWLEDGMENT

Both authors thank Dan Golomb from the University of Massachusetts Lowell (USA), for providing data on use and cost of lubricants and for his very useful suggestions.

REFERENCES

Angelino, G.(1967). Perspectives for the liquid phase compression Gas Turbines. *J. of Engng for Pow.*, 229-237

Blok K., W.C. Turkenburg, C.A. Hendriks and M. Steinberg (1992). Proc. of the First Int. Conf. on CO₂ Removal and Vol.33, No 5-8 of *Energy Conv.& Management*, Amsterdam

Iantovski E.(1994). *Energy and Exergy Currents*, NOVA Science Publ., NY

Iantovski E., K. Zwagolsky, V. Gavrilenko (1994). The COOPERATE-demo power cycle. *Proc. of the 2nd ICCDR*, Tokyo

Iantovski E., K. Zwagolsky, V. Gavrilenko (1994). The COOPERATE power cycle. AES-Vol.33, *Intern. Mech. Engng Congress*, ASME, pp 105-112, Chicago

Iantovski E., V. Schwets, N. Akinfiev, A. Kelebejev, V. Pimenov (1995). Fuel-fired power plant with the sequestering of liquid carbon dioxide in a deep aquifer. *Proc. Intern Conf. Efficiency, Cost. Optimization and Simulation of Energy Systems*, ECOS '95, Istanbul

Iantovski E., Ph. Mathieu, D. Golomb (1996). *Int. Conf. on analysis and utilization of oily wastes*, AUZO'96, Gdànsk (Poland)

Mathieu Ph., P.J. Dechamps, N. Pirard (1994). The use of CO₂ in an existing industrial gas turbine. *49 th ATI National Congress*, Perugia (Italy)

Mathieu Ph. (1995). Combined Cycle Project: CO₂ Mitigation through CO₂/steam/argon gas turbine cycles and CO₂/steam/argon gasification. *Final Report of the Combined Cycle Project*, European Joule II Programme

Riemer P.W.F. (1993), *Proc. of the Int. Energy Agency CO₂ Disposal Symposium* and vol. 34, No 9-11 of Energy Conversion and Management, Oxford

Riemer P., H. Audus, A. Smith (1993). *Carbon dioxide capture from power stations*, IEA Greenhouse Gas Programme, Report

Riemer P.W.F. and A.Y. Smith (1995). *Proc. of the Int. Energy Agency Greenhouse Gases: Mitigation Options* and Vol. 37, No 6-8 of Energy conversion and Management, London

Tamaura Y., K. Okazeki, M. Tsuji, S. Mirai (1993). *Proc. on the Int. Symposium on CO₂ Fixation and Efficient Utilization of Energy*, Tokyo

Energy Convers. Mgmt Vol. 38, Suppl., pp. S147–S152, 1997
© 1997 Elsevier Science Ltd. All rights reserved
Printed in Great Britain
0196-8904/97 $17.00 + 0.00

Pergamon

PII: S0196-8904(96)00261-0

Combining Cryogenic Flue Gas Emission Remediation with a CO₂/O₂ Combustion Cycle

Z. Meratla

CDS Research Ltd, 20 Brooksbank Avenue, North Vancouver, BC V7J 2B8, Canada

ABSTRACT

An alternate approach to conventional flue gas emission remediation, using well established cryogenic principles, is described. The design focuses on the following design objectives: low energy consumption, full heat recovery, complete water removal, high recovery efficiency and capture of NO_x, SO_x and CO_2 as commercial grade NO_2, SO_2 and CO_2 where needed. Heavy metals may also be extracted. The patented process yields commercial grade feed stock for secondary industries producing acids, fertilizers, polymers, methanol, etc., as well as CO_2 for enhanced oil recovery and sequestration, and lends itself to total system energy integration. Conversion of the recovered products may be effected on site or off site. The application of the cryogenic process to a coal fired power plant with a CO_2/O_2 combustion cycle, using by-product oxygen from a hydrogen production facility, is described. © 1997 Elsevier Science Ltd

KEYWORDS

Cryogenic, flue gas, CO_2,

INTRODUCTION

Most of the current commercial wet or dry type flue gas remediation technologies applied to power plants and other single point air emission sources generate a waste residue that creates its own waste disposal or storage problem. These technologies are generally designed to strip a specific effluent, typically NO_x, VOC's or SO_x but not all at the same time. Other potentially harmful effluents, such as, heavy metals, dioxin, furans, arsenic, etc., do not normally lend themselves to treatment by these processes. A list of some of the environmental technologies in use is given in reference [1].

The application of cryogenic processes to flue gas emission remediation, is still at an early stage of application. Most of the studies carried out to date have focused on the capture and sequestration of carbon dioxide. Reference [2] reviewed a number of CO_2 recovery and disposal options.
For small applications, cryogenic condensation and recovery of VOC's has been demonstrated, as illustrated in reference [3]. The use of liquid nitrogen, however, imposes a limitation on the size of VOC recovery systems and represents a significant energy penalty due to producing and delivering liquid nitrogen. For large recovery systems, there are more efficient closed cycle refrigeration techniques that need to be considered, such as, in situ refrigeration using mixed component refrigerants or other cooling systems suitable for the specific application, including recovering the latent heat of cryogenic fuels as described in reference [4].

The process described herein falls within the category of remediation technologies that reduces and recycles harmful effluents [5] and is capable of recovering a broad range of effluents at the same time as well as being applicable to most emission sources. The technology is applicable to conventional combustion cycles using air-fuel mixture or advanced combustion process using CO_2/O_2 -fuel mixture. The focus of this paper is on the advanced combustion process which is gaining increasing interest.

DESCRIPTION OF THE PROCESS

The basic process scheme for a CO_2 recycle combustion process is illustrated in fig. 1 [2]. This scheme was originally developed to facilitate CO_2 recovery at power plants for enhanced oil recovery and has subsequently gained interest as a potential means of CO_2 capture due to its the high CO_2 concentration in the flue gas stream. Particulate matter is removed using conventional methods. A small amount of combustion NO_X is produced from atmospheric entrained nitrogen with the coal and/or present in the fuel itself. Without flue gas treatment, it can be seen that the impurities, including combustion water, are recycled into the boiler.

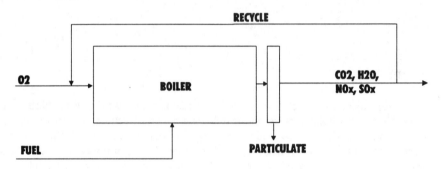

Figure 1 - Basic CO2/O2 - Fuel Combustion Cycle

The cryogenic flue gas remediation technology described herein removes water, and NO_x and SO_x as NO_2 and SO_2. The quantity of CO_2 needed for conversion into other products is treated to remove O_2 and small amounts of nitrogen where present. The residual stream is recycled into the boiler with the excess CO_2 used for enhanced oil recovery and sequestration. The process illustrated in fig. 2 recovers the waste heat and effects a low pressure separation in order to yield a low energy penalty. Residual fine particulate matter is precipitated with controlled condensing of the water produced by the combustion process and present in the fuel.

Typically, 99% of the water content of the flue gas stream is recovered in two steps followed by a final moisture removal step to prevent water freezing in the cold section of the recovery plant. The absence of free water in the cold section of the plant inhibits chemical corrosion.   Established deriming procedures are integrated into the plant operating procedures, thus, preventing corrosion of the equipment when the plant is warmed up during maintenance. In order to meet the specific corrosion prevention requirements of each emission source and the flue gas maximum temperature, a broad range of materials and coatings have been identified and others are under appraisal for the cryogenic technology.
Water and chemicals are not added to the flue gas stream, thus, minimizing the amount of CO_2, NO_2, SO_2 and SO_3 that dissolves in and/or reacts with water to produce highly acidic solutions with the consequent adverse corrosion impact on the equipment and the problem of managing the waste sludge produced.

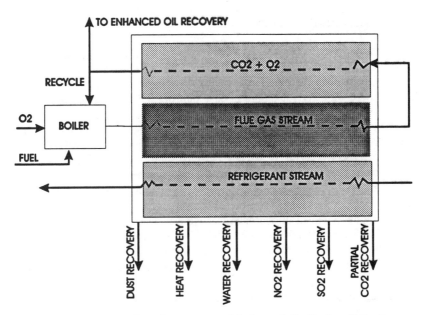

Figure 2 - Schematic Arrangement of the Cryogenic Flue Gas Remediation Process

Extensive experience in the liquefied natural gas industry has shown that traditional technologies for the removal of hydrates and CO_2 are not sufficient to arrest the flow of heavy metals, such as mercury. The cryogenic technology lends itself to installing traps for heavy metal and other air toxics at locations in the process stream where they are immune to clogging and poisoning. This approach also facilitates the handling and treatment of contaminated traps.

In order to enhance the conversion of NO_x and SO_x in the flue gas stream into NO_2 and SO_2 so as to maximize the recovery of acid rain producing gases, oxidation is effected after water removal. Both existing and new oxidation processes have been evaluated. NO_2 and SO_2 lends themselves to recovery with minimal energy penalty.

After the targeted effluents in the flue gas stream have been recovered, the clean residual stream that has been cooled down is circulated in counter flow to the flue gas stream being processed to recover the sensible heat, as shown in fig. 2, and leaves the treatment plant at a temperature of 20 to 25° C. Thus, apart from the thermal losses in the process plant, the refrigeration load needed is essentially dictated by the recovered products only.

POTENTIAL APPLICATIONS

The economic justification for the cryogenic flue gas remediation process in conjunction with the CO_2/O_2 cycle results from its competitiveness with emerging recovery technologies and recovering products with a commercial value, and the ability to accommodate low grade fuels without environmental penalties. This approach moves away from traditional techniques aimed at converting one form of waste into another with the attendant waste handling and storage problems. As the move towards recovering re-usable products gains momentum [5], there will be a need to find additional innovative applications that will serve as outlets for the anticipated quantities of recovered feed stock in the form of NO_2, SO_2 and CO_2.

OXYGEN PRODUCTION

Up to now the focus on oxygen supply for the CO_2/O_2 combustion cycle has been directed towards air separation plants[2]. For many sites, air separation plants offer one of the best options, given their current capacity. However, there does not seem to be a universal solution that is applicable to all sites. To that end, Canada is endowed with large quantities of coal, oil from wells as well as tar sand, natural gas and hydro-electric power potential. The combination of a small population, relatively low growth in energy demand and a preference for certain type of fuels, such as natural gas, has resulted in a vast hydro-electric power potential, in the order of tens of GWe, remaining untapped. To benefit from this clean and renewable source of energy, Canada is actively pursuing international hydrogen export opportunities based on hydro-electric power. This initiative offers the following benefits:

- it provides large scale hydrogen supply from an energy source completely free of greenhouse gas emissions,
- it allows the early development of hydrogen technologies necessary for next century environmental solutions, such as, fuel cell stationary power generation, fuel cell powered vehicles, hydrogen fueled submarines, hydrogen fueled gas turbines, aircraft powered with hydrogen [7], etc.,
- it permits the development of CO_2 recycling processes such as, converting CO_2 to polymers, methanol, etc.,
- it starts displacing some of the hydrocarbon fuels whilst alternate clean sources of energy are being developed,
- it provides very large quantities of by-product oxygen.

The oxygen may be sold in gaseous form to power plants fueled with hydrocarbons in conjunction with a CO_2/O_2 cycle or used on a coal fired power plant integrated with the electrolysis plant, as shown in fig. 3. The latter option offers an efficient vehicle to recycle some of the flue gas CO_2 in the form methanol and/or polymers. There is, also, the potential to effect a high level of energy integration.

In terms of capacity, a 1000 MWe hydro-electric power plant in conjunction with an electrolysis plant produces approximately:
- 480 ton per day of hydrogen
- 3840 ton per day of oxygen

Thus, assuming it takes 22.5 ton/day of oxygen to produce 1 MWe output from a coal fired plant [2], considered a conservative estimate for the process conditions described herein, a 1000 MWe hydro-electric power plant producing hydrogen can support a 170 MWe coal fired power plant provided with a CO_2/O_2 cycle. Although the electrolysis of water is energy intensive, the by-product oxygen offers reduced electrolysis energy penalty and eliminates the capital cost of a sizable air separation plant.

ADVANCED COMBUSTION PROCESS

The cryogenic flue gas remediation technology is ideally suited to advanced combustion processes of the type illustrated in fig. 3. It can be seen that some of the hydrogen produced in the electrolysis plant is used to convert part of the CO_2 produced by the coal fired power plant into polymers and methanol. The SO_2 recovered is either sold "As Is" for de-inking newspapers or as a feed stock, or converted on site into sulfuric acid or ammonium sulfate [5]. Sufficient CO_2 is recycled with the oxygen produced in the electrolysis plant to maintain the heat transfer characteristics of coal/air inside the boilers. Since all the CO_2 processed in the cryogenic flue gas treatment plant is dry, there is no water in the recycled CO_2 stream. The cryogenic flue gas treatment extracts from the flue gas stream SO_2, combustion water, a small amount of NO_2 due to the nitrogen dissolved in the coal and heavy metals, if present. The residual

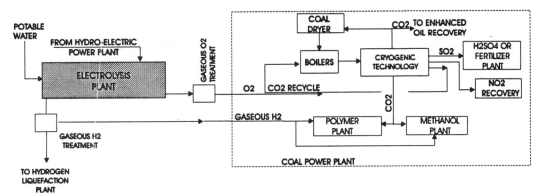

Figure 3 -
ILUSTRATION OF AN INTEGRATED ENERGY SYSTEM
COMBINING A CO2/O2 CYCLE COAL FIRED POWER PLANT
AND A HYDROGEN PRODUCTION FACILITY

stream is recycled "As Is" since it consists essentially of CO_2 with traces of O_2. In order to increase the efficiency of the boilers, most of the latent and sensible heat normally used to heat and vaporize water contained in the coal is eliminated by using hot CO_2 to dry the coal being fed into the boilers and displace the free air inside it which may produce NO_x. This also minimizes the amount of water exhausted from the boilers and the quantity of gaseous effluents that react with it to produce acids.

CONCLUSION

The cryogenic flue gas remediation technology under development offers a new approach for mitigating air pollution from flue gas stacks. It focuses on recycling and re-using the recovered acid rain producing gaseous effluents.

When used in conjunction with an electrolysis plant aimed at producing hydrogen for export, the cryogenic process provides an effective treatment of harmful effluents exhausted by a coal fired power plant using a CO_2/O_2 combustion cycle with all or part of the oxygen derived from the electrolysis plant.

ACKNOWLEDGMENT

The concept for the cryogenic flue gas remediation technology was originally developed by scientists working at Canada's sub-atomic research facility, TRIUMF, located in Vancouver, B.C.
Initial funding for this technology was provided by the National Research Council of Canada.

References
1.   Environmental technologies: solutions for the future,
        Hydrocarbon Processing, August 1995, pp 111-120.
2.   A research needs assessment for the capture, utilization and disposal of carbon dioxide from
        fossil fuel-fired power plants, DOE/ER-30194, Vol 2 of 2, July 1993.
3.   Hender, Gregory W, Sameshima, Glenn T.
        Cryogenic vapor recovery of VOC emissions, 93-TP-51-02, Air & Waste Management
        Association,   Denver, June 1993.

4.   Meratla, Z., (CDS Research Ltd) <u>US Patent 08/293,297</u>, 1995
        International patents pending.

5.   Flue gas chemicals hitting the market, <u>Chemical Marketing Reporter</u>, October 1994, pp 3-4.

Pergamon

Energy Convers. Mgmt Vol. 38, Suppl., pp. S153–S158, 1997
© 1997 Elsevier Science Ltd. All rights reserved
Printed in Great Britain
0196-8904/97 $17.00 + 0.00

PII: S0196-8904(96)00262-2

THE ROLE OF IGCC IN CO₂ ABATEMENT

R. PRUSCHEK and G. OELJEKLAUS
University of Essen, Universitätsstraße 2-17, D-45141 Essen, Germany

G. HAUPT and G. ZIMMERMANN
Siemens AG Power Generation Group (KWU), P.O.B. 3220, D-91050 Erlangen, Germany

D. JANSEN and J.S. RIBBERINK
Netherlands Energy Research Foundation (ECN), P.O.B. 1, NL-1755 ZG Petten, The Netherlands

ABSTRACT

IGCC technology per se involves the potential of highest efficiencies, thus reducing the CO_2 output accordingly. Moreover, the intermediate stage of synthesis gas makes it possible to remove most of the carbon compounds before combustion with acceptable additional auxiliary power demand. The separated CO_2 stream is of highest purity and therefore suited for disposal e.g. in the deep sea or for reuse in chemical syntheses. So, methanol synthesis based on power plant CO_2 has been investigated.

This contribution presents the results of a pre-basic design for a coal-fired 300 MW-class IGCC power plant with methanol production using an external H_2 source. Based on a Siemens Model V94.3A gas turbine-generator, the standard IGCC has been equipped with plant components including CO shift reactors, CO_2 scrubber, methanol synthesis reactors and distillation unit; additional investment costs amount to approx. 25 %. This concept is based solely on proven process engineering methods.

Primary energy utilization as well as the resulting methanol production costs based on appropiate generating costs are discussed. Comparative CO_2 emission figures make the advantage of such a coproduction process regarding this perfectly clear. © 1997 Elsevier Science Ltd

KEYWORDS

CO_2 removal; IGCC power plant; CO shift reactor; Rectisol wash; efficiency calculations; coproduction; methanol synthesis; hydrogen source; CO_2 emission figures.

INTRODUCTION

The observed increase of CO_2 concentration in the atmosphere gives cause for concern about man-made climate change. In view of recent and future developments in the field of IGCC expressed by increasing overall efficiencies (Pruschek *et al.*, 1996), a significant CO_2 emissions reduction becomes possible solely by replacing old and low-efficient stations with advanced IGCC power plants

Fig. 1 shows comparative net efficiencies for large-scale units including a prognosis for mid-term future improvements by further enhancing the turbine inlet temperature. In case of IGCC, a level close to 50 % has been achieved with gas turbine-generators available today and for well-gasifiable coals. First estimates give cause for hope to exceed 53 % in the next two years leading to CO_2 emissions reductions of 25-30 % within this area.

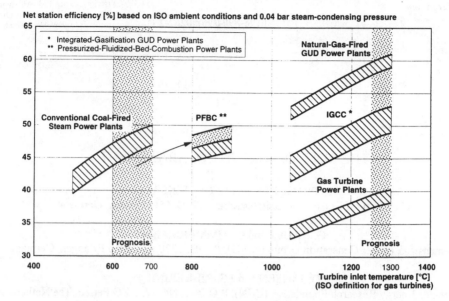

Fig. 1: Thermal Performance of Different 600 MW-Class Power Plants

Advisory panels, however, have recommended to reduce the CO$_2$ emission by 60 or even 80 % by the middle of the next century related to the emission of 1987. Thus, limiting the environmental pollution and CO$_2$ emission caused by energy consumption is one of the greatest challenges to science and technology. Neither efficiency enhancement as discussed above, prime measure for CO$_2$ emission abatement of power plants, nor substitution of fossil fuels by carbon-free energy will comply with this requirement. If coal has to be used under such circumstances, the ultimate measure, viz. CO$_2$ removal, has to be considered as an option. However, the essential problem of such a strategy is the disposal of extracted CO$_2$. If it could be stored separate from atmosphere e.g. in deep sea or underground, CO$_2$ emission from IGCC would be reduced by 88 % by the removal system investigated here. Another option, less efficient with respect to CO$_2$ emission abatement and more sophisticated and costly, is reusing CO$_2$ for methanol synthesis.

IGCC WITH CO$_2$ REMOVAL

The pre-basic design described afterwards is based on a 300 MW-class IGCC power plant with a Siemens Model V94.3A gas turbine-generator with 1190 °C ISO inlet temperature. Gas generation and treatment include (Pruschek *et al.*, 1994):

- Pittsburgh No. 8 coal is gasified in a pressurized entrained-flow (PRENFLO) oxygen-blown gasifier.
- The sensible heat of coal-derived gas is used for steam generation by raw gas cooling.
- Dust is primarily removed by cyclone and candle filter arrangements.
- Alkali compounds and residual fine dust are separated in a Venturi (water) wash.
- CO and H$_2$O are converted into CO$_2$ and H$_2$ in a water gas shift reactor system.
- Sulfur compounds (H$_2$S, COS) and CO$_2$ are simultaneously removed by Rectisol wash.

By this means, around 88 % of CO$_2$ (related to C input by coal) are removed and extracted in gas phase with acceptable amount of energy and investment costs. Table 1 compares salient data of IGCC with CO$_2$ removal and a standard IGCC as reference case. The incorporation of CO$_2$ separation results in a 6 percentage points lower net efficiency.

Table 1: Salient Data of IGCC with CO_2 Removal
Compared to the Reference Case

	Reference Case	With CO_2 Removal
CO_2 Emission (Stack) Absolute	72.9 kg/s	8.4 kg/s
Specific	0.69 kg/kWh	0.09 kg/kWh
Coal Heat Input	811.2 MJ/s	876.1 MJ/s
Gross Power Output Gas Turbine (V94.3A)	238.8 MW	234.1 MW
Steam Turbine	177.7 MW	170.2 MW
Auxiliary Power Requirement	37.8 MW	49.1 MW
Gas Treatment from This	3 %	24 %
Net Power Output	378.6 MW	355.2 MW
Net Efficiency	46.7 %	40.5 %

IGCC WITH COPRODUCTION OF ELECTRICITY AND METHANOL

The recovered CO_2 fraction from IGCC as described above has to be either disposed or further utilized. Oceans represent the largest CO_2 storage facility compared to other conceivable possibilities such as aquifers or empty gas fields. However, environmental impact cannot be excluded. In addition, ocean disposal of CO_2 consumes extra energy for transportation, liquefaction and possibly freezing, further deteriorating net efficiency of the overall system.

Annual world-wide CO_2 emissions from energetic use of fossil fuels are in the order of magnitude of 22 Gt (1988), corresponding to a carbon inventory of some 6 Gt C. About 20 % of them come from electric power generation in coal-fired power plants representing 1.2 Gt C in the fuel. In contrast, carbon used world-wide for raw materials of the chemical industry such as ethylene, propylene, methanol amounts to only 0.09 Gt C. These figures show that only a small fraction of the CO_2 emitted by power plants would be necessary as a feedstock for the production of chemicals, which would not justify demonstration of CO_2 reuse technology on technical scale. The only potential worth analyzing more closely is substitution of mineral-oil-based fuels in the transportation sector by methanol, as the major part of the carbon requirement of 1.2 Gt C could be met from power plant emissions. For this purpose, CO_2 and H_2 are converted into the liquid energy carrier methanol which can be stored more easily and which is therefore more suitable to application in today's transportation sector infrastructure. Bound carbon would then be ultimately released into atmosphere by subsequent combustion. Required H_2 must of course be produced using a carbon-free or low-carbon energy source (e.g. by hydropower-driven water electrolysis) in order to reduce CO_2 emissions of the overall system.

Stoichiometric conversion of hydrogen and carbon dioxide to methanol following the chemical reaction equation

$$3 H_2 + CO_2 = CH_3OH + H_2O + 49.9 \text{ kJ/mol}$$

requires 2.1 m^3 H_2/kg CH_3OH. Assuming e.g. the whole fuel demand of Germany's traffic sector to be covered by methanol, about 220 billion m^3 H_2/a are required, which is eleven times the present annual H_2 production in Germany.

Fig. 2 shows how a CO_2-based methanol synthesis plant (Göhna and König, 1994) can be integrated into an IGCC. They are linked through CO_2 fraction removed, purge and off-gas, and imported and exported steam as well. This integration of the methanol system is considerably influencing the design of IGCC's water/steam cycle, e.g. some of the compressors being driven by back pressure turbines operated with high pressure steam. Hydrogen as reactant to be imported is assumed to be available at plant limits at 100 % purity and 66 bar.

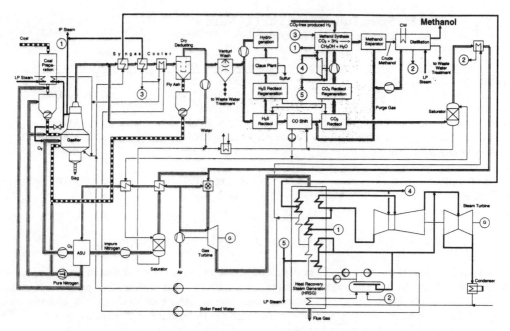

Fig. 2: Integrated PRENFLO Gasification GUD Power Plant with CO Shift,
CO₂ Removal and Utilization for Methanol Synthesis
(GUD is a registered trade mark and an acronym for the the German
words "Gas und Dampf" meaning gas and steam)

Salient data of the coproduction system are:

- Coal input   2300   tonnes/d
- Hydrogen input   780   tonnes/d
- CO₂ (intermediate product)   5500   tonnes/d
- Methanol product (>99.85 wt%)   3800   tonnes/d
- Total gross power production   354   MW
    Gas turbine   234 MW
    Steam turbine   120 MW
- Total net power output   310   MW

An independent evaluation of both processes, power generation and methanol production, is difficult and
not highly conclusive due to the many process interfaces. If input energy carriers coal and hydrogen as well
as methanol produced are evaluated with their lower heating values (LHV), the relative energy yield of the
overall system amounts to appr. 67 %. Methanol can be further converted to synthetic gasoline via MTG
(Methanol To Gasoline) process of Mobil Oil.

ECONOMICAL CONSIDERATIONS

Total plant investment costs for coproduction of power and methanol/gasoline comprise costs of IGCC with
CO₂ removal and of methanol synthesis/MTG plant. Specific costs of coproduced methanol/gasoline
strongly depend on hydrogen price. CO₂ from IGCC could be considered as for free in case extra costs
caused by CO₂ removal are allocated to generating costs. In the present case, a simplified cost calculation
model has been used which is solely based on investment costs for the additional equipment of the methanol

system and hydrogen production costs depending on the different sources. Taking into account that the CO_2 stream has been provided by adding a removal plant to a standard IGCC, the specific methanol production costs estimated as described above are penalized by thereby higher generating costs according to the ratio of electricity and coproduced methanol/gasoline.

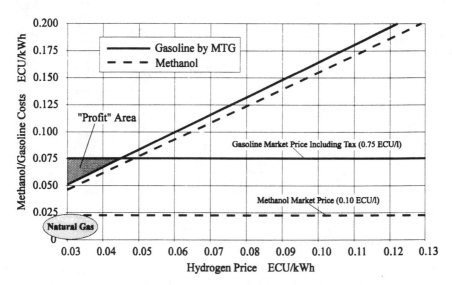

Fig. 3: Methanol/Gasoline Cost vs. Hydrogen Price

In Fig. 3, methanol/gasoline production costs are plotted against hydrogen price (1 ECU = 1.25 US$). General target should be that methanol costs are below market price and gasoline costs are below market price including tax. Both comparative figures are indicated in the diagram showing that even with today's cheapest hydrogen produced from natural gas by steam reforming (not CO₂-free!), methanol is not competitive. Synthetic gasoline is only below market price if based on this hydrogen source and not charged with tax.

COMPARATIVE CO₂ EMISSIONS

Finally, the described coproduction process and separate generation of electric power and methanol, which should be used for transportation in correspondingly modified gasoline engines, have to be compared as regards total CO₂ output of the systems. Primary energy sources are then coal and a CO₂-free energy source such as hydropower also supplying hydrogen via water electrolysis wherever necessary. In Fig. 4, primary energy requirements and accompanying CO₂ emissions for different cases of interest are collected. It turns out that only power generation via IGCC and parallel use of cars fuelled with hydrogen from a CO₂-free source (Case 2) show a (slightly) lower CO₂ output than IGCC with coproduction of electric power and methanol as substitute for oil-based gasoline (Base Case).

CONCLUSIONS

CO₂ removal from IGCC makes drastic CO₂ reduction possible as required by the IPCC in the next century if it can be disposed of. A conceivable CO₂ disposal strategy apart from deep sea dumping and underground storage is reutilization of extracted CO₂, which is of course a more expensive way, and CO₂ emission

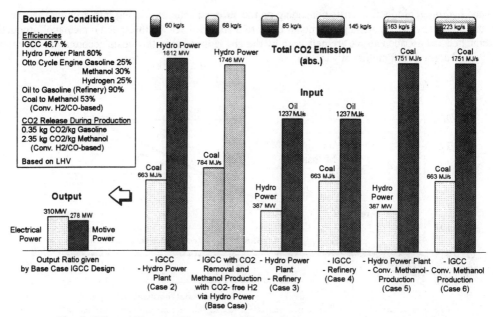

Fig. 4: Primary Energy Demand and CO_2 Emission
for Combined Electrical and Motive Power

abatement of the overall system is less pronounced. In principle, a couple of routes are available for reuse of CO_2's C atom as a "feedstock" in chemical industry. Significant quantities of CO_2, as e.g. emitted from fossil-fired power plants, are only required if methanol/synthetic gasoline can substitute oil-based fuels used in traffic sector today. Ultimately, retained CO_2 only serves to store the energy carrier H_2 in a liquid fuel. A net CO_2 reduction, of course, is only achieved if necessary H_2 is based on CO_2-free primary energy sources. Depending on the respective hydrogen source, coproduction of electricity and methanol/gasoline via IGCC ends up with production costs around present gasoline market price including tax. These results will have to be compared with alternatives available in the middle of the next century if CO_2 reduction up to 80 % would be required. As emphasized at the beginning, such scenarios have to be seen in the context of stringent regulation on CO_2 emission reduction which cannot be fulfilled solely by rational use of energy and by use of C-free energy resources e.g. due to acceptance problems in the nuclear sector.

REFERENCES

Göhna, H. and P. König (1994). Producing methanol from CO_2. *CHEMTECH*, JUNE 1994, 36-39.

Pruschek, R., G. Oeljeklaus, V. Brand, G. Haupt and G. Zimmermann (1994). GUD Power Plant with Integrated Coal Gasification, CO Shift and CO_2 Washing. In: *POWER-GEN EUROPE '94* (Conference Papers), Book III, Vol. 6, pp. 205-225. Cologne/D, 17-19 May 1994.

Pruschek, R., G. Oeljeklaus, D. Boeddicker, G. Haupt and G. Zimmermann (1996). Enhancement Potentials of Combined Cycles with Integrated Coal Gasification - Efficiency, Cost Effectiveness and Availability. In: *POWER-GEN EUROPE '96* (Conference Proceedings), Vol. II, pp. 103-126. Budapest/H, 26-28 June 1996.

ACKNOWLEDGEMENT

This study was supported in part by the European Union (EU) within the framework of the JOULE II programme.

Pergamon

Energy Convers. Mgmt Vol. 38, Suppl., pp. S159–S164, 1997
© 1997 Elsevier Science Ltd. All rights reserved
Printed in Great Britain
0196-8904/97 $17.00 + 0.00

PII: S0196-8904(96)00263-4

Water gas shift membrane reactor for CO_2 control in IGCC systems: techno-economic feasibility study

M. Bracht[1], P.T. Alderliesten[1], R. Kloster[2], R. Pruschek[2], G. Haupt[3], E. Xue[4], J.R.H. Ross[4], M.K. Koukou[5] and N. Papayannakos[5]

1. Netherlands Energy Research Foundation (ECN) P.O. Box 1, 1755 ZG Petten, The Netherlands
2. University of Essen, Germany
3. Siemens AG/KWU, Germany
4. University of Limerick, Ireland
5. National Technical University of Athens, Greece

ABSTRACT

A novel reactor concept, the water gas shift membrane reactor (WGS-MR) for CO_2 removal in IGCC systems has been investigated. In order to establish full insight in the possibilities of the application of such a reactor, a multidisciplinary feasibility study has been carried out comprising system integration studies, catalyst research, membrane research, membrane reactor modelling and bench scale membrane reactor experiments. The application of the WGS-MR concept in IGCC systems is an attractive future option for CO_2 removal as compared to conventional options. The net efficiency of the IGCC process with integrated WGS-MR is 42.8 % (LHV) with CO_2 recovery (80 % based on coal input). This figure has to be compared with 46.7 % (LHV) of an IGCC without CO_2 recovery and based on the same components, and with 40.5 % (LHV) of an IGCC with conventional CO_2 removal. Moreover, an economic analysis indicates favourable investment and operational costs. The development of the process is considered to be technically feasible. However, it became clear that the technology of inorganic high selective gas separation membrane manufacturing and high temperature ceramic materials engineering is not yet mature and that further development in this area remains necessary. © 1997 Elsevier Science Ltd

KEYWORDS

CO_2-removal, IGCC, Power generation, Cycle calculations, Water gas shift, Water gas shift Catalysts, Ceramic membranes, Membrane reactor, Membrane reactor model, Computational fluid dynamics

INTRODUCTION

In a no regret scenario, it is of utmost importance to have cost effective methods available for carbon dioxide emission reduction for fossil fuelled power production systems. CO_2 removal might be an option which has to be used ultimately. Possibilities to control the emission of carbon dioxide from an IGCC system are believed to compare favourably to other large coal-based combined cycles due to the relatively high efficiency and the relative ease of CO_2 removal from the high pressure fuel gas. Conventional process schemes for CO_2 removal for oxygen blown entrained bed gasifiers comprise low temperature gas cleaning, a separate Water Gas Shift (WGS) conversion to convert the CO with steam in the fuel gas mixture into CO_2 and H_2, followed by CO_2 removal for example by scrubbing with solvents. The extent of the shift conversion determines the final CO concentration in the fuel gas and thus the maximum level of CO_2-control. The thermodynamic equilibrium conversion of CO decreases as temperature increases. In order to achieve high CO conversion, a two-step catalytic process is usually employed.

One of the methods to circumvent conversion limitations is to enforce an equilibrium displacement to the product side. From an energy-efficiency viewpoint, this should be achieved by continuous removal of one of the product components directly at the place where it is formed.

A promising approach to reach this, is the implementation of a catalytic membrane reactor, in which the WGS reaction is combined with H_2 separation from the reaction mixture in one reactor, using ceramic (inorganic) membranes selectively permeable to hydrogen. The application of inorganic membranes is necessary because of the temperature level at which the reaction occurs (typically 300-500°C). In order to establish full insight in the possibilities of the application of such a reactor, a multidisciplinary feasibility study has been carried out comprising system integration studies, catalyst research, membrane research, membrane reactor modelling and bench scale membrane reactor experiments.

WATER GAS SHIFT MEMBRANE REACTOR

A schematic of the membrane reactor considered here is shown in fig.1. The membrane has been placed inside the reactor shell and the catalyst outside the membrane tube. The WGS takes place on the catalyst and hydrogen permeates selectively through the membrane to the so-called permeate side. Removing hydrogen from the reaction zone forces the chemical equilibrium of the reaction to the product side. The sweep gas dilutes the product and raises the trans membrane partial pressure difference resulting in a higher driving force. After reaction the feed stream leaves the reactor as the retentate stream, which is enriched with CO_2.

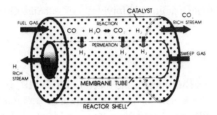

Fig. 1. Schematic of the water gas shift membrane reactor

Membranes

A summary of inorganic membrane development has been prepared by Hsieh (1989). Several materials seem to have good potential for industrial gas separation applications, however, the development of most materials is still on laboratory scale with typical sample sizes of several cm^2. For large scale applications tubular systems are thought to have the best possibilities when the state of the art of production technology is considered in connection with the attainable surface area and module construction aspects (Sarraco et al., 1994). Porous membranes are currently thought to have good opportunities because of their relatively high permeability which translates directly to the (limited) size of the installation. The porous materials considered in this study have been scaled up and are available in tubular geometry with a surface area upto several hundreds cm^2. For this study defect-free tubular Knudsen diffusion membranes have been made and exploratory research on scaled up tubular high-selective microporous silica membranes has been performed. The results of the membrane research work are summarized in table 1.

The characteristics of the microporous silica membranes have been used for further evaluation of the concept, although it is recognised that the stability of the microporous silica membranes is still worrisome. These membranes show a combination of an acceptable permeability with an acceptable selectivity, as will be shown below. Further research must be directed towards improving the stability of these membranes or towards other more stable materials with simular characteristics. Membrane reactor tests have been postponed until the stability problems have been solved.

Table 1. Results of membrane research.

	Microporous silica membranes	Knudsen diffusion membranes
Permselectivity H_2/CO_2	15	max. 4.77
Permselectivity H_2/CH_4	47	max. 2.83
H_2 permeability [mol / m^2 s Pa]	$2\ 10^{-6}$	$\sim 1\ 10^{-5}$
Stability in WGS condition	material degrades to Knudsen diffusion properties	good

Catalysts

For the process a catalyst is required which is sufficiently active and selective for the WGS reaction at the optimum process conditions: low steam to CO ratio's at the optimum process temperatures and pressures.

Furthermore, the catalyst should be resistant to the contaminants present in the fuel gas after gas cleaning. Many catalysts have been screened. Based on the activity, selectivity and sulphur resistivity Fe-Cr and Pt on ZrO_2 have been selected as most promising. A disadvantage of the commercially available Fe-Cr catalyst is the low activity at the selected operation temperature, there is no activity below 300°C. Pt/ZrO_2 shows promising results. It shows a higher activity, a better S-resistivity than a Fe-Cr catalyst and a very good selectivity. However, Pt/ZrO_2 is more expensive than the Fe-Cr catalyst and is still under development. In the process considered in this study a lower steam/CO ratio than generally employed in a WGS is being used (see System integration). Experiments showed that this did not lead to undesirable side products as carbon or CH_4 for the Fe-Cr or Pt/ZrO_2 catalysts.

Reactor model

Several models have been developed to gain insight in the various heat and mass transfer processes that take place in a membrane reactor (Bracht et al., 1995; Koukou et al). Various concepts have been evaluated and a laboratory reactor and a full scale reactor have been designed using these models.
The phenomena that have been taken into account in the different membrane reactor models are: permeation, WGS reaction, gas flow along the membrane, axial mass dispersion, radial mass dispersion and heat effects. The membrane permeability is relatively so large that radial dispersion effects play an important (detrimental) role in the overall reactor performance and have to be taken into account. Models including the dispersion effects have been implemented using the PHOENICS® computational fluid dynamics (CFD) software package. The models have been validated with experimental results from bench scale experiments. With the use of the models reactor designs can be made in which the detrimental dispersion effects have been minimised. The reactor performance parameters and their targeted values (see System Integration) are shown in table 2.

Table 2. Membrane reactor performance parameters and their values

MR Parameter	definition	target value
CO conversion	CO converted/CO entering	90%
H_2 recovery	H_2 in permeate/H_2+CO entering	80%
CO+CO₂ recovery	CO+CO_2 in retentate/CO+CO_2 entering	80%

SYSTEM INTEGRATION

The integration of the WGS-MR for CO_2 recovery in an IGCC requires some adaptation with respect to conventional IGCC schemes. The required reactor boundary conditions, inlet and outlet stream quantities have to be optimised to meet the WGS-MR performance targets and to obtain optimum process integration for best utilization of energy. The WGS-MR requires a higher reactor inlet temperature for high performance. A hot gas cleaning operating at higher temperatures, as an alternative to the conventional low temperature wet gas cleaning, is a promising option that has been investigated too.
Important boundary conditions that have been set concern:
- Gasification process: PRENFLO, entrained-flow, oxygen-blown gasification with dry coal dust.
- 2 options for synthesis gas cleaning system: • Conventional wet gas cleaning (low temperature)
  • Advanced dry gas cleaning (elevated temperatures)
- 2 options for CO_2 delivery conditions: • gaseous state, 15°C, 1bar
  • liquid state, 15°C, 150 bar

Main differences to a conventional IGCC concern the additional components for CO_2 recovery and CO_2 treatment and the gas cleaning in case of hot gas cleaning option. The basic process scheme is shown in figure 2. The hot gas cleaning showed to be advantageous mainly because it avoids a clean gas heat-up. Based on considerations of overall IGCC performance and realistic membrane reactor performance the target performance of the WGS-MR has been defined as shown in table 2. To enhance the hydrogen recovery of the WGS-MR, a sweep gas is fed at the low pressure side of the membranes in countercurrent flow mode. The sweep gas consists of the nitrogen fraction from the ASU, saturated with water to increase the available mass flow and as a measure for NO_x suppression in the gas turbine combustion chamber. The unseparated gas stream from the WGS-MR still contains unconverted and unseparated fuel (CO, H_2) which is utilized by burning the gas stream with oxygen from the ASU in a catalytic burner (CO + H_2 concentration is below the flammability limits). The hot exhaust gas from the burners is cooled in an additional pressurized heat recovery steam generator (HRSG) producing HP live steam for the steam turbines.

A H_2O/CO ratio of 1.3 has been set for the WGS-MR feed gas as a mean value between high amounts of steam losses in case of a higher ratio and too low CO conversion in case of a lower ratio. A too low a value would also be irrealistic because of the problem of carbon formation. The feed side and permeate side pressure are resp. 36 and 21 bar. The 15 bar pressure drop across the membrane is necessary in order to have

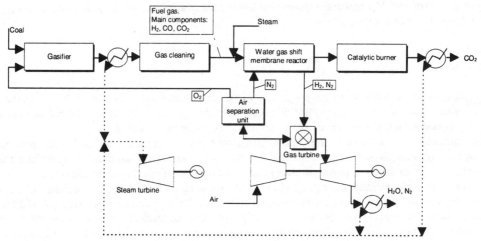

Fig. 2. Basic process scheme of IGCC with integrated WGS-MR system for CO_2 recovery

enough driving force for hydrogen transport through the membrane. The inlet temperature of the reactor has been set to 325°C based on considerations of best IGCC performance, activity window of the catalyst and thermodynamic WGS conversion equilibrium level.

Membrane reactor configuration

A two step adiabatic WGS-MR system (as opposed to a one step isothermal reactor) is envisioned to be the most feasible option currently from a technical point of view. A layout of the system is shown in figure 3. In the first shift reactor the bulk of the conversion, accompanied with the largest temperature rise will occur. After cooling, the conversion is being pulled over the thermodynamic equilibrium conversion by hydrogen extraction from the reaction zone with the membranes in the membrane reactor. The temperature rise in the second step will be moderate. In these circumstances it is more certain that membranes will be able to operate satisfactorily and fulfill lifetime requirements. From a heat management point of view the two reactors are relatively simple devices, because no heat has to be carried away in the reactor itself, being adiabatic systems.

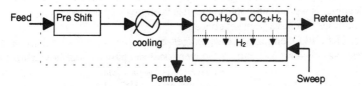

Fig. 3. The water gas shift membrane reactor system

Cycle calculations

Table 3 summarises the results of the cycle calculations. The application of the WGS-MR concept in IGCC systems is an attractive future option for CO_2 removal as compared to conventional options. Although the rate of CO_2 removal is a little lower than in the conventional case, the efficiency of the novel system is significantly higher. By improvement of the WGS-MR reactor performance, by using more selective membranes, the overall IGCC/WGS-MR efficiency can be raised even further.

Table 3. Results of cycle calculations

IGCC configuration	Net electric efficiency [%]	Overall CO_2 recovery [%][1]
No CO_2 removal	46.7	-
Conventional CO_2 removal[2]	40.5	88
WGS-MR + conventional gas cleaning	42.8	80
WGS-MR + dry gas cleaning	43.4	80

1: based on coal input;   2: CO_2 removal by Rectisol wash as elaborated by Pruschek et al. (1996).

Feasibility

Design and dimensions of the membrane reactor have been established by closely evaluating the necessary membrane surface area and catalyst volume, the possibilities for membrane fabrication, the possibilities for membrane sealing, the size of the total system and mass transfer effects. The main design characteristics for the reactor are:

- Membrane tube length   2 m
- Total membrane surface area   2176 m^2
- Total number of membrane tubes 3300
- Membrane tube diameter   10.4 cm
- Total catalyst volume   25.3 m^3

Based on these starting points a reactor has been designed which dimensions are compatible with the size of other IGCC components. Large tubes are necessary to reduce the number of (expensive) sealings. A dead end tube configuration is used in order to avoid unacceptable thermal compression and expansion stresses. This necessitates the insertion of a tube inside the membrane tubes for gas supply to the dead end side. With respect to the technical feasibility of the WGS-MR integrated in an IGCC, three levels of development can be distinguished for parts of the system:

1) *IGCC*. IGCC itself has been demonstrated in several projects around the world. In Europe two major projects are currently going on: Buggenum, the Netherlands and Puertollano, Spain. These projects illustrate the interest for and possibilities of IGCC as power generation option. The IGCC still has a great potential for efficiency improvements

2) *Dry gas cleaning (in the option with dry gas cleaning)*. Dry gas cleaning at elevated temperatures is a technology that is being developed. The application will lead to further efficiency improvements as shown also by the work in this project. Currently, options are already being tested on bench/pilot scale.

3) *Membrane reactor*. Obviously the membrane reactor is the element in this new system that is, at least technically, the least mature. Prospects from laboratory scale experience and from model calculations look very good. However, in scaling up the technology to bench and pilot scale a number of hurdles have to be taken. The issue in membrane development is connecting sufficient selectivity and permeability with a good stability of the membranes. The permeability of the membranes should not be much smaller than the value measured and used here. This would lead to larger necessary surface areas and an unacceptable size of the reactor system.

The evaluation of the investment costs is based on a detailed economic analysis for a conventional IGCC.

Table 4. Results of the economic evaluation

IGCC case	base-case	conventional CO_2 removal	WGS-MR CO_2 removal with conventional gas cleaning
CO_2 recovery[1]	-	88	80
CO_2 state	-	gaseous (1.3 bar, 38 °C)	gaseous (1.0 bar 74 °C)
Net power [MW]	379	355	433
Net efficiency [% LHV]	46.7	40.5	42.8
Specific investment[2] [ECU/ kW]	1560	1869	1594
COE[3] (6000 h/a) [ECU / kWh]	0.0787	0.0867	0.0835

1: based on coal input;   2: European Currency Unit, 1ECU=1.27 US$, 30 July 1996;
3: Cost of electricity generation.

Additional components that are required and included in the economic analysis are: saturation system, CO pre shift reactor, WGS-MR, catalytic burners, 2nd HRSG for retentate gas and gas expansion turbine for retentate gas. Results of the economic analysis and comparison with a system with conventional CO_2 removal are summarised in table 4. The efficiency of the system with a WGS MR is higher than a system with conventional CO_2 removal, moreover the investment costs are lower. This shows the economic feasiblility of the novel reactor concept.

CONCLUSIONS

The water gas shift membrane reactor concept in an IGCC as an option for CO_2 mitigation is attractive. From system optimization and cycle efficiency calculations it has been shown that the system proposed has a higher efficiency connected with lower costs compared to conventional options for CO_2 removal. Further efficiency improvement in the process can be obtained by:
* using catalysts that allow lower steam/CO rations
* improving the membrane ractor performance by using more selective membranes than that are currently at hand

By preparing a full scale reactor pre-design it has been shown that a reactor with a reasonable size emerges. It became clear, however, that especially the technolgy of inorganic high-selective membrane and module manufacturing is far from mature and that further development is required.

ACKNOWLEDGEMENT

The research work described in this paper has been funded in part by the COMMISSION OF THE EUROPEAN UNION within the framework of the JOULE II programme (1992-1995), Sub programme: COAL. The authors would like to thank: G. Oeljeklaus (University of Essen), G. Zimmermann (Siemens, KWU), M. O'Keeffe (University of Limerick), N.C. Markatos (National Technical University of Athens), A. Bos, K. Hemmes, A.B.J. Oudhuis and P.P.A.C. Pex (ECN) for their substantial contributions to the project

REFERENCES

Alderliesten P.T. and M. Bracht (1996), An attractive option for CO_2 control in IGCC systems: Water gas shift Integrated H_2/CO_2 Separation (WIHYS) process. Phase I: Proof of principle. Final Report CEC project JOU2 CT92-0158.

Bracht M., A. Bos, P.P.A.C. Pex, van H.M. Veen, P.T. Alderliesten, (1995). Water gas shift membrane reactor. In: *Proceedings of Euromembrane '95, Bath* (W.R. Bowen et al. ed) Vol.2, 425-430.

Hsieh H.P. (1989). Inorganic Membranes. *Am. Inst. Chem. Eng.Symposium Series, No. 261*, 84, 1-18.

Hsieh H.P. (1989). Inorganic Membrane Reactors. *Am. Inst. Chem. Eng. Symposium Series, No. 268*, 85, 53-67.

Jansen D, A.B.J. Oudhuis and H.M. van Veen (1992). CO_2 reduction potential of future coal gasification based power generation technologies. Energy Convers. Mgmt., 33, 365-372.

Koukou M.K., N. Papayannakos, N.C. Markatos Dispersion effects on membrane reactor performance. accepted for publication in AIChE Journal.

Pex P.P.A.C. et al. (1995). Materials aspects of microporous inorganic gas separation membrane manufacturing. In: *Proceedings of Euromembrane '95, Bath* (Bowen W.R. et al. ed) Vol.1, 295-300.

Pruschek R., G. Oeljeklaus, V. Brant, G. Haupt, G. Zimmermann and J.S. Ribberink (1994). Combined cycle power plant with integrated coal gasification, CO shift and CO_2 washing. In: *Proceedings of the conference on Carbondioxide Removal*, Kyoto, Pergamon Press.

Pruschek R. (1996), Coal-fired multicycle power generation systems for minimum noxious gas emission, CO_2 control and CO_2 disposal. Final Report CEC project JOU2-CT92-0185.

Saracco G., G.F. Versteeg and van W.P.M. Swaaij (1994). Current hurdles to the succes of high temperature membrane reactors, *J. Membrane Science*, 95 , 105-123.

van Veen H.M., M. Bracht, E. Hamoen and P.T. Alderliesten (1996). Feasibility of the application of porous inorganic gas separation membranes in some large scale chemical processes. In: Fundamentels of Inorganic Membrane Science and Technology (A.J. Burggraaf, L. Cot eds) pp 641-680, Elsevier.

Xue E., M. O'Keeffe, J.R.H. Ross, Catalysis Today 30 (1-3) 107-118.

Energy Convers. Mgmt Vol. 38, Suppl., pp. S165–S172, 1997
© 1997 Elsevier Science Ltd. All rights reserved
Printed in Great Britain
0196-8904/97 $17.00 + 0.00

Pergamon

PII: S0196-8904(96)00264-6

NATURAL GAS UTILISATION

WITHOUT CO_2 EMISSIONS

Bjørn Gaudernack, Institute for Energy Technology, Norway.
Steinar Lynum, Kværner Engineering, Norway.

ABSTRACT

To reduce CO_2 emissions into the atmosphere, alternatives for "CO_2-free" utilisation of natural gas should be developed. Hydrogen production by conventional technology with CO_2 sequestration, or by newly developed technology for pyrolytic cracking of natural gas, are promising alternatives. The state of the art of such processes is briefly reviewed, and some scenarios for possible applications are indicated.

© 1997 Elsevier Science Ltd

KEYWORDS

Natural gas, Hydrogen, Steam reforming, CO_2 sequestration, Pyrolytic cracking, Carbon black.

INTRODUCTION

Natural gas (NG) is recognised as the fossil fuel causing least damage to the environment. This is because it is clean, has a low carbon ratio, and can be used efficiently in GTCC plants for (combined heat and) power production. However, even this type of plant will emit substantial amounts of CO_2, contributing to the increased contents of greenhouse gases in the atmosphere. Alternatives for NG utilisation without release of CO_2 to the atmosphere should therefore be developed. One possibility is to convert NG into hydrogen and/or hydrogen-containing products.

"CO_2-free" production of hydrogen from NG can be accomplished by two main routes:

1. Hydrogen from NG by conventional technology (e.g. steam reforming) with CO_2 sequestration.
2. High temperature pyrolysis of NG, yielding pure hydrogen and carbon black.

Technologies for industrial scale realisation of these options have been developed and evaluated in Norway, which is a large producer and exporter of NG.

STEAM REFORMING OF NG WITH CO_2 SEQUESTRATION

Advances in Steam Reforming

Steam reforming (SMR) of natural gas is the dominant method for hydrogen production, and the technology is well proven. Nevertheless, developments and improvements are taking place. Conventional steam reformers are large and awkward pieces of equipment, operating at high temperatures with poor heat transfer and emissions of CO_2 and NO_x. They are being replaced by more compact and efficient reformers of the heat exchanger type. It has also proven advantageous to carry out the reforming in two stages, adding oxygen for autothermal reforming in one of the stages. These features are incorporated in ICI Katalco's Leading Concept Hydrogen (LCH) process, using the specially designed Gas-Heated Reformer (GHR) in the first stage. This results in a much more compact plant (plot area reduced by 25-30 %) and reduced emissions. It is also claimed that investments and operating costs will be considerably reduced (Abbishaw and Cromarty, 1996).

The possibilities of incorporating selective sorbents or membranes into steam reformers or shift reactors are being investigated. Selective removal of one of the reaction products (H_2 or CO_2) will shift the equilibrium and create more favourable thermodynamic conditions for the process, permitting lower operating temperatures. An example is the Sorption Enhanced Reaction Process (SERP) being developed by Air Products and Chemicals, where a special CO_2-selective sorbent has been developed. This is expected to result in: lower operating temperatures, reduced downstream purification, less expensive construction materials (Anand et al., 1996). An alternative is to remove hydrogen selectively through H_2-permeable membranes. Thin supported membranes of a Pd-Ag alloy have been investigated with promising results (Lægsgaard Jørgensen et al., 1995); ceramic membranes are also being developed (Aihara et al., 1996).

Capture and Disposal of CO_2

The state of the art of CO_2 capture and disposal will be adequately described in other presentations at this conference. What should be emphasised here is the greater ease and lower cost at which CO_2 can be separated from the steam reformer/shift reactor product, compared to separation from flue gases from NG-fired power plants. The cost of CO_2 avoidance at a GTCC plant has been estimated at 80 USD/t CO_2 (Holloway, 1996), whereas Audus et al. (1996) calculated the cost of CO_2 capture in a hydrogen plant to be approximately 20 USD/ t CO_2 (resulting in a 25 % increase in the costs of the hydrogen produced).

Underground disposal of CO_2 is the method considered to be best proven and closest to practical large-scale application. In October 1996, the Norwegian company Statoil will start a 1Mt/a CO_2 disposal operation at the Sleipner field off-shore Norway (Audus et al., 1996). The potential storage capacity for CO_2 in underground formations off-shore Norway has been estimated to be very large (Holloway, 1996). This study also shows that the costs of transportation and injection of the CO_2 - depending, obviously, on volumes and distances - will be small relative to the costs of separating CO_2 from flue gases.

Scenario I: Hydrogen for Export

This scenario, as sketched in fig. 1, assumes a large hydrogen plant at one of the NG terminals on the western coast of Norway, e.g. the Kårstø terminal. The boundary conditions would be quite similar to those used in the study made by Foster Wheeler for Statoil and the IEA Greenhouse Gas R&D Programme (Audus et al., 1996). The relevant case is a plant with separation and disposal of CO_2 (~5000 t/d) in an off-shore aquifer, consuming 2.45 MNm^3/d of NG and producing 6.75 MNm^3/d of hydrogen. With a "centre price" of 3 USD/GJ for NG the cost of the hydrogen produced will be 7 USD/GJ.

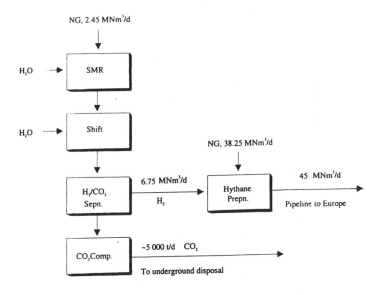

Fig. 1: Scenario I: Hydrogen for Export

From a site at a NG terminal, export of hydrogen to the European continent by pipeline is conceivable. A special pipeline for hydrogen would be expensive, but a feasible alternative is to export hydrogen as a mixture with NG, so-called Hythane. 6.75 MNm^3/d of hydrogen would give 45 MNm^3/d of Hythane containing 15 vol % of H_2. This would require a pipeline capacity of 16 billion Nm^3/year, which is a fairly normal size for a NG pipeline. The 15 % Hythane can be handled in a standard pipeline, and the cost of transportation should not deviate much from the cost for NG transportation (Öney *et al.*, 1994). At 3 USD/GJ for NG and 7 USD/GJ for hydrogen, the cost of the 15 % Hythane would be 3,2 USD/GJ. The advantages of Hythane as a clean fuel may be worth this price increase. There is considerable interest in Europe for Hythane as fuel for vehicles and other applications. If neat hydrogen is required, it could be separated from the Hythane by a PSA or membrane process. In this scenario 1,8 Mt/y of CO_2 would be avoided, at a cost of approximately 56 USD/t.

Scenario II: Integrated Industrial Complex

Hydrogen is an excellent energy carrier, but it is also a main ingredient in many useful products. Examples are methanol (important raw material for chemicals, and potential transportation fuel), ammonia (important fertiliser component), hydrogen peroxide (environmentally benign bleaching agent, replacing chlorine compounds). In the scenario sketched in fig. 2, production of these commodities as well as liquid hydrogen and electric power is envisaged. This integrated industrial complex would require a NG feed of 5 MNm^3/d, twice that of the hydrogen plant in the previous scenario. The sizes of the methanol and ammonia plants are representative of present-day commercial projects. The size of the power plant (350 MW) is equal to that of the NG-fired power plants planned in Norway, and the electric efficiency is assumed to be the same (58 %). The power demands for oxygen production and hydrogen liquefaction have been estimated at 55 MW and 11 MW, respectively. This is based on specific consumption of 0,29 kWh/kg O_2 (Bolland, 1993) and 0,95 kWh/l of LH_2 (Bracha *et al.*, 1992). Total requirements of the complex can therefore be expected to be in the order of 100 MW, leaving some 250 MW available for export.

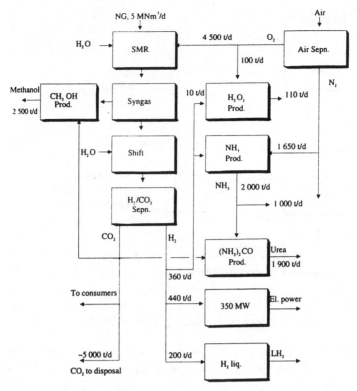

Fig. 2: Scenario II: Integrated Industrial Complex

Both liquid hydrogen and methanol could potentially be used as fuels in the transport sector, LH$_2$ for aviation and heavy-duty fleet vehicles, methanol for light fuel cell powered vehicles. If the total production envisaged from this industrial complex were to be introduced in the Norwegian transport sector, very substantial reductions in national CO$_2$ emissions would be achieved. An estimate shows that a total of 2,65 Mt/y of CO$_2$ can be avoided in this scenario.

PYROLYTIC CRACKING OF NATURAL GAS

Advances in Pyrolytic Cracking

Pyrolytic cracking of hydrocarbons implies heating to a temperature where the hydrocarbon decomposes into its elementary components, i.e. carbon and hydrogen. Industrial emphasis has until recently been on the production of a carbon product, so-called carbon black (CB). This is traditionally produced in the Furnace Black Process, using heavy oil as a feedstock and NG for heating. This process is characterised by low feedstock utilisation, high emissions and a gaseous by-product of low value. In recent years developments have been made to improve these conditions, and emphasise production of pure hydrogen as a valuable by-product. Two main lines have been followed in pyrolysis of NG:

1. High temperature pyrolysis in plasma arcs, and
2. Catalytic pyrolysis at lower temperatures.

The latter approach has been studied a.o. in Russia (Kuvshinov *et al.*, 1996) and the USA (Muradov, 1996). Pyrolysis of NG in plasma arcs has been investigated several places, a.o. in Russia (Parmon *et al.*, 1996) and France (Fulcheri and Schwob, 1995). These studies are of a theoretical character and have not resulted in any industrial process so far. The Norwegian company Kværner, however, has developed such technology for industrial production of carbon black and hydrogen. The process is briefly described below.

The Kværner CB & H Process

In 1990 Kværner started development of the innovative CB & H (Carbon Black and Hydrogen) process. Since 1992 a full-scale pilot plant has been operated, and the development has resulted in technology ready for commercialisation. The process is based upon a plasma torch, integrated in a reactor system in such a way that the basic pyrolytic process parameters can be controlled and adjusted to allow production of various types of carbon black to desired specifications. The plasma gas is hydrogen, which is recirculated in the process. Thus, apart from the feedstock and the electricity demand for the plasma torch, the process is self-sufficient.

The process flow-sheet is schematically shown in fig. 3. Any hydrocarbon can be used as feedstock. With NG feed, the electric energy requirement is about 1 kWh/Nm^3 of hydrogen produced. Purity of the hydrogen is >99,7 %, and feedstock utilisation is close to 100 %. The process is virtually emission-free. It is modular, based on 8 MW plasma torch units. High efficiency and good economy can be achieved also in small and medium scale plants.

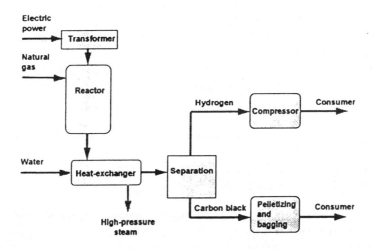

Fig. 3: Flow Diagram for the CB & H Process

The cost of hydrogen produced will depend upon the price obtainable for the carbon black and the NG and electricity prices. A relationship is shown in fig. 4. It appears that in a large plant with upper level NG and electricity prices, a CB price of 140 USD/t (which is low in relation to current market prices) will give a hydrogen cost of 0.075 USD/Nm^3. This is equal to the cost derived for NG decarbonisation by Audus *et al.* (1996). Thus, the CB & H process must be considered as a competitive alternative for NG decarbonisation. An added advantage is the ability to operate economically on a small scale. The Kværner CB & H process was first presented in 1994 (Lynum, 1994) and has later been described by Lynum *et al.* (1996) and Gaudernack and Lynum (1996).

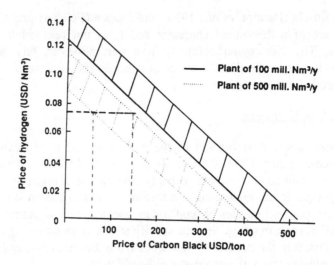

Fig. 4: Hydrogen - CB Price Relationship for the CB & H Process.
(Band width reflects gas and electricity price range.)

Markets for Carbon Black

The traditional main market for carbon black is the rubber industry. Of the world annual production of approximately 6 Mt[1], about 90 % is consumed by this industry.

The metallurgical industry is identified as an interesting, new CB market. Pyrolytically produced CB is expected to exhibit several advantages compared to petrol coke, which is the traditional carbon material used for carbonisation and reduction. Pyrolytic CB is very clean, highly reactive and pulverised. Its introduction in the metallurgical industry will have several positive effects, such as reduced sulphur emissions and a reduced carbon consumption, which in turn will lead to reduced CO_2 emissions. Use of a clean carbon material may also lead to low impurity levels in products, as well as reduced energy demands.

The following applications of CB in the metallurgical industry have been evaluated as interesting:
• carbon additive/carburiser for the steel and foundry industry
• reduction material for production of SiC (Si, FeSi).

The total consumption of carbon additives in Europe is approximately 300.000 t/y. The European production of SiC is approximately 130 000 t/y, with a consumption of reduction material of about 150 000 t/y. CB produced by the Kværner CB & H process has been tested for this application and has shown excellent properties.

Scenario III: Integrated Concept for CO_2 Reduction

Combinations of the Kværner CB & H process with other industrial processes, utilising the hydrogen and CB produced, may result in reduced CO_2 emissions. An example is illustrated in fig. 5, showing a NG fired power plant in combination with the CB & H process and a metallurgical process (for SiC production).

---

[1] Assuming 60 % feedstock utilisation, this means that 4 Mt of C, i.e. 15 Mt of CO_2 are emitted. By switching to the CB & H process, this emission would be avoided.

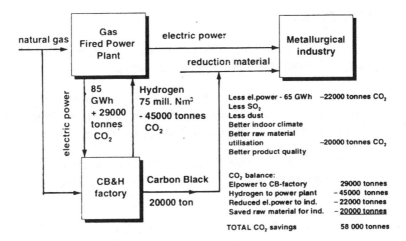

Fig. 5. Scenario III: CO_2 Reduction by Combining CB & H with other Processes

Carbon Black as an Energy Carrier

Hitherto, pyrolytic cracking of natural gas has mainly been considered for meeting the demands of carbon black, regarding hydrogen as a by-product. The idea of using the process for large-scale production of hydrogen has been launched, but is commonly met with the following objections: the CB markets will soon be saturated, then there will be huge amounts of a carbon by-product causing storage problems, and the energetic value of the carbon will be wasted. This is not valid if CB is regarded as an energy carrier. It is certainly easier to store than CO_2, causing no environmental damage. Its energetic value will remain indefinitely. Being very clean and reactive, CB will have excellent properties as a fuel, e.g. for an IGCC power plant. The elimination of ash removal and sulphur clean-up steps would reduce investment and operating costs substantially. Sequestration of CO_2 from an IGCC plant can be carried out relatively easily and inexpensively (Holloway, 1996).

It is also likely that CB can be combusted efficiently in a mixture of oxygen and recycled CO_2, for possible incorporation in a GTCC plant. In this case the flue gas would be virtually pure CO_2, easily disposable. Thus, it seems reasonable to assign a minimum value to CB at least corresponding to the fuel value of coal. If this is assumed to be 2 USD/GJ (Audus *et al.*, 1996) and the LHV of CB is assumed to be 32,5 GJ/t, the fuel value of CB will be 65 USD/t. From figure 4 it can be seen that at this CB price, hydrogen can still be produced at 0,075 USD/Nm^3 by the Kværner CB & H process, provided that the lower level prices of NG and electricity are applicable.

Conclusions

CO_2-free production of hydrogen is an efficient and environment-friendly way of utilising natural gas. It will also be the cheapest way of hydrogen production in the near and middle term. There are two main options: conventional hydrogen production with CO_2 sequestration, and pyrolytic cracking of natural gas to hydrogen and carbon black. Both options are industrially feasible and can produce hydrogen at 7 USD/GJ or less. If the conventional markets for carbon black become saturated, it can be easily stored as an excellent energy carrier. The Kværner CB & H process will make it possible to meet increasing demands for carbon black and hydrogen in the short and medium term, as well as having potential for large scale, CO_2-free utilisation of natural gas in the long term.

REFERENCES

Abbishaw, J. and B. Cromarty (1996). NPRA Paper no. AM-96-62, National Petroleum Refiners Association Annual Meeting, San Antonio, Texas.

Aihara, M., H. Ohashi, S. Semonova, H. Ohya and Y. Negishi (1996). High Temperature Corrosion-Resistant Separation Membranes for Thermochemical Processes. *Proc. of 11th World Hydrogen Energy Conference (WHEC)*, **1**, 861-866.

Anand, M., J.R. Hufton, S.G. Mayorga, S. Nataraj, S. Sircar, T. Gaffney (1996). Sorption Enhanced Reaction Process (SERP) for the Production of Hydrogen. USDOE Hydrogen Program Review, Miami, Florida.

Audus, H., O. Kårstad and M. Kowal (1996). Decarbonisation of Fossil Fuels: Hydrogen as an Energy Carrier. *Proc. 11th WHEC*, **1**, 525-534.

Bolland, O., SINTEF, Norway (1993). Pers. comm.

Bracha, M., G. Lorenz, A. Patzelt, M. Wanner (1992). Large-Scale Hydrogen Liquefaction in Germany. *Proc. 9th WHEC*, **2**, 1001-1010.

Fulcheri, L. and Y. Schwob (1995). From Methane to Hydrogen, Carbon Black and Water. *Int. Journal of Hydrogen Energy*, **20**, 3, 197-202.

Gaudernack, B. and S. Lynum (1996). Hydrogen from Natural Gas without Release of CO_2 to the Atmosphere, *Proc. 11th WHEC*, **1**, 511-523.

Holloway, S. (Ed.) (1996). The Underground Disposal of Carbon Dioxide. Final Report of JOULE II Project No. CT92-0031.

Kuvshinov, G.G., Y.J. Mogilnykh, D.G. Kuvshinov, S.G. Zavarukhin, V.N. Parmon (1996). New Ecologically Sound Technology to Produce Hydrogen and New Carbon Material via Low Temperature Catalytic Pyrolysis. *Proc. 11th WHEC*, **1**, 655-660.

Lynum, S. (1994). CO_2-free Hydrogen from Hydrocarbons. The Kværner CB & H process. 5th Annual US Hydrogen Meeting, National Hydrogen Association, Washington D.C.

Lynum, S., J. Hugdahl, R. Hildrum (1996). The Kværner CB & H process. Carbon Black World, Nice.

Lægsgaard Jørgensen, S., P.E. Højlund Nielsen, P. Lehrmann (1995). Steam Reforming of Methane in a Membrane Reactor. *Catalysis Today*, **25**, 303-307.

Muradov, N.Z. (1996). Hydrogen Production by Catalytic Cracking of Natural Gas. *Proc. 11th WHEC*, **1**, 697-702.

Parmon, V.N., G.G. Kuvshinov, V.A. Sobyanin (1996). Innovative Processes for Hydrogen Production from Natural Gas and other Hydrocarbons. *Proc. 11th WHEC*, **3**, 2439-2448.

Öney, F., T.N. Veziroglu, Z. Dülger (1994). Evaluation of Pipeline Transportation of Hydrogen and Natural Gas Mixtures. *Int. Journal of Hydrogen Energy*, **19**, 10, 813.

Pergamon

Energy Convers. Mgmt Vol. 38, Suppl., pp. S173–S178, 1997
© 1997 Elsevier Science Ltd. All rights reserved
Printed in Great Britain
0196-8904/97 $17.00 + 0.00

PII: S0196-8904(96)00265-8

COMPARISON OF CO_2 REMOVAL SYSTEMS FOR FOSSIL-FUELLED POWER PLANT PROCESSES

G. Göttlicher and R. Pruschek
Universität GH Essen
Universitätsstr. 2-17
D-45141 Essen, Germany

ABSTRACT

Around 300 articles on CO_2 removal from fossil-fuelled power generation systems have been reviewed. The technical approach and the published performance data differ widely. A survey has been prepared comprising a data collection of around 60 variants of power plants with CO_2 removal. For better comparison, the efficiencies were computed for a standardized CO_2 pressure. Thermodynamic cycles and CO_2 separation techniques, such as CO_2 wet absorption, dry adsorption, membrane separation, CO shift, air separation, CO_2 gas turbines, CO_2-blown gasification are described, and a literature database regarding general aspects of CO_2 emissions was compiled. If coal is to be used, the IGCC combined with CO shift conversion and physical absorption was found to be the most appropriate option which could be built on the basis of today's technology. Further progress in membrane separation and in high-temperature fuel cells could improve efficiency and economy in the long term. © 1997 Elsevier Science Ltd

KEYWORDS

CO_2 removal, power plant, gas turbine combined cycle (GTCC), integrated gasification combined cycle (IGCC), absorption, adsorption, membrane, fuel cell, CO_2 turbine, efficiency, costs of CO_2 avoidance.

INTRODUCTION

There is a large number of publications on carbon dioxide removal from fossil-fuelled power generating systems. Some are based on fundamental considerations, others even report on results of basic engineering design. The performance data given in the articles differ not only due to specific design features and removal options but also in the degree of design analyses and due to different assumptions on boundary conditions. Although at this stage of development it is difficult to finally comment on the advantages and disadvantages of the individual proposals, a survey including analysis and evaluation of conceivable power generating systems with CO_2 removal was prepared (Pruschek and Göttlicher, in course of printing). For the evaluation of the published efficiencies, costs and technical maturity a common reference basis is required. Furthermore, it seems to be useful to classify the removal options and essential process components to gain a better insight into their performance. This is outlined in the following sections. Criteria observed among others are the specific energy expenditure for CO_2 removal, the rate of CO_2 retained related to the carbon input and the purity of the removed CO_2 fraction at standardized boundary conditions. Also the technical maturity of the proposed removal processes, such as absorption, adsorption, membrane separation, cryogenic separation, air separation, and of components like CO shift converter or CO_2 turbine has been taken into account.

CLASSIFICATION OF PROCESSES - FUNDAMENTAL CONSIDERATIONS ON GAS SEPARATION TECHNOLOGIES - COMPARISON OF ENERGY REQUIREMENTS

The CO_2 removal systems can be grouped into five process families:

Process family I comprises processes with CO shift or steam reforming before CO_2 removal. The hydrogen-rich fuel gas can be combusted with air after H_2 /CO_2 separation. Also the option of H_2 /CO separation without CO shift is possible. Process family II covers processes where fuel is combusted in an atmosphere of oxygen mixed with recycled CO_2 or steam. Process family III includes all kinds of fossil-fuel-fired power generation systems in which CO_2 is removed from the flue gas after combustion at the exhaust end of the plant. Process family IV includes the so-called Hydrocarb processes whereby carbon is separated from the fuel prior to combustion. Process family V deals with CO_2 separation in fuel cells suitable for the use of fossil-fuel-derived gases.

Table 1 shows the reversible work w_r to separate carbon dioxide, hydrogen or oxygen from flue gas, fuel gas or air. The reversible separation work is lowest when carbon dioxide is separated from a shifted coal gas; it is two to three times as high, when carbon dioxide is separated from flue gas. The energy demand in technical separation processes is definitely much higher.

Absorption is the most commonly used CO_2 separation technology attaining high purities and removal rates. The energy demand of physical absorption processes is predominantly caused by compression and pumping of solvent. It can be as low as 0.03 kWh per kg of carbon dioxide removed from a gas under elevated pressure (Condorelli et al., 1991). Chemical absorption processes need heat for regeneration, which strongly depends on solvent concentration in the aqueous solution. For comparison, the heat demand for regeneration is expressed as loss of turbine work of the bled off steam. At high pressure (pressurized gasification, reforming) chemical absorption requires around 0.11 kWh/kg CO_2 (Shell, 1990), at lower pressure (flue gas) around 0.34 kWh/kg CO_2 (Smelser et al., 1991). CO_2 removal from a pressurized synthesis gas additionally reduces the power output of the gas turbine combined cycle by approximately 0.03 kWh per kg of CO_2.

Adsorption is also proposed for this kind of CO_2 removal using pressure swing (PSA) or a combined pressure temperature swing (PTSA) for regeneration. In the chemical industry, adsorption processes are mostly employed in small-scale applications. High removal rates and purities imply sophisticated equipment. Cited energy requirements are in the range of 0.16 to 0.18 kWh/kg CO_2 at CO_2 concentrations of 28 to 34 mole % in the feed gas (Wakamura et al., 1992) and 0.55 to 0.7 kWh/kg CO_2 at a CO_2 concentration of 10 to 11.5 mole % (Ishibashi et al., 1995).

Tab. 1: Reversible separation work w_r and energy demand of existing gas separation processes. The energy demand is expressed as electrical energy.

Feed gas	Shifted coal gas	Air	Flue gas: Coal-fired steam power plant	Flue gas: IGCC	Flue gas: GTCC
mole % CO_2	40.5	0.032	11.1	8.1	3.2
wt. % CO_2	87.1	0.049	17.1	11.7	4.9
Reversible Separation Work	$w_r = \varepsilon_i R_i T_0 \ln p_{tot} / p_i$				
ε_i mass fraction of the separated species i; R_i individual gas constant; p_i partial pressure, p_{tot} total pressure					
CO_2 Separation [kWh/kg CO_2]	0.026	0.142	0.048	0.055	0.069
H_2 Separation [kWh/kg H_2]	0.4354 ($\triangleq$ 0.0266 kWh/kg CO_2)				
O_2 Separation [kWh/kg O_2]		0.0336	-	-	-
Energy Demand of Gas Separation Processes					
Phys. Absorption: ≈ 0.09 kWh/kg CO_2 (assuming equal pressure of feed and product) Additional losses by CO shift: exerg. eff. 95 to 97% ($\triangleq$ 0.1-0.15 kWh/kg CO_2)		Air Separation (Two-Column Distillation): ≈ 0.27 kWh/kg O_2 (1 bar)	Chem. Absorption: ≈ 0.34 kWh/kg CO_2 Cryogenic Distillation/Absorption 0.6 to 1.0 kWh/kg CO_2 (not demonstrated on technical scale)		

Membranes for CO_2 separation from flue gases should have a minimum CO_2/N_2 selectivity of 200 (van der Sluijs, 1992). At present the best polyimid membranes attain a CO_2/N_2 selectivity of 30. Hydrogen can be removed from synthesis gases by polymer membranes, which presently attain CO_2/H_2 selectivities up to 10. Ceramic membranes achieve up to 15 on laboratory scale, expensive palladium metal membranes even 100 at temperatures between 300 and 400°C. Energy consumption is determined by the pressure ratio across the membrane. In the case of a shifted coal-derived fuel gas the energy demand is in the range of 0.04 to 0.07 kWh/kg CO_2, however, only low removal rates and insufficient purities of CO_2 are obtained. Membrane reactors combined with CO shift (Bracht et al., 1996) could reduce the energy losses due to lower additional steam demand.

Cryogenic CO_2 separation can be achieved by sublimation, condensation or, as generally applied, by cryogenic supported absorption. The latter comes close to the physical absorption by Rectisol with a comparable energy demand. In case of pressurized coal-derived synthesis gases it is calculated to be around 0.04 kWh/kg CO_2 or 0.1 kWh/kg CO_2, respectively, including energy for compression of CO_2 to the feed gas pressure, and around 0.6 to 1 kWh/kg CO_2 for separation from flue gas (Golomb et al., 1989).

EFFICIENCY REDUCTION DUE TO CO₂ REMOVAL

CO_2 separation processes applied to a fossil-fuel-fired power plant give rise to additional energy consumption and to the direct reduction of power output due to reduced flow rates.

CO₂ Compression. Assuming a pipeline pressure of 110 bar, the energy requirement for liquefaction by intercooled 5-stage compression starting from 1 bar amounts to around 0.12 kWh/kg CO_2. Thus, in case of 90% CO_2 removal the liquefaction of CO_2 reduces the efficiency of coal-fuelled power plants by 3.2 to 5.1 percentage points. The efficiency of natural-gas-fired power plants is reduced by 2.2 to 2.8 percentage points. These values depend on the carbon content and heating value of the fuel and the CO_2 removal rate. The following figures on efficiency reduction are related to CO_2 removal at 1 bar (gaseous state).

Process Family I: The efficiency reduction due to CO_2 removal is predominantly caused by:
- exergy destruction due to steam reforming or gasification and CO shift (exergy efficiency of CO shift:
   $\approx$ 95 to 97%, corresponding to an overall efficiency reduction of 2.5 to 5 percentage points)
- energy demand of the gas separation process, i.e. regeneration, compression, pumping (efficiency reduction
   $\approx$1 percentage point)
- volume displacement of the separated carbon dioxide, which is not converted into useful work but dissipated
   (efficiency reduction $\approx$1 percentage point).
Exergy losses by CO shift, steam reforming or gasification depend, among others, on the amount of steam required. Membranes incorporating CO shift reaction (WIHYS process) have the lowest steam demand. The efficiency reductions for Process Family I are in the range of 4 to 7.4 percentage points for an IGCC with CO shift, when 80 to 90% of the CO_2 are removed, and around 14.5 percentage points for a natural-gas-fired combined cycle (GTCC) with steam reforming, CO shift and absorption by MDEA at a removal rate below 60%.

Process Family II: If a CO_2-rich flue gas is produced by combustion in a mixture of oxygen and recycled flue gas or steam, the efficiency reduction not only depends on the carbon content of the fuel, but, in contrast to other processes, also on the content of hydrogen and oxygen and especially on the efficiency of the air separation unit (ASU). In natural-gas-fired power plants much of the energy expenditure for oxygen production is consumed for oxidation of hydrogen. This kind of CO_2 removal is less favourable than flue gas scrubbing. The energy demand of air separation related to the energy of the fuel directly reduces the overall plant efficiency. Further deterioration of the efficiency has to be expected if a CO_2 gas turbine is to be used, which in principle applies higher pressure ratios. Considering long-term options, it should be mentioned that decomposition of CO_2 in front of MHD generators would also impair the performance. On the other hand, CO_2 -blown gasification could have a positive overall effect on efficiency. The efficiency reductions due to O_2/CO_2 firing cited in the literature and derived from in-house studies are in the range of 4.8 to 8.5 percentage points concerning IGCC power plants and around 6 percentage points concerning natural-gas-fired power plants.

<u>Process Family III</u>:   In this process scheme the bulk of proposals apply flue gas scrubbing with amine-based chemical sorbents. Such removal processes consume up to two thirds of steam as bled off steam (2-5 bar) for solvent regeneration. Efficiency reduction figures cited in the literature are in the range of 8 to 11 percentage points, when 80 to 90% of the CO_2 are removed from coal-fired power plants, or 5.5 to 11 percentage points when removed from GTCCs.

<u>Process Family IV</u>:   Processes like "Hydrocarb" (Steinberg *et al.*, 1991) aim at the separation of carbon from the fuel. This option, however, is a reasonable approach only for a fuel mix consisting of biomass and natural gas or oil, where the CO_2 released is generated from the biomass and the sequestered fossil carbon is disposed. According to Steinberg, the conversion efficiency (energy of produced methanol divided by energy input of natural gas and biomass input, LHV) amounts to 60.9%. To compare this process with the other CO_2 removal options as discussed before, one has to multiply this conversion efficiency with a cycle or plant efficiency (reciprocating engines, steam and gas cycles).

<u>Process Family V</u>:   Phosphoric acid fuel cells (PAFC), molten carbonate fuel cells (MCFC) or solid oxide fuel cells (SOFC) can be operated with fossil-fuel-derived gases. The efficiency of power plants with PAFC is below the efficiencies of other combined cycles. MCFC or SOFC combined cycles are expected to achieve higher efficiencies. In all three cases the extra energy demand in the range of 0.05 to 0.11 kWh/ kg CO_2 is required for the necessary separation of residual fuel and CO_2 from the anode exhaust. This energy demand reduces the overall power plant efficiency by around 1.5 to 4.3 percentage points if coal is used as a fuel and 1.0 to 2.4 percentage points in natural-gas-fired plants. In case of the MCFC, a part of the CO_2 has to be recycled to the cathode. Therefore, the attainable CO_2 removal rate is probably limited to around 80% with specific CO_2 emissions of 0.15 to 0.17 kg of CO_2 per kWh.

EFFICIENCIES AND COSTS

In the near future, IGCCs could possibly achieve efficiencies around 55% with gas turbine inlet temperatures of 1250°C (ISO), GTCCs are supposed to achieve around 60%. Higher efficiencies are conceivable in advanced energy conversion systems, such as MHD combined cycles and fuel cell combined cycles, which are long-term options.

Efficiencies of power plants incorporating CO_2 removal based on today's available technology (efficiencies without CO_2 removal: IGCC: 50%, GTCC: 58%) and respective CO_2 emissions are depicted in Fig. 1 and listed in more detail in Tab. 2. These data are derived from in-house studies and the evaluation of published data. For a complete list of references the reader is referred to the complete report of this study (Pruschek and Göttlicher, in course of printing).

Processes with O_2/CO_2 firing in IGCCs attain CO_2 removal rates of almost 100%. The CO_2 gas turbine needed is, however, not available on the market. Similar efficiencies are obtained by IGCCs with CO shift and physical absorption removing around 90% of the CO_2 (Pruschek *et al.*, 1996), using exclusively commercially available components. The physical CO_2 wash could be potentially replaced by an H_2 separation membrane system which might lead to an efficiency enhancement especially when combined with hot gas cleaning (Bracht *et al.*, 1996). Power generation from coal gas by means of high-

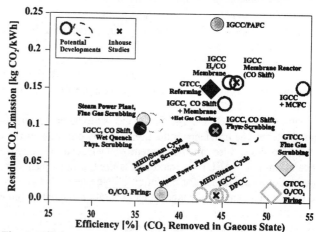

Fig. 1: Efficiency and specific CO_2 emission of power plants with CO_2 removal.

Tab. 2: Estimate of efficiencies, specific CO_2 emissions and costs based on present status of technology.

Processes With CO₂ Removal (liquid CO₂, 110 bar)					Reference Processes (Similar Process Type)				Cost of CO₂ Avoided
η = net efficiency, LHV (*) cost assumed by the authors	DM/ kW	kg CO₂ /kWh	η [%]	DM/ kWh	DM/ kW	kg CO₂/ kWh	η [%]	DM/ kWh	[DM/t CO₂ liq.]
PROCESS FAMILY I: CO₂ CAPTURE/REMOVAL FROM FUEL GAS PRIOR TO COMBUSTION									
IGCC-CO-Shift, Wet Quench	3300	0.108	31.4	0.134	2350	0.83	38.9	0.095	48
IGCC-CO-Shift, Phys. Abs.	3595	0.102	40.2	0.134	2700	0.64	50.0	0.102	58
IGCC-CO-Shift, Sea Water (*)	3330	0.095	41.8	0.125	2700	0.64	50.0	0.102	43
IGCC-CO-Shift, Chem. Abs.	3720	0.055	39.2	0.138	2700	0.64	50.0	0.102	61
IGCC-CO-Shift, Membrane, HG Clean-up	3400	0.110	41.1	0.127	2700	0.64	50.0	0.102	48
IGCC-CO-Shift , Membrane Reactor	3360	0.175	42.6	0.125	2700	0.64	50.0	0.102	49
IGCC-CO-Shift, Cryo. Distillation (*)	3600	0.121	39.7	0.134	2700	0.64	50.0	0.102	61
IGCC-CO-Shift, Adsorption (*)	4800	0.136	29.8	0.179	2700	0.64	50.0	0.102	151
Air-Blown IGCC,CO-Shift,Chem.Abs.(*)	3440	0.179	34.4	0.135	2300	0.69	46.7	0.093	82
IGCC, H₂ /CO Membrane (*)	3500	0.060	41.8	0.130	2700	0.64	50.0	0.102	47
Natural-Gas-Fired Processes									
Steam Power Plant, Reformer (*)	2660	0.053	33.3	0.182	2000	0.46	45.0	0.136	115
GTCC, Reformer	1600	0.156	42.0	0.130	940	0.36	57.5	0.089	207
PROCESS FAMILY II: O₂/CO₂ FIRING									
Steam Power Plants	3660	0.007	33.6	0142	2400	0.72	45.0	0.096	64
Direct Coal-Fired Combined Cycle (*)	3090	0.007	40.6	0.119	2000	0.61	52.5	0.081	63
MHD Combined Cycle	3110	0.007	38.2	0.122	2000	0.64	50.7	0.082	63
IGCC + PFBC (Partial Gasification)	3690	0.007	34.8	0.141	2300	0.69	46.7	0.093	71
IGCC - O₂/H₂O-Blown Gasification	3900	0.007	39.9	0.142	2700	0.64	50.0	0.102	63
IG-STIG O₂/H₂O-Blown Gasification (*)	4210	0.007	28.2	0.165	2700	0.84	38.4	0.110	66
IGCC - CO₂/H₂O-Blown Gasification	3890	0.007	40.1	0.142	2700	0.64	50.0	0.102	63
Natural-Gas-Fired Processes									
Steam Power Plants (*)	2580	0.007	37.1	0.169	2000	0.46	45.0	0.136	74
GTCC	1590	0.007	48.6	0.118	940	0.36	57.5	0.089	84
Evaporative Cycle (*)	1540	0.007	45.8	0.121	900	0.37	56.0	0.089	89
Internal Combustion Rankine Cycle	3620	0.007	34.1	0.207	940	0.36	57.5	0.089	337
PROCESS FAMILY III: CO₂ CAPTURE/REMOVAL FROM FLUE GAS AFTER COMBUSTION									
Steam Power Plant, Freezing	4000	0.190	33.8	0.151	2400	0.783	45.0	0.096	92
Steam Power Plant, Chem. Abs.	3415	0.113	32.0	0.137	2400	0.783	45.0	0.096	60
Steam Power Plant, Membrane (*)	3930	0.514	28.0	0.157	2400	0.783	45.0	0.096	225
MHD Combined Cycle, Chem. Abs.	2850	0.082	41.0	0.112	2000	0.783	50.7	0.082	43
IGCC + PFBC, Chem. Abs.	3550	0.110	33.8	0.112	2300	0.69	46.7	0.093	79
Natural-Gas-Fired Processes									
Steam Power Plants (*)	2750	0.042	36.3	0.176	2000	0.46	45.0	0.136	97
GTCC, Chem. Abs.	1370	0.052	50.3	0.110	940	0.36	57.5	0.089	69
HAT, Chem. Abs. (*)	1370	0.088	50.8	0.014	900	0.36	57.5	0.087	74
PROCESS FAMILY V: CO₂ CAPTURE/REMOVAL USING FUEL CELLS									
IG-PAFC, Membrane (Present Cost)	7510	0.279	40.6	0.244	5710	0.783	50.5	0.187	113
IG-PAFC, Membrane (Predicted Cost)	3700	0.279	40.6	0.136	2650	0.783	50.5	0.100	71

Running hours: 7000 h/a; interest: 8%/a; depreciation time: 20 a; planning+construction: 5 a; inflation: 3.5%/a; owner's contribution: 5%; insurance + tax + maintenance: 4.7%/a; coal price: DM 3/GJ; nat. gas price: DM 8.33/GJ temperature fuel cell combined cycles is a promising long-term option. As far as natural gas cycles are concerned, the most favourable choice is flue gas scrubbing.

The data on specific investments as shown in Tab. 2 are derived from literature and from data obtained from the power industry. Electricity generating costs and costs of CO_2 avoidance were calculated for removal of CO_2 in the liquid state. Cost of CO_2 avoidance is based on comparison of CO_2 emissions and electricity generating costs. Considering natural-gas-fired GTCCs, flue gas scrubbing turns out to be the most economic option. As far as coal is concerned, cost estimations of IGCC with CO shift and physical absorption as well as of conventional steam power plants with flue gas scrubbing are based on market prices. The IGCC turns out to be the more economical solution. Cost calculations based on technologies which are not state-of-the-art are uncertain. In this context one should realize that there has been an unexpected decrease in the costs of commercially available power plants in the past few years, which has not been considered in the before-mentioned calculations.

ACKNOWLEDGEMENT

The research work presented in this paper has been carried out within the framework of the Community Research Programme with a financial contribution from the European Commission.

REFERENCES

Bracht, M., P.T. Alderliesten, R. Kloster, R. Pruschek, G. Haupt, E. Xue, J. Ross, M. Koukou and N. Papayanakos (1996): Water gas shift membrane reactor CO_2 Control in IGCC systems: techno-economic feasibility study. Paper presented at *ICCDR-3*, Boston, 1996.

Condorelli, P., S.C. Smelser, G.J. Mc Cleary, G.S. Booras and R.J. Stuart (1991): *Engineering and Economic Evaluation of CO_2 Removal from Fossil-Fuel-Fired Power Plants*. Volume 2: Coal Gasification Combined-Cycle Power Plants. EPRI IE-7365, Vol. 2. 1991.

Golomb, D., H. Herzog, J. Tester, D. White and S. Zemba (1989): *Feasibility, Modelling and Economics of Sequestering Power Plant CO_2 Emissions In the Deep Ocean*. Massachusetts Institute of Technology, Energy Laboratory, Dec. 1989, MIT-EL 89-003.

Ishibashi, M., H. Ohta, N. Akutsu, S. Umeda, M. Tajlka, J. Izumi, A. Yasutake, T. Kabata and Y. Kageyama (1995): Technology for Removing Carbon Dioxide from power Plant Flue Gas by the Physical Adsorption Method. *Energy Convers. Mgmt*. Vol. 37, No. 6-8, pp. 929-933.

Pruschek, R. and G. Göttlicher, (in course of printing): *Concepts of CO_2 Removal from Fossil-Fuel-Based Power Generation Systems*.

Pruschek, R., G. Oeljeklaus, G. Haupt, G. Zimmermann, D. Jansen and J.S. Ribberink (1996): Role of IGCC in CO_2 Abatement. Paper presented at *ICCDR-3*, Boston, 1996.

Shell International Petroleum Maatschappij and Koninklijke/Shell Explorati en Produktie Laboratorium (1990): *Carbon dioxide disposal from coal based combined cycle power stations in depleted gas fields in the Netherlands*. Publikatiereeks Lucht nr. 91, Ministry of Housing, Physical Planning and Environment, Air Directorate, Leidschenham, NL, 1990.

van der Sluijs, J. P., C.A. Hendriks and K. Blok (1992): Feasibility of Polymer Membranes for Carbon Dioxide Recovery from Flue Gases. *Energy Convers. Mgmt*. Vol. 33, No. 5-8, Pergamon Press, Oxford, 1992, pp. 429-436.

Smelser, S.C., R.M. Stock, G.J. Mc Cleary, G.S. Booras and R.J. Stuart (1991): *Engineering and Economic Evaluation of CO_2 Removal from Fossil-Fuel-Fired Power Plants*. Volume 1: Pulverized Coal-Fired Power Plants. EPRI IE-7365, Vol. 1, 1991.

Steinberg, M. and E.W. Grohse (1991): *A feasibility study for the coprocessing of fossil fuels with biomass by the Hydrocarb process*. Research report US EPA-600/7-91-007 (NTIS DE91-011971), 1991.

Wakamura, O., K. Shibamura, K. Uenoyama (1992): Development of PSA Plant for Manufacture of Carbon Dioxide from Combustion Waste Gas. *Nippon Steel Technical Report No. 55*, October 1992, pp. 51-55.

Energy Convers. Mgmt Vol. 38, Suppl., pp. S179–S186, 1997
© 1997 Elsevier Science Ltd. All rights reserved
Printed in Great Britain
0196-8904/97 $17.00 + 0.00

 Pergamon

PII: S0196-8904(96)00266-X

THERMODYNAMIC AND ENVIRONMENTAL ASSESSMENT OF INTEGRATED GASIFICATION AND METHANOL SYNTHESIS *(IGMS)* ENERGY SYSTEMS WITH CO_2 REMOVAL

GIORGIO CAU
Department of Mechanical Engineering,
University of Cagliari, Italy

ROBERTO CARAPELLUCCI, DANIELE COCCO
Department of Energetics,
University of L'Aquila, Italy

ABSTRACT

Environmental and technical concerns arise from the extensive utilization of the huge world coal reserves. They can nowadays be overcome only resorting to advanced energy conversion technologies. Integrated coal gasification and methanol synthesis power plants can perform a load-following service, by producing and storing liquid methanol, and reduce the carbon dioxide emissions, by removing CO_2 and disposing it outside the atmosphere. In this paper, a comparative performance analysis among the different *IGMS* power plants has been carried out, by evaluating the influence of the coal gasification and methanol synthesis technologies on the main performance characteristics, such as the methanol production, the specific CO_2 emissions and the overall efficiency. © 1997 Elsevier Science Ltd

KEYWORDS

Coal Gasification; Methanol Synthesis; CO_2 Removal; Energy Storage Systems; Load Modulation

INTRODUCTION

The extensive utilization of the huge world coal reserves can nowadays be conceived only resorting to advanced energy conversion technologies with high efficiency and low environmental interactions. In the recent years, in particular, an increasing attention has been devoted towards the study of the carbon dioxide effects on the global climate change and of the options of reducing the CO_2 emissions from coal based power plants.

A reduction of the CO_2 emissions can be obtained by increasing the overall efficiency of the power plants, but a more significant reduction can only be achieved by removing and storing the CO_2 outside the atmosphere. Among the different options, the carbon dioxide removal based on coal gasification appears to be very attractive, due to its lower energy losses and costs (ICCDR-1, 1992; ICCDR-2, 1994; Blok, 1993).

The conventional integrated coal gasification combined cycle *(IGCC)* power plants are most suitable for a base-load service, mainly due to their operating characteristics and high capital costs. In order to utilize an *IGCC* power plant for a load-following service, some of the energy derived by the coal gas must be stored during periods of low electrical demand and, subsequently, utilized for the peak-load generation. The integrated coal gasification and methanol synthesis *(IGMS)* power plants can perform a load-following service by converting a portion of the syngas to methanol; both syngas and methanol are produced at a steady rate, and the latter is burned during periods of high electrical demand. The capabilities of these integrated power plants to operate the gasifier at a high capacity factor can allow for a gasifier size and cost reduction (Brown *et al.*, 1988; Paffenbarger and Eustis, 1990; Frenduto *et al.*, 1993; Cau *et al.*, 1995; Carapellucci *et al.*, 1995).

Besides the load-following service, the *IGMS* power plants allow to remove a portion of the carbon dioxide from the coal gas and dispose it outside the atmosphere, due to the peculiar characteristics of the coal gasification and methanol synthesis processes.

In the paper, a comparative performance analysis among the different *IGMS* power plants, with and without CO_2 removal, has been carried out. The influence of the coal gasification and methanol synthesis technologies and of the shift conversion and carbon dioxide removal process characteristics on the methanol production, carbon dioxide removal and efficiency of the integrated power plant has also been evaluated.

IGMS POWER PLANTS

If a coal gasification power plant is integrated with a methanol synthesis process, the resulting *IGMS* power plant can perform a load-following service: a portion of the produced coal gas feeds the power plant for the base-load generation, while the remainder is converted to liquid methanol, which is stored and burned during periods of high electrical demand. An *IGMS* power plant (figure 1) includes three main sections: a coal gasification section (*CGS*), a methanol synthesis section (*MSS*) and a power generation section (*PGS*).

The *CGS* produces the raw fuel gas from coal, oxygen (produced in the air separation unit) and steam (diverted from the *PGS*), by means of entrained, fixed or fluidized bed gasifiers. It also produces steam, by cooling the raw gas and the gasifier. The raw gas is then cleaned, heated and usually saturated; a portion of it feeds the methanol synthesis section.

The *MSS* produces methanol, purge gas, that is the unreacted clean gas, and saturated steam, by cooling the methanol reactor or the reactor exit gas. Purge gas and saturated steam are utilized, with the clean gas directly fed to the *PGS*, for the base-load generation, while the stored methanol is utilized for the peak-load generation, in the same *PGS* or in external power plants. The methanol is produced by means of *GPH* or *LPH* processes. The conventional *GPH* process requires a feed gas with optimum $H_2/CO/CO_2$ molar ratios, in order to attain a high feed gas conversion (Supp, 1990; Lee, 1990). Hence, the clean gas composition must be suitably modified by means of a conditioning system, which includes

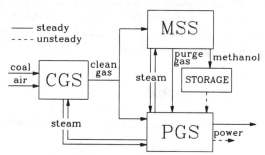

Fig. 1. Functional scheme of the IGMS power plant.

the shift conversion and the carbon dioxide removal processes. In the shift reactor, the *CO* reacts with H_2O forming H_2 and CO_2; the H_2O is introduced as liquid water, in a saturator, and as steam, diverted from the *PGS*, in the shift reactor. A portion of the carbon dioxide is then removed from the shifted gas by means of a carbon dioxide removal system; in particular, in the following analysis, the latter is based on a Rectisol process. The advanced *LPH* process does not require any clean gas treatment nor saturation, since the methanol production is favoured by a low H_2O content in the feed gas. Figures 2a,b show the simplified schemes of the *MSS*, respectively based on *GPH* and *LPH* processes.

Owing to the availability and the requirement of remarkable amounts of steam with different characteristics, as well as to the production of gaseous and liquid fuels suitable for gas turbines, the combined cycle (*CC*) power plants are the most suitable solutions for the *PGS*. In particular, the *CC* power plant considered in the present study is based on a heavy-duty gas turbine and on a two pressure levels heat recovery steam generator (*HRSG*).

Carbon Dioxide Emissions in thermal power plants

Table 1 reports the CO_2 emissions of some conventional thermal power plants, fed by different primary fuels (methane, methanol, coal). The thermal power plants considered in table 1 are a *CC* power plant, with a two pressure levels *HRSG*, a *PCF* power plant, with flue gas treatment, and three *IGCC* power plants, respectively based on dry-feed entrained bed (*ETD*), fluidized bed (*FLD*) and slagging fixed bed (*FXS*) gasifiers. All the *IGCC* power plants are based on a combined cycle with the same main characteristics of the *CC* power plant. The coal is the "Pittsburgh N. 8", with a *LHV* equal to 27.5 *MJ/kg*

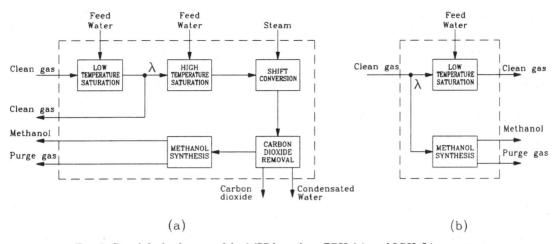

Fig. 2. Simplified schemes of the MSS based on GPH (a) and LPH (b) processes.

and a *C/H* mass ratio equal to 14.67. The efficiency of the *PCF* power plant has been assumed equal to 38%, whereas the efficiencies of the *CC* and *IGCC* power plants have been evaluated, with reference to the operating parameters reported in table 2, by means of a modular simulation model (Carapellucci and Cau, 1992; Cau *et al.*, 1993; Cau and Cocco, 1993).

The specific CO_2 emissions are reported with reference to the mass of primary fuel ($M^f_{CO_2}$), its chemical energy ($M^c_{CO_2}$), the electrical output ($M^e_{CO_2}$). As the most significant $M^e_{CO_2}$ emissions are concerned, the two *CC* power plants, fed by methane and methanol, reveal values respectively equal to about 45% and 55% of the corresponding values of the reference *PCF* power plant. Moreover, owing to their higher efficiency, the *IGCC* power plants allow for a slight reduction of the $M^e_{CO_2}$ emissions with respect to the *PCF* power plant fed by the same coal.

Table 1. Carbon dioxide emissions of some conventional thermal power plants.

		CC		PCF	IGCC		
					ETD	FLD	FXS
Fuel		Methane	Methanol	Coal	Coal	Coal	Coal
H_i	[MJ/kg]	50.0	19.96	27.5	27.5	27.5	27.5
$M^f_{CO_2}$	[kg/kgf]	2.75	1.375	2.585	2.585	2.585	2.585
$M^c_{CO_2}$	[g/MJ]	55.0	68.9	94.0	94.0	94.0	94.0
η	[%]	50.0	50.7	38.0	41.1	43.4	41.8
$M^e_{CO_2}$	[g/kWh]	396.0	489.2	889.2	823.4	779.8	809.6

Table 2. Operating parameters of the IGMS sections.

Gasification process				Combined Cycle				Synthesis process		
	ETD	FLD	FXS	Steam Turbine		Gas Turbine			GPH	LPH
p_G [bar]	25	25	25	p_{HP} [bar]	80	β	12.1	p_M [bar]	60	60
T_G [°C]	1400	1000	400	T_{HP} [°C]	500	T_{max} [°C]	1100	$T_{M,IN}$ [°C]	220	220
α [kg$_{H_2O}$/kg$_C$]	0.86	0.60	0.65	p_{LP} [bar]	7			$T_{M,OUT}$ [°C]	260	230
μ [kg$_{O_2}$/kg$_C$]	0.10	0.30	0.30	T_{LP} [°C]	200			T_{shift} [°C]	500	-

A more remarkable reduction of the carbon dioxide emissions in *IGCC* power plants can be obtained by removing and storing the CO_2 outside the atmosphere. The CO_2 removal options proposed in the literature are essentially based on the coal gas shift conversion process and carbon dioxide removal by means of absorption processes (Schutz *et al.*, 1992; Jansen *et al.*, 1992), or on the coal gas combustion with oxygen in semi-closed gas turbine cycles using CO_2 as working fluid (Hendriks and Blok, 1992; Mathieu and De Ruyck, 1993). The first option appears to be very interesting if performed in *IGCC* power plants integrated with methanol synthesis processes, allowing to reduce the CO_2 emissions and to perform the load following service, at the same time.

The capabilities of the *IGMS* power plants to perform a load-following service were evaluated in a previous analysis (Cau *et al.*, 1995; Carapellucci *et al.*, 1995), considering different coal gasification and methanol synthesis processes. In particular, the analysis pointed out the possibility of reducing the CO_2 emissions in the case of the *IGMS* power plants based on the *GPH* process, due to the CO_2 removal performed by the coal gas conditioning system included in the *MSS*. The CO_2 removal assumes remarkable values in the case of a high load modulation, being it directly related to the methanol production. Table 3 reports the main performances of the *IGMS* power plants in the case of the maximum methanol production, that is for the maximum *MSS* size; it results from the maximum achievable value of the syngas repartition ratio λ between the syngas fed to the *MSS* and that produced by the *CGS*. This λ_{max} corresponds to the minimum amount of gaseous fuels (clean gas and purge gas) that must feed the *PGS* in order to assure the minimum waste heat to the $HRSG$[1]. The *IGMS* power plant performances have been evaluated with reference to the "Pittsburgh N. 8" coal and to the operating parameters summarized in table 2, by assuming a *CO* conversion ratio $(CO)_C$ in the shift reactor and a CO_2 absorption ratio $(CO_2)_A$ in the removal tower both equal to 90%.

As shown, the *IGMS* power plants based on the *GPH* process are characterized by λ_{max} values lower than 1. Moreover, the methanol (M_M) produced by the *GPH* process results about 2.0-2.5 times higher than that produced by the *LPH* process, with a corresponding lower purge gas production (M_{PG}). If the methanol is utilized in an external *CC* power plant $(\eta_{CC} = 50.7\%$, table 1), the mean overall efficiency of the *IGMS* power plant (η) is about 30-32% (*GPH*) and 37-39% (*LPH*). The low efficiency in the case of the *GPH* process is mainly due to the energy losses in the clean gas conditioning system, which requires a great amount of steam (M_S) to be diverted from the *PGS*. If the *IGMS* power plants are not specifically conceived to the carbon dioxide removal, the CO_2 removed by the absorption tower M_{CO_2} is wasted to the atmosphere, giving $M^e{}_{CO_2}$ emissions of about 30-38% (*GPH*) and 10-12% (*LPH*) higher than those of the corresponding *IGCC* power plants.

The specific $M^e{}_{CO_2}$ of the *IGMS* power plants emissions can be reduced only in the case of the *GPH* process, due to the presence of the clean gas conditioning system, if the removed carbon dioxide M_{CO_2} is compressed and stored outside the atmosphere. For example, if the CO_2 is compressed up to 60 *bar* by means of an intercooled two stage compressor, the specific $(M^e{}_{CO_2})_R$ emissions result about 10-20% lower than those of the corresponding *IGCC* power plants, although the overall efficiency ($)_R$ decreases.

Table 3. Performances of the IGMS power plants in the case of λ_{max}

		GPH			LPH		
		ETD	*FLD*	*FXS*	*ETD*	*FLD*	*FXS*
$M_{coal\,gas}$	$[kg/kg_{coal}]$	1.726	1.629	1.666	1.726	1.629	1.666
λ_{max}		0.850	0.920	0.755	1.0	1.0	1.0
M_M	$[kg/kg_{coal}]$	0.767	0.738	0.623	0.305	0.360	0.329
M_{PG}	$[kg/kg_{coal}]$	0.041	0.147	0.122	1.422	1.269	1.338
M_S	$[kg/kg_{coal}]$	1.193	1.228	1.017	-	-	-
M_{CO2}	$[kg/kg_{coal}]$	1.145	1.025	0.909	-	-	-
η	[%]	29.84	31.90	31.90	37.17	39.31	37.39
$M^e{}_{CO2}$	$[g/kWh]$	1134.0	1060.8	1060.8	910.4	860.8	905.0
$(\eta)_R$	[%]	28.43	30.63	30.78	-	-	-
$(M^e{}_{CO2})_R$	$[g/kWh]$	663.2	666.9	713.0	-	-	-

IGMS POWER PLANTS WITH CO_2 REMOVAL

As already mentioned, the *IGMS* power plants based on the *GPH* process can allow to reduce the carbon dioxide emissions. However, the CO_2 removal, being directly related to λ, assumes significant values only for a high methanol production, with noticeable penalties of the overall efficiency. A remarkable CO_2 removal, for any methanol production and with higher efficiencies, can be achieved if the *IGMS* power

---

[1] It is needed for preheating and superheating the saturated steam, produced by the gasifier and the methanol reactor, and for producing the superheated steam, required by the gasification and shift conversion processes.

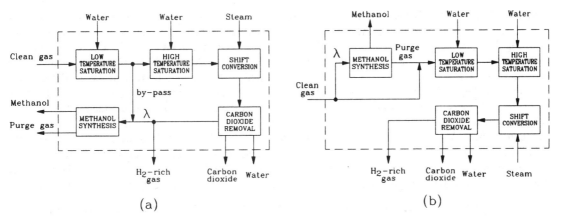

Fig. 3. *Simplified schemes of the MSS based on GPH (a) and LPH (b) processes,*
in IGMS power plants with CO₂ removal.

plants, based on both the *GPH* and the *LPH* processes, are suitably arranged in order to feed the *PGS* with an H_2-rich gas.

Figures 3a,b report the simplified schemes of the *MSS* in *IGMS* power plants with CO_2 removal. In the case of the *GPH* process, all the clean gas produced by the *CGS* feeds the shift conversion and carbon dioxide removal processes. The H_2-rich gas produced is then distributed between the *PGS* and the methanol reactor, according to the value of λ. The portion of the H_2-rich gas fed to the methanol reactor is mixed with a suitable amount of the clean gas, in order to optimize the $H_2/CO/CO_2$ molar ratios. In the case of the *LPH* process, a *CO* shift conversion and a CO_2 removal processes are introduced just upstream from the *PGS*. The gas conditioning system, which is not needed if the *LPH* process is utilized in *IGMS* power plants without CO_2 removal, is fed by the purge gas and, if λ is lower than 1, by a portion of the clean gas.

Among the operating parameters of the *IGMS* power plants with CO_2 removal, the *CO* conversion ratio in the shift reactor $(CO)_C$ appears to be the most significant. In fact, it remarkably influences both the overall efficiency, through the steam required by the shift reactor, and the CO_2 emissions, through the H_2-rich gas composition.

Figures 4a,b show, for given values of the methanol production (M_M), the influence of $(CO)_C$ on the shift reactor steam (M_S) and on the ratio R_{CO_2} between the removed CO_2 and the overall CO_2 produced by the coal. In particular, they refer to *IGMS* power plants based on *FLD* gasifiers and on both the *GPH* and *LPH* methanol synthesis processes. The CO_2 absorption ratio in the removal tower $(CO_2)_A$ has been always

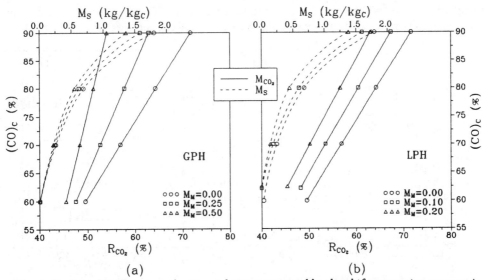

Fig. 4. *Carbon dioxide removal ratio and steam required by the shift conversion process in*
function of the CO conversion for IGMS power plants with CO₂ removal.

assumed equal to 90%. If the shift conversion process is performed without feeding any steam (M_S=0), $(CO)_C$ assumes its minimum value (about 60-62%, depending on the water content of the saturated gas), allowing for R_{CO_2} values in the range 45-50%. Higher R_{CO_2} values are achievable if $(CO)_C$ is increased by introducing steam in the shift reactor. Considering a limit value of $(CO)_C$ equal to 90%, R_{CO_2} results in the range 55-75% (GPH) and 65-75% (LPH). For $(CO)_C$ equal to 90%, about 30-40% of the available steam produced by the PGS must be fed to the shift reactor. On the other hand, even if all the available steam feeds the shift reactor, the maximum values of $(CO)_C$ are in the range 92-94%, being the shift conversion reaction governed by equilibrium conditions. For a given value of $(CO)_C$, increasing the methanol production, both M_S and R_{CO_2} decrease. In the case of the ETD and FXS gasification processes, the influence of $(CO)_C$ and M_M on M_S and R_{CO_2} reveals a behaviour similar to that showed by figures 4a,b. As a result of the previous analysis, a $(CO)_C$ value equal to 80% has been always assumed for the following study, in order to obtain a significant carbon dioxide removal without remarkable plant efficiency losses.

Figures 5a,b report, in the case of a FLD gasification process, the influence of λ on the mass flows across the boundary of the control volume showed in figures 3a,b. The entering mass flows, referred to the mass of coal, are the coal gas produced by the CGS (M_{CG}), the steam fed to the shift reactor (M_S) and the water utilized in the saturators (M_{FW}). The exiting mass flows are the H_2-rich gas fed to the PGS (M_{H_2}), the methanol (M_M), the purge gas (M_{PG}), the removed carbon dioxide (M_{CO_2}) and the condensed water (M_{CW}). In the case of the LPH process (figure 5b), the mass flow M_{PG} is absent, since the purge gas is mixed to the clean gas and fed to the conditioning system. Increasing λ, the mass flows M_M and M_{PG} increase, whereas the mass flows M_{H_2}, M_{CO_2}, M_S, M_{FW} and M_{CW} decrease. As the methanol production is concerned, these IGMS power plants reveal comparable values with respect to the corresponding IGMS without CO_2 removal, allowing for a maximum methanol production of about 0.8 kg/kg_c (GPH) and 0.3 kg/kg_c (LPH). However, the IGMS power plants with CO_2 removal reveal remarkable values of M_{CO_2} for any methanol production, both in the case of GPH and LPH processes. In particular, M_{CO_2} results equal to about 1.7 kg/kg_c for λ=0 and 1.2 kg/kg_c for λ=λ_{max}; this behaviour is mainly due to the carbon fixed in the methanol. In the case of the ETD and FXS gasification processes, the influence of λ on the mentioned mass flows reveals a behaviour similar to that showed by figures 5a,b, with differences in the range 5-10% for the maximum methanol production and 10-15% for the removed carbon dioxide.

As already mentioned, the IGMS power plants with CO_2 removal can perform, at the same time, the load-following service and the carbon dioxide removal. Consequently, the most significant performances are the methanol production M_M, the specific CO_2 emissions $M^e_{CO_2}$ and the overall efficiency η. These performances are reported in figures 6a,b for the different coal gasification and methanol synthesis processes here considered. The performances of the IGMS power plants have been evaluated by assuming that the methanol is utilized in an external CC power plant with an efficiency equal to 50.7% (table 1) and that the removed carbon dioxide is compressed up to 60 *bar* through an intercooled two stage compressor, for the final disposal outside the atmosphere.

In the limit case of M_M =0, the IGMS power plants reduce to IGCC power plants with coal gas shift conversion and CO_2 removal, being absent the methanol synthesis process. In this case, the overall

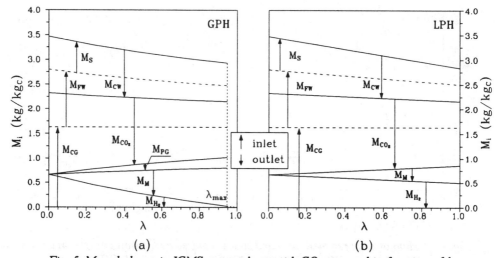

Fig. 5. Mass balance in IGMS power plants with CO_2 removal in function of λ.

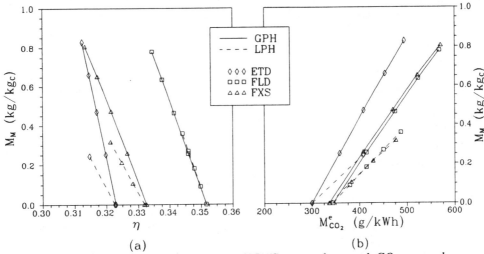

Fig. 6. Performance and emissions of IGMS power plants with CO_2 removal.

efficiency is about 20% lower than that of the corresponding *IGCC* power plants without any carbon dioxide removal, but with a reduction of the $M^e_{CO_2}$ emissions equal about to 60% (table 1). For low M_M values, the *IGMS* power plants with CO_2 removal are characterized by η values in the range 32-35%, which are comparable with those reported in the literature for *IGCC* power plants with CO_2 removal based on shift conversion (Schutz *et al.*, 1992; Jansen *et al.*, 1992) and for advanced *IGCC* power plants based on CO_2 gas turbines (Hendriks and Blok, 1992; Mathieu and De Ruyck, 1993). As the CO_2 emissions are concerned, the *IGMS* power plants with CO_2 removal reveal $M^e_{CO_2}$ values higher than those achievable using CO_2 gas turbines. However, producing liquid methanol, they allow for a load modulation, as well as for the use of the methanol as a strategic fuel.

Increasing M_M, the overall efficiency decreases, due to the energy losses of the methanol synthesis process, resulting an increase of the specific CO_2 emissions. For a given methanol production, the solutions based on the *GPH* process reveal higher η values and lower $M^e_{CO_2}$ values than those based on the *LPH* process; moreover, these solutions allow for higher values of the maximum methanol production. Consequently, if a high load modulation is required, the *IGMS* power plants based on the *GPH* process allow for a methanol production up to about 0.8 kg/kg_C, with slight penalties of the overall efficiency, even if the specific $M^e_{CO_2}$ emissions remarkably increase. As the coal gasification technologies are concerned, the solutions based on *ETD* gasifiers reveal the lowest $M^e_{CO_2}$ emissions, while those based on *FLD* gasifiers are characterized by the highest overall efficiencies.

CONCLUSIONS

The capabilities of the *IGMS* power plants to perform, at the same time, the load-following service and the CO_2 removal, have been evaluated considering different coal gasification and methanol synthesis processes. The *IGMS* power plants allow to operate the load-following service by producing and storing liquid methanol, but they are characterized by higher CO_2 emissions with respect to the conventional coal based power plants. However, due to their peculiar characteristics, the *IGMS* power plants can be suitably arranged in order to reduce the CO_2 emissions.

The analysis carried out in the paper reveals that the removed CO_2 and the overall efficiency of the *IGMS* power plants with CO_2 removal remarkably depend on the CO conversion ratio, due to the large amount of steam required by the shift conversion process. Moreover, the solutions based on the *GPH* synthesis process reveal performances slightly higher than those of the corresponding solutions based on the *LPH* process; they also allow for higher values of the maximum methanol production.

As pointed out by the analysis, the *IGMS* power plants with CO_2 removal reveal interesting capabilities of producing a strategic fuel, such as the liquid methanol, and of reducing the carbon dioxide emissions. In particular, these integrated power plants are characterized by overall efficiencies comparable with those of the more advanced power plants with CO_2 gas turbines, even if their CO_2 emissions are higher. However, the *IGMS* power plants allow for a chemical energy storage and are based on well-established CO_2 removal and methanol synthesis technologies. Moreover, their performances, both in terms of overall efficiency and CO_2 emissions, can be further improved, for a given methanol production, by optimizing the integration among the coal gasification, CO_2 removal, methanol synthesis and power generation sections.

ACKNOWLEDGEMENTS

This work was supported by the "Ministero dell'Università e della Ricerca Scientifica e Tecnologica".

REFERENCES

Blok, K. (1993). Final report on the integrated research programme on carbon dioxide recovery and storage. Utrecht University, January.

Brown, W.R., R.B. Moore and J. Klosek (1988). Coproduction of electricity and methanol. *Eighth Annual EPRI Conference on Coal Gasification*, Palo Alto, California, October 19-20.

Carapellucci, R. and G. Cau (1992). Un sistema di simulazione modulare per la valutazione delle prestazioni dei sistemi energetici. *VI Convegno Gruppi combinati: prospettive tecniche ed economiche*, Genova, Italy, November 12-13.

Carapellucci, R., G. Cau and D. Cocco (1995). Integrated coal gasification and methanol synthesis power plants. Part II. Parametric analysis and performance evaluation. *Twelfth Annual International Pittsburgh Coal Conference*, Pittsburgh, Pennsylvania, September 11-15.

Cau, G.,D. Cocco and T. Pilloni (1993). Equilibrium model for predicting performances of coal and low grade fuel gasification systems. *2nd International Conference "Energy and Environment towards the year 2000"*, Capri, Italy, June 3-5.

Cau, G., and D. Cocco (1993). Performance prediction and evaluation of coal gasification plants: conceptual development of simulation models. Dep. of Mech. Eng., University of Cagliari, *Internal Report*, Italy, December.

Cau, G., R. Carapellucci and D. Cocco (1995). Integrated coal gasification and methanol synthesis power plants. Part I. Assessment of the coal-to-methanol conversion process. *Twelfth Annual International Pittsburgh Coal Conference*, Pittsburgh, Pennsylvania, September 11-15.

Frenduto, F.S., J. Klosek and E. Osterstock (1993). Liquid phase methanol energy storage with gasification combined-cycle power generation. *ASME/IEEE Power Generation Conference*, Kansas City, USA, October 17-22.

Hendriks, C.A. and K. Blok (1992). Carbon dioxide recovery using a dual gas turbine IGCC plant. *ICCDR-1, Energy Conversion Management*, Vol. 33, n. 5-8, pp. 387-396.

ICCDR-1 (1992). Proc. of the *First Int. Conf. on Carbon Dioxide Removal*, Amsterdam, March 4-6.

ICCDR-2 (1994). Proc. of the *Second Int. Conf. on Carbon Dioxide Removal*, Kyoto, October.

Jansen, D., A.B.J. Oudhuis and H.M. van Veen (1992). CO_2 reduction potential of future coal gasification based power generation technologies. *ICCDR-1, Energy Conversion Management*, Vol. 33, n. 5-8, pp. 365-372.

Lee, S. (1990). *Methanol synthesis technology.* CRC Press, Boca Raton.

Mathieu, P. and J De Ruyck (1993). CO_2 capture in CC and IGCC power plants using a CO_2 gas turbine. IGTI-Vol 8, ASME Cogen Turbo Power Conf., Bournemouth, UK, September 21-23.

Paffenbarger, J.A. and R.H. Eustis (1990). A coal gasification combined cycle power plant with methanol storage. *Report EPRI GS/ER-6665.*

Schutz, M., M. Daun, P.M. Weinspach, M. Krumbeck and K.R.G. Hein (1992). Study on the CO_2 recovery from an ICGCC-plant. *ICCDR-1, Energy Conversion Management*, Vol. 33, n. 5-8, pp. 357-363.

Supp, E. (1990). *How to produce methanol from coal.* Springer-Verlag, Berlin.

Energy Convers. Mgmt Vol. 38, Suppl., pp. S187 S192, 1997
© 1997 Elsevier Science Ltd. All rights reserved
Printed in Great Britain
0196-8904/97 $17.00 + 0.00

Pergamon

PII: S0196-8904(96)00267-1

CO₂ RECOVERY IN A POWER PLANT
WITH CHEMICAL LOOPING COMBUSTION

M. Ishida and H. Jin

Research Laboratory of Resources Utilization, Tokyo Institute of Technology
4259 Nagatsuta, Midoriku, Yokohama, Japan 226

ABSTRACT

This paper points out the features of CO_2 recovery in a novel power plant by employing a chemical-looping combustor. This chemical-looping combustor consists of the two reactors; fuel reacts with metal oxide, the reduced metal reacts with oxygen in air, and both reactions proceed with no visible flame. For this reason, the proposed LNG-fueled power generation system with the chemical-looping combustor has the following major advantages: (a) Since the thermal efficiency of this plant may reach a new-generation level, i.e., higher than 60% (based on LHV), the generation of CO_2 per kWh electricity can significantly be decreased to 0.33 kg-CO_2/kWh; (b) This combustor does not need pure oxygen which requires high power consumption for O_2 generation, say for O_2/CO_2 combustor; (c) Due to the presence of only CO_2 gas and water vapor in the exhaust gas from the fuel reactor, CO_2 is highly concentrated and may easily be recovered by condensation of the water vapor without any extra cost which is generally large in conventional separation processes based on absorption (MEA, Selexol), adsorption (PSA, TSA), membranes, or cryogenics etc. © 1997 Elsevier Science Ltd

KEYWORDS: CO_2 capture, flue gas of power plant, chemical-looping combustion, looping material, reaction kinetics, carbon deposition.

INTRODUCTION

Concerns about suppressing greenhouse impact have led many nations to adopt strategies to limit CO_2 emissions, initially only through improved energy efficiency and changes in fuels (Lundberg 1990; Riemer 1996). For another option of the CO_2 recovery and sequestration, so far, there have been many technologies such as absorption (MEA, Selexol), adsorption (PSA, TSA), membranes, or cryogenics etc. Most of these technologies, however, require a large amount of energy. For example, if CO_2 will be recovered from the exhaust gas in power plants, which are the largest and the most concentrated CO_2 generation source of many industries, it gives rise to the relative decrease of the thermal efficiency from 9% - 27% and increase in the power generation cost between 1.3 to 2.3 times (Akai et al., 1995; Kimura et al., 1995).

Hence, the situation we confronted is how to deal with an important but complicated subject: CO_2 recovery together with lower energy penalty, instead of consumption of large amount of energy. To tackle this difficulty, Ishida and Jin (1994a) have investigated and proposed a novel power generation plant that allows CO_2 to easily be recovered and gives a new-advanced-level thermal efficiency for power generation by combining a chemical-looping combustor with saturation for air in a gas turbine cycle (CLSA).

The aim of this paper is to develop the chemical-looping combustor in the CLSA system. We have experimentally examined the kinetics and life of several looping materials such as Fe₂O₃/YSZ,

NiO/Al₂O₃, and NiO/YSZ (yttria-stabilized ziroconia) by means of a thermogravimetric analyzer (TGA) and scanning electron microscopy (SEM).   At the same time, the effect of carbon deposition in reduction was also investigated by the energy-dispersive spectrometry (EDS).

CHEMICAL-LOOPING COMBUSTION

The chemical-looping combustor is one of the key components in the proposed power generation system (CLSA).   It consists of two continuous reactions as shown in the following formulae (1) and (2).   The reduction product (M) of a metallic oxide (MO) obtained according to the formula (1) is utilized in the oxidation reaction of the formula (2).   Therefore, it is a chemical-looping reaction with MO as an oxygen carrier.   The reaction of the formula (1) is an endothermic reaction of MO and a fuel (RH) with low-level energy absorption in a low-temperature region (about 600 - 1,000 K) and the reaction of the second step is an exothermic oxidation of the reaction product (M) of the first step in a high-temperature region (about 800 - 1,700 K).   A high-temperature exhaust gas is produced by the heat of the reaction and is utilized for driving a gas turbine.

$$RH + MO \longrightarrow m\,CO_2 + n\,H_2O + M \quad (1)$$
$$M + 0.5\,O_2 \longrightarrow MO \quad\quad\quad\quad\quad (2)$$

where M represents a reduction product of MO and is exemplified by metals such as iron (divalent, trivalent), nickel, copper, and manganese, etc.

EXPERIMENTS

In our previous study (Ishida and Jin 1994b), we pointed out the fact that the nickel oxide particles mixed with yttria-stabilized zirconia (YSZ, stabilized by addition of 8% Y₂O₃), NiO/YSZ (weight ratio of 3:2), have good reaction kinetics and durability.   Here, we examine the kinetic behavior of reaction by comparing the NiO/YSZ particle with the other kinds of looping materials, i.e., Fe₂O₃/YSZ and NiO/Al₂O₃.   All of these particles were prepared by the same procedure as before.

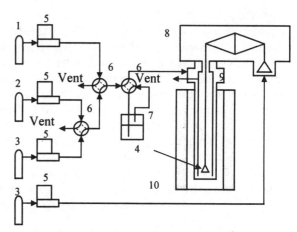

1 Air cylinder, 2 CH₄ cylinder, 3 N₂ cylinder, 4 Platinum pan
5 Mass flow controller 6 Four-port switch valve, 7 Saturator
8 Micro-balance, 9 Reaction tube, 10 Electric furnace

Fig. 1 Schematic diagram of experimental apparatus

The particle form we used here, instead of powder form, is mainly to easily recycle the looping material between the two reactors and to avoid the presence of dust from the product gas of the reactors to meet the condition of gas expansion process in gas turbines.

Reactivity measurements for both reduction and oxidation experiments were conducted by means of a thermogravimetric analysis (TGA) system, shown schematically in **Fig. 1**.   The main part of this system is a CAHN 2000 electrobalance (8) housed in a two-port jar with one port connected to the reactor tube.   It can handle weights up to 3.5 g and is sensitive to weight changes as small as 0.1 µg. The hangdown tube consists of two concentric quartz tubes, the inner of 14 mm ID, the outer of 22 mm ID, and length of 460 mm.   This length is capable of sufficiently preheating the reactant gases to the specified temperature.   The reactant gases which have passed through a saturator (7) enter the annular region (9), and then go into the inner tube, in which they react with the solid reactant.   The solid sample is placed on a 5.5 mm platinum pan (4) suspended from the sample arm of the balance with thin quartz wire.   The effluent gases are then purged through a side port of the reactor tube. Inert gas (N₂) through another port was used to prevent corrosive gases from reaching the weighing unit.

When the temperature reached the desired value by an electric furnace (11), two four-port switching valves (7) were used to switch gaseous streams from inert nitrogen to reactant gases. The weight of the solid particle and the reaction temperature were recorded continuously. In all experiments hereafter the gas flow rate of fuel and air was fixed at 5.0 ml/s. Pure hydrogen was used as a reducing gas in this stage.

RESULTS AND DISCUSSION

Kinetic features of three kinds of particles

Figures 2 and **3** show the comparison of the reduction and oxidation rates for the above three kinds of materials.   The fractional oxidation X is plotted against the reaction time, where X is defined as $X = 1 - [(W_i-W)/(W_{oxd}-W_{red})]$.   Hence, $X = 0$ and $X = 1$, respectively, correspond to M and MO. The reduction was conducted at 873 K and the oxidation at 1273 K. The spherical particle was nearly 2.0 mm in diameter. It is found in Fig. 2 that only NiO/YSZ particle (curve A) has a high initial reduction rate. But both NiO/Al₂O₃ particle (curve B) and Fe₂O₃/YSZ particle (curve C) show relatively low initial rates and their change of X with time t was different from that of NiO/YSZ particle (A). Another behavior to be pointed out is that NiO/YSZ particle (A) has not only high reaction rate but also high conversion in completing reduction, compared to NiO/Al₂O₃ particle (curve B), in which reduction of 60% took place.   For oxidation at 1273 K we observed from Fig. 3 that NiO/Al₂O₃ particle has a very high initial oxidation rate, higher than that of NiO/YSZ particle, and Fe₂O₃/YSZ particle has rather low oxidation rate and low reaction conversion, giving rise to incomplete oxidation.   From the engineering viewpoint, NiO/YSZ particle is quite good but NiO/Al₂O₃ particle has a good advantage in material cost and must be investigated in more detail.

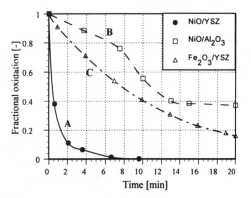

Fig. 2 Comparison of reduction rates

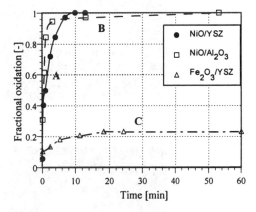

Fig. 3 Comparison of oxidation rates

The fact that the NiO/YSZ particle (curve A in Fig. 3) has extraordinary high reduction and oxidation rates points out that the YSZ plays a quite significant role in the NiO/YSZ particles.   The addition of the YSZ to NiO not only promotes reaction kinetics due to its high solid diffusivity for oxygen at high temperature, but also increases physical strength of particle.   Hence, the particle has an advantage of no shrinkage during reaction and it is favorable for its cyclic use.   However, the effectiveness of YSZ is not always true for other kinds of materials.   For example, as we have represented in Figs. 2 and 3, the addition of YSZ to Fe₂O₃ does not provide the above advantages.   When the addition of YSZ to Fe₂O₃, the reaction rates both in reduction and in oxidation were low, especially in oxidation, reaction took place only on the outside surface of particles.   This behavior is completely different from that of NiO/YSZ.

We measured the changes of structure of particles and grains inside the particles by a scanning electron microscopy (SEM), as shown in **Photo 1**.   C0 (fresh sample) and C1 (after a cycle), respectively, show the cross section of Fe_2O_3/YSZ (3:2) particle in Photo 1.   It can be noted from comparison of C0 and C1 Photos that Fe_2O_3/YSZ particle has quite different micro-structure between the states before and after reaction.   In more detail, the size of pore in the particle became smaller as the reaction proceeded, and at the same time, a thin layer was formed and fine cracks were generated on the outer surface of Fe_2O_3/YSZ (3:2) particles.   The thin dense layer provided a big resistance to gas diffusion as so to result in pore close and incomplete reaction.   This observation of a thin layer formed was in accordance with the result in Fig. 3 that oxidation took place only on the outer surface.

On the basis of the overall reactivity by TGA and the internal phenomena of particles by SEM, we may say that the Fe_2O_3/YSZ particle is incapable of being used in a practical reactor, since the cracked layer on the outer surface may generate dust.   On the other hand, NiO/YSZ particle has high reaction rates both in reduction of NiO and in oxidation of Ni together with high conversion, and a good physical strength for durability.

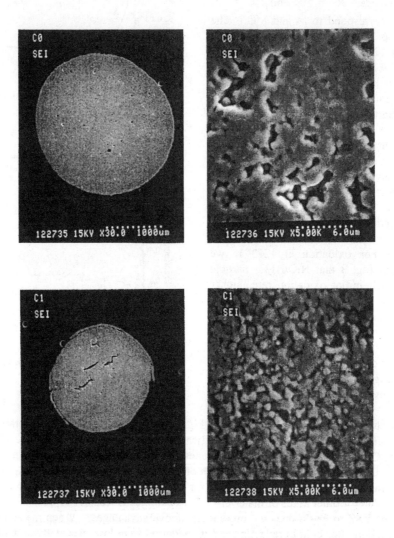

Photo 1   Cross section of Fe_2O_3/YSZ (3:2) particle by SEM
C0 before reaction; C1 after a cycle of reaction

<u>Avoidance of carbon deposition</u>
Another problem we have to solve is the carbon deposition in reduction stage, which will strongly degrade the physical strength of particles.   Carbon deposition is also observed in fuel reforming for fuel cell or reduction of metal oxide in hot gas cleanup in IGCC etc.   Carbon deposition can be usually avoided by addition of water vapor in fuel with high ratio of H_2O/CH_4 in fuel (2-3 times).   In addition, the carbon deposition can be prevented also by the addition of some kinds of material in particle.

Photo 2 illustrates the distribution of Ni, Zr, O, and C maps over the cross section of the NiO/YSZ (3:2) particle measured by EDS (energy-dispersive spectrometry).   When CH_4 was used as fuel, carbon deposition was observed for the NiO/YSZ particle.   It is found in photo 2 that the map of carbon element matches with Zr in YSZ, instead of the Ni layer which is previously confirmed by Sakai and Chida (1988).   It means that the increase in the amount of YSZ addition would not allow the prevention of carbon deposition in this material.   We, however, have confirmed that when we use the NiO/YSZ particle the carbon deposition can be controlled at very low ratios of H_2O/CH_4 (mole ratio), about 10% at the temperature of 1023 K and 30% at 973 K.

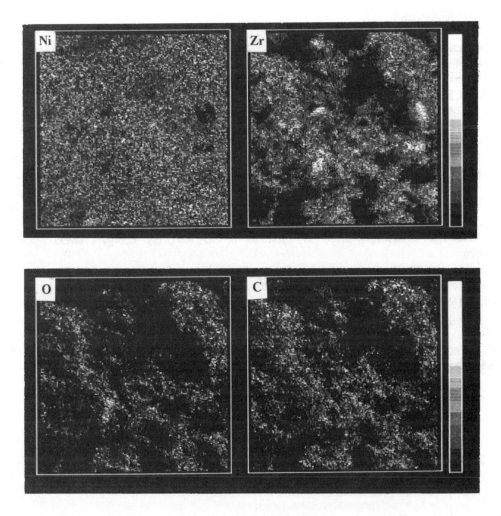

**Photo 2   Distribution map of Ni, Zr, O, and C elements
over the cross section on the NiO/YSZ particle (×2000)**

CONCLUSIONS

1. Significant increase in thermal efficiency results in a CO_2 emission rate of 0.33 kg/kWh in the proposed CLSA system.   This value is much lower than that in a conventional power plant.

2. The CO_2 emitted at very lower rate can be effectively recovered without any separation process, i.e., no extra energy requirement for CO_2 separation.

3. The addition of YSZ (yttria-stabilized ziroconia) to NiO particle show a good property, but it shows no good matching with Fe_2O_3.   From engineering viewpoint, the NiO/Al_2O_3 particle is perhaps suitable to reduce material cost.

4. The carbon deposition in reduction can be avoided by adding a small amount of water vapor to fuel.

5. This process will become one of good candidates to effectively recover CO_2 from a power generation plant.

ACKNOWLEDGMENT

This research has been supported by New Energy and Industrial Technology Development Organization (NEDO) in Japan, as a project of "Proposal-Based Advanced Industrial Technology R & D Program".

REFERENCES

Akai, M., Kagajo, T. and M. Inoue (1995). Performance Evaluation of Fossil Power Plant with CO_2 Recovery and Sequestering System. Energy Convers. Mgmt **36**, 801-804.

Ishida, M. and H. Jin (1994a). A New Advanced Power-generation System Using Chemical-looping Combustion. Energy - The International Journal **19**, 415-422.

Ishida, M. and H. Jin (1994b). A Novel Combustor Based on Chemical-looping Reactions and its Reaction Kinetics. Chemical Engineering of Japan **27**, 296-301.

Kimura, N., Omata, K., Kiga, T., Takano, S., and S. Shikisima (1995). The Characteristics of Pulverized Coal Combustion in O_2/CO_2 Mixtures for CO_2 Recovery. Energy Convers. Mgmt. **36**, 805-808.

Lundberg, G. (1990). The Future of Fossil Fuel in Power Plants. Proceedings of the Florence World Energy Research Symposium Firenze, Italy, 3-20.

Riemer, P. (1996). Greenhouse Gas Mitigation Technologies, an Overview of the CO_2 Capture, Storage and Future Activities of the IEA Greenhouse Gas R&D Programme.   Energy Convers Mgmt. **37**, 665-670.

Sakai, N. and T. Chida (1988). Reduction of Nickel Oxide with Carbon Monoxide Accompanied by Carbon Deposition. J. of Chemical Engineering Society in Japan (in Japanese), **14**, 368-373.

SECTION 3

GEOLOGICAL STORAGE

Pergamon

Energy Convers. Mgmt Vol. 38, Suppl., pp. S193–S198, 1997
© 1997 Elsevier Science Ltd. All rights reserved
Printed in Great Britain
0196-8904/97 $17.00 + 0.00

PII: S0196-8904(96)00268-3

AN OVERVIEW OF THE UNDERGROUND DISPOSAL OF CARBON DIOXIDE

SAM HOLLOWAY

British Geological Survey, Keyworth, Nottingham NG12 5GG, UK

ABSTRACT

The underground disposal of industrial quantities of CO_2 is entirely feasible. Cost is the main barrier to implementation. The preferred concept is disposal into porous and permeable reservoirs capped by a low permeability seal, ideally, but not necessarily, at depths of around 800 metres or more, where the CO_2 will be in a dense phase. New concepts and refined reservoir models are continually emerging. As more regional estimates are carried out it appears that there will be ample underground storage capacity in the worlds sedimentary basins. Storage will be stable over geological timescales. The (remote) possibility of an escape of CO_2 from a storage reservoir onshore merits further investigation and modelling. It would be highly desirable to learn as much as possible from the operators of the new CO_2 disposal schemes arising from natural gas processing in offshore gas fields, as few such opportunities may arise. © 1997 Elsevier Science Ltd

KEYWORDS

Carbon dioxide, underground disposal, storage concepts, safety, stability, overview

BACKGROUND

The fundamental reason for considering CO_2 disposal underground is that it could lessen the environmental damage caused to the planet by disposing of all anthropogenic CO_2 into the atmosphere - without causing profound changes to our current way of life.

At present it is not possible to say what role the underground disposal of CO_2 will play in the range of measures which could be used to ameliorate global warming. But, given the long timescale necessary to implement CO_2 disposal (e.g. conversion or rebuilding of power plant), now is the time to evaluate options.

For geographical reasons, both underground and ocean disposal would likely have a role to play if CO_2 disposal were implemented. Underground disposal would be particularly suitable for areas of the world far from deep oceans, where the continental shelf or land area are underlain by sedimentary basins. Examples include much of northern Europe (Haugen & Eide, 1996) and Western Canada (Law & Bachu, 1996). Ocean disposal might be more suitable for countries adjacent to ocean trenches and with less suitable underground reservoirs; Japan might perhaps be an example. A combination of the two methods - disposal at relatively shallow depths below the deep ocean floor has also been proposed (Koide *et al.*, 1996).

FEASIBILITY

Underground disposal is a perfectly feasible way of disposing of the very large quantities of CO_2 produced by fossil fuel fired power plants.

It is very important to realise that underground disposal of CO$_2$ is already a reality, not just a theoretical concept. CO$_2$ is obtained from natural underground CO$_2$ fields in the USA, transported long distances under pressure in pipelines and injected underground into onshore oilfields to be used in enhanced oil recovery (EOR) projects (Taber, 1993). This has been an everyday operation for years. Furthermore, although these EOR operations are on a smaller scale than envisaged for disposal from power stations, they are not considered to involve any undue risks to man or the natural environment.

The first of two large offshore gas field developments in which the underground disposal of CO$_2$ is a major feature; the Sleipner Vest project, operated by Statoil (Korbul & Kaddour, 1995) started in late August 1996. A second, even larger, scheme is planned as part of the development of the Natuna gas field, in the South China Sea, by Pertamina and Esso Exploration and Production Natuna, Inc. (Anon., 1996). In these schemes CO$_2$ will be stripped from the hydrocarbon gases on the platform before disposal. In both cases, the purpose of underground disposal is to prevent the CO$_2$ entering the atmosphere, where it acts as a greenhouse gas. Thus these projects are even closer analogies to CO$_2$ disposal from power plants than the EOR industry, and they will further validate the concepts of CO$_2$ disposal from fossil fuel fired power plants.

OBSTACLES TO IMPLEMENTATION

The main obstacle to the underground disposal of CO$_2$ from fossil fuel fired power plant is its high cost. Most of the cost is incurred by the necessity to separate the CO$_2$ from, or concentrate it in, the flue gas, e.g. Doherty & Harrison (1996). This cost varies widely, depending on the type of power plant and separation technology considered, but is in the range \$27-\$65 per tonne CO$_2$ avoided (Doherty & Harrison, 1996). A realistic cost of \$40 per tonne of CO$_2$ avoided would increase the cost of electricity generation by at least 2 cents/kWh, 40% above current levels (Riemer, 1996).

Total CO$_2$ underground disposal costs (including capture, transport and injection underground) have been estimated at around \$52/tonne CO$_2$ in Alberta (Gunter et al., 1996). Cost factors associated with disposal in depleted oil and gas reservoirs in Texas are discussed by Bergman et al. (1996).

When viewed in isolation, the costs of CO$_2$ disposal may seem very high, but new work on acid gas reinjection from gas processing plants (Chakma, 1996) emphasises that the cost may be acceptable when examined in combination with other measures. For example, CO$_2$ disposal combined with offshore power production could be economic in Norway (Holt & Lindeberg, 1996). If a proportion of any CO$_2$ captured from power plant could be used for EOR this would have a very positive impact on the costs, e.g. DeMomtigny et al. (1996).

STORAGE CONCEPTS

The preferred underground storage concept is injection via wells into deep reservoir rocks capped by very low permeability seals such as shales or claystones. This could be into fluid traps, but reservoir modelling suggests that CO$_2$ could simply be injected into certain large, essentially horizontal aquifers without the need for the conventional structural or stratigraphic fluid traps of the sort in which oil and natural gas are found, e.g. Gunter et al. (1993). This vastly increases the underground storage potential of sedimentary basins. It is important that reservoir modelling, both of injection and the subsequent migration of CO$_2$ continues, as advances are continuously being made, e.g. Lindeberg (1996). Novel storage concepts, such as the substitution of CO$_2$ for methane in coal reservoirs continue to be proposed (Gunter et al., 1996).

Much of the CO$_2$ will be stored in a free state, or dissolved in the formation water of the reservoir. However, a natural mineral trapping mechanism exists in certain sandstone reservoirs. CO$_2$/water mixtures will react with basic aluminosilicate minerals within the rock matrix and either precipitate calcite or other carbonates, or produce bicarbonate brines. Either route will result in the fixing a proportion of the carbon for geological timescales e.g. Gunter et al. (1993).

STABILITY OF STORAGE

Advantages of underground disposal include retention of the CO_2 for timescales of tens of thousands to millions of years. For example, the CO_2 in the Pisgah Anticline in Central Mississippi, USA, is thought to have originated by thermal metamorphism of Jurassic carbonates by the Jackson Dome igneous intrusion (Studlick *et al.*, 1990) during late Cretaceous times, which ended some 65 million years ago.

SAFETY

The use of supercritical CO_2 in the EOR industry is probably the best pointer to the low risks associated with the underground disposal of CO_2. Risks associated with the pipeline and surface injection facilities of any CO_2 disposal scheme will be analogous to those at EOR sites. Legal and regulatory issues are discussed by Bergman *et al.* (1996), while Van der Meer & Summerfield in Cox *et al.* (1996) point out that risks can be minimized by appropriate choice of materials, and designed so that in the event of a system failure, only small amounts of CO_2 can escape. The probablity (Van der Meer & Summerfield in Cox *et al.*, 1996) and consequences (Kryse & Tekeila, 1996) of a pipeline rupture have been analysed.

Given that deaths have occurred as a result of natural emissions of CO_2 in volcanic regions (Le Guern & Sigvaldason, 1989), it is important to consider the various kinds of leak which could theoretically occur from a CO_2 storage reservoir. Theoretically, leaks could result from the failure of an injection well or occur via an unidentified migration pathway from the reservoir, such as a geological fault (Cox *et al.*, 1996). A certain amount of information about the likelihood of such leaks could perhaps be gained from the track record of natural gas storage schemes in aquifers and abandoned hydrocarbon fields.

In reality, the potential effects of leaks from the storage reservoir would have to be assessed on a site-specific basis. The effects of a slow but persistent leak of CO_2 into confined spaces, such as housing, should be considered, as should the distribution and dispersal of a large gas cloud caused by a sudden release of CO_2 such as might be caused by a blowout in an injection well.

In coastal areas, risks to man and the onshore natural environment associated with leaks from a storage reservoir could be further minimised by disposal offshore (Holloway, 1996).

STORAGE CAPACITY

There are two main problems associated with estimating the global underground storage capacity for CO_2. Firstly, storage concepts for CO_2 have evolved over the last few years and, secondly, little is known about even the basic parameters (e.g. distribution, thickness, porosity, permeability) of deep reservoirs in the subsurface outside the major petroleum provinces. It is far from certain how much CO_2 could be stored in individual aquifers with well known parameters, never mind those which are poorly described. Storage capacity can be modelled using reservoir simulators. However, storage efficiency, the fraction of the reservoir which can be filled with CO_2, varies considerably between reservoirs (Van der Meer, 1995), and a 'global average' figure is elusive, not least because there is no global average reservoir. These problems mean that regional or global studies of storage capacity, e.g. Koide *et al.* (1992), Riemer (1996), Bachu *et al.* (1994) have commonly been based on totally different assumptions; firstly about the storage concept, and secondly about the volumes of storage space available for each storage concept.

For example, the storage capacity of natural gas fields is sometimes estimated to be higher than the storage capacity of aquifers e.g. Riemer (1996). This seems paradoxical at first sight, as a natural gas field forms only a small part of the aquifer in which it occurs. However, a very high proportion of the pore space in a depleted gas field can sometimes be available for CO_2 storage, because in many gas fields there is little water invasion during production. When production from such fields ends, the pore space is occupied largely by low pressure methane. This is highly compressible, leaving a high proportion of the pore space to be occupied by injected CO_2. The apparent paradox is completed if the conservative assumption is made that CO_2 can be stored only in conventional fluid traps in the aquifer (i.e. spaces analogous to oil and gas fields, but filled with water rather than oil or gas). Reservoir modelling suggests that if a realistic injection rate is assumed, only a small proportion of the pore space in such a structure can be occupied by injected CO_2 before gravity override and viscous fingering mean that it begins to escape from the trap (van der Meer, 1995).

Undoubtedly the greatest contrast in storage volumes comes in the aquifer estimates. The upper estimates consider that a small fraction of the whole pore volume of a saline aquifer could be used for CO_2 storage, whereas the lower estimates consider that only a small fraction of the pore space which makes up the conventional fluid traps present in a saline aquifer could be used for CO_2 storage. It no longer seems realistic to assume that only conventional fluid traps can be used. For example, the CO_2 which is currently being injected into the Utsira Formation at the Sleipner West gas field is not being injected into a conventional fluid trap. When it is appreciated that CO2 can simply be injected into certain large, essentially horizontal aquifers without the need for conventional fluid traps, calculated storage volumes commonly change from space for tens of years of emissions in the studied area to space for hundreds of years of emissions, or even more. For example, van der Straaten (1996) estimated the theoretical underground CO_2 storage capacity of the European Union and Norway. The total storage volume of the oil and gas fields in this area was calculated to be 6 Gt CO_2 and 27 Gt CO_2 respectively. The trap storage potential of the aquifers was calculated to be 19.5 Gt CO_2 offshore and 10.7 Gt CO_2 onshore. The storage potential in aquifers not confined to traps was estimated to be 57 Gt CO_2 onshore and 716 Gt CO_2 offshore. These estimates represent trap storage potential for 67 years and untrapped storage potential for 810 years of 1990 CO_2 emissions from power plant in the European Union (0.95 Gt/yr). Of course, these figures are highly speculative, and a proportion of this theoretically available storage space will prove unsuitable for a wide variety of geological, economic and political reasons. Nevertheless, it is clear that if even a small fraction of the untrapped storage potential is used there will be enough storage space for all likely CO_2 disposal projects in Western Europe.

It is worth noting that all estimates of the storage capacity of oil and gas fields are also likely to be too high, as they assume that all such fields (in some cases over a certain size) will be available and suitable for storage: in practice this will not prove to be the case.

Koide et al. (1992) used a totally different, areal method to estimate global underground CO_2 storage capacity. They assumed that 1% of the area of the onshore sedimentary basins of the world could be used for CO_2 storage and pointed out that if offshore basins were also considered, the storage capacity would increase appreciably. On the assumption that the CO_2 would be stored dissolved in the formation water, they calculated that it could be stored at a rate of 492,000 tons CO_2/km^2. Their estimate of global storage capacity using this method was 320 GtCO_2, 13.6 Gt of which was in the Western European onshore area.

The majority, if not all, subsequent estimates have been higher. For example, the IEA Greenhouse Gas R & D Programme (Riemer, 1996) estimated that the global storage capacity of oil and natural gas fields alone is at least a further 180 Gt C (approx. 660 Gt CO_2). They estimate the global storage capacity of saline aquifers as >100 Gt C (approx. 367 Gt CO_2). This figure is low because they only counted the fluid traps in the aquifers and did not consider using a proportion of the pore space of the whole aquifer for CO_2 storage.

Bachu et al. (1994) who based their estimates on using a proportion of the pore space of entire aquifers, point out that the total underground storage capacity of the Alberta Basin (Canada) is 20 Gt CO_2. This alone is 16% of the global figure estimated by Koide et al. (1992).

WHERE TO FROM HERE?

It would be highly desirable to cooperate as closely as possible with the operators of CO_2 disposal schemes, in order to monitor and learn as much as possible from them. This is perhaps the most immediate priority, as few such practical demonstration opportunities may occur.

In the field of safety and stability of storage, the track record of leaks from gas storage schemes and natural gas seeps should be investigated. Further modelling of the movement and dispersion of CO_2 clouds in varying terrains and weather conditions should be attempted.

Storage concepts are still evolving and new concepts will affect storage capacity estimates. Further reservoir simulation and further geochemical experiments will be required, to provide data for geochemical modelling.

More natural CO_2 accumulations should be studied, as analogues of man-made storage facilities, to determine the longer term mineralogical reactions caused by the introduction of CO_2 into reservoir rock.

The major barrier to implementation of the underground disposal of CO_2 is cost. No matter where it is to be disposed of, reduction of the costs of CO_2 separation or concentration is a research priority if widespread CO_2 disposal from power plants is to become a reality.

ACKNOWLEDGEMENT

This paper is published with the approval of the Director, British Geological Survey (NERC).

REFERENCES

Anon. (1996). *Greenhouse Issues*, (Newsletter of the IEA Greenhouse Gas R&D Programme), Number 22, January 1996.

Bachu, S, Gunter, W D and Perkins, E H. (1994). Aquifer disposal of CO_2: hydrodynamical and mineral trapping. Energy Conversion and Management, Volume 35, p.264-279.

Bergman, P D, Drummond, C J, Winter, E M & Chen, Z-Y (1996). Disposal of power plant CO2 in depleted oil and gas reservoirs in Texas. *Proceedings of the Third International Conference on Carbon Dioxide Removal*, Massachussetts Institute of Technology, Cambridge, MA, USA, 9-11 September, 1996.

Chakma, A (1996). Acid gas re-injection - a practical way to eliminate CO2 emissions from gas processing plants. *Proceedings of the Third International Conference on Carbon Dioxide Removal*, Massachussetts Institute of Technology, Cambridge, MA, USA, 9-11 September, 1996.

Cox, H, Heederik, J P, van der Meer, L G H, van der Straaten, R, Holloway, S, Metcalfe R Fabriol H and Summerfield, I (1996). Safety and stability of underground storage, pp.116-162 in: Holloway, S (Ed.)*The Underground Disposal of Carbon Dioxide* Final Report of Joule II project No. CT92-0031; '

DeMomtigny, D, Gelowitz, D, Kritpiphat, W, & Tontiwatchwuthikul, P (1996). Simultaneous Production of Electricity, Steam and CO_2 from Small Gas-fired Cogeneration Plants for Enhanced Oil Recovery: A Feasibility Study for Western Canada. *Proceedings of the Third International Conference on Carbon Dioxide Removal*, Massachussetts Institute of Technology, Cambridge, MA, USA, 9-11 September, 1996.

Doherty, P & Harrison R (1996). Techno-economic modelling of the underground disposal of carbon dioxide. Chapter 8 in: Holloway, S (Ed.), *The Underground Disposal of Carbon Dioxide*. Final Report of Joule II project No. CT92-0031.

Freund, P & Ormerod, W (1996). Progress towards storage of CO_2. *Proceedings of the Third International Conference on Carbon Dioxide Removal*, Massachussetts Institute of Technology, Cambridge, MA, USA, 9-11 September, 1996.

Gunter, W D, Bachu, S, Law D H-S, Marwaha, V, Drysdale, D L, Macdonald, D E & McCann, T J (1996). Technical and economic feasibility of CO_2 disposal in aquifers within the Alberta sedimentary basin. *Energy Conversion and Management*, Volume 37, Nos 6-8, pp.1135-1142.

Gunter, W D, Gentzis, T, Rottenfusser, B A & Richardson, R J H (1996). Deep coalbed methane in Alberta, Canada: A fossil fuel resource with the potential of zero greenhouse gas emissions. *Proceedings of the Third International Conference on Carbon Dioxide Removal*, Massachussetts Institute of Technology, Cambridge, MA, USA, 9-11 September, 1996.

Gunter W D, Perkins, E H & McCann T J (1993). Aquifer disposal of CO₂-rich gases: reaction design for added capacity. *Energy Conversion and Management*, Volume 34, pp. 941-948.

Haugen, H A & Eide L I (1996). CO₂ Capture and Disposal: The realism of large scale scenarios. *Energy Conversion and Management*, Volume 37, Nos 6-8, pp. 1061-1066.

Holloway, S (1996). An overview of the Joule II project 'The Underground Disposal of Carbon Dioxide'. *Energy Conversion and Management*, Volume 37, Nos 6-8, pp.1149-1154.

Koide, H, Shindo, Y, Tazaki, Y, Iijima, M, Ito, K, Kimura, N & Omata, K (1996). Deep sub-seabed disposal of CO2 - The most protective storage. *Proceedings of the Third International Conference on Carbon Dioxide Removal*, Massachussetts Institute of Technology, Cambridge, MA, USA, 9-11 September,

Koide, H, Tazaki, Y, Nogichi, Y, Nakayama, S, Iijima, M, Ito, K & Shindo, Y (1992). Subterranean containment and long term storage of carbon dioxide in unused aquifers and in depleted natural gas reservoirs. *Energy Conversion and Management*, Volume 34, pp.619-626.

Korbul, R & Kaddour, A (1995). Sleipner Vest CO₂ disposal - injection of removed CO₂ into the Utsira Formation. *Energy Conversion and Management*, Volume 36, Nos 6-9, pp. 509-512.

Kruse, H & Tekeila, M (1996). Calculating the consequences of a CO₂-pipeline rupture. *Energy Conversion and Management*, Volume 37, Nos 6-8, pp.1013-1018.

Lackner, K S, Butt, D P & Wendt, C H (1996). Progress on binding CO2 in mineral substrates. *Proceedings of the Third International Conference on Carbon Dioxide Removal*, Massachussetts Institute of Technology, Cambridge, MA, USA, 9-11 September, 1996.

Law, D H-S & Bachu, S (1996). Hydrogeological and numerical analysis of CO₂ disposal in deep aquifers in the Alberta sedimentary basin. *Energy Conversion and Management*, Volume 37, Nos 6-8, pp.1167-1174.

Le Guern, F & Sigvaldason, G E (Eds) (1989). The Lake Nyos event and natural CO₂ degassing. Special Issue of *Journal of Vulcanology and Geothermal Research*, Volume 39, Nos 2-3.

Lindeberg, E (1996) vertical convection in an aquifer column under a CO2 gas cap. *Proceedings of the Third International Conference on Carbon Dioxide Removal*, Massachussetts Institute of Technology, Cambridge, MA, USA, 9-11 September, 1996.

Riemer, P (1996). Greenhouse Gas Mitigation Technologies, an Overview of the CO₂ Capture, Storage and Future Activities of the IEA Greenhouse Gas R&D Programme. *Energy Conversion and Management*, Volume 37, Nos 6-8, 665-670.

Studlick, J R J, Shew, R D, Basye, G L and Ray, J R (1990). A giant carbon dioxide accumulation in the Norphlet Formation, Pisgah Anticline, Mississippi. In: *Sandstone Petroleum Reservoirs* (J Barwis, McPherson & J. R J Studlick, Eds), pp. 181-203. Springer Verlag, New York.

Taber, J J (1993). The use of supercritical CO₂ in enhanced oil recovery. In: *Proceedings of the Second International Conference on Carbon Dioxide Utilisation*, University of Bari, Bari, Italy.

Van der Meer, L G H (1995). The CO₂ storage efficiency of Aquifers. *Energy Conversion and Management*, Volume 36, No. 6-9, 513-518.

Van der Straaten, R (1996). Inventory of the theoretical CO₂ storage capacity of the European Union and Norway: Results and Conclusions. Pp.105-110 in: Holloway, S (Ed), Final Report of the Joule II project No. CT92-0031: 'The Underground Disposal of Carbon Dioxide'

Energy Convers. Mgmt Vol. 38, Suppl., pp. S199–S204, 1997
© 1997 Elsevier Science Ltd. All rights reserved
Printed in Great Britain
0196-8904/97 $17.00 + 0.00

Pergamon

PII: S0196-8904(96)00269-5

PROGRESS TOWARD STORAGE OF CARBON DIOXIDE

P. FREUND and W. G. ORMEROD

IEA Greenhouse Gas R&D Programme, CRE Group Ltd, Cheltenham, Glos. GL52 4RZ, U.K.

ABSTRACT

Abatement of carbon dioxide emissions by capturing CO_2 from flue gases and other process streams can be achieved using available technology although at substantial cost. Having captured the gas, it must be sequestered in such a way that it does not reach the atmosphere for several hundred years. There are many options available for storing CO_2; assessments by the IEA Greenhouse Gas R&D Programme and others have shown that many of these are relatively inexpensive, compared with the cost of capturing the gas in the first place. However, there are significant uncertainties involved in most of these schemes, indicating the need for further research and development in order to reduce the risks of their application.

In this paper, a summary will be given of recent progress in CO_2 storage, including an overview of work on aquifer storage, ocean storage and the storage role of forests. External influences on decisions about CO_2 storage will be discussed. Opportunities will be examined for further research to progress these technologies. Future possibilities for practical work on storage of CO_2 will be discussed.

© 1997 Elsevier Science Ltd

KEYWORDS

Carbon dioxide; storage of carbon dioxide; ocean storage, underground storage; above-ground storage; aquifer storage.

INTRODUCTION

Capture and storage of carbon dioxide has been seriously considered as an option for reducing greenhouse gas emissions only in recent years. The process of capturing CO_2 from flue gases can be achieved using technology which is available now or is in the advanced stages of development. The cost of capture and compression is substantial - a minimum of $100/t C (Riemer and Ormerod, 1994) - which sets a challenge for further technical development. Having captured the gas, in order to make a useful contribution to mitigation of climate change, it must be stored for several hundred years. However, this part of the process is relatively poorly understood. Work is underway in various fields to learn more about CO_2 storage and the purpose of this paper is to review these developments.

OPTIONS FOR STORAGE OF CARBON DIOXIDE

There are many options available for storing CO_2; some involve adaptation of existing techniques; others are quite new and radical concepts (IEA GHG 1994a). Because of the large quantities that need to be stored, the options most often discussed are those using natural reservoirs underground or in the deep ocean. It is also conceivable that CO_2 could be stored on land, either through the natural growth of plants, or in an artificial store. Each of these options will be discussed briefly below.

Given the scale of anthropogenic emissions (6 GtC/y from fossil fuels), a worthwhile storage option must offer global capacity of 10's to 100's of GtC, at least, and the cost must be competitive with other mitigation options. Security of storage is another key issue. At a minimum, it must be possible to store CO_2 for several hundred years, with confidence that almost all of it will stay where it is put. These factors form the basis for distinguishing between the various technical possibilities.

Storage underground

Use of disused oil and gas fields to store CO_2 is attractive, not only because of the conceptual tidiness of using the space left by removal of hydrocarbons, but also because the knowledge gained about such fields from their previous use (e.g. detailed geology, existence of a geological seal) gives confidence in their use for storing CO_2. The storage capacity will vary depending on the nature of the reservoir - the best would be a closed reservoir which had not been invaded by water when the hydrocarbons were removed. More likely some invasion of water has taken place and dense phase CO_2 will rise to the top of the trap and displace only a limited amount of the water in the formation. In the absence of a detailed assessment of existing oil and gas fields, the range of estimates for capacity is quite wide (IEA GHG 1994a). The global capacity for storage in disused gas fields could be as much as 140GtC, and in disused oil fields perhaps 40GtC (because of typically lower recovery of oil than gas). A hypothetical scheme for storage in an abandoned on-shore gas field has been costed at \$8.2/tC (IEA GHG 1994a).

Aquifers are another possibility for underground storage. Lack of information on their structure, especially the existence of a suitable seals, has led to wide ranges of estimates for their capacity. The IEA GHG Programme estimated the global CO_2 storage capacity of aquifers to be at least 100Gt C, but others (Holloway *et al*, 1995) have suggested that capacity in Europe alone may be 200Gt C, but with more conservative assumptions this would reduce to 30Gt C. Interaction with reservoir rock introduces another factor which may increase long-term capacity (Gunter *et al*, 1993). A hypothetical scheme for injection into an onshore aquifer indicated costs of about \$4.7/tC (IEA GHG 1994b). Wide availability of these reservoirs has, however, led to them being targeted for the first CO_2 storage schemes, now underway or being planned.   The Sleipner Vest natural gas field contains 9.5% CO_2; this gas is being separated and re-injected into the Utsira formation under the North Sea, in the world's first commercial-scale CO_2 storage in an aquifer.   Other such projects are under consideration, including the massive Natuna field, offshore Indonesia, which Exxon and Pertamina are planning to develop. This field contains 71% CO_2.

Further storage options arise from use of CO_2 to enhance oil or gas recovery. Use of natural CO_2 for EOR is already practised and this also provides an opportunity for storage of CO_2. It has been estimated that the global capacity for storage in this way may amount to 65GtC, at a cost not very different from the other underground options (IEA GHG, 1994c). It is expected that EOR schemes using CO_2 captured from fossil fuel combustion will be undertaken within the next few years. In an analogous way it is proposed, in a paper to this conference (Gunter *et al* 1996), that CO_2 could be used to enhance the recovery of coal bed methane, the CO_2 displacing the sorbed CH_4. In both cases, the extra revenue helps to offset some of the cost of storage.

Storage in the deep ocean

Since the original proposal by Marchetti (1977), storage in the oceans has provoked various innovative concepts. Many of these have attractive features, not least in recognising the fact that the oceans are the ultimate natural sink for atmospheric carbon. The capacity of the oceans for storing CO_2 has been estimated to be upwards of 1000 GtC (IEA GHG, 1994a). Injecting dense-phase CO_2 into deep water could delay return to the atmosphere by several hundred years, if it is done in the appropriate location (Bacastow *et al*, 1995) However, the energy costs of compressing and moving CO_2 sufficiently far offshore to access the deep water, have stimulated investigation of other approaches.   An alternative is to inject at shallower depths (perhaps <500m) and form a dense CO_2-seawater plume which would sink and carry the CO_2 to deeper waters (Drange and Haugan, 1992)). Depth of discharge and length of pipeline are two of the key influences on the cost of ocean storage, which has been estimated at \$4.1/tC for shallow depths rising to \$21/tC for deeper schemes (IEA GHG, 1994a) The environmental impact of such schemes may raise concern but this might be handled by frequently changing the point of discharge (Nakashiki *et al*, 1995). Other ideas (Watson *et al*, 1994); Jones, 1996) involve artificial fertilisation of some parts of the oceans, to enhance uptake of atmospheric CO_2. The international implications of ocean storage scheme is another of the unknowns; further research is essential in order to answer key questions.

Storage on land

Natural terrestrial carbon stores (e.g. forests) are widely discussed (IEA GHG, 1995) - absorbing CO_2 from the atmosphere using the energy in sunlight to drive the photosynthesis process is a relatively cheap method of sequestering carbon.   Estimated costs can be as low as \$1/tC, but depend greatly on the cost of land. Providing there is adequate management of the plantations, forests can remain in place for hundreds of years and it has been estimated that efficient management of the worlds forests could provide an additional storage capacity of 1.2GtC/year (Winjum at al, 1992). This makes forestry a very popular approach, especially for international co-operations such as embodied in Activities Implemented Jointly (AIJ). Another terrestrial-based concept envisages use of thermally insulated artificial stores containing solid CO_2 (Seifritz, 1992). However a store with capacity of 10^8 m^3 (sufficient to hold the lifetime output of one 500Mw$_e$ power station) would cost around \$500/tC (IEA GHG, 1994).

Interim Conclusions

For most of these options, the cost of storing a tonne of carbon is much less than the cost of capturing it in the first place. Hence cost reduction is not the major target for advancing the technology of CO_2 storage. Questions of performance, environmental impact and jurisdiction are ones where there is need for further research and these issues will now be examined in further detail for each of the main storage options.

ANALYSIS OF STORAGE OPTIONS

Ocean Storage

The IEA Greenhouse Gas R&D Programme has held a series of specialist workshops (IEA GHG, 1996a, 1996b) in the last year to review the state of knowledge in key aspects of ocean storage of CO_2, to identify research needs and also to examine issues associated with practical implementation. The first two workshops addressed the basic scientific issues of performance, namely the timescale of CO_2 returning to the atmosphere and the potential impact on the marine environment; a third workshop examined legal and jurisdictional aspects.

Ocean general circulation models (OGCMs) provide the only route for estimating the time taken for CO_2, injected in the deep ocean, to resurface. The most recent OGCMs are 3-dimensional and were developed to track the movement of CO_2 dissolved in surface waters down to deep waters and the eventual return of most of that CO_2 back to the surface. A timescale of around 1000 years is normally associated with the sinking of dense cold Arctic waters and the reappearance in surface waters in other parts of the world ocean system. Superimposed on this physical circulation is the ocean carbon cycle which is responsible for the net removal of around 2Gt C/year from the atmosphere to the ocean. A principal aim of this modelling effort is to establish an accurate estimate of the oceanic uptake of CO_2 for various IPCC emission scenarios; currently, the best model-based estimate of oceanic uptake has a ±30% uncertainty (Siegenthaler *et al*, 1993). One OGCM has been reconfigured to track CO_2 from point sources in the deep ocean back to surface waters (Bacastow *et al*, 1995) and the results from this exercise are very revealing, the model predicting that the time taken for CO_2 to resurface from deep water is not only a function of starting depth but also of geographic location.

A major concern associated with using the ocean as a store for CO_2 is the impact this may have on the marine environment. Injection of CO_2 into seawater will create, in the vicinity of the injection point, a volume of low pH water and it is known that marine organisms will only tolerate small pH changes (i.e. <1pH unit). Injection techniques and engineering design may have the capability to reduce mortality rates among marine organisms to very low levels, but there is concern that sub-lethal effects (e.g. growth and reproduction rates), which are very difficult to quantify, may be more widespread (IEA GHG 1996b) In particular the presence of a strata of low pH water could act as a barrier to the vertical migration of marine organisms. This can occur on both a 24 hour cycle and on a seasonal cycle, and can be part of the reproductive cycle. This has serious implications because these migrations can link activity in surface waters to water at depths of 1000m and, on the shelf break, to depth of 1600m. These considerations would indicate that discharge depths would have to exceed 1600m, which although possible in principle, is outside present-day experience. Another key aspect is that the activity of deep sea benthic organisms is also linked to the productivity of surface waters, and bacteria, responsible for remineralisation of organic carbon back to inorganic carbon, make up a large proportion of the benthic biomass. Isolation techniques, designed to contain CO_2 as a liquid on the seabed, may therefore also interfere with the fluxes involved in the ocean carbon cycle.

Significant legal and jurisdictional issues may also have to be faced over use of the deep ocean for storage of CO_2. There is a trend, with waste disposal issues, towards the Precautionary Principle which puts the onus on the party involved to demonstrate that no environmental damage will ensue and that a problem is not being transferred from one part of the environment to another. Moreover, legislators concerned with use of the oceans for waste disposal (e.g. 1972 London Agreement) are responding to international pressure to protect the deep oceans and are progressively tightening the legislation.

Underground Storage

More confidence can currently be associated with the performance of underground storage of CO_2 and this has resulted in the first commercial scale CO_2 storage application in the Utsira aquifer formation associated with the Sleipner Vest gas field. Retention times in aquifers are estimated to be orders of magnitude higher than for ocean storage (thousands of years or higher) and only a catastrophic release would render their performance unsatisfactory. Both aquifers and depleted oil/gas wells therefore represent a more conventional storage concept where only leakages would result in a return to atmosphere. Confidence can also be drawn from artificial natural gas stores and examples of underground CO_2 stores that provide natural analogues where CO_2 has been contained over geological timescales (Pearce et al, 1996). Storage of CO_2 in depleted gas/oil reservoirs or in deep aquifers does not therefore generate fundamental scientific questions over the validity of the concept, but there are environmental and some safety concerns. These concerns relate to the possibility of a catastrophic release at an injection well, or smaller leakage from fingering or geological faults.

In principle, use of depleted gas or oil reservoirs would seem to represent the most logical storage option and to have advantages over storage in aquifers. Exploration costs should be relatively small, traps are proven, reservoir properties are well known and some of the infrastructure might be re-used. Furthermore, expertise in transporting and injecting CO_2 is available from EOR schemes. However, as with ocean storage, legal and jurisdictional issues may present a significant barrier to commercial application. The current examples of commercial application are 'self-contained' in that a gas or oil producer also wishes to store CO_2 or use it for EOR - there is no conflict of interest in terms of the ownership of the store of CO_2. Environmental/safety problems are also reduced by it being offshore and away from any population centre. It may be questioned whether the position would be as satisfactory for an independent CO_2 emitter. In this respect, use of aquifers would seem to represent a simpler situation than use of depleted oil and gas fields and may account for the absence, so far, of any commercial scheme involving such fields.

Storage on Land

The world's forests hold 90% of all above-ground terrestrial carbon and 40% of all below-ground carbon (Winjum *et al.* 1992). Deforestation, to clear land for agriculture, is currently estimated to account for emissions of 1.6 ± 1.0Gt C/y to atmosphere (IPCC, 1994). Much of the land cleared by deforestation has proved to be poor quality agricultural land and this provides the opportunity for reforestation, to reverse the trend of a decreasing terrestrial carbon store.

There are several examples of CO₂ emitters in industrial countries funding forestry projects, usually in developing countries, to offset the CO₂ emissions of a specific fossil fuel-fired power plant (e.g. FACE 1995). This is an alternative, indirect route which achieves the same result as a CO₂ capture and storage scheme, but has the advantages of not reducing the energy conversion efficiency of the power plant, appearing to have better environmental credentials and, at present, being lower in cost. The disadvantage of the method is the large area of land required (a 500MW$_e$ coal-fired power plant with a 35 year life, would require a forestry scheme covering 1400km², maintained for several hundred years). This large requirement for land has many economic, social and political implications which suggest that, if implemented on any significant scale, costs would rise.

Offset forestry schemes are low in cost, at present, as they are hand-picked; sufficient funding is provided to subsidise a forestry scheme which has good credentials, not only to sequester carbon but also to provide forestry products. Moreover, most schemes are in developing countries where labour costs are low. With increasing implementation, suitable land will become more scarce and the market for forest products may saturate. The IEA GHG Programme is sponsoring a study, in a tropical country, to examine the cost of large scale forestration for C-sequestration, and, more specifically, to try to quantify the impact on cost of an increasing uptake of land for forestry.

FUTURE DIRECTION

Significant progress has been made over the last few years, not only in our overall understanding of the concepts associated with the various CO₂ storage options but also in identifying the research needs associated with the various options. Political decisions to restrict CO₂ emissions have also helped to bring about the first commercial proposals for utilising aquifers for storage of CO₂ and these provide unique opportunities to monitor performance, validate predictive models and improve the level of our understanding. The IEA GHG Programme is proposing to hold an expert workshop to develop a performance monitoring strategy that could be applied to a commercial aquifer CO₂ storage scheme.

Expert workshops, such as those sponsored by the IEA GHG Programme on ocean storage, can prove valuable in highlighting research work in related fields. This links such work more directly to the concept of CO₂ storage which then becomes a more integral part of some of the large international science programmes on climate change (IGBP, WCRP etc.). Through such workshops, research needs can be identified more closely, results and ideas are shared internationally and plans can be formulated for international experiments.

The concept of Activities Implemented Jointly, as a method of encouraging schemes for reducing greenhouse gas emissions, has resulted in many proposals for off-set forestry schemes in developing tropical countries. Some schemes are already in operation. The developing countries will play an important part in limiting greenhouse gas emissions in the future, and forestry schemes offer a way of both enhancing the terrestrial carbon sink and assisting the economic development of these countries.

The future direction in all cases must be towards practical work - experiments, demonstrations, commercial schemes. A high level of funding is required for such work and international collaboration can help to share the cost. The IEA GHG Programme provides a mechanism for such collaboration.

REFERENCES

Bacastow R B and Dewey R K (1996). Effectiveness of CO_2 sequestration in the post-industrial ocean. *Energy Convers. Mgmt.*, 37, 1079-1086.

Drange H and Haugan P M (1992). Sequestration of CO2 in the deep ocean by shallow injection. Nature, 357, 318-320.

FACE (1995). Face Foundation in Practice. Face Foundation, Utrechtseweg 310, NL-6800 AN Arnhem, The Netherlands.

Gunter W D, Perkins E H, McCann T J (1993). Aquifer disposal of CO2 rich gases: reaction design for added capacity. *Energy Convers. Mgmt.*, 34, 941-948.

Gunter W D, Gentiz T, Rottenfusser B A, Richardson R J H (1996). Deep coalbed methane in Alberta, Canada: a fossil-fuel resource with the potential of zero greenhouse gas emissions. *Paper to be presented at ICCDR-3 Conference, MIT, Cambridge, MA.*

Holloway S, Heederick J P, van der Meer L G H, Czernichowski-Lauriol I, Harrison R, Lindeberg E, Summerfield I R, Rochelle C, Schwarzkopf T, Kaarstad O, Berger B (1996). The underground disposal of carbon dioxide - summary report. *European Communities Joule II Programme*. British Geological Survey, Keyworth, Nottingham, UK.

IPCC (1994) Radiative forcing of climate change. Report of the scientific assessment working group.

IEA GHG (1994a). Carbon dioxide capture from power stations. *IEA Greenhouse Gas R&D Programme*, Cheltenham, UK.

IEA GHG (1994b). Carbon dioxide disposal from power stations. *IEA Greenhouse Gas R&D Programme*, Cheltenham, UK.

IEA GHG (1994c). Carbon dioxide utilisation. *IEA Greenhouse Gas R&D Programme*, Cheltenham, UK.

IEA GHG (1995). Global warming damage and the benefits of mitigation. *IEA Greenhouse Gas R&D Programme*, Cheltenham, UK.

IEA GHG (1996a). Ocean storage of CO2, workshop 1 - ocean circulation. *IEA Greenhouse Gas R&D Programme*, Cheltenham, UK.

IEA GHG (1996b). Ocean storage of CO2, workshop 2 - environmental impact. *IEA Greenhouse Gas R&D Programme*, Cheltenham, UK.

Jones I S F (1996). Enhanced carbon dioxide uptake by the world's oceans. *Energy Convers. Mgmt.*, 34, 1049-1052.

Marchetti C (1977). On geoengineering and the CO2 problem. *Climate Change*, 1, 59-68.

Nakashiki N, Ohsumi T, Katano N (1995). Technical view on CO2 transportation onto the deep ocean floor and dispersion at intermediate depths. In: *Direct Ocean Disposal of Carbon Dioxide*, (N Handa and T Ohsumi eds.) Terra Scientific Publishing Company, Tokyo.

Pearce J M, Holloway S, Wacker H, Nelis M K, Bateman K (1996). Natural occurrences as analogues for the geological disposal of carbon dioxide. *Energy Convers. Mgmt.*, 37, 1123-1128.

Riemer P W F and Ormerod W G (1995) International perspectives and results of carbon dioxide captur, disposal and utilisation studies. *Energy Convers. Mgmt.*, 36, 813-818.

Seifritz W (1992). The terrestrial storage of CO2-ice as a means to mitigate the greenhouse effect. In: *Hydrogen Energy Progress IX* (Pottier C D J and Veziroglu T N, eds.).

Siegenthaler U and Sarmiento J L (1993). Atmospheric carbon dioxide and the ocean. *Nature*, 365, 119-125.

Watson A J, Law C S, Van Scoy K A, Milero F J, Yao W, Friederich G E, Lidicoat M I, Wanninkhof R H, Barber R T, Coale K H (1994). Minimal effect of iron fertilisation on sea-surface carbon dioxide concentrations. *Nature*, 371, 143-145.

Winjum J K, Dixon R K, Schroeder P E (1992). Estimating the global potential of forest and agroforest management practices to sequester carbon. Water, Air and Soil Pollution.

Energy Convers. Mgmt Vol. 38, Suppl., pp. S205–S209, 1997
© 1997 Elsevier Science Ltd. All rights reserved
Printed in Great Britain
0196-8904/97 $17.00 + 0.00

Pergamon

PII: S0196-8904(96)00270-1

ACID GAS RE-INJECTION - A PRACTICAL WAY TO ELIMINATE CO2 EMMISSIONS FROM GAS PROCESSING PLANTS

Amit Chakma

Environmental Systems Engineering, Faculty of Engineering

The University of Regina, Regina, Saskatchewan, Canada

ABSTRACT

Natural gas processing plants separating CO_2 and H_2S from natural gas streams are major contributors to CO_2 emissions in various gas producing regions of the world. The present practice is to convert the separated H_2S into elemental sulfur through a well know chemical conversion process, the Claus process and to vent the CO_2 into the atmosphere. In some cases it might be possible to re-inject the acid gases into either a depleted petroleum reservoir or into aquifers thus reducing CO_2 emissions from natural gas plants. A number of such acid gas re-injection projects are now operational in Alberta. This paper provides a background leading to the development of these projects. © 1997 Elsevier Science Ltd

KEY WORDS

CO_2 Emissions, Acid Gas Injection, Depleted Oil and Gas Reservoirs

INTRODUCTION

While there are many large point sources for CO_2 generation, power plants have been singled out as one of the sources from which CO_2 can be extracted in sufficiently large quantities, so that the cost of its separation can be kept as low as possible. Even with such economy of scale advantage, in the absence of any regulatory requirement, separation of CO_2 from power plant flue gases using existing technology is considered to be prohibitively expensive.

In some special regions, there might be other sources of CO_2, that can be extracted in a much more cost effective manner compared to extraction from power plants. In the case of regions producing natural gas containing CO_2 as an impurity, the incremental cost of CO_2 separation and disposal can be significantly lower. Most of the gases produced from geological formations contain varying amounts of acidic components such as H_2S and CO_2 as impurities. These impurities need to be removed from the natural gas stream to minimize health hazards, corrosion and environmental problems before the gas can be transported to the markets for end-use. H_2S and CO_2 thus removed from gas streams need to be disposed of in a safe and environmentally acceptable manner. The conventional practice is to convert H_2S into elemental sulphur and to emit the CO_2 into the atmosphere. The process of converting H_2S into elemental sulphur is an expensive one. In addition, lower sulfur prices in recent times makes this conventional option economically unattractive.

In view of the above, during the last few years, there has been a growing interest in the re-injection of acid gases containing H_2S and CO_2 into geological formations instead of converting H_2S into elemental sulfur. The acid gases separated from natural gas streams (typically using an amine based process) are compressed and re-injected into an appropriate geological formation. These formations can be depleted oil and gas reservoir, oil reservoir, aquifers etc. Currently, there are over 12 acid-gas re-injection projects in operation throughout Alberta. A review of these acid-gas re-injection projects and their impact in minimizing the emission of CO_2 into the atmosphere will be provided. Although CO_2 injection into a geological formation for the mitigation of green house gas emission is considered to be an expensive proposition, in all the projects to be described, it is found to be a cheaper option compared to the conversion of H_2S into elemental sulfur. Although, the intention of the project operators is to dispose of H_2S, the side benefit of the process is less emission of CO_2 into the environment. The lesson learned from this innovative approach of solving the H_2S disposal problem is very clear - CO_2 separation and disposal may appear to be prohibitively expensive when examined in isolation, but the cost may prove to be acceptable in many cases when examined in combination with other applications.

NATURAL GAS PROCESSING

Acid gases such as CO_2 and H_2S are separated from natural gas streams usually by contacting them with a chemical solvent such as alkanolamine. Acid gases are chemically absorbed by the solvent in an absorber and natural gas with much reduced acid gas content is thus obtained. The chemical solvent containing the absorbed acid gases is regenerated in a regenerator with the aid of stripping vapor to be used again in the absorption process. This regeneration process produces an acid gas rich stream that requires either further treatment or disposal.

Figure 1 shows a block diagram of the acid gas removal and disposal process. As shown, acid gases are normally sent to a sulphur recovery unit consisting of a Claus Plant where H_2S is converted to elemental sulphur. Depending on the sulphur recovery levels required, an additional tail gas clean-up unit may have to be added and finally the residual H_2S must be incinerated. All these processing steps add to the cost of H_2S separation and disposal. Unless elemental sulphur can be sold at a reasonable price, the gas producer often has to subsidize the sulphur recovery operation from the sales of natural gas and other liquid products extracted from it.

The degree of sulphur recovery in Alberta is dictated by the total amount of sulphur present in the raw natural gas stream. When sulphur content of the gas entering the plant exceeds 2000 tons/day, then gas processors are required to recover 99.9% of the sulphur from the feed gas. On the other hand, if the sulphur content is less, the level of recovery is reduced. Prior to 1988, when the sulphur content of the feed gas was less than 10 tons/day, the sour gas could be flared. Since 1988, this practice is no longer allowed and sulphur recovery is required. For plants with sulphur content of 1 to 5 tons/day, sulphur recovery requirement is 70%. Such low capacity sulphur recovery processes are not cost effective. In such cases, re-injection into an underground formation appear to be more attractive option especially if a depleted oil/gas reservoir is available.

RAW NATURAL GAS

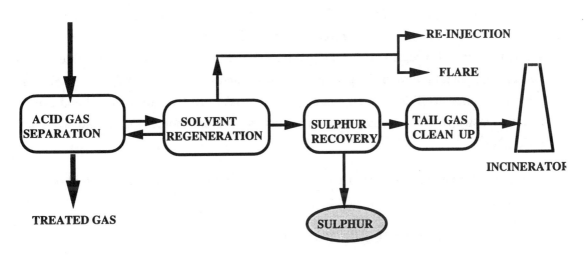

Figure 1: Schematic block diagram of acid gas removal and disposal process.

INJECTION OPTIONS

A number of possibilities exist for the injection of acid gases into geological formations. They include injection into a depleted oil or gas reservoirs, a different acid gas bearing zone within the same reservoir, or an aquifer.

Among all these options, injection into a depleted oil or gas reservoir is the most attractive one from a cost effectiveness point of view. In such a case part of the basic infrastructure including the injection wells and pipelines are already in place from previous production operations.

When a different acid gas bearing zone in and around the same formation is available, re-injection into such zones may also be contemplated.

Aquifer disposal of acid gas requires that the gas be injected at a depth where the water is considered to be non-potable because of its high salinity. The gas can either be injected directly into the aquifer or it can be pre-mixed with water at the surface and the sour water can then be re-injected. Among these two options, the direct injection option is the favoured one.

SAFETY ISSUES

Safety concerns must be satisfied before acid gas injection projects can be contemplated. The primary concern is related to the leakage of injected gases which can be lethal considering the presence of highly toxic H_2S. The injected acid gas must be contained within the formation. When the acid gas is re-injected into a depleted reservoir, the mechanical integrity of the injection wells as well as piping must be examined. If the gas is re-injected into the original formation, then migration of the re-injected gas to the producing wells need to be considered as well. However, petroleum reservoirs either depleted or under production, have proven ability to contain hydrocarbons including the acidic gases and thus are very reliable from a containment point of view, as long as the mechanical and structural integrity of the wells and other production facilities can be maintained. The risks involved are minimal and can be handled easily if precautionary measures such as routine monitoring of the wells are undertaken.

The issue of containment in aquifers is a bit more complicated than that in hydrocarbon reservoirs.

ACID GAS RE-INJECTION PROJECTS

Examples of a number of acid gas re-injection projects in Alberta are given in Tables 1 and 2. Both depleted oil and gas reservoirs as well as aquifers have been used for acid gas re-injection. Typically, gases are injected at formation depths of over 1500 m. The injection pressure is a direct function of the depth of the formation. Although the main reason for disposal is H_2S, the gases also contain significant amounts of CO_2.

Table 1:   Example of Re-Injection into Depleted Petroleum Reservoirs

Formation	Year Started	Injection Depth (m)	Injection Pressure (kPa)	%H_2S	%CO_2
Acheson	1989	1,231	9,300	15	83
Zama	1995	1,478	14,479	38	60

Table 2:   Example of Re-Injection into Aquifers

Formation	Year Started	Injection Depth (m)	Injection Pressure (kPa)	%H2S	%CO2
Provost	1994	1,427	13,500	9	91
Pembina	1994	2,865	28,800	68	30

RE-INJECTION COST

Re-injection cost primarily includes the cost of compression of the acid gas streams from essentially atmospheric pressure to the formation pressure, which typically varies with the depth and type of the formation. There is also the cost of pipelining the gas from the gas plant to the disposal site, which again varies from case to case and depends on the distance through which the gas must be transported.

If the gas is to be injected into a depleted oil/gas reservoir, the cost of the infrastructure including the drilling of injection wells, installing of the pipeline system etc., is negligible. As was shown earlier, injection into both depleted oil and gas reservoirs with existing infrastructure as well as injection into aquifers with no existing infrastructure have been undertaken and found to be cost effective.

Expressed in terms of CO_2, compression cost can hover around $ 2 - $ 3 per ton of CO_2 disposed. The cost of compression also depends on the size of the plant. Larger plants require lower costs due to economy of scale. If new injection wells are to be drilled, and pipelines are to be constructed, the cost may double. Even this cost level is much less than what is required for the separation/disposal of CO_2 from power plant flue gas streams.

RE-INJECTION OF CO_2 ONLY GASES??

From the above discussion, it is clear that acid gases can be re-injected, and the cost of such operations are not prohibitive when disposal is required by law. The question arises whether disposal of CO_2 produced from natural gas plants can be carried out in a cost effective manner, especially if depleted petroleum reservoirs are available with existing facilities. The answer must be in the affirmative. If it can be done for H_2S, it can also be done for CO_2. However, no one is contemplating re-injection of CO_2 only gases in Alberta - because its is not required by law. However, the Sleipner experience certainly proves that where there is a will there is a way.

Energy Convers. Mgmt Vol. 38, Suppl., pp. S211–S216, 1997
Published by Elsevier Science Ltd
Printed in Great Britain
0196-8904/97 $17.00 + 0.00

Pergamon

PII: S0196-8904(96)00271-3

DISPOSAL OF POWER PLANT CO_2 IN DEPLETED OIL AND GAS RESERVOIRS IN TEXAS

Perry D. Bergman
USDOE, Pittsburgh Energy Technology Center,
P.O. Box 10940, Pittsburgh, PA 15236-0940

Edward M. Winter
Burns and Roe Services Corporation,
P.O. Box 18288, Pittsburgh, PA 15236

Zhong-Ying Chen
Science Applications International Corporation,
11251 Roger Bacon Drive, Reston, VA 22090

ABSTRACT

Texas accounts for more than 10 percent of carbon dioxide (CO_2) emissions from power plants in the U.S. and has a large concentration of oil and gas reservoirs which could serve as disposal sites. Two of the largest power plants, accounting for 20 percent of the state's CO_2 emissions, have adequate reservoir volumes within 50-100 miles to allow CO_2 disposal for 30 years or more. However, the majority of these fields are still producing and will not be sufficiently depleted and ready for CO_2 injection for 10 years or more.

Oil reservoirs are economically preferred to gas reservoirs since operators are willing to buy the CO_2 for enhanced oil recovery. For retrofit to existing power plants, estimated costs are \$50-\$65/ton CO_2 for disposal in nearby gas reservoirs and \$40-\$55/ton CO_2 for disposal in oil reservoirs. For advanced technologies in future power plants estimated costs are \$15-\$25/ton CO_2 for disposal in gas reservoirs, and \$5-\$15/ton CO_2 for disposal in oil reservoirs. Avoided costs are approximately 50 % higher. Comparative cost differences between power plants using coal or natural gas are discussed. Technical and regulatory factors which may limit the use of some categories of reservoirs as disposal sites are identified. Published by Elsevier Science Ltd

KEYWORDS

CO_2 disposal; oil gas reservoirs; costs; limitations

INTRODUCTION

Underground disposal of carbon dioxide (CO_2) has been identified as one high priority area for research and development relating to global climate change (Herzog et al., 1993). A study of disposal of CO_2 from utility power plants into abandoned oil and gas reservoirs in the U.S. showed that the logistics for full-scale disposal are not favorable (Winter and Bergman, 1994). Most of the potential reservoir volumes ultimately available are concentrated in the south-central U.S. while power plant sources of CO_2 emissions are predominantly in the upper midwest and southeast U.S., 500-1500 miles away from disposal sites. The high cost of long-distance pipelining of CO_2 would make large-scale disposal from these plants unattractive.

However, some CO_2 might be disposed of in abandoned reservoirs, in places where power plants are near the reservoirs. Texas is a prime candidate, since it accounts for over 10 percent of the national CO_2 emissions from power plants and has the greatest concentration of reservoirs. This study describes preliminary evaluations of technical, economic, and regulatory factors limiting CO_2 disposal in on-shore oil and gas reservoirs in Texas.

RESULTS AND DISCUSSION

Natural gas and oil reservoirs needed to be considered separately for two reasons - the actual disposal processes would be quite different, and the two options fall under different regulatory agencies in Texas and need to meet different requirements. In the former, CO_2 could be permanently stored in the depleted gas reservoirs as waste. In oil reservoirs, CO_2 could be injected to recover additional oil, with CO_2 remaining in the rock structures afterwards.

The economics of disposal in oil reservoirs is more favorable than disposal in gas reservoirs because of the by-product oil credit. The CO_2 has value to oil producers for enhancing oil recovery. However, injection rates, storage efficiency, and overall reservoir capacities are lower for oil reservoirs than for gas reservoirs. Also, regulatory restrictions relating to CO_2 purity are apt to be more stringent for gas reservoirs than oil reservoirs.

One economic limitation for both gas and oil reservoir disposal is distance from power plants to disposal sites and the associated cost of transporting the CO_2 by pipeline. Pipeline costs, based on one published cost estimate are about \$0.02/mile/ton CO_2 (ETSU, 1994) . For a 100 mile pipeline, the charges would be of the order of \$2-\$3/ton CO_2 transported.

Mapping of the power plant sources of CO_2 in Texas showed that one third of the CO_2 emissions comes from four power plants. Only one of the power plants has enough abandoned or nearly depleted reservoir capacity within about 50-100 miles to store CO_2 for a significant period of time. This is the W. A. Parish power plant south of Houston, which accounts for about 10 percent of the power plant CO_2 emissions in Texas (about one percent of the U.S. total). One inactive reservoir, within 100 miles, can be used to dispose of the CO_2 from this plant for almost 30 years. As other nearby reservoirs become depleted, more CO_2 can be sequestered, and the time period will be extended. The other three power plants do not have a large abandoned reservoir volume nearby. The Martin Lake power plant in Rusk County, which also releases about 10 percent of Texas power plant CO_2, is close to many large producing oil and gas reservoirs. When these reservoirs are depleted sometime in the future, they have the capacity to accept the CO_2 from this plant for more than 75 years. Therefore, these two power plants could ultimately be used for disposal of about 20 percent of Texas CO_2 emissions from power plants, about 2 percent of the U.S. total.

Economics

On the basis of economics, oil reservoirs are preferred over gas reservoirs for CO_2 storage, and coal flue gas is preferred over flue gas from natural gas turbines.

CO_2 Capture and Avoided Costs. Processes exist for CO_2 capture using amine scrubbing. For example, there are three coal-fired power plants currently using the Kerr-McGee process, licensed by ABB Lummus. Two of the plants sell chemical-grade CO_2 and the third sells food-grade CO_2. Based on published data from ABB Lummus, the cost of CO_2 captured by retrofitting an amine scrubber to an existing power plant is \$35-\$71/ton CO_2 captured (ABB Lummus Crest) (See Table 1.) The high end of the range is for food-grade liquid CO_2, extracted from gas turbine flue gas, while the low end of the range is for chemical-grade gaseous CO_2 from coal-fired boilers. Two other estimates are in the range \$30-\$40/ton CO_2. The first is for a coal-fired boiler (Tabor, 1994) and the second for a gas-fired power plant (St. Clair and Simister, 1983).

The costs in Table 1 are on the basis of CO_2 captured from power plant flue gas. Since CO_2 capture requires a significant expenditure of energy, with the production of additional CO_2 if fossil fuels are used for power generation, not all of the CO_2 captured is CO_2 avoided, that is, the original amount of CO_2 that would have been emitted into the atmosphere without CO_2 capture. According to an estimate by Fluor Daniel for a pulverized coal power plant (Booras and Smelser, 1991), a 35 percent reduction in electrical power output can be expected when the plant is retrofitted for CO_2 capture and disposal. Therefore, the total CO_2 captured is greater than the avoided

Table 1. Costs of Retrofit CO_2 Capture

	Coal Boiler		Gas Turbine Plant	
	Food Grade (liquid)	Chemical Grade (gaseous)	Food Grade (liquid)	Chemical Grade (gaseous)
Cents/kWh	6.2	3.9	4.1	2.9
$/ton CO_2	$54	$35	$71	$52

CO_2 by a factor of approximately 1.5. A cost of $35/ton of CO_2 captured translates into $53/ton of CO_2 avoided.

Comparison of Costs for Coal-Fired and Natural Gas-Fired Plants The cost of capturing CO_2 from a gas plant is higher than from a coal boiler because of the lower concentration of CO_2 in the flue gas from gas turbines. The concentration of CO_2 in the flue gas from coal fired boilers is typically 15 percent and oxygen 3.5 percent, while the concentration of CO_2 in gas turbine exhaust is typically 3 percent and oxygen 15 percent, because of the higher excess air used. The cost of removing a pound of CO_2 from a gas turbine flue gas will be significantly higher than from coal boilers because a much larger volume of gas must be processed, making the equipment larger and more expensive to operate. On the other hand, less CO_2 is given off per unit of energy in a gas turbine plant than in a coal boiler.

The net result is that, on the basis of a unit of CO_2 removed ($/ton CO_2), it is more expensive to capture CO_2 from a gas turbine power plant than from a coal fired plant. Based on the values in Table 1 and $15/ton for liquefaction and disposal, the cost of CO_2 capture and disposal from a coal plant would be approximately $50/ton and from a gas turbine plant approximately $65/ton CO_2. Avoided costs would be $75 and $100/ton CO_2, respectively. On the basis of a unit of electricity generated (cents/kWh), the lower CO_2 production in the gas turbine plant more than offsets the higher cost of CO_2 capture.

CO_2 for EOR is currently selling for $10-$17/ton (Tabor, 1994). Therefore, disposal in oil reservoirs would cost about $10-$15/ton CO_2 less than disposal in gas reservoirs or $40-$55/ton CO_2 captured ($60-$83/ton CO_2 avoided). (See Table 2.)

Table 2. CO_2 Disposal Costs

$/ton CO_2 captured	Current Retrofit Technology		Future Integrated Technologies
	Coal-Fired Plant	Gas Turbine Plant	
Into Oil Reservoirs	$40	$55	$5-$15
Into Gas Reservoirs	$50	$65	$15-$25

Note: Transportation (pipeline) costs not included.

CO_2 capture in new, integrated plants, based on advanced technologies, will be significantly less expensive, estimated in the range of $15-$25/ton CO_2 captured (ETSU, 1994). This would be the cost for CO_2 disposal in gas reservoirs. The net cost of disposal in oil reservoirs would be $5-$15/ton CO_2 captured.

Gas Reservoir Purchase and Development. Before CO_2 can be injected into an abandoned gas reservoir, the exhausted reservoir would first have to be purchased and prepared for CO_2 disposal. Requirements would be expected to be similar to natural gas storage reservoirs. A recent estimate for the cost of developing a gas storage reservoir with a capacity of 125 Bcf of natural gas was $24,000,000 (Tejas, 1996). This included purchasing the abandoned reservoir as a storage container, locating all the plugged wells in the reservoir, examining all the well records and replugging any wells that don't meet standards, testing for reservoir integrity either by pressure or tracer, buying right-of-ways for pipelines, and constructing all the infrastructure needed at the site, including compressors, metering and monitoring instrumentation, gas conditioning equipment, etc. The storage field would have 13 wells, which could transfer up to 80,000,000 cf/day of natural gas. Purchase and development of a gas reservoir this size for CO$_2$ disposal would add about $2-$3/ton CO$_2$ to the cost of disposal.

One of the gas reservoirs near the W. A. Parish power plant has had cumulative natural gas production of 1.3 Tcf, 10 times larger than the storage reservoir discussed above. Development cost would be considerably greater. The CO$_2$ output from the W. A. Parish power plant would be 1 Bcf/day, 14 times greater than above. Therefore, possibly 80 to 100 injection wells might be required.

Technical Factors

A number of site-specific factors will limit the number of reservoirs suitable for storage of CO$_2$ (Tabor et al., 1996). Reservoir size, depth, chemical reactivity of the minerals, permeability, reservoir structure, well spacing, oil density and viscosity are all factors that need to be considered.

Corrosion of Equipment. Corrosion will be a potential problem regardless of whether disposal is into oil or into gas reservoirs. Carbon dioxide containing moisture can be very corrosive to mild steel equipment, as experienced in many EOR projects. Corrosion might be a problem in two areas: 1) at the injection wells, and 2) at the point where old, abandoned wells enter the underground formation into which the CO$_2$ is being injected.

There are numerous examples of the use of CO$_2$ for EOR, where corrosion is under control. About 50 oil fields in the U.S. are currently being injected with CO$_2$ for EOR (Tabor, 1994). These use naturally occurring CO$_2$, which is relatively dry. In a few cases in the past, CO$_2$ has been captured from flue gases and used for EOR. For example, Lubbock Power and Light supplied CO$_2$ from one of its natural gas boilers for enhanced oil recovery at a nearby reservoir without corrosion problems (St. Clair and Simister, 1983). Also, some natural gas reservoirs contain CO$_2$, and corrosion of the producing well tubing and the pipeline is routinely controlled by gas drying, use of inhibitors, CO$_2$ scrubbing, or by a combination of these operations.

Therefore, injection well corrosion is considered an economic problem rather than a technical barrier. Likewise, since many CO$_2$ EOR projects are being conducted and no CO$_2$ release has been reported, corrosive leakage up old well bores is not considered a technical barrier to CO$_2$ disposal. It may, however, be an economic factor, increasing the cost of preparing a reservoir for CO$_2$ storage.

Chemical Reactivity of the Rock Structure. The natural CO$_2$ deposits in New Mexico, Colorado, and Wyoming are found in sandstone and carbonate formations. Even though slow chemical reactions do occur in these formations, they would not be expected to have any serious effect (Pearce et al., 1996). Other types of formations will have to be tested for adverse reactions.

Depth. Depth will be a deciding factor in a number of cases. For CO$_2$ disposal, the pressure of the reservoir determines the density of the CO$_2$ and, therefore, the amount of CO$_2$ that can be stored. Depth is also a factor in enhanced oil recovery. The efficiency of EOR depends on pressure and, thus, on reservoir depth. Carbon dioxide is miscible with reservoir oil at high pressure, but not at low pressures. Greater miscibility has cost benefits associated with increased oil recovery. The threshold pressure, above which miscibility occurs, is called the Minimum Miscibility Pressure (MMP), which corresponds roughly to a depth of 2500 feet. Miscible CO$_2$ displacement results in approximately 22 percent higher recovery, while immiscible displacement achieves approximately 10 percent higher recovery (Tabor, 1994). Therefore, there is a greater payback for miscible displacement, and deeper reservoirs are preferred, where pressures are above the MMP.

Oil Properties. The MMP also depends on the composition of the oil. Higher density oils, higher viscosity oils, and oils with more multiple aromatic ring structures have a higher MMP (Tabor, 1994). Miscible displacement can be achieved only at greater pressures (greater depths) than with lighter oils. Historically, CO$_2$ EOR has only been used on oils with densities below about 0.92 g/cc (API gravities greater than about 22) and viscosities lower than about 10 cp, because of greater miscibility and higher recovery efficiencies. Therefore, reservoirs containing light crude oils are preferred. Increased oil production also provides more capacity, for sequestering CO$_2$.

Ultimate Capacity in Oil vs Gas Reservoirs. The ultimate demand for CO$_2$ in enhanced recovery in oil reservoirs in the U.S. has been estimated at approximately 50 Tcf (Tabor, 1994). Since total recycle of CO$_2$ was assumed in this estimate, this amount is presumably the CO$_2$ that would remain in the reservoir after EOR. Much of this 50 Tcf reservoir volume is in western Texas, eastern New Mexico and the Rocky Mountain area. By way of comparison, the CO$_2$ disposal capacity of gas reservoirs in Texas, based on cumulative gas production, is about 250 Tcf CO$_2$. If the ultimate gas production is twice the current cumulative production, gas reservoirs in Texas have an ultimate CO$_2$ disposal capacity more than 10 times the capacity of oil reservoirs. If all Texas power plant CO$_2$ emissions were captured and used for EOR in oil reservoirs, these reservoirs would fill in about 10 years. Gas reservoirs would be filled in about 100 years.

Timing. Not every reservoir will be immediately ready for injection of CO$_2$. One strategy for oil reservoirs is to delay using CO$_2$ for enhanced recovery until after secondary recovery with water injection has been used. There is an optimum time to begin water flooding and then, later on, an optimum time to start tertiary recovery with CO$_2$. The availability of reservoirs for CO$_2$ disposal will be primarily determined by the economics of oil and gas production. While timing is not an absolute barrier to disposal, the uncertainties associated with not knowing when reservoirs will be ready for CO$_2$ injection will greatly influence any overall CO$_2$ disposal strategy.

Regulatory Concerns

Two regulatory agencies in Texas have overlapping responsibility for underground projects. The Texas Railroad Commission (RRC), Oil and Gas Division, has responsibility for oil and gas production and all matters relating to this, including injection of known or new and proposed substances for the purpose of enhancing recovery of oil or gas. The Texas Natural Resource Conservation Commission (TNRCC) has primary responsibility for the safety and protection of drinking water supplies. Thus, the RRC has the role of overseeing and issuing permits for oil and gas production, which is important to Texas' economic and financial well being. The RRC's position seems to be almost to promote - certainly to try to avoid hindering - production. On the other hand, the role of TNRCC is to protect the environment and the natural resources of Texas. TNRCC's rulings are seen as more stringent and restrictive. It appears that disposal of CO$_2$ from power plant flue gases in a depleted gas reservoir would fall under the jurisdiction of TNRCC, while injection of the same CO$_2$ into an oil reservoir for the purpose of enhancing oil recovery would fall under RRC.

Contaminants in CO$_2$. An important issue is what classification the waste CO$_2$ would fall into for gas reservoir disposal. If contaminants in the power plant flue gas, such as SO$_2$, NOx, trace heavy metals (from coal ash), and toxic hydrocarbons, end up in the concentrated CO$_2$ stream in significant quantities, then this stream would probably have to be classified as a hazardous, Class 1 (EPA classification), material. This would make it much more difficult, although not impossible, to obtain underground disposal permits. Permitting procedures for hazardous wastes take much longer and require much more justification. If the injectant has no hazardous compounds, it can be classified as Class 5, and permitting is much more lenient. If CO$_2$ is used for enhancing oil recovery, then it can be classified as a Class 2 material and permitting will be handled much more easily, by RRC.

Technology does exist to purify the CO$_2$ sufficiently to avoid the Class 1 hazardous waste classification. For example, Applied Energy Systems currently operates a coal-fired power plant in Poteau, OK, which incorporates an amine scrubber capturing 200 tons/day of CO$_2$ from flue gas and producing food-grade liquid CO$_2$. Therefore, underground disposal is ultimately a matter of economics and the urgency of CO$_2$ reduction.

CONCLUSIONS

No technological barriers were identified which would make CO_2 disposal in oil or gas reservoirs inherently infeasible. There are factors, however, which may limit disposal to certain types of reservoirs based on geological characteristics, location, and oil properties.

Gas reservoirs in Texas have more than 10 times the capacity for CO_2 disposal than oil reservoirs.

Capturing and disposing of CO_2 from power plant flue gas will be expensive. The cost for a retrofit installation has been estimated as \$50/ton CO_2 captured for coal-fired and \$65/ton for gas-fired power plants. For disposal in depleted gas reservoirs this cost must be increased by the cost of developing the disposal site.

If the CO_2 is sold for EOR, the price paid by the oil producers will partially offset the cost of disposal. The net difference in cost between disposal in gas reservoirs and oil reservoirs will be roughly the sales price of naturally occurring CO_2 for EOR, or currently \$10-\$17/ton CO_2. The net cost of disposal in oil reservoirs for EOR is in the range \$40-\$55/ton CO_2 captured. The CO_2 sales price depends on the market price of oil. If oil prices rise, producers can afford to pay more for CO_2, and the cost of disposal in oil reservoirs would decrease.

Future power plants based on advanced technology would have higher efficiencies and lower CO_2 emissions per unit of energy generated. They could also be designed for more efficient CO_2 capture. Estimated disposal costs are \$15-\$25/ton CO_2 captured for disposal in gas reservoirs and net \$5-\$10/ton CO_2 captured for oil reservoirs.

REFERENCES

ABB Lummus Crest, Inc. CO_2 Recovery from Flue Gas. Technical Brochure.

Booras, G.S., S. C. Smelser (1991). Engineering and Economic Evaluation of CO_2 Removal from Fossil-Fuel-Fired Power Plants. *Energy,* 16, No. 11/12, pp 1295-1305.

ETSU (1994). Full Fuel Cycle Study on Power Generation Schemes Incorporating the Capture and Disposal of Carbon Dioxide, Vol 4. Report prepared for IEA Greenhouse Gas R&D Programme.

Herzog, H., E. Drake, J. Tester (1993). A Research Needs Assessment for the Capture, Utilization and Disposal of Carbon Dioxide From Fossil Fuel-Fired Power Plants. Report by Energy Laboratory of Massachusetts Institute of Technology for USDOE.

Pearce, J. M., S. Holloway, H. Wacker, M. K. Nelis, C. Rochelle, and K. Bateman (1996). Natural Occurrences as Analogues for the Geological Disposal of Carbon Dioxide. *Energy Convers, Mgmt,* 37, Nos 6-8, pp 1123-1128.

St.Clair, J. H., W. F. Simister (1983). Process to Recover CO_2 from Flue Gas Gets First Large-scale Tryout in Texas. *Oil & Gas Journal,* Technology, Feb 14, 1983, pp 109-113.

Tejas Gas Corporation (1996). Personal communication.

Tabor, J. J. (1994). Enhanced Oil Recovery. Report prepared for IEA Greenhouse Gas R&D Programme, Publ OE17.

Tabor, J.J., F. D. Martin, R. S. Seright (1996). EOR Screening Revisited. SPE/DOE Tenth Symposium on Improved Oil Recovery, Tulsa, OK.

Winter, E. M., P. D. Bergman (1994). Carbon Dioxide Disposal in Abandoned Oil and Gas Reservoirs. Air & Waste Management Association 87th Annual Meeting & Exhibition, Cincinnati, OH.

Energy Convers. Mgmt Vol. 38, Suppl., pp. S217–S222, 1997
© 1997 Elsevier Science Ltd. All rights reserved
Printed in Great Britain
0196-8904/97 $17.00 + 0.00

Pergamon

PII: S0196-8904(96)00272-5

DEEP COALBED METHANE IN ALBERTA, CANADA:
A FUEL RESOURCE WITH THE POTENTIAL OF ZERO GREENHOUSE GAS EMISSIONS

W.D. Gunter[1], T. Gentzis[1], B.A. Rottenfusser[2] and R.J.H. Richardson[3]

[1]Alberta Research Council, PO Box 8330, Edmonton, Alberta, T6H 5X2, Canada
[2]Redfoot Enterprises, PO Box 538, Devon, Alberta, T0C 1EO, Canada
[3]Alberta Geological Survey, 9945 108St., Edmonton, Alberta, T5K 2G6, Canada

ABSTRACT

A huge incentive to develop Alberta's deep coalbed methane (CBM) resources is found in evolving enhanced gas recovery (EGR) technologies using CO_2 injection, somewhat similar to enhanced oil recovery (EOR). However, unlike CO_2-EOR where CO_2-breakthru eventually occurs, the injected CO_2 is sequestered in the reservoir by sorption to the coal surface. The mechanism is that the CO_2 displaces the sorbed CH_4 from the coal surface, two molecules of CO_2 being trapped for every molecule of CH_4 released. Already this CO_2-EGR process, although in an embryonic stage, has shown increased yields of produced CH_4 over conventional CBM recovery. Future successful tuning of the process may allow the design of efficient null-greenhouse-gas-emission power plants which are fuelled by CBM from deep coal beds. In this closed CO_2-cycle process, the waste CO_2 produced by CH_4-fuelled power plants is injected into CBM reservoirs to produce more CH_4. Other scenarios are possible using offsets. A simple mass balance argument, based on a 2 to 1 coal-sorption-selectivity for CO_2 over CH_4, supports the feasibility of building and operating fossil-fuelled **green** power plants. © 1997 Elsevier Science Ltd

KEYWORDS

Coalbed methane production; carbon dioxide disposal; greenhouse gases; power plants; Alberta basin.

CARBON DIOXIDE SINKS

An important source of anthopogenic carbon dioxide (CO_2) emissions results from fossil fuel combustion for electric power generation and from other industrial activities. Alberta produced 127 million tons of CO_2 in 1990, 30 % of which originated from coal-fired thermal power stations. Still, CO_2 emissions are steadily rising. Canada is committed to reducing greenhouse gas emissions to 1990 levels by the year 2000. Therefore, large environmentally acceptable "sinks" for long-term storage of CO_2 have to be identified that are economically acceptable.

In Canada, the bulk of the fossil fuels are found in the Alberta Basin, a huge sedimentary basin occupying a volume of approximately 2 million km^3 (Mossop and Shetsen, 1994). Hitchon (1996) showed that the deep aquifers in this basin are capable of trapping huge amounts of waste CO_2. The capacity of these traps are so large that no other sinks would be needed to safely dispose of all of Alberta's CO_2 emissions over the next century. On the other hand, this method of CO_2 disposal is expensive, of the order of $50/tonne if flue gas separation costs are included (Gunter *et al.*, 1995). A synergy is needed between the production of the fossil fuels in the Alberta Basin and the disposal of their gaseous emissions created during the energy conversion process. There are two, which have been or are being investigated. Enhanced oil recovery (injecting CO_2 into a reservoir to produce more oil) is done on a regular basis and is considered mature technology (Bailey and McDonald, 1993). In contrast, production of CBM by enhanced recovery techniques utilizing injection of CO_2, the subject of this paper, is in an embryonic stage.

COALBED METHANE PRODUCTION

In Canada, the bulk of the coalbed methane resources lies in the Alberta Basin. There is coalbed methane production from several coal basins in the US, but around 96% comes from just two -- the San Juan Basin, located in southwestern Colorado and northwestern New Mexico, and the Black Warrior basin in Alabama (Dawson, 1995). Estimates of the CBM resource in North America vary greatly, but by all reasonable estimates, a huge amount exists. In the United States, estimates of CBM resources range from 275 to 649 trillion cubic feet (TCF); in Canada there is considerably more uncertainty with estimates ranging over an order of magnitude, from 200 to 3000 TCF (Nikols and Rottenfusser, 1991). Total CBM production in the U.S. exceeded 700 billion cubic feet (BCF) in 1993. There has been very little production in Canada.

Canada's conventional natural gas reserves are declining, estimated at 53 TCF (with an annual consumption of 2 TCF) in 1995. Consequently, new sources of gas are needed for the future. Approximately 5% of the CBM resources are recoverable under present-day economics utilizing primary production technology. North America's CBM resources, particularly Canada's, lie largely undeveloped because the necessary technology to do so is currently not available. Enhanced gas recovery methods are being developed which may allow more of the CBM resource to be targeted for exploitation.

General Properties of Coalbed Methane

Coal forms by the compaction of plant material that accumulated in swamps, generally under tropical or semi-tropical conditions, similar to the Florida Everglades today. The plant material transforms to coal by the action of both pressure and temperature as the coal swamp is buried under younger sediments. Coals are classified according to the proportion of carbon they contain, which increases as the degree of maturation increases. Lignite has 70% carbon, bituminous coals vary between 80 and 90% carbon, and anthracite has more than 90% carbon. Gases are generated during the conversion of the plant material to coal, and they are either absorbed onto the coal or are dispersed into the pore spaces around the coal seam. The amount of gas formed depends on the rank of coal. Methane is the dominant gas (about 95%), with the remaining gases including ethane, carbon dioxide, nitrogen, helium and hydrogen; hydrogen sulfide is found only in trace amounts, even in high-sulfur coals. In addition to the gas formed, large amounts of water are released during the maturation process. This process of maturation results in coalbeds that are generally naturally-fractured, low-pressure, low-permeability, water-saturated gas reservoirs.

Most coals contain two forms of porosity: microporosity and macroporosity. The average microporosity is contained in the coal matrix, is generally less than 1%, and is rank-dependent. The majority of coalbed methane is present in the sorbed state, attached to the coal surface in these micropores. The primary controlling factors for the amount of gas stored in coal are the confining pressure and the surface area contained within the coal micropore system. The surface area of the coal on which the gas is adsorbed is huge, ranging from 20 to 200 sq. meters/gram. Consequently, saturated coalbed methane reservoirs can easily have up to 5 times the amount of gas contained in a conventional gas reservoir of comparable size, temperature and pressure. The macroporosity of coal is within fractures called cleats, which are commonly found at right angles to bedding surfaces. Cleats are of two types: the most prominent set is 'face' cleats and the secondary set is 'butt' cleats (Kendall and Briggs, 1933). Face cleats are much more continuous than butt cleats and control the flow of gas to the wellbore. On the other hand, the butt cleats are associated with the diffusion of gas from the coal matrix to the face cleats. Cleating in coal is determined by at least two diagenetic factors: the progressive compaction of organic matter due to coalification, and tectonic forces acting on the coal (McCulloch *et al.*, 1974). Permeability in coal beds is controlled by the cleat system in the coalbeds (Jones *et al.* 1988) and is generally low (< 10 md).

Methane, although present in coal beds in a sorbed state in the micropore system, can also be stored either as free gas or dissolved in water in the cleat space. In order for a free gas phase to exist, the pressure exerted by the gas phase has to be equal to that of the water. Normally, the water pressure is greater than the gas pressure in the virgin methane reservoir. As the water is removed from the cleat system, the pressure in the coal is reduced until the water pressure equals the gas pressure (termed desorption pressure); then methane gas is desorbed from the coal matrix (i.e. the micropore system) to the adjacent cleats (the macropore system). Usually, large quantities of water have to be extracted from the coalbed before the desorption pressure is reached - the point at which a two phase fluid exists, gas and water. Degasification of coal is a two-step process: first, desorption of the gas from the coal matrix followed by flow of the gas and water through the cleats to the production well (Seidle and Arri, 1990).

If the coalbeds have developed a well-cleated system, the methane gas present in the coal matrix will desorb into the cleats as soon as the pressure in the cleats decreases sufficiently. If the rate of gas desorption from the coal matrix and diffusion through the butt cleats to the face cleats is much higher than the rate of fluid flow in the face cleats, then gas production is flow-limited. On the other hand, if the rate of gas desorption from the matrix and diffusion in the butt cleats is very slow compared to the rate of fluid transport in the face cleats, then gas production is diffusion-limited. Which process is controlling production depends on the properties of the coalbed.

Coal Gas Sorption

Experimentally-measured adsorption isotherms for binary and ternary mixtures of CO_2, CH_4 and N_2 show that the equilibrium gas and adsorbate phase compositions differ considerably, and that the total amount of gas mixture adsorbed is strongly dependent on composition and system pressure (Stevenson *et al*, 1991). CO_2 is the most strongly adsorbed gas , then CH_4, with N_2 being the least adsorbed. The approximate adsorption ratios are 4:2:1; that is 4 molecules of CO_2 are adsorbed for 2 molecules of CH_4 and for every 1 molecule of N_2, when comparing pure gases at the same temperature and pressure.

Enhanced Gas Recovery (EGR) Techniques for Coalbed Methane

When produced, coalbed methane is commonly recovered by means of reservoir pressure depletion. While this method is simple and effective, it is not efficient. Reduction in reservoir pressure deprives the fluids of the energy necessary to flow to the wellbore. Furthermore, when the resevoir is water saturated, there are disadvantages associated with the long delay in methane gas production and the large quantities of water produced. This may be overcome by injection of a gas into the coal seams. Due to the presence of a gas phase, methane gas recovery can be enhanced either by reducing the partial pressure of CH_4 through the introduction of a lower-adsorbing gas such as N_2 (Puri and Yee, 1990) or displacement by the introduction of a higher-adsorbing gas such as CO_2 (Arri *et al.* 1992). Injecting gas at the start of the recovery process not only allows reservoir pressure to be maintained but also has the potential to produce methane gas quickly.

Such enhanced methods of coalbed methane recovery consist of flooding the coal bed with a relatively inert gas (e.g. nitrogen or argon, Yee *et al.*, 1995), with a strongly adsorbing gas (e.g. carbon dioxide), oxygen-depleted stream of air (pressure swing adsorption method described in Puri and Pendergraft, 1995, and Yee and Puri, 1995) or gas mixtures. Furthermore, a two-step injection process involving a strongly-adsorbable fluid (e.g. carbon dioxide) followed by a weakly-adsorbable gas (e.g. nitrogen) can be used to stimulate the release of residual methane from the coal bed. The strong adsorber, CO_2, displaces and desorbs the CH_4, while the inert gas, N_2, forces the excess CO_2 to move through the coal bed. As the fluid moves through the coal bed it desorbs more CH_4 from the coal matrix and sweeps it to the production well. Alternatively, complete and rapid demethanation has been reported in the laboratory when injection of CO_2 was done in a cyclical fashion (three cycles over 90 days) (Fulton *et al.*, 1980) at pressures up to 800 psig (Reznik *et al.*, 1984).

PRODUCTION OF CH_4 FROM AND DISPOSAL OF CO_2 INTO COAL BEDS

The EGR process for coals has already been tested in the field. In the late 1980s, Amoco Corporation through a series of patents based on experiments and numerical simulations, demonstrated the potential of EGR for coalbed methane. In 1993, Amoco was the first to successfully demonstrate the process in the field for a small N_2 pilot in the San Juan Basin, Colorado. This was followed by a CO_2 pilot, started in December, 1993. In 1995, Meridian, started their first CO_2 pilot, also located in the San Juan Basin. The performance results for these two CO_2 pilots have not been released.

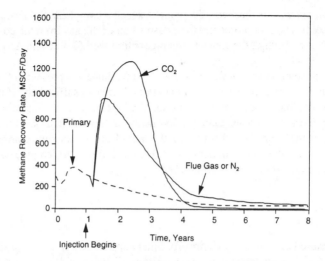

Fig. 1. Comparison of methane recovery rates for a fully methane-saturated coalbed for primary pressure depletion, for pure CO_2 injection, for pure N_2 injection, and for flue gas injection, where injection and production wells are arranged in a five spot pattern. An injection pressure of 2000 psia, a reservoir pressure of 1500 psia, a permeability of 10 md, a porosity of 0.5%, a coalbed thickness of 10 feet and a drainage area of 46 acres were assumed. Modified after Chaback et al. (1995).

Ignoring water, Chaback *et al.* (1995) have shown by modelling, that injected CO_2 is stripped chromatographically as it flows through a homogenous coalbed, displacing the CH_4 and pushing it to the producing well (Figure 1). Very little of the CO_2 shows up at the production well until most of the CH_4 has been produced. In contrast, when N_2 injection is used, the N_2 appears at the production well almost immediately. With a simulated flue gas (i.e. 15% CO_2 and 85% N_2), N_2 appears at the production well immediately and the methane production profile is very similar to the pure N_2 injection case. However, CH_4 production is strongly retarded compared to the pure CO_2 injection case. In all three cases, cumulative methane production is enhanced by more than a factor of two, and the bulk of the methane is produced much earlier compared to primary pressure depletion methods (Figure 1). When water is present, production of water occurs along with the gas. Water production is also enhanced compared to the primary case, but this is compensated for by the much higher production rates for CH_4 and the better sweep of the reservoir. EGR techniques utilizing CO_2 and N_2 injection should be able to compete economically with primary production of CBM, and at the same time, trap the CO_2 in the coalbeds.

CAPACITY OF CO_2 DEEP COALBED SINKS IN ALBERTA

We believe that the large amounts of CO_2 sequestered in coalbeds would result in a significant reduction in greenhouse gas emissions and should have a very positive environmental impact. Even using the lower estimates of 200 TCF for the coalbed methane resources in Alberta, this number equates to a storage capacity for CO_2 of 20 gigatonnes in the Alberta Basin; equivalent to disposing of all the CO_2 emissions of Alberta coal-fired power plants for more than 500 years. This capacity of Alberta's coalbeds is similar to the capacity estimated for the Alberta Basin aquifers by Bachu *et al.* (1994).

SYNERGY WITH POWER PLANTS

The possibly synergy on a large scale between the production of CH_4 and capture of CO_2 emissions during production of electricity should also be considered. The idea is for coalbed methane, produced following injection of waste CO_2 or flue gas into deep coal seams, to fuel a typical 500 MW coal-fired power plant in central Alberta (e.g., Edmonton Power's Genesee plant) or a series of smaller capacity natural gas-fired power plants, such as Edmonton Power's Rosedale and Clover Bar plants, both located in urban areas. A study is needed to address whether it is possible to operate a **"closed loop"** system by which zero CO_2 is released into the atmosphere in the power cycle. In other words, the aim is to reduce global warming while generating electrical power from **green** power plants. This closed loop process would allow utilities to access an untapped source of natural gas in Alberta at a time when there is a decline in the conventional natural gas reserves in Canada.

Air is normally used as the oxidant for the fuel in a power plant, the oxygen being converted to CO_2 during combustion. Consequently, the flue emissions of power plants are low in CO_2 (10 to 14%) with the majority of the gas being N_2. Only a portion of this CO_2-poor flue gas could be injected into coalbeds to recover methane, and a zero CO_2 emissions scenario could not be maintained without first separating the CO_2 from the N_2 followed by injection of the CO_2-rich stream. Alternatively, it would not be necessary to purify the flue gas if the Argonne O_2-fired method is used. In this case, air is separated cryogenically into O_2 and N_2 before combustion. The feed (coal or natural gas) is burnt in oxygen diluted with CO_2 rather than in the presence of air. The resulting combustion gas "product" contains more than 95% CO_2 with the remainder being SO_x and NO_x compounds. It is not anticipated that the latter two gases would make their way through the coalbeds to the production well. If this scheme is used, make-up CO_2 may be required for the injection side of the cycle because, ideally, two molecules of CO_2 would adsorb to the coal, releasing one molecule of CH_4.

If the production of methane is decoupled from disposal of CO_2, barren unsaturated coalbeds can be used for scrubbing of flue gas. In this case, the coalbed will strip out the CO_2 and deliver a N_2 stream to the production well containing less than 0.01% CO_2 as long as the coalbed does not become gas-saturated (Chaback *et al.*, 1995). However, the proximity of deep methane-saturated coalbeds to coal-or gas-fired power plants, coupled with a sound pipeline infrastructure in Alberta should make the former options more attractive.

Other scenarios may be developed using offsets: for example, coal may be used as the fuel for the power plant; the CO_2 may originate from a gas plant; and the CH_4 may be sold for profit. Whatever makes economic sense fits as long as CO_2 emission offsets, based on using CH_4 as fuel are maintained!

CONCLUSIONS

Some technological problems remain to be solved. From the production side, the EGR methods pioneered by Amoco for coalbed methane exploitation require further refinement. Specifically the problem of low permeability resulting in poor injectivity and communication in the coalbeds has to be overcome. Delaying or minimizing water production is also important. CBM production must be designed to provide a steady flow of CH_4 to the power plant. From the power plant side, the efficiency of oxygen-fired power plants has to be increased and the technology improved. These problems will be solved in the future.

We believe that development of CO_2-EGR technology to exploit Alberta's CBM resources will lead to a **synergy between an increased supply of fossil fuel and decreased global warming**. Recovery of CBM will lead to increased total gas reserves in Alberta. Utilization of this CBM in fuelling power plants and disposal of the waste CO_2 in deep coal reservoirs which act as a huge geological sinks leads to reduction in emitted CO_2, the major culprit of global warming.

ACKNOWLEDGEMENTS

Bill Kaiser of the Texas Bureau of Economic Geology thoroughly reviewed the paper. Don Macdonald of Alberta Energy provided encouragement for the past 5 years, through all our CO_2 mitigation work. In 1991, Don discussed with us the idea of using Alberta's coal beds to sequester CO_2. In 1996, a $120,000 'Proof of Concept' study was started, supported by government and industry, to firm up the conclusions of this paper.

REFERENCES

Arri, L.E., Yee, D., Morgan, W.D., and Jeansonne, M.W. (1992) Modeling coalbed methane production with binary gas sorption. *Society of Petroleum Engineers Paper* No. *24363,* presented at the SPE Rocky Mountain Regional Meeting, Casper, Wyoming, U.S.A., May 18-21, 1992.

Bachu, S., Gunter, W.D., and Perkins, E.P. (1994) Aquifer disposal of CO_2: Hydrodynamic and mineral trapping. *Energy Convers. Mgmt. 35,* p. 269-279.

Bailey, R.T. and McDonald, M.M. (1993) Carbon dioxide capture and use for EOR in Western Canada 1. General overview. *Energy Convers. Mgmt. 34,* p. 1145-1150.

Chaback, J.J., Yee, D., Volz, R.F., Seidle, J.P., and Puri, R. (1995) Method for treating a mixture of gaseous fluids within a solid carbonaceous subterranean formation. *Amoco Corporation U.S. Patent No. 5,439,054*

Dawson, F.M. (1995) Coalbed methane: A comparison between Canada and the United States. *Geological Survey of Canada Bulletin 489*, 60 pp.

Fulton, P.F., Parente, C.A., Rogers, B.A., Shah, N., and Reznik, A.A. (1980) A laboratory investigation of enhanced recovery of methane from coal by carbon dioxide injection. *Society of Petroleum Engineers Journal*, p. 65-72.

Gunter, W.D., Bachu, S., Law, D., Marwaha, V., Drysdale, D.L., Macdonald, D.E., and McCann, T.J. (1995) Technical and economic feasibilty of CO_2 disposal in aquifers within the Alberta Sedimentary Basin. Canada. *Energy Convers. Mgmt. 37*, p. 1135-1142.

Hitchon, B. (editor) (1996) *Aquifer Disposal of Carbon Dioxide: hydrodynamic and mineral trapping - proof of concept.* Geoscience Publishing Ltd. (Box 79088, 1020 Sherwood Drive, Sherwood Park, Alberta T8A 5S3, Canada) 165 pp.

Jones, A.H., Bell, G.J., and Schraufnagel, R.A. (1988) A review of the physical and mechanical properties of coal with implications for coal-bed methane well completion and production. In: Fassett, J.E., ed., *Geology and coal-bed methane resources of the northern San Juan Basin, Colorado and New Mexico.* Rocky Mountain Association of Geologists Guide Book, p. 169-181.

Kendall, P.F. and Briggs, H. (1933) The formation of rock joints and the cleat of coal. *Proc. Royal Society of Edinburg 53,* p. 164-187.

McCulloch, C.M., Deul, M., and Jerran, P.W. (1974) Cleat in bituminous coalbeds. *U.S. Bureau of Mines Report of Investigations 7910*, 25 pp.

Mossop, G. and Shetsen, I. (compilers)(1994) *Geological Atlas of the Western Canada Sedimentary Basin.* Cdn. Soc. Petrol Geol. and Alberta Research Council, Calgary, Alberta, 510pp.

Nikols, D.J. and Rottenfusser, B.A. (1991) Coalbed methane - A Canadian resource for the 1990's. Coalbed Methane - 1991, *Rocky Mountain Association of Geologists*, p. 249-253

Puri, R., and Yee, D. (1990) Enhanced coalbed methane recovery. *Society of Petroleum Engineers Paper* No. *20732*, presented at the 65th Annual Technical Conference and Exhibition, New Orleans, U.S.A., September 23-26, 1990.

Puri, R., and Pendergraft, P.T. (1995) Method for the recovery of coalbed methane. *Amoco Corporation, Patent PCT/US94/11672.*

Reznik, A.A., Singh, P.K., and Foley, W.L. (1984) An analysis of the effect of CO_2 injection on the recovery of in-situ methane from bituminous coal: An experimental simulation. *Society of Petroleum Engineers Journal*, October 1984, p. 521-528.

Seidle, J.P., and Arri, L.E.(1990) Use of conventional reservoir models for coalbed methane simulation. *Canadian Institute of Mining/Society of Petroleum Engineers Paper 90-118*, presented at the CIM/SPE International Technical Meeting, Calgary, Alberta, Canada, June 10-13, 1990.

Stevenson, M.D., W.V. Pinczewski, Somers, M.L., and Bagio, S.E. (1991) Adsorption/desorption of multicomponent gas mixtures at in-seam conditions. *Society of Petroleum Engineers, 23026.* Proceedings of the SPE Asia-Pacific conference in Perth, Western Australia, p. 741-756.

Yee, D., and Puri, R.(1995) Method for the recovery of coalbed methane. *Amoco Corporation Patent PCT/US94/11819.*

Yee, D., Seidle, J.P., and Puri, R. (1995) Method for the recovery of coalbed methane with reduced inert gas fraction in produced gas. *Amoco Corporation Patent PCT/US94/11673.*

 Pergamon

Energy Convers. Mgmt Vol. 38, Suppl., pp. S223–S228, 1997
© 1997 Elsevier Science Ltd. All rights reserved
Printed in Great Britain
0196-8904/97 $17.00 + 0.00

PII: S0196-8904(96)00273-7

SIMULTANEOUS PRODUCTION OF ELECTRICITY, STEAM, AND CO₂ FROM SMALL GAS-FIRED COGENERATION PLANTS FOR ENHANCED OIL RECOVERY

David deMontigny[#], Weerapong Kritpiphat[#], Don Gelowitz[#], and Paitoon Tontiwachwuthikul[#*]

[#] Process Systems Laboratory, Industrial Systems Engineering Division
Faculty of Engineering / Energy Research Unit, The University of Regina
Regina, Saskatchewan, S4S 0A2, CANADA; Tel. (306) 585-4726 Fax. (306) 585-4855

ABSTRACT

In recent years, global warming has been blamed on the so called "greenhouse effect" and has caught the attention of scientists and politicians throughout the world. There is an increasing concern surrounding the emission levels of greenhouse gases, particularly carbon dioxide (CO_2). This paper is an extension of earlier work[1] to show how cogeneration concepts can be used to reduce production costs by simultaneously producing electricity, CO_2 and steam for enhanced oil recovery (EOR) applications.

© 1997 Elsevier Science Ltd

KEYWORDS

Carbon Dioxide Recovery; Cogeneration; Enhanced Oil Recovery; Greenhouse Effect

INTRODUCTION

Primary production methods for oil are becoming less effective and as a result oil companies are beginning to invest money in enhanced oil recovery (EOR) applications. Traditional flooding agents for oil reservoirs include water, steam, and light end gases. Recently, carbon dioxide (CO_2) has become a popular flooding agent. A variety of industrial sources produce CO_2 as a byproduct but the cost of capturing and transporting CO_2 for EOR has been quite high. One method for overcoming this economic challenge is the use of cogeneration concepts for on site CO_2 recovery and injection into an oil reservoir.

Cogeneration is the simultaneous production of electrical or mechanical power and usable thermal energy, such as steam or hot air, to be used for heating, drying, or absorption refrigeration.[2] Cogeneration systems are becoming popular because they utilize less fuel than traditional plants and operate more efficiently. This is because heat and electricity are produced simultaneously in the same system, rather than in two separate systems. A typical power generating station, based on condensing steam turbine technology, converts between 30 and 35 percent of the fuel energy in its boilers into electricity[2] with the remainder being rejected to the environment. In cogeneration, this waste heat is utilized as a sequential input into another process and as a result they can convert 60 to 80 percent of the available fuel energy into electricity and useful heat.

STEAM-CO₂ INJECTION

Efforts to improve EOR techniques have led to steam-CO_2 reservoir injection studies. The search for an improved flooding agent is being driven by two factors: (1) the demand for traditional flooding agents such as ethane and propane has increased, and (2) industry continually strives to improve the *sweep efficiency* of flooding agents. Steam-CO_2 injection schemes are based on simple concepts and produce very good results. The technique works best in an oil producing region where there are at least two horizontal wellbores, one of which is deeper than the other. The procedure[3] begins with the injection of liquid CO_2 into the reservoir through the deepest wellbore. This brings the CO_2 into an initial contact with the underground oil. After some time has passed, steam is injected into the reservoir via the same deep wellbore. When the CO_2 is contacted by the steam it heats up and expands into a gaseous state. The pressure from the steam injection causes (i) the CO_2 gas to dissolve into the petroleum and (ii) the steam to condense into water. The resulting mixture of petroleum, CO_2, and water is less viscous and less dense than the petroleum liquid alone and the mixture flows upward with greater ease. A second horizontal wellbore above this action receives the mixture and channels it to the surface.

* Author to whom correspondence should be addressed.

DESIGN CONCEPT AND SPECIFICATIONS

This study originated out of an idea to localize CO_2 production to the actual site where it will be used. On site production is desirable because it eliminates transportation costs. Cogeneration is used to produce steam and CO_2 which can be immediately injected into the reservoir for EOR purposes.

Figure 1 illustrates the general layout of the cogeneration plant. The pumpjack pumps oil from the reservoir to the surface. The oil is treated and prepared for sales. Natural gas is brought on site and burned in a turbine. The turbine turns a generator and electricity is produced for on site consumption and sale to the electric grid. The flue gas from the turbine is treated in the CO_2 recovery plant where the CO_2 is stripped from the flue gas and the solvent monoethanolamine (MEA) is regenerated. The CO_2 is then compressed and injected into the reservoir. Cogeneration produces steam which is used on site in various processes. Excess steam is injected into the reservoir according to the steam-CO_2 injection scheme.

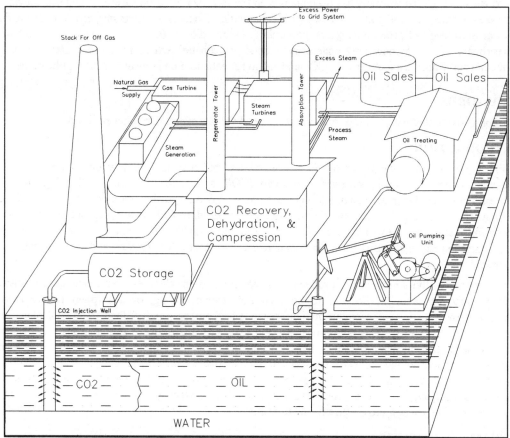

Figure 1: Simplified Cogeneration Plant Layout for EOR Project

Note that the plant operates continuously, always producing steam and CO_2 for injection. The steam-CO_2 procedure requires that CO_2 and steam be injected separately. To deal with this continuous supply of flooding agents it is necessary to have a well established oil field where some wellbores are used for CO_2 injection and others are used for steam injection all at the same time. It would be possible to have five spot or nine spot drilling arrangements in an oil producing field operating under a steam-CO_2 injection schedule. Such systems would certainly improve the overall sweep efficiency of flooding agents and improve the overall EOR project. The elevation restrictions that apply to injection and production wellbores would still apply whereby injection wellbores are deeper.

This study involved the simulation and economic analysis of a cogeneration based gas turbine power plant that produces steam and CO_2 flooding agents for EOR. Three different gas turbine models were compared: the 251B12 @ 49.1 MW; the LM2500 @ 22.8 MW; and the MARST14000 @ 10 MW. Comparisons were based on the highest operating efficiency at lowest capital cost ($/kW). The turbines were referenced from the 1992-93 Gas Turbine World Handbook[4]. All turbines are natural gas fired, assume no duct losses, and are rated at ISO base load conditions. Design specifications for the turbines (fuel consumption, pressure ratio, etc.) were entered into ASPEN PLUS Release 9.1, an advanced computer process simulator developed by ASPEN© Technology Inc. This process simulator was used for all individual combined-cycle cogeneration models. All gas streams and steam lines were assumed to behave under Ideal Gas Law relationships.

A summary of the simulation results is presented in Table 1 for each gas turbine modeled. The assumptions concerning the percentage of CO_2 recovered and the heat duty required for solvent regeneration are typical for the MEA solvent.[6,7] In this study, the surplus steam available is low compared to typical cogeneration systems. This is because some flue gas energy is used in a heat exchanger to heat the off gas from the absorption tower prior to exiting back to the stack. This in effect would eliminate the need to provide additional energy in order to force the gases up the stack.

The power generated and CO_2 produced is presented in Figure 2 for each gas turbine modeled in the study. The size of power plant and recovery system is determined by the amount of steam and CO_2 required for injection. All incremental oil production is governed by the available CO_2 produced for injection into the reservoir. The average industry CO_2 injection rate is approximately 6 mscf/Bbl oil recovered. A range of 1-10 mscf/Bbl was analyzed. Studies out of the University of Wyoming[8] have shown that the optimized ratio of CO_2 over steam is roughly 8 liters of CO_2 (std. conditions) per 10 cc of steam (cold water equivalent). In most cases the CO_2 available produced more incremental oil

Table 1: Design Output Specifications

Power Plant	MarsT14000	LM2500	251B12
Electric Power Generated by Gas Turbine (MW)	10.000	22.800	49.100
Electric Power Generated by HP Steam Turbine (MW)	1.210	2.474	6.034
Electric Power Generated by MP Steam Turbine (MW)	0.878	1.797	4.382
Electric Power Generated by LP Steam Turbine (MW)	0.937	1.920	4.680
Total Electricity (MW)	13.025	28.991	64.197
Maximum CO_2 Produced (tonnes/day)	147.430	269.268	653.493
Natural Gas Consumption (MMSCFD)	2.779	5.072	12.316
Steam Circulation (lb/hr)	28812	57020	139671
Useful Heat from Steam Condensation (Btu/hr)	29,470,000	59,180,000	148,760,000
Electric Efficiency (%)	38.40	46.82	42.69
Steam Efficiency (%)	25.45	28.00	28.99
Power/Steam Ratio	1.51	1.67	1.47
CO_2 Recovery Plant			
Percentage of CO_2 Removal (%)	90.0	90.0	90.0
Heat Required for Regenerating Solvent (Btu/lbmole CO_2)	70,000	70,000	70,000
Heat Required for Regenerating Solvent (Btu/hr)	19,350,198	35,341,463	85,770,909
Electricity Consumption (MW)	0.845	1.543	3.745
Product			
Net ISO-Rated Electricity (MW)	12.180	27.448	60.451
CO_2 Recovery Rate (MMSCFD)	2.514	4.592	11.145
(tonnes/day)	132.687	242.341	588.143
Excess Steam (MMBtu/day)	191.198	449.852	1,273.281
Water In/Steam Out (lb/hr)	1,200	2,000	30,000

than the steam available, therefore the amount of CO_2 available became the fixed parameter in the steam-CO_2 ratio. Ideally, the steam required to satisfy this ratio can be obtained as excess steam from the cogeneration plant itself, otherwise it must be purchased from outside sources. Figure 3 illustrates the incremental oil production based on various CO_2 injection rates versus the net power generated. Simulations revealed that only the turbine model 251B12 had enough excess steam available to satisfy the steam-CO_2 ratio. The other two systems required that steam be purchased to satisfy the steam-CO_2 ratio.

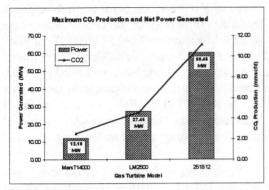

Figure 2: CO₂ Production and Power Produced

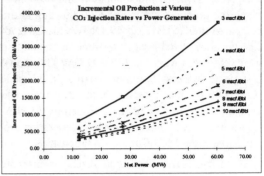

Figure 3: Incremental Oil Production

ECONOMICS

The installation of a cogeneration power plant with CO_2 recovery at an existing oil producing facility justifies the economics because extra revenues are generated from incremental oil production and the sale of excess electrical power to the grid.

The annual net cashflow in Table 2 is calculated based on no taxes paid, fixed oil price and fuel costs. The internal rate of return (IRR) is then determined from these parameters. Calculations were based on a cost of \$2.00/mscf natural gas, \$21.00/Bbl oil, electricity revenues of 50 mils/KWh, royalty oil volume at 20%, oil operating costs of \$4.50/Bbl, make-up water at \$1.55/1000 US gal, import steam at \$3/MMBtu, all Canadian dollars at Nov. 1992, 15 year project life, and 10% salvage value on capital after 15 years. Figure 4 summarizes before-tax and after-tax economics.

One of the major obstacles facing the economic stability of this study is the ability to sell electricity to the local utility. From Table 1, we see that internal electricity consumption for each model represents roughly 6% of

Table 2: Economic Summary

	MarsT 14000	LM2500	251B12
November 1992 Total Installed Cost (Million$)			
Power Plant	16.89	32.47	47.38
CO₂ Plant	7.53	13.75	33.37
CO₂ Compression	0.00	0.00	0.00
CO₂ Injection Well Conversion (Four Wells)	0.16	0.16	0.16
Total	24.58	46.38	80.91
Annual Revenues (Million$/yr)			
Oil @ $21.00/Bbl ($CD)	2.78	5.07	12.30
Electricity @ 50 mils/KWh	4.80	10.82	23.83
Steam @ $3/MMBtu	0.00	0.00	0.00
Total	7.58	15.89	36.13
Annual Operating Cost (Million$/yr)			
Power Plant			
O & M, Power Plant	0.80	1.47	2.95
Fuel, Power Plant @ $2.00/mscf	1.83	3.33	8.09
Cooling Water, Power Plant @ $0.23/1000 US gal	0.08	0.18	0.51
Total	2.70	4.98	11.55
CO₂ Plant			
O & M, CO₂ Plant	0.30	0.31	0.36
Cooling Water, Power Plant @ $0.23/1000 US gal	0.02	0.04	0.10
Total	0.32	0.36	0.46
Oil Production Operating Costs ($4.50/Bbl)	0.74	1.36	3.30
Chemical Costs	0.07	0.12	0.30
Waste Disposal Costs	0.06	0.11	0.27
Property Taxes (1% of Capital Costs/yr)	0.02	0.03	0.05
Operating Cost Total	3.91	6.96	15.93
Annual Gross Profit (Million$/yr)	3.66	8.93	20.20
Depreciation Cost	1.47	2.78	4.85
Annual Taxable Income (Million$/yr)	2.19	6.14	15.35
Taxes	0.00	0.00	0.00
Net Income after Taxes (Million$/yr)	2.19	6.14	15.35
Annual Net Cash Flow (Million$/yr)	3.66	8.93	20.20
Estimated IRR after Taxes	13.00%	17.45%	23.80%
ROI (%)	15.15%	19.42%	25.07%
Payback Period (yrs)	6.60	5.15	3.99
CO₂ Price Before Taxes ($/mscf)	3.05	1.74	0.68
CO₂ Price After Taxes ($/mscf)	3.51	2.21	1.01

the total power generated. Therefore, at 50 mils/KWh, the sale of electricity will contribute to a major portion of annual revenues. This study assumes that electricity is sold and purchased at 50 mils/KWh in order to eliminate the need to estimate the power requirements for incremental oil production as well as power consumed on an existing facility.

In this study an exchange rate of $73 US for every $100 CD was used. The steam turbines, waste heat boiler, superheaters, economizer, condensate pump, and CO_2 recovery system costs were based on previous studies and adjusted utilizing an equation from Peters and Timmerhaus.[9,10] Operating and maintenance costs, including salaries and fringes, are based on previous studies and adjusted to Canadian context.[2,6,11] It is assumed that 1 person will operate the power plant and 1 person will operate the CO_2 recovery facility both on a 24 hour basis. Power plant maintenance is estimated at 4.67 mils/KWh, and at 3.50% of capital/yr for the CO_2 plant. Cooling water

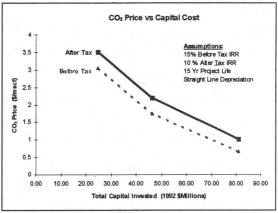

Figure 4: CO_2 Price vs Capital Cost

costs for the plant depend on the use of excess steam. The cost of the cooling water is estimated at $0.23/1000 US gallons as raw water price since water treatment is assumed already in place. Chemical costs include chemicals required to treat the cooling water at $0.0053/1000 US gallons, MEA makeup at $1.45/tonne of CO_2 and triethyleneglycol (TEG) makeup at $0.7/lb.[6,11] Other additional costs include waste disposal costs at $1.40/tonne of CO_2, insurance costs, which are included in the capital, and property taxes at 1.00% of capital costs/year.

PRICE SENSITIVITIES

In this study, it is assumed electricity is sold, therefore, the investment risk is relatively low. Based on this major assumption, the IRR is fixed to 15% before tax and 10% after tax in order to determine price sensitivities. The total amount of CO_2 produced from each turbine model dictates the amount of incremental oil that will be produced. Depending on the steam available to satisfy the steam-CO_2 ratio, steam is either bought or sold. The price sensitivity in $/Bbl is based on the amount of incremental oil produced for every mscf of CO_2 and steam injected. From this, the necessary oil revenue sensitivity from the CO_2 and steam injection rates is determined for each turbine model. Figure 5 represents the necessary revenues for the 251B12 turbine for the after tax scenario. The tax rate is assumed to be 45%.

In conjunction with an EOR project, the construction of a cogeneration combined cycle CO_2 recovery plant is dependent on the combined economic feasibility. The amount of incremental oil produced and the price of the oil dictates the allowable price of CO_2 at the wellhead. Previously, in Table 2, as well as in Figure 4, estimated CO_2 prices in $/mscf are determined for both before tax and after tax based on the economics of this study. The CO_2 price is high for the smaller sized turbine models, however, for model 251B12, the CO_2 price becomes economical at $0.68/mscf ($12.93/tonne) before tax and $1.01/mscf ($19.22/tonne) after tax based on necessary revenue at 5 mscf/bbl.

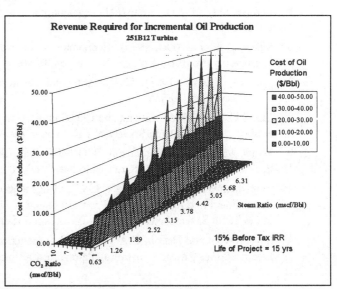

Figure 5: Revenue Required Before Tax

CONCLUSION

Industry is trying to reduce the recovery cost of CO_2 to near \$1.25/mscf. This paper has demonstrated the feasibility of achieving low price CO_2 through cogeneration techniques by placing a combined-cycle cogeneration CO_2 and steam recovery plant in an EOR field.   Such a system has numerous advantages.[6] Some of these include:

- Production of electricity for EOR operations and sale to the local grid.
- The combined-cycle plant provides the EOR field operator a means by which shut-in natural gas can be used to produce CO_2, electricity, and oil, thereby finding a marketplace for the shut-in gas.
- The Rankine cycle can be modified to export steam for thermal EOR or used to energize reboilers or operate absorption refrigeration units.
- Eliminates CO_2 transportation costs (both via semi-trailer or pipeline).
- The large volumes of CO_2 needed by EOR formations tend to favor the larger gas turbines and CO_2 plants, therefore benefiting from the large economy of scale needed to deliver low cost CO_2.

As well the environment benefits through the reduction of greenhouse gases and the economy benefits because the construction of such projects create short and long term jobs and utilize the area's own resources. Cogeneration techniques are becoming increasingly popular, however, the market for excess energy in the form of electricity is a major limiting factor.

REFERENCES

1. Tontiwachwuthikul, P. et al. (June, 1996). Carbon Dioxide Production from Cogeneration for Enhanced Oil Recovery: An Economic Evaluation. The Journal of Canadian Petroleum Technology, 35, pp. 27-33.

2. Solar Turbines Incorporated. Cogeneration Systems. San Diego, California., pp. 8.

3. Jennings, A.R. (Jan., 1991). Use of CO_2/Steam to Enhance Floods in Horizontal Wellbores. United States Patent, #4,982,786.

4. 1992-1993 Electric Power and Budget Pricing for Basic Gas Turbine Packages. 1992-1993 Gas Turbine World Handbook, pp. 1-20, 1-24 and 3-6 to 3-14.

5. Limaye, D.R. (1987). Industrial Cogeneration Applications, pp.118. Fairmount Press, Lilburn, Georgia.

6. Miller, D.B. et al. (Oct. 1986). Economics of Recovering CO_2 from Exhaust Gases. Chemical Engineering Progress, Vol. 82, No. 10, pp.38-46.

7. Astarita, G. et al. (1983). Gas Treating with Chemical Solvents, 1st Edition.   John Wiley and Sons, New York, U.S.A.

8. Liu, S.C. et al (1995).   Steam and CO_2 Combination Flooding of Fractured Cores: Experimental Studies.   The Petroleum Society of CIM, Paper # 95-80.

9. Peters, M.S. and Timmerhaus, K.D. (1991).   Plant Design and Economics for Chemical Engineers, 4th Edition. McGraw-Hill, New York, U.S.A.

10. Tontiwachwuthikul, P. et al (Oct. 1993).   Cogeneration Concepts for Recovering CO_2 From Fossil-Fuel Fired Power Plants. Paper presented at The 5th Saskatchewan Petroleum Conference of the South Saskatchewan Section. Petroleum Society of CIM and CANMET.

11. Snyder W.G. and Depew, C.A. (1986). Co-production of Carbon Dioxide (CO_2) and Electricity. Electric Power Research Institute. Palo Alto, California.

Energy Convers. Mgmt Vol. 38, Suppl., pp. S229–S234, 1997
© 1997 Elsevier Science Ltd. All rights reserved
Printed in Great Britain
0196-8904/97 $17.00 + 0.00

Pergamon

PII: S0196-8904(96)00274-9

VERTICAL CONVECTION IN AN AQUIFER COLUMN UNDER A GAS CAP OF CO$_2$

ERIK LINDEBERG & DAG WESSEL-BERG

IKU Petroleum Research
N-7034 Trondheim, Norway

ABSTRACT

The basic equation for volume, heat and CO$_2$ flux in a porous medium which is subject to both a temperature field and molecular diffusion have been analysed with respect to the stability criteria for convectional vertical flow in a porous medium. This analysis reveals under what condition vertical convection may occur, which is important for the total storage capacity of CO$_2$ in aquifers. © 1997 Elsevier Science Ltd

KEYWORDS

CO$_2$, aquifer, vertical convection, molecular diffusion, Rayleigh convection.

BACKGROUND

When CO$_2$ is injected into an aquifer there are several physical forces affecting the transport in the porous medium. The viscous, gravity and capillary forces determine the flow pattern on a short time scale. In a large (width >> height) homogeneous reservoir the displacement mechanism will typically be dominated by gravity rather than viscous forces except for close to the injection well. This will result in a CO$_2$ phase overlaying the water phase. The naturally occurring temperature gradient will only partially be compensated by the pressure gradient and therefore contribute to destabilise the water column. The problem of stability is further complicated by the CO$_2$ diffusing into the water column which will modify the density gradient. In a density gradient being introduced by heat flow and molecular diffusion, the on set of convection is only conditional and will depend on several fluid and reservoir properties. If vertical convective currents occur, this might increase the total storage capacity of the aquifer significantly since deep water undersaturated with respect to CO$_2$ will be transported to the top of the column and contacted by the CO$_2$ phase. This will increase diffusion rate. Standard reservoir simulators being used to predict the distribution of CO$_2$ in aquifers (Holt *et al.* 1994, Weir *et al.* 1994) do not model any conditions to set up a vertical flow and this phenomenon must therefore be studied separately to investigate if these reservoir simulators give reliable predictions.

MOLECULAR DIFFUSION

When CO$_2$ is injected into an aquifer the viscous forces will dominate close to the injection well where the rates are high. The CO$_2$ will partially mix with water and the water contacted by CO$_2$

in this region will be saturated by CO_2. If the distribution of gas is not controlled by high permeable zones (thief zones) or semi-permeable horizontal layers (Lindeberg 1996), the CO_2 will, however, accumulate under the cap rock seal. Only the top of the water column will then be contacted by CO_2 and the dissolution of CO_2 will be controlled by molecular diffusion if no vertical convection occurs. This is a slow process as illustrated in Figure 1 where the dissolved amount of CO_2 in an infinitely high 1 m^2 water saturated reservoir column under a CO_2 gas cap is plotted as function of time. The porosity is 30%. The amount CO_2 dissolved due to diffusion during a injection period

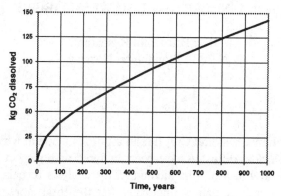

Fig. 1 CO_2 dissolved in a 1 m^2, infinitely high water column due to diffusion from a gas cap. The reservoir has a porosity of 0.3.

of 25 years will be small compared to CO_2 in the gas phase (Holt *et al.* 1994). If CO_2 is injected under a horizontal seal with open boundaries, however, the fate of the CO_2 will be of interest in a much longer perspective of time if the feasibility of a particular reservoir shall be evaluated. After 1000 years 143 kg CO_2 has been dissolved in the column, corresponding to a gas cap height of 0.6 m at the same conditions. This is a significant contribution to the storage potential. There is, however, another effect of the diffusion which might be more important for the storage potential. When CO_2 dissolves in water the density of the water increases. The water on the top of the column will therefore be denser than the water below. This will contribute to destabilise the column with respect to convective flow. In Figure 2 a density profile due to temperature and pressure is compared to a profile resulting after diffusion of CO_2 into the column. The effect of these gradients will be analysed.

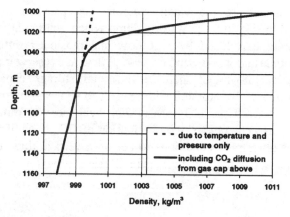

Fig. 2   Density gradients in a water column due to temperature and pressure compared to the density gradient where the combined effect of pressure, temperature and diffusion.

CONVECTION DUE TO THE TEMPERATURE GRADIENT ONLY

The natural occurring vertical temperature gradient (the geothermal gradient) will destabilise the water column because the warmer water on the bottom of reservoir will be less dense than the colder water on the top of the column. (This effect is only partially compensated by the pressure gradient that is changing the density in reverse direction.) To study the stability condition in detail, the basic equations describing flow and fluid properties must be analysed. These equations are: Darcy equation for steady state gives the volume flux:

$$\mathbf{j} = -\frac{k}{\mu}(\nabla p - \rho \mathbf{g}) \tag{1}$$

Equation of heat conduction

$$\Gamma \frac{dT}{dt} + \nabla T \mathbf{j} = \kappa \nabla^2 T \tag{2}$$

Equation of continuity

$$\nabla \cdot \mathbf{j} = 0 \tag{3}$$

The energy equation (equation of state) gives the density:

$$\rho = \rho_o \left[1 - \beta (T - T_o) \right] \tag{4}$$

where $\mathbf{g}$ is the gravity field, p is the pressure, k is the rock permeability, μ is the viscosity of water, Γ is a constant containing the specific heat capacitance of fluid and rock, T is temperature, κ is the thermal diffusivity, φ is the porosity of the rock and β is the coefficient of thermal expansion of water.

Equations (1) to (4) have been analysed for a boundary condition corresponding to the conditions found in a homogeneous reservoir (Horton and Rogers 1945). The horizontal boundaries were assumed to be impermeable and perfect heat conductors. It is found that the conduction state remains stable if the Rayleigh number,

$$Ra = \frac{kgh\beta\Delta T \rho_0}{\mu\kappa} < 4\pi^2 = Ra_c \tag{5}$$

Instability occurs as convection when the Rayleigh number exceeds the critical Rayleigh number, Ra_c. Convection appears in form of cellular motion with horizontal wave number π, as illustrated in Figure 3.

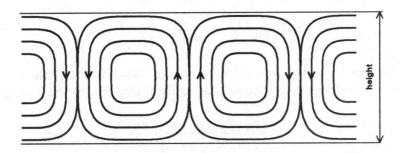

Fig. 3   Stream lines for Rayleigh convection in a porous medium induced
by instability due to a negative density gradient.

The Rayleigh number has been computed for some high permeable reservoirs at conditions typical for the Norwegian continental shelf. The conditions and results are given in Table 1. It is found that the Rayleigh number is of same order of magnitude as the critical Rayleigh number, but in all four examples it is smaller ($Ra < Ra_c$), indicating that the fluid columns are stable. Convection will therefore typically not occur under these conditions. The calculations were based on the assumption of porous medium homogenous in the height direction. Most reservoirs are horizontally divided by layers of lower permeability. The presence of any horizontally low permeable layer will further decrease the possibility for convectional currents (Bjørlykke et al. 1988), and most likely all aquifers are initially stable.

CONVECTION DUE TO BOTH TEMPERATURE AND CONCENTRATION GRADIENT

The interface of an undersaturated water column below a gas cap of CO_2 will immediately be saturated with CO_2. Gradually the CO_2 will invade the column but due to a decreasing gradient the flux will also gradually decrease. Since the water density increases with CO_2 concentration the CO_2 diffusion will further destabilise the column. This is illustrated in Figure 2 where the density gradient due to temperature gradient is compared to the density gradient due to the combined effect of temperature and concentration gradient. This problem requires two more relations, the equation of diffusion:

$$\frac{\partial \rho}{\partial t} + \frac{\mathbf{j}}{\varphi} \nabla \rho = \nabla \cdot \left(\frac{D}{\varphi} \nabla \rho \right) \tag{6}$$

where D is the molecular diffusion coefficient and φ is the rock porosity, while equation (4) must be modified to also take the concentration dependence of the density into account

$$\rho = \rho_o \left[1 - \beta (T - T_o) + \gamma c \right] \tag{7}$$

review where c is the CO_2 concentration in water and γ is the partial derivative of density with respect to concentration. This equation system (1 - 3) and (6 - 7) is more complex than the previous since the density gradient is time dependent. The density gradient will vary with time, but to get a first approach to the problem both the temperature and concentration gradients has been linearized and the system analysed by introducing small perturbations on both the temperature and concentration gradient analogous to the procedure of Horton and Rogers (1945). This gives a stability criteria for the combined effect of temperature and concentration on the column, S:

$$S = \frac{kgh\beta\Delta T \rho_0}{\mu\kappa} + \frac{kgh\gamma\Delta c \rho_0}{\mu D} < 4\pi^2 = Ra_c \tag{8}$$

If $\gamma \Delta c \rightarrow 0$ or $D \rightarrow \infty$, S becomes equal to Ra (5), which is consistent with the physical representation. The two density contributions can easily be identified by rearranging the equation:

$$S = \left(\frac{\Delta \rho_T}{\kappa} + \frac{\Delta \rho_c}{D} \right) \frac{kgh}{\mu} \tag{9}$$

Also the S-numbers were calculated in Table 1. While the Rayleigh numbers were close to the critical Rayleigh number, the S-numbers are two to three orders of magnitude larger, mainly due to the fact that the thermal diffusivity is in the order of two magnitudes larger than molecular diffusion. The conditions for convective currents are accordingly typically satisfied.

Table 1 The Rayleigh number and the S-number have been calculated for various typical North Sea reservoir conditions.

Properties:					
Temperature	$t =$	30	50	70	90 °C
Pressure	$p =$	100	150	200	300 bar
Porosity	$\Phi =$	0.3	0.3	0.3	0.3
Permeability	$k =$	2000	1000	500	100 mD
Permeability	$k =$	$1.97 \cdot 10^{-12}$	$9.87 \cdot 10^{-13}$	$4.93 \cdot 10^{-13}$	$9.87 \cdot 10^{-14}$ m²
Acceleration of gravity	$g =$	9.81	9.81	9.81	9.81 m/s²
Density of 3.5% brine @ p and T	$\rho =$	1036.0	1030.1	1021.8	1013.5 kg/m³
Coefficient of thermal expansion	$\beta =$	0.000351	0.000351	0.000351	0.000351 1/K
Temperature difference	$\Delta T =$	8.0	8.0	8.0	8.0 K
Height	$h =$	160	160	160	160 m
Thermal diffusivity	$\kappa =$	$3.67 \cdot 10^{-7}$	$3.67 \cdot 10^{-7}$	$3.67 \cdot 10^{-7}$	$3.67E \cdot 10^{-7}$ m²/s
Molecular diffusivity	$D =$	$2.20 \cdot 10^{-9}$	$3.40 \cdot 10^{-9}$	$4.90 \cdot 10^{-9}$	$6.30 \cdot 10^{-9}$ m²/s
Viscosity	$\mu =$	$8.4 \cdot 10^{-4}$	$5.8 \cdot 10^{-4}$	$4.3 \cdot 10^{-4}$	$3.4 \cdot 10^{-4}$ Pa s
Geothermal gradient	$dT/dh =$	0.05	0.05	0.05	0.05 K/m
Density difference due to T	$\Delta\rho_T =$	2.910	2.893	2.870	2.847 kg/m³
Density difference due to c	$\Delta\rho_c =$	14.42	14.42	14.42	14.42 kg/m³
Rayleigh number	$Ra =$	29.3	21.1	14.1	3.53
S-number from (9)	$S =$	24204	11348	5315	1046
Rayleigh number, critical $= 4\pi^2$	$Ra_c =$	39.5	39.5	39.5	39.5

DISCUSSION

The results must be used with some precaution due to the simplified approach. The largest uncertainty is introduced by the linearization of the concentration gradient. This assumption gives a significant deviation from an actual concentration gradient in the beginning, when only a fraction in the top of the column has been affected by diffusing CO_2. On an intermediate time scale, when the CO_2 has penetrated deeper into the column, and before the concentration at the bottom has significantly been changed, this assumption is best. By calculating the standard deviations between the linear density profile and various density profiles computed at different times for pure molecular diffusion in a 160 m high aquifer column, this corresponds to a time of approximately 40000 years. After 40000 year the deviation from the linear gradient will increase and will be dominated by errors in the bottom of the column.

A more comprehensive analysis should include a numerical solution of the original non linear equations which would yield a stability analysis where the transient diffusion behaviour at short periods of times is better taken care of. This is of particular importance because the shorter time scale (< 3000 years) is the most interesting in a CO_2 storage strategy (Lindeberg 1996).

Possible coupling between heat diffusion and molecular diffusion (the Sorret effect) should also be considered because it could either enhance or inhibit the penetration of CO_2 into the column. This is, however, only a second order effect and should be taken into account only when an improved handling of the transient behaviour is available.

CONCLUSIONS

Simplified analysis of CO_2 diffusing from a gas cap into a water column indicates that density gradients will set up convectional currents. The temperature gradient, enhancing this convection, can however be neglected. Existing reservoir simulators with provisions to model both Darcy flows and molecular diffusion can not handle conditionally unstable states. They will therefore always give convection in a fluid column with a negative density gradient, even if the condition for convective flow are not met. However, for the CO_2/water system the results above show that this is not a serious defect because an unstable column will typically set up convection in high permeable aquifers.

REFERENCES

Bjørlykke, K., Mo, A. and Palm, E. (1988). Modelling of thermal convection in sedimentary basins and its relevance to diagenetic reactions. *Marine and Petroleum Geology*, Vol. 5, November, 338-351.

Holt, T., Jensen, J.-I. and Lindeberg, E. (1994). Underground storage of CO_2 in aquifers and oil reservoirs. *Energy Convers. Mgmt* Vol. **36** No. 6-9, 535-538.

Horton, C.W. and Rogers, F.T., Jr. (1945). Convection currents in a porous media. Journal of Applied Physics, Vol. **16**, June, 367-370.

Lindeberg, E.G.B. (1996). Escape of CO_2 from aquifers. Paper presented at ICCDR-3, Boston 9-11. August 1996. Accepted for publishing in *Energy Convers. Mgmt.*

Weir, G.j., White, S.P. and Kissling, W.M. (1994). Reservoir storage and containment of greenhouse gases. *Energy Convers. Mgmt* Vol. **36** No. 6-9, 531-534.

Energy Convers. Mgmt Vol. 38, Suppl., pp. S235–S240, 1997
© 1997 Elsevier Science Ltd. All rights reserved
Printed in Great Britain
0196-8904/97 $17.00 + 0.00

Pergamon

PII: S0196-8904(96)00275-0

ESCAPE OF CO₂ FROM AQUIFERS

ERIK LINDEBERG

IKU Petroleum Research
N-7034 Trondheim, Norway

ABSTRACT

Many large aquifers consist of wide structures only confined by a horizontal cap rock with no distinct traps. By numerical simulations the distribution of injected CO_2 into this type of aquifers has been analysed. The results show that the distribution of CO_2 is controlled by gravity forces and the horizontal permeability just below the cap rock. The escape rate through a possible open periphery or fractures 8000 m from an injection well was simulated. Some of the injected CO_2 will escape from a high permeable aquifer, but the long residence times showed that an effective storage will still be provided in a strategy to combat accumulation of atmospheric CO_2.

© 1997 Elsevier Science Ltd

KEYWORDS

Underground CO_2 storage, aquifers, residence time, escape rate.

INTRODUCTION

CO_2 disposal into gas and oil reservoirs is usually considered as a safe and permanent storage option. Aquifers does not have a similar proved sealing capacity. The integrity of the cap rock is more uncertain unless it maintain a non-hydrostatic pressure gradient. Disposal of 1 million tonnes industrial CO_2 per year into an aquifer with hydrostatic pressure started September 1996 (Utsira, North Sea). The permanency of this storage option, is a question of the time scales considered. Even from gas and oil reservoirs hydrocarbons are continuously escaping up into overlaying sediments and finally enter into the atmosphere or ocean although the residence times are millions of years. Molecular diffusion and migration will contribute to the transport. The storage capacity for any underground disposal project must therefore be related not only to the quantity of CO_2 that can be stored, but also to the residence time for the injected CO_2. In this respect underground disposal of CO_2 resembles ocean disposal where also a limited retention of CO_2 is expected. In previous studies when aquifer disposal potential has been studied in geological traps (van der Meer 1992, Holt and Lindeberg 1994), the storage capacity was defined as the amount CO_2 injected when the CO_2 phase reaches the spill point. This criterion will be satisfactory for a dipping trap or an anticline. For storage under a large, but finite horizontal cap rock which is considered as open in the periphery through a spill point or an open fracture, this type of definition is in not applicable. To evaluate non-dipping aquifers for deposition, a more comprehensive analysis must be performed. Criteria must be formulated based on escape rates, and the residence time compared to a project life time of the fossil fuel era and the time scale of natural climate changes. If aquifer CO_2 disposal is also to be applied in large scale on land, escaping CO_2 could also constitute a safety hazard (Holloway 1996).

DISTRIBUTION OF THE CO_2 PHASE IN A HOMOGENOUS RESERVOIR

The distribution of injected CO_2 in a horizontal aquifer was modelled by Darcy's equation for transport of the two phases, water and CO_2, (White et al. 1994) and Fick's law for the molecular diffusion of CO_2 in the water phase. The fluid properties for the CO_2/brine system (density and solubility) have been well described by respectively Duan et al (1992) and Enick and Klara (1990). The relative permeabilities used in the model were based on previous measurements (Holt and Lindeberg 1994). The problem was solved numerically by use of the Eclipse 100 reservoir simulator from Interra. The gird was represented in cylindrical geometry with the injection well in the centre. CO_2 was injected in the bottom of a 160 m thick aquifer at a rate of 6 million tonnes/year. To minimise numerical dispersion, a high spatial vertical resolution was used. The optimal vertical spacing of the grid was found to be 22 grid blocks varying from 20 m thickness in the bottom of the reservoir to 0.08 m near the top as illustrated in Figure 1, left, where the vertical grid lines represents the block boundaries. The thinnest blocks near the top are not visible due to the scale. A coarser grid will give too high numerical dispersion resulting in an underestimation of the free gas volume. The horizontal (radial) axis was divided into 250 blocks of 50 m length. Initially the aquifer had no dissolved CO_2 and was at hydrostatic pressure. The pressure in the periphery of the model was maintained at hydrostatic pressure during the simulations, corresponding to hydraulic communication to the surface water. The rock porosity was 0.3 and the aquifer temperature was 30°C in all simulations. This aquifer can resemble e.g. the Utsira formation in the North Sea.

Figure 1, left, shows the profile of the reservoir and the CO_2 front at different times after start of injection . The injection was stopped after 25 years. CO_2 phase will accumulate as a thin layer under the cap rock slowly migrating radially while the thickness of the bubble continuously decreases. The radial migration rate will gradually decrease because the driving force is the hydrostatic head due the density difference between CO_2 and water and the thickness of the gas cap. The spreading will not continue indefinitely due to dissolution of CO_2 in the water phase by diffusion and convection (Lindeberg and Wessel-Berg, 1996). The bubble will reach a maximum radius after which it will diminish and finally disappear when all the CO_2 has been dissolved in the water column below. This maximum bubble radius and the dissolution rate will depend on the horizontal permeability as illustrated in Figure 1, right. The vertical permeability was one tenth of the horizontal permeability in all simulations.

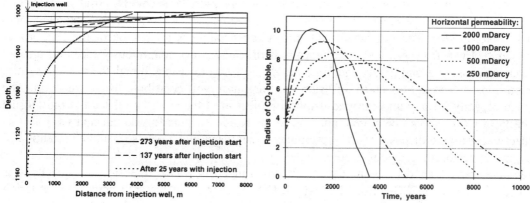

Fig. 1   Left: The curves represent the CO_2 front in a vertical cross section of the reservoir with 2000 mDarcy permeability. The areas left of the curves are the free CO_2 phase at different time after injection start. When injection ceases after 25 years the CO_2 phase will continue to migrate and accumulates just below the cap rock. Right: The radius of the gas phase versus time for four different horizontal permeabilities.

ESCAPE OF CO₂ FROM AN AQUIFER

If the cap rock has an open boundary or an open fracture at a distance 8000 m from the injection well it can be seen from Figure 1, right, that no CO_2 can escape from the aquifer with 250 mDarcy permeability because the bubble starts to retreat before the gas phase reaches the spill point. For the higher permeabilities, some CO_2 will escape in these examples. A new set of simulations was performed with a similar reservoir model as above, but this time with a spill point 8000 m from the injection well. The CO_2 reaching the spill point was continuously removed from the model and the CO_2 transport through the boundary was monitored. The results are illustrated in Figure 2. The escape curves depend strongly on the permeability of the top layers where the most important transport goes on. With a permeability of 4000 mDarcy the first CO_2 returns to the atmosphere after 250 years and during the next 1200 years 11% of the injected CO_2 is lost. After that time, the losses are insignificant. In an even less favourable situation the cap rock is dipping upwards from the injection well to the spill point with a dip of 1:1000. This gives a loss of almost 20% during the following 2000 years. This represents the worst case of all scenarios. If all the anthropogenic CO_2 is deposited in similar rather poor aquifers, instead of being emitted into the atmosphere directly, this will still be a significant contribution in the CO_2 emission abatement strategy as illustrated in Figure 2, right, were the resulting escape emission profile is compared to the "business as usual" scenario based on a total available $7.1 \cdot 10^9$ tonne fossil carbon. The worst case escape scenario will give a maximum CO_2 emission in year 3100 of one third of the present emission. The strong peak in CO_2 emissions in the reference scenario is effectively avoided imposing only a minor environmental load probably disguised by other natural variation.

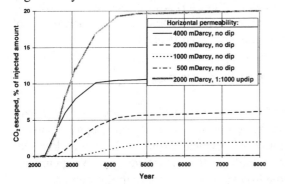 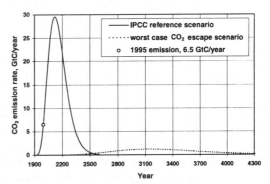

Fig. 2   Escape of CO_2 from an aquifer where a spill point is located 8000 m from the injector. Left: The cumulative escape of CO_2 from aquifers with different permeabilities and cap rock dip. Right: The escape rate from the worst case scenario (1:1000 up dip) if all the future CO_2 is injected in similar leaky reservoirs, compared to the IPCC reference scenario.

DISTRIBUTION OF CO₂ IN AN AQUIFER WITH LAYERED HETEROGENEITIES

Due to the sedimentary origin of underground aquifers these will often contain stratified layers consisting of sands with varying permeability or low permeable shales. A 20 m thick zone with 10 000 mDarcy permeability was introduced 100 m below the cap rock into a similar model as above with a permeability of 500 mDarcy. The numerical vertical grid was refined in the thief zone to minimise numerical dispersion. The result of the simulation showed that the distribution of CO_2 did not differ significantly from the distribution in the homogeneous case. In the next simulation a low permeable shale (0.1 m thick with a vertical permeability of 0.01 mDarcy), was placed on top of the high permeable zone. The simulated distribution of CO_2 is illustrated in Figure 3. The CO_2 was penetrating deeper into the thief zone for some time, but after injection stopped, the CO_2 migrated through the shale and accumulated under the cap rock. During the migration from the thief zone up into the reservoir above, the CO_2 did, however, contact large volumes of water and much CO_2 was dissolved as illustrated in Figure 3 (lower set). In this case the thief zone increased the storing capacity, because it retained the CO_2 deeper in the aquifer for some time, resulting in less CO_2 accumulating under the cap rock. Another type of heterogeneity was also investigated. An aquifer was divided in four by three low permeable horizontal shales (each 0.1 m thick with a vertical permeability of 0.01 mDarcy) and was subject to the same injection scheme as above. The results are illustrated in Figure 4. A similar effect can be observed as in previous example. The migration of CO_2 to the cap rock was retained by the shales. The presence of semipermeable horizontal shales appears to increase the vertical sweep and accordingly also the dissolution of CO_2.

To achieve a sufficient vertical retention in the examples above, a shale permeability, k as low 0.01 mDarcy was needed for a 0.1 m thick shale. By combining Darcy's and Poiseuilles's equations, the permeability can be expressed as a function of a characteristic pore radius, r as $k = r^2 / 8$. This gives a $r \approx 10^{-9}$ m. The surface tension between CO_2 and water is typically 35 mN/m at these conditions, and the required pressure difference for a CO_2 bubble to enter a water wet shale pore with radius r can be calculated from $\Delta p = 2 \sigma / r \approx 70$ MPa. The hydrostatic pressure from a 10 meter CO_2 column in water will only give a $\Delta p = 0.03$ MPa. The CO_2 penetration through the shales in the simulations above is therefore an artefact due to the capillary pressure for the shale was not taken into account. Any shale with a permeability of less than 5 mDarcy will actually be an effective cap rock for a reservoir of this height. The layered sand in the examples above should be treated as isolated formations with respect to migration of free CO_2. A shale with a sufficiently large pore to allow a CO_2 bubble to enter, will have so large permeability that it will not affect the distribution pattern significantly.

MIGRATION OF CO_2 UNDER A NOT PERFECTLY FLAT CAP ROCK

In practice the term "flat cap rock" is an approximation because there are usually small anticlines, inverted peaks and valleys on all cap rocks. From the cap rock topography obtained by a seismic survey on the Utsira formation (Statoil 1994), typical peaks on the cap rock could be 30 m high (or deep) and 6250 m wide. A cap rock with attics corresponding to this geometry was included in an aquifer model which was studied by simulation under similar conditions as the homogeneous aquifers, but with an injection rate six times as large (33 million tonne CO_2/year, corresponding to the CO_2 emissions from a 3000 MW coal fired power plant or the total Norwegian CO_2 emission). The CO_2 was injected below a peak in the cap rock as illustrated in Figure 5. The first attic is filled up completely and only a fraction of the CO_2 will spill over into the next attic.

In this example the attic formed a trap encircling the injection well due to the cylindrical geometry of the model. If this roughness of the cap rock formed a continuous inverted valley with a slope reaching a spill point, it would be important to answer the question of what velocity CO_2 will be transported. The front velocity can be estimated by Darcy's equation using a steady state approach assuming that the CO_2 river will reach a constant rate. This is in accordance with an assumption that the velocity is controlled by the slope and not by the injection rate. The *depth* of the river and accordingly the CO_2 *rate* will, however, still depend on the injection rate. An upward slope of $S = 1:1000$ and a permeability of $k = 2000$ mDarcy were used in the analytical solution. Only the gas viscosity $\mu = 0.065$ mPa s entered the equation because a worst case assumption of unity gas saturation was applied. The velocity (equal to the CO_2 flux) can be estimated from

$$j = -\frac{k}{\mu}\Delta\rho gS$$

where $\Delta\rho$ is the density difference between the two phases (250 kg/m^3). This gives a velocity of 2.5 m/year or 3200 year to reach a spill point 8000 from the injection well. The agreement with the solution of the transient problem presented in Figure 2 is striking.

CONCLUSIONS

The distribution of CO_2 in high permeable aquifers will be controlled by gravity forces and the permeability in the uppermost part of the aquifer, as this is where the CO_2 will accumulate as a spreading phase. In a long term perspective, the solubility of CO_2 in the water column below, will contribute to inhibit the spreading of the CO_2 phase. If a spill point exists within several 1000 meters from the injection well this can result in loss of CO_2 into the atmosphere. The aquifers will still be useful as deposits for CO_2 in an abatement strategy against climate change, due to the fact that only a fraction of the CO_2 will escape, and that a significant residence time can be achieved if some simple measures regarding disposal site and injected quantities are taken. Any concavities of the cap rock would effectively trap CO_2 unless they form a continuous system of inverted sloping valleys reaching a spill point.

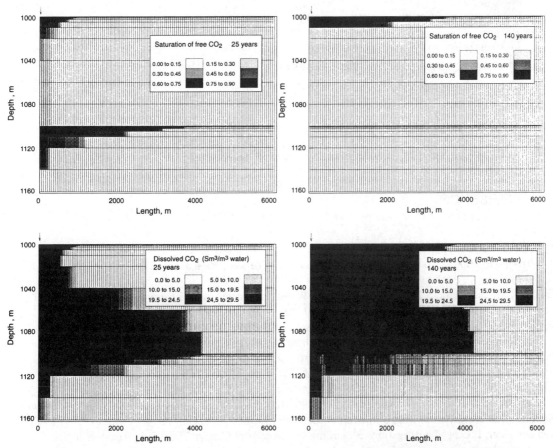

Fig. 3   Vertical cross section of an aquifer with a high permeable zone under a low permeable layer. The upper figures show the saturation of free CO_2 and the lower figures show the CO_2 concentration in water.

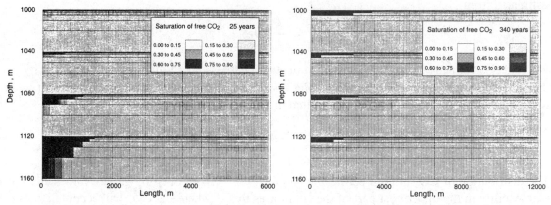

Fig. 4   Vertical cross section of an aquifer divided by three layers of low permeability (2000 mDarcy, 0.1 m thick) 25 years of continuos injection and at a time 340 years after injection start.

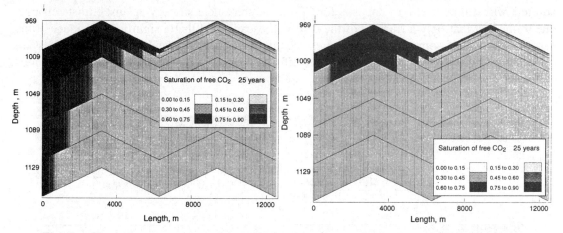

Fig. 5   Vertical cross section of a reservoir with a not perfectly flat cap rock. The attics are 31 m high and 6250 m wide. Injection rate is 33 million tonne CO_2/year.

REFERENCES

Duan, Z, Møller, N. and Weare, J.: An Equation of State for the CH_4-CO_2-H_2O System: I. Pure Systems from 0 to 1000°C and 0 to 8000 bar. *Geochimica et Cosmochimica Acta*, Vol. 56, pp. 2605-2617, 1992.

Enick, R.M. and Klara, S.M.: CO_2 Solubility in Water and Brine Under Reservoir Conditions, *Chem.Eng.Comm.*, Vol. 90, pp. 23-33, 1990.

Holloway, S.: Safety of the Underground Disposal of Carbon Dioxide, Presented at ICCDR-3, September 9-11, 1996. Accepted for publishing in *Energy Convers. Mgmt.*

Holt, T., Jensen, J.-I. and Lindeberg, E.: Underground Storage of CO_2 in Aquifers and Oil Reservoirs. *Energy Convers. Mgmt.* Vol. 36, No. 6-9, pp. 535-538, 1995.

Lindeberg, E., Wessel-Berg, D.: Vertical convection in an aquifer column under a CO2 gas cap. Presented at ICCDR-3, September 9-11, 1996. Accepted for publishing in *Energy Convers. Mgmt.*

Statoil: Structural Contour Map of the top of the Utsira formation, 6. Oct. 1994.

van der Meer, L.G.H.: Investigations Regarding the Storage of Carbon Dioxide in Aquifers in the Netherlands. *Energy Convers. Mgmt.* Vol. 33, No. 5-8, pp. 611-618, 1992.

Weir, G.J., White, S.P. and Kissling, W.M.: Reservoir Storage and Containment of Greenhouse Gases. *Energy Convers. Mgmt.* Vol. 36, No. 6-9, pp. 531-534, 1995.

Pergamon

PII: S0196-8904(96)00276-2

Energy Convers. Mgmt Vol. 38, Suppl., pp. S241–S245, 1997
© 1997 Elsevier Science Ltd. All rights reserved
Printed in Great Britain
0196-8904/97 $17.00 + 0.00

SAFETY OF THE UNDERGROUND DISPOSAL OF CARBON DIOXIDE

SAM HOLLOWAY

British Geological Survey, Keyworth, Nottingham NG12 5GG, UK.

ABSTRACT

The risks associated with the transport and injection of carbon dioxide are reasonably well understood and already borne in the USA. There is a remote possibility that CO_2 disposed of underground could leak from a storage reservoir, either through an unidentified migration pathway or as the result of a well failure. The kind of threat that this might represent may be judged by comparison with naturally occurring volcanic CO_2 emissions. Diffuse CO_2 emissions through the soil or via carbonated springs in volcanic areas do not appear to represent a threat as long as the CO_2 is able to disperse into the atmosphere. However, when CO_2 is able to build up in enclosed spaces it poses a definite threat. Large CO_2 clouds associated with sudden emissions from volcanic vents or craters also pose a lethal threat. However, there appears to be little analogy between such events and any possible leak from a storage reservoir via a natural unidentified migration pathway. Modelling of the development, migration and subsequent dispersal of any CO_2 cloud which might arise from a well failure is recommended. © 1997 Elsevier Science Ltd

KEYWORDS

Carbon dioxide, underground disposal, safety, volcanos

INTRODUCTION

The safety and stability of the underground storage of CO_2 are fundamental to its social acceptance (Cox *et al.*, 1996). The risks associated with the pipeline transport of CO_2 and with surface injection facilities are reasonably well understood (van der Meer & Summerfield in Cox *et al.*, 1996) and are already borne by the enhanced oil recovery industry in the USA. Both the likelihood (van der Meer & Summerfield in Cox *et al.*, 1996) and the possible consequences (Kruse & Tekeila, 1996) of a CO_2 pipeline rupture have been analysed. However, one of the questions which must be answered if the large scale underground disposal of CO_2 is to become publicly acceptable is: 'However remote the risk of an escape, what will happen if CO_2 leaks from an underground storage reservoir and makes its way to the ground surface or sea bed?'. This question has already been addressed using reservoir simulation (van der Meer in Cox *et al.*, 1996). The following brief review of some of the literature on natural emissions of CO_2 at the ground surface is intended to throw some light on this question from a different angle.

It should be noted that this paper does not in any way attempt to assess the chances of an escape of CO_2 from a storage reservoir. This could perhaps be assessed from the track record of underground natural gas storage schemes and the enhanced oil recovery industry.

The most likely concepts for an underground CO_2 storage reservoir are:

1. A volume of CO_2 equivalent to at least the lifetime emissions of a single 500 MW (e) coal-fired power

plant will be confined within a conventional subsurface fluid trap, probably at depths of around 800 metres or more below the surface (e.g. Holloway & van der Straaten, 1995). Such a trap would be directly analogous to the natural gas (methane) fields found throughout the oil and gas producing regions of the world. The amount of CO_2 stored would be at least 100 million tonnes, and in the reservoir the CO_2 would most likely be in the supercritical phase.

2. Similar, or larger, volumes of CO_2 could be disposed of in large, essentially horizontal, reservoirs confined only by a cap rock. In such a case, modelling suggests that much of the CO_2 would simply react with aluminosilicate minerals in the reservoir rock or dissolve in the formation water of the reservoir. The remainder would migrate only very slowly, with gravity or aquifer flow the only forces acting to move it towards the basin margins e.g. Gunter *et al.* (1993), Lindeberg & van der Meer (1996).

In both these cases there appear to be two main types of leak that could occur. Firstly, there is a remote possibility that CO_2 could be released from the reservoir due to the failure of an injection well. If the reservoir pressure was high, this could result in a very localised, high velocity emission of CO_2 to the surface (Cox *et al.*, 1996). Secondly, there is the possibility that a more diffuse, or lower volume and lower pressure, emission of CO_2 to the surface could occur as a result of migration to the outcrop of the reservoir rock or through other unidentified migration pathways such as faults. This type of leak could result in the emission of CO_2 through the soil and/or in carbonated springs.

NATURAL CO_2 EMISSIONS

Most natural emissions of CO_2 are associated either directly or indirectly with volcanic activity. The CO_2 arises from degassing of the magma (Farrar *et al.*, 1995). CO_2 emissions from volcanos are only rarely dangerous because the vast majority of the CO_2 is emitted at high temperatures, and often high pressures, from fumaroles or during explosions. So the CO_2 rises well above ground level and becomes mixed with, and diluted in, the atmosphere (Stupfel & Le Guern, 1989).

Diffuse Emissions of CO_2

Apart from emissions from vents and fumaroles, the most commonly observed emissions of CO_2 are those of naturally carbonated springs. These discharge water which has become saturated with CO_2 of volcanic origin underground at pressures higher than atmospheric (Sigvaldason, 1989). They are not generally regarded as hazardous, except where found in caves. However, in the 1948 eruption of Hekla (Iceland), 8 sheep and a fox were asphyxiated when entering invisible ponds of CO_2. These ponds formed on calm nights in depressions in the ground surface. The CO_2 is thought to have originated from degassing of carbonated groundwater; when it emerged it had the same temperature as the groundwater. The ponds contained a 2 m thick layer of air mixed with 40% CO_2 (Thorarinson, 1950).

Additionally, there are areas of the world where volcanic CO_2 emerges through the soil, or forms invisible pockets in depressions in the ground or caves. Some of these are briefly described below.

Mammoth Mountain. Mammoth Mountain, in eastern California, is a volcano which was last active several hundred years ago. It forms part of a larger volcanic basin known as the Long Valley Caldera. It is a popular ski resort, because of the deep snow that accumulates on this eastern margin of the Sierra Nevada mountain chain (Williams, 1995).

Areas of dead and dying trees are found on Mammoth Mountain. Until recently the cause of their death was a mystery. But it is now known that the trees are being killed by large volumes of CO_2 being emitted through the soil from magma deep within the mountain (Farrar *et al.*, 1995). Farrar *et al.* (1995) report that in areas of tree kill > 30 hectares in size, the CO_2 flux through the soil is approximately 1200 tonnes per day, i.e. about 40 tonnes per hectare per day. It is noteworthy that there is little fumarolic activity on Mammoth Mountain and the CO_2 is all emerging through the soil.

The flux of CO_2 first became apparent in March 1990, when a US Forest Ranger reported symptoms of asphyxia when entering a small snow-covered cabin near Horseshoe Lake. Between 1990 and 1994 other, similar incidents were reported by persons entering confined spaces, including subterranean utility vaults. A preliminary survey of confined spaces in areas of known high CO_2 discharge near Horseshoe Lake found CO_2 levels >1% in campsite lavatories and small tents, 25% in a small cabin, and 89% in a utility vault of 0.6 m diameter and 1 m depth. All of these values exceed US health standards, and the last two would be quickly lethal. Thus it is clear that CO_2 flux through the soil can be dangerous, because the gas builds up in confined spaces.

Mount Etna. Mount Etna, on the island of Sicily, releases measurable amounts of CO_2 from its outer flanks, in areas where no visible emissions take place. On the upper flanks of the volcano, a persistent CO_2 excess of 20 - 100 ppm occurs at 1.5 m above ground level in the ambient air around the volcano, well outside the plume and far from the craters. An airborne survey, flown in a circle centred 15 km from the summit at an altitude of 3200 m, indicated that a CO_2 anomaly of 10 ppm or more persists kilometres away from both sides of the plume, forming a dome of CO_2 over the volcano (Allard et al., 1991). This dome cannot be due to CO_2 from the plume; it seeps upward through the soil. This CO_2 flux does not appear to be regarded as a hazard.

Discussion. The flux of CO_2 through the soil at Mammoth Mountain and Mount Etna does not appear to be a danger to human life when it simply emerges through the ground and disperses into the atmosphere. However, it can become dangerous when it builds up to form high concentrations in confined spaces. Above 5% concentration the gas causes an increase in respiration rate along with a number of other symptoms, such as headache, breathing difficulty, palpitation, dizziness and weakness. At 10% concentration, CO_2 causes instantaneous unconsciousness and will quickly be lethal. If the concentration is 20% the gas is instantaneously fatal (Kruse & Tekeila, 1996). Thus it seems reasonable to assume that emission of CO_2 through the soil could be a major problem in housing, especially therefore in urban areas. It could also be a danger in topographic hollows, such as quarries.

It is extremely difficult to imagine circumstances in which emissions of CO_2 by seepage through the sea bed could pose a risk to human life. For technological reasons, underground disposal of CO_2 offshore is likely to occur in fairly shallow water depths; probably less than 200 m. At these depths, escaping CO_2 could pose a risk to marine ecosystems. The seriousness of this risk would perhaps depend mainly on the flux of CO_2 through the sea bed.

Emissions of CO_2 from volcanic craters, crater lakes and vents

Recorded emissions of this type consist of large volumes of initially nearly pure CO_2 which are emitted almost instantaneously. They can form dense, ground-hugging clouds which, at worst, can lead to almost instantaneous deaths. Some examples are given below:

Dieng (Central Java, Indonesia). The following account is based on the work of Allard et al. (1989). On 20 February 1979, during the eruption of the Dieng volcano, a large outflow of pure CO_2 was exuded from an existing volcanic vent (Sigludug Crater) and a reopened extinct eruptive fracture. As a result, sheets of the dense gas flowed down from the volcano onto the nearby Batur Plain. 142 people who fled from the eruption down a path from the village of Kepucakan towards the village of Batur were engulfed by the gas sheets and killed almost instantly. The gas emitted was nearly pure carbon dioxide with subordinate amounts of methane and sulphur compounds. It is thought that magmatic CO_2, accumulated beneath the Dieng volcano, was the source of the lethal gas, the effusion of which was triggered by the pressure release generated by the phreatic eruption. The total discharge of the 1979 Dieng event might have approached 0.1 km³, i.e. close to the lower output estimated for the Lake Nyos catastrophe.

There are persistent CO_2 emissions on the Dieng Plateau, which destroy the surrounding vegetation, and local people are aware of 'Death Valleys'. Four workers were killed by CO_2 at the Dieng 13 geothermal well in February, 1988.

Lake Nyos, Cameroon. During a period of about 4.5 hours in the late evening of August 21, 1986, a huge mass of concentrated CO_2, was emitted from Lake Nyos, a volcanic crater lake in Cameroon (Sigvaldason, 1989). The mass of CO_2 emitted has been estimated at $1-3 \times 10^8$ kg CO (Zhang, 1996). A lethal concentration of the gas reached a height of 120 m above the lake surface, and the total volume of the lethal gas cloud may have been up to 1.3 km^3 (Kanari, 1989). It flowed out of the spillway at the northwest end of the lake and down the topographic slope, along two valleys. It killed more than 1700 people in a thinly populated area, and all animal life along its course as far as 14 km from the crater. Two hypotheses have been put forward to account for this disaster. The first, (Tazieff, 1989) considers that there was a phreatic eruption in the lake, which exuded a very large amount of CO$_2$. The second, and most widely accepted, considers that it was caused by a sudden and violent release of CO$_2$ caused by the overturn of the 220 m deep lake, the hypolimnium of which became oversaturated with CO$_2$ of volcanic origin, caused by a slow leak of CO$_2$ into the lake waters from below (e.g. Sigvaldason, 1989).

It should be noted that, at least as far as can be judged from a review of the literature, the topography around Lake Nyos appears to provide ideal conditions for the emitted CO$_2$-rich gas cloud to remain concentrated rather than disperse. High crater walls surround the lake on east and west sides, and the natural water spillway in the northwest corner of the lake provides a natural outlet for the CO$_2$ into a valley system where it would remain confined.

Lake Monoun, Cameroon. A very similar CO$_2$ gas burst occurred at approximately 11.30 pm on August 15, 1984 at Lake Monoun in Cameroon, causing 37 casualties (Sigurdsson *et al.*, 1987). The origin of this gas burst was assigned to a landslide on the lake margin causing the lake waters to overturn and emit dissolved CO$_2$. The most likely reconstruction of events is as follows: "The resulting dense CO -dominated cloud drifted with the prevailing wind a few hundred metres to the east and settled in the depression along the nearby Panke river. Although a loud noise was heard in the village of Njindoun about 11.30 the previous night, people travelling along the road across the Panke River before daybreak the following morning were unaware of the danger, and were asphyxiated as they entered the cloud. The cloud claimed its last two victims about 5.00 am, but had dissipated by about 10.30 am, when the local doctor and police commandant were able to approach the victims. It is clear from eyewitness accounts that the cloud was greyish and visible in daylight and that it smelled and tasted bitter." (Sigurdsson *et al.*, 1987).

Discussion. It seems extremely unlikely that the sudden emission of a large ground-hugging cloud of CO$_2$ from a volcanic vent, such as occurred at the Dieng volcano, could actually be in any way analogous to CO$_2$ escaping via a natural migration pathway from a storage reservoir. The author is unaware of any closer analogies, e.g. recorded instances of the sudden rupturing of underground oil fields or natural gas fields resulting in the emission of large quantities of gas or oil at the ground surface.

The sudden emissions of concentrated CO$_2$ from crater lakes in the Cameroon are the result of slow emissions of carbon dioxide into deep, stratified lakes. It would be relatively simple to determine whether any such lakes occur in the vicinity of a proposed CO$_2$ disposal site. Thus the possibility of an analogous event resulting from the leakage of CO$_2$ from a storage reservoir could easily be excluded.

Nevertheless, the disasters on the Dieng Plateau, Lake Nyos and Lake Monoun indicate that large clouds of concentrated CO$_2$ can be almost instantaneously lethal. It appears possible that if CO$_2$ was emitted from an underground storage reservoir as a result of an injection well failure, a cloud of dense CO$_2$-rich gas could develop. Further research into the chances of such a well failure occurring, how a failed well could be sealed, the CO$_2$ emissions which might result, and modelling of the movement and dispersion of any CO$_2$-rich cloud that might develop, is recommended.

ACKNOWLEDGEMENT

This paper is published with the permission of the Director, British Geological Survey (NERC).

REFERENCES

Allard, P, Carbonnelle, J, Dajlevic, D, Le Bronec, J, Morel, P, Robe M C, Maurenas, J M, Faivre-Pierret, R, Martin, D, Sabroux, J C & Zettwoog, P. 1991. Eruptive and diffuse emissions of CO$_2$ from Mount Etna. *Nature*, **351**, 387-391, 30 May 1991.

Allard, P, Dajlevic, D & Delarue, C. 1989. Origin of carbon dioxide emanation from the 1979 Dieng eruption, Indonesia: Implications for the origin of the 1986 Nyos catastrophe. *Journal of Volcanology and Geothermal Research,* **39** (2-3), 195-206.

Baxter, P J & Kapila, M. 1989. Acute health impact of the gas release at Lake Nyos, Cameroon, 1986. *Journal of Volcanology and Geothermal Research,* **39** (2-3), 265-275.

Cox, H, Heederik, J P, Van Der Meer, L G H, Van Der Straaten, R, Holloway, S, Metcalfe, R, Fabriol, H & Summerfield, I. 1996. Safety and stability of underground storage. Pp.116-162 in: *The Underground Disposal of Carbon Dioxide.* Final Report of Joule II project No. CT92-0031.

Farrar, C D, Sorey, M L, Evans, W C, Howle, J F, Kerr, B D, Kennedy, B M, King, B-Y & Southon, J R. 1995. Forest-killing diffuse CO_2 emission at Mammoth Mountain as a sign of magmatic unrest. *Nature,* **376,** 675-677, 24 August 1995.

Gunter W D, Perkins, E H & McCann T J. 1993. Aquifer disposal of CO2-rich gases: Reaction design for added capacity. *Energy Conversion and Management,* Volume 34, pp. 941-948.

Holloway, S & Van Der Straaten, R. 1995. The Joule II project 'The Underground Disposal of Carbon Dioxide'. *Energy Conversion and Management,* **36** (6-9), 519-522.

Kanari, S. 1989. An inference on the process of gas outburst from Lake Nyos, Cameroon. *Journal of Volcanology and Geothermal Research,* **39** (2-3), 135-150.

Kruse, H & Tekeila, M. 1996. Calculating the consequences of a CO_2-pipeline rupture. *Energy Conversion and Management,* Volume 37, Nos 6-8, pp.1013-1018.

Lindeberg, E & Van der Meer. L G H. 1996. Reservoir modelling and enhanced hydrocarbon recovery. Pp. 163-182 in:The Underground Disposal of Carbon Dioxide. Final Report of Joule II project No. CT92-0031.

Sigurdsson, H., Devine, J.D., Tchoua, F.M., Presser, T.S., Pringle, M.K.W.. & Evans, W.C. 1987. Origin of the lethal gas burst from Lake Monoun, Cameroun. *Journal of Volcanology and Geothermal Research,* **31,** 1-16.

Sigvaldason, G E. 1989. International Conference on Lake Nyos Disaster, Yaounde, Cameroon 16-20 March, 1987: Conclusions and Recommendations. *Journal of Volcanology and Geothermal Research,* **39** (2-3), 97-107.

Stupfel, M & Le Guern, F. 1989. Are there biomedical criteria to assess an acute carbon dioxide intoxication by a volcanic emission? *Journal of Volcanology and Geothermal Research,***39** (2-3), 247-264.

Tazieff, H. 1989. Mechanisms of the Nyos carbon dioxide disaster and so-called phreatic steam eruptions. *Journal of Volcanology and Geothermal Research,* **39** (2-3), 109-116.

Tazieff, H, Faivre-Pierret, R X & Le Guern, F. 1986. La catastrophe de Nyos, (Cameroun), 21 Aout, 1986. Ministere de la Cooperation, Paris, Contribution C.F.R. no 811, CNRS-CEA, Gif sur Yvette, 79pp.

Thorarinson, S. 1950. The eruption of Mt Hekla 1947-1948. *Bulletin Volcanologique, Series II,* Volume 10, 157-168.

Williams, S N. 1995. Dead trees tell tales. *Nature,* **376,** 644, 24 August 1995.

Zhang, Y. 1996. Dynamics of CO_2-driven lake eruptions. *Nature,* v. 379, p.57, 4th January, 1996.

Energy Convers. Mgmt Vol. 38, Suppl., pp. S247–S252, 1997
© 1997 Elsevier Science Ltd. All rights reserved
Printed in Great Britain
0196-8904/97 $17.00 + 0.00

Pergamon

PII: S0196-8904(96)00277-4

GAS POWER WITH CO_2 DEPOSITION LOCATED ON ABANDONED PLATFORMS

TORLEIF HOLT & ERIK LINDEBERG

IKU Petroleum Research
N-7034 Trondheim, Norway

ABSTRACT

A pre-study of a concept for power production offshore on abandoned platforms shows that a net delivered 300 MWe combined cycle power plant and a CO_2 sequestering plant can be located on a single Condeep platform. By replacing the present gas turbine power production on neighbouring oil fields, the CO_2 emissions will be reduced by 1.6 Mt per year, corresponding to 4.4% of the total Norwegian emissions. The concept appears to be both technically and economically feasible. © 1997 Elsevier Science Ltd

KEYWORDS

Offshore power production, CO_2 emission, CO_2 deposition, Abandoned platforms, Economy.

INTRODUCTION

Abandonment of platforms in the North Sea is fast becoming an important issue as an increasing number of oil and gas fields are approaching the end of their productive life, and operators consider how to dispose of the redundant installations. At the end of 1995 there were more than 420 producing fields in the Norwegian, UK, Danish, Dutch and Irish waters (Terdre *et al.* 1996). Eventually, all will cease production and the facilities installed upon them will have to be decommissioned, creating a great challenge to the technical and scientific community to find environmentally and economically acceptable solutions for their abandonment. The legislative framework is still in the process of formation, and the result will depend on international negotiations.

Approximately 70 installations are placed on the Norwegian continental shelf. The cost of abandoning these structures has been estimated to a total of 50 billion NOK (NPD 1994). Many of the structures are relatively small units (less than 40 000 tonnes), typically on a steel jacket, while others are floaters. These are easy to remove completely and recycle for a moderate cost. About ten of the Norwegian platforms are, however, large concrete gravity base structures consisting of more than 500 000 tonnes of steel and concrete with the base pressed into the sea-floor. The abandonment of these large units provides the real challenge.

It has previously been suggested that CO_2 from a gas power plant could be sequestered and used as injection gas in oil reservoirs for improved oil recovery (Holt and Lindeberg 1988, Holt and Lindeberg 1993). In an economic environment where the credit could be taken both on a CO_2 tax saving, the sale of electric power and enhanced oil recovery (EOR), this concept was proven to be economically interesting. In addition to the power and CO_2 separation plant, the major investment costs were a high voltage direct current transport system for power and a large offshore carrier if the plant was to be located offshore near the oil field. For an onshore location of the power plant, an extra CO_2 pipeline from the shore to the injection wells was required.

In the concept presented here, all these extra investments can be omitted compared with previous studies, because the power plant is assumed located on an abandoned oil platform, and the power is distributed to nearby platforms on less expensive AC cables. A combined cycle power unit with a gross generating capacity of 360 MWe is considered. When CO_2 separation is combined with this power plant, the net energy output will be approximately 300 MWe. The power is sold at a competitive price compared to the present cost for offshore production. The power cost offshore is high due to the CO_2 tax, and low generating efficiency compared to a combined cycle plant.

The sequestered CO_2, approximately one million tonnes per year, will be injected into some of the existing wells, either into an oil reservoir or into an over or under laying aquifer. The existence of aquifers in the vicinity of oil reservoirs makes the concept flexible in respect to treatment of breakthrough CO_2 from petroleum reservoirs. No EOR credit is accounted for this injection in the base case economic calculations, however.

This paper is based on a pre study evaluating of the use of old petroleum platforms for power production offshore. The study (Holt and Lindeberg 1996) was performed for the environmental foundation Bellona.

SITE OF POWER PLANT

Both at the Ekofisk, Frigg and Heimdal fields there are large platforms that will be taken out of use in the near future. These platforms are, however, ruled out as potential sites for offshore power plants due to subsidence of the field or to small power markets, respectively. The next really large platform that goes out of production is the Statfjord A platform. This Condeep type platform was one of the last of this type which was not designed to be refloated. Removal may therefore be particularly difficult. It is located on the Tampen bank 220 km north west of Bergen with major producing fields nearby. The production on these fields will also decline, but a more than 300 MWe power demand is expected in this oil province for at least 15 years out in the next millennium.

TECHNICAL DESCRIPTION

A simplified flow diagram for the concept gas power without release of CO_2 is shown in Figure 1. A brief description of each main element also including the carrier is given below.

The Statfjord A platform is a large concrete, gravity base structure consisting of several hundred thousand tonnes of steel and concrete with the base pressed into the sea-floor at 146 m depth. When the present process equipment is removed it offers a 90m × 70m base construction area with a total carrying capacity of 50 000 tonnes. For this concept the power and the CO_2 separation plants represent the main mass loads.

The placement of several plants on a limited area will require a different plant layout compared to ordinary land based installations. In this work it is assumed that it is possible to place the equipment on the available space. This assumption is justified by the fact that the weight carrying capacity of the platform is almost twice the weight of the process equipment. The possibility of

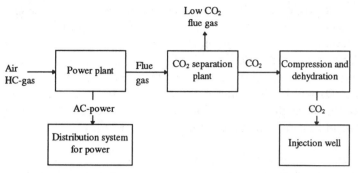

Fig. 1   Power production with CO_2 separation and deposition. Simplified flow diagram of the concept.

reducing the space requirements of a standard plant is exemplified by the EPOS (Electric Power On Sea) power plant (Jeffs 1981). This 375 MWe plant was designed to be built on 2520 m^2 area compared to 6000-8000 m^2 plant area for the standard 360 MWe plants. Based on drawings of the layout of the EPOS-concept it is likely that the construction area of this plant could have been compacted further.

A modified standard combined cycle power plant with a net output of 360 MWe forms the basis for the power plant calculations. The modifications, compared to a standard plant, are that low pressure steam is extracted from the process to feed the MEA-reboiler in the CO_2 separation plant (see below). Further, a fraction of the exhaust gas from the steam boiler is cooled and recirculated to the inlet of the gas turbine compressor where it is mixed with air before compression. Based on previous work, 40% of the exhaust gas are assumed recirculated (Bolland *et al*. 1991, Kværner 1995). This increases the CO_2 concentration in the exhaust gas from approximately 3% to 6%, and thus allows a reduction in the size of the CO_2 separation plant.

Due to the extraction of low pressure steam from the steam turbine, the efficiency of the power plant is reduced. Further, mechanical work and electricity are supplied to the CO_2 separation plant, to the CO_2 compressors and other auxiliaries. The net effect delivered from the plant is thus reduced to 300 MWe. The calculations of gas consumption and CO_2 production are based on 360 MWe power, produced at a thermal efficiency of 60%. The mass of a 360 MWe combined cycle power plant is estimated to 13 200 tonnes.

CO_2 can be removed from power plant exhaust gas commercially by means of several separation processes, but here only chemical active absorption based on use of monoethanolamine (MEA) is considered. The optimal amount of CO_2 that can be removed from the exhaust depends on the size of the absorption unit and the concentration of CO_2 in the exhaust. For a standard plant the economical recovery limit is approximately 85% for 3% CO_2 in the exhaust and 90-92% for 8%. Here a plant with 90% CO_2 recovery was used in the calculations. The light weight and smaller sized process developed by Kværner Water Systems was considered to be taken into use, but was not chosen due to the higher costs, as size and weight limitations were not to be critical for the present carrier. The mass and area of a standard separation plant needed for the 360 MWe power plant were estimated to 12 500 tonnes and 110m × 55m, respectively. It is assumed that the area requirements can be greatly reduced by a more compact lay out. The height of the absorption tower will typically be 45m. The CO_2 from the separation plant is cooled, compressed and dried before it is piped to the injection well. The necessary equipment, which includes a gas scrubber, multistage compressors and a drying plant.

The dried and compressed CO_2 leaves the platform and is injected into an underground formation. The mass to be injected corresponds to 914 000 tonnes per year (90% of the CO_2

formed during power production). This is comparable to the amount to be injected into the Utsira formation in the Sleipner Vest CO_2 disposal project, where only one injection well is to be used (Korbøl and Kaddour 1995). At the end of the petroleum production period there will be a large number of available wells on the platform. It is assumed that one existing well can be chosen for injection purposes, eventually after shutting off the well bore at the desired position and re-perforating. The CO_2 will be deposited in a location where there is no danger for the fluid to migrate to any of the petroleum producing reservoirs in the region. The optimal location for CO_2 deposition may be in an aquifer or an isolated part of an oil reservoir.

It will be necessary to lay two major AC cable links from the generating platform to the two neighbouring fields. It will also be necessary to connect Statfjord B and C to Statfjord A, but these cables are much shorter. Gullfaks A, B and C are already inter-connected, but it is possible that this line has to be upgraded.

REDUCTION IN CO_2 EMISSIONS

By substituting local power produced with gas turbines with power from a central combined cycle plant the CO_2 emissions will decrease even without CO_2 deposition due to the higher thermal efficiency of the combined cycle process. The present CO_2 emissions and emissions from a combined plant, both with and without CO_2 recovery and deposition, are summarised in Table 1. In the calculation some emissions from four remaining gas turbines kept for back up are also included, assuming that each turbine is in operation three months per year. The figures are compared to the present Norwegian CO_2 emissions of 36.4 million tonnes per year.

Table 1 CO_2 emissions for various power production scenarios.

Case	CO_2 emissions (10^6 ton/year)	Part of Norwegian emissions
Use of present installed gas turbines	1.80	4.9 %
Central plant with CO_2 deposition	0.173	0.5 %
Central plant without CO_2 deposition	1.073	2.9 %

ECONOMIC ANALYSES

The investment costs for the various elements are summarised in Table 2. In addition to the basic equipment costs for the separate elements, come interest in the building period, engineering costs, administration costs, training, testing and start up costs and contingency. The total addition amounts to 22% of the total equipment costs. By taking the petroleum production platform in use as carrier for the power plant, the abandonment and the corresponding costs are postponed. These costs can be regarded as an income or a negative investment cost at the start of the project, but enter as a cost at the end of the project period.

Following removal of topside equipment, the main platform construction may either be dumped in deep water or demolished and the steel recovered. It has not been possible to acquire reliable cost data for these operations. One reason for this is that abandonment is still far ahead for most installations in the North Sea. In a study by the Norwegian Petroleum Directory (1994) the average cost of abandoning each of the about 70 large and small installations placed on the Norwegian Continental Shelf was 108 million USD (MUSD). Since the Statfjord A platform belongs to the small group of large installations, and is not designed to be refloated, the cost of abandonment will be considerable larger than the average. In this work it will be assumed that 153.5 MUSD can be saved if only the platform is cleared for existing process equipment compared to complete abandonment. This is an uncertain, but most likely conservative estimate. For comparison, the costs of abandonment of 92 000 tonnes steel gravity base, and 19 000 tonnes topside weight platform at Maureen (UK) is estimated to 69 MUSD or 100 MUSD, for dumping or full recovery, respectively. This platform is designed to be refloated (Terdre *et al.* 1996).

Table 2   Calculation of total capital investment.

Component	Amount (10^6 USD)
1. Power plant	265.5
2. CO_2 separation plant	136.8
3. Compressors and Dehydration plant	14.9
4. Power distribution system	69.7
5. Injection well	12.3
6. Total costs of equipment	499.2
7. Interests	35.5
8. Additional costs (22 % of 6)	109.6
9. Carrier	-153.5
10. Total capital investment	490.8

The costs of operation and maintenance (running costs) of the installations amount to 72.5 MUSD. Costs of the gas feed to the power plant (39.1 MUSD), running and maintenance of the power plant (21.1 MUSD) and CO_2 separation plant (9.8 MUSD), are the main cost elements. Costs and incomes to the project are based on a yearly running time of 8000 hours.

The incomes to the project are due to the delivery of electric power to the petroleum installations in the region. The price of electric power is estimated to 0.071 USD/kWh based on the gas price, CO_2 tax and running costs of the gas turbines presently in operation. The incomes from sales of electric power amount to 169.6 MUSD. The possible incomes from sale of CO_2 for enhanced oil recovery (EOR) in the neighbouring oil fields are estimated to 18.9 MUSD. This possible income is not included in the base case calculations.

The net present value (NPV) and the internal rate of return (IRR) for the project is determined to 340 MUSD and 17.6%, respectively for the base case calculations with 15 years project time and 7% discount rate. No income taxes are included in the calculations. This implies that no attempts are made to divide the profits between the state and the owners of the project. The sensitivity on NPV on changes in some of the economic factors is summarised in Figure 2.

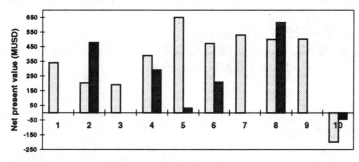

Fig. 2.   Sensitivity analyses.1) Base case NPV, 2) investment cost varied ± 30%, 3) change to small size and light weigh CO_2 recovery plant, 4) platform abandonment cost varied ± 50%, 5) sales price of electric power varied ± 20%, 6) running costs varied ± 20%, 7) sales of CO_2 for EOR included, 8) increased lifetime to 20 and 25 years, 9) full CO_2 release to atmosphere, 10) transport of power to land, 20 and 25 years project lifetime.

CONCLUSIONS

One single Condeep platform has a carrying capacity for building a net 300 MWe combined cycle power plan with CO_2 removal. This power production is sufficient to replace the existing power production from gas turbines on the Norwegian oil fields Statfjord, Gullfaks and Snorre.

Power production with CO_2 disposal from a 300 MWe central unit will reduce the emissions of CO_2 with 1.6 Mt per year compared to the present use of gas turbines. These reductions represent 4.4% of the total Norwegian CO_2 emissions.

The present concept is technically feasible. The internal rate of return is estimated to 17.6%. The net present value is 2211 MNOK from a total capital investment of 3190 MNOK.

Central power production without CO_2 removal gives and internal rate of return of 32.2% provided that all the power (360 MWe) can be sold offshore. This solution reduces the CO_2 emissions with 0.7 Mt, which corresponds to 2.0% of the present Norwegian emissions.

REFERENCES

BOLLAND, O., SÆTHER, S., HYLLSETH, M. and LUNDE, O. 1991: Gas Fired Power Plant with Reduced Emissions of Carbon Dioxide. Summary Report. STF report no. STF15 A91070, Trondheim.

FIGVED, J.O. 1992: Høgspent - likestrømsforsyning fra land. Foredrag ved NIF-seminar om el-kraftforsyning offshore,10-12 juni, Stavanger (in Norwegian).

HOLT, T. and LINDEBERG, E. 1988: Miljøvennlig gasskraft kombinert med økt oljeutvinning - en forstudie. IKU report no. 34.2776.00/01/88.

HOLT, T. and LINDEBERG, E. 1993: CO_2 som injeksjonsgass. En forstudie over muligheter for å øke oljeutvinning og å redusere utslippene av klimagasser ved å bruke CO_2 som injeksjonsgass. Hovedrapport. IKU report no. 34.2899.00/06/92, Trondheim (in Norwegian).

HOLT, T. and LINDEBERG, E. 1996: Offshore gas power with CO2 removal supplied from a plant located on an abandoned platform. A pre-study. IKU report no. 54.5051.00/01/96.

JEFFS, E. 1981: EPOS gas energy from marginal offshore fields, Modern Power Systems (February) pp. 25-27.

KORBØL, R. and KADDOUR, A. 1995: Sleipner Vest CO_2 Disposal - Injection of Removed CO_2 into The Utsira Formation." *Energy Covners. Mgmt*. Vol. 36, No. 6-9, pp. 509-512.

Kværner 1995: CO_2-separasjon fra gassturbiner offshore. Kværner Water Systems report no. 3578-0000, 38 pp.

NPD Annual Report 1979 - 1994. Stavanger.

TERDRE, N., CORCORAN, M. and STEINNES, H. 1996: The North Sea ABANDONMENT Handbook 1996. The TCS Partnership, Surrey, England.

Energy Convers. Mgmt Vol. 38, Suppl., pp. S253–S258, 1997
© 1997 Elsevier Science Ltd. All rights reserved
Printed in Great Britain
0196-8904/97 $17.00 + 0.00

Pergamon

PII: S0196-8904(96)00278-6

DEEP SUB-SEABED DISPOSAL OF CO_2
- THE MOST PROTECTIVE STORAGE -

H. KOIDE,[*1] Y. SHINDO,[*2] Y. TAZAKI,[*3] M. IIJIMA,[*4]
K. ITO,[*4] N. KIMURA,[*5] and K. OMATA[*5]

[*1] Geological Survey of Japan,AIST,MITI,
1-1-3 Higashi,Tsukuba,305 Japan,
[*2] National Institute of Materials and Chemical Research,AIST,MITI,
1-1 Higashi,Tsukuba,305 Japan,
[*3] Kanto Natural Gas Dev. Co.,
3-1-20 Nihonbashi Muromachi,Tokyo,103 Japan,
[*4] Mitsubishi Heavy Industries,Ltd.,
3-3-1 Minatomirai,Nishi-ku,Yokohama,220 Japan,
[*5] Electric Power Development Co.,
6-15-1 Ginza,Chuo,Tokyo,104 Japan

ABSTRACT

The sub-seabed disposal of CO_2 is safer than the disposal of CO_2 in inland aquifers. Even if small amounts of CO_2 seeped out of sea floor, CO_2 would disperse and dissolve into sea water. On the surface of the sea, there exist no depressions where CO_2 may concentrate. Sediments under deep sea floor are very cool because the deep oceanic water is usually at a few degrees centigrade. CO_2 hydrate is formed in sediments under wide areas of ocean floor deeper than about 300m. Virtually complete isolation of huge amounts of CO_2 is possible by the deep sub-seabed disposal. Liquid CO_2 with heavy suspension intrudes laterally under light unconsolidated sediments at sea floor deeper than about 3700m. Lateral intrusion technique for the super-deep sub-seabed disposal of CO_2 can protect the ecology on the sea floor. © 1997 Elsevier Science Ltd

KEYWORDS

CO_2; sub-seabed disposal; CO_2 hydrate; aquifer; unconsolidated sediments

INTRODUCTION

The principal cause of global warming is the huge amount of CO_2 emitted by burning fossil fuels extracted from underground deposits. Although the carbon dioxide is not very poisonous to animals and is even useful for plants, the enormous volume of manmade CO_2 emissions makes the development of effective mitigation technology extremely difficult. About 60% of CO_2 emissions in Japan are from fixed massive sources such as electric power plants, mills, etc. While intensive efforts are being made for the development of CO_2 recovery technology from flue gas of power plants and mills, we should find suitable spaces for the disposal of separated CO_2. Oceanic disposal of CO_2 is still questionable due to the risk to the marine environment.

Depleted natural gas and oil fields have porous underground reservoir rocks and trap structures that can contain gas and liquids safely over several million years. Some natural gas deposits contain a large fraction of CO_2. It is possible to use depleted underground natural gas and oil reservoirs as storage sites for CO_2 because impermeable caprocks prevent the leakage of CO_2. Simple carbon dioxide is stored as a supercritical fluid in reservoirs deeper than about 800m. Cooler reservoirs having a low geothermal gradient can store more carbon dioxide than warmer reservoirs having a high geothermal gradient(Koide *et al.*, 1994).

Prediction of the fate of underground CO_2 is necessary for the safe subterranean disposal of CO_2. About 3 MPa of pressure at about 5°C is sufficient to form CO_2 hydrate in aquifers. Formation of CO_2 hydrate in the pores of rocks virtually completely prevents leakage of carbon dioxide(Koide *et al.*, 1995). Carbon dioxide injected in deep reservoirs may migrate upward in aquifer systems and forms a CO_2 hydrate cap at the depth range of 200m-400m below the water surface in cool sedimentary basins in high latitudes and in wide areas of cool marine sedimentary basins deeper than two or three hundred meters(Koide *et al.*, 1995).

In and around the Japanese islands, the Engineering Advancement Association of Japan (ENAA) estimated (Tanaka *et al.*,1994) that as much as 2 billion tons of CO_2 can be stored as a supercritical fluid in oil and gas reservoirs, as much as 1.5 billion tons in aquifers with trap structures, as much as 16 billion tons as solution in groundwater in other aquifers on land and as much as 72 billion tons in offshore aquifers. Koide *et al.*(1992) estimated that 320 billion tons of CO_2, at least, can be stored in aquifers in sedimentary basins in the world.

Artificial emission of CO_2 concentrates in urban areas and in industrial centers. However, it is very difficult in densely populated areas to find lands available for CO_2 injection-transportation

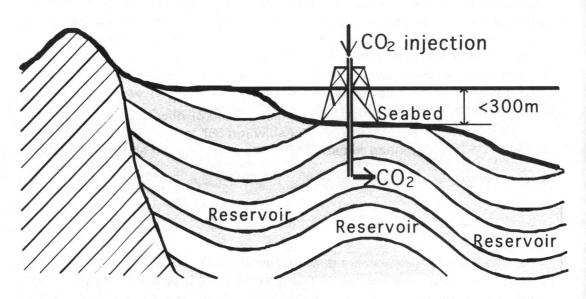

Fig.1 Shallow sub-seabed disposal of CO_2

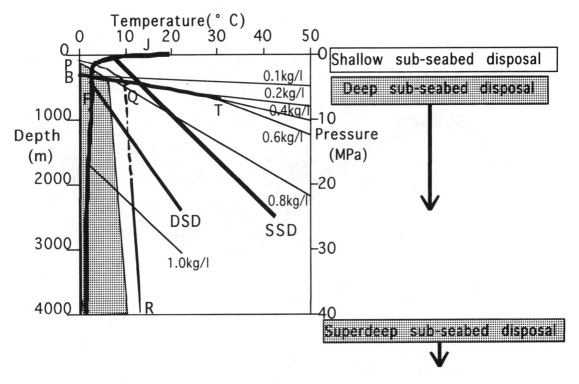

Fig.2 Sub-seabed disposal and state of CO_2 at temperatures and pressures below the seabed. B-Q-T is the boiling curve of CO_2, while T denotes the critical point at 31.1°C,7.39MPa. P-Q-R is the decomposition criterion of CO_2 hydrate. The dotted area is the temperature-pressure range where CO_2 hydrate may be formed in marine sediments. J-F-K is a temperature profile in the northern Pacific Ocean. The lines SSD and DSD are simplified examples of possible temperature-pressure increase with depth in marine sediments in cases of shallow and deep sub-seabed disposal, respectively.

facilities. The sub-seabed disposal of CO_2 has great advantage for safety and for accessibility in Japan that consists of small islands.

SHALLOW SUB-SEABED DISPOSAL OF CO_2

The sub-seabed disposal of CO_2(Fig.1) is technically and economically feasible for submarine aquifers below shallow seabed(Korbøl and Kaddour, 1995). The cost for construction of injection facilities is reduced where offshore platforms for oil or gas production are available. The disposal of CO_2 in submarine aquifers is safer than the disposal of CO_2 in inland aquifers. Even if small amounts of CO_2 seeped out of sea floor, CO_2 would disperse and dissolve into sea water. On the surface of the sea, there exist no depressions where CO_2 may concentrate. CO_2 disposal in aquifers under sea floors shallower than 200 or 300m is safe and economically reasonable technology for CO_2 mitigation. Huge volumes of CO_2 storage are expected in sedimentary basins under continental shelves in the world.

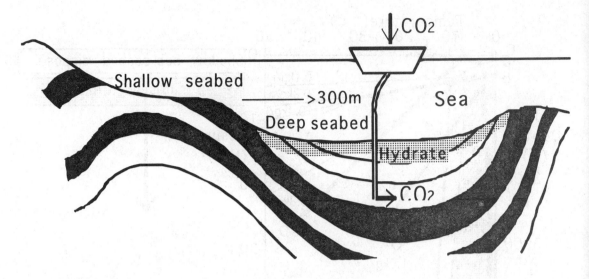

Fig.3 Deep sub-seabed disposal of CO_2.

DEEP SUB-SEABED DISPOSAL OF CO_2

Sediments under deep sea floor are very cool because the deep oceanic water is usually at only a few degrees centigrade. CO_2 hydrate is formed in sediments under wide areas of ocean floor deeper than about 300m where the temperature of sediments is lower than about 5°C(Fig.2). CO_2 injected in submarine aquifers migrates and forms CO_2 hydrate cap in cool sediments just below deep ocean floor(Fig.3). Virtually complete isolation of huge amounts of CO_2 is possible by the deep sub-seabed disposal in sedimentary basins under sea floor deeper than about 300m. Deep drilling at deep seabed is technically difficult and expensive. Deep sub-seabed aquifers provide the most protective but less accessible storage for CO_2.

SUPERDEEP SUB-SEABED DISPOSAL OF CO_2

Shindo *et al.*(1993) proposed long-term storage of liquid CO_2 in depressions of very deep seabed. Thin film of CO_2 hydrate and layer of CO_2 saturated sea water protect heavy CO_2 liquid in depressions of seabed deeper than 3700m(Fig.4). However, some ecological impacts are inevitable even at very deep seabed.

Thick unconsolidated sediments fill some depressions of seabed. Effective density of some shallow unconsolidated sediments is only a little higher than seawater. Effective density of liquid CO_2 can easily be made higher with mixture of heavy suspension such as clay or ash than shallow unconsolidated sediments under seabed deeper than 3700m. Falling steel penetrator can easily penetrate some 10 meters into unconsolidated sediments under deep sea floor. Liquid CO_2 with heavy suspension can be injected laterally under lighter unconsolidated sediments without serious damages to sea floor(Fig.5). Lateral intrusion technique for the super-deep sub-seabed disposal of CO_2 hardly affect the ecology on the seabed.

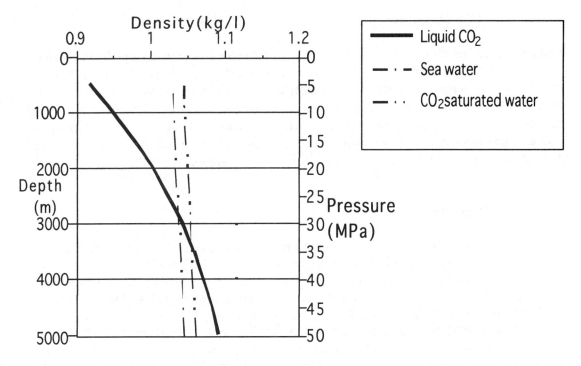

Fig.4. Approximate relation between pressure and density of CO_2, sea water and CO_2 saturated sea water at 3°C.

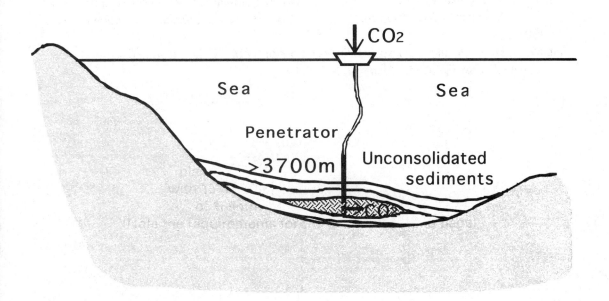

Fig.5 Superdeep sub-seabed disposal of CO_2

CONCLUSION

Shallow sub-seabed disposal of CO_2 in aquifers under seabeds shallower than 200 or 300m is technically and economically feasible and safer than inland aquifer disposal. Deep sub-seabed disposal of CO_2 in aquifers under seabeds deeper than 300m provide the most protective storage for CO_2 thanks to formation of CO_2 hydrate in sub-seabed sediments. Superdeep sub-seabed disposal of CO_2 under light unconsolidated sediments also hardly affect the ecology on the seabed. Three types of sub-seabed disposal of CO_2 - shallow, deep and super-deep sub-seabed disposal - can realize CO_2 emission free fossil fuel power generation.

REFERENCES

Fujioka, Y., M. Ozaki, K. Takeuchi, Y. Shindo and Y. Yanagisawa (1995). Ocean sequestration at the depths larger than 3700m. *Energy Convers. Mgmt,* 36, 551-554.

Koide, H., M. Takahashi, H. Tsukamoto and Y. Shindo (1995). Self-trapping mechanisms of carbon dioxide in the aquifer disposal. *Energy Convers. Mgmt,* 36, 505-508.

Koide, H., M. Takahashi, Y. Shindo, Y. Noguchi, S. Nakayama, M. Iijima, K. Ito and Y. Tazaki (1994). Subterranean disposal of carbon dioxide at cool formation temperature, *The Conference Proceedings from CLEAN AIR'94,* 63-72.

Koide, H., Y. Tazaki, Y. Noguchi, S. Nakayama, M. Iijima, K. Ito, and Y. Shindo (1992). Subterranean containment and long-term storage of carbon dioxide in unused aquifers and in depleted natural gas reservoirs. *Energy Convers. Mgmt,* 33, 619-626.

Korbøl, R. and A. Kaddour (1995) Sleipner Vest CO_2 disposal - Injection of removed CO_2 into the Utsira formation. *Energy Convers. Mgmt,* 36, 509-512.

Shindo, Y., Y. Fujioka, M. Ozaki, K. Takeuchi and H. Komiyama (1993). New concept of deep sea CO_2 sequestration. *Proc. Int. Symp. on CO_2 Fixation & Efficient Utilization of Energy,* 307-314.

Tanaka, S., H. Koide and A. Sasagawa (1994). Possibility of CO_2 underground sequestration in Japan. *Energy Convers. Mgmt,* 36, 527-530.

Energy Convers. Mgmt Vol. 38, Suppl., pp. S259–S264, 1997
Published by Elsevier Science Ltd
Printed in Great Britain
0196-8904/97 $17.00 + 0.00

Pergamon

PII: S0196-8904(96)00279-8

PROGRESS ON BINDING CO$_2$ IN MINERAL SUBSTRATES

Klaus S. Lackner, Darryl P. Butt, Christopher H. Wendt

Los Alamos National Laboratory, Los Alamos, New Mexico 87545

ABSTRACT

Based on current estimates of reserves, coal could satisfy even a very much increased world energy demand for centuries, if only the emission of CO$_2$ into the atmosphere could be curtailed. Here we present a method of CO$_2$ disposal that is based on combining CO$_2$ chemically with abundant raw materials to form stable carbonate minerals. A major advantage of this method is that the resulting waste product is thermodynamically stable and environmentally neutral. It is therefore possible to store large quantities permanently with minimal environmental impact and without the danger of an accidental release of CO$_2$ which has proven fatal in quantities far smaller than contemplated here. The raw materials to bind CO$_2$ exist in nature in large quantities in ultramafic rocks. They are readily accessible and far exceed what would be required to bind all CO$_2$ that could possibly be generated by burning the entire fossil fuel reserves. In this paper we outline a specific process that we are currently investigating. Our initial rough cost estimate of about 3¢/kWh is encouraging. The availability of a CO$_2$ fixation technology would serve as insurance in case global warming, or the perception of global warming, would cause severe restrictions on CO$_2$ emissions. If the increased energy demand of a growing world population is to be satisfied from coal, the implementation of such a technology would be unavoidable.   Published by Elsevier Science Ltd

KEYWORDS

Carbon Dioxide Disposal; Carbonate Mineral; Peridotite; Serpentinite; Magnesium Oxide; Calcium Oxide; Greenhouse Effect; Power Plant Emission; Emission Control.

INTRODUCTION

An economic method of CO$_2$-disposal in solid form would eliminate the potential danger of climate changes due to greenhouse gas emissions from burning fossil fuels. The availability of such technology would allow the large coal reserves (estimated at 10,000 Gt (United Nations, 1993)) to satisfy the growing demand for world-wide energy without an impact on atmosphere and climate. More immediately, such a technology could protect a large investment in energy infrastructure in case that strict limits on future CO$_2$ emissions are deemed necessary.

In this paper we report on the current state of our effort to develop a method of binding CO$_2$ as chemically inert mineral carbonate. We outline a process that we consider viable and we discuss a preliminary cost estimate. Our discussion will leave out the extraction of CO$_2$ from flue gases and their transportation to the disposal site which has been studied by others and is much better understood than the disposal (Audus *et al.*, 1995). Also, these steps would be in common with many other disposal methods (Booras & Smelser, 1991; Hendriks *et al.*, 1991; Herzog *et al.*, 1993; Steinberg & Cheng, 1985).

There are two major reasons for considering disposal of CO$_2$ as a carbonate mineral. First, the oxides of calcium and magnesium are available in nature in quantities vastly exceeding those of fossil fuels, even if one limits oneself to those that are not already carbonated. Secondly the carbonation reaction, that binds CO$_2$ to calcium oxide or magnesium oxide, is thermodynamically favored and exothermic even if one starts from natural minerals in which these oxides are chemically bound into a silicate structure. Thus, at least in principle, there is no need to consume energy which suggests that the cost could be held low. The widespread presence of natural carbonates, which is explained by their thermodynamic stability, demonstrates that the waste product we generate is environmentally benign and stable.

A major advantage of mineral carbonate disposal is that it avoids long term storage of CO_2 as a free phase. Storing gigatons of a substance that is gaseous at ambient surface conditions is fraught with risk. Just as in nuclear waste disposal, any storage system would have to be guaranteed as safe for centuries against the possibility of an accidental release. A natural disaster in 1986 at Lake Nyos, Cameroon, has demonstrated these hazards (Kling et al., 1987). The waters deep in the crater lake had been saturated with CO_2 from underwater volcanic vents. Due to an instability in the layering, the lake turned over causing a sudden release of a CO_2 bubble on the order of $0.1\,km^3$ (Freeth, 1994) or about the weekly output of a single gigawatt power plant. The gas overflowed the crater rim and, because it is heavier than air, flowed down a valley killing 1700 people. The danger of a sudden gas release from the storage site is completely eliminated in our approach, because the mineral carbonate once formed is stable and the reaction cannot be readily reversed.

RESOURCES

The only common oxides that readily form stable mineral carbonates are calcium and magnesium oxides. The prototypical reactions are those of the pure oxides which at $179\,kJ\,mol^{-1}$ and $118\,kJ\,mol^{-1}$ respectively are very exothermic. The heat release may be compared with the enthalpy of reaction in burning carbon to CO_2 which is $394\,kJ\,mol^{-1}$.

Pure CaO and MgO are rare in nature, but many common minerals may be regarded as a combinations of these oxides with others forming a silicate matrix. The large scale of fossil fuel consumption, which amounts to 6 Gt of carbon per year world-wide, requires that mineral deposits used in the disposal must be very large. Fortunately there are very large deposits rich in magnesium or calcium, or both. Of course, we exclude deposits that are already carbonated, like limestone. Instead, we consider igneous deposits that are essentially free of carbonates.

Ultramafic igneous rocks are rich in magnesium oxide. The richest is dunite. It belongs to the peridotite group and contains ~50% magnesium oxide by weight as forsterite. In its pure form forsterite contains 60% magnesium oxide. Peridotites in general have high magnesium content, as do serpentinites that are frequently associated with them. In North America the largest deposits are in the West. Large areas of California are covered by serpentine or serpentinized peridotites. An exposed peridotite slab in the Klamath mountains of Northern California with a size of about $1000\,km^2$ by 1 km or about $3,000\,Gt$ (Irwin, 1977; Lindsley-Griffin, 1977) demonstrates the scale of the deposits. Other large deposits are found in the Northeast. There are many more smaller deposits, for example in Ontario, Maine, Virginia and Texas. Peridotite can be found all over the world (Coleman, 1977; Nicolas, 1989). In the context of CO_2 disposal, one should mention Scandinavia and Japan. The total amount of magnesium oxides found in such readily accessible deposits far exceeds the worldwide coal reserves which are estimated to about 10,000 Gt.

The deposits of calcium oxides in basalts are even larger and more widespread. The Columbia River flood basalt is an example of a large deposit. It forms a sheet about 1 km thick with a volume of $200,000\,km^3$ (Hooper, 1982). This sheet alone could absorb all the CO_2 generated from the estimated world coal reserves 2.5 times over. There are even larger flood basalt sheets, e.g., the Dekka flats in India. However, large deposits containing calcium oxide are rarely as rich as magnesium bearing deposits. In most cases one must settle for calcium oxide concentrations between 9 and 13%. From the standpoint of material processing, this gives a clear advantage to magnesium bearing minerals.

Table 1. Abundant rock types rich in magnesium and calcium oxide. The table is taken from Lackner et al. (1995). R_C is the mass ratio of rock needed for CO_2 fixation to carbon burned. R_{CO_2} is the corresponding mass ratio of rock to CO_2.

	Rock Type	MgO, wt%	CaO, wt%	R_C	R_{CO_2}
Peridotites	Dunite	49.5	0.3	6.8	1.8
	Harzburgite	45.4	0.7	7.3	2.0
	Lherzolite	28.1	7.3	10.1	2.7
Serpentinite		~40	~0	~8.4	~2.3
Gabbro		~10	~13	~17	~4.7
Basalt	Continental tholeiite	6.2	9.4	26	7.1

Binding the CO_2 derived from burning one ton of carbon requires 4.7 t of CaO or 3.3 t of MgO. In the table we give the equivalent amounts for the different types of rocks. Magnesium rich deposits have a big advantage in the amount of material that needs to be processed, which turns out to be about 7 to 10 times the amount of carbon burned. The excavation of magnesium bearing minerals would actually be smaller in scope than most coal mining, because one avoids the overburden typical for coal mining. In terms of mass ratios, these overburdens typically amount to a factor of 20.

ECONOMICS

In the absence of a detailed design we compare an outline of the carbonation disposal process to the cost of existing industrial processes. From this comparison we conclude that a cost of $15 per ton of raw mineral is not an unreasonable goal. For peridotite, this corresponds to $30/t of CO_2 which compares favorably to the Scandinavian CO_2 tax of $50/t (Kaarstad, 1995). At a conversion efficiency of 33%, complete disposal would add 3 ¢ to the kWh. The cost of CO_2 collection and pipelining may add another 3 ¢. At a base price of electricity of 3 ¢/kWh, coal would still be competitive with nuclear energy.

Here we give a simple outline of our estimate. For additional detail see Lackner et al. (1995) and references therein. We consider a process in which raw material is dug up, crushed and ground. Thereafter a small number of processes, e.g. leaching, distilling and reverse calcination are performed to bind CO_2. The final step is the disposal of the resulting materials. For purposes of this discussion, we assume that none of the resulting products have any economic value.

Digging, crushing and grinding of rock costs about $4/t. In copper mining this even includes a flotation process to separate out the copper sulfide. The entire cost of copper refining can be estimated conservatively from a recent low in copper prices to be below $9/t of ore. This cost includes not only the refinement process but also the disposal of waste tailings. Copper mining also operates on a very large scale (measured in ore throughput) and thus is likely to take advantage of economies of scale. The operations performed directly on the ore are simpler than what we are proposing. If we neglect the costs that arise in the processing of the much smaller amounts of copper sulfide and raw copper, this comparison sets a lower limit. Other comparisons can be made to the production of lime, magnesia and alumina. Expressed in terms of the cost per ton of raw material input, all these processes range between $20 to $50 per ton, i.e., 1.5 to 3 times more expensive than the goal we have set ourselves. On the other hand these processes are at least as complex as those we are considering. They consume energy, and it can be shown that they have not fully made use of the available economies of scale. Indeed the lime and cement making which has the largest throughput comes closest to our goal, whereas the manufacture of magnesia, which is performed at a scale at least three orders of magnitude too small, is the most expensive. Therefore, we consider these numbers as loose upper limits that quite likely can be significantly reduced.

CHEMISTRY

We have found (Lackner et al., 1995) that for all common calcium and magnesium bearing minerals the carbonate reaction is exothermic and under ambient conditions it is favored thermodynamically. Because of the high entropy of gaseous CO_2, the equilibrium shifts towards free CO_2 as the temperature increases. This results in an upper limit to the process temperature which is quite high for the carbonation of calcium oxide ($P_{CO_2} = 1$ bar at 890°C), but which is relatively low for the carbonation of magnesium oxide (410°C). For the most common minerals, which we have investigated, this temperature limit ranges from to 170°C to 410°C.

Since the carbonate is the thermodynamically favored state, the goal is to identify a reaction path that with minimal losses in energy can achieve this ground state at reasonable reaction rates. The simplest approach would be direct carbonation of a mineral powder in a gas solid reaction. This would have the advantage that the heat of the reaction would be released at high temperatures and with minimum dilution. The problem of this approach is that the reaction kinetics for calcium and magnesium silicates tends to be too slow and that raising the temperature in order to speed up the reaction kinetics is prohibited by thermodynamic constraints.

We exposed a number of minerals ground to a grain size of 50 to 100 μm to a CO$_2$ atmosphere (0.78 bar) at various temperatures (140 to 300°C) below the calcination point. Carbonation in most cases was insignificant even when the exposure lasted for days. The direct carbonation of minerals is therefore likely to require higher pressure or pretreatment. At high pressures (340 bar) we succeeded in carbonating a serpentinite to about 30% of the stoichiometric limit by flowing carbon dioxide through a packed bed of serpentinized powder. We are currently exploring pretreatment options for serpentinite that have been suggested in the literature and which could improve the efficacy of this process (Drägulescu et al., 1972).

An alternative approach is carbonation of pure hydroxides or oxides, which would have to be extracted from the minerals in a prior processing step. We found from the literature (Bhatia & Perlmutter, 1983) as well as from our own experiments that the carbonation of CaO and Ca(OH)$_2$ progresses rapidly. One can achieve near completion in minutes. However, the high reactivity of calcium oxides and hydroxides is counterbalanced by their low concentrations in naturally occurring minerals. The measured rate of carbonation of MgO under atmospheric pressure turned out to be far too slow to be of practical interest. The reaction rate of Mg(OH)$_2$ on the other hand appears to be very promising and we are developing a processing scheme based on this reaction (Lackner et al., 1996).

In our first experiments with Mg(OH)$_2$, we obtained significant carbonation by slowly flowing CO$_2$ ($P_{CO_2} \sim 0.5$ bar) over a bed of Mg(OH)$_2$ powder in runs that extended over 12 hours (Butt et al., 1996). The average grain size was 20 μm and about 8% of the stoichiometric maximum carbonation was achieved. In more recent experiments with similar powders we obtained 8% of the stoichiometric maximum in about 10 minutes ($P_{CO_2} = 0.78$ bar). We are still investigating the details of the kinetics which is governed by a complicated interaction between the water being driven off the magnesium hydroxide and the CO$_2$ that enters the powder. Published data concerning the water release vary and it appears that complications like crack formation in the grain can greatly affect the outcome of the reaction.

Further experiments on Mg(OH)$_2$ are investigating the effects of varying grain size, gas flow, temperature and pressure on the overall kinetics. Below the thermodynamic maximum temperature, an increase in temperature drastically increases the rate of the initial uptake of CO$_2$. It also raises the level of carbonation that is achieved before rates level off. We have observed that slight pressure increases of 0.2 bar raise the rate of uptake and the maximum uptake for a given temperature. However, we expect that the major advantage of increased pressure will come from raising the operating temperature. Based on thermodynamic arguments, doubling the pressure would allow one to raise the temperature by more than 30°C which based on current data should drastically increase the level of carbonation. Currently we are working on experiments in an intermediate pressure range (10 to 100 bar). This pressure regime is of great interest because it can be contained in an industrial setting. Furthermore, pipeline delivery of CO$_2$ to the disposal site would already have precompressed the gas to these pressures so there would be no additional energy penalty. At even higher pressures (340 bar and 500°C), we have already shown that the reaction goes to completion in less than 2 hours. These results will be reported in more detail at a later date.

The cost of the process is critically affected by the reaction time, and simple estimates suggest that the reaction needs to go to completion in less than one hour. It is likely that the dehydroxylation and concurrent carbonation of magnesium hydroxide can be performed well within this constraint. However, the final experiments demonstrating this point are still outstanding.

To extract magnesium hydroxide from mineral rock we are studying a well known aqueous process for magnesium extraction in order to adapt it to our situation. In summary, forsterite and serpentinite rocks are decomposed in hydrochloric acid, yielding magnesium chloride in solution. The hydrochloric acid can then be recovered in a series of steps leading to hydrochloric acid, water and magnesium hydroxide. The magnesium hydroxide is then carbonated in a gas-solid reaction at elevated pressure and temperature (see Fig. 1). The advantage of the gas-solid reaction is that the reaction heat can be captured for use in the earlier steps of the process. If the efficiency of the process is not sufficient to actually make use of this heat energy, we have already shown that the carbonation of magnesium hydroxide can also be performed in an aqueous environment where reaction rates can be fast even at low pressures, but where heat recovery is simply not feasible. This reaction was completed in less than 30 minutes when using 20 μm magnesium hydroxide powder.

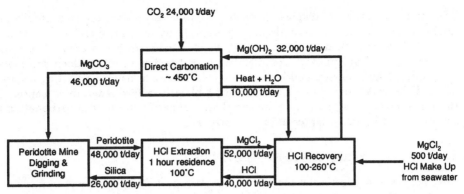

Fig. 1. A diagram of the material flows in CO$_2$ disposal. The absolute rates of processing are geared to match the CO$_2$ output of a power plant with 33% conversion efficiency that produces 1 GW of electric power.

We have reproduced results obtained in the forties (Houston, 1945) and fifties (Barnes *et al.*, 1950) on the extraction of magnesium from forsterite and serpentinite. The process can be performed rapidly at about 100°C. It is slightly exothermic and can maintain its own operating temperature. The end-product is magnesium chloride in aqueous solution. The silica can be precipitated as silica gel, by gradually raising the pH of the solution, for example by adding MgCl(OH) which is produced in a later step of the process.

For a successful recovery of the hydrochloric acid, it is important that the formation of soluble chlorides is minimized. Fortunately, all peridotite rocks are remarkably free of alkali ions, which contribute less than 1% to the mineral mass. Therefore the irreversible formation of alkali chlorides is not a serious problem. A frequent and unavoidable contamination in all peridotites is iron oxide which is also extracted in the acid leaching as iron chloride. Fortunately, neutralizing the solution with MgCl(OH) will cause the complete precipitation of iron oxides and hydroxides while forming additional magnesium chloride. It has been shown by Houston that a straightforward unoptimized design will readily recover 95% of the hydrochloric acid (Houston, 1945).

The acid recovery begins with concentration of the magnesium chloride solution to form hydrated magnesium chloride crystals of composition MgCl$_2 \cdot n$H$_2$O with $n = 6$. With gradual heating, the water is driven of in steps moving from $n = 6$, to approximately $n = 1$. Further heating of the $n = 1$ powder leads mostly to the formation of HCl and MgCl(OH) rather than additional water release (Smith & Veazey, 1932). MgCl(OH) is readily disassociated into MgCl$_2$ and Mg(OH)$_2$ by reintroducing it into aqueous solution. Thus we have a complete cycle for the recovery of the hydrochloric acid which is akin to the process used in the manufacture of magnesium hydroxide from seawater (Copp, 1994).

We have performed the individual steps in the laboratory and have found that without optimization for high reaction rates the process can be performed with residence times of about 30 minutes which puts it well within our goal. Future work will aim to improve the speed of the reaction and most importantly we will try to develop, at least in a simulation, a process that minimizes heat losses. Given the large amount of water that is repeatedly recycled through these steps a careful design that avoids heat loss is absolutely critical making this process economically viable.

CONCLUSIONS

We have identified the outline of a viable process for disposal of CO$_2$ in the form of carbonate minerals. Based on our first cost estimates we are encouraged to pursue this line of research further.

At this point we are favoring a specific implementation using Mg(OH)$_2$ extracted from peridotite or serpentinite rock. However, the direct solid gas reactions with minerals at higher pressure are still under study and may reveal practical alternatives using these same minerals or others such as basalt. Another possibility, which is related in principle, is the direct carbonation of underground calcium bearing minerals by injection of CO$_2$ at high pressure into suitable reservoirs (Bachu *et al.*, 1994).

The major advantage of disposal in carbonate minerals is that the environmental impact is minimal. Carbonate minerals are already common in nature and therefore the disposal is safe. The most obvious impact arises from mining which is comparable in scope to the mining of coal in the first place. Our approach completely avoids the danger of a sudden accidental release of the CO$_2$ into the atmosphere, and it does not require safeguards for preventing a slow re-introduction of the CO$_2$ which could cause a climate problem in a few generations hence. The stability of the waste product is guaranteed by the thermodynamic stability of carbonate minerals and it is demonstrated by much larger natural deposits.

ACKNOWLEDGEMENTS

This work was supported by the U. S. Department of Energy.

REFERENCES

Audus, H., P. W. F. Riemer & W. G. Ormerod (1995). Greenhouse Gas Mitigation Technology Results of CO$_2$ Capture & Disposal Studies. In *Proceedings of the 20th International Technical Conference on Coal Utilization, & Fuel Systems*, Coal & Slurry Technology Association and the U.S. Department of Energy's Pittsburgh Energy Technology Center, Clearwater, Florida, pp. 349–358.

Bachu, S., W. D. Gunter & E. H. Perkins (1994). Aquifer Disposal of CO$_2$: Hydrodynamic and Mineral Trapping. *Energy Convers. Mgmt.* **35**, 269–279.

Barnes, V. E., D. A. Shock & W. A. Cunningham (1950). Utilization of Texas Serpentine. The University of Texas, No. 5020, Bureau of Economic Geology.

Bhatia, S. K. & D. D. Perlmutter (1983). Effect of the Product Layer on the Kinetics of the CO$_2$-Lime Reaction. *AIChE Journal* **29**, 79–86.

Booras, G. S. & S. C. Smelser (1991). An Engineering and Economic Evaluation of CO$_2$ Removal from Fossil-fuel-fired Power Plants. *Energy* **16**, 1295–1305.

Butt, D. P., K. S. Lackner, C. H. Wendt, S. D. Conzone, H. Kung, Y-C. Lu & J. K. Bremser (1996). Kinetics of Thermal Dehydroxylation and Carbonation of Magnesium Hydroxide. *J. Am. Ceram. Soc.* **79**, 1892–1898.

Coleman, R. G. (1977). *Ophiolites: Ancient Oceanic Lithosphere?* Springer-Verlag, Berlin.

Copp, A. N. (June 1994). Magnesia/Magnesite. *American Ceramic Society Bulletin* **73**, 111–113.

Drăgulescu, C., P. Tribunescu & O. Gogu (1972). Lösungsgleichgewicht von MgO aus Serpentinen durch Einwirkung von CO$_2$ und Wasser. *Revue Roumaine de Chimie* **17**, 1517–1524.

Freeth, S. J. (1994). Lake Nyos: can another disaster be avoided? *Geochemical Journal* **28**, 163–172.

Hendriks, C. A., K. Blok & W. C. Turkenburg (1991). Technology and Cost of Recovering and Storing Carbon Dioxide From an Integrated-Gasifier, Combined-Cycle Plant. *Energy* **16**, 1277–1293.

Herzog, H., E. Drake & J. Tester, eds. (1993). *A Research Needs Assessment for the Capture, Utilization and Disposal of Carbon Dioxide from Fossil Fuel-Fired Power Plants.* Energy Laboratory, Massachusetts Institute of Technology, Cambridge, MA 02139-4307.

Hooper, P. R. (1982). The Columbia River Basalts. *Science* **215**, 1463–1468.

Houston, E. C. (1945). Magnesium from Olivine. American Institute of Mining and Metallurgical Engineers, Technical Publication No. 1828, Class D, Nonferrous Metallurgy, No. 85.

Irwin, W. P. (1977). Ophiolitic terranes of California, Oregon and Nevada. In *North American Ophiolites*, (ed. Coleman, R. G. & W. P. Irwin), State of Oregon Department of Geology and Mineral Industries, Bulletin 95.

Kaarstad, O. (1995). Norwegian Carbon Taxes and Their Implication for Fossil Fuels. In *Proceedings of the 20th International Technical Conference on Coal Utilization, & Fuel Systems*, Coal & Slurry Technology Association and the U.S. Department of Energy's Pittsburgh Energy Technology Center, Clearwater, Florida, pp. 359–370.

Kling, G. W., M. A. Clark, H. R. Compton, J. D. Devine, W. C. Evans, A. M. Humphrey, E. J. Koenigsberg, J. P. Lockwood, M. L. Tuttle & G. N. Wagner (1987). The 1986 Lake Nyos Gas Disaster in Cameroon, West Africa. *Science* **236**, 169–175.

Lackner, K. S., C. H. Wendt, D. P. Butt, E. L. Joyce & D. H. Sharp (1995). Carbon Dioxide Disposal in Carbonate Minerals. *Energy* **20**, 1153–1170.

Lackner, K. S., D. P. Butt, C. H. Wendt & D. H. Sharp (1996). Carbon Dioxide Disposal in Solid Form. In *The Proceedings of the 21st International Technical Conference on Coal Utilization & Fuel Systems*, Coal & Slurry Technology Association and the U.S. Department of Energy's Pittsburgh Energy Technology Center, Clearwater, Florida, pp. 133–144.

Lindsley-Griffin, N. (1977). The Trinity Ophiolite, Klamath Mountains, California. In *North American Ophiolites*, (ed. Coleman, R. G. & W. P. Irwin), State of Oregon Department of Geology and Mineral Industries, Bulletin 95.

Nicolas, A. (1989). *Structures of Ophiolites and Dynamics of Oceanic Lithosphere.* Kluwer Academic Publishers, Dordrecht.

Smith, A. K. & W. R. Veazey (1932). Dehydration of Magnesium Chloride. United States Patent Office, Patent No. 1,874,373.

Steinberg, M. & H. C. Cheng (February 1985). *A Systems Study for the Removal, Recovery, and Disposal of Carbon Dioxide from Fossil Fuel Power Plants in the U.S.* U. S. Department of Energy, Washington, DC.

United Nations (1993). *1991 Energy Statistics Yearbook.* New York.

SECTION 4

OCEAN STORAGE

Energy Convers. Mgmt Vol. 38, Suppl., pp. S265–S271, 1997
© 1997 Elsevier Science Ltd. All rights reserved
Printed in Great Britain
0196-8904/97 $17.00 + 0.00

Pergamon

PII: S0196-8904(96)00280-4

OCEAN SYSTEMS FOR MANAGING
THE GLOBAL CARBON CYCLE

DWAIN F. SPENCER, PRINCIPAL, SIMTECHE
WHEELER J. NORTH, PROFESSOR EMERITUS, CALTECH

24 Fairway Place, Half Moon Bay, CA 94019

Kerckhoff Marine Laboratory, Caltech
101 Dahlia Street, Corona Del Mar, CA 92625

ABSTRACT

Carbon dioxide is formed in all processes utilizing fossil fuels. Controlling the emissions of CO_2 from a number of processes by forming CO_2 hydrates (clathrates), may be an effective approach for both absorbing CO_2 from multicomponent gas streams (Flue gases, Anaerobic digester gases, etc.) and sequestering CO_2 in the deep oceans. Further, ocean marine farms may be an effective process for extracting CO_2 from the atmosphere and forming both valuable products and rejecting excess CO_2, in the form of clathrates, to the deep ocean. Preliminary engineering analyses indicates that clathrate formation for controlling both conventional fossil fuel gaseous CO_2 emissions and those associated with marine farm anaerobic digester gases may provide a meaningful control strategy for CO_2. © 1997 Elsevier Science Ltd

KEYWORDS

Carbon dioxide clathrates; extraction from multicomponent gases; oceanic sequestration; marine farm systems.

INTRODUCTION

Over the last 20 years, increasing global emissions of carbon dioxide from fossil resources, and the potential impacts on the earths' atmosphere, have become a subject of intense scientific investigation. Major emphasis has been placed on assessing the range of impact of these increased emissions and methods to reduce or restrict the increasing use of fossil fuels.

Much less emphasis is being placed on the research, development, and demonstration of engineered systems to manage and control CO_2 emissions, particularily from stationary fossil sources, such as powerplants, chemical plants, petroleum refineries, etc.

The purpose of this paper is to summarize key new fundamental research on the formation of CO_2 clathrates from pure CO_2 gas streams and to discuss the potential for CO_2 clathrate formation from multi component gas streams. Typical emission streams including powerplant stack gases, coal or petroleum coke synthesis gases, natural gas/carbon dioxide mixtures, etc. are all potential sources for carbon dioxide extraction and fixation, in the form of CO_2 clathrates.

Once formed, these solid clathrates can be injected, in slurry form, into the ocean or deep aquifers for CO_2 sequestration. Finally, an active CO_2 management and control system will be discussed which a) extracts CO_2 from the atmosphere, at a distributed level, i.e. non-point source, b) converts the marine carbon to natural gas and carbon dioxide, c) sequesters the carbon dioxide in the deep ocean, and d) provides liquified natural gas, as well as other high valued products, as articles of commerce.

FORMATION OF CARBON DIOXIDE CLATHRATES

It is well known that liquid or gaseous carbon dioxide when mixed with water at temperatures below 10°C and pressures in the range of 10-70 bars, form solid clathrates (Takenouchi and Kennedy, 1965). These clathrates typically have 2 to 8 CO_2 molecules, bound into a matrix of 46 water molecules. If full lattice occupancy is achieved, 8 CO_2 molecules are trapped within the matrix of 46 water molecules. This produces a mole fraction of CO_2 of only 0.148, but a weight fraction of 0.36.

The purpose of a research program, undertaken at Caltech and sponsored by the Electric Power Research Institute, was to determine how such clathrates could be formed under minimal pressure conditions; thus reducing compression requirements for CO_2, or CO_2 containing gaseous mixtures. This work was reported in a number of EPRI progress reports (North, et al., 1993; North, et al., 1995) and in a paper (North, et al., 1993).

Initial experiments were conducted in a batch chamber, at temperatures of 4-6°C and pressures of 65 to 200 bars. Through a series of experiments, stable CO_2 clathrates with greater than 25% by weight can now be formed at -1 to 0°C (salt water) and pressures of 12 to 14 bars. These greatly reduced formation conditions have been achieved by developing an empirical model of CO_2 clathrate formation, which indicates three distinct phases in the development of stable clathrate formation, namely a) CO_2 dissolving in water, b) hydrate precursors forming i.e. structured water, and c) final hydrate formation with the introduction of additional CO_2.

The research then moved to the use of two interconnected semi continuous flow bubbling reactors for clathrate formation and formation pressures were reduced to the 12 to 14 bar levels. These reduced formation conditions are achieved by stepwise formation/decomposition of the clathrate. Each subsequent clathrate formation step occurs at a lower pressure, indicating that the nucleated water has a "memory" from the previous formation. Stable formation pressures of 13 bar were achieved, at reactor temperatures slightly below zero centigrade.

Finally, two continuous flow reactors, designed by D. Johnson, a project subcontractor, were operated in late 1995. Our previous studies had shown that CO_2 hydrate formation at 12-14 bars pressure would most likely require two or three reactor stages, to ensure that an adequate quantity of precursors occurred in the stream of CO_2-water entering the final stage.

The Johnson Fluidic Venturi consists of two coverging cylindrical chambers, the eductor tube and plenum, discharging into a common terminal space, the tail tube. Two Fluidic Venturis were constructed. The second unit produced excellent gaseous CO_2, water mixing at flow rates of 0.36 liters per second. Unit 2 could accommodate about 120 grams of CO_2 per second and produce 1.7 metric tons of hydrates per hour. This represent near full saturation of the lattice, as the CO_2 weight fraction is 0.33.

Optimal operation occurred when the water pressure in the plenum was approximately 6.7 bars greater than the CO_2 gas pressure i.e. the water educts the CO_2 into the Fluidic Venturi. Therefore, it may be possible to further reduce the CO_2 compression requirements for second stage injection, if the Fluidic Venturi is used as the clathrate formation reactor. In a fully engineered system, it is likely that a two stage reactor system would be employed. The first to form the clathrate precursors, and the second for clathrate production. Approximately 25% of the CO_2 is absorbed in the precursor stage, so this portion might be compressed to 20 bars. The remaining 75% would be compressed only to 12-14 bars, and perhaps less, if the eductor tube provides 6-7 bars of suction. With these low pressures for stable clathrate formation, one can consider hydrate formation from atmospheric pressure CO_2 streams, such as flue gases, as well as pressurized streams, such as found in synthesis gas production from natural gas, coal, petroleum, coke, etc.

CLATHRATE FORMATION FROM MULTI COMPONENT GAS STREAMS

All of the clathrate formation research performed to date has been with pure CO_2 gases. In order to obtain a pure CO_2 stream from a multi component gas, conventional separaton techniques subject the multi-component gas to a number of energy consuming steps, namely (1) absorption of the CO_2 from the gaseous stream by a host solvent e.g. monoethanolamine or Selexol, (2) removal of CO_2 from the host solvent e.g. by steam stripping, and (3) compression and cooling of the stripped CO_2 for clathrate formation or alternatively, (4) compressing or liquifying the pure CO_2 gas stream for pipelining or injection into the ocean.

Recently, one of us (Spencer), has conceived an approach for CO_2 clathrate formation from multicomponent gas streams, including

a) Powerplant Flue Gases, b) Turbo Charged Boiler Flue Gases, c) Coal Gasification Product Gases, d) Shifted Coal Gasification Product Gas, and e) Anaerobic Digester Product Gas. In these applications, the exhaust (flue) gas from processes a and e are at atmospheric pressure, and processes b,c, and d at pressures from 10 to 60 bars. Mole fractions of CO_2 in the multi component gas streams vary from 0.02 to 0.50 over the range of these applications.

Due to the extremely high preferred solubility of carbon dioxide in water, and even higher preferred CO_2 solubility in nucleated water, a single step process for extracting CO_2 from these gaseous streams, forming CO_2 clathrates, and clathrate slurries, has been designed. The CO_2 is now ready for injection as a CO_2-water slurry into either the ocean or a deep aquifer system. A U.S. patent application has been filed (Spencer,1996), earlier this year.

At this point, only theoretical separation efficiencies have been developed. Comparisons of the energy efficiency for separating and pipelining CO_2 using conventional amine scrubbing/stripping techniques e.g.monoethanolamine, with the new process of forming CO_2 clathrates directly, have been developed for both a 500 Mwe conventional coal fired powerplant and utilizing Selexol for CO_2 separation in an integrated coal gasification combined cycle (IGCC) powerplant. The base energy losses for both of these systems are taken from Smelser (1991). The energy loss associated with separation and pipelining of CO_2 using conventional technology from a conventional coal fired powerplant is estimated to consume 35 percent of the power output. The similar loss for an integrated coal gasification combined cycle powerplant is 12 percent.

Theoretical estimates of these losses, if the new clathrate formation process is utilized for both separation and slurry pipelining, are 9.0 and 4.4 percent respectively for the conventional coal plant or the IGCC powerplant. Thus, the energy losses, as well as process simplification, are much more attractive utilizing the new clathrate separation process.

Hidy and Spencer have submitted a grant application to the U.S. Environmental Protection Agency for a 3 year research program entitled, "Exploratory Design and Evaluation of a New CO_2 Removal Process" (Hidy, <u>et al.</u>, 1996). This work is planned to performed at the University of Riverside Center for Environmental Research and Technology, in conjunction with Dr. North of Caltech. At least, two fundamentally different reactor configurations will be tested, one a con-current flow reactor, the second, a countercurrent flow reactor.

CLATHRATE SEQUESTRATION IN THE OCEAN

As discussed above, it is envisaged that slurries of CO_2 clathrate will be formed for injection into the ocean or deep aquifers. With a slurry concentration of 50 percent by weight clathrate, the CO_2 content of the slurries is approximately 16-

17 percent. The specific gravity of the clathrates are approximately 1.16, so that the hydrate should slowly settle through sea water after injection and will either redisolve or sink to the ocean floor. (Cole, et al., 1993) have shown that the deep ocean can be a repository for at least 1200 gigatons of CO_2, with minimal change in oceanic acidity. This amount is equivalent to approximately 2 pre-industrial atmospheric CO_2 contents.

If the clathrates reach the sediments, it is estimated that an additional 2800 gigatons of carbon could be sequestered. Of course, these estimates assume uniform dispersal over the sea floor or injection in a deep advection zone such as the North Atlantic Deep Water flow. In any case, there is clearly a large repository potential in the deep oceans. Pumping clathrate slurries to depth of 700 meters or so and discharging the slurries in a downward flow direction should allow complete sequestration of the CO_2 for many centuries, if not permanently.

MARINE FARM SYSTEMS FOR GLOBAL CARBON MANAGEMENT

One of us, (Spencer,1991), has proposed large scale floating macroalgal systems for CO_2 fixation from the atmosphere. These "so called" marine farms, each some 100,000 acres in extent, would not only fix airborne CO_2, but also, produce valuable byproducts. By harvesting and processing species such as Macrocystis Pyrifera, e.g. by anaerobically digesting these species to form approximately a 50:50 mixture of methane and carbon dioxide, a portion of the carbon is used to produce clean fuels and the remainder of the carbon can be sequestered. The carbon dioxide, representing approximately 50 percent of the anaerobic digester gas, would be stripped from the methane in the form of CO_2 clathrates and injected as a slurry into the ocean. The methane would be liquified and become an article of commerce. Other coproducts can also be produced from these marine farms.

A 100,000 acre marine farm fixes approximately 1 million tons of carbon per year, of which 50 percent is recycled as fuel (methane) and 50 percent is sequestered as CO_2 clathrates in the ocean. Thus; 1000 such marine farms would be necessary for managing 1 gigaton of carbon annually.

By utilizing anaerobic digestion of the harvested macro algal species, such as Macrocystis Pyrifera, 90-95 percent of the nutrients (nitrogen and phosphorus) are retained in the digester liquors and can be recycled to the farm to provide the necessary nutrients. Upwelled seawater would provide the "makeup" nutrients necessary for equilibrium macro algal production.

Overall system and process designs have been developed (SIMTECHE Proprietary Information); however a full preliminary design of the system has not yet been conducted. Very preliminary performance and costs indicate that such a system could be ecomically viable, if a broad product slate of high value products can be established.

Such a multi-component product slate could include a) Food-algae, fish and invertebrate polycultures, b) Biopolymers, such as agar, algin, carogens, c) Organics, such as acetone, and organic acids, d) Fuels, such as alcohols, in addition to methane, etc. This type of floating, tended farm system could serve many useful purposes, and assist in the distribution of CO_2 clathrates being injected over large ocean areas.

SUMMARY

This paper summarized some recent experimental work focused on controlling CO_2 emissions from stationary sources. It has shown that systematic study of the properties of CO_2-H_2O mixtures and conditions, can greatly reduce the energy necessary to fix and sequester CO_2.

Further, recent advances have shown that a) stable, high mole fraction CO_2 clathrates can be formed in continuous reactors, b) CO_2 clathrate formation from a range of multi component gaseous streams is feasible and for some gases, very easily achieved, and c) system concepts which include global CO_2 management can be integrated with productive marine farm systems for fuels, chemicals, etc. and clathrate injection of the CO_2 waste stream.

Although this is only the beginning of the search for meaningful engineered systems to manage CO_2, the efforts to date, are both promising and exciting. Although the cost effectiveness of these systems must yet be determined, the fundamental reductions of energy losses associated with these systems, compared to conventional techniques, indicate that this cost effectiveness should be achieved, as engineering and operational data are obtained.

REFERENCES

Cole, K.H., Stegen, G.R., and Spencer, D.F., The capacity of the deep oceans to absorb carbon dioxide, Presented at IEA International Conference on Climate Change, Oxford, U.K. (1993).

Hidy, G.M., and Spencer, D.F., Grant application for Exploratory design and evaluation of a new CO_2 removal process, Research proposal submitted to U.S. Environment Protection Agency, (1996).

North, W.J., Morgan, J.J., Johnson, D.E., Spencer, D.F., Use of hydrate for sequestering CO_2 in the deep ocean , Second U.S./Japan workshop on mitigation and adaptation technologies related to climate change. Honolulu, Hawaii (1993).

North, W.J. and Morgan, J.J., Investigation of CO_2 hydrate formation and dissolution, EPRI Progress Report for 1 October 1991 to 30 June, 1993.(1993)

North, W.J. and Morgan, J.J., Investigations of CO_2 hydrate formation and dissolution, EPRI Progress Report for 1 July 1993 to 31 December, 1995 (In Publication).(1995)

Smelser, S.C., Stock, R.M., and McCleary, G.J., Engineering and Economic Evaluation of CO_2 Removal from Fossil-Fuel-Fired Power Plants-Vols 1,2, EPRI IE-7365, June 1991.

Spencer, D.F., Open ocean macroalgal farms for CO_2 mitigation and energy production, Proceedings of IEA International Conference on Technology Responses to Global Environmental Challenges, Kyoto, Japan (1991).

Spencer, D.F., Methods of selectively separating CO_2 from a multicomponent gaseous stream, U.S. Patent Application No. 081643,151, 30 April 1996.(1996)

Takenouchi, S. and Kennedy, G.C., Dissociation pressures of the phase CO_2-5 3/4 H_2O, J. Geol., 73, 383-390, (1965).

Pergamon

Energy Convers. Mgmt Vol. 38, Suppl., pp. S273–S277, 1997
© 1997 Elsevier Science Ltd. All rights reserved
Printed in Great Britain
0196-8904/97 $17.00 + 0.00

PII: S0196-8904(96)00281-6

Cost Comparison in Various CO₂ Ocean Disposal Options

YUICHI FUJIOKA *, MASAHIKO OZAKI*, KAZUHISA TAKEUCHI* ,
YUJI SHINDO** and HOWARD J. HERZOG***

*Mitsubishi Heavy Industries, LTD.
5-717-1 Fukahori-machi, Nagasaki 851-03, Japan
**Former member of Agency of Industrial Science and Technology, MITI Japan
*** Massachusetts Institute of Technology

ABSTRACT

CO_2 ocean disposal is one of several ways proposed to mitigate the greenhouse gas effect. In this paper, we estimate the cost of various CO_2 disposal options. The system consists of a conventional pulverized coal power plant, CO_2 separation using an amine scrubber, liquefaction and transportation. In this investigation, the power plant and the CO_2 separation plant are common. The liquefaction process and transportation systems are different. We evaluate the costs of pipeline, liquid CO_2 tanker for CO_2 lake, liquid CO_2 tanker dispersion, dry ice tanker and CO_2 dispersion by OTEC. The effect of distance between a power plant and the disposal site, plant size, and plant efficiency on the cost for each case is evaluated. Costs of ocean disposal options are estimated about 90 to 180 per tonne of CO_2 avoided. The least cost transportation option is dependent on the distance between power plant and site. © 1997 Elsevier Science Ltd

INTRODUCTION

Increase of CO_2 in the atmosphere may cause extreme changes to the global environment. If we decide to continue to use fossil fuels and want to avoid emitting large amounts of CO_2 into the atmosphere, we think that CO_2 ocean disposal is one promising method to consider. It is unclear which is the best CO_2 ocean disposal strategy among the many proposed options. The cost estimation of ocean CO_2 disposal will show the way of research and development of CO_2 ocean disposal. CO_2 ocean disposal is a temporary technology. Renewable energies without the fossil fuels are desirable, however it takes much time, expense and personnel to develop such new energies. The cost estimation in power plant with a CO_2 ocean disposal show comparable cost of the electricity versus alternative energy.

OCEAN DISPOSAL OPTIONS

In this study, five options for the CO_2 ocean disposal shown in Fig.1 are compared. It is assumed that all options have the same plant on land that consists of two 554MW units of pulverized coal fired boilers Smelser (1991), and a CO_2 separation unit which can capture 93% of CO_2 in the flue gas by an amine scrubber. The gross efficiency of this plant is 37.6% and the net efficiency is 35% before capture and the maximum handling CO_2 is about 20,400 tons/day. Additional production processes of high pressure liquid CO_2 for pipeline transportation, or low temperature liquid CO_2 for tanker transportation, or dry ice, as shown in Fig.2, are on land.

① Fig. 1 shows liquid CO_2 transported via a pipeline and dispersed in the sea at depths from 1,000 to 2,000 m. Golomb (1993), Adams et al. (1994) proposed the mixing device at the outlet of pipeline to promote dissolving CO_2 to seawater. Haugan et al. (1995) estimated the behavior of CO_2 plume in deep sea. On land liquid CO_2 for EOR has been transported by pipelines for more than 25 years. The CO_2 pipeline is operated at 90 to 120 bar to prevent vaporization at a temperature of 60 C. The CO_2 pipeline needs compressors to recover pressure drop at 300-kilometer intervals.

② Fig.1 shows liquid CO_2 transported with a tanker and CO_2 lake at the bottom of the sea at depths under 3700 m. The density of liquid CO_2 is heavier than that of CO_2 dissolved seawater and the density of CO_2 dissolved water is heavier than that of ambient seawater. If eddies are mild near sea bottom, there will be density stratification over the CO_2 lake that refrain CO_2 diffusion and stabilized CO_2 lake (Ohsumi, 1993: Fujioka et al., 1995). In order to estimate the cost of the transporting system, we have initially designed a tanker for liquid CO_2 at a temperature of -50 C and a pressure of 6.83 bar. Number of tankers, tanker size and service velocity can be planned according to the given conditions such as CO_2 transportation rate and distance from land to the site. For example, two tankers that have 4 spherical tanks of 20 meter diameter and service velocity of 12 to 15 knots can transport 17,000 m³/day of CO_2 as far as 400 kilometers

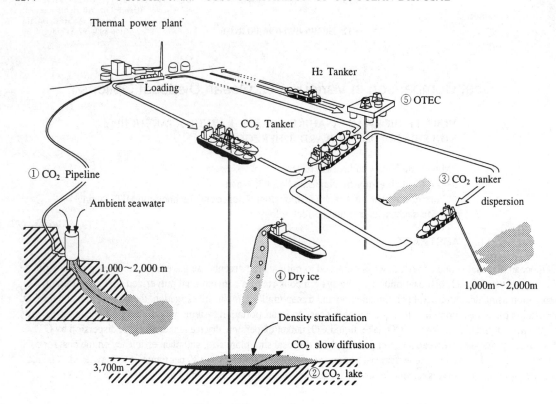

Fig. 1. CO_2 ocean disposal options

constantly. At the site over the CO_2 lake, transported liquid CO_2 is sifted from the tanker to the floating base that may be similar to CO_2 tanker, and CO_2 is injected into the deep water with a pipe suspended under the floating base.

③ Fig.1 shows tanker for CO_2 dispersion into the sea at depths from 1,000 to 2,000 meters. It has a long pipe that has many small holes. The eddies that are produced by the pipe stir the spouted liquid CO_2. If the pipe at a diameter of 0.263 meter is moved at a velocity of 0.3 knot and the 100 kilograms/s of CO_2 is spouted from the holes, it takes about 44 seconds to dilute liquid CO_2 into 1/3200 CO_2 concentration.

④ Fig.1 shows disposal of dry ice. The dry ice is transported bulk carrier and is dumped into the sea. This option is the lowest environmental impact (Herzog et al., 1996). Liquid CO_2 is adiabatically expanded to make dry ice. CO_2 gas was also produced at expansion process for dry ice. The CO_2 gas is recycled to the compressor that is used to make liquid CO_2.

⑤ Fig.1 shows a ocean thermal energy conversion system (OTEC), a CO_2 dissolution unit and a H_2O electrolysis unit on a floating platform. Pumping up deep seawater cools down condensers in OTEC and is used to dissolve CO_2 and returns to deep sea. The electricity produced by OTEC is converted to H_2 with electrolysis in the ocean. The H_2 is transported to land by H_2 tankers.

The annual expenses of units such as power plants, separation unit, CO_2 liquefaction unit and CO_2 transportation unit, energy and personnels were calculated. The definition of the CO_2 disposal cost is as follows. CO_2 disposal cost = (Total electricity cost with CO_2 disposal - Total electricity cost without CO_2 disposal) / (Amount of avoided CO_2). Avoided CO_2 equals exhausted CO_2 without CO_2 disposal minus exhausted CO_2 with CO_2 disposal.

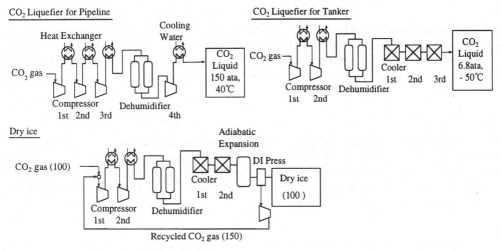

Fig. 2. CO$_2$ liquefaction and production of dry ice processes

RESULTS AND DISCUSSION

Comparison of ocean CO$_2$ disposal options

Fig. 3 shows the electricity cost. Electricity cost without CO$_2$ disposal is 53 mills / kWh. Electricity cost with CO$_2$ pipeline disposal is 2.3 times higher than that without CO$_2$. Electricity cost with CO$_2$ lake and CO$_2$ tanker dispersion are 2.5 times higher. Electricity cost with dry ice is 3.5 times. Electricity cost with OTEC is 2.8 times. CO$_2$ disposal costs, of which CO$_2$ transportation is 200 kilometers long, are also shown in Fig. 3.

The disposal cost by CO$_2$ pipeline is the lowest and \$90 per ton CO$_2$ avoided. The cost with the CO$_2$ tanker is higher than that with the CO$_2$ pipeline because the production of low temperature CO$_2$ needs more complicated processing than that of high pressure CO$_2$. Decrease the cost for cooling liquid CO$_2$ can reduce the cost of CO$_2$ tanker transportation. Compressing 2.5 times larger CO$_2$ gas in dry ice production than that in liquid CO$_2$ makes high cost of the dry ice disposal.

The costs of decreasing CO$_2$ with various natural energies such as OTEC are higher than that of CO$_2$ ocean disposal via pipeline. It is obvious that the electricity cost produced by the natural energies is higher than the cost produced by fossil fuels for the present.

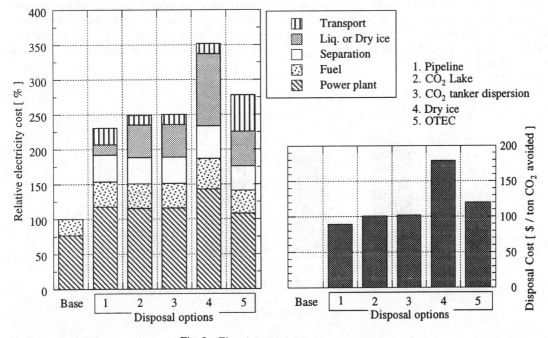

Fig. 3. Electricity and CO$_2$ Disposal Cost

Distance

If the CO_2 disposal site is near land and the distances are 50 to 100 kilometers long, CO_2 pipeline disposal is preferable. Cost of CO_2 pipeline disposal is proportional to the transportation distance. Cost of CO_2 transportation by CO_2 tanker does not significantly depend on the transport distance. In the case of transporting more than 300 kilometers, the total cost of the CO_2 pipeline disposal is higher than that of the CO_2 lake and the CO_2 tanker dispersion as in shown in Fig. 4.

Plant size

For a maximum diameter of the CO_2 pipeline of 60 centimeters, the CO_2 pipeline can transport the amount of CO_2 that is exhausted by 4,000 MW power plant. If the number of power plants increase and the total amount of liquid CO_2 of the ocean disposal is greater than the CO_2 from 4,000MW, another CO2 pipeline will be needed. It is suggested that CO_2 pipelines be used for transportation from the power plants to a port and that CO_2 tankers be used for transportation from the port to a site of CO_2 disposal.

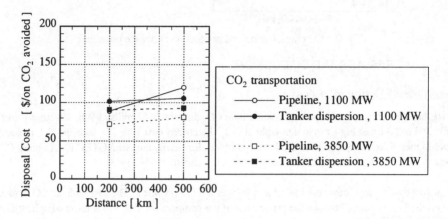

Fig. 4.   Effect of transportation distance and plant size on CO₂ disposal costs

Plant efficiency

New types of power plants such as pressurized fluid bed combustion (PFBC) or integrated gasification combined cycle (IGCC) will yield higher plant efficiencies. As seen in Fig. 5, this will substantially reduce the cost of CO_2 capture, and disposal assuming that the cost of the base power plant remains the same. Even if the price of the base power plant increases significantly with increased efficiency, it still benefits CO_2 capture and disposal. For example, in Fig. 6, CO_2 capture from a 37.6% efficient plant increased electricity costs by a factor of 2.3. However, capture from a 47% efficient plant that costs twice as much as the 37.6% efficient plant, raises the electricity cost only by a factor of about 1.9.

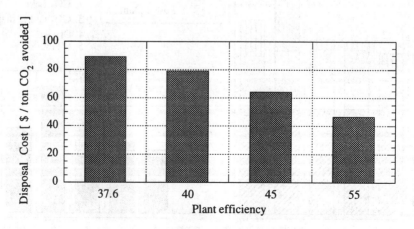

Fig.5.   CO₂ disposal cost vs plant gross efficiency

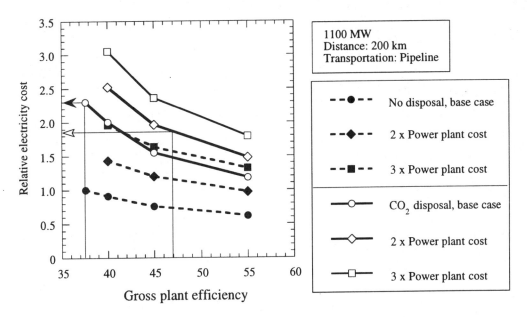

Fig.6. Effect of net plant efficiency and power plant cost CO_2 disposal costs

CONCLUSIONS

1) The estimated cost of CO_2 disposal is about $90 ~ $180/ton-avoided CO_2. The CO_2 ocean disposal increases the electricity cost by a factor of 2 ~ 2.5.

2) In view of the cost, CO_2 pipeline disposal, CO_2 lake and CO_2 tanker dispersion are preferable. The cost of CO_2 ocean disposal depends on the distance from a power plant to a disposal site. The cost of CO_2 disposal using pipelines is the lowest for the case that a disposal point near a power plant.

3) If the distance from a power plant to a disposal site is long or the amount of CO_2 is large, the transportation by a tanker decreases CO_2 disposal cost.

REFERENCES

Adams, E.E., D.S. Golomb and H.J. Herzog, (1995). Ocean Disposal of CO_2 at Intermediate Depths, Energy Convers. Mgmt, 36, 447- 452

Golomb, D., (1993). Ocean Disposal of CO_2 : Feasibility, Economics and Effects, Energy Convers. Mgmt, 34 (9-11), 967-976

Fujioka Y., M. Ozaki, K. Takeuchi Y. Shindo and H.Y. Yanagisawa, (1995). Ocean CO_2 Sequestration at the Depths larger than 3700 m, Energy Convers. Mgmt, 36, 551- 554

Haugan P. M. (1995). F. Thorkildsen, G. Alendal, Dissolution of CO_2 in the ocean, Energy Convers. Mgmt, 36 461-466

Herzog H.J., E. Adams, D. Auerbach and J. Caulfield, (1996), Environmental Impacts of Ocean Disposal, Energy Convers. Mgmt, 36, 999 - 1005

Ohsumi, T., (1993). Prediction of Solute Carbon Dioxide Behavior around a Liquid Carbon Dioxide Pool on Deep Ocean Basins, Energy Convers. Mgmt, 34 (9-11), 1059-1064

Smelser S.C., R. M. Stock, G. J. McCleary (1991). Engineering and economics evaluation of CO_2 Removal from fossil-fuel-fired power plants, Vol. 1. ERPI IE-7365, vol.1, project 2999-10

Pergamon

Energy Convers. Mgmt Vol. 38, Suppl., pp. S279–S286, 1997
© 1997 Elsevier Science Ltd. All rights reserved
Printed in Great Britain
0196-8904/97 $17.00 + 0.00

PII: S0196-8904(96)00282-8

TRANSPORT SYSTEMS FOR OCEAN DISPOSAL OF CO_2 AND THEIR ENVIRONMENTAL EFFECTS

DAN GOLOMB
University of Massachusetts Lowell
Lowell, Massachusetts 01854, USA

ABSTRACT

After removal from the power plant flue gas, the CO_2 has to be transported from the power plant to the deep ocean. Because of the vast quantities that need to be transported, this is a formidable engineering task. The transport systems will add significantly to the cost of CO_2 disposal, and they may pose an environmental and safety risk. Considering that a single 1000 MW bituminous coal-fired power plant equipped with CO_2 removal system will deliver between 7.2-8.2E6 T/y of CO_2 (228-260 kg/s), the only practical transport systems appear to be pipes and large tankers in which the CO_2 is transported as a liquid. The environmental impact of offshore pipes will occur mainly during the construction phase. In USA territorial waters pipes need to be buried to a depth of 61 m. This means blasting and digging through sediments and reefs. After construction, the underwater habitat will be restored in a few years. Tanker transport requires major port and docking facilities, which could lead to perturbations of coastal habitat, and contamination due to leakages, spills and effluents. As CO_2 is stored and transported as a pressurized liquid, there is a risk of rupture and release of large quantities of CO_2. At concentrations in air above 10%, CO_2 is an asphyxiant, and at smaller concentrations it causes nausea, vomiting, diarrhea and skin lesions. However the environmental impacts and safety hazards of CO_2 transport systems can be minimized or altogether prevented with prudent management and exercise of precaution. © 1997 Elsevier Science Ltd

KEYWORDS

CO_2 disposal; CO_2 transport systems; CO_2 pipes; CO_2 tankers; CO_2 environmental impact; CO_2 health risk.

INTRODUCTION

In order to make a dent in the greenhouse effect, a significant portion of the anthropogenic CO_2 emissions need to be disposed in a manner that the CO_2 will not reimerge into the atmosphere for prolonged times. Disposing CO_2 into the deep ocean, 1000 m or deeper, is considered a possible stowage medium from which CO_2 will not resurface for at least several hundred years (Dewey *et al.*, 1996). Also, disposal at depth is assessed to cause only minimal and localized environmental impact (Auerbach, 1996; Caulfield *et al.*, 1996). However, getting the CO_2 from the emission sources to the deep ocean presents a formidable engineering task, and it will add significantly to the cost of CO_2 disposal. Furthermore, the transport systems may disturb the environment, and cause a safety and health risk.

The most likely sources from which the emissions will be transported to the ocean are power plants. Power plants emit vast quantities of CO_2 in a concentrated manner. About one third of global CO_2 emissions are due to power plants (Keeling, 1994), and power production is highly centralized. For ocean disposal, coastal power plants will be the likely candidates, since that involves only offshore transport. Golomb (1994) identified several coastal zones on the industrial continents from whence the offshore distance to the 1000 m depth is less than 200 km. Most likely, new advanced cycle coastal power plants will have to be built in which the CO_2 removal and disposal systems are integrated into the plant design in order to minimize the energy losses. The quantity of CO_2 emitted by a power plant depends on the characteristics of the fuel (carbon content and heat content), thermal efficiency or heat rate of the plant, and average load capacity of the plant. Taking Illinois No. 6 coal as an example, with carbon content 70.8%, heat content 25.8 MJ/kg, a power plant with rated capacity of 1000 MW that works at 90% of its capacity, will have the emission rates of CO_2 as a function of thermal efficiency listed in Table 1. For the following discussion we shall assume that an advanced cycle 1000 MW power plant with CO_2 removal will have a net thermal efficiency of 35%, and thus emit 260 kg/s of CO_2.

Table 1. CO_2 emission rates for a 1000 MW power plant

Thermal eff. (%)	25	30	35	40	45
E_{CO2} (10^6 T/y)	11.5	9.6	8.2	7.2	6.4
E_{CO2} (kg/s)	364	304	260	228	203

PHYSICAL PROPERTIES OF CO_2

In order to comprehend the technical difficulties, and the potential environmental and safety hazards of CO_2 transport, it is useful to review the physical properties of CO_2. Figure 1 presents the phase diagram of CO_2. The triple point of CO_2 is 216.8K. At the triple point solid, liquid and gaseous CO_2 coexist. At the triple point the vapor pressure is 5.18 bar. Solid CO_2 has a vapor pressure of 1 atm at 194.6K . At a temperature of 298K liquid CO_2 has a vapor pressure of 60 bar. The critical temperature is 304.6K, the critical pressure is 73.8 bar. Liquid CO_2 has a density of 914 kg/m³ at 273K. At the critical point the density is 467 kg/m³; and at the triple point solid CO_2 has a density of 1512 kg/m³. The physical properties of CO_2 dictate the choice of a transport system. For maximum throughput, and ease of loading and unloading, CO_2 ought to be in the liquid or supercritical

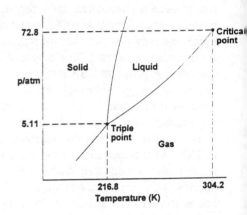

Figure 1. The Phase Diagram of Carbon Dioxide

phase. However, at normal temperatures of 10-25°C, liquid CO_2 has a vapor pressure of 40-60 bar. Thus, pipes, storage tanks and vessels for liquid and supercritical CO_2 need to be constructed with thick walls in order to withstand such pressures. For transport of solid CO_2 ("dry ice"), the problem is that at normal temperatures the solid directly vaporizes (sublimates) without going through the liquid phase. The transport of solid CO_2 also demands specially insulated and pressurized vessels.

TRANSPORT SYSTEMS OF CO_2

CO_2 can be transported as a gas, liquid, or solid, in pipes, tankers, cargo vessels, barges, trucks or railroad cars. For transport of large quantities of CO_2, such as produced by coal fired power plants, transport as a liquid in pipes and tankers appears to be the most economical and practical.

Pipe Transport. The quantity of CO_2 that can be transported in a pipe depends on the pipe diameter and the state of the CO_2, that is, its density and temperature. In order to avoid a two-phase flow, and for maximum throughput, the CO_2 must be in the supercritical ("dense") phase. Skovholt (1993) estimated the minimum pipe diameters necessary for transporting supercritical CO_2 in onshore and offshore pipes. For the transport of 8.2 million T/y of CO_2 (1000 MW coal fueled plant), the minimum diameter is 50 cm (20 inches) for onshore pipes, and 45 cm (18 inches) for offshore pipes. We recalculated the minimum diameter for *offshore* pipes for various throughputs using the pipe flow model developed by Uhl et al. (1965). The repumping pressure P, the pumping energy E(MW), and the number of pumping stations PS, required to transport supercritical CO_2 to the deep ocean over a 100, 200 and 500 km distance, respectively, has also been calculated. In the model, the pipe is submerged in water at an average ambient temperature of 10°C. The initial pressure of CO_2 is 140 bar. Repumping is deemed necessary whenever the pressure falls below 80 bar in order to avoid a two-phase flow. Table 2 summarizes the results. For a 1000 MW plant (throughput 260 kg/s), pipe diameter 24 inches, over a linear distance of 100 to 200 km, the pressure drop is negligible, and no repumping is necessary. Over a distance of 500 km, one repumping station is necessary, requiring an energy of 0.5 MW. However, for the same plant, pipe diameter 16 inches, even over a distance of 100 km, one repumping station would be necessary with an energy requirement of 1.3 MW. For 200 km, three repumping stations would be necessary (E = 2.7 MW), and for 500 km, seven stations (E = 6.6 MW). Clearly, large diameter pipes are necessary if submerged repumping stations are to be avoided. If a pipe is used to discharge the output of ten 500 MW power plants over a linear distance of 100-200 km, a diameter of 40 inches may be necessary. Because of the large diameter of the pipe, and the great depth to be reached (minimum 1000 m) for CO_2 disposal, the most likely pipelaying method will be the so-called tow-lay method. In this method, long lengths (up to 10 km) of pipe are assembled onshore or in shallow water, towed by a lay barge, and the segments are connected under water. Using this method, pipes have been laid to offshore oil and gas wells in the Gulf of Mexico, the North Sea and other locations to a depth of about 900 m (Rayborn, 1996).

Table 2. Pipe Diameters, Repumping Stations and Energy Consumption for Offshore Pipes

Length of pipeline		100 km			200 km			500 km		
D (inch)	m (kg/s)	ΔP (bar)	E (MW)	PS	ΔP (bar)	E (MW)	PS	ΔP (bar)	E (MW)	PS
12	130	45	0.9	1	90	1.8	3	245	4.4	7
16	130	0	0	0	2	0.4	1	68	1.3	1
16	260	34	1.3	1	68	2.7	3	170	6.6	7
24	260	0	0	0	0	0	0	13	0.5	1
24	390	0	0	0	70	0.4	1	73	4.3	1
32	650	0	0	0	0	0	0	19	1.9	1
40	1300	0	0	0	0	0	0	46	9	1

Floating Platforms and Vertical Pipes. If the disposal of CO_2 is to occur far offshore, transport by tankers to a floating platform with an attached vertical discharge pipe will be necessary. Offshore platforms are usually divided into two general categories, fixed and compliant. Fixed types fully extend to the sea bed and remain in place by a combination of the weight and piles driven into the sea bed. Little or no motion of such structures is experienced. Compliant type platforms are more responsive to external effects. Their movements are controlled by mooring systems which typically consist of chains, cables, ropes and anchors. Alternatively, the compliant platform may be dynamically positioned by means of computer controlled thrusters or propellers, and maintain position within prescribed limits of set location (Mazurkiewicz, 1987; Ranney, 1979). For deep ocean disposal of CO_2, a compliant, semi-submersible platform seems to be most suitable. Usually, the platform is supported by columns connected to large underwater displacement hulls. Because of the large weight of the vertical free dangling pipe attached to the platform, the underwater displacement hulls of a CO_2 disposal platform must be much larger than that of conventional oil exploration platforms. Today's semi-submersible platforms are designed for operation in water depths up to 1000 m. The mooring system usually consists of eight anchors placed in a spread pattern and connected to the hull by chain or wire rope. These platforms are used for oil drilling and pumping from the ocean floor. Large diameter free dangling pipes may pose great technical difficulties. Present offshore oil drilling and pumping facilities use pipes that do not exceed 16 inches in diameter, and the pipe is fastened to the sea floor by a template. The Ocean Thermal Energy Conversion (OTEC) facility off the shore of Hawaii uses a sea floor mounted pipe rather than a vertical pipe dangling from a platform, because previous experience showed that dangling pipes are prone to breaking (Nihous, 1995). This pipe is made of polyethylene and reaches a depth of 700 m within a distance of about 5 km from the shore. A free dangling pipe would experience enormous forces due to gravity, pressure differential inside/outside, thermal expansion and hydrodynamic forces due to water motion. The connection of the pipe to the platform must be flexible to withstand the mutual motions of the pipe and platform of surge, sway, yaw, heave, pitch and roll (Demirbilek, 1989). An offshore platform with an attached vertical pipe probably would serve several coastal power plants. We expect the pipe diameter to be at least 36 inches, serving six to eight 1000 MW power plants. Srikandarajah and Mahendran (1987) estimated the wall thickness and submerged weight of various diameter pipes as a function of length. For a 36 inch diameter pipe reaching to 1000 m depth the wall thickness is 38 mm, and the submerged weight is 200 Tonnes. For a 3000 m vertical pipe the wall thickness would be 57 mm, and the submerged weight 2500 Tonnes.

Tanker Transport. Transportation of liquid CO_2 in a tanker requires pressurization and refrigeration, or full insulation. An optimal regime would be a temperature of -50°C to -20°C, at which pressures the liquid has a vapor pressure of 5-20 bar. Such vessels need to be built with heavily constructed holding tanks. According to Fujioka (1966), Mitsubishi Heavy Industries builds semi-refrigerated tankers for transport of pressurized liquid chemicals, such as LPG and ammonia. The standard design tanker has a volume of 22,000 m³ with 4 tanks, 5,500 m³ each, capable of transporting a liquid at a temperature of about -50°C, pressure about 6 bar. Such a tanker could accept one day's load of CO_2 from a 1000 MW plant. Assuming that such a tanker has a speed of 15 knots, that it takes 6 hours each for loading and unloading, and the time of the return journey, the tanker can travel about 170 km to a disposal platform with one day's load of CO_2. For greater distances, proportionally more tankers will be necessary. Larger tankers, up to 125,000 m³, for the transport of LNG are being built today, however they are not pressurized, and therefore not suitable for the transport of liquid CO_2.

Dry Ice Transport. For dry ice transport, special cargo vessels must be used. Solid CO_2 has a vapor pressure of 1 bar at 194.6K. Therefore, the holds of the vessel need to be refrigerated or well-insulated to prevent evaporation during voyage. Solid CO_2 has a density that is approximately 1.5 times the density of liquid CO_2, but this still requires enormous volumes to transport the output of large power plants. A 1000 MW plant will require one vessel per day with a volume of about 16,000 m³. For rapid sinking and slow dissolution purposes it is desirable to dump the solid CO_2 in large blocks. The loading onto the vessel, storage, and unloading on the open seas of large blocks of solid CO_2 will be a formidable task. One should mention also the extra energy requirement for producing solid CO_2. After separation from the flue gas, solidifying CO_2 requires approximately twice the energy compared to liquefying it. If a power plant with CO_2 removal and liquefaction would have a net thermal efficiency of 35%, that with solidification would have an efficiency of about 31%.

Cost of Transport Systems. The cost of laying a pipe, excluding the cost of the pipe itself, is about $1M/km in shallow water, and $2M/km in deep water, with an average cost of $1.5 M/km (Rayborn, 1996). The cost of a 22,000 m³ tanker is about $40-50M (Fujioka, 1996). The cost of a near-shore floating platform for oil and gas exploitation is about $80-100M. The cost of a mid-ocean platform anchored to the deep bottom, or kept in position by dynamic positioning is in the order of $200-250M (Daniels, 1996). To this one should add the cost of mounting a vertical discharge pipe descending to at least 1000 m, or preferably deeper. Let us assume that the total cost of a distant floating platform with a storage tank and vertical pipe is about $250M. Accordingly, the crossover balance of costs is about at 250 km. That is, up to 250 km it would be cheaper to lay a sea floor mounted pipe to the desired depth ($375M); at distances greater than 250 km, it would be cheaper to employ 2 tankers ($50M each) shuttling to and from a floating platform ($250M).

ENVIRONMENTAL EFFECTS OF TRANSPORT SYSTEMS

The major environmental effect of pipe transport will occur during the construction phase. Inland pipelines may cross open water, wetland and upland habitats. New pipeline canals through wetlands are typically 3 m wide, which is necessary for the push-ditch method of pipeline construction. In the USA, since 1970, backfilling of newly dredged pipeline canals has been required by permitting agencies. In the Gulf of Mexico, offshore and onshore oil explorations typically disturbed 0.7 hectares of wetlands per km of pipeline (Turner and Cahoun, 1988). Backfilling newly dredged pipeline canals is typically more successful in freshwater wetlands with higher inorganic content in the soils than in salt or brackish wetlands (DOI, 1994). Restoration and revegetation of the habitat after construction will be required under the Resource Conservation and Recovery Act. During the laying of submerged pipes, sediment displacement (resuspension of sediments) and coral reef disturbance will be caused by pipeline trenching, which in the USA is required by law in water depths to 61 m . Such trenching and burying is required in shallow depths as an engineering precaution in order to reduce movement of the pipe during periods of high currents and storms. It is estimated that 5,000 cubic meters of sediment will be resuspended for each kilometer of pipeline trenched. Pipelaying barges use an array of eight 9,000 kg anchors to both position the barge and to move it forward along the pipeline route. The anchors are continually moved as the pipelaying operation proceeds. The area actually affected by anchors will depend on water depth, wind, currents, chain length, and the size of the anchor and chain. Service vessel anchoring is assumed not to occur in water depths greater than 150 m. In such water depths the vessels will always tie up to a platform or buoy (DOI, 1994). Upon completion of the pipelaying operations, underwater trenches will have to be refilled. Probably, it will take several years before the underwater habitat is restored to its previous state. Thereafter, no further disturbances are expected, as maintenance actions of underwater pipelines will be very infrequent.

The environmental impacts of tanker transport include (a) accidental releases from valves and ruptures with possible injury to the crew, and (b) release into the atmosphere of fuel combustion products. Both diesel and steam propelled tankers release into the marine atmosphere copious quantities of particles (flyash and soot), SO$_x$ and NO$_x$, and of course CO$_2$. It has been reported that along the most frequently used shipping channels in the Atlantic and Pacific Oceans, high altitude artificial clouds are formed due to ship emissions (Coakley et al., 1987). If tanker transport is used for deep sea disposal, floating platforms must be used for unloading the CO$_2$ from the tanker and pumping it into a vertical pipe extending from the platform. The floating platform and the vertical pipe will cause minimal impact to the marine environment. The operating crew on the platform will be exposed to storms and other environmental hazards, as well as to possible accidental releases of CO$_2$ during unloading and vessel rupture. The platform may cause a navigational hazard to ships and the vertical pipe to submarines. If tankers are used to transport CO$_2$ from a coastal power plant or a coastal terminal of inland pipelines, a port facility has to be provided for the tankers. The facility will include storage tanks, pumping stations and docks, as well housing for the workers and tanker crew. Constructing the facility will require coastal land acquisition and some disruption of coastal habitat. The facility's presence, along with the associated access routes, alters the natural hydrology and geography of the area over time, resulting in increased runoff from the land. Saltwater intrusion may occur if the facility is built near wetlands. If existing vegetation is modified, greater erosion and land loss around the facility may take place. Increases of nonpoint source pollution due to runoff at support facilities may contribute particulate matter, heavy metals, oil and grease, fecal coliform and high nutrient loadings to local streams, estuaries and bays, causing elevation of the contaminants in the surrounding waters, low dissolved oxygen levels, and high turbidity. Supporting onshore infrastructure also contributes to routine and accidental point source discharges to the surrounding bodies of water. Service bases and marine terminals contribute contaminants through their use of anti-fouling marine paints, discharge of domestic waste, and through the release of heavy metals and contaminated sludge (DOI, 1994).

<u>Dispersion of accidental releases of CO₂.</u> An accidental release of pressurized CO_2 in transport systems may arise because of mechanical failures, human error and natural disasters. The released CO_2 will cause a hazard to the surrounding population, workers at the facility and the surrounding fauna. CO_2 is known to be toxic at concentrations of about 6%, resulting in nausea, vomiting, diarrhea, irritation to mucous membranes, skin lesions and sweating. At about 10% concentrations, CO_2 is fatal due to asphyxiation (Sharp, 1972). The 1986 Lake Nyos, Cameroon, disaster caused death to about 1700 people and 3000 cattle. In that event magmatic CO_2 which accumulated at the lake bottom suddenly was released into the atmosphere. Due to a yet undetermined perturbation the lake rolled over, bringing up the CO_2 from the hypolimnion first to the surface layer, then vented into the atmosphere and blown toward the nearby village (Kling et al., 1987).

At ambient temperatures, CO_2 gas is heavier than air and will disperse mainly horizontally. The dispersion rate depends on the CO_2 concentration in the cloud and ambient conditions, such as winds, atmospheric stability and temperature. The principal dilution mechanism is entrainment of air through the top of the cloud. Using the model of Fay (1987), we estimate the dispersion of CO_2 from the rupture of a pipe or storage tank. The results are given in Tables 3 and 4, and illustrated in Figures 2 and 3. Here the dilution time means the time at which the concentration of CO_2 in the air χ decreases below 10%. The distance R is the radius from the rupture point to the given concentration, and v is the velocity at which the cloud expands. For the pipeline, a capacity of 390 Tonnes corresponds to a release of CO_2 from a 500 m segment of a 40 inch pipeline, 1170 Tonnes 1600 m , 2340 Tonnes 3200 m, 3510 Tonnes 4800 m, and 4680 Tonnes 6500 m pipeline. The segments have been selected with the assumption that CO_2 can be shut off with check valves at those distances upstream of the rupture point. Taking a midrange, 2340 Tonnes, a fatal concentration $\chi = 10\%$ will be reached in 2 minutes at a radius from the rupture point R = 580 m. For the storage tank rupture, a capacity of 28,000 Tonnes corresponds to an output of a 1000 MW power plant in 30 hours. Each subsequent capacity is a doubling of the number of hours. The rupture of the 28,000 Tonnes tank would cause fatalities ($\chi \geq$ 10%) within a radius of 1.3 km up to 3 minutes. Similar results have been obtained by Kruse and Tekiela (1995). While an accidental release from a pipe or tank rupture is highly unlikely, the risk cannot be dismissed. Leaks from tanker holds, solid CO_2 transport vessels, pumps, loading and unloading pipes may occur more frequently. While their radius and time of impact may be shorter than that of pipelines and storage tanks, such leaks may endanger the operating crew, and proper precautions need to be exercised, such as wearing of gas masks and protective clothes.

Table 3. Dispersion from pipeline rupture

Capacity (1000 ton)		Dilution Time (min)							
		0.5	1	1.5	2	3	4	6	10
0.39	R (m)	187	263	321	370	450	523	640	739
	χ (%)	100	21.9	10.1	6.6	3.9	2.8	1.7	1.3
	v (m/s)	6.3	4.4	3.6	3.1	2.5	2.2	1.8	1.5
1.17	R (m)	245	346	422	487	595	687	842	971
	χ (%)	100	35.7	13.9	8.6	4.9	3.4	2.1	1.6
	v (m/s)	82	5.8	4.7	4.1	3.3	2.9	2.3	2
2.34	R (m)	294	411	502	580	709	818	1001	1155
	χ (%)	100	54.6	17.3	10.3	5.7	3.9	2.4	1.8
	v (m/s)	9.8	6.9	5.6	4.8	3.9	3.4	2.8	2.4
3.51	R (m)	325	456	556	641	784	905	1108	1279
	χ (%)	100	75.9	20	11.5	6.24	4.28	2.6	1.9
	v (m/s)	10.9	7.6	6.2	5.4	4.4	3.8	3.1	2.7
4.68	R (m)	351	491	599	670	843	973	1191	1374
	χ (%)	100	100	22.3	12.5	6.7	4.5	2.8	2
	v (m/s)	11.7	8.2	6.7	5.8	4.7	4.1	3.3	2.9

Table 4. Dispersion from storage tank rupture

Capacity (1000 ton)		Dilution Time (min)						
		0.5	1	1.5	2	3	4	6
28	R (m)	553	771	939	1082	1322	1524	1865
	χ (%)	100	100	56.9	22.9	10.5	6.7	4
	v (m/s)	18.4	12.8	10.4	9	7.3	6.4	5.2
56	R (m)	660	918	1118	1288	1573	1814	2218
	χ (%)	100	100	100	31	12.7	8	4.6
	v (m/s)	22	15.3	12.4	10.7	8.7	7.6	6.2
112	R (m)	787	1094	1332	1533	1872	2158	2639
	χ (%)	100	100	100	45.3	15.8	9.6	5.3
	v (m/s)	26.2	18.2	14.8	12.8	10.4	9	7.3
224	R (m)	940	1304	1586	1825	2227	1567	3139
	χ (%)	100	100	100	76.5	20.1	11.5	6.2
	v (m/s)	31.3	21.7	17.6	15.2	12.4	10.7	8.7

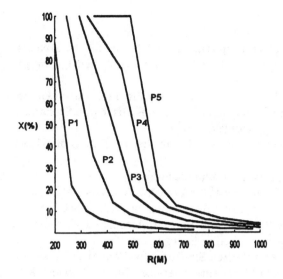

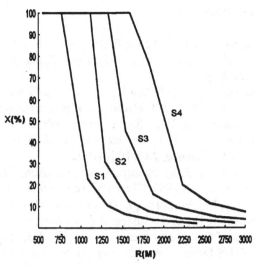

Figure 2. Dispersion From Pipeline Rupture
P1 is release of 390 T, P2 = 1170 T, P3 = 2340 T, P4 = 3510 T, P5 = 4680 T

Figure 3. Dispersion From Storage Tank Rupture
S1 is release of 28 kT, S2 = 56 kT, S3 = 112 kT, S4 = 224 kT

CONCLUSIONS

While liquid CO_2 can be transported in trucks, railroad cars and barges, practical transport systems include only pipes and large tankers. Onshore and offshore pipelines will cause some environmental disturbances during construction. Tankers and floating platform may not cause significant environmental disturbances, but docking facilities may alter the natural hydrology and geography of the area, saltwater intrusion over wetlands, beach erosion, and point and nonpoint source pollution. Accidental releases of CO_2 from pipes, storage tanks, vessel holds, connecting pipes and valves may endanger the workers and surrounding population. However, the environmental impacts and safety hazards of CO_2 transport systems can be minimized or altogether prevented with prudent management and exercise of precaution.

ACKNOWLEDGMENT

This work was performed with the indirect support of the US Department of Energy, Pittsburgh Energy Research Center, under a consulting agreement with the Energy Laboratory, Massachusetts Institute of Technology. The guidance and discussions of Howard Herzog are gratefully acknowledged. Yulin Shao and. Herbert Sinnock were helpful in gathering data and performing calculations.

REFERENCES

Auerbach, D. (!996) Near Field Impacts of Ocean CO_2 Disposal, Part I: A Model for the Toxicity to Marine Organisms of Lowered pH, Paper presented at ICCDR-3, to be published in *Energy Convers. Mgmt.*

Caulfield, J.A., D.I. Auerbach, E.E. Adams and H.J. Herzog (1996) Near Field Impacts of Ocean CO_2 Disposal, Part II: Probabilistic Plume Exposure Simulations for Various Discharge Scenarios, Paper presented at ICCDR-3, to be published in *Energy Convers. Mgmt.*

Coakley, J. A., R. L. Bernstein and P. H. Durkee (1987) Effect of Ship-Stack Effluents on Cloud Reflectivity, *Science*, **237**, 1020-1022.

Daniels, J. (1996) Ensearch Corp., Houston, TX, private communication.

Demirbilek, Z. (1989) "Tension Leg Platforms: an Overview of the Concept, Analysis and Design" in Tension Leg Platforms, Z. Demirbilek, ed. Amer. Soc. Civ. Eng. publication, New York, NY.

Dewey, R.K., G.R. Stegen and R.Bacastow (1996) Far Field Impacts Associated with Ocean Disposal of CO_2, Paper presented at ICCDR-3, to be published in *Energy Convers. Mgmt.*

DOI (US Department of Interior) (1994) Environmental Impact Statement for Central and Western Planning Areas, OCS EIS/EA MMS 94-0019, Minerals Management Service , Gulf of Mexico OCS Region, New Orleans, LA.

Fay, J. A. (1987) Modeling the Lake Nyos Disaster, *Atmospheric Environment*, **21**, No.3,

Fujioka, Y. (1996) Mitsubishi Heavy Industries, Ltd., Nagasaki R&D Center, private communication.

Golomb, D., H. Herzog, J. Tester, D. White and S. Zemba (1989) Feasibility Modeling and Economics of Sequestering Power Plant CO_2 Emissions in the Deep Ocean, Energy Laboratory, Massachusetts Institute of Technology, MIT-EL 89-003, Cambridge, MA 02139.

Keeling, C.D. (!994) Global Historical CO_2 Emissions, in Trends '93: A Compendium of Data on Global Change, ORNL/CDIAC-65, Oak Ridge National Laboratory, Oak Ridge, TN.

Kling, G. W., M. A. Clark, H. R. Compton, J. D. Devine, W. C. Evans, A. M. Humphrey, E. J. Koenigsberg, J. P. Lockwood, M. L. Tuttle and G. N. Wagner (1987) The 1986 Lake Nyos Gas Disaster in Cameroon, West Africa, *Science*, **236**, 169-175.

Kruse, H. and M. Tekiela (1995) Calculating the Consequences of a CO_2 Pipeline Rupture, *Energy Conversion and Mgmt.* **37**, 1013-1018..

Mazurkiewicz, B. K. (1987) "Offshore Platforms General" in Offshore Platforms and Pipelines, B. K. Mazurkiewicz, ed. Trans Tech Publications, Germany.

Nihous, G. C. (1995) Private Communication, Pacific International Center for High Technology Research, Honolulu, HI 96822.

Ranney, M. W. (1979) Offshore Oil Technology, Noyes Data Corporation, Park Ridge, NJ.

Rayborn, E. (1996) Kvaerner R.J. Brown & Assoc., Houston, TX, private communication.

Sharp, G. R. (1972) in Aviation Medicine: Physiology and Human Factors, G. Dhenin, ed., pp. 202-203, Tri-med Books, London, UK.

Skovholt, O. (1993) CO_2 Transportation System, *Energy Conversion and Mgmt.*, **34**, 1095-1105.

Sriskandarajah, T. and I. K. Mahendran (1987) "Parametric Considerations of Design and Installation of Deepwater Pipelines" in Offshore Oil and Gas Pipeline Technology, IBC Technical Services, Ltd., London, UK.

Turner, R. E. and D. R. Cahoun (1987) Causes of Wetland Loss in the Coastal Central Gulf of Mexico, OCS Study MMS 90-0009, prepared for the US Department of Interior, Minerals Management Service, Gulf of Mexico OCS Region, New Orleans, LA.

Uhl, A.E. (1965) Steady Flow in Gas Pipelines, Report prepared for the American Gas Association, Arlington, VA.

Pergamon

Energy Convers. Mgmt Vol. 38, Suppl., pp. S287–S294, 1997
ⓒ 1997 Elsevier Science Ltd. All rights reserved
Printed in Great Britain
0196-8904 97 $17.00 + 0.00

PII: S0196-8904(96)00283-X

Marine Carbonate Formations: Their Role In Mediating Long -Term
Ocean-Atmosphere Carbon Dioxide Fluxes - A Review.

C.N. Murray and *T.R.S. Wilson

Space Applications Institute,
Joint Research Centre, 21020 Ispra (VA), Italy.

*University of Liverpool, Laboratory of Oceanography,
PO Box 147, Bedford Street North, Liverpool, L69 3BX, UK.

ABSTRACT

Deep marine sediment formations far from tectonic plate edges represent stable regions which have not been sufficiently considered for their potential permanent storage capacity of carbon dioxide. The purpose of this paper is to review mechanisms involving carbon dioxide which may be significant on various geological time scales, and to consider how these might relate to the geological evidence of past surface conditions. There is evidence that large fluctuations in atmospheric carbon dioxide have occurred in the past. In order to predict even the broad outline of this adjustment, it is necessary to consider the behaviour of carbon dioxide in the atmosphere, the oceans, the sediments and in the deep lithosphere. The ultimate adjustment of the system to anthropogenic inputs of fossil fuel carbon will take many tens of thousands of years, and the time constants of this process cannot be predicted with accuracy at the present state of knowledge. However, it is possible to draw inferences from what is known, and the conclusions have relevance to discussion of the question of disposal of carbon dioxide within deep sedimentary systems. ⓒ 1997 Elsevier Science Ltd

KEYWORDS

Carbon dioxide disposal, sedimentary formations, permanent storage, ocean-atmosphere interaction.

GEOLOGICAL CARBON CYCLES

Climate varies naturally on all time scales from hundreds of millions of years to a few years (e.g. Folland et al. 1990). Broecker and Denton (1990), among others, have argued that glacial cycles are driven by massive reorganisations of the ocean-atmosphere system that link cyclic changes in the earth's orbit to the advance and retreat of ice sheets. Prominent in recent Earth's history have been the 100,000 year Pleistocene glacial-interglacial cycles when climate was mostly cooler than at present (Imbrie and Imbrie 1979). This period began about 2 million years before present (BP) and was proceeded by a warmer epoch having only limited glaciation, mainly over Antarctica, called the Pliocene (5-2 Myr BP). Global surface temperatures have typically varied by 5-7 °C through the Pleistocene ice age cycles, with large changes in ice volume and sea level, and temperature variations as great as 10-15 °C in some middle and high latitude regions of the Northern Hemisphere. Since the beginning of the current interglacial epoch (about 10,000 years BP) global temperatures have fluctuated within a much smaller range. Some fluctuations have nevertheless lasted several centuries, including the Little Ice-Age which ended in the nineteenth century and which was global in extent.

Emiliani (1955) produced the first complete record of variations in past glaciations on the basis of foraminifera (single cell marine organisms) which produce calcium carbonate shells. The ratio of ^{18}O to ^{16}O in the tests (carbonate) reflect the ratio of the two oxygen isotopes in the seawater in which they developed. It has been shown that this ratio is closely linked to the proportion of the world's water that is incorporated into glaciers and ice sheets. The processes of evaporation and precipitation tend to favour the return of ^{18}O over ^{16}O to the oceans (by a very small factor due to the relative difference in molecular weight). Thus water falling as snow on ice sheets becomes relatively depleted of ^{18}O. As glaciers and ice sheets grow, so ^{16}O becomes proportionally lower in seawater.

On the basis of cores from seafloor sediments Emiliani found that the isotopic ratio varied in general with the astronomical cycles as predicted by Milankovic (1879-1958). Milankovic identified three components affecting the earth's astronomical behaviour, two that change the intensity of the seasons (tilt, and shape of earth's orbit) and a third that affects the interaction of the other two. The tilt of earth's spin axis varies between 21.5 to 24.5 degrees (presently 23.50) and varies over a cycle of 41 thousand years. As the tilt increases the more intense are the seasons in both hemispheres. The second, weaker factor is the shape of the earth's orbit. Over a period of 100 thousand years the orbit stretches into a more eccentric ellipse and then grows more nearly circular again. The affect is that as eccentricity increases the difference in distance from the sun of the nearest and farthest point of the earth's orbit changes, intensifying the season in one hemisphere and moderating them in the other. The third factor is the precession, or wobble, of the earth's spin axis which varies with a cycle of 23 thousand years. Precession determines whether summer in a given hemisphere falls at a near or far point in the orbit, thus enhancing or weakening the effect of orbit eccentricity.

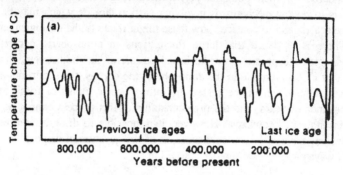

Fig 1. Global temperature variation over the last million years (scale 2^0C per vertical unit).

Figure 1 (ICCP-1990) shows the mean global temperature fluctuations over the last million years. The variations of temperature estimated from various environmental parameters largely coincide with the astronomical cycles. We do not know exactly why these factors initiate global climate changes. It seems likely that the mechanism is indirect, involving reorganisation of some part of the planetary fluid circulation (Broecker and Dalton 1990). The calcium carbonate of deep-ocean sediments reflects these glacial- interglacial climatic modifications through variations in their oxygen isotope ratio, a fact which points to the potential significance of this component in the global carbon cycle.

What is known about the concentration of carbon dioxide in the Earth's atmosphere during earlier ages? Gammon et al (1985) reviewed the available data for a period up to 100 million years BP, and concluded that very large changes in carbon dioxide occurred between 10^8-10^7 years BP. Recent studies have shown that on the basis of the pH dependent fractionation of boron isotopes between sea water and precipitated carbonate minerals (Spivack et al 1993) that sea water boron concentrations and isotopic composition have been constant during the past 21 Myr. They reported that 21 Myr ago, surface ocean pH was only 7.4 ± 0.2, but it then increased to 8.2 ± 0.2 (roughly the present value) about 7.5 Myr ago. Such results are consistent with the suggestion that carbon dioxide concentrations may have been much higher before 21 Myr than today (figure 2, from Spivack et al 1993).

Sanyal et al (1995) have presented boron isotope data indicating changes in ocean pH between the last glacial and the Holocene period. They have estimated that the deep Atlantic and Pacific oceans had a pH 0.3 ± 0.1 units higher during the last glaciation and suggest that the accompanying change in carbonate ion concentration is sufficient to account for the decrease in atmospheric pCO2 during the glacial period (from 280 ppmv to 200 ppmv, e.g. Barnola et al, 1987). Sanyal et al propose that these results are consistent with the hypothesis (Archer and Meyer-Reimer 1994) that low carbon dioxide content of the glacial atmosphere was caused by an increase in the ratio of organic carbon to carbonate carbon reaching the seafloor, which led to an increase in carbonate ion concentration (and thus in pH) and to a corresponding decrease in pCO$_2$.

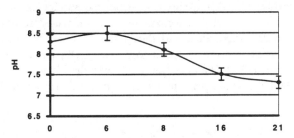

Fig 2. Calculated pH values of surface seawater as a function of age, 10^6 years.
(from Spivack et al 1993)

Thus from an initial concentration, perhaps as high as several thousand ppmv in Cretaceous (136-65 BP) times, the atmospheric level fell towards much lower values varying in the range of 200-300 ppmv characteristic of the glacial-interglacial cycles of the past two million years or so. This was followed by a post glacial period of relatively stable carbon dioxide concentrations varying between 260-290 ppmv which lasted from about 10 thousand years BP until the beginning of the nineteenth century. The third stage, from 1800 BP to the present, is one in which human impact on the global carbon cycle becomes clearly measurable in the atmosphere. It has been estimated that the increase in atmospheric levels of carbon dioxide during the last deglaciation was comparable in magnitude to the increase from the beginning of the industrial revolution to the present day.

The quantity of anthropogenic induced emissions since the beginning of the industrial revolution (fossil fuel burning plus land use changes) has caused an increase in atmospheric carbon dioxide concentrations from 280 ppmv (~ 600 Gtons C) to 355 ppmv (~ 750 Gtons C). This is equivalent to a net overall increase in atmospheric carbon dioxide of some 550 Gtons CO$_2$ and has taken place in about 200 years compared to the 5-8 thousand years of the last glacial to interglacial climate changeover. However, whereas the oceans appear to have acted as a slow source of carbon dioxide during deglaciation, present day oceans are acting as a relatively rapid overall sink. This contrast may have significant implications for our understanding of possible future climate fluctuations, and some likely mechanisms are explored below.

Among others Berger (1976) has estimated that about half of the deep-sea floor is covered by biogenically produced pelagic sediments made up of calcareous oozes (~ 50%) with concentrations of calcite varying between 10-90%, siliceous oozes (~15%), and by yellow, brown and reddish deep sea clays (~ 35%). The ocean is typically supersaturated with calcite at shallow and intermediate depths and under saturated in deep waters as its solubility increases with pressure. As Archer and Meyer-Reimer (1994) point out any imbalance in the sources (terrestrial weathering and alteration) and sinks (deep sea and shallow water depositions) for CaCO$_3$ will change the ocean dissolved carbonate ion concentration (CO$_3^{--}$) in the direction of restoring this balance.

Berner (1994) has drawn together the available information on these processes in order to estimate the relative change in atmospheric carbon dioxide over the past 600 million years. He concludes that present pCO$_2$ levels are lower than at any previous period having been approximately one order larger during the early history of the Earth. Berner concludes that this fall in atmospheric pCO$_2$ appears to have been compensated for by an increase of about 5% in solar energy output, so that the surface temperature of the Earth has remained remarkably constant.

THE INFLUENCE OF OCEAN SEDIMENTATION PROCESSES

Since the carbon deposited within sediments must be derived from the surface ocean which is in equilibrium with the atmosphere, it seems obvious that the atmospheric carbon dioxide concentration must be influenced by the fluctuations in the rate of deposition of carbon noted above. The linking processes by which this influence can be exerted are rather complex. It is surprising, for instance, that the deposition of large quantities of calcium carbonate from the surface ocean tends to raise the atmospheric pCO2. If the calcareous rocks so formed are later weathered sub-aerially, the process removes carbon dioxide from the atmosphere and thus tends to lower the pCO2. Similarly, the deposition of organic carbon tends to decrease the atmospheric pCO2, just as its later oxidation or combustion will have the opposite effect. The important parameter, therefore, includes not only the rate of deposition of carbon within the sediments, but the ratio between organic and inorganic carbon preserved in the sediments.

How is this ratio controlled? All phytoplankton produce organic carbon compounds from dissolved carbon dioxide. Some organisms also secrete calcium carbonate to form their skeletal or shell components (using carbon from dissolved carbonate ions), while others do not have such hard components, or make them of other material such as silica. Even though the latter group of organisms is less effective at removing carbon from solution, they are more effective (per gram of carbon fixed as protoplasm) at lowering pCO$_2$ because they do not simultaneously precipitate calcium carbonate. Such precipitation changes the alkalinity of the surface water and lowers the pH thus tending to increase the atmospheric pCO$_2$. In the modern ocean, it is known that the growth of coral provides a significant removal pathway for calcium in the shallow tropical ocean (Wilson, 1975): insofar as most of the associated organic carbon is respired before burial while an appreciable fraction of the calcite can be preserved by burial, this process actually tends to increase atmospheric pCO$_2$ rather than decreasing it.

The first possible control is therefore ecological. Any process which tends to favour the growth of silica-secreting organisms such as diatoms, at the expense of calcite-secretors such as the coccolithophorids, will tend to lower the atmospheric pCO$_2$ even if the total organic biomass formed in the surface layer and sinking from that layer remains constant. Conversely, an ecological shift favouring calcite (or aragonite) secretors would tend to raise the atmospheric pCO$_2$. The factors which control the species composition of the phytoplankton include temperature, nutrient levels and light availability, but the more subtle indirect factors are not yet understood.

Evidence exists that such indirect links can be significant in climate control. Curry and Lohmann (1985) reported on a study of carbonate and inorganic carbon deposition in the Atlantic over the past 160,000 years. They were able to show that the accumulation of carbonate in shallow sediments decreased by one half during glacial maxima, while greater remineralisation of organic carbon occurred below 3750m. These observations imply that compared to present conditions organic carbon productivity was relatively more important than carbonate mineral formation during the last glacial maximum. This change could be expected to produce a lower pCO$_2$. It is known that lower atmospheric pCO$_2$ characteristic of glacial periods can account for at least an appreciable fraction of the global surface temperature change during these times. These observations support the contention that ecological shift within the phytoplankton could be implicated in natural climate variation.

It follows that the changes induced by fossil fuel burning might well change the species composition of the planktonic biota, but it is not yet possible to predict the magnitude or even the sign of this impact. This in itself is a powerful argument for minimising the release of carbon dioxide to the atmosphere and surface ocean. The time scale involved in any such feedback loop active in the surface ocean would be of the order of a few decades to centuries, and thus potentially of consequence for strategic decision making.

In the longer term, the biospheric pCO$_2$ control is mediated through processes in the interior of the ocean. It is known that the oxidation of organic carbon is a fairly rapid process: after sinking out of the euphotic zone, much of the organic carbon vertical flux is oxidised before reaching the sea floor. By contrast, the inorganic skeletal components are dissolved at or near the sea bed, before or during burial. It follows that more rapid burial, occurring perhaps because of an enhanced terrigenous component from riverine run-off or aeolian transport, will tend to remove carbonate minerals from contact with the ocean before they can dissolve. Such a change would tend to produce an ocean of lower pH and therefore a higher atmospheric pCO_2.

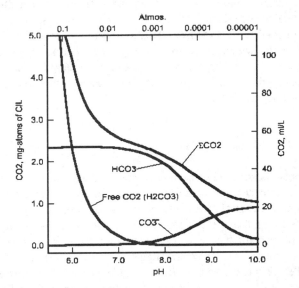

Fig 3. Variations in total CO_2, free CO_2, HCO_3^- and CO_3^{--}, with hydrogen ion concentration in sea water (Sverdrup et al., 1946).

The most significant deep-ocean feedback loop is however mediated through the influence of pressure and temperature on the chemistry of sea-water. The solubility of both calcite and aragonite increases with increase in pressure or decrease in temperature. Consequently, both these minerals dissolve far more efficiently in deep, cold water, and the deepest sediments are often almost completely devoid of inorganic carbon. The effect of increased atmospheric carbon dioxide is to decrease the pH of ocean water (figure 3), and this favours solution of carbonates. In fact, the depth at which carbonates begin to dissolve rapidly becomes more shallow within the ocean. This exposes a larger area of seafloor to under- saturated seawater, so that a larger fraction of the vertical flux of that mineral will be dissolved before burial. In other words, the effect of fossil-fuel burning is to increase the ratio of organic to inorganic carbon preserved in the sediments, and thus to increase the efficiency with which the oceans can remove carbon dioxide from the atmosphere. In effect, the extra carbon dioxide is taken up by the dissolution of inorganic carbonate in a kind of global deep ocean titration process.

Farrell and Prell (1989) showed that the depth at which sediments of 90% calcite are found has oscillated with a period of 100kyr, calcite-rich sediments being found at greater depth during glacial intervals than during interglacials. This is consistent with the outline above, a less corrosive deep water composition resulting from lower atmospheric carbon dioxide concentration during glacials. However, Oxburgh and Broecker (1993) have shown that the sediments of the deep Pacific are out of equilibrium with respect to the water chemistry. These sediments respond relatively slowly to changes in water chemistry, the bioturbated layer of the deep sediments taking several kiloyears to equilibrate with the Glacial-Holocene transition. It follows, by analogy, that the homeostatic mechanism outlined above cannot, therefore, be capable of controlling atmospheric carbon dioxide composition over shorter periods.

Since the deep ocean and the atmosphere are in reasonably good contact on this time scale, release of carbon dioxide to the deep ocean, while avoiding the large atmospheric excursions consequent on atmospheric disposal, will lead to an eventual significant increase over periods of several hundred years to a few kiloyears. Because the mixing rate of the deep ocean is very slow, the titration process outlined above will then reduce the atmospheric concentration over tens of kiloyears thereafter. It follows that the release of carbon dioxide to the deep ocean, although much less undesirable than release directly to the atmosphere, is still associated with considerable long-term climate uncertainties.

Because carbon dioxide forms stable ice-like hydrates under deep ocean conditions, it can be argued (Murray et al 1996) that calcareous sediment deeper than a kilometre or so within the ocean might offer the possibility of almost complete isolation from the atmosphere. Geologically, there is an encouraging analogy with methane hydrates which appear to form a quantitative trap for methane diffusing upward toward the sediment water interface. Direct observation of carbon dioxide behaviour on the deep ocean floor after release from hydrothermal systems is also encouraging. The practical problems of emplacement are, however, daunting.

LONG-TERM CONSIDERATIONS

The early proto-atmosphere is believed to have contained high carbon dioxide concentrations, and it required the development of eucaryotic organisms to produce an atmosphere sufficiently depleted in carbon dioxide that appreciable glaciation could occur. It has been known since the early work of Svante Arrhenius (1859-1927) that the geologic record holds much more carbon dioxide than any conceivable early atmosphere could have contained. It follows that there must be a means of supplying carbon dioxide to the atmosphere, and Arrhenius suggested that this was vulcanism. The later work of Chamberlain (see Revelle 1985) extended this idea to include the role of weathering in removing carbon dioxide from the atmosphere during the breakdown of continental crust; Garrells & Mackenzie (1972) and Berner et al (1983) have more recently developed this model in the light of recent knowledge.

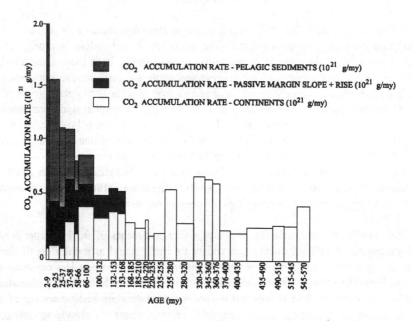

Fig.4 Accumulation of CO2 as carbonate (1021 g per million years) on the continents (Ronov 1980), in passive margin slopes and rise deposits and in pelagic (Hay 1985) sediments over the past ~ 600My.

Since the rate of weathering is not strongly dependent upon atmospheric carbon dioxide concentration, the system is not well-constrained with respect to changes in the global intensity of volcanic activity. The rate of accumulation of carbonate rock (figure 4) has varied significantly over the past 600 million years (Ronov, 1980, and references cited therein, Hay, 1985), and this may well represent the response of the ocean carbonate system to changes in the rate of supply of vulcanogenic carbon dioxide and of bicarbonate ion resulting from weathering. This picture is complicated by the fact that an appreciable fraction of volcanic carbon dioxide is recycled from carbonate sediments subducted at ocean margins. Long-term ecological change is also significant: the development of the oceanic calcareous plankton since the Mesozoic (225-65 Myr BP) may be responsible for the apparent shift from shallow sea accumulation to the deep ocean shown in figure 4 (Hay, 1985).

Sundquist (1985) has reviewed the time-constants appropriate to the various reaction-systems which influence atmospheric carbon dioxide concentration. By means of a series of model systems, he demonstrated that the sediments must be taken into account for time scales greater than a few hundred years, while continental weathering effects are significant at time scales greater than one hundred thousand years. It would be reasonable to consider the natural system as containing a hierarchy of control systems of successively increasing capacity and of decreasing sensitivity. Thus, small perturbations may produce a series of quite subtle feedback adjustments, while a more determined assault on the stability of the system may result in considerable transient disturbance until the long-term homeostasis asserts itself.

In view of the changes in solar luminosity which have occurred, and in the face of considerable fluctuations in the intensity of vulcanism during the break-up and movement of tectonic plates, it is remarkable that the surface temperature of the earth has apparently remained in the liquid water range since the evolution of an oxygenated atmosphere around 2Gyr BP. However, within this broad stability, appreciable fluctuations have occurred, and the geological record shows evidence of the reorganisations and adjustments which resulted. In the face of this evidence of variability over long time periods, it is difficult to argue that the homeostatic mechanisms, which undoubtedly exist in the natural system, are capable of very close control. It would be more prudent to accept that considerable short-term perturbation of surface conditions is possible, with homeostasis only occurring reliably (if at all) on time scales one or two orders of magnitude longer than the time span of recorded human history.

CONCLUSIONS

Disposal of the products of combustion directly to the atmosphere has the advantage of low cost, but the disadvantage that a large transient atmospheric concentration spike is created. This is because the deep ocean equilibrates with the atmosphere on a time scale of hundreds to thousands of years, while present day anthropogenic releases are much more rapid.

Clearly, any proposal which seeks to prevent a rapid atmospheric transient while maintaining the rate of fossil fuel utilisation must depend on the capture of carbon dioxide before release to the atmosphere, and the subsequent isolation of this captured material from the atmosphere. While release to the deep ocean waters would produce a worthwhile benefit (e.g. Wilson 1983), a much better degree of isolation would be obtained by disposal within deep ocean sediments. Natural analogues indicate that clathrate formation could confer a very high degree of isolation, eventual return occurring by subduction and subsequent vulcanism over hundreds of millions of years. Spread over such a time scale, a release which would produce serious effects over a few hundred years would be reduced in impact by several orders of magnitude, an obviously desirable outcome. Further, such an option would allow the continued use of abundant, inexpensive carbon rich fuel resources over the next few hundred years.

REFERENCES

Archer, D. and Meyer-Reimer, E. (1994). Effects of deep-sea sedimentary calcite preservation on atmospheric CO_2 concentration. *Nature*, **367**, 260-263.

Barnola, J.M., Raynaud, D., Korotkevitch, Y.S. and Lorius, C. (1987). Vostok ice core provides 160,000 year record of atmospheric carbon dioxide. *Nature*, **239**, 408-414.

Berger, W.H. (1976). Biogenous deep-sea sediments: production, preservation and interpretation. In: *Chemical Oceanography* (J. P. Riley and R. Chester Eds.), **5**, 29, 265-372, Academic Press.

Berner, R.A. (1994). 3GEOCARB II: A revised model of atmospheric CO_2 over Phanerozoic time. *American Journal of Science,* **294**, 57-91.

Berner, R.A., Lasaga, A.C. and Garrels, R.M. (1983). The carbonate-silicate geochemical cycle and its effect on atmospheric carbon dioxide over the last 100 million years. *American Journal of Science,* **283**, 641-683.

Broecker, W.S. and Peng, T-H. (1982). *Tracers in the sea.* Lamont-Doherty Geological Observatory, Columbia University, New York, 689 pages.

Broecker, W.S. and Denton, G.H. (1990). What drives glacial cycles? *Scientific American,* **262**, 1, 42-50.

Curry, W.B., and Lohmann, G.P. (1985). Carbon deposition rates and deep water residence time in the equatorial Atlantic ocean throughout the last 160,000 years. In: *The carbon cycle and atmospheric carbon dioxide: Natural variations Archaen to present* (E. T. Sundquist and W.S. Broecke, Eds.), AGU Geophysical Monograph No. 32, 285-302.

Emiliani, C. (1955). Pleistocene temperatures. *J. Geol.*, **63**, 538-578.

Emiliani, C. (1978). The cause of the ice ages. *Earth Planet. Sci. Lett.*, **37**, 349-352.

Folland, C.K., Karl, T.R., and Vinnikov K. Ya. (1990). Observed climate variations and change. In: *Scientific Assessment of Climate Change.* Report IPCC, 195-242.

Farrell, J.W. and Prell, W.L. (1989). Climatic change and $CaCO_3$ preservation: an 800,000 year bathymetric reconstruction from the central Pacific ocean. *Palaeoceanography,* **4**, (4), 447-466.

Gamon, R.H., Sunquist, E.T., and Fraser, P.J. (1985). History of carbon dioxide in the atmosphere. In: *Atmospheric carbon dioxide and the global carbon cycle* (J.R. Trabalka, Ed.), Oak Ridge National Laboratory Report DOE/ER-0239, 26-62.

Garrels, R.M. and Macenzie, F.T. (1972). A quantitative model for the sedimentary rock cycle. *Marine Chemistry,* **1**, 27-41.

Hay, T.W. (1985) Potential errors in estimates of carbonate rock accumulating through geologic time. In: *The carbon cycle and atmospheric carbon dioxide: Natural variations Archaen to present* (E.T. Sundquist and W.S. Broecker, Eds.), AGU Geophysic. Monograph. 32, 627 pages.

Imbrie, J., Imbrie, K.P. (1979*). Ice Ages - solving the mystery.* Macmillan, London, 224 pages.

Murray, C.N., Visintini, L., Bidoglio, G., Henry, B. (1996). Permanent storage of carbon dioxide in the marine environment: the solid CO_2 penetrator. *Energy Convers. Mgmt.* **37**, 6-8, 1067-1072.

Oxburgh, R., Broecker, W.S. (1993). Pacific carbonate dissolution revisited. *Palaeogeography, Palaeoceanography, Palaeoecology,* **103**, (1-2), 31-39.

Revell, R. (1985). The scientific history of carbon dioxide. In: *The carbon cycle and atmospheric carbon dioxide: Natural variations Archaen to present* (E.T. Sundquist and W.S. Broecker, Eds.), AGU Geophysical Monograph No. 32, 1-4.

Ronov, A.B. (1980). The Earth's sedimentary shell: quantitative patterns of its structures, compositions and evolution (in Russian), (A.A. Yaroshivskiy, Ed.) Nauka, Moscow.

Spivack, A.J., You, C-F., and Smith H.J. (1993). Foraminiferal boron isotope ratios as a proxy for surface ocean pH over the last 21 Myr. *Nature,* **363**, 149-151.

Sanyal, A., Hemming, N.G., Hanson, G. N., Broecker, W.S. (1995). Evidence for a higher pH in the glacial ocean from boron isotopes in foraminifera. *Nature,* **373**, 234-236.

Sundquist E.T. (1985). Geological perspectives on carbon dioxide and the carbon cycle. In: *The carbon cycle and atmospheric carbon dioxide: Natural variations Archaen to present* (E.T. Sundquist and W.S. Broecker, Eds.) AGU Geophysical Monograph No. 32, 627 pages.

Sverdrup, H.U., Johnson, M.W., Fleming, R.H. (1946). *The Oceans their Physics, Chemistry and General Biology.* Prentice Hall, New York, 1087 pages.

Wilson, T.R.S. (1975). Salinity and the major elements of sea water. In: *Chemical Oceanography,* (Riley J.P. and G. Skirrow, Eds.), **1**, 6, 365-413, Academic Press.

Wilson, T.R.S. (1983). The deep ocean disposal of carbon dioxide. *Energy Convers. Mgmt.* **33**, 5, 627-634.

Energy Convers. Mgmt Vol. 38, Suppl., pp. S295–S300, 1997
© 1997 Elsevier Science Ltd. All rights reserved
Printed in Great Britain
0196-8904/97 $17.00 + 0.00

Pergamon

PII: S0196-8904(96)00284-1

TESTING THE WATERS:
AN ANALYTICAL FRAMEWORK FOR TESTING THE POLITICAL FEASIBILITY OF SCENARIO-BASED PROPOSALS FOR DISPOSING OF CO_2 IN THE OCEANS

Professor Judith KILDOW

Department of Ocean Engineering, Massachusetts Institute of Technology,
Cambridge, MA 02139

ABSTRACT

Industrial nations need an array of arrangements to meet obligations for the Rio Treaty CO_2 emissions standards for the year 2000. Ocean storage of power plant CO_2 emissions is one option.

Two strategies seem feasible: 1) Transfer CO_2 directly from the power plant by pipeline to the appropriate ocean depth and release it through diffusers; 2) transport contained CO_2 to an offshore site and diffuse it by pipe to appropriate depths. However, jurisdictional location of the disposal site, the environmental consequences of the diffused CO_2 on water quality and living resources, and public understanding and acceptance will ultimately influence decisions. Legality will be tested by The London Dumping Convention, The Law of the Sea Convention, the RIO Treaty and Agenda 21. Public acceptance will depend on public trust.

This paper presents a decision framework based on the two ocean disposal scenarios above. It allows scientists and policy-makers to examine these strategies using legal and socio-political baseline parameters. Both disposal strategies, if determined feasible, would have added costs necessary to assure the public of the integrity of the disposal activity. This decision framework helps policy makers formulate questions, determine trade-offs, and ultimately decide the practicality of the CO_2 Ocean Disposal strategy. © 1997 Elsevier Science Ltd

KEYWORDS

ocean disposal; CO_2; ocean policy; public acceptance; decision-making frame; hidden policy costs.

FRAMEWORK

Industrial nations will need an array of arrangements to meet their Rio Treaty obligations from the UN Conference on Environment and Development to limit CO_2 emissions to the standards set for the year 2000. One strategy for which environmental impact tests are being undertaken is ocean disposal of power plant CO_2 emissions before they reach the atmosphere. This solution has some drawbacks, even if technically feasible: 1) it relies on the oceans as a waste sink at a time when the oceans are under scrutiny for protection by international governmental organizations and non-governmental public interest groups; 2) it transfers CO_2 to another medium instead of reducing global production of CO_2, violating the spirit of the RIO Treaty; 3) it is probably only feasible where power plants are near to the coast, thereby affecting a small fraction of the CO_2 produced globally.

The design of a decision framework to determine the socio-political feasibility of the ocean option requires certain assumptions to establish the parameters. The first set of assumptions refers to the generic category of power plant capture and ocean disposal of CO_2, not to the specific disposal options. These follow:

1. The evidence for climate change is clear enough to convince the public that new and possibly riskier measures must be taken to address the problem.

2. International and national imperatives have resulted in incentives that convince the power industry to invest in measures to curb CO$_2$ emissions, e.g. carbon taxes or mandated legal measures.

3. Other available options are either not enough to offset the necessary reductions, using the ocean option as a supplement, or they are not comparable in risk and other costs.

A second set of assumptions refers to the specific options. They include:

4. Scientific evidence for each ocean disposal option clearly indicates socially acceptable impacts.

5. The "public" and its designated decision makers are persuaded that a particular disposal option is safe, and that its operators will carry it out reliably and with integrity.

This decision framework focuses on two ocean disposal scenarios described below. Its purpose is to 1) logically elaborate the socio-political and legal elements that scientists and policy makers should consider to test the waters for a CO$_2$ ocean disposal option; 2) demonstrate the potential influence of these considerations over the outcome of technical and economic decisions, particularly "hidden" costs; 3) provide benchmarks from past cases as a measure of acceptability and to identify trade-offs.

The first option assumes use of a pipeline from an onshore power plant to an offshore disposal site. One of the most efficient ways to transport CO$_2$ is as a liquid under pressure through a pipe. This assumes that there is a feasible location for long-term storage. The deeper the location and the farther offshore the greater the costs and risks of rupture or damage to the pipe. Scientists estimate a minimum of 500 meters depth is needed with locations no more than a few hundred kilometers offshore. These and other combinations of requirements—involvement of shipping routes or environmentally sensitive areas—limit this application to only a few places.

Laying pipelines under water is well-understood technology. If the pipe must be buried, the question of dredge spoils disposal as well as turbidity and toxicity must be weighed, as must potential environmental damage if blasting is necessary. If the pipe rests on the sea floor, there are risks of interference with fishing boats and shipping. There is ample precedent for analyzing some of these risks. However, even though engineers can solve the technical problems, public perceptions of risks will influence governing bodies that must permit the activity.

The second option—to transport the pressurized liquid CO$_2$ via pressurized, insulated tankers to a floating platform with a dangling, pressurized pipeline with diffusers, placed over an appropriate disposal site—utilizes new technologies and involves complex logistical operations. This option may appear more economic at first glance, but socio-political aspects may change the equation significantly. This method has risks that must be made acceptable to the public. Its costs depend upon the degree of safety and monitoring systems required by permitters. There are a number of logistical problems. The interdependence of each step demands smooth phasing; possibilities of snags at any one phase pose legal, political, technical and economic problems. The transfer of CO$_2$ from the power plant to the tanker, the safety of the transport route for the ship, and the transfer to the platform, down the pipeline and into the water column must all occur with precision. The possibility of bad weather requires backup storage capacity and affects the safety of ship transport. Without an effective monitoring system, some less-than-honest ship operators might dump their cargo to save on fuel and other costs, dump early to avoid bad weather or dump unintentionally in areas where detrimental effects would be unacceptable. While these risks may be low, they must be accounted for in a public forum. This will have added costs, which will be discussed later.

Past cases of attempts at similar logistically complicated enterprises have been highly scrutinized, and difficult to achieve. The U.S. never permitted offshore ship incineration because the public feared an "out of sight, out of mind" scenario. Deep sea mining of manganese nodules requires complex logistical timing and technical reliability to avoid polluting and to avoid huge economic losses. Its costs have never made it viable. Seabed disposal of high level radioactive wastes also posed logistical and phasing problems that affected public perceptions of its risks. Those and other problems, including monitoring demands, halted that option—thus far.

Both CO$_2$ disposal options will need to have "convincing" air-tight monitoring systems that guarantee the integrity of the operation—from construction through operations to ocean impacts of the process. These will all have added costs, leaving aside their feasibility.

The public must be assured that the oceans will be protected from harm. How that occurs is critical. Socio-political feasibility depends upon the "public's trust," and on scientific evidence supporting both the climate change problem and the ocean disposal options, particularly the uncertainties regarding the impacts of CO$_2$ disposal on ocean systems.

THE PUBLIC TRUST

The common property nature of the ocean makes it vulnerable, lacking the checks and balances of a private enterprise market system where the incentives for appropriate uses and disincentives for abuse are clearer. The public interest related to ocean activities requires protection from governments, both international and national, and participation in the regulatory process by public interest groups and other non-governmental organizations that provide market-type, but less tangible incentive to "do the right thing," as illustrated by the response to over fishing and pollution.

Controls or restrictions can take different political, legal and economic forms. Both disposal options raise international legal and political questions about national obligations, as well as considerations of national policies and laws. However, if the activity and its impacts fall outside national jurisdiction, international conventions will likely become an issue. The International Maritime Organization (IMO) presides over the London Dumping Convention and its Five Annexes, prohibiting a host of substances from being placed in the ocean environment. The Group of Experts on the Scientific Aspects of Marine Pollution advises the IMO on these matters by classifying substances according to their toxicity to ocean life; IMO restricts them accordingly, with the consensus of its members. If CO$_2$ were to be disposed of in international waters, most likely the London Dumping Convention would have to be amended. If CO$_2$ is placed in national waters, adherents to the Convention would still be subject to its conditions, and most nations considering this option are signatories to the Convention. Other international agreements posing political barriers are the UN Convention on the Law of the Sea, The Rio Treaty and Agenda 21 from the UN Conference on Environment and Development, which establish standards of conduct for nations to maintain the integrity of ocean systems. There are also national programs in almost every coastal nation that either advise or actually restrict nations from disposing of most substances in the oceans. If the entire activity and its impacts occur within national waters, national policies and laws will dominate the decision. However, pressures from the international community cannot be overlooked.

The case of ocean disposal may pose special economic considerations. Even if the option is found acceptable under certain conditions, it probably would be considered a "short-term" insurance policy or bridge to more long-term options. However, because of the apparent large expenditures required to launch this enterprise, the public would want to a guarantee that it would not evolve into a long-term option—and the exclusive domain of a few wealthy nations. Growth of the practice over time to include many nations could change the parameters of the environmental impacts. Carbon taxes, disposal taxes, subsidies and other mechanisms could be used to restrict expansion. Additional investments in other more viable alternatives would be essential.

The second element of "public trust" that is important to consider is the legacy of environmental decisions from both the private sector and governments that has left the public cynical of corporate and government officials when it comes to health, safety and the environment. The coastal zones of the world are currently under pressure from development and suffer from pollution, over exploitation of natural resources, and other environmental problems. Health problems and economic losses draw increasing attention, and the "protectors" of the ocean are growing more vocal and powerful everywhere. International agreements call for a legal concept known as "the precautionary principle," advising decision makers to err on the side of caution when considering environmental risks.

UNCERTAINTIES

The combination of the distrust of past decisions by governments and industry, and the public's interest in presenting itself as a surrogate constituency to protect the oceans, represents non-trivial barriers to implementation of the ocean disposal option. These conditions demand large degrees of certainty about the safety of such an enterprise. This in turn raises the other element of socio-political considerations, the uncertainties surrounding the evidence for climate change and the remaining large areas of uncertainty related to both the options of ocean disposal considered here.

First, uncertainties about public health and survival—The possibility that CO$_2$ in the atmosphere is forcing climate changes influences public perceptions of risks from climate change. While much of the evidence points toward the validity of this hypothesis, its seriousness and the need to incur other risks to solve the problem must be weighed in terms of the public's perception of which risks are acceptable and which are not. If climate change does not seem too serious, it is unlikely the public will accept disposal of CO$_2$ in the oceans, without first attempting other more palatable and less risky options.

The second uncertainty—Public perceptions of risks associated with disposing of CO$_2$ in the oceans. The certainty of evidence addressing the extent to which CO$_2$ will change ocean pH, and what, if any, harm may be caused to ecosystems, will need to be clearly demonstrated. Even if the evidence indicates low risk, the public will need clear, complete information to understand its implications. Using the oceans for waste disposal, even if for a relatively benign substance, will not easily gain public support. Two cases illustrate this point. Recently, the Acoustical Thermal Ocean project set out to take the temperature of the oceans with long distance, underwater acoustical soundings, but was met with a restraining order as a result of not adequately informing the public of the importance or details of the project, and did not counter misleading information from opponents who convinced the public that noise generated by the project would harm marine mammals. The project leaders had also overlooked legal details that were used to temporarily stop the project. When the public became better informed, and legal obstacles mediated, the project proceeded, but incurred large financial and other losses.

In the second case, the federal government, enforcing the U.S. ban on all offshore dumping, closed the so-called 106 Mile dump site off New York City although there was no evidence of harm to marine life. Although there was scientific information to justify keeping the site open, that did not change the decision, demonstrating that availability of some information does not guarantee good decisions; more is always desirable. Public understanding happens through rational methods to inform stakeholders and their surrogates, governmental and non-governmental groups, through the media and through public involvement. The better the information available, the more accurate the public's perception of risk.

The uncertainty issue goes beyond convincing the public of the acceptability of risks. The public will demand guarantees that its perceptions are correct. Disposal operators will need to monitor both the operation of the activity and the impacts on the oceans. Guaranteeing that the disposal takes place as advertised will be important. Monitoring the entry, fate and effects of the CO$_2$ when it enters the ocean may be required to assuage public fears that the impacts might accumulate and become harmful over time.

COSTS

A summary of the socio-political considerations regarding the feasibility of carrying out the ocean disposal option for the storage of CO$_2$ reveals some of the hidden costs of the enterprise that might not have been estimated in the original engineering analyses. Total costs must include:

- capital costs of the technology,
- associated operating costs,
- costs of back-up systems or buffers for contingencies,
- costs for monitoring the operation,
- costs for monitoring the impacts of disposal,
- public information and education costs,
- costs of funding other options that would ultimately replace the need for this option.

There are certain to be other costs as well, not covered here.

Lawyers and policy makers will have to determine whether to risk the CO_2 ocean disposal strategy. It is a case of risk and necessity. This framework can provide them with baseline parameters and analogies with which to formulate questions, determine hidden costs and trade-offs and make final decisions. The costs of taking the perceived risks must be weighed against the perceived urgency of the situation.

REFERENCES

Global Strategies for Marine Environmental Protection. (1991). *GESAMP Reports and Studies No. 45*. London.

Handa, N. and T. Ohsumi, ed. (1995). *Direct Ocean Disposal of Carbon Dioxide*. Terrapub, Tokyo.

Herzog, H., E. Adams, D. Auerbach and J. Caulfield. (1995). *Technology Assessment of CO₂ Ocean Disposal*. MIT-EL 95-001. Massachusetts Institute of Technology Energy Laboratory, Cambridge, Massachusetts.

Kildow, J., ed. (1980). *Deepsea Mining*. MIT Press, Cambridge, Massachusetts.

Kildow, J. (1982). *Socio-Political Aspects of a Subseabed Disposal Program for High Level Nuclear Waste*. Report for Sandia Laboratory, Contract No. 74-2545.

National Academy of Sciences. (1984). *Disposal of Industrial and Domestic Wastes: Land and Sea Alternatives*. National Academy Press, Washington D. C.

Omerod, W. and M. Angel. (1996). *Ocean Storage of CO₂, Workshop 1, Ocean Circulation*. IEA Greenhouse Gas R&D Programme.

Omerod, W. (1996). *Ocean Storage of CO₂, Workshop 2, Environmental Impact*. IEA Greenhouse Gas R& D Programme.

Decision Framework

Implementation Costs =

Technical
+
Logistics Monitoring
+
Impacts Monitoring
+
Buffers Against Uncertainties
+
Public Information & Education
+
Funding of Other Options

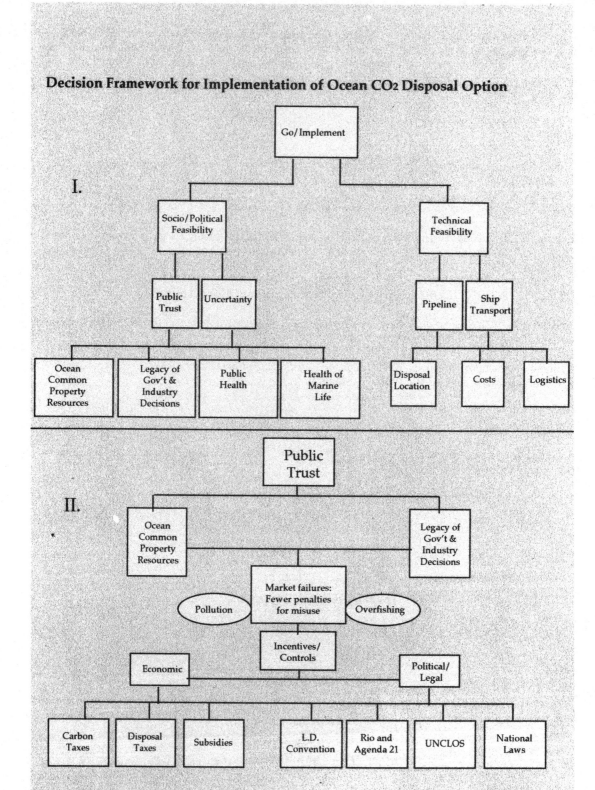

Decision Framework for Implementation of Ocean CO₂ Disposal Option

Pergamon

Energy Convers. Mgmt Vol. 38, Suppl., pp. S301–S306, 1997
© 1997 Elsevier Science Ltd. All rights reserved
Printed in Great Britain
0196-8904/97 $17.00 + 0.00

PII: S0196-8904(96)00285-3

CO₂ CLATHRATE-HYDRATE FORMATION AND ITS MECHANISM
BY MOLECULAR DYNAMICS SIMULATION

S. HIRAI, K. OKAZAKI, Y. TABE and K. KAWAMURA[††]

Research Center for Carbon Recycling and Utilization,
[††] Department of Earth and Planetary Sciences, Faculty of Science,
Tokyo Institute of Technology , Ohokayama, Meguroku, Tokyo, 152, Japan

ABSTRACT

Molecular dynamics simulation has been conducted in order to obtain the fundamental understanding for the formation mechanism of CO_2 clathrate-hydrate that suppresses the dissolution of liquid CO_2 isolated at deep ocean floor. It was demonstrated that the H_2O molecules formed a characteristic cage structure of type I clathrate around the CO_2 guest molecules after 260 ps from the initial condition of H_2O molecules at pressurized water state. CO_2 clathrate-hydrate formation kinetics has elucidated that the interactions between the CO_2 guest molecules would form a low potential region, which has an effect to suppress the H_2O molecules motions in a two-dimensional plane and assist to form cage structures consisted of 5 and 6 membered rings. © 1997 Elsevier Science Ltd

KEYWORDS

CO_2 sequestration in ocean, CO_2 clathrate-hydrate, Molecular dynamics simulation

INTRODUCTION

CO_2 clathrate-hydrate has recently been receiving special attention in the global warming mitigation strategy of CO_2 sequestration in ocean (Marchetti, C., 1977 and Steinberg, M. *et al.*, 1984), which is removal of CO_2 from fossil fuel fired power plant and injection of liquid CO_2 in the ocean. CO_2 clathrate-hydrate film, formed at the interface between liquid CO_2 and sea water, has an effect to suppress the dissolution of CO_2 into the sea water. Quantitative estimations for the effect of CO_2 clathrate-hydrate on the liquid CO_2 dissolution rate, which are those of the dissolution behavior of clathrate-hydrate covered CO_2 droplets and time-scale for CO_2 isolation at a seabed, was made by the authors (Hirai et al, 1996A). The authors also proposed a clathrate-hydrate model that CO_2 molecules directly permeate through the imperfect structure of clathrate-hydrate layer (Hirai et al, 1996B). The understanding of CO_2 clathrate-hydrate from both macro- and microscopic viewpoints is important in both fundamental and applied aspects of CO_2 sequestration in ocean.

CO_2 clathrate-hydrate is, on the molecular point of view, a solid-like crystalline where water molecules (host molecules) are linked through hydrogen bonding making lattice structure with cavities in which CO_2 molecules (guest molecules) are included. CO_2 clathrate-hydrate belongs to type I clathrates and complete crystal structure of type I clathrate in a unit cell consists of 6 large cages and 2 small cages, where these 8 cages are consisted of 46 H_2O molecules. CO_2 molecules are only contained in the large cages due to its large molecular size.

The mechanism for the kinetics of clathrate-hydrate formation is the point of special interest. Formation kinetics, the formation of clusters of water molecules around molecules and joining these clusters to create a hydrate nucleus, was proposed by Christiansen and Sloan (Christiansen and Sloan, 1994). In author's knowledge, computer simulation study for the growth of natural gas hydrate made by Baez and Clancy are the first attempt for the formation of clathrate-hydrate using molecular dynamics, where it was shown that gas atoms readily diffused from the liquid to hydrate crystal and water molecules formed pentagons, timers and tetramers to complete partial cavities or form new ones (Baez and Clancy, 1994). The molecular dynamics simulation study for the formation of Argon clathrate-hydrate was also recently conducted by the authors (Hirai et al, 1996C). A complete formation of clathrate-hydrate characteristic cage structure was demonstrated and the effect of Argon guest molecules to form the cage structure was clarified.

On the other hand, the effect of guest molecules, Ar and CO_2, on the structural stability is elucidated to be different by the authors (Hirai et al, 1996D) using MD simulation which was not that of clathrate formation. It was made clear that the effects of a CO_2 molecule shape and the Coulomb forces acting between the CO_2 guest molecules and H_2O host molecules have destabilizing effect on the clathrate-hydrate structure.

In the present paper, investigation using molecular dynamics simulation has been conducted in order to clarify the mechanism and to obtain the fundamental understanding of CO_2 clathrate-hydrate formation. Calculation started from the initial condition that 48 CO_2 molecules were placed in 368 H_2O molecules in pressurized equilibrium water state. After around 270 ps, the H_2O molecules formed a cage-like structure. A reasonable interpretation for the formation of CO_2 clathrate-hydrate structure is presented based on the interactions between guest and host molecular forces and cage structure formation.

THE INTERATOMIC POTENTIAL MODEL AND THE SIMULATION METHOD

The present calculation applies a potential model which treats H_2O systems as an assemblage of hydrogen and oxygen atoms (Kumagai *et al.*, 1994). The model expresses the inter- and intramolecular interactions in terms of a single interatomic potential function and motions of H and O atoms have a total freedom.

The interatomic potential function is made of two body and three body terms. The two body terms are expressed by eq. (1) for pairs of atoms

$$U_{ij} = \frac{z_i z_j e^2}{r_{ij}} - \frac{c_i c_j}{r_{ij}^6} + f_o\left(b_i + b_j\right)\exp\left[\frac{\left(a_i + a_j - r_{ij}\right)}{b_i + b_j}\right]$$

$$+ f_o D_{ij}\left\{\exp\left[-2\beta_{ij}\left(r_{ij} - r_{ij}^*\right)\right] - 2\exp\left[-\beta_{ij}\left(r_{ij} - r_{ij}^*\right)\right]\right\}. \tag{1}$$

Here, r_{ij} is a interatomic distance, f_o a constant for unit adaptations for these terms. The first, the second and the third terms represent Coulomb, short range repulsions and van der Waals interaction terms, respectively. The fourth term (the Morse function) is only employed to estimate the interactions between O-H atoms and C-O atoms. The three body part is only employed for H-O-H groups as

$$U_{HOH}\left(\theta_{HOH}\right) = -2f_k\left\{\cos\left[2\left(\theta_{HOH} - \theta_o\right)\right] - 1\right\}\sqrt{k_1 k_2}, \tag{2}$$

Table 1. Parameters of interatomic potential models

Two body term	H₂O		CO₂	
	O	H	O	C
z	-0.78	0.39	-1.00	2.00
a [Å]	1.841	0.037	1.885	0.690
b [Å]	0.131	0.057	0.150	0.100
c [kJ$^{1/2}$ Å^{3}mol$^{-1/2}$]	49.09	0.0	40.92	0.0
D [Å]	75.0		69.0	
β [Å^{-1}]	2.74		2.00	
r* [Å]	0.82		1.10	
Three body term	H-O-H			
f_k [J]	1.1×10^{-11}			
θ_0 [degree]	99.5			
r_m [Å]	1.40			
g_r [Å^{-1}]	7.0			

where θ_{HOH} is the angle of H-O-H and and define the effective range of three body potential by the following equation

$$k_i = 1 \Big/ \Big\{ \exp\big[g_r \big(r_{OH}(i) - r_m \big) \big] + 1 \Big\}. \tag{3}$$

The three body forces apply perpendicular to the O-H bond and its magnitude is given by

$$FH(i) = -2 f_k \sin\big[2 \big(\theta_{HOH} - \theta_0 \big) \big] \sqrt{k_1 k_2} \Big/ r_{OH}(i). \tag{4}$$

The parameters that appear in above equations are listed in Table 1.

The calculation cell is a cubic one of 24Å³ where 8 basic cells (12Å³) of type I clathrate-hydrate are included. In a complete clathrate hydrate structure, 48 large cages and 16 small cages are formed. In the first stage, 368 H_2O molecules were calculated at the temperature condition of 300K with fixed calculation volume until it reached an equilibrium state. In the second stage, 48 CO_2 molecules were placed in the calculation cell, where the coordinates of 48 CO_2 molecules were center of the cages whose positions are obtained by the neutron-ray diffraction experiments (Hollander *et al.*, 1977). Note that H_2O molecules did not form a cage structure at this initial condition and CO_2 molecules are placed at the position of center of the cage under the assumption that the cage is constructed. 12 H_2O molecules are removed when the CO_2 molecules are placed in the cell because they were near the CO_2 molecules. The calculation started from the initial coordinates stated above and the temperature and the pressure were kept constant during the calculation, which are those of 270K and 9.0MPa, respectively. The positions of the 48 C atoms in CO_2 molecules are fixed during the calculation because CO_2 molecules gather and condense if they move freely with H_2O molecules. The time integration was carried out using time step of 0.4 fs.

RESULTS AND DISCUSSIONS

The clathrate hydrate of type I consists of large cages and small cages, which include 12 pentagons and 2 hexagons ($5^{12}6^2$) and 12 pentagons (5^{12}), respectively. Fig.1 illustrates the time variation of number of 5 and 6 membered rings (pentagons and hexagons, respectively) for the case of CO_2 as guest molecules. The two dotted lines in the figure represent the number of 5 and 6 membered rings when the complete hydrate clathrate structure is formed, which are those of 384 and 48, respectively. 5 membered rings increases and 6 membered rings decreases and they reach near the numbers of 300 and 50, respectively, after 270 ps, longer time to reach an equilibrium state than that of Ar (Hirai *et al.*, 1996C).

Fig. 2 depicts the profile of pair-correlation function (P.C.F.(r)) between C atoms (in CO_2) and O atoms (in H_2O). P.C.F.(r) could be calculated by the following equation

$$P.C.F.(r) = n_{ij} \Big/ \Big[4\pi \big(N_i N_j / V \big) r_m^2 \Delta r \Big]. \tag{5}$$

Here, n_{ij} is the number of pairs of O atoms that exists in the distance from $r_m - \Delta r/2$ to $r_m + \Delta r/2$. N_i and N_j are the number of O atoms of H_2O in the calculation cell and V is the volume of the calculation cell. If H_2O molecules make a cage-like structure around the CO_2 molecules, distances between CO_2 and H_2O become nearly constant and P.C.F.(r) should show a higher peak profile. The profile of t=265 ps shows higher peak of P.C.F.(r) than that of 12, 48 and 168 ps, which suggests that a cage-like structure is forming

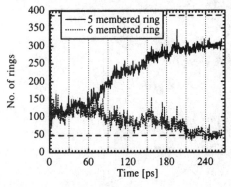

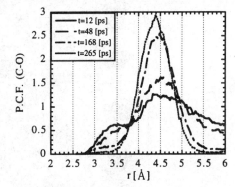

Fig. 1. Time variation of number of 5 and 6 membered rings

Fig. 2. Time variation of P.C.F.(r) between C atoms (in CO_2) and O atoms (in H_2O)

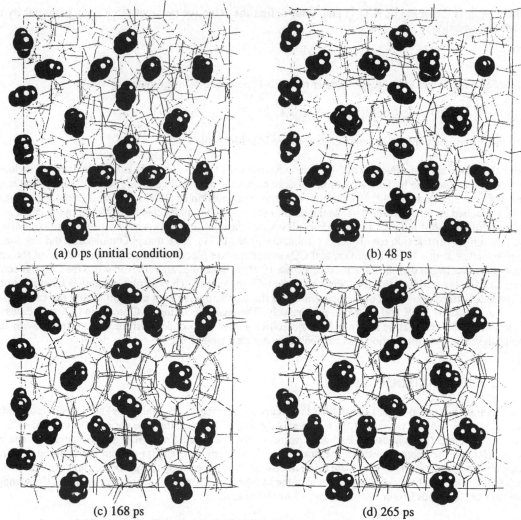

(a) 0 ps (initial condition)   (b) 48 ps

(c) 168 ps   (d) 265 ps

Fig. 3. Time variation of network of H$_2$O at the time of (a) 0 ps (initial
condition), (b) 48 ps, (c) 168 ps and (d) 265 ps

with time.

In Figs.3 (a)-(d), the network of H$_2$O at the time of 0 ps (initial condition), 48 ps, 168 ps and 265 ps are

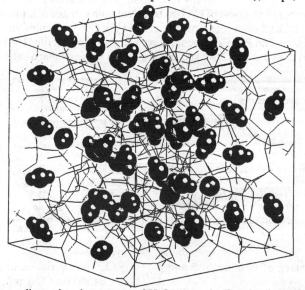

Fig. 4. Three dimensional structure of H$_2$O network of at the time of 265 ps

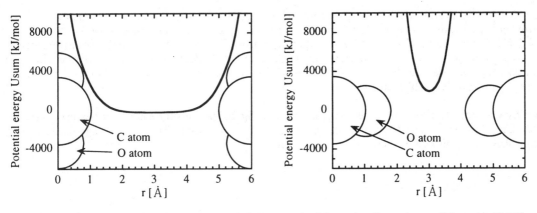

(a) CO₂ molecules axis perpendicular to the C-C line　　(b) CO₂ molecules axis parallel to the C-C line

Fig. 5. Potential energy profile U_{sum} an O atom in H_2O
molecule is place between the two CO_2 molecules

shown for whole calculation cell. The lines represent the hydrogen bonds connecting the coordinates between the O atoms in H_2O molecules. It can be clearly shown by the figure that hydrogen bonds connecting the H_2O molecules show successive pentagons and hexagons and forms a cage-like structure in Fig.3 (d), which is not observed at the initial condition (Fig.3 (a)). Fig.4 shows the cage structure at the time of 265 ps from the different angle, to see the three dimensional structure, which also shows the nearly complete clathrate-hydrate crystal structure.

Thus, it is demonstrated that CO_2 clathrate-hydrate structure is formed by the present molecular dynamic simulation. We make a consideration for the mechanism of clathrate-hydrate formation. Formation of clathrate-hydrate structure is making 5 and 6 membered rings and rearrangement of these rings forms a cage structure. Now, we pay attention on the formation of the 6 membered rings which is formed between the two CO_2 guest molecules. The faces of the 6 membered rings, when they are formed, are perpendicular to the line connected between the two C atoms in CO_2 guest molecules (hereafter, C-C line). Therefore, in order to investigate the formation mechanism of the CO_2 clathrate-hydrate structure, we investigate the forces and the motion of H_2O molecules placed between two CO_2 guest molecules as a simple system. We here consider the motion of an O atom in a H_2O molecule placed between the two CO_2 molecules placed at a distance 6Å, which is the distance between C atoms in two CO_2 molecules. Potential energy profile formed between the CO_2 molecules when an O atom in H_2O molecule is place between them is shown in Fig. 5. Figs. 5 (a) and (b) describe the potential energy profile when axis of CO_2 molecules is perpendicular to the C-C line and parallel to it, respectively. U_{sum} is the summation of Coulomb, short range repulsions and van der Waals potential energy described in eq. (1). Fig. 5 (a) shows that, between two CO_2 molecules, a low potential region is formed and its width where the potential profile is flat is approximately 2Å. Therefore,

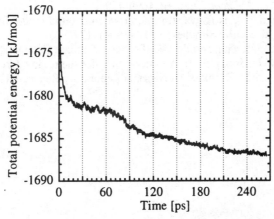

Fig. 6. Time variation of the total potential energy

the low potential profile region where O atoms in H_2O molecule could freely move is restricted to the narrow potential region. Its effect is pronounced when the axis of CO_2 molecules is parallel to C-C line (Fig.5 (b)). The low potential region becomes extremely narrow. Therefore, when a number of H_2O molecules are placed between the CO_2 guest molecules, the O atoms in H_2O molecules motions are restricted in a two dimensional plane perpendicular to the C-C line and H_2O molecules would rearrange and make 6 membered ring and construct a cage-like structure.

The time variation of the total potential energy in shown in Fig. 6. The potential energy reaches an equilibrium state when the number of 5 and 6 membered rings becomes nearly constant. Therefore, the cage-like formation is a stable state of low potential energy in this system.

CONCLUDING REMARKS

The present study demonstrated the formation of CO_2 clathrate-hydrate structure, that is H_2O molecules reaches from pressurized water state to a cage-like formation, can be simulated by molecular dynamics simulation.   It was revealed from the simulation that a low potential region formed by the interaction between the guest molecules assist the formation of 5 and 6 membered rings and to form a cage structure, which is a stable state of low potential energy. The position of the guest CO_2 molecules are fixed and the CO_2 concentration is supersaturated in the present calculation condition. The growth and propagation of hydrate-clathrate in actual case including its mechanism, should be made in future research. We plan to continue this work on CO_2 clathrate-hydrate growth including its nucleation.

REFERENCES

Baez, L. U. and P. Clancy (1994). Computer Simulation of the Crystal Growth and Dissolution of Natural Gas Hydrates, *First International Conference on Natural Gas Hydrates*, Sloan, E. D., Hoppel, J. and Hrafow, M. A. Ed., Vol. **715** of the Annals of the New york Academy of Science, 177-186.

Christiansen R. L. and E. D. Sloan (1994). Mechanism and Kinetics of Hydrate Formation, *First International Conference on Natural Gas Hydrates*, Sloan, E. D., Hoppel, J. and Hrafow, M. A. Ed., Vol. **715** of the Annals of the New york Academy of Science, 283-305.

Hirai, S., K. Okazaki, Y. Tabe, K. Hijikata and Y. Mori (1996A). Dissolution Rate of Liquid CO_2 in Pressurized Water Flow and Effect of Clathrate Film, *Energy Int. J.*, (to be published)

Hirai, S., K. Okazaki, N. Araki, H. Yazawa, H. Ito and K. Hijikata (1996B). Transport Phenomena of Liquid CO_2 in Pressurized Water Flow with Clathrate-Hydrate at the Interface, *Energy Conv. Mgmt.*, **37**,1073-1078.

Hirai, S., K. Okazaki, S. Kuraoka and K. Kawamura (1996C). Molecular Dynamics Simulation for the Formation of Argon Clathrate-Hydrate Structure, *submitted to US-Japan Meeting "Molecular and Microscale Transport Phenomena," Aug. 7-10, Santa Barbara, CA.*

Hirai, S., K. Okazaki, S. Kuraoka and K. Kawamura (1996D). Study for the Stability of CO_2 Clathrate-Hydrate Using Molecular Dynamics Simulation, *Energy Conv. Mgmt.*, **37**,1087-1092.

Hollander, F. and G. A. Jeffrey (1977). Neutron Diffraction Study of the Crystal Structure of Ethylene Oxide deuterohydrate at 80 K, *J. Chem. Phys.*, **66**,4699-4705.

Kumagai, N., K. Kawamura and T. Yokokawa, An interatomic potential model for H_2O :applications to water and ice polymorphs, Molecular Simulation., **12** (1994), 177-186.

Marchetti, C. (1977). On Geoengineering and the CO_2 Problem, Climate Change ,**1**, 59-68.

Steinberg, M., H. C. Cheng and F. Horn (1984). A system study for the removal, recovery and disposal of CO_2 from fossil fuel power plants, Brookhaven National Laboratory Report, OE/CH/00016.

Pergamon

Energy Convers. Mgmt Vol. 38, Suppl., pp. S307–S312, 1997
© 1997 Elsevier Science Ltd. All rights reserved
Printed in Great Britain
0196-8904/97 $17.00 + 0.00

PII: S0196-8904(96)00286-5

DISSOLUTION MECHANISMS OF CO₂ MOLECULES IN WATER CONTAINING CO₂ HYDRATES

T. UCHIDA[1], A. TAKAGI[2], S. MAE[2] and J. KAWABATA[1]

[1] Hokkaido Nation. Ind. Res. Inst.; 2-17 Tsukisamu-higashi, Sapporo 062, Japan
[2] Dept. of App. Phys., Fac. Eng., Hokkaido Univ.; N13W8, Sapporo 060, Japan

ABSTRACT

In situ Raman spectroscopic measurements were performed in a $CO_2 + H_2O$ system. When aqueous CO_2 coexisted with the hydrate, CO_2 spectra were different from those obtained when the hydrate was absent. Spectral analysis indicated that the water structure surrounding a CO_2 molecule in solutions with hydrates resembled conditions in the hydrate. © 1997 Elsevier Science Ltd

KEYWORDS

CO_2, water, CO_2 hydrate, Raman spectroscopy, CO_2 solubility

INTRODUCTION

Ocean disposal of CO_2 is a promising option for controlling emissions of this species. CO_2 released in the ocean at depths below about 500 m will form a CO_2 hydrate layer at the interface with sea water. This hydrate layer is expected to control dissolution of CO_2 in sea water. However, recent studies on the stability of the CO_2 hydrate have revealed that the hydrate layer is stable only if the sea water contacting it is saturated with CO_2 (Aya *et al.*, 1993; Hirai *et al.*, 1995). It is important, therefore, to understand the solubility of CO_2 in the presence of CO_2 hydrates.

Yamane and Aya (1995) determined that the solubility of CO_2 in water decreased with decreasing temperature over a range where CO_2 hydrates form. This behavior was similar to the observed temperature dependence of the dissolution rate of a CO_2 droplet in water, hence, they concluded that solubility is a dominant factor in the dissolution of a CO_2 droplet covered with a hydrate layer.

The temperature dependence of CO_2 solubility affects not only CO_2 hydrate stability but also physical properties of CO_2 hydrates measured by Raman spectroscopy. Uchida *et al.* (1995) estimated the hydration number of CO_2 hydrates using CO_2 solubility deduced from information obtained at non-hydrate forming temperatures. If the temperature dependence of solubility is affected by the presence of the hydrate phase, then this estimate of the hydrate number may include a significant error.

In this study the vibration of CO_2 molecules in solution was monitored via Raman spectroscopy for the cases in which CO_2 hydrates were present and absent. The vibration of CO_2 molecules should indicate if the CO_2 solution condition changes when hydrates form. Results suggested that the water structure surrounding a CO_2 molecule is affected by the presence of the CO_2 hydrate phase.

EXPERIMENTAL PROCEDURES

The experimental set up is nearly identical to that employed in a previous study described by Uchida *et al.* (1995). The primary components comprise a pressure vessel with silica glass windows and a laser Raman spectrometer. A cooling rod was inserted into the center of the vessel to form the CO_2 hydrate. Transparent hydrates grow slowly only in the region around the rod. This arrangement prevents uncontrolled scattering of the laser beam and allows Raman spectra of both CO_2 hydrate and CO_2 in solution to be obtained. An Argon ion laser (λ=488 nm) and JEOL-JRS-400T type spectrometer with a triple monochromator were used to conduct the Raman spectroscopic analysis. Scattered light was monitored with a photomultipler (Hamamatsu TV R585) whose output was recorded with a computer.

About 4 cm^3 of pure water was initially introduced into the vessel, and temperatures of the vessel and the cooling rod were set at T_v = 282.2 ($\pm$0.1) K and T_{cr} = 281.2 K, respectively. The vessel was then pressurized with CO_2 gas to P = 3.78 ($\pm$0.01) MPa, which is the dissociation pressure of CO_2 hydrate at 282.2 K. Following pressurization, CO_2 began to dissolve in the water. Raman spectra of CO_2 molecules in solution were monitored during the dissolution process. When the water was saturated with CO_2, experimental conditions were changed to T_v = 278.4 K, T_{cr} = 277.4 K and P = 4.16 MPa to form CO_2 hydrates on the cooling rod. After the hydrate formed, T_{cr} was set equal to T_v. Finally, experimental conditions were reset to T_v = T_{cr} = 282.2 K and P = 3.78 MPa and Raman spectra of both CO_2 in solution and CO_2 hydrate were measured.

RESULTS

Raman spectra of CO₂ molecules during dissolution

Figure 1 shows the Raman spectra of CO_2 molecules in solution. Each curve corresponds to CO_2 in solution at time after pressurization (shown in regend) during the dissolution process. Two spectral peaks are observed between 1250 and 1450 cm^{-1} ($v_l \approx$ 1274 cm^{-1} and $v_u \approx$ 1383 cm^{-1}). These peaks are Fermi resonance, which are mixtures of CO stretch (v_1) and OCO bend ($2v_2$). Raman spectra of CO_2 molecules were measured every hour following pressurization. The experimental condition was T_v = 282.2 K, T_{cr} = 281.2 K and P = 3.78 MPa. However, CO_2 hydrate was formed neither in the vessel nor on the cooling rod at this condition.

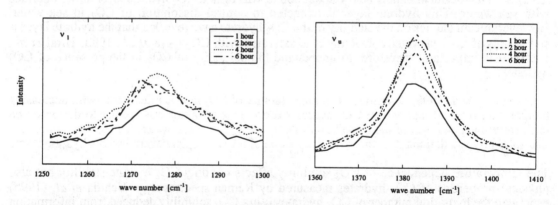

Fig. 1:   Two peaks (v_l: left and v_u: right) of Raman spectra of CO_2 molecules dissolved in water as a function of time.

This figure demonstrates that the Raman intensity of both peaks gradually increase with time, attaining an equilibrium level. This indicates that saturation is achieved after several hours. On the other hand, the wave number of these peaks did not change significantly during the dissolution.

Raman spectra of CO₂ molecules in solution and in the hydrate

Raman spectra of CO_2 molecules in solution were measured both in the presence and absence of

CO_2 hydrates. Figure 2 presents typical spectra of CO_2 molecules in aqueous solutions and in the hydrate phase. This shows that the Raman intensities of CO_2 in solution were similar in both cases.

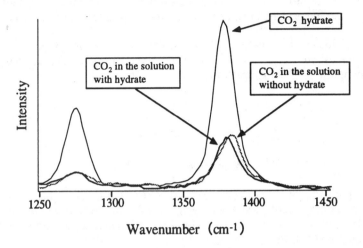

Fig. 2:   Raman spectra of CO_2 molecules in the hydrate (thin solid line);
          in solution with hydrate (thick solid line); and in solution
          without hydrate (thick dotted line).

Since Raman intensity depends upon the molecular concentration, the similarity of Raman intensities for CO_2 in the solutions indicates that the CO_2 concentration in the solution with hydrate present at the equilibrium condition is almost the same as when hydrates do not exist.

Table 1 provides the Wave numbers, v_1 and v_u, for both conditions (it also includes wave numbers of the CO_2 hydrate). Although the experimental uncertainty is estimated as ± 1 cm⁻¹, the measured wave numbers were reproduced within this uncertainty.

Table 1. Wave numbers of Fermi resonance of CO_2 in solution
         and in the hydrate

conditions	v_1 [cm⁻¹]	v_u [cm⁻¹]
CO_2 in solution without hydrate	1274	1383
CO_2 in solution with hydrate	1275	1381
CO_2 in the hydrate	1277	1380

These results demonstrate that Raman shifts of CO_2 molecules in solution coexisting with CO_2 hydrates were slightly different from those in solution without CO_2 hydrates, and those shifts were intermediate between those in solution without CO_2 hydrates and in CO_2 hydrate phase.

DISCUSSION

CO₂ solubility and CO₂ vibration in water

The comparison of CO_2 spectra in solutions with and without a hydrate phase indicated that the CO_2 concentration in water is similar for both cases at the present experimental conditions. Yamane and Aya (1995) measured the temperature dependence of CO_2 solubility and determined that as temperature decreases at 30 MPa, the difference in solubility of solutions with and without hydrate increases. At equilibrium, however, this difference disappeared. The results of the present study are consistent with this result even though the pressure was much lower than 30 MPa.

Although solubility appeared to the same, Raman spectra indicated that the vibration of CO_2 molecules in solution with hydrates was intermediate between the vibration of CO_2 in the solution without hydrates and CO_2 in the hydrate phase. This may indicate that hydrate can affect the dissolution mechanism. In order to discuss the dissolution mechanism of CO_2 molecules in water, we need to estimate the unperturbed frequencies (v_1 and $2v_2$) of the Fermi resonance.

Amat and Pimbert (1965) and Howard-Lock and Stoicheff (1971) developed a model to estimate v_1 and v_2 from perturbed frequencies (v_1 and v_u). According to this model, v_1 and v_2 are calculated, employing the following equations:

$$v_1 = 0.5 (v_l + v_u) - 0.5 \Delta \tag{1}$$

$$v_2 = 0.5 (0.5 (v_l + v_u) + 0.5 \Delta) \tag{2}$$

$$\Delta^2 = X^2 - 4W_{10°0-02°0}{}^2 \tag{3}$$

where Δ is the separation of the unperturbed vibration levels, X is the observed frequency separation of the two components and $W_{10°0-02°0}$ is the Fermi coupling constant. If the value of $W_{10°0-02°0}$ is estimated to be -50.98 cm⁻¹ (Montero, 1983), then we obtain the unperturbed frequencies presented in Table 2.

Table 2. Frequencies of normal vibration estimated from Fermi resonance

conditions	X [cm⁻¹]	Δ	v_1 [cm⁻¹]	v_2 [cm⁻¹]
CO_2 in solution without hydrate	109	38.5	1309	674
CO_2 in solution with hydrate	106	29.0	1314	671
CO_2 in the hydrate	103	17.8	1319	661

The calculation indicates that v_1 increases and v_2 decreases as CO_2 moves from an aqueous solution to a solution with hydrates, and finally to the hydrate phase.

The vibrational frequency may be affected by both interactions between CO_2 and H_2O and CO_2 and CO_2. Since CO_2 concentration in solution at 30 MPa is approximately 0.075 (Yamane and Aya, 1995), CO_2-CO_2 interactions should be less prominent than CO_2-H_2O interactions. Also, CO_2-H_2O interactions in the solution with hydrates may be less pronounced than when hydrates are absent because v_1 and v_2 in solution with hydrate are closer to those in the hydrate phase. This arises from the fact that the intermolecular interaction in the hydrate cage is negligible small (van der Waals and Platteeuw, 1959).

Estimation of average spacing of CO_2 and H_2O molecules and CO_2 dissolution mechanism

Changing in the degree of intermolecular interactions between CO_2 and H_2O molecules may result from differences in the average spacing of CO_2 and H_2O in the solutions. In order to estimate this spacing, we applied a cavity model for gas hydrates developed by Nakahara *et al.* (1988).

The simplified model considers a CO_2 molecule trapped in a spherical cavity of diameter $2a$. The cavity is assumed to be surrounded by a dielectric medium having the same permittivity ε'. The carbon atom in a CO_2 molecule exists at the center of the cavity. The periodicity of the cavity in the hydrate is neglected. With these assumptions, the vibration frequency v' in the cavity can be related to that in vacuum, v_0, by the equations:

$$v'/v_0 = (1/\varepsilon^*)^2 \tag{4}$$

$$\varepsilon^* = \frac{1}{c}\int_0^c \frac{db}{1-\left(\dfrac{\varepsilon'-1}{\varepsilon'-0.5}\right)\left(\dfrac{b}{a}\right)^3} \tag{5}$$

where b is an arbitrary radius from the center of the cavity ($b<a$) and $c = 1.160$ Å is the interatomic distance between carbon and oxygen in CO_2. When we consider the CO stretching mode (v_1), $v_{10} = 1332$ cm^{-1} and ε' is either the permittivity of water ($\varepsilon' = 1.78$ for $v_{10} = 1332$ cm^{-1}) or of the hydrate ($\varepsilon' = 1.82$). We then calculate the value of a and the partial molar volume v listed in Table 3.

Table 3. Average distance between CO_2 and H_2O and partial molar volume

conditions	v_1 [cm^{-1}]	ε'	a [Å]	v [cm^{3}mol^{-1}]
CO_2 in solution without hydrate	1309	1.78	2.50	39.4
CO_2 in solution with hydrate	1314	1.78	2.67	48.0
CO_2 in the hydrate	1319	1.82	3.07	73.0

Table 3 indicates that $a = 3.07$ Å in the CO_2 hydrate which coincides with the radius of the average free space of the large cage in structure I hydrates, 2.91 ($\pm$0.29) Å (Sloan, 1990). On the other hand, $a = 2.50$ Å in the CO_2 solution without hydrate is almost the same as the van der Waals radius of the CO_2 molecule, 2.56 Å (Sloan, 1990). This suggests that CO_2 molecules in the solution without hydrates are surrounded closely by water molecules. Formation of hydrates increases the average distance between a CO_2 molecule and surrounding H_2O molecules. This value is intermediate between the intermolecular spacing in the solution with no hydrates and the spacing in the hydrate phase. These results suggest that, when CO_2 in solution coexists with the hydrate phase, the structure of water surrounding the dissolved CO_2 molecule changes slightly, and begins to resemble the condition in the hydrate.

It also should be noted that the predicted partial molar volume of CO_2 in the solution without hydrates, $v = 39.4$ cm^{3}mol^{-1}, is close to the value obtained from density measurements of CO_2 solutions, $v = 30.5$ cm^{3}mol^{-1} (Ohsumi, 1993). The average partial molar volume in the hydrate can be calculated based on crystal structure arguments to be about 90 cm^{3}mol^{-1}. This does not differ significantly from $v = 73.0$ cm^{3}mol^{-1} calculated with the cavity model.

CO_2 solubility in water is determined by equilibrium between partial pressures in the gas phase and in water phase. However, if a hydrate phase coexists with the gas and solution phases, then CO_2 solubility may be determined by a competition between gas-solution equilibrium and solution-solid equilibrium. In the present study, it was determined that the structure of water surrounding a CO_2 molecule in solutions with hydrates is similar to the hydrate structure. Therefore, it is concluded that CO_2 solubility in the solution with hydrate is controlled by solution-solid equilibrium. In this case, solubility may increase with temperature. This behavior agrees with the observations of Yamane and Aya (1995).

CONCLUSIONS

In situ Raman spectroscopic measurements were performed in solutions for cases in which CO_2 hydrates were present and absent. Raman intensity of CO_2 gradually increased with time following pressurization, attaining an equilibrium level. This indicated that saturation is achieved after several hours for this experimental set up. When aqueous CO_2 coexisted with the hydrate, the vibration of CO_2 molecules was intermediate between the vibration of CO_2 without hydrates and CO_2 in the hydrate phase. Theoretical models suggested that the structure of water surrounding a CO_2

molecule in solutions with hydrates is resembled conditions in the hydrate. It is therefore concluded that CO_2 solubility in the solution with hydrates is controlled by solution-solid equilibrium whereas that without hydrates is determined by gas-water equilibrium.

ACKNOWLEDGMENTS

The authors thank I. Aya, K. Yamane, T. Ohsumi, S. Hirai, T. Ebinuma and H. Narita for their fruitful discussions about these experiments. They also thank M. Mori for her help in the preparation of the figures.

REFERENCES

Amat, G. and M. Pimbert (1965). On Fermi Resonance in Carbon Dioxide. *J. Molecular Spectroscopy*, **16**, 278-290.

Aya, I., K. Yamane and N. Yamada (1993). Effect of CO_2 concentration in water on the dissolution rate of its clathrate. *Proc. of the Int. Symp. on CO₂ Fixation and Efficient Utilization of Energy, RCCRU of Tokyo Inst. of Tech.*, 351-360.

Hirai, S., K. Okazaki, N. Araki, H. Yazawa, H. Ito and K. Hijikata (1995). Dissolution and diffusion phenomena of liquid CO_2 in pressurized water flow with clathrate-hydrate at the interface. *Proc. of Int. Conf. on Tech. for Marine Env. Preserv. (MARIENV'95)*, vol. **2**, 901-905.

Howard-Lock, H. E. and B. P. Stoicheff (1971). Raman Intensity Measurements of the Fermi Diad v_1, $2v_2$ in $^{12}CO_2$ and $^{13}CO_2$. *J. Molecular Spectroscopy*, **37**, 321-326.

Montero, S. (1983). Raman intensities of Fermi diads. I. Overtones in resonance with nondegenerate fundamentals. *J. Chem. Phys.*, **79**, 4091-4100.

Nakahara, J., Y. Shigesato, A. Higashi, T. Hondoh and C. C. Langway, Jr. (1988). Raman spectra of natural clathrates in deep ice cores. *Phil. Mag.*, **57**, 421-430.

Ohsumi, T. (1993). Prediction of Solute Carbon Dioxide Behavior around a Liquid Carbon Dioxide Pool on Deep Ocean Basins. *Energy Conves. Mgmt*, **34**, 1059-1064.

Sloan, E. D., Jr. (1990). *Clathrate Hydrates of Natural Gases*. Marcel Dekker, Inc., New York.

Uchida , T., A. Takagi, J. Kawabata, S. Mae and T. Hondoh (1995). Raman spectroscopic analyses on the growth process of CO_2 hydrates. *Energy Convers. Mgmt.*, **36**, 547-550.

van der Waals, J. H. and J. C. Platteeuw (1959). Clathrate Solutions. *Adv. Chem. Phys.*, **2**, 1-57

Yamane, K. and I. Aya (1995). Solubility of carbon dioxide in hydrate region at 30MPa. *Proc. of Int. Conf. on Tech. for Marine Env. Preserv. (MARIENV'95)*, vol. **2**, 911-917.

Pergamon

Energy Convers. Mgmt Vol. 38, Suppl., pp. S313–S318, 1997
© 1997 Elsevier Science Ltd. All rights reserved
Printed in Great Britain
0196-8904/97 $17.00 + 0.00

PII: S0196-8904(96)00287-7

NUMERICAL SIMULATION FOR DISSOLUTION OF LIQUID CO2 DROPLETS COVERED WITH CLATHRATE FILM IN INTERMEDIATE DEPTH OF OCEAN

S. HIRAI, K. OKAZAKI, Y. TABE and K. HIJIKATA*

Research Center for Carbon Recycling and Utilization
* Department of Mechano-Aerospace Engineering
Tokyo Institute of Technology
2-12-1, O-okayama, Meguro-ku, Tokyo 152, Japan

ABSTRACT

Dissolution of liquid CO_2 droplets at intermediate depth of ocean is strongly effected by the release methods of liquid CO_2 and hydrate formation. The present paper presents numerical simulations for the dissolution behavior of CO_2 droplets released from (1) pipeline outlet which is a fixed point at intermediate ocean depth and (2) pipe of moving ship. Clathrate-hydrate formation on the CO_2 droplet surface was fully included for both cases (1) and (2). External flow conditions around the CO_2 droplets are different between (1) and (2). A rising plume water flow is formed for case (1) and a free stream turbulence induced by the wake behind the pipe is for case (2). It was indicated that, for case (1), the released droplet diameter is required to be controlled to be less than 0.8cm to obtain the complete dissolution for the travel distance less than 1000m. The large travel distance is caused by the rising motion of plume water flow. Even if the CO_2 is released from 16 separated branches, the travel distance is reduced to 30%. In addition, increase of ambient CO_2 concentration drastically decreases the dissolution. On the other hand, for case (2), it was demonstrated that easier released droplet size condition to obtain complete dissolution is acceptable when liquid CO_2 is released from a pipe of moving ship, which also possesses a high potential that the ambient CO_2 concentration would not increase. © 1997 Elsevier Science Ltd

KEYWORDS

Dissolution of liquid CO_2, CO_2 clathrate-hydrate, Plume, Wake

INTRODUCTION

Liquid CO_2 dissolution in the intermediate depth of the ocean is a mitigation strategy for global warning, which stands against deep ocean storage of CO_2 at a seabed. Liquid CO_2 droplets released at the intermediate sea depth dissolve in the process of rising upward due to buoyancy in the sea water. In this process, CO_2 clathrate hydrate film is present at the liquid CO_2 droplet surface, which has an effect to reduce the CO_2 dissolution rate.

Studies for the dissolution behavior of CO_2 droplets covered with clathrate film was first made in quiescent pressurized water (Aya *et al.*, 1993 and Shindo *et al.*, 1993). The dissolution rate of hydrate-covered CO_2 droplets where the flow field around the droplet is present was investigated at various temperature, pressure and velocity conditions (Hirai *et al.*, 1996A). The surface concentration of clathrate-covered CO_2 droplet, which represents the resistance for dissolution due to the existence of hydrate, has been presented, which would be a valuable database for the precise simulation of CO_2 droplets dissolution (Hirai *et al.*, 1996B).

In order to estimate the strategy of liquid CO_2 dissolution in the intermediate depth of the ocean, technical conditions that exist to obtain the complete dissolution of CO_2 droplets must be clarified. It is strongly related to the existence of CO_2 clathrate hydrate and also depends on the external flow conditions around the CO_2 droplets, which is different according to release methods. To release CO_2 droplets from pipeline introduced from power plant lies on a fairly established technology and it is also a significantly economical strategy. CO_2 is released from a fixed pipeline outlet at intermediate ocean depth, and a successive release of a large amount of CO_2 droplet forms a rising plume. The plume increases the CO_2 droplet rise velocity, which is directly accompanied by the decrease of effective dissolution

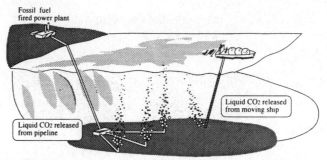

Fig. 1. Dissolution of liquid CO₂ droplets at intermediate depth of ocean. CO₂ droplets released from pipeline outlet with formation of a plume and from a pipe hanged from a moving ship.

rate of liquid CO₂. Numerical simulation study of dissolving CO₂ droplets in a plume suggests that released droplet size is a dominating factor to obtain dissolution (Haugan *et al.*, 1995), but the effects of hydrate were not considered in the simulation. Hydrate would force a longer rising distance for CO₂ dissolution. Decrease of effective dissolution rate in the presence of plume could be avoided when CO₂ droplets are released from a pipe of moving ship (Nakashiki *et al.*, 1995). CO₂ dissolves in a free stream turbulence induced by the wake behind the pipe hanged from a moving ship. Dissolved CO₂ would be diluted because the release point moves with ship. These two methods are schematically shown in Fig. 1. The understanding of dissolution behavior for these two cases must be made clear, which also makes environmental impact estimation to be possible.

The present paper presents numerical simulations for the dissolution behavior of CO₂ droplets released by the two cases mentioned above. The dominating factors of rising travel distance for two cases were evaluated precisely and the conditions in order to obtain complete dissolution of hydrate covered liquid CO₂ droplet was made clear. It was elucidated that plume water flow has an effect to enforce large travel distance for dissolution. Increasing the numbers of ports to decrease plume velocity has a little effect to reduce the travel distance and travel distance increases drastically with increase of ambient CO₂ concentration of water. On the other hand, it was demonstrated that to released CO₂ from a moving ship has a large potential for easier dissolution that travel distance reduces to 30% as compared with that in a plume and the ambient CO₂ concentration would not increase.

SIMULATION OF CO₂ DROPLETS DISSOLUTION

Dissolution of CO₂ droplets in a plume

The simulation presents CO₂ droplets dissolution behavior released from pipeline outlet of 1m diameter placed at 1500m ocean depth. The CO₂ flow rate is 185kg/s that is exhausted from 1GW coal-fired power plant. We assume that the outlet of pipeline is consisted of one or many exhaust ports and the released CO₂ droplet size could be controlled by the orifices. CO₂ droplets rise upward due to buoyancy in sea water and water plume is formed by a continuous release of droplets. The conservation equations of mass and momentum of two phases (liquid CO₂ and water) in an axisymmetric cylindrical coordinate are integrated to the radial direction assuming profiles given by the equations, $W(r) = W_m f(r/b)$, $C(r) = C_m f(r/\lambda_1 b)$, $\rho_w = \rho_\infty + \Delta\rho_w f(r/\lambda_2 b)$. Here, $W(r)$ and W_m : plume velocity and maximum plume velocity at the plume axis, respectively, $C(r)$ and C_m : CO₂ volume fraction in plume and maximum CO₂ volume fraction in plume, respectively, b : HWHM (half width at half maximum) of plume velocity, r : coordinate in the radial direction, ρ_w : water density, ρ_∞ : ambient water density, $\Delta\rho_w$: maximum water density minus ρ_∞. The function f is assumed to take the form $f(x) = 1/(1 + ax^2)$ with $a = \sqrt{2} - 1$ (Rajaratnam, 1976). λ_1 and λ_2 are constants and take the values 0.8 and 1.25, respectively (Liro *et al.*, 1992). The integrated conservation equations are

$$\frac{d}{dz}(Q_d + Q_w) = 2\pi\rho_\infty bEW_m. \tag{1}$$

$$\frac{d}{dz}(M_d + M_w) = \pi b^2 [s_1 C_m(\rho_\infty - \rho_{CO_2}) - s_6 \Delta\rho_w]g. \tag{2}$$

where

$$Q_d = \pi b^2 \rho_{CO_2} C_m (s_1 U + s_2 W_m),$$ (3)

$$Q_w = \pi b^2 \rho_\infty (s_3 - s_5 C_m) W_m,$$ (4)

$$M_d = \pi b^2 \rho_{CO_2} C_m (s_1 U^2 + 2s_2 U W_m + s_5 W_m^2),$$ (5)

$$M_w = \pi b^2 \rho_\infty (s_4 - s_5 C_m) W_m^2.$$ (6)

The function f is different from Milgram (Milgram, 1983), but the derivation of integrated equations is similar to that one. Here, Q_d:mass flux of CO_2 droplets, Q_w:mass flux of water, U: slip velocity of CO_2 droplet, E : entrainment coefficient(=0.1) (Haugan *et al.*, 1995), M_d:momentum flux of CO_2 droplets, M_w:momentum flux of water, z : coordinate in the vertical direction, g : gravitational acceleration, ρ_{CO_2}: liquid CO_2 density, $s_i (i=1\text{-}6)$: constants and take the values, $s_1=1.55$, $s_2=0.634$, $s_3=2.41$, $s_4=0.805$, $s_5=0.409$, $s_6=3.77$. Constants s are derived automatically in the process of integration.

Dissolution behavior of liquid CO_2 droplets is described by eq. (7) (Hirai *et al.*, 1996A). The mass transfer coefficient k in eq. (7) was estimated using eqs. (8) ,(9) for high Schmidt number flows (Clift *et al.*, 1978).

$$\frac{d(d_{CO_2})}{dz} = -\frac{2k(C_o - \overline{C_w})}{\rho_{CO_2}(U + \overline{W})} - \frac{d(\rho_{CO_2})}{dz} \frac{d_{CO_2}}{3\rho_{CO_2}},$$ (7)

$$Sh = 1 + Re^{0.41} Sc^{1/3} \qquad (Re \leq 100),$$ (8)

$$Sh = 1 + 0.752 Re^{0.472} Sc^{1/3} \qquad (Re \geq 100).$$ (9)

Here , C_o and $\overline{C_w}$: the surface concentration of CO_2 droplet and CO_2 concentration in the plume averaged over the cross section, respectively, Sh : the Sherwood number (= $k\, d_{CO_2}/D$), Re : the Reynolds number (= $U d_{CO_2}/\nu$), Sc : the Schmidt number (= ν/D), ν : the kinematic viscosity, d_{CO_2} : the CO_2 droplet diameter, D : the diffusion coefficient, $\overline{W}$: plume velocity averaged over the cross section, evaluated by the reduced equation of $\overline{W} = s_2 / s_1 W_m$.

The droplet slip velocity U was estimated by eq. (10) (Clift *et al.*, 1978 and Hirai *et al.*, 1996B).

$$U = \sqrt{\frac{4 d_{CO_2} (\overline{\rho}_{plume} - \rho_{CO_2}) g}{3 C_D \overline{\rho}_{plume}}},$$

$$\log_{10}[C_D Re/ 24 - 1] = -0.7133 + 0.6305w \qquad (26 \leq Re \leq 260),$$

$$\log_{10} C_D = 1.6435 - 1.1242w + 0.1558w^2 \qquad (260 \leq Re \leq 1,500),$$

$$\log_{10} C_D = -2.4571 + 2.5558w - 0.9295w^2 + 0.1049w^3 \quad (1,500 \leq Re \leq 12,000).$$ (10)

Here, C_D is drag coefficient, $w = \log_{10} Re$ and $\overline{\rho}_{plume}$ is the water density averaged over the cross section of plume.

CO_2 droplet dissolving behavior in a plume can be obtained precisely by calculating eqs. (1) -(10), employing the

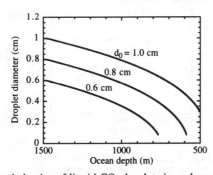

Fig. 2. Dissolution behavior of liquid CO_2 droplets in a plume. CO_2 droplets are released at 1500m depth of ocean,varying the released droplet diamater d_0=1.0, 0.8 and 0.6.

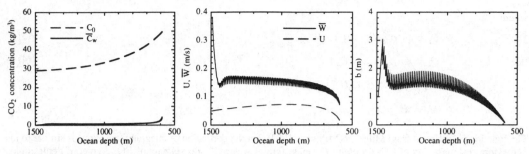

Fig. 3. The profile of the surface concentration C_0 and plume CO_2 concentration $\overline{C_w}$ for d_0=0.8cm.

Fig. 4. The profile of the droplet slip velocity U and plume velocity $\overline{W}$ for d_0=0.8cm.

Fig. 5. The profile of plume radius b (FWHM) for d_0=0.8cm.

experimental surface concentration C_0 (Hirai et al., 1996B) which describes the effect of hydrate to reduce dissolution. The diffusion coefficient D in eqs. (8) ,(9) is evaluated from the equation by Wilke and Chang (Wilke and Chang, 1955), which shows a good agreement in a high pressure range (Hirai et al., 1996C). In the simulation, the measured temperature profile in the ocean (Nishiyama, 1987) was used, which is approximated by $T(℃)=46291/z^{1.3781}$(m). The peeling model employed in the present calculation is identical to the one by Haugan et al. (Haugan et al., 1995).

In Fig. 2, the dissolution behavior of CO_2 droplet is shown varying the released initial droplet size. When the initial released diameter d_0 is 0.8 cm, the dissolution finishes in the rising distance of 900 m, whereas when it is 1.0 cm, the dissolution does not finish before CO_2 droplet becomes gaseous at a 500 m depth. The released droplet diameter is a dominating factor for the dissolution, which was also indicated previously (Haugan et al., 1995). The present calculation shows extremely longer rising distance than that of Haugan et al. (Haugan et al., 1995). It is caused by the fact that the effect of hydrate to reduce dissolution rate was not included (Haugan et al., 1995), which has two different points in the plume model. One is the mass transfer coefficient of fluid sphere used in the simulation (Haugan et al., 1995) is much larger than that of rigid sphere employed in eq. (8) and (9) due to the existence of hydrate at the droplet surface. The other is decrease of surface concentration C_0 due to the existence of hydrate. One of the dominant driving force for dissolution is the difference between the surface concentration C_0 and plume CO_2 concentration $\overline{C_w}$, $C_0 - \overline{C_w}$ in eq. (7). The profile of C_0 and $\overline{C_w}$ versus ocean depth is depicted in Fig. 3 in the case of released droplet size d_0=0.8cm. $C_e - \overline{C_w}$ becomes to be approximately 1/2 of $C_{sat} - \overline{C_w}$, which represents that hydrate has a large suppression effect on dissolution. The profile of droplet slip velocity U, plume velocity $\overline{W}$ and formation of plume described in terms of its HWHM b is shown in Fig. 4 and 5, respectively, for the case of d_0=0.8cm. The plume velocity $\overline{W}$ is 2-3 times larger than droplet slip velocity U, which extremely reduces the effective dissolution rate. The plume width decreases gradually with decrease of ocean depth and 115 times of peelings are observed.

One of the method to reduce the effect of retardation of the dissolution due to plume is to release liquid CO_2 from separated ports, in order to decrease the flow rate. Its effect is calculated and shown in Fig. 6, varying numbers of ports , 1,4 and 16 in the case d_0=0.8cm. It can be seen from the figure that increasing the number of ports does not drastically decreases the travel distances, i.e., 20-30%. Therefore, to increase number of ports possesses considerably small effect to shorten the dissolution distance and to permit a large released droplet size.

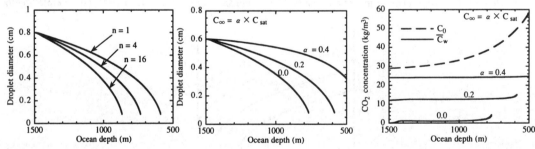

Fig. 6. Effect of number of exhaust ports n on the dissoluion behavior of CO_2 droplets

Fig. 7. Effect of ambient CO_2 concentration on the dissoluion behavior of CO_2 droplets

Fig. 8. Effect of ambient CO_2 concentration on the surface concentration C_0 and plume CO_2 concentration $\overline{C_w}$

The CO_2 concentration of ambient water would increase (1) when the mixing between CO_2 dissolved water and fresh water is slow as compared with CO_2 dissolution into sea water and (2) due to the peeling of CO_2 dissolved water. The effect of CO_2 concentration of ambient water, C_∞ on the dissolution behavior is shown in Fig. 7 in the case of $d_0=0.6$cm, $C_\infty=0\%, 20\%$ and 40% of saturated concentration. It can be seen from the figure that increase of ambient CO_2 concentration drastically reduces the dissolution. It is caused by the decrease of the driving force for dissolution, $C_o - \overline{C_w}$ in eq. (7) by the increase of $\overline{C_w}$, due to the increase of ambient CO_2 concentration (Fig. 8).

Dissolution of CO_2 droplets released from moving ship

The plume is formed by a successive release of CO_2 droplets from a fixed points and it was shown that plume has a large effect to decrease the effective dissolution rate of CO_2 droplets. One of the method to avoid the plume formation is to release CO_2 droplets from a pipe hanged from a moving ship (Nakashiki *et al.*, 1995). CO_2 droplets rise upward in the wake flow induced by the pipe of moving ship. It also has a favorable effect that ambient CO_2 concentration would not increase. The simulation described in this section is the dissolution behavior of CO_2 droplets that rise upward in the wake behind the pipe. The ship is assumed to move in the speed of 6 knot (3.09m/s) and 185kg/s of liquid CO_2 is released from the tip of the pipe. The pipe diameter is assumed to be 1m where many orifices are attached to control the released droplet size. In this case, the Reynolds number, based on the pipe diameter and moving pipe velocity is approximately 3×10^6, which indicates that homogeneous turbulence is formed behind the moving pipe.

The rise velocity (especially C_D in eq. (10)) and mass transfer coefficient k is affected by the homogeneous free stream turbulence, which effect is represented by the parameter, I_R, by the following equation (Clift *et al.*, 1978),

$$I_R = \sqrt{u'^2} / U \qquad (11)$$

U is the CO_2 droplet rise velocity and $\sqrt{u'^2}$ is root mean square of velocity fluctuation induced by the wake of pipe hanged from moving ship. The ratio of $\sqrt{u'^2}$ to U_{wake-m}, maximum velocity of the wake, takes the following value (Townsend, 1976),

$$\sqrt{u'^2} / U_{wake-m} = 0.28 \qquad (12)$$

which is independent to the distance from the pipe. U_{wake-m} is expressed in terms of moving pipe velocity U_{pipe} and dimensionless distance from the pipe x/d_{pipe}, (where d_{pipe} is the pipe diameter) as follows (Schlichting, 1979),

$$U_{wake-m} = U_{pipe} C_r \sqrt{x/d_{pipe}} \qquad (13)$$

with $C_r = C_{D-pipe} \sqrt{U_{pipe} d_{pipe} / \varepsilon_0} / 4\sqrt{\pi}$ and $C_{D-pipe} = 0.7$ for $Re = 3 \times 10^6$ (Roshko, 1961) and $\varepsilon_0 = 0.0222(C_{D-pipe} U_{pipe} d_{pipe})$. C_D in equation (10) becomes the function of I_R by eqs. (14),(15) in the range $0.07 < I_R < 0.5$ (Clift *et al.*, 1976). C_D in the range $I_R < 0.07$ was estimated by the interpolation between C_D substituting $I_R = 0.07$ in eqs. (14) , (15) and C_D used in quiescent water (in eq. (10)).

$$C_D = 162 I_R^3 / Re \qquad (Re \le 50) \qquad (14)$$

$$C_D = 0.133(1 + 150 / Re)^{1.565} + 4 I_R \qquad (50 \le Re \le 700) \qquad (15)$$

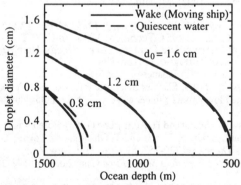

Fig. 9. Dissolution behavior of liquid CO_2 droplets released from moving ship. CO_2 droplets, varying the released droplet diamater $d_0=1.6$, 1.2, and 0.8cm. CO_2 droplets, in the wake flow induced by the pipe. Dashed lines represents dissolution behavior of liquid CO_2 droplets in a quiescent water.

Employing these equations (11)-(15), the parameter I_R is evaluated and the CO_2 droplet rise velocity could be estimated using eq.(10) replacing ρ_{plume} to ρ_w and using eqs. (14) and (15). The CO_2 droplet dissolution behavior could be calculated by eq. (7) with $\overline{W} = 0$ and $\overline{C_w} = 0$. The ship movement is implemented by the distance x in eq. (13). The effect of turbulence induced by the wake on mass transfer coefficient k is negligibly small at the present case of small I_R (Clift et al., 1976).

In Fig. 9, the dissolution behavior of CO_2 droplets in wake is presented along with the case when CO_2 droplet dissolves in a quiescent water. It can be seen from the figure that turbulence induced by the wake hardly effect the dissolution behavior as compared with that in quiescent water. The CO_2 droplets rise velocity is around 0.5cm/s (0.005m/s) and pipe hanged from the ship moves in the speed of 3.09m/s. The ship moves 20000m (20km) while CO_2 droplet rises 400m. The ship moves far away and the turbulence induced by the moving pipe becomes negligibly small. Thus, turbulence induced by the moving ship hardly affect the CO_2 droplet dissolution behavior and it is nearly same as that in quiescent water. Note that the rising travel distance of CO_2 droplet released from moving ship is 200m (Fig. 9) for $d_0=0.8$cm, while it is 900m for the same d_0 in plume (Fig. 2).

CONCLUDING REMARKS

The simulations of CO_2 droplet dissolution in a plume and in a wake were conducted to obtain the fundamental understanding associated with released methods. The effect of hydrate is fully included in the simulation. A plume has an extremely strong effect to decrease the dissolution of CO_2 droplets even if CO_2 is released from separated branches, whereas turbulence induced by a wake of moving pipe has negligibly small effect. The travel distance in a wake is less than 30% of that in a plume. In addition, CO_2 droplet dissolution in a plume strongly affected by the CO_2 concentration to cause unfavorable effect that retardation of CO_2 dissolution would be pronounced. On the other hand, CO_2 droplet dissolution by a moving ship is not an established method but it posses a high potential that CO_2 droplet could be easily dissolve and ambient CO_2 concentration would not increase.

Acknowledgement—The authors would like to thank Mr. H. Takao (Nitto Koatsu Co.) for assistance with the present study.

REFERENCES

Aya, I., K. Yamane and N. Yamada (1993). Effect of CO_2 Concentration in Water on the Dissolution Rate of its Clathrate, *Proc. Int. Symp. on CO_2 Fixation & Efficient Utilization of Energy*, Tokyo, 351-360.
Clift, R., J. R. Grace and M. E. Weber (1978). *Bubbles, Drops and Particles*, pp. 97-284, Academic Press, New York
Haugan, P. M., F. Thorkildsen and G. Alendal (1995). Dissolution of CO_2 in the Ocean, *Energy Conv. Mgmt.*, **36**, 461-466.
Hirai, S., K. Okazaki, N. Araki, H. Yazawa, H. Ito and K. Hijikata (1996A). Transport Phenomena of Liquid CO_2 in Pressurized Water Flow with Clathrate-Hydrate at the Interface, *Energy Conv. Mgmt.*, **37**,1073-1078.
Hirai, S., K. Okazaki, Y. Tabe, K. Hijikata and M. Yasuo (1996B). Dissolution Rate of Liquid CO_2 in Pressurized Water Flow and Effect of Clathrate Film, *Energy Int. J.*, (to be published).
Hirai, S., K. Okazaki, H. Yazawa, H. Ito, Y. Tabe and K. Hijikata (1996C). Measurements of CO_2 Diffusion Coefficient and Application of LIF in Pressurized Water, *Energy Int. J.*, (to be published).
Liro, C. R., E. E. Adams and H. J. Herzog (1992). Modeling the Release of CO_2 in the Deep Ocean, *Energy Conv. Mgmt.*, **33**,667-674.
Milgram, J. H. (1983). Mean flow in round bubble plumes, *J. Fluid Mech.*, **133**, 345-376.
Nakashiki, N., T. Ohsumi and N. Katano (1995). Technical View on CO2 Transportation onto the Deep Ocean Floor and Dispersion at Intermediate Depths, *Direct Ocean Disposal of Carbon Dioxide*, N. Handa and T. Ohsumi, eds. Tokyo; Terrapub, pp.183-194.
Nishiyama, K. (1987). *Umi to Anzen*, pp. 9-11 (in Japanese).
Rajaratnam, N. (1976). *Turbulent Jets*, pp. 27-49, Elsevier Scientific Pub. Co., New York.
Roshko, A. (1961). Experiments on the Flow Past a Cylinder at Very High Reynolds Number, *J. Fluid Mech.*, **10**, 345-356.
Shindo, Y., Y. Fujioka, Y. Yanagisawa, T. Hakuta and H. Komiyama (1993). Formation and Stability of CO_2 Hydrate, Direct Ocean Disposal of Carbon Dioxide, N. Handa and T. Ohsumi, eds. Tokyo; Terrapub, pp. 217-231.
Shlichting, H. (1979). *Boudary Layer Theory*, pp.729-757, McGraw Hill, New York.
Townsend, A. A. (1976). *The Structure of Turbulent Shear Flow* (2nd. Ed.), pp. 188-258, Cambridge University Press, Cambridge
Wilke, C. R. and P. Chang (1955). Correlation of Diffusion Coefficient in Dilute Solutions *A.I.Ch.E. J.* **1**, 264 -270

Energy Convers. Mgmt Vol. 38, Suppl., pp. S319–S324, 1997
© 1997 Elsevier Science Ltd. All rights reserved
Printed in Great Britain
0196-8904/97 $17.00 + 0.00

PII: S0196-8904(96)00288-9

Pergamon

DISPERSION OF CO_2 DROPLETS IN THE DEEP OCEAN

H. Teng[1], S.M. Masutani[2] and C.M. Kinoshita[2]

[1]National Institute of Materials and Chemical Research
1-1 Higashi, Tsukuba, Ibaraki 305 JAPAN

[2]Hawaii Natural Energy Institute, University of Hawaii at Manoa
2540 Dole Street, Holmes 246, Honolulu, HI 96822 USA

ABSTRACT

Dispersion of CO_2 droplets in the deep ocean under the influence of currents and turbulence has been investigated. It is shown that: (1) the concentration of the CO_2 effluent can be described by Gaussian distributions under distorted coordinates; (2) ocean turbulence does not influence the direction of motion of the CO_2 droplets; however, it can exert a significant influence on effective droplet velocities; and (3) deep ocean currents alter the direction of droplet motion, distorting the droplet cloud. Both deep currents and marine turbulence can favorably affect CO_2 discharge in the ocean by enhancing dissolution and limiting local environmental impacts. © 1997 Elsevier Science Ltd

KEYWORDS

CO_2; contaminant; droplet; ocean; dispersion.

INTRODUCTION

Ocean disposal and sequestration of CO_2 captured from fossil fuel combustors is being evaluated as a means to slow its accumulation in the atmosphere. The majority of scenarios proposed to date call for discharge of pure, liquefied CO_2 at depths below the thermocline (i.e., below about 500 m). The behavior of the effluent following release in the deep ocean needs to be confirmed via experiment and modeling studies, since it will impact profoundly the extent of the hazard posed to the marine environment and the effectiveness of the concept as a greenhouse gas emissions countermeasure.

The CO_2/seawater system is hydrodynamically unstable; hence liquid CO_2 effluent will break up into a dispersed droplet phase (Masutani *et al.*, 1995; Nishikawa *et al.*, 1995). At ocean pressures greater than 44.5 bar and temperatures less than 283 K, a thin solid hydrate layer will rapidly encase the CO_2 droplets (Aya *et al.*, 1992; Teng *et al.*, 1995), impeding dissolution. Break up of the effluent and hydrate effects have been afforded inconsistent treatment in earlier studies of the fate of CO_2 discharged in the deep ocean (Golomb *et al.*, 1989; Golomb *et al.*, 1992; Liro *et al.*, 1992; Morishita *et al.*, 1993; Haugen *et al.*, 1995; Holder *et al.*, 1995; Kobayashi, 1995). The liquid effluent has been modeled both as a continuous phase (no break up) and a droplet plume; hydration of the CO_2 has been predicted to proceed to completion, been confined to a thin film at the seawater-CO_2 interface, or ignored.

In this study, it is assumed that liquid CO_2 discharged below 500 m will exist initially as a droplet cloud that can be assigned a "contaminant" concentration, C. The droplets comprising the contaminant phase are buoyant at depths above about 3000 m and dissolve slowly (due to the hydrate film) as they move through the water. While the behavior of dissolving CO_2 droplets has been modeled previously (see, for example, Golomb *et al.*, 1989; Liro *et al.*, 1992; Holder *et al.*, 1959; Haugen *et al.*, 1995), the effects of deep ocean currents and fluid turbulence have not been considered directly. If the droplets are less than 10 mm in diameter, then their buoyant velocities are of the same order of magnitude as the velocities typically associated with ocean eddies and deep currents (Robinson, 1983). This suggests that the dispersion of the droplets may be influenced significantly by ocean hydrodynamics.

To investigate the effects of ocean currents and turbulence on the transport of CO_2 released in the deep ocean, an analysis was conducted of the advection and dispersion of a contaminant droplet phase. The analysis considers a line release of liquid CO_2 that breaks up immediately to form a cloud of small, monodispersed droplets which are assumed not to agglomerate nor coalesce. To facilitate the analysis, the contaminant concentration is taken to represent all carbon initially derived from the effluent.

ADVECTION-DISPERSION WITH NEUTRAL BUOYANCY

To avoid buoyant rise of the CO_2 effluent, discharge at depths between 3000 and 3500 m has been proposed (Aya *et al.*, 1992; Ohsumi *et al.*, 1992; Lund *et al.*, 1994). At the temperatures and pressures which prevail at these depths, the difference in the densities of the liquid CO_2 contaminant and seawater are small; the CO_2 droplets may be considered neutrally buoyant and, therefore, amenable to classic dispersion analysis. Although this line of inquiry will not provide information on the motion of individual droplets, it can predict the spatial transport characteristics of the droplet ensemble.

In the deep ocean, stable vertical stratification leads to predominant horizontal (steady or unsteady) flows characterized by length scales much larger than those in the vertical direction. The following advection-dispersion equation describes the evolution of the contaminant CO_2 (or carbon) concentration:

$$\frac{\partial C}{\partial t} + u(z,t)\frac{\partial C}{\partial x} = k_x \frac{\partial^2 C}{\partial x^2} + k_z \frac{\partial^2 C}{\partial z^2} \ . \tag{1}$$

Here, u is the (horizontal) ocean current velocity, and k_x and k_z are, respectively, the horizontal and vertical dispersion coefficients. The point x = 0; z = 0 corresponds to the release site.

The ocean current velocity may be modeled as

$$u = u_0(t) + \alpha(t)z \ . \tag{2}$$

The linear shear distorts the droplet cloud. To account for this deformation, we introduce a distorted coordinate system, (X,Z,t), in which Z = z and

$$X = x - \int_0^t u_0(t')dt' - zG_x(t) \ , \tag{3}$$

where G_x is a distortion parameter that must be determined. Under the new coordinates, (1) becomes

$$\frac{\partial C}{\partial t} + (\alpha - \frac{dG_x}{dt})z\frac{\partial C}{\partial X} = (k_x + k_z G_x^2)\frac{\partial^2 C}{\partial X^2} - 2G_x k_z \frac{\partial^2 C}{\partial X \partial z} + k_z \frac{\partial^2 C}{\partial z^2} \ . \tag{4}$$

To simplify the analysis, boundary conditions are assumed to be time-independent. If G_x is taken to be

$$G_x = \int_0^t \alpha(t')dt' \ , \tag{5}$$

then (1) takes the form of a time-dependent dispersion equation (Young *et al.*, 1982; Smith, 1982):

$$\frac{\partial C}{\partial t} = K_x \frac{\partial^2 C}{\partial X^2} \ . \tag{6}$$

The dispersion coefficient, K_x, is $K_x \equiv k_x + k_z G_x^2$. A Laplace transform yields a solution for (6):

$$C(X,t) = \frac{M}{\sqrt{2\pi\sigma_x^2}}\exp(-\frac{X^2}{2\sigma_x^2}) \ , \tag{7}$$

where M is the source strength of the discharge and σ_x^2 is a variance that satisfies $d\sigma_x^2/dt = 2K_x$. Under distorted coordinates, (7) indicates that dispersion of a CO_2 droplet cloud produces a Gaussian concentration distribution. The distortion parameter G_x depends on the ocean current: for steady currents, the shear factor α is constant and $G_x = \alpha t$ and $\sigma_x^2 = 2k_x t + (2/3)\alpha^2 k_z t^3$; for temporally periodic currents, $\alpha = \alpha'\cos(\omega t)$, leading to $G_x = (\alpha'/\omega)\sin(\omega t)$ and

$$\sigma_x^2 = 2k_x t + \frac{k_z}{2\omega}(\frac{\alpha'}{\omega})^2 (2\omega t - \sin 2\omega t) \ , $$

where α' and ω are the amplitude and frequency of the current, respectively.

ADVECTION-DISPERSION WITH POSITIVE BUOYANCY

Disposal system costs and limitations on submerged pipeline technology recommend discharge of liquid CO_2 at moderate depths between 500 and 1500 m (Herzog *et al.*, 1993). In this region, buoyancy of the contaminant must be considered. Under the constraints of the dispersion analysis, a vertical current is employed to account for buoyant transport. The functional form of this pseudo-current is inferred by considering the terminal rise velocity, V, of a CO_2 droplet, which is attained when fluid drag balances buoyancy:

$$g\Delta\rho \frac{4}{3}\pi r^3 = \frac{1}{2}C_D \pi r^2 \rho_w V^2 . \tag{8}$$

In this expression, g is the gravitational acceleration, $\Delta\rho$ the density difference between seawater and CO_2, r the droplet radius , C_D the drag coefficient, and ρ_w the seawater density. An expression for the droplet velocity is derived from (8):

$$V = (\frac{8g\Delta\rho r}{3C_D\rho_w})^{1/2} . \tag{9}$$

The drag coefficient may be estimated via the Stokes' relationship (White, 1991): $C_D = (12\mu_w)/(rV\rho_w)$. The droplet velocity then is given by $V = (2/9)(g\Delta\rho r^2/\mu_w)$, where μ_w is the seawater viscosity. In order to derive an expression for V as a function of location and time, consider the vertical displacement of the buoyant CO_2 droplets:

$$z = \int_0^t V dt' = \frac{2g\Delta\rho}{9\mu_w} \int_0^t r^2(t') dt' . \tag{10}$$

Here, it is assumed that the upper limit of integration, t, is less than the time required for complete dissolution of the droplets (at which point, buoyant rise is irrelevant). In addition, changes in fluid properties with depth are ignored. Since experiments suggest that the shrinkage rate of a CO_2 droplet in seawater, $A \equiv |dr/dt|$ is approximately constant (Aya *et al.*, 1992; Shindo *et al.*, 1995), (10) becomes

$$z = \frac{2g\Delta\rho}{27\mu_w A}(r_0^3 - r^3) , \tag{11}$$

leading to

$$r^2 = r_0^2(1 - \frac{27\mu_w A}{2g\Delta\rho r_0^3}z)^{2/3} ,$$

where r_0 is the initial radius of the droplets. It follows that

$$V = \frac{2g\Delta\rho r_0^2}{9\mu_w}(1 - \frac{27\mu_w A}{2g\Delta\rho r_0^3}z)^{2/3} . \tag{12}$$

Since A is very small (Aya *et al.*, 1992; Shindo *et al.*, 1995), (12) may be approximated as

$$V \approx V_0 + \gamma z , \tag{13}$$

where $V_0 = (2/9)(g\Delta\rho r_0)/\mu_w$ and $\gamma = -2A/r_0$.

As indicated by (13), the vertical velocity of the contaminant CO_2 phase depends only on position, decreasing linearly with increasing z. The evolution of the contaminant concentration, C, may, therefore, be described by an advection-dispersion equation

$$\frac{\partial C}{\partial t} + u(z,t)\frac{\partial C}{\partial x} + w(z)\frac{\partial C}{\partial z} = k_x \frac{\partial^2 C}{\partial x^2} + k_z \frac{\partial^2 C}{\partial z^2} , \tag{14}$$

where the velocity of the vertical pseudo-current, which accounts for buoyant transport, is modeled after (13) to be $w(z) \equiv w_0 + \beta z$. Once again, $u(z,t) = u_0(t) + \alpha(t)z$. The parameters α and β depend on the hydrodynamic conditions at the disposal site. Letting $C(x,z,t) = C_1(x,t)C_2(z,t)$ yields

$$\frac{\partial C_1}{\partial t} + u(z,t)\frac{\partial C_1}{\partial x} = k_x \frac{\partial^2 C_1}{\partial x^2} \tag{15}$$

and

$$\frac{\partial C_2}{\partial t} + w(z)\frac{\partial C_2}{\partial z} = k_z \frac{\partial^2 C_1}{\partial z^2} \ . \tag{16}$$

Applying a procedure similar to that used in the neutral buoyancy case, (15) and (16) are solved yielding

$$C_1 = \frac{M_X}{\sqrt{2\pi\sigma_X^2}} \exp(-\frac{X^2}{2\sigma_X^2}) \ , \tag{17}$$

and

$$C_2 = \frac{M_Z}{\sqrt{2\pi\sigma_Z^2}} \exp(-\frac{Z^2}{2\sigma_Z^2}) \ , \tag{18}$$

which leads to

$$C(X,Z,t) = \frac{M}{2\pi\sigma_X\sigma_Z} \exp\left[-(\frac{X^2}{2\sigma_X^2} + \frac{Z^2}{2\sigma_Z^2})\right] , \tag{19}$$

where $M = M_X M_Z$ is the source strength and X and Z are distorted coordinates given by

$$X = x - \int_0^t u_0(t')dt' - z\int_0^t \alpha(t')dt' \qquad (z \neq 0) \ ,$$

$$Z = z - \int_0^t w_0 dt' - z[(\beta z / w) + \exp(-wt / z)] \qquad (w \neq 0) \ .$$

The variances are $\sigma_X^2 = 2K_x t$ (where $K_X \equiv (k_x + k_z G_x^2)|_{kz} \to 0 \cong k_x$) and

$$\sigma_Z^2 = \int_0^t 2K_z dt' = 2k_z \int_0^t (1 - G_z)^2 dt' \ ,$$

with $K_Z = k_z(1-G_z)^2$ and $G_z = \exp(-wt/z) + (\beta z/w)$. Equation (19) indicates that the dispersion of buoyant CO_2 droplets also produces a Gaussian distribution in the distorted coordinate frame. Noting that stable vertical stratification of the deep ocean produces $k_x >> k_z$, and considering the definitions of the variances, it is obvious that the effect of ocean turbulence on the dispersion of CO_2 effluent is manifested primarily through σ_X. As expected, strong turbulence (i.e., large values of k_x) results in large σ_X, enhanced spatial dispersion, and a small peak value of effluent concentration, C.

The concentration profile of the droplet cloud at x = 0, i.e., on the axis of the discharge, is of interest. At this locale, (19) becomes

$$C|_{x=0} = \frac{M}{2\pi\sigma_X^2\sigma_Z^2} \exp\left[-((\int_0^t u dt')^2 / 2\sigma_X^2 + Z^2 / 2\sigma_Z^2)\right] . \tag{20}$$

Tilting of the axis of the droplet cloud from vertical therefore is induced through a non-zero value of

$$(\int_0^t u dt')^2 / (2\sigma_X^2) \ ;$$

i.e., by the ocean current. For a steady ocean current,

$$(\int_0^t u dt')^2 / (2\sigma_x^2) = \frac{u^2 t}{4 k_x} \; ;$$ (21)

for a periodic ocean current,

$$(\int_0^t u dt')^2 / (2\sigma_x^2) = (\int_0^t u_0(t')dt' + \frac{\alpha'}{\omega} z \sin \omega t)^2 / 4 k_x t \; .$$ (22)

DISCUSSION AND CONCLUSIONS

While the preceding analysis is not relevant to individual CO_2 droplets, it demonstrates clearly that currents and ocean turbulence can exercise significant influence on the motion of the droplet ensemble. If buoyancy can be neglected, as, for example, in some of the very deep ocean disposal scenarios, then from (7), the concentration of the effluent assumes a one-dimensional Gaussian distribution under distorted coordinates. Transport effects are manifested primarily through the variance term. For a steady ocean current, the variance can be expressed as $\sigma_x^2 = 2 k_x t + (2/3)\alpha^2 k_z t^3$. Since the growth of σ_x is nonlinear with respect to time, the dispersion of the droplets is not purely horizontal. Shearing by the ocean current may induce a tilting of the axis of the droplet cloud. In general, for a temporally periodic current, the variance also grows nonlinearly with time; however, at large values of t, $2\omega t >> \sin 2\omega t$, and, thus, $\sigma_x^2 = 2 K_x t$, where $K_x = k_x + (k_x/2)(\alpha'/\omega)^2$. The parameter α'/ω can be quite large at some locations (Gargett *et al.*, 1981), resulting in the effective dispersion coefficient K_x exceeding k_x by a significant amount.

The CO_2 effluent concentration also assumes a (two-dimensional) Gaussian distribution under distorted coordinates when droplet buoyancy is non-negligible. Examination of (19) and (20) indicates that turbulence effects, which are manifested through the dispersion coefficients, k_x and k_z, combine with currents to alter the effective droplet velocities and reduce peak values of C (through the variances). Turbulence, however, does not deflect the droplet cloud; the tilting of the cloud axis arises solely through the action of currents.

The conclusions of the present study may be summarized as follows: (1) analytical solutions can be obtained for the concentration of a CO_2 contaminant being dispersed and advected by deep ocean turbulence and currents through the application of a coordinate transformation; the contaminant concentration is described by one- (neutral buoyancy) or two-dimensional (buoyant) Gaussian distributions in the distorted coordinate frame; (2) turbulence does not influence the direction of the droplet motion but may alter effective droplet velocities during dispersion; (3) currents induce changes in the direction of the droplet motion, tilting the axis of the droplet cloud; and (4) these induced changes in the effective velocities and direction of the droplet cloud can enhance mass transfer and limit local impacts on the marine environment; hence, ocean turbulence and currents may favorably influence marine disposal of CO_2.

ACKNOWLEDGMENTS

This study was funded in part by U.S. Department of Energy Grant No. DE-FG22-95PC95206. This support does not constitute an endorsement by the Department of Energy of the views expressed herein.

REFERENCES

Aya, I., K. Yamane and N. Yamada, (1992). Stability of Clathrate-Hydrate of Carbon Dioxide in Highly Pressurized Water. HTD-Vol. 215, pp. 17-22, ASME, Fairfield.

Gargett, A.E., P.J. Hendricks, T.B. Sanford, T.R. Osborn and A.J. William (1981). A Composite Spectrum of Vertical Shear in the Upper Ocean. *J. Phys. Oceanogr.* **11**, 1258-1271.

Golomb, D., H. Herzog, J. Tester, D. White and S. Zemba (1989). Feasibility, Modeling and Economics of Sequestering Power Plant Emissions in the Deep Ocean. Report No. 89-003, Energy Laboratory, MIT, Cambridge.

Golomb, D.S., S.G. Zemba, J.W.H. Dacey and A.F. Michaels (1992). The Fate of CO₂ Sequestered in the Deep Ocean. *Energy Convers. Mgmt* **33**, 675-683.

Haugan, P.M., F. Thorkildsen and G. Alendal (1995). Dissolution of CO₂ in the Ocean. *Energy Convers. Mgmt* **36**, 461.

Herzog, H., E. Drake and J. Tester (1993). A Research Needs Assessment for the Capture, Utilization and Disposal of Carbon Dioxide from Fossil Fuel-Fired Power Plants, DOE/ER-30194.

Holder, G.D., A.V. Cugini and R.P. Warzinski (1995). Modeling Clathrate Hydrate Formation During Carbon Dioxide Injection into the Ocean. *Environ. Sci. Technol.* **29**, 276.

Kobayashi, Y. (1995). Physical Behavior of Liquid CO_2 in the Ocean. In: *Direct Ocean Disposal of Carbon Dioxide* (N. Handa & T. Ohsumi eds.), pp. 165-181, Terrapub, Tokyo.

Liro, C.R., E.E. Adams and H.J. Herzog (1992). Modeling the Release of CO_2 in the Deep Ocean. *Energy Convers. Mgmt* **33**, 667-674.

Lund, P.C., Y. Shindo, Y. Fujioka and H. Komiyama (1994). Study of the Pseudo-Steady-State Kinetics of CO_2 Hydrate Formation and Stability. *Int. J. Chem. Kinetics* **26**, 289.

Masutani, S.M., C.M. Kinoshita, G.C. Nihous, H. Teng, L.A.Vega and S.K. Sharma (1995). Laboratory Experiments of CO_2 Injection into the Ocean. In: *Direct Ocean Disposal of Carbon Dioxide* (N. Handa & T. Ohsumi eds.), pp. 239-252, Terrapub, Tokyo .

Morishita, M., K.H. Cole, G. R. Stegen and H. Shibuya, (1993). Dissolution and Dispersion of a Carbon Dioxide Jet in the Deep Ocean. *Energy Convers. Mgmt* **34**, 841.

Nishikawa, N., M. Ishibashi, H. Ohta, N. Akutsu, M. Tajika, T. Sugitani, R. Hiraoka, H. Kimuro and T. Shiota (1995). *Energy Convers. Mgmt* **36**, 489.

Ohsumi, T., N. Nakashiki, K. Shitashima and K. Hirama (1992). Density Changes in Water Due to Dissolution of Carbon Dioxide and Near-Field Behavior of CO_2 from a Source on the Deep-Sea Floor. *Energy Convers. Mgmt* **33**, 685.

Robinson, A.R. (1983). *Eddies in Marine Science*, Springer-Verlag, Berlin.

Shindo, Y., Y. Fujioka, Y. Yanagisawa, T. Hakuta and H. Komiyama (1995). Formation and Stability of CO_2 Hydrate. In: *Direct Ocean Disposal of Carbon Dioxide* (N. Handa & T. Ohsumi eds.), pp. 217-231, Terrapub, Tokyo.

Smith, R. (1982). Dispersion of Tracers in the Deep Ocean. *J. Fluid Mech.* **123**, 131-142.

Teng, H., C.M. Kinoshita and S.M. Masutani (1995). Hydrate Formation on the Surface of a CO_2 Droplet in High-Pressure/Low-Temperature Water. *Chem. Eng. Sci.* **50**, 559-564.

White, F.M. (1991). *Viscous Fluid Flow,* 2nd ed., McGraw-Hill, Inc., New York.

Young, W.R., P.B. Rhines, and G.J.R. Garrett (1982). Shear-Flow Dispersion, Internal Waves and Horizontal Mixing in the Ocean. *J. Phys. Oceanogr.* **12**, 515-527.

Pergamon

PII: S0196-8904(96)00289-0

Energy Convers. Mgmt Vol. 38, Suppl., pp. S325–S329, 1997
© 1997 Elsevier Science Ltd. All rights reserved
Printed in Great Britain
0196-8904/97 $17.00 + 0.00

INFLUENCE OF DISPOSAL DEPTH ON THE SIZE OF CO_2 DROPLETS PRODUCED FROM A CIRCULAR ORIFICE

H. TENG, A. YAMASAKI and Y. SHINDO

National Institute of Materials and Chemical Research
Agency of Industrial Science and Technology, MITI
1-1 Higashi, Tsukuba, Ibaraki 305, Japan

ABSTRACT

Disposal of liquid, anthropogenic CO_2 in intermediate-depth ocean waters has been accepted as a means to reduce atmospheric concentrations of this greenhouse species and mitigate global warming. Since the CO_2-seawater system is hydrodynamically unstable, liquid CO_2 effluent in the ocean will break up into droplets. This paper discusses the influence of both the discharge orifice and properties of CO_2 and seawater on the size of CO_2 droplets produced. It is found that the droplet size is affected strongly by disposal depth. This suggests that determination of the behavior of CO_2 discharged in the ocean must consider not only the orifice configuration, but also fluid properties at the disposal site. © 1997 Elsevier Science Ltd

KEYWORDS

CO_2; ocean disposal; disposal depth; droplet size.

INTRODUCTION

Since the primary natural removal process of atmospheric CO_2 is uptake by the ocean, disposal and sequestration of anthropogenic CO_2 in the ocean has been considered as a potential strategy for counteracting rising levels of this greenhouse gas. The majority of disposal scenarios proposed to date call for discharge of liquefied CO_2 through submerged pipelines at depths below the mixed layer of the ocean, either in intermediate-depth waters (between 500 and 1500 m) or in the abyssal zone (depths >3000 m). Since CO_2 is only slightly miscible in seawater, the CO_2-seawater system is hydrodynamically unstable and liquid CO_2 effluent issuing from these pipelines (i.e., liquid CO_2 jets in seawater) will break up into droplets.

At a given disposal depth, the pattern of droplet formation from a circular orifice depends principally on orifice diameter and jet velocity: large orifices and high jet velocities will result in non-uniform, multiple-droplet formation; small orifices and low jet velocities will produce a train of uniformly-sized droplets (Teng *et al.*, 1995a). If liquid CO_2 is discharged in seawater at pressures greater than 44.5 bar and temperatures less than 10°C (conditions satisfied at depths >500 m), a thin, solid hydrate layer will form rapidly on the surface of CO_2 droplets (Teng *et al.*, 1995b). Multiple-droplet formation promotes droplet interactions following jet breakup. Because the surface hydrate layers are reactive, on contact, CO_2 droplets can easily agglomerate to form droplet clusters through hydrogen bonding (Teng *et al.*, 1996). This has been observed in laboratory simulations (Aya *et al.*, 1992; Nishikawa *et al.*, 1995). To avoid unwanted droplet agglomeration, the jet velocity should be kept low and the orifice should not be large (Teng *et al.*, 1996).

At low jet velocities, droplets form at the orifice. Droplet formation comprises two major steps. The first is the creation of a precursor droplet at the orifice; the droplet undergoes a quasi-static growth and resides on the orifice. The second step is liftoff, which occurs when droplet buoyancy exceeds retarding forces and the droplet detaches the orifice. This second step is a dynamic process and a "necking" phenomenon can be observed during detachment. Breakup of the neck, which connects the precursor droplet to the orifice, will add volume to the droplet (Clift *et al.*, 1978). Since necking is induced by buoyancy, the size of droplets produced will be influenced strongly by the relative densities of CO_2 and the ambient seawater.

In comparison with seawater, liquid CO_2 is much more compressible: at 7°C, the density of liquid CO_2 is 893 kg/m^3 at 50 bar and increases to 1031 kg/m^3 at 300 bar; seawater density changes from about 1020 kg/m^3 at the surface to 1040 kg/m^3 at 3000 m ($\sim$ 301 bar). Thus, buoyancy of a CO_2 droplet in the ocean varies greatly with disposal depth. This, in turn, implies that the size of CO_2 droplets produced by the same orifice can differ significantly.

Above 3000 m, the CO_2 droplets which form from the breakup of the effluent will have positive buoyancy and will ascend toward the ocean surface as they dissolve. Since the hydrate layer formed on the CO_2 droplets is very thin, it is not expected to influence buoyant motion of the droplets (Masutani *et al.*, 1993; Teng *et al.*, 1995b). However, hydrate formation greatly hinders mass transfer of CO_2 from the droplets into seawater (Aya *et al.*, 1992; Teng *et al.*, 1995b), which results in longer traveling distances before full dissolution. Thus, successful sequestration of CO_2 in the ocean requires that the initial size of CO_2 droplets be carefully controlled (Golomb *et al.*, 1989; Herzog *et al.*, 1991; Holder *et al.*, 1995).

To model accurately the behavior of CO_2 released in the ocean, the influence of CO_2 and seawater properties, which vary with depth, on the initial droplet size must be understood. This work describes the results of an analytical study of the effects of disposal depth on the size of CO_2 droplets produced with a circular orifice.

PREDICTION OF DROPLET SIZE FROM HYDRODYNAMIC INSTABILITY THEORY

When liquid CO_2 is injected through a circular orifice into seawater, droplets will form at the orifice or at the end of the jet as result of hydrodynamic instability. In order to avoid droplet agglomeration following jet breakup, the discharge velocity should not be high. At low velocities, the discharge jet can be modeled assuming laminar flow and temporal instability (actually, no jet will be observed at low velocities; here, to be consistent with the terminology employed in hydrodynamic instability theory, the CO_2 effluent is termed a jet). The jet instability can then be described by the following characteristic equation (Teng *et al.*, 1995a):

$$[1 + \frac{1}{2}\frac{\rho_{seawater}}{\rho_{CO_2}}\eta\frac{K_0(\eta)}{K_1(\eta)}]\beta^2 + 2Z\eta^2\beta = \eta^2(1-\eta^2), \tag{1}$$

where $\eta \equiv kr_0$, k is the dimensional wavenumber, and r_0 is the orifice radius; K_0 and K_1 are zeroth- and first-order modified Bessel functions of the second kind, $\beta \equiv \omega(2\rho_{CO_2}r_0^3/\sigma)^{1/2}$, and ω is the dimensional growth rate in disturbance amplitude; $Z \equiv (3\mu_{CO_2} + \mu_{seawater})/(2r_0\rho_{CO_2}\sigma)^{1/2}$ is a modified Ohnesorge number, ρ the density, σ the interfacial tension, and μ the viscosity.

According to Rayleigh's classical theory (Rayleigh, 1945), breakup of the jet is induced by the "most-unstable surface wave," which experiences a maximum-growth rate in amplitude. Since, at low velocities, jet breakup results in a train of uniformly-sized droplets, it is reasonable to assume that each most-unstable wave produces a single droplet. Under this assumption, an expression for droplet size can be derived (Teng *et al.*, 1995a):

$$\frac{d_{drop}}{d_0} = (\frac{3\pi}{2\eta_m})^{1/3}, \tag{2}$$

where d_{drop} is the droplet diameter, d_0 is the orifice diameter, and η_m is the most-unstable wavenumber. The jet is unstable whenever $\beta > 0$ which requires that $\eta < 1$. Under this condition, $\eta K_0(\eta)/K_1(\eta) \ll 1$. Noting that $\rho_{seawater}/\rho_{CO_2} < 2$, (1) can be approximated as

$$\beta^2 + 2Z\eta^2\beta = \eta^2(1-\eta^2). \tag{3}$$

The most-unstable wavenumber can be obtained by applying the condition $d\beta/d\eta|_{\eta = \eta_m} = 0$ to (3):

$$\eta_m = (2 + 2Z)^{-1/2}. \tag{4}$$

Substituting (4) into (2) and rearranging yields the following relationship for droplet diameter

$$\frac{d_{drop}}{d_0} = (\frac{3\pi}{\sqrt{2}})^{1/3}(1 + Z)^{1/6}. \tag{5}$$

The necking effect can be accounted for by including a modifier to (5)

$$\frac{d_{drop}}{d_0} = \alpha(\frac{3\pi}{\sqrt{2}})^{1/3}(1 + Z)^{1/6}, \tag{6}$$

where α is an experimentally-determined parameter.

Important factors affecting the droplet detachment include: droplet buoyancy, interfacial tension force at the orifice, momentum of the effluent, inertia of the droplet, and drag force. If both the effluent and ambient velocities are low, then only the buoyancy force ($\pi/6d_{drop}^3\Delta\rho g$), which acts to remove the droplet from the orifice, and the interfacial tension force at the orifice ($\pi d_0\sigma$), which acts to hold the droplet to the orifice, need to be considered. It may be assumed then that $\alpha=\alpha(Bo)$, where $Bo\equiv d_0^2\Delta\rho g/\sigma$ is the Bond number, $\Delta\rho$ is the difference in the local densities of seawater and CO_2, and g is the gravitational acceleration. Equation (6) is a general expression that is not restricted to the CO_2-seawater system. Since no experimental data on depth dependence of the size of CO_2 droplets has been reported, α is estimated from the literature on seven liquid-in-liquid systems where the jet fluid is either kerosene, n-heptane, or a mixture of liquid paraffin and kerosene (all with positive buoyancy), and the ambient fluid comprises either water or an aqueous solution of starch syrup (Teng *et al.*, 1995a). The parameter α inferred from these experimental data is

$$\alpha=0.831+2.48Bo. \qquad (7)$$

Including the factor α in (5) significantly improves the agreement between experiment and predictions: the maximum difference between predicted and measured droplet diameters for the seven liquid-in-liquid systems of 18% [using (5)] reduces to 3.8% when (6) and (7) are employed.

INFLUENCE OF DISPOSAL DEPTH ON CO_2 DROPLET SIZE

From (6) and (7), it is seen that dimensionless droplet diameter is a function of Bond number and Ohnesorge number. Since both Bo and Z depend on the orifice diameter and fluid properties at the disposal site, the size of CO_2 droplets produced from a given orifice will vary with disposal depth. The influence of disposal depth on the size of droplets produced by a 5 mm diameter orifice is shown in Table 1. Fluid properties used to calculate Bo and Z were taken from Golomb *et al.* (1989). These results indicate that the size of CO_2 droplets varies significantly with disposal depth.

Table 1. Influence of disposal depth (h) on droplet size

h (m)	Bo	Z	$(d_{drop}/d_0)_{d_0=5\ mm}$
500	0.659	2.71×10^{-3}	4.64
600	0.550	2.86×10^{-3}	4.13
700	0.430	3.02×10^{-3}	3.57
800	0.355	3.16×10^{-3}	3.22
900	0.297	3.28×10^{-3}	2.95
1000	0.248	3.35×10^{-3}	2.72
1100	0.221	3.38×10^{-3}	2.60
1200	0.193	3.40×10^{-3}	2.46
1300	0.173	3.41×10^{-3}	2.37
1400	0.152	3.42×10^{-3}	2.27
1500	0.131	3.43×10^{-3}	2.18

Since both seawater and liquid CO_2 are low-viscosity fluids, the influence of Ohnesorge number on the droplet diameter is negligible. Furthermore, interfacial tension changes only slightly in the ocean (Golomb *et al.*, 1989). Thus, CO_2 droplet size is basically determined by $\Delta\rho$ and d_0. At depths exceeding 2000 m, droplet buoyancy is negligibly small; in this case, the parameter α should be set to unity because the necking effect becomes negligible. Since Bo is proportional to d_0^2, the dimensionless droplet diameter also varies with changes in orifice diameter. The dependence of droplet size on orifice diameter is shown in Table 2. The dimensionless droplet size as a function of orifice diameter and ocean depth is plotted in Fig.1.

Results presented in Table 2 and Fig.1 indicate that CO_2 droplets decrease in size with disposal depth for a given orifice, and that dimensionless droplet size increases with orifice diameter at a given disposal depth. These trends largely reflect the necking effect; in general, an increase in orifice diameter or droplet buoyancy enhances the necking effect and, therefore, yields larger droplets. Tables 1 and 2 (as well as Fig.1) show clearly that the size of CO_2 droplets is influenced strongly by disposal depth in the intermediate-depth disposal. Hence, both orifice configuration and local fluid properties need to be considered for successful implementation of the CO_2 disposal concept.

Table 2. Dependence of droplet size on orifice diameter

h (m)	$(d_{drop}/d_0)_{d_0=2\ mm}$	$(d_{drop}/d_0)_{d_0=3\ mm}$	$(d_{drop}/d_0)_{d_0=4\ mm}$
500	2.06	2.67	3.53
600	1.97	2.49	3.21
700	1.89	2.29	2.85
800	1.83	2.16	2.62
900	1.79	2.06	2.45
1000	1.75	1.98	2.31
1100	1.73	1.94	2.22
1200	1.71	1.89	2.14
1300	1.69	1.85	2.08
1400	1.68	1.82	2.02
1500	1.66	1.78	1.96

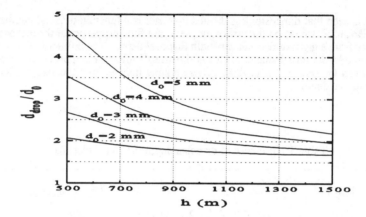

Fig.1. Dependence of dimensionless droplet size on orifice diameter and ocean depth.

Another phenomenon encountered in low-discharge-rate disposal is orifice blockage by hydrates, which has been observed in some laboratory studies (Aya *et al.* 1992; Hirai *et al.*, 1995). In deep waters, droplet buoyancy becomes small; thus, if droplets cannot be removed quickly from the orifice, hydrates may cover the entire droplet surface, causing it to stick to the orifice (note that the orifice material can accelerate hydrate formation). However, orifice blockage may be avoided in intermediate-depth waters where droplet buoyancy is significant; buoyancy will remove the droplet from the orifice before blockage occurs (Masutani *et al.*, 1993; Teng *et al.*, 1995b, 1996). Hydrodynamic forces in the ocean can also remove droplets from the orifice.

SUMMARY

In order to avoid unwanted droplet agglomeration following breakup of CO_2 effluent in the ocean, the discharge velocity of the liquid CO_2 should be low. This will produce a train of uniformly-sized droplets. At intermediate depths, these droplets will be buoyant and will rise toward the ocean surface as they dissolve slowly due to the inhibiting effect of a surface hydrate layer. The travel distance of droplets may affect the term of sequestration of the CO_2 as well as the extent of impact on the marine environment. Since this travel distance depends on the initial diameter of the droplets, the droplet formation mechanism at the discharge orifice warrants study. The present investigation has identified critical factors that influence droplet size for low-velocity discharge where necking induced by CO_2 buoyancy may be significant. It was determined that both orifice diameter and fluid properties (i.e., depth of discharge) are important. At a given depth, the dimensionless droplet diameter would increase with increasing orifice diameter. However, for a given discharge orifice, droplet diameter would decease with depth as a result of diminishing CO_2 buoyancy. These results suggest that both orifice configuration and disposal depth should be selected appropriately to ensure successful implementation of the ocean disposal concept.

Acknowledgement — The authors wish to express their gratitude to Dr. S.M. Masutani of the Hawaii Natural Energy Institute for his valuable suggestions on improving this paper.

REFERENCES

Aya, I., K. Yamane and N. Yamada (1992). Stability of clathrate-hydrate of carbon dioxide in highly pressurized water. In: *Fundamentals of Phase Change: Freezing, Melting, and Sublimation* (P.G. Kroeger and Y. Bayazitoglu, Ed.), ASME, pp17-22.

Clift, R., J.R. Grace and M.F. Weber (1978). *Bubbles, Drops, and Particles.* Academic Press, New York.

Golomb, D., H. Herzog, J. Tester, D. White and S. Zemba (1989). *Feasibility, Modeling and Economics of Sequestering Power Plant CO₂ Emissions in the Deep Ocean.* MIT Energy Laboratory Report, MIT-EL 89-003.

Herzog, H., D. Golomb and S. Zemba (1991). Feasibility, Modeling and Economics of Sequestering Power Plant CO₂ Emissions in the Deep Ocean. *Environ. Prog.*, **10**, 64-74.

Hirai, S., K. Okazaki, N. Araki, K. Yoshimoto, H. Ito and K. Hijikata (1995). Experiments for dynamic behavior of carbon dioxide in deep sea. *Energy Convers. Mgmt*, **36**, 471-474.

Holder, G.D., A.V. Cugini and R.P. Warzinski (1995). Modeling clathrate hydrate formation during carbon dioxide injection into the ocean. *Environ. Sci. Technol.*, **29**, 276-278.

Masutani, S.M., C.M. Kinoshita, G.C. Nihous, H. Teng and L.A. Vega (1993). An experiment to simulate ocean disposal of carbon dioxide. *Energy Convers. Mgmt*, **34**, 865-872.

Nishikawa, N., M. Ishibashi, H. Ohta, N. Akutsu, M. Tajika, T. Sugitani, R. Hiraoka, H. Kimuro and M. Moritoki (1995). Stability of liquid CO₂ spheres covered with clathrate film when exposed to environment simulating the deep sea. *Energy Convers. Mgmt*, **36**, 489-492.

Rayleigh, Lord (1945). *Theory of Sound.* Dover, New York.

Teng, H., C.M. Kinoshita and S.M. Masutani (1995a). Prediction of droplet size from the breakup of cylindrical liquid jets. *Int. J. Multiphase Flow*, **21**, 129-136.

Teng, H., C.M. Kinoshita and S.M. Masutani (1995b). Hydrate formation on the surface of a CO₂ droplet in high-pressure, low-temperature water. *Chem. Eng. Sci.*, **50**, 559-564.

Teng, H., A. Yamasaki and Y. Shindo (1996). The fate of liquid CO₂ disposed of in the ocean. *Energy*, **21**, 756-774.

Energy Convers. Mgmt Vol. 38, Suppl., pp. S331–S336, 1997
© 1997 Elsevier Science Ltd. All rights reserved
Printed in Great Britain
0196-8904/97 $17.00 + 0.00

Pergamon

PII: S0196-8904(96)00290-7

LES STUDY OF CO_2 ENRICHED GRAVITY CURRENTS

Guttorm Alendal

Nansen Environmental and Remote Sensing Center
Edv. Griegsvei 3A, N–5037 Solheimsviken, NORWAY

ABSTRACT

Earlier integrated gravity current models have been used to study both natural existing ocean currents and artificially generated currents with CO_2 enriched seawater. The latter being a possible option for ocean disposal of waste CO_2 from powerplants or other large point sources. These types of models are not resolving the interior of the current but only describes the mean velocity and density over cross-sections. The models parameterise friction and entrainment, and earlier studies have shown a high sensitivity on the friction parameterisation and thereby the entrainment of ambient water. In this study a Navier Stokes solver connected with a Large Eddy Simulation (LES), to account for the cascade of energy to subgrid scale eddies, is used to model the current close to the outlet. The code resolves the interior of the current and no entrainment or friction parameterisations are needed. Only the first 250 meters after the outlet are modelled and the results shows high entrainment. The numerical results are used together with a simple time dependent integrated gravity current model to estimate entrainment and friction parameters.

© 1997 Elsevier Science Ltd

KEYWORDS

Ocean disposal, bottom gravity current, CO_2 injection, entrainment and friction parameterisation.

INTRODUCTION

Marchetti (1977) was the first to suggest sequestration of fossil fuel CO_2 into the ocean by injecting it into the outflow of dense, saline water through the Gibraltar Strait, which would distribute the CO_2 at intermediate depths over the northern Atlantic. Later Haugan and Drange (1992) suggested to create a gravity current by dissolving the CO_2 into seawater hence increase the density. To study this option an integrated gravity model was used (Drange and Haugan, 1992; Drange et al., 1993). This model was sensitive to the parameterisation of friction and entrainment (Alendal et al., 1994), and these parameterisations are essential for how deep the current will penetrate before the density difference vanishes. The model also used an assumption on the width over height ratio that, at its best, is questionable. Earth rotation will bend the current, forcing it into near geostrophic balance, i.e. near horizontal flow direction, hence a long distance has to be travelled before reaching the deep ocean. To avoid the effect from rotation and the height over width assumption, a study of a current flowing in a canyon or duct was made (Haugan et al., 1996), but still the uncertainties in friction and entrainment parameterisation make it hard to conclude on the effectiveness of this injection option.

This paper reports on a study of gravity currents of CO_2 enriched water using a Navier Stokes solver. The interior of the current is resolved and the earlier mentioned parameterisations are no longer needed. The code uses a local eddy viscosity based on Renormalisation Group Theory (Yakhot and Orszag, 1986) to account for the cascade of turbulent energy from resolved scales to subgridscale eddies. The viscosity increases with higher strain and no wall function seems to be needed. Several testings of the code have been carried out (Alendal, 1996) against Direct Numerical Simulation of open channel flow (Nagaosa and Saito, 1996), and three lock-release experiments (Hacker et al., 1996).

MODEL DESCRIPTION AND SOLUTION METHODS

The model solves the following equations:

$$\frac{\partial}{\partial t}\mathbf{u} + \mathbf{u} \cdot \nabla \mathbf{u} = -\nabla p + Ri\theta\mathbf{g} + \nabla \cdot \frac{1}{Re_\tau}\tilde{S}, \tag{1}$$

$$\nabla \cdot \mathbf{u} = 0, \tag{2}$$

$$\frac{\partial}{\partial t}\theta + \mathbf{u} \cdot \nabla\theta = \frac{1}{RePr}\nabla^2\theta, \tag{3}$$

where $\mathbf{u}$ is the velocity vector, p pressure, $\mathbf{g}$ is a directional vector for the gravity force dependent on the inclination of the bottom, and θ is a scalar. $\tilde{S}$ is the stress tensor with components

$$S_{ij} = \frac{1}{2}\left(\frac{\partial u_i}{\partial x_j} + \frac{\partial u_j}{\partial x_i}\right). \tag{4}$$

All equations have been nondimensionalised with a characteristic velocity U and length L, giving a characteristic time scale $L\,U^{-1}$. CO_2 concentration, C, has been nondimensionalised against a reference value, C_0, and a characteristic difference, ΔC, according to

$$\theta = \frac{C - C_0}{\Delta C}.$$

This scalar is used as a dynamic scalar making it possible to create stratified environment and density differences. The Boussinesq approximation and a linear equation of state with an expansion coefficient β have been used,

$$\rho = \rho_0\left(1 + \beta(C - C_0)\right). \tag{5}$$

CO_2 concentration is the only tracer included in the code so there is no stratification in the ambient fluid. The non-dimensional numbers in Eq. (1)–(3) are;

$$Re = \frac{U\,L}{\nu_0}, \quad Ri = \frac{g\,\beta\,\Delta C\,L}{U^2}, \quad Pr = \frac{\nu_0}{\kappa},$$

which are the Reynolds number, the Richardson number and the Prandtl number, respectively, ν_0 $(m^2 s^{-1})$ is molecular viscosity and κ $(m^2 s^{-1})$ the diffusion parameter for the scalar.

In Large Eddy Simulation (LES) all variables and equations are filtered giving rise to extra terms in the momentum equations due to the nonlinearities in Eqs. (1) and (3). For the momentum equations, these terms are usually divided into a Leonard stress term and subgridscale stresses, the latter often assumed to be dependent on the resolved strain and an eddy viscosity. The Leonard stress can be of importance for the backscattering of energy from subgridscale eddies to the resolved scales. However, Antonopoulos-Domis (1981) claimed that for staggered grid better results can be obtained if the Leonard stress is neglected. For this reason and for the sake of simplicity this term is left out of the present calculations.

The local eddy viscosity, ν, $(m^2 s^{-1})$ is based on a result from Renormalisation Group Theory (Yakhot and Orszag, 1986) which involves solving the third order equation

$$\nu = \nu_0 \left[1 + H \left(\frac{a \Delta^4}{2(2\pi)^4 \nu_0^3} \nu \left(\frac{\partial v_i}{\partial x_j} + \frac{\partial v_j}{\partial x_i} \right)^2 - C \right) \right]^{1/3}, \tag{6}$$

for ν. Here $a = 0.12$, $C = 75$, Δ is the filtering scale, and the function H is defined

$$H(x) = \begin{cases} x, & \text{for } x > 0 \\ 0, & \text{for } x < 0. \end{cases} \tag{7}$$

For large stress Eq. (6) gives a Smagorinski type eddy viscosity, while the other extreme with low stress the eddy viscosity returns to molecular viscosity. The local turbulent Reynolds number appearing in Eq. (1) follows

$$Re_\tau = \frac{UL}{\nu}. \tag{8}$$

Yakhot and Orszag (1986) also supply a turbulent diffusion coefficient for turbulent transport, but the code has shown to be slightly too dissipative when tested against a lock-release experiment, so only molecular dissipation is used in the scalar transport equation (Alendal, 1996).

The momentum equations are solved using a fractional timestep method (Gresho, 1991). The first step solves the momentum equation by neglecting the pressure gradient, the second step solves an elliptic equation to find the pressure which is used in the final step to adjust the velocity. Eq. (2) is implicitly solved in the process. The two last steps involve a transformation of the intermediate velocity field, which is not solenoidal, to the wanted solenoidal velocity field. The elliptic pressure field problem is solved with black-red ordering SOR (Golub and Van Loan, 1989) updating the relaxation parameter for each half step to accelerate convergence (Press et al., 1992). For the transport equation, Eq. (3), an upwind scheme for the advection has been used to avoid false extremal values that are likely to occur for ordinary central difference discretisation. This introduces extra diffusivity into the code.

RESULTS

To study the onset of a gravity current generated when CO_2 enriched water is released on an inclined slope, with an inclination angle of 0.01 rad, the code is set up for simulation of a flow with Richardson number set to 30 and with a Reynolds number of 10^6. The Prandtl number for CO_2 in water is ~ 700. Initially the enriched water has no velocity and only the first 250 meters of the current is modelled.

The beauty of non-dimensional modelling is that several currents can be modelled with the same simulation as long as they have the same Richardson, Reynolds and Prantl numbers, but with an expansion coefficient for CO_2 fixed to 10^{-5} $(kg\ m^{-3}\ mol^{-1})$, a characteristic length of 10 m and velocity $0.1\ m\ s^{-1}$ the simulation correspond to an excess CO_2 concentration $\sim 300\ mol\ m^{-3}$. This is approximately 25 % of the maximum CO_2 that can be resolved in seawater (Drange et al., 1993), and gives an excess density $\sim 2.5\ kg\ m^{-3}$. All results given later use these characteristics on the current.

The averaged velocity over the cross-section and the maximum velocity are shown in Fig. (1). The current experiences a rapid increase in velocity close to the outlet whereafter the maximum velocity decreases while the average velocity remains at a constant value. The maximum velocity seems to be constant with time, this is due to the rate of entrainment of ambient water with zero momentum. Notice the big difference in peak value compared to average velocity. This suggests another uncertainty in the entrainment parameterisation in integrated gravity current modelling; the entrainment is assumed to be dependent on the velocity of the current but when the velocity

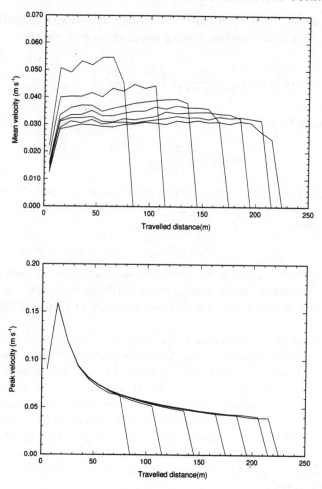

Fig. 1. Average and maximum along-stream velocity taken over cross-
sections perpendicular to the main flow direction. The velocities
are taken after each 1000 seconds. Notice how the average velocity
decreases with time but stays nearly constant in spatial direction.
The peak velocity remains constant in time but decreases in the
along-stream direction.

in the current varies as much as shown the question of correct choice of velocity to use arises. The
correct choice of velocity to use in entrainment parameterisation is near the plume edge, but this
velocity is not available for integrated gravity current models. The front velocity decreases with
time from ~ 0.03 m s^{-1} to 0.01 m s^{-1}.

In an attempt to compare the numerical results with an integrated gravity current the following
equations where solved for the entrainment parameter E and friction parameter C_D:

$$\frac{\partial}{\partial t}\left(\rho A\right) + \frac{\partial}{\partial x}\left(\rho A U\right) = E\rho_e bU \tag{9}$$

$$\frac{\partial}{\partial t}\left(\rho A U\right) + \frac{\partial}{\partial x}\left(\rho A U^2\right) = g\left(\rho - \rho_0\right) A sin\phi - C_D \rho bU^2 \tag{10}$$

here A (m^2), U (m s^{-1}) and ρ (kg m^{-3}) denotes the area, average velocity and density over cross-
sections of the current. x is the along stream direction and ρ_0 (kg m^{-3}) is the ambient density.
The inclination is denoted ϕ. In Fig. (2) the resulting entrainent and friction parameters are shown

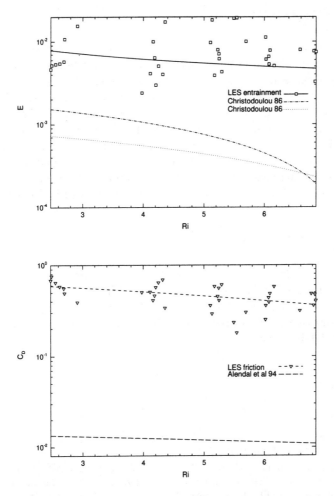

Fig. 2. Calculated entrainment and friction parameters resulting from Eqs.
(9) and (10). Each symbol represent a data point and equations for
the Least Square curves are given in Eq. (11). Also shown are the
entrainment parameterisations given by Christodoulou (1986) and
the Richardson dependent friction parameter developed in Alendal
et al. (1994).

as function of the Richardson number. Each symbol in Fig. (2) represent a data point, and a Least
Square method is used to find equations for the Richardson dependent entrainent and friction
parameters

$$E = 0.0127 \, Ri^{-0.58}, \qquad C_D = -0.05 \, Ri + 0.71. \tag{11}$$

In Fig. (2) the entrainent parameter given by Christodoulou (1986) and the Richardson dependent
friction parameter from Alendal *et al.* (1994) are also shown for comparison. The resulting entrainment parameter are higher than given by Christodoulou (1986) which is partially caused by the
fact that the code has shown to be too diffusive when tested against the lock-release experiments,
smearing out large gradients in CO_2 concentration (Alendal, 1996). High entrainment might also
be responsible for higher friction, the friction parameter found here are order of magnitude higher
than the Richardson dependent friction parameter developed in Alendal *et al.* (1994).

CONCLUSIVE REMARKS

An LES code has been used to model a release of heavy CO_2 enriched seawater. Presently the current have only been modelled close to the outlet. It is clear that entrainment is very important for the dynamics of the current, and the difference in maximum compared with average velocity shows that there are a wide range of velocities in the current. If the order of magnitude of the friction parameter found is correct it is encouraging in view of the results from integrated gravity current modelling. However, this study cannot say whether a release of CO_2 enriched water on a sloping ocean floor will be an option for sequestration into the deep oceans. Further studies with models that resolve the interior on larger scale will be necessary to predict how far the current will go before the density difference vanishes. Earth rotation are important for gravity currents on bigger scales and the effect from the Coriolis force should be included in further studies.

ACKNOWNLEDGMENTS

This work has been supported by the Norwegian Research Council. I would like to thank Dr. Ryuichi Nagaosa at National Institute for Resources and Environment in Japan for letting me bring his Navier Stokes solver with me when I returned from a visit to his laboratory. Finn Thork-ildsen and Dr. Helge Drange at NERSC have been of great help in the preparation of the code and this paper through useful discussions.

REFERENCES

Alendal, G. (1996), Implementation and testing of an RNG-based subgridscale eddie viscosity in a Navier Stokes solver, *J. Comp. Phys.*, Submitted.

Alendal, G., Drange, H., and Haugan, P. M. (1994), Modelling of deep-sea gravity currents using an integrated plume model, in *The Polar Oceans and Their Role in Shaping the Global Environment: The Nansen Centennial Volume*, edited by Johannessen, O. M., Muench, R. D., and Overland, J. E., volume 85 of *AGU Geophysical Monograph*, pp. 237–246, American Geophysical Union.

Antonopoulos-Domis, M. (1981), Large-eddy simulation of a passive scalar in isotropic turbulence, *J. Fluid Mech.*, **104**:55–79.

Christodoulou, G. C. (1986), Interfacial mixing in stratified flows, *Journ. of Hydr. Res.*, **24**(2):77–92.

Drange, H., Alendal, G., and Haugan, P. M. (1993), A bottom gravity current model for carbon-dioxide enriched sinking seawater, *Energy Conv. Mgmt.*, **34**(9-11):1065–1072.

Drange, H. and Haugan, P. M. (1992), Carbon dioxide sequestration in the ocean: The possibility of injection in shallow water, *Energy Convers. Mgmt.*, **33**(5-8):697–704.

Golub, G. H. and Van Loan, C. F., editors (1989), *Matrix computations*, Number 3 in Johns Hopkins Series in the Mathematical Sciences, Johns Hopkins, London, second edition.

Gresho, P. M. (1991), Some current CFD issues relevant to the incompressible Navier-Stokes equations, *Comp. Methods in Appl. Mech. Eng.*, **87**:201–252.

Hacker, J., Linden, P. F., and Dalziel, S. B. (1996), Mixing in lock-release gravity currents, *Dyn. Atmosph. and Ocean*, **24**:183–195.

Haugan, P. M., Alendal, G., and Drange, H. (1996), Transport of dissolved CO_2 in the ocean by gravity currents, *J. Marine Systems*, submitted.

Haugan, P. M. and Drange, H. (1992), Sequestration of CO_2 in the deep ocean by shallow injection, *Nature*, **357**:318–320.

Marchetti, C. (1977), On geoengineering and the CO_2 problem, *Climatic Change*, 1:59–68.

Nagaosa, R. and Saito, T. (1996), Direct Numerical Simulation of turbulence structure and heat transfer across a free surface in stably stratified open–channel flows, *Int. J. Mass and Heat Transfer*, In press.

Press, W. H., Teukolsky, S. A., Vetterling, W. T., and Flannery, B. P. (1992), *Numerical Recipies in FORTRAN*, Cambridge University press, Cambridge, second edition.

Yakhot, V. and Orszag, S. A. (1986), Renormalization group analysis of turbulence. I. Basic theory, *J. Scient. Comp.*, **1**(1):3–51.

Energy Convers. Mgmt Vol. 38, Suppl., pp. S337–S341, 1997
© 1997 Elsevier Science Ltd. All rights reserved
Printed in Great Britain
0196-8904/97 $17.00 + 0.00

PII: S0196-8904(96)00291-9

Impacts of High Concentration of CO2 on Marine Organisms; a Modification of CO2 Ocean Sequestration

K. Takeuchi, Y. Fujioka, Y. Kawasaki*, and Y. Shirayama**

Nagasaki R & D Center, Mitsubishi Heavy Industries, Ltd.
5-717-1, Fukahori, Nagasaki 851-03, JAPAN

* Central Research Institute of Electric Power Industry
1-16-1,Ohte-machi, Chiyoda-ku, Tokyo 100, JAPAN

**Division of Marine Ecology, Ocean Research Institute, University of Tokyo
1-15-1, Minami-dai, Nakano-ku, Tokyo 164, JAPAN

ABSTRACT

Experimental studies regarding impacts of high concentration of CO_2, as a modification of CO_2 ocean sequestration, on marine organisms were carried out. As representatives of the marine organisms bacteria and nematodes were investigated. The former is not only the major decomposer but also the most important potential genetic resource, and the latter the most abundant in number and the highest in species diversity in the benthic ecosystem. Both nematodes and bacteria were so resistant to the high concentration of CO_2 that drastic impacts were observed only under the condition of pH 5.5~6 or less. In addition to the experiment using a monospecific material, however, further studies are needed taking interactions of variety of marine species into account. © 1997 Elsevier Science Ltd

KEY WORDS

CO_2; biological impact; bacteria; benthos; high-pressure vessel

INTRODUCTION

To decide the mitigation scenario for the rapid increase of atmospheric concentration of CO_2 due to the anthropogenic activities, it is necessary to compare the global impacts between the cases of CO_2 sequestration and direct release of CO_2 to the atmosphere. The cumulative decrease in the ocean surface pH due to the increase of atmospheric CO_2 from the pre industrial level of 280 ppm to present 360 ppm is found to be almost 0.1 pH unit in cold waters and slightly less than 0.09 pH units in the warmest surface water.(Haugan and Drange, 1996). Such a change of pH in the ocean surface may have affect on the biological activities in the euphonic zone where primary production occurs.

One of the feasible option for the CO_2 mitigation is to capture the CO_2 exhausted from power plants and to sequester it into the ocean. Three designs have been proposed as the method of ocean sequestration, namely deep-sea sequestration, the mid-water dispersion and the shallow water injection (Shindo *et al.*,1995; Adams *et al.*, 1995; Haugan and Drange, 1992). On the basis of detailed investigations regarding the three methods, it has been widely recognized that the CO_2 sequestration in the ocean is a promising option from the point of view of its capacity, its economy and its effectiveness. Many research are being conducted using the high pressure vessel and the numerical simulation to clear uncertainties about the behavior of CO_2 in the process of dissolution and dispersion (Shindo *et al.*,1995; Nakashiki *et al.*,1995). One of the most important research subjects regarding the behavior of CO_2 after the sequestration is to clarify the affects of seawater with high concentration of CO_2 on the marine organisms. The increase of partial pressure of CO_2 and decrease of pH of the seawater may have an impact on marine organism directly. Moreover the change of the specific density, the ionic strength, the temperature, and the heavy metal concentration of the seawater may have an impact on marine organisms indirectly. As a result, nektons, that have high swimming ability, may avoid from the impacted area. On the other hand, bacteria, plankton and benthos, that lack the mobility, may decline in number due to the impacts mentioned above. In addition to the acute impacts, genetic impacts may be present in the long term.

There are few studies concerned with the impacts of the high concentration of CO2 on the marine organisms. Several studies reported the effects of low-pH seawater on the marine organisms. Herzog(1996) summarized potential environmental impacts by the ocean disposal of CO2 and mentioned that impacts around the releasing point are inevitable, however the size and severity of the impacted area will depend on the exact release technology.

If CO2 are sequestered to the ocean, impacts are estimated to be the largest at the coastal zone and the ocean floor because the biomass is high in these areas. Whichever option of the ocean sequestration is chosen, the seawater with high concentration of CO2 will diffuse and sink onto the ocean floor. Therefore it is prerequisite to estimate the CO2 impacts on the marine organisms living in the benthic realm. However, enough information is not available to estimate impacts concerned with CO2 sequestration on deep-sea benthic organisms. In this paper, we carried out experiments regarding the impact of high concentration of CO2 on marine organisms to provide information necessary for estimating the environmental impact due to ocean disposal of CO2.

MATERIALS AND METHODS

As representatives of marine organisms, bacteria and nematodes were selected. The former is not only the major decomposer of organic material in the ocean but also the most important potential genetic resource, and the latter the most abundant in number and the most high in species diversity in metazoa of the benthic ecosystem. Nematodes were sorted out from the sands collected in the subtidal zone off the Otsuchi Marine Research Center, Ocean Research Institute, The University of Tokyo and transferred to the laboratory of MHI Nagasaki R&D Center keeping the temperature cool (under 5 ℃). Three nematode species, that are abundant and readily identical under the dissecting microscope, i.e. , *Metachromadora* sp. , *Symplocostoma* sp. and *Mesacanthion* sp. were used in the experiments.

Eleven bacteria species isolated from marine habitats (seawater, sediment, etc.) were selected from ATCC (American Type Culture Collection; see Table 1). Among them, *Pseudomonas bathycetes* was the species that had been isolated from the deep-sea sediment.

Table 1. The list of marine bacteria used in the present study.

	ATCC No.		Temp.	Sample
Type Strain	29570	*Alteromonas rubra*	26 ℃	surface seawater
Type Strain	33046	*Alteromonas aurantia*	25 ℃	surface seawater
Type Strain	19260	*Flavobacterium marinotypicum*	18 ℃	
Type Strain	25915	*Photobacterium angustum*	26 ℃	seawater
	23597	*Pseudomonas bathycetes*	26 ℃	deep sea sediment
Type Strain	27130	*Pseudomonas stanieri*	26 ℃	seawater
Type Strain	27135	*Oceanospirillum jannaschii*	26 ℃	seawater
Type Strain	27119	*Marinomonas vega*	26 ℃	seawater
Type Strain	15381	*Vibrio marinus*	18 ℃	seawater
Type Strain	7744	*Vibrio fischeri*	26 ℃	
Type Strain	17749	*Vibrio alginolyticus*	37 ℃	spoiled horse mackerel

To study the affects of CO2 on the survival rates of nematodes, individuals were put in the wells of microplates and each well was filled with 1 ml of seawater that was prepared at a given concentration of CO2, adjusted using a gas mixed with standard CO2, N2, and O2 gases. The mixed gases were bubbled into seawater, and the equilibrium state were confirmed by checking the pH using a pH electrode. The headspace of the wells of microplates were replaced with the same mixed gas as that used to adjust the CO2 concentration of the seawater. These microplates were placed in the chambers, where the temperature was controlled at 20 ℃ and the CO2 concentrations was kept at the given level. Five replicated experiments were done for each condition and 10 individuals were put in each microplate. Their survival rates were determined by observing nematodes continuously under a dissecting microscope.

The impacts of CO2 on the growth of the bacteria were assessed at both atmospheric and high pressures. Stock culture of 11 species listed in Table 1 were maintained by weekly transfer to the growth medium (Difco Marine Broth 2216) . Subcultures prepared by inoculating the growth medium with all strains were incubated for 12 hrs at given temperature. To evaluate the affects of CO2, its concentrations in the growth media were adjusted using the similar method as that used in the nematodes experiment. In the experiments using bacteria, however, to achieve the sterile conditions, mixed gases were supplied after filtering through

0.2 μm membrane filter. The 5 ml of growth media at adjusted CO_2 concentrations were dispensed in 18-mm sterile test tubes. To each tube, 100 μl of the subcultures were inoculated and incubated at given temperature with an aid of a shaker (120 rpm). To keep the CO_2 concentration constant, CO_2 mixed gases passed through a 0.2 μm sterile membrane filters were aerated to the headspace of the test tubes. As the indicator of the bacterial growth, optical densities of the growth mediums at 660 nm were monitored continuously using photometer. Five replicates were taken for each experimental condition.

The impacts of CO_2 on the growth of the bacteria were compared among the atmospheric- and the high-pressure (350 atm) conditions using two marine bacteria species that could grow under the high-pressure condition (*Pseudomonas bathycetes* and *Marinomonas vega*). Subcultures were prepared in the way mentioned above, and they were inoculated into the growth medium so as to be the concentration of approximately 10^4 cells per ml. In this experiment, 1/5 Marine Broth 2216 (Difco Marine Broth 2216 diluted by seawater) were used as the growth medium. The inoculated medium was dispensed into syringes modified for the use in the high-pressure vessel. The syringes, each containing a Teflon ball, were placed in the pressure vessel and incubated at the high pressures. Pressure was applied to the entire system by means of a equipped hydraulic pump. Hydrostatic pressure was transmitted through the piston. The pressure in the vessel was monitored using a pressure gauge through the experiment. The vessel was submerged in water kept at 26 ± 0.1 ℃. As control media, syringes were placed in the incubator at the same temperature but at atmospheric pressure. The growth media sampled continuously were plated on agar plates (Difco Marine Agar 2216) and incubated for 72 hours at atmospheric pressure to count the number of viable colonies.

RESULTS AND DISCUSSION

Effects of CO_2 on the survival of nematodes.

Even nematodes under the natural (control) condition decreased gradually with days in the present experiments, because the experiments were carried out under starvation conditions. When the survival curves of nematodes at lower pH condition were compared with those at pH 8.0 (control), no significant difference was found in the experiments at pH 7.0 and 6.2 (Fig. 1). At pH 5.4, however, survival rates of nematodes were significantly lower than in the control. Among the three species studied, either *Mesacanthion* sp. or *Symplocostoma* sp. were more sensitive than *Metachromadora* sp. The former two species are more active than the last species. These active species are considered to be stressed more strongly in the present starved condition and are probably easier to be damaged by the environmental change.

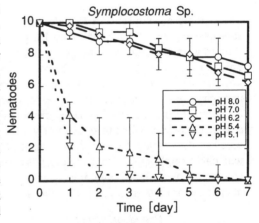

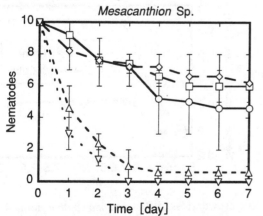

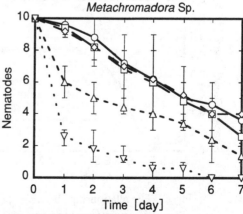

Fig. 1. Survival curve of three nematode species under different pH conditions. Plots denote the mean of five experiments and vertical bars the range of survivals.

Effects of CO2 on the growth of bacteria

The growth of bacteria after 9 hrs of culture was strongly affected and decreased to less than 50% of the control medium (pH=7.7) under the conditions of less than 6.0 in pH (Fig. 2) except for *Alteromonas rubra*. The species was more sensitive than other species and even at pH=6.5 it could not grow more than 50% of the control. The experiment was carried out under eutrophic conditions. The in situ condition of the ocean is however more oligotrophic. In such a condition, bacteria might be more sensitive to the change of pH, because they are already stressed by the severe trophic conditions.

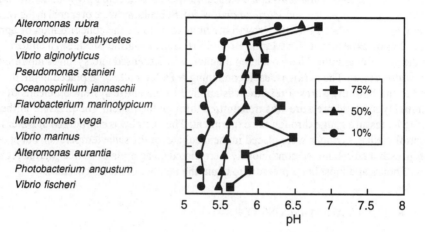

Fig. 2. The growth of 11 marine bacteria species after 9 hrs as a ratio of the control experiment.

The specific growth rates of *Pseudomonas bathycetes* and *Marinomonas vega* tended to be higher at atmospheric pressure than at 350 atm. The pH dependence of specific growth rates for these species however showed a similar pattern at the two pressure conditions. In the case of the former species, the specific growth rates were comparable to the control (pH=8.2) at pH=6.9, slight low at pH=6.2, and significantly lower at pH=5.5 and 5.1. The latter species grew at the comparable rate to the control condition (pH=8.2) through pH=6.2, but significantly lower at condition of pH=5.5 and 5.1. The present results did not support the general idea that deep-sea species are more sensitive than the shallow-water species to the environmental change because they live in an extremely stable condition. Further experiments using variety of deep-sea species are necessary to confirm the findings obtained in the present study. The present study however dealt with only a single species. In the future, it is necessary to study the impacts of CO2 onto the marine biological community as a whole taking interspecific relations into account. For this purpose, experiments using a microcosm or an in situ benthic chamber should be realized in the future.

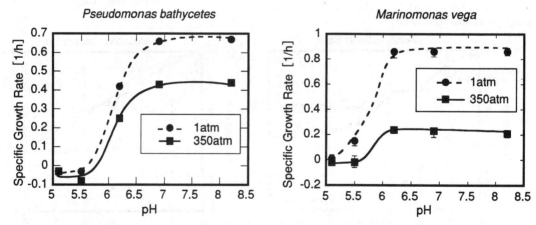

Fig. 3. The relationships between the specific growth rates of *Pseudomonas bathycetes* and *Marinomonas vega*, and pH. These curves were obtained under atmospheric and high-pressure conditions.

CONCLUSION

The present experiments aimed to evaluate the impacts of high concentration of CO_2, as a modification of CO_2 ocean sequestration, on marine organisms. The results revealed that distinct impacts on the survival of both nematodes and bacteria existed under the conditions of pH=5.5~6 or lower but no acute affects were observed at conditions of higher pH.

ACKNOWLEDGEMENT

This research was commissioned by Ministry of International Trade and Industry, Japan.

REFERENCES

Haugan, P. M. and H. Drange, (1996). *Energy Convers. Mgmt.*, 37, 1019-1022

Herzog, H. J., E. E. Adams, D. Auerbach, J. Caulfield, (1996). *Energy Convers. Mgmt.*, 37, 995-1005

Adams, E. E., D. S. Golomb, H. J. Herzog,(1995). *Energy Convers. Mgmt.*, 36, 447-452

Shindo, Y., T. Hakuta, Y. Fujioka, K. Takeuchi, H. Komiyama,(1995). *Energy Convers. Mgmt.*, 36, 479-484

Nakashiki, N., T. Hikita,(1995). *Energy Convers. Mgmt.*, 36, 453-456

Haugan, P. M. and H. Drange, (1992). *Nature*, 357, 318-320

Energy Convers. Mgmt Vol. 38, Suppl., pp. S343–S348, 1997
© 1997 Elsevier Science Ltd. All rights reserved
Printed in Great Britain
0196-8904/97 $17.00 + 0.00

Pergamon

PII: S0196-8904(96)00292-0

NEAR FIELD IMPACTS OF REDUCED pH FROM OCEAN CO$_2$ DISPOSAL

JENNIFER A. CAULFIELD, DAVID I. AUERBACH, E. ERIC ADAMS, HOWARD J. HERZOG

Massachusetts Institute of Technology, 77 Massachusetts Avenue, Cambridge, MA 02139 USA

ABSTRACT

A methodology has been developed to quantify mortality suffered by marine zooplankton passing through a CO$_2$-enriched sea water plume. Predicted impact depends on the mode of injection with scenarios which disperse the CO$_2$ showing least impact. Benthic impacts also depend on injection mode, with localized effects expected for any scenario in which the plume contacts the bottom. Effects of scale were found whereby the predicted impact from 10 power plants is greater than 10 times the impact of one plant at the same site. Based on available data, our modeling suggests that mortality associated with exposure to low pH can be avoided by properly dispersing the CO$_2$ and keeping the plume off of the seabed. © 1997 Elsevier Science Ltd

APPROACH

Ocean disposal of CO$_2$ will only make sense if the environmental impacts to the ocean are significantly less than the avoided impacts of atmospheric release. Here we address the near field impacts associated with decreased pH, and compare results for different methods of CO$_2$ injection. Our work is part of an MIT study performed for US DOE (Adams and Herzog, 1996), and more information can be obtained in that report or in Auerbach et al. (1996) and Caulfield et al. (1996).

Near Field Perturbations

The most significant impact from ocean disposal of CO$_2$ will derive from the lowering of pH which results from the reaction of CO$_2$ with water. Near the discharge, impacts will depend on the mode of CO$_2$ discharge as well as the loading. We have evaluated five discharge scenarios:

- dry ice injected at the ocean surface from a stationary ship.

- liquid CO$_2$ injected between depths of 1000 and 1500 m from a pipe towed by a moving ship and forming a droplet plume.

- liquid CO$_2$ injected at a depth of 1000 m from a bottom pipe and forming a droplet plume.

- dense CO$_2$-seawater mixture injected at a depth of about 700 m forming a bottom gravity current which ultimately places the CO$_2$-enriched seawater at a depth of 1000 m.

- liquid CO$_2$ introduced to a seafloor depression forming a "deep lake" at a depth of about 4000 m.

Mathematical models were developed to simulate the near-field physical-chemical perturbations caused by each of these discharge scenarios. Results have been referenced to a conventional 500 MW_e coal-fired station with 90% capture efficiency and a 20% energy penalty, which results in CO_2 emissions to the ocean of 130 kg/s. The models are summarized below.

For the *dry ice scenario*, each standard plant would produce a cube, 3 m on a side, every 5 minutes; cubes were assumed to be released continuously from a stationary ship in 4000 m of water into a uniform current of 5 cm/s. Cube fall velocity and dissolution rate were based on measurements by Nakashiki *et al.* (1991) and a wake-induced diffusivity was added to ambient horizontal diffusion based on Okubo (1971). Undissolved dry ice reaching the seabed was assumed to melt and then behave like a deep lake.

A *towed pipe scenario* has been suggested by a number of Japanese researchers (Ohsumi, 1995). We assumed that liquid CO_2 would be released continuously over a depth of 1000 to 1500 m from a ship traveling at 5 m/s. Near-field mixing due to pipe-induced turbulence was added to ambient horizontal diffusion.

For the *droplet plume scenario* we considered CO_2 injection from ten closely-spaced nozzles per plant emanating from a bottom-mounted pipe at a depth of 1000 m in an ambient current of 5 cm/s. Near-field mixing from each nozzle was simulated by modifying the droplet plume model of Liro *et al.* (1992). Lateral spreading due to plume intrusion in the stratified ocean was included, and ambient diffusion was simulated based on modification of Brooks' model (1960).

The *dense plume scenario* follows the suggestion of Haugan and Drange (1992) and presumes that a method is provided to concentrate the CO_2, e.g., through use of a mixing vessel (Adams *et al.*, 1995) or a gas lift system (Kajishima *et al.*, 1995), such that the resulting CO_2-enriched seawater sinks as a gravity current. Our gravity current simulations, based on a modification of Drange *et al.* (1993) for application to submarine canyons, suggest that a concentrated release at a depth of 700 to 800 m could reach 1000 m, at which depth the enriched seawater would lift off the bottom and behave similarly to a droplet plume.

The *deep lake scenario* envisions liquid CO_2, with density slightly greater than ambient seawater, forming a stable "lake" within a seafloor depression. We simulated the Nishinosima Trough, located off the coast of Japan, using topographic data provided by T. Oshumi (CRIEPI, Abiko, Japan, personal communication). Mass transfer through the hydrate film expected to form at the lake interface was simulated using experimental results from Hirai *et al.* (1996) based on an overlying current speed of 3 cm/s. Both scale-dependent horizontal diffusion and a constant vertical diffusion of 100 cm^2/s (based on Nakashiki and Ohsumi, 1993) were simulated using a modification of Brooks' model. Initially, the lake filling rate would exceed the rate of CO_2 dissolution, meaning the flux of CO_2 to the water column would be less than for the other scenarios (i.e., the CO_2 would be partially isolated). We evaluated conditions after 50 years, when the simulated rate of dissolution is about 90% of the filling rate for one plant and a little less than 50% for ten plants.

CO_2 concentration distributions predicted for each scenario were converted to near field maps of absolute pH and pH change relative to ambient. Figure 1 illustrates the former for the one plant droplet plume. The near field perturbations, indicated by the minimum pH outside of the immediate plume area, and the volume of water with pH < 7 are summarized in Figure 2. Results clearly indicate that those scenarios designed to disperse the CO_2 during injection, e.g., the dry ice and the towed pipe, have the least impact. By contrast, the droplet and dense plumes show greater impact because the diluted seawater is forced to intrude into a relatively narrow (about 20 m) layer in the stratified ambient. Indeed the dense plume, because it must rely on a *concentrated* CO_2-seawater solution, has the greatest impact. It should be emphasized that these results pertain to base case parameter assumptions and that a wide range of results have been obtained with different parameter values. For example, in the droplet plume scenario, it is possible to disperse the liquid CO_2 by releasing it over an extended portion of the pipe, thereby lowering its impact.

Biological Impacts

Available literature on impacts of low pH on marine organisms were reviewed. We focused on zooplankton and benthos, excluding nekton (because we assumed they would avoid regions of stress) and phytoplankton (because the CO_2 from most scenarios would be sequestered at depths below significant phytoplankton abundance). Figure 3 summarizes data on mortality, indicating strong dependence on both pH and exposure time. Data from this figure were used to fit isomortality curves (e.g., an LC_{50} curve connects data for 50% mortality). To account for reproductive and sublethal effects, and the fact that deep-water organisms may be more sensitive to pH than the shallow organisms used in most studies, the curves were lowered by 0.25 pH units.

Since organisms in the field would experience a time-varying pH, we discretized the varying experience into time intervals of constant pH which could be added on the assumption that discrete experiences producing the same individual mortality are equivalent. The relative position of an organism to the centerline of plume concentration will also vary with time due to ambient turbulence. This variability was handled through a Monte Carlo approach in which individual organism paths were simulated with a random walk routine based on the same scale-dependent diffusion (Okubo, 1971) used to describe plume spreading. To account for the fact that relatively unstressed organisms will reproduce, allowing depleted populations to recover once the stress is removed, a logistic equation for density-limited population growth was used based on a growth rate for copepods. Population recovery was allowed to begin when the pH reached 7.35, with full recovery at pH 7.5. As an example, Figure 4 shows a spatial map of zooplankton mortality for the one plant droplet plume. Results for all scenarios are summarized in Figure 5 which displays integrated mortality (the equivalent volume of seawater with 100% mortality) and the peak spatial deficit. Note the non-linear scale dependence: the magnitude of the mortality from ten plants can be more than 100 times greater than that from one plant, again implying that near-field impacts can be reduced by dispersing the emissions. Qualitatively, the results concerning mortality reinforce those made based on physical perturbation -- i.e., those scenarios that disperse the CO_2 have least impact -- but the quantitative comparisons are different.

In addition to water column effects, the deep lake, dry ice and dense plume scenarios will impact benthic communities. The towed pipe scenario targets a range of water depths and hence is able to avoid impact at the ocean floor. As modeled in this study, the droplet plume also distributes CO_2 only to intermediate water depths; however, depending on design and location, a sinking droplet plume could impact the ocean floor. We assumed that lowered pH will affect benthic organisms according to Figure 3. Because benthic organisms are stationary, the toxicities were evaluated by extrapolating the curves to 1000 hours. Results are displayed in Figure 6. Areas of bottom impact shown in Figure 6 are based on an integration of area as a function of pH weighted by the fractional mortality of organisms at that pH. In addition, for the deep lake, the bottom area of the lake is also included.

CONCLUDING COMMENTS

The major environmental impact of ocean CO_2 disposal would derive from the lowering of pH caused by the reaction of CO_2 with seawater. Depending on the method of release, pH can be reduced to as little as about 4 very near the injection point from its ambient value of about 8.

Impacts would be felt primarily by those organisms which reside at depths of 1000 m or greater and are unable to avoid regions of low pH near the point of injection (e.g., zooplankton, bacteria and benthos). In comparison with the size of the world's oceans, the areas of acute impact would be very small for all injection scenarios. Nonetheless, there would be significant differences among scenarios, with the least impact predicted from those methods which disperse the CO_2 as it is being injected and which keep it off the seabed. Based on available data, it appears possible to design a discharge system which effectively eliminates mortality by spreading the CO_2 sufficiently in either the vertical or horizontal directions, e.g., by releasing it from either an inclined pipe towed by a moving ship or from multiple ports distributed along a fixed pipe laid on the seafloor. The precautionary principle suggests that future R&D efforts be focused on such methods.

An important aspect not addressed in this paper is the far field impact of ocean CO_2 disposal. At basin to global scales, the impacts are less dependent on the type of release, but may reflect the release location (Dewey, *et al.*, 1997) Even with business-as-usual emissions of CO_2 to the atmosphere, the average pH of the world's oceans may ultimately decline by about one-half of a pH unit as our reserves of fossil fuels are consumed (Herzog *et al.*, 1995). Direct injection of CO_2 will affect these perturbations by slightly increasing the pH change at depth, while slightly mitigating the change in the surface waters (Haugan and Drange, 1995). While such perturbations are below thresholds of acute toxicity identified in our near field studies, they could lead to chronic impacts involving species reproduction, community structure, etc., which may be difficult to quantify.

ACKNOWLEDGMENT

This study was supported by the U.S. Dept. of Energy, Pittsburgh Energy Technology Center, under Grant No. DE-FG22-94PC227. The authors are grateful for the assistance provided by Perry Bergman and Janice Murphy of PETC.

REFERENCES

Adams, E. and H. Herzog. (1996)" Environmental Impacts of Ocean Disposal of CO_2", Final Report, DOE Grant No. DE-FG22-94PC227., MIT Energy Lab.

Adams, E., D. Golomb, X. Zhang and H. Herzog (1995). "Confined release of CO_2 into shallow seawater" In: *Direct Ocean Disposal of Carbon Dioxide* (Handa and Ohsumi, ed.) pp153-164.

Auerbach, D. (1996) "Global warming mitigation via ocean disposal of power plant-generated CO_2: an environmental and political analysis" MS thesis, Dept. of Civil and Environ. Engrg., MIT, Cambridge, MA.

Brooks, N. (1960) "Diffusion of sewage effluent in an ocean current" In: *1st Int'l Conf on Waste Disposal in the Marine Environment,* (Pergaman Press) pp. 246-267.

Caulfield, J. (1996) "Environmental impacts of carbon dioxide ocean disposal: plume predictions and time dependent organism experience", MS thesis, Dept. of Civil and Environ. Engrg., MIT, Cambridge, MA.

Dewey, R.K., G.R. Stegen and R. Bacastow (1997) "Far-field impacts associated with ocean disposal of CO_2" *Energy Convers. Mgmt* (this volume).

Drange, H., G. Alendal and P. Haugan (1993) "A bottom gravity current model for CO_2-enriched seawater", *Energy Convers. Mgmt* 34:1065-1072.

Haugen P.M. and H. Drange, "Sequestration of CO_2 in the Deep Ocean by Shallow Injection," *Nature*, 357, pp. 318-20 (1992).

Haugan, P. and H. Drange (1995) "Effects of CO_2 on the ocean environment", *Energy Convers. Mgmt* 37:1019-1022.

Herzog, H., E. Adams, D. Auerbach and J. Caulfield (1995) "Technology assessment of CO_2 ocean disposal", Report MIT-EL 95-001, MIT Energy Lab., Cambridge, MA.

Hirai, S., K. Okazaki, T. Yutaka, K. Hijikata and Y. Mori (1996) "Dissolution rate of liquid CO_2 in pressurized water flow and effect of clathrate film", *Energy Int'l Journal* (in press).

Kajishima, T., T. Saito, R. Nagaosa and H. Hatano (1995) "A gas-lift system for CO_2 release into shallow seawater". *Energy Convers. Mgmt* 36:467-470.

Liro, C., E. Adams and H. Herzog (1992) "Modeling the release of CO_2 in the deep ocean", *Energy Convers. Mgmt* 33:667-674.

Nakashiki, N. and T. Ohsumi (1993) "Simulation of the near field behavior of CO_2 from a source on the deep-sea floor. In: *Proc Int'l Symp.on CO_2 Fixation and Efficient Utilization of Energy*: 331-336.

Nakashiki, N., T. Ohsumi and K. Shitashima (1991) "Sequestering of in a deep-ocean--fall velocity and dissolution rate of solid CO$_2$ in the ocean" CRIEPI Report (EU 91003).

Ohsumi, T. (1995) "CO$_2$ disposal options in the deep sea", *Marine Technology Society Journal*, 29(3): 58-66.

Okubo, A. (1971). "Oceanic diffusion diagrams", *Deep Sea Research* 18: 789-802.

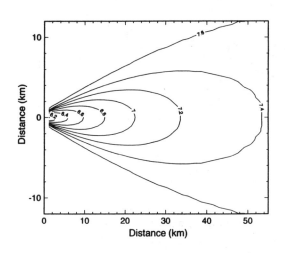

Fig. 1: pH distribution at 1000 m for one-plant droplet plume

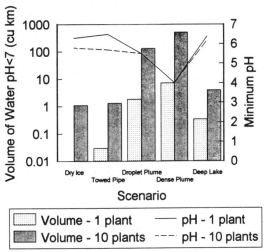

Fig. 2: Summary of near field perturbations

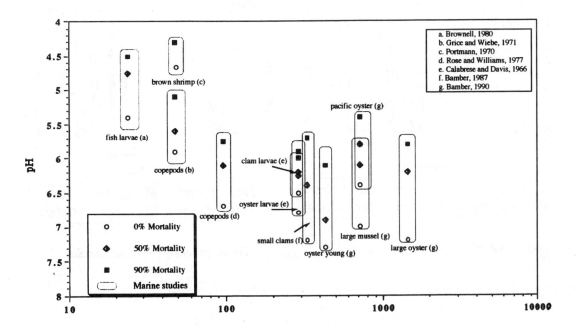

Fig. 3: Mortality of marine organisms (zooplankton and benthos) due to low pH

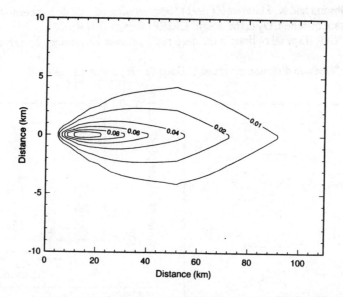

Fig. 4: Zooplankton deficit due to mortality for one-plant droplet plume

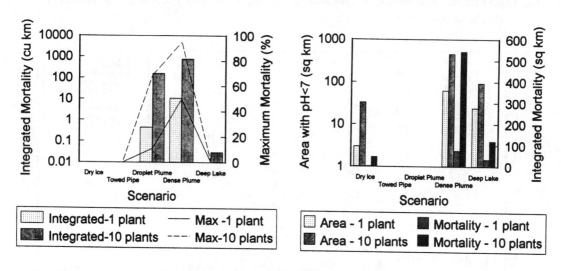

Fig. 5: Summary of water column mortality   Fig. 6: Summary of benthic mortality

Pergamon

Energy Convers. Mgmt Vol. 38, Suppl., pp. S349–S354, 1997
© 1997 Elsevier Science Ltd. All rights reserved
Printed in Great Britain
0196-8904/97 $17.00 + 0.00

PII: S0196-8904(96)00293-2

Far-Field Impacts Associated With Ocean Disposal of CO$_2$

Richard K. Dewey[1], Gilbert R. Stegen[2] and Robert Bacastow[3]

[1]CEOR, University of Victoria, P.O. Box 1700, BC, Canada V8W 2Y2
[2]SAIC, 13400B Northup Way #36, Bellevue WA 98005
[3]Scripps Institution of Oceanography, La Jolla, CA 92093

ABSTRACT

We present predictions of the far-field and long-term environmental impacts that might be associated with ocean disposal of CO$_2$ using two detailed ocean circulation models. The Mesoscale Ocean Dispersion Model (MODM) simulates the transport and dispersion of injected CO$_2$ at spatial scales between 25-2000 km, and over periods between days to 10 years. The Global Carbon Cycle Model (GCCM) predicts both the ocean distribution of CO$_2$ and the atmospheric concentration beyond several centuries. The models have been used to evaluate differences between injection locations with regard to the short- and long-term sequestration of the injected CO$_2$. The results presented here suggest that for injection sites along the east coast of the United States and south east of Japan, the Gulf Stream and Kuroshio, respectively, dominate the initial fate of the injected CO$_2$, while long-term sequestration efficiencies are characterized by deep-water formation in the north Atlantic and slow diffusive upwelling in the North Pacific, respectively. © 1997 Elsevier Science Ltd

Keywords
Carbon Dioxide, CO$_2$, ocean sequestration, injection, disposal, impacts, carbon cycle, dispersion, modelling.

Introduction

The modelling analyses presented here will attempt to predict changes in the local concentration of CO$_2$ in the ocean and the global concentration of atmospheric CO$_2$, including contributions from other chemical compounds important to the carbon cycle, from which ecosystem and climate analyses may be inferred. In particular, the concentration of CO$_2$ in both the ocean and atmosphere will be evaluated for various ocean disposal scenarios, from which first order assessments of the long-term and far-field efficiency of ocean disposal of CO$_2$ can be made. Two models will be described, one designed to simulate the advection and dispersion of CO$_2$ at oceanic mesoscales (25-2000 km), the second to simulate the influence of ocean circulations and carbon cycling of CO$_2$ over global scales and intervals exceeding several centuries. Additional results and analyses are presented in Adams and Herzog (1996).

Mesoscale Ocean Dispersion Modelling

A wide range of oceanographic and hydrodynamic processes influence the advection and dispersion of material discharged into the stratified ocean. The temporal and spatial scales associated with oceanic dispersion vary from seconds to decades, and from centimeters to thousands of kilometers, respectively. To first order, the problem can be broken down into two dominant processes: advection by the virtually unbounded mean flow and diffusion by turbulence and eddies (non-mean flow components). The development of a global, or even regional ocean model which attempts to include numerical solutions to both the large and small scale processes affecting the dispersion of material is a difficult problem to both formulate and execute computationally. Our approach is to utilize an existing 3-dimensional ocean velocity field from the advanced Parallel Ocean Circulation Model (POCM) of the Naval Postgraduate School (Monterey CA) and NCAR (Semtner and Chervin, 1992), and combine this "mean flow field" with a stochastically driven turbulent velocity field. Within this turbulent, 3-dimensional, time-varying ocean we track neutrally buoyant particles representing clouds of dissolved CO$_2$. Injection sites throughout the global ocean can be investigated, and the resulting CO$_2$ plume(s) can be monitored and evaluated.

The Mesoscale Ocean Dispersion Model (MODM) is ideally suited for determining the intermediate environmental impacts associated with ocean sequestration of CO$_2$ at time scales of days to years and spatial scales between 25 and

2000 km (meso-scales). "Clouds" of CO$_2$ are continuously released at user specified injection site(s) and tracked as they are advected and mixed by a seasonally varying turbulent ocean. The distribution and concentration of injected CO$_2$ is monitored. The monthly POCM data has a $\frac{1}{2}°$x$\frac{1}{2}°$ horizontal resolution in longitude and latitude, with 20 vertical layers, more closely spaced near the ocean surface. We integrate these POCM velocity fields with locally scaled stochastic turbulent velocities (u',v',w'). Neutrally buoyant clouds of dissolved CO$_2$ are tracked as they are advected by the velocity interpolated to an exact geographic location by an oceanic current [u,v,w]=[U+u',V+v',W+w'], where [U, V, W] are the spatially and temporally interpolated velocities derived from the monthly POCM velocity data, and [u',v',w'] are the local stochastic turbulent velocities. Although the "mean" velocity field is gridded at the resolution set by the POCM model, the velocity of each CO$_2$ cloud is spatially interpolated to an exact position, so CO$_2$ is free to move anywhere in the ocean. A limited model domain (approximately 2500km x 2500km) is established around each injection site for computational purposes.

The turbulent velocities represent the influence of sub-grid scale processes not resolved by the POCM. The horizontal turbulent velocities (u',v') are modeled as a first order Markov process (Sawford, 1984; Bennett, 1987) using the Langevin equation,

$$u'(x,y,z,t+\Delta t) = \left(1 - \frac{\Delta t}{T_L(y)}\right) u'(x,y,z,t) + \sigma_u(x,y,z,t) \left(\frac{2\Delta t}{T_L(y)}\right)^{1/2} \xi, \tag{1}$$

where Δt is the dispersion model time step (4 hours), $T_L(y)$ is the local Lagrangian integral time scale of the turbulence (=1/2 local inertial period), σ_u^2 is the local turbulent kinetic energy, and ξ is a random number drawn independently from a Gaussian distribution with zero mean and unit variance. The Langevin equation can result in the "false" accumulation of particles in regions of the domain were both the mean and turbulent velocities are very weak, trapping the particles indefinitely. To alleviate this, a small horizontal up-gradient "drift" velocity is added to the turbulent components in regions of very weak mean flow (< 1 cm/s) to avoid false accumulation of material. The local turbulent kinetic energy (TKE) is parameterized to be a fraction (i.e. 5 %) of the local mean kinetic energy. The vertical component of the turbulent velocity field (w') is derived separately by scaling the square root of the TKE by the local Richardson number, $w'(x,y,z,t)=\sigma_u (\xi/4Ri)$, where ξ is a new random number and $Ri(x,y,z,t)$ is calculated from the local density and mean velocity fields.

A primary feature of the CO$_2$ plumes generated using the particle tracking technique is the apparent confinement of material by the dominant flow characteristics of the mean circulation. Unlike finite difference gradient diffusion models, where material is forced to diffusion uniformly in all directions, the Lagrangian formulation constrains the effective dispersion to realistic path-lines in the flow. The stochastic dispersion given by (1), in good agreement with observed oceanic dispersion (Ledwell, *et al.* 1993), primarily acts to migrate material across path-lines, which may subsequently diverge or converge. The dominant path-lines (time varying streamlines) of the flow act to limit the dispersion to specific regions and patterns, much as ocean drifter experiments have found that Lagrangian advection tends to preferentially sample convergent regions of the real ocean (Davis, 1991). Various model validation tests have been performed which indicate the model is well behaved statistically and reproduces realistic oceanic plumes. It is assumed that the injected liquid CO$_2$ has dissolved into solution with the ambient seawater, and that the resulting solution has reached a depth of neutral buoyancy. The injection rate at each site is 1.119×10^{-3} GtC/year, or 1.26 m^3/s of liquid CO$_2$. This amount of CO$_2$ represents the CO$_2$ from one 500 MWe power plant configured to capture CO$_2$ with a 90% efficiency.

Injection sites east of New York, Washington DC, Cape Hateras, and Miami were evaluated, with injection depths of 1000, 1500, 1700, and 2000m. The injected CO$_2$ was advected by several dominant ocean circulations in the region, and the resulting plumes share many characteristics. The most significant oceanic feature near these injection sites was the Gulf Stream circulation, with it's strong horizontal currents and regions of upwelling. The CO$_2$ from near-shore and shallower sites eventually enters into regions of local upwelling and portions rise towards the upper layers of the ocean. Examination of the Gulf Stream structure and historical data (Stommel, 1965) reveals that the isopycnals are tilted significantly upwards towards the northwest. Vertical velocities in the model (represented by mean flow along tilted isopycnal surfaces) in the vicinity of these injection sites are significant, but realistic (Hall, 1986), with magnitudes as large as 2.5 m/day (2.8×10^{-3} cm/s). Consequently, a fraction (few %) of the CO$_2$ plume from most injection sites rises towards the upper ocean. For deeper injections (i.e. 2000m) only traces (1%) of the CO$_2$ would enter into the upwelling regions, the majority would enter a southward undercurrent and vent out of the southern wall of the model domain into the Equatorial Atlantic. A characteristic of this deep, steady (3 cm/s) southward stream was the confinement of the CO$_2$ to a narrow filament, only a few tens of kilometers wide but over

2000 kilometers long. Despite the "turbulent" dispersion apparent in the more divergent upper circulation, this deep steady flow exhibits very little dispersion. Similar non-dispersive regions of the ocean are more plentiful at depth (i.e. > 2000m) away from boundaries and convective circulations.

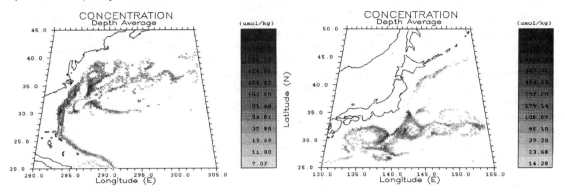

Figure 1 a) Example CO₂ plume (depth averaged) from injections east of Cape Hateras at depths of 1500, 1700 and 2000m after 3 years. b) Injection of CO₂ at 1000m south of Tokyo (32N, 135E) after 3 years. Injection rate for both runs was 1.26 m³/s.

Shown in Fig. 1a is a depth averaged CO₂ plume from an injection seaward of Cape Hatteras, where the continental slope is nearest to the coast. For this simulation, the total injection rate of 1.26 m³CO₂/s was divided into three separate injections and allowed to diffuse from progressively deeper sites in a line east of Cape Hatteras (at 1500, 1700, and 2000m). After three years, both eastward and southward plume extensions are evident, with an approximate partition of 1/3 of the CO₂ heading towards the northeast, 1/3 towards the south (within the under current) and 1/3 remaining trapped within the "slowly" dispersion eddy fields associated with meanders in the Gulf Stream. The minimum and maximum depths of penetration of injected CO₂ were 178 and 2208 m, respectively.

For comparison, an injection site was selected south of Japan, near the Kuroshio current, in water 2750 m deep (Fig 1b). The injected CO₂ is advected and recirculates within a complex, time-varying flow. Plume separations during the seasonal shifts in the Kuroshio result in isolated patches and detached filaments of CO₂. Here, portions of the CO₂ are advected eastward within the main Kuroshio extension, while significant amounts of CO₂ recirculate on an annual time scale back towards the initial injection longitude and latitude. For this injection site and depth (1000 m) minimum and maximum plume depths were 518 and 1620 m, respectively.

Global Carbon Cycle Modelling

The Global Carbon Cycle Model (GCCM), with a realistic circulation ocean, a well mixed atmosphere, and various oceanic carbon cycle elements was applied to evaluate the far-field environmental impacts associated with ocean disposal of CO₂ on time scales out to 500 years. The GCCM is based on the ocean circulation model of Mairer-Riemer and Hasselmann (1987). The addition of annual biological carbon cycle components is described by Bacastow and Mairer-Riemer (1990). The model consists of two primary components; the ocean circulation model driven by seasonal climatology (wind, temperature, salinity) and thermohaline dynamics, and the oceanic carbon cycle components which integrate the sequestered CO₂ as dissolved inorganic carbon. The global distribution of CO₂ is most influenced by the natural ability of the oceans to advect, convect, and mix carbon and nutrients, and to establish fertile biological regions for primary productivity. The model consists of a 72 x 72 grid covering the entire globe, divided into 15 vertical layers. Average grid spacing is 2.5° x 5° in latitude and longitude, and vertical bins range between 100 and 1000m in depth, with thinner layers near the surface of the ocean. The basic carbon cycle is divided into 7 main elements:

1) Anthropogenic CO₂ emitted into the atmosphere and ocean (during sequestration).
2) Air/sea exchange of CO₂ based on local partial pressures, location, and ice cover.
3) Biological assimilation (photosynthesis) of carbon and nutrients and the generation of local concentrations of dissolved organic carbon (DOC), particulate organic carbon (POC) and calcium carbonate CaCO₃.
4) Particulate (POC and CaCO3) sinking from the upper oceanic layers.
5) Ocean advection and convection of all dissolved components (including nutrients/upwelling).

6) Remineralization of organic carbon (DOC and POC) below the surface layers.

7) $CaCO_3$ dissolution at depths of undersaturation.

The advection, convection and mixing of carbon by ocean currents depends on the magnitude, direction and structure of the 3-dimensional current field which, in turn, is driven by atmospheric and thermohaline forcing. Biological assimilation of carbon occurs primarily in the near surface euphotic layer, and is a function of light intensity and nutrient availability. Light intensity is a function of latitude and season, while nutrient availability depends on previous biological uptake (Redfield ratios), remineralization, and oceanic transport of carbon and nutrients. Experience with the model indicates that much of the structure and variability in the global carbon cycle system can be reproduced (Bacastow and Maier-Reimer, 1990). A known limitation of the present model formulation is the omission of dissolution of $CaCO_3$ rich sediments, although dissolution of in situ $CaCO_3$ during photosynthesis is included. Inclusion would allow additional CO_2 buffering in the deep ocean, and improve the sequestration predictions. This limitation becomes important at longer time scales (i.e. 100 years), and would be more important in the Atlantic than the Pacific due to the deep convection in the Atlantic which brings CO_2 in contact with the $CaCO_3$ rich Atlantic sediments.

The simulations described here track two primary components. First is the global anthropogenic CO_2 production, most of which is assumed to be discharged directly into the atmosphere, and second, the CO_2 associated with the sequestration process. The global production curve q(t) is matched to the observed anthropogenic CO_2 production up to 1990 and also to the rate of production during 1990, and is modelled as a logistics function. The total anthropogenic CO_2 production has been set to 4240 GtC is consistent with a symmetric production curve as hypothesized by Hidy and Spencer (1994). The sequestration schedule starts in the year 2000, and runs 100 years. The global carbon cycle model was initiated from a pre-industrial state, taken to be the year 1740, up to the year 2000. Ocean sequestration is assumed to begin in the year 2000 at a constant rate. For the following simulations, the CO_2 captured from multiple (50) coal burning 500 MWe electric power plants is assumed to be the source for the sequestered CO_2, with an ocean injection rate of 5.597×10^{-2} GtC/year (2.051×10^{11} kg CO_2/year) at two separate oceanic locations (off New York and Tokyo). The total CO_2 sequestered in the ocean is then 1.1194×10^{-1} GtC/year (4.102×10^{11} kg CO_2/year). For comparison, the estimated 1990 global production of CO_2 was 6.01 GtC/year (2.2×10^{13} kg CO_2/year). The sequestered amount of CO_2 represents less than 2% (1.86%) of the global production in 1990, and by the year 2090 when global production is predicted to reach levels of ~32 GtC/year (1.17×10^{14} kg CO_2/year), this sequestration level represents only 0.34% of the global CO_2 production. Therefore, it should be noted that the present simulations sequester a very small fraction of the global CO_2 emissions, and that any real sequestration program would likely inject an increasing amount, or fraction (%) of the anthropogenic CO_2.

Three model simulations were conducted: 1) No ocean sequestration, all anthropogenic CO_2 is discharged into the atmosphere; 2) Ocean sequestration off New York; and 3) Ocean sequestration off Tokyo. Due to the relatively small amount of CO_2 sequestered, a combined ocean sequestration is obtained by summing the New York and Tokyo results. The depth of ocean sequestration was set at 1000m, which represents a GCCM model layer with interfaces at 900 and 1500 m. Due to the energy expended capturing and transporting the CO_2 to the sequestration sites, the "no ocean sequestration" simulation generates less CO_2 per MWe, and therefore actually discharges less CO_2 into the ocean/atmosphere system. Consequently, the "no ocean sequestration" simulation tracks 26% less "sequestration" CO_2 than do the "ocean sequestration" simulations.

Shown in Fig. 2a are the time histories of the amount of ocean sequestered CO_2 returned to the atmosphere in units of ppm (parts per million) for the control (no sequestration) and ocean sequestration simulations. The control shows a linear increase in atmospheric concentration of the captured CO_2 during the sequestration interval (2000 - 2099), with a net increase of approximately 4.15 ppm. Once the sequestration is stopped, the steady decline in the control case reveals the natural up-take of atmospheric CO_2 by the ocean. The ocean sequestration curve shows a much more gradual increase during the first 100 years, due partly to the small fraction of CO_2 discharged directly into the atmosphere plus CO_2 exchanged from the ocean surface. After year 2100 the ocean sequestration curve increases more slowly, revealing the buffering efficiency of the deep ocean. However, CO_2 continues to escape from the ocean into the atmosphere, and after 500 years, the two curves are only 0.6 ppm apart. At even longer time scales, when the ocean/atmosphere system has reached equilibrium (12% atmosphere/ 88% ocean), the curves will have crossed and the ocean sequestration case will result in more net CO_2 reaching the atmosphere. This is because capturing CO2 generates 26% more CO_2 per Mwe.

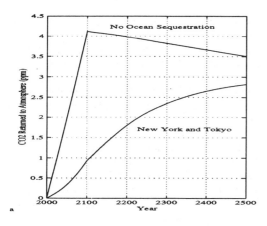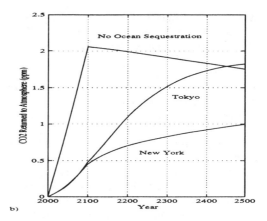

Figure 2 (a) Time histories of the injected CO_2 returned to the atmosphere (ppm) for the control run (no sequestration) and combined ocean injection simulations, and (b) each site (New York and Tokyo) separately.

Differences between New York and Tokyo become more apparent when the amount of CO_2 returned to the atmosphere from each site is examined independently (Fig. 2b). Although similar during the first 100 years, the period of actual ocean injection, the curves representing the amount of CO_2 returned to the atmosphere for the ocean sequestration cases for New York and Tokyo depart significantly after year 2100 (when ocean sequestration ceases). The diverging curves reveal differences between the local circulation and effective buffering capabilities of the north Atlantic and north Pacific. These curves clearly indicate that New York (north Atlantic) is a more efficient sequestration site than Tokyo (north Pacific) at time scales out to 500 years. However, it should be noted that after 200 years (year 2200) both injection sites buffer at least 55% of the CO_2 from returning the atmosphere, and in the case of New York, the return is less than 35%. In these simulations, the global production of CO_2 has nearly ceased by the year 2200. It is therefore more important to evaluate the buffering capabilities during this early period, when global CO_2 production is most vigorous.

The CO_2 injected at 1000 m off of New York enters into the Gulf Stream system, which advects most of the CO_2 into the central north Atlantic. Once the CO_2 reaches the high latitude regions of the Atlantic, a portion of the sequestered CO_2 is transported to great depth by deep water convection, a process resulting from sea-surface cooling. However, the nature of vertical convection (mixing) acts to simultaneously transport CO_2 both towards the surface and the bottom of the ocean. CO_2 that reaches the ocean surface is then allowed to exchange with the atmosphere, and after a year, is globally distributed and may return to the ocean through local air/sea exchange. The deep water convection transports a significant amount of CO_2 down to the bottom layers of the ocean, where it is advected by the deep (3000 m), southward thermohaline circulation. The CO_2 which enters the deep (3000 m) thermohaline circulation is efficiently buffered from the atmosphere, and is the primary reason for the difference between the amount of CO_2 returned to the atmosphere for New York and Tokyo.

The CO_2 that is injected off Tokyo enters another western boundary current, the Kuroshio. The Kuroshio advects and diffuses the CO_2 into the central north Pacific, much in the same way that the Gulf Stream transports "New York" CO_2 into the central north Atlantic. However, the Pacific does not extend as far north as does the Atlantic, and there are no regions of deep water formation and convection. In addition, the Pacific represents the "end" of the thermohaline conveyor belt (which initiates in the north Atlantic), and the general trend in the north Pacific is a gradual diffusion upwards. After 60 years the upper layers (< 1000 m) of the north Pacific contain most of the injected CO_2. Very little of the CO_2 is penetrating into deep regions of the ocean, except for small traces in the high latitude convective regions in the north and south Atlantic. After 300 years, 200 years after injection has stopped, the upper layers of the entire global ocean contain Tokyo sequestered CO_2. The CO_2 concentrations are lowest in the deep (3000m) Pacific, further indicating the presence of persistent upwelling and vertical diffusion there. The upward diffusion gradually transports the injected CO_2 towards the surface waters of the Pacific, where it vents into the atmosphere and causes an increase in atmospheric CO_2 concentrations. Ultimately, after 500 years the amount of CO_2 reaching the atmosphere from the Tokyo injection site exceeds that for an equivalent electric power generation without sequestration due to the energy penalty introduced by the capture and sequestration processes which consume energy. Despite this long term trend, the buffering capacity of the north Pacific after 200 years (year 2200) is still 45% better than the case of no sequestration, thereby alleviating some of the short-term impacts associated with pure atmospheric discharge of CO_2.

After 60 years, CO_2 injected off of New York has resulted in a global ocean surface layer increase of 0.01 μmole/kg. At depth, large horizontal gradients in CO_2 still exist, with maximum concentrations remaining near the injection site. After 300 years, global ocean surface layer concentrations have increased to 0.5 μmole/kg, while concentrations in the deep Atlantic are on the order 1.0 μmole/kg. Virtually no sequestered CO_2 reaches the deep (3000m) Pacific, where concentrations remain below 0.001 μmole/kg. For the injection simulation off of Tokyo, marginally higher surface layer concentrations persist after 300 years (0.6 to 0.9 μmole/kg), while deep Pacific concentration are still low (0.001 μmole/kg). Changes in the pH due to these concentration increases are predicted to be small (-0.1) even near the source where concentrations increase by 10 μmole/kg.

The participation of the injected CO_2 in the model's carbon cycle enhances the buffering predictions over non-carbon cycling models (i.e. MODM). Biological assimilation of the CO_2 through photosynthesis in the ocean surface layers is the single most important component of the carbon cycle in reducing atmospheric concentrations. Ocean biological production is governed by the availability of nutrients, which are replenished from below by diffusion and upwelling. Both dissolved and particulate organic carbon (DOC and POC) are formed and are advected away. In the case of POC, the material is allowed to sink below the surface layer, where it is remineralized. Approximately ⅓ of the flux is associated with sinking POC, while ⅔ is advected and diffused away as DOC. An important chemical change to sea water in response to elevated CO_2 levels is the loss of saturation conditions with respect to biogenic calcite (calcium carbonate: $CaCO_3$). Additional buffering would be realized with the inclusion of sediment ($CaCO_3$) dissolution in the model.

Conclusions

The model simulations presented here suggest that although deeper injection sites are more efficient at keeping the injected CO_2 away from the atmosphere, injections near strong circulation features such as the Gulf stream and Kuroshio may permit the CO_2 to entrain and up-well towards the surface layers of the ocean. Generally, CO_2 that is sequestered into the deep ocean (> 2000m), whether directly or indirectly through processes such as deep water formation, is effectively sequestered from contact with the atmosphere for many centuries.

References

Adams, E. and H.Herzog, 1996: Environmental Impacts of Ocean Disposal of CO2: Volume II, Chapter IV Far-Field Impacts, Department of Energy, (in press).

Bacastow, R. and E.Mairer-Reimer, 1990; Ocean-circulation model of the carbon cycle., Climate Dyn., **4**, 95-125.

Bennett, A. F., 1987; A Lagrangian analysis of turbulent diffusion, Reviews of Geophys., Vol **24**(4), 799-822.

Davis, R.E., 1991; Lagrangian ocean studies, Annu. Rev. Fluid Mech., **23**, 43-64.

Hall, M.M., 1986: Horizontal and vertical structure of the Gulf Stream velocity field at 68° W., J. Phys. Oceano., **16**, 1814-1828.

Hidy, G.M. and D.F.Spencer, 1994; Climate alteration: A global issue for the electric power industry in the 21st century., Energy Policy, **22**(12), 1005-1027.

Ledwell, J.R., A.J.Watson, and C.S.Law, 1993; Evidence of slow mixing across the pycnocline from an open-ocean tracer-release experiment., Nature, **364**, 701-703.

Mairer-Reimer, E. and K.Hasselmann, 1987; Transport and storage of CO_2 in the ocean - An inorganic ocean-circulation carbon cycle model, Climate Dyn., **2**, 63-90.

Sawford, B.L., 1984; The basis for, and some limitations of, the Langevin equation in atmospheric relative dispersion modelling, Atmospheric Environ., Vol **18**(11), 2405-2411.

Semtner, A. J., and R. M. Chervin, 1992; Ocean general circulation from a global eddy-resolving model., J. Geophys. Res., Vol **97**(C4), 5493-5550.

Stommel, H., 1965; *The Gulf Stream: A physical and dynamics description.*, 2nd Edition, Uni.of Califonia Press, Berkeley.

Pergamon

Energy Convers. Mgmt Vol. 38, Suppl., pp. S355–S360, 1997
© 1997 Elsevier Science Ltd. All rights reserved
Printed in Great Britain
0196-8904/97 $17.00 + 0.00

PII: S0196-8904(96)00294-4

Dispersion of CO_2 Injected into the Ocean at the Intermediate Depth

Norikazu Nakashiki and Takashi Ohsumi

Abiko Research Laboratory, Central Research Institute of Electric Power Industry,
1646 Abiko, Abiko-City, Chiba 270-11, Japan
(nakasiki@criepi.denken.or.jp, ohsumi@criepi.denken.or.jp)

ABSTRACT - The biological impact on the ambient marine environment should be assessed in the CO_2 injection at the intermediate depth. The long term dispersion (several decades) will also have a potential impact on the marine ecosystem in the wide area. The velocity field and dispersion of the substance in the North Pacific ocean at the intermediate depth were computed using an OGCM, and discussed with the comparison with the observed temperature, salinity and CFCs profiles. A model was applied to estimate the diffusion of the artificially injected CO_2. © 1997 Elsevier Science Ltd

1. INTRODUCTION

Ocean disposal of the collected CO_2 is a promising option for isolating the large amount of anthropogenic CO_2 from the atmosphere. The effectiveness of isolation (i.e., retention time) for the CO_2 injected at the North Pacific intermediate deep water was discussed by Nakashiki and Hikita (1994). The long-term transportation of the water mass containing the CO_2 injected at the North Pacific intermediate water was studied using an ocean general circulation model (OGCM), which suggested that the CO_2 storage at the intermediate depth could provide an effective measure of isolation of the collected CO_2 for at least 200 years.

The biological impact in the ambient marine environment also should be assessed on the CO_2 injection at the intermediate depth. Although dispersed by the oceanic turbulence, the CO_2 remains in the water mass of the injection point with a relatively high concentration. Figure 1 shows the schematic diagram of the relation between time, space (i.e.,diffusion area) scales and the modeling approaches. For the prediction of CO_2 concentration in the sub-grid scale near the injection point, the particle trace method with the random walk process could be applied to on the simulated OGCM results of the oceanic flow obtained in the coarse grid. The long term dispersion (several decades) will make the concentration of injected CO_2 lower, but it could have potential impact on the marine ecosystem in large area. It is difficult to apply the particle trace method on the long term dispersion of CO_2, because the huge number of the particle is needed to properly estimate the low concentration area. A regional model nested with OGCM seems to be useful for the estimation for such long term diffusion of CO_2. The nested model approach between a regional model and an OGCM has been presented by Nakashiki (1995) on the computation of the flow fields.

In this study, we mainly concern with the computation of the dispersion of substance at the intermediate depth in the ocean. First, an OGCM with isopycnal diffusion is applied to calculate the profiles of chlorofluorocarbon (CFC-11 and CFC-12) which are the artificial and stable tracer permeating through the sea surface into the ocean. The model performance was discussed in the comparison between the observation and the computational result. Second, concentrations of CO_2 injected North Pacific Ocean was calculated over 20 years. Computed distribution of CO_2 injected at the intermediate depth in the ocean, were compared with the concentration and variation of the dissolved CO_2 under the natural conditions, to discuss the potential impact of CO_2 storage into the ocean.

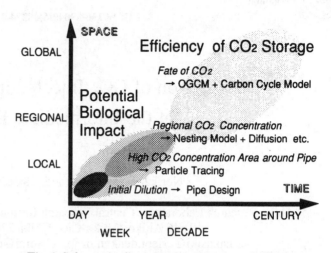

Fig.1 Schematic diagram showing the relation between time, space scales and the modeling approaches

2. MODEL DESCRIPTION AND COMPUTATIONAL RESULTS

The prognostic OGCM (Nakashiki, 1995) was employed to calculate the annual mean flow fields, using the temperature, salinity data (NOAA WOA, 1994) and wind field (Hellerman and Rosenstein, 1983). The resolution of the computational mesh were $2^0 \times 2^0$ and 10 - 500m in the horizontal and vertical directions, respectively. In this study, the isopycnal mixing effect (Redi, 1982) where substances are diffused and transported along the density surface, was adopted to compute CFCs or CO_2 dispersion at the intermediate depth of the ocean. The diffusivity tensor was written as follows.

$$K^g = \frac{A_H}{\left(1+\delta^2\right)} \times \begin{bmatrix} 1+\dfrac{\rho_y^2 + \varepsilon\rho_x^2}{\rho_z^2} & (\varepsilon-1)\dfrac{\rho_x\rho_y}{\rho_z^2} & (\varepsilon-1)\dfrac{\rho_x}{\rho_z} \\[2ex] (\varepsilon-1)\dfrac{\rho_x\rho_y}{\rho_z^2} & 1+\dfrac{\rho_x^2 + \varepsilon\rho_y^2}{\rho_z^2} & (\varepsilon-1)\dfrac{\rho_y}{\rho_z} \\[2ex] (\varepsilon-1)\dfrac{\rho_x}{\rho_z} & (\varepsilon-1)\dfrac{\rho_y}{\rho_z} & \varepsilon+\delta^2 \end{bmatrix} , \quad where \quad \delta = \frac{\left(\rho_x^2+\rho_y^2\right)^{1/2}}{\rho_z} \qquad - (1)$$

In Eq.(1), the subscripts x, y and z denote the density gradient in horizontal and vertical directions. A_H is the isopycnal diffusivity, and $\varepsilon = K_H/A_H$ (K_H: the diapycnal diffusivity) is a measure of the relatively weak vertical cross-isopycnal mixing. In this study, A_H and ε were employed as 5×10^5 cm²/s and 2×10^{-6}, respectively.

2.1 Calculation of the CFCs Distribution

To discuss the validity of the model, the dispersion of CFCs in the ocean was computed, where CFCs are the artificial stable passive tracer, and dissolved into the ocean only through the sea surface. Atmospheric concentrations of CFCs were well understood (for example, Haine and Richards, 1995), and the uptake of CFCs was estimated using "piston velocity" based on the sea surface wind and solubility of CFCs (Liss and Merlivat, 1986).

$$Flux = k \times \left(F\chi' - C\right) , \quad k = (2.85U - 9.65)\sqrt{\frac{S_c\left(CO_2, 20°C\right)}{S_c\left(CFCs, T\right)}} \qquad - (2)$$

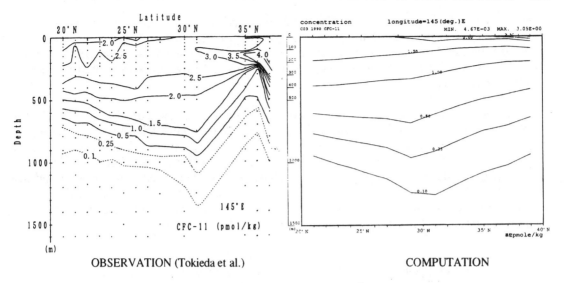

OBSERVATION (Tokieda et al.)   COMPUTATION

Fig. 2-1 CFC-11 profiles at $145°$E section

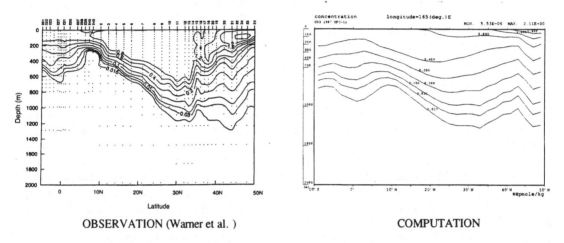

OBSERVATION (Warner et al.)   COMPUTATION

Fig. 2-2 CFC-12 profiles at $165°$E section

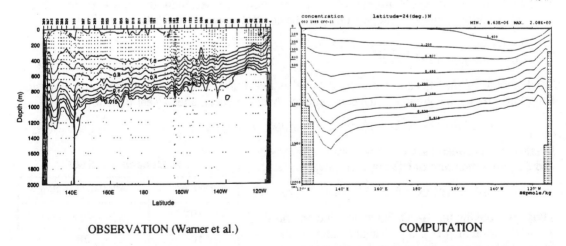

OBSERVATION (Warner et al.)   COMPUTATION

Fig. 2-3 CFC-11 profiles at $24°$N section

where F is a solubility function of CFCs (Warner and Weiss, 1985), χ' is atmospheric concentration, C is a concentration at sea surface, U is a wind speed and S_C is a Schmidt number. In this study, the observed annual mean wind field (Hellerman and Rosenstein, 1983) and the computed results of the annual mean temperature, salinity and flow fields were used to calculate the CFCs in the ocean.

Figure 2-1 shows the calculated distribution and observation (Tokieda *et al.*, 1996) of CFC-11 in the meridional section at 145° E. Figures 2-2 and 2-3 denote the computational results of the CFCs dispersion and observations (Warner, *et al.*) in the meridional section at 165° E and zonal section at 24°N, respectively. In the computation, CFCs uptake rate in the surface mixing layer is smaller and penetration at the equational region is deeper than the observation. However, Figures indicate that the computed CFCs distributions including the penetration depth of CFCs qualitatively agree with the observation, which suggests that the model can predicts the short-term dispersion of CO_2 at the intermediate depth of the North Pacific Ocean with high accuracy.

2.2 Dispersion of CO_2 Injected at the Intermediate Depth of the Ocean

Then, the model is applied to estimate the potential impact of the CO_2 injection at the intermediate depth of the ocean. The injection point is the same point as in the previous study (Nakashiki and Hikita, 1994) which investigated the effectiveness of ocean storage in the North Pacific ocean. The continuous point source shown below was given at the intermediate depth of the ocean, and the dispersion of CO_2 was computed over 20 years.

    CO_2 injection point   :   *$25°$ N, $150°$ E, 950m depth*

Figures 3-1, 3-2 and 3-3 show the calculated dispersion of CO_2 in horizontal and vertical sections, after the continuous injection of 5 years. The solid lines in the figures denote the dilution defined by the concentration normalized to the source concentration. The hatched area shows the region where the dilution is under 10^{-3} (i.e., high CO_2 concentration). Figure 3-1 shows the horizontal section at the depth of 950m (injection depth). Figures 3-2 and 3-3 show the meridional section at 150° E and the zonal section at 25° N, respectively. These figures show that the high concentration area was limited around the injection point. Figures 4-1, 4-2 and 4-3 show the dilution after 20 years. The injected CO_2 was transported horizontally by the sub-arctic gyre and sub-tropical gyre, and diffused to the downward in the western North Pacific area.

To investigate the potential impact of CO_2 injection, we assumed the concentration at the continuous source, mixed homogeneously in the computational source cell.

    Source cell   :   *$2° \times 2° \times 100\,m$*
    Concentration of the source   :   *$2.3 \times 10^3\,mole\,/\,l$*   - (3)

Total amount of the injection was found to be as two third of the total anthropogenic emission from Japan by the integration of the concentration over the ocean after 1 year injection.

    Amount of the CO_2 injection :   *820,000,000 ton / year*   - (4)

Relationship between the CO_2 concentration increment and the partial pressure of CO_2 is as follows:

       1 (μ atm) of $\Delta pCO_2 = 6 \times 10^{-9}$ mole/l   - (5)

Thus, we obtain the pCO_2 anomaly due to the artificial injection of CO_2. Table 1 shows the corresponding anomaly of partial pressure of CO_2 to the computed dilution based on the assumption

Table 1   Dilution vs. ΔpCO_2

Dilution	Anomaly (μ atm)
10^{-3}	380
10^{-4}	38
10^{-5}	3.8

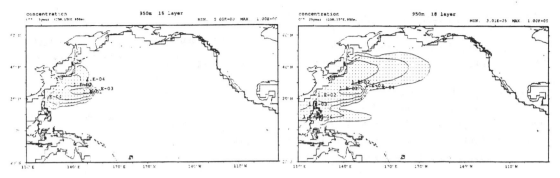

Fig. 3-1 CO_2 dispersion at 950m depth section

(Injection over 5 years)

Fig. 4-1 CO_2 dispersion at 950m depth section

(Injection over 20 years)

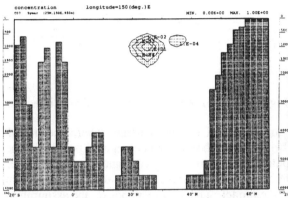

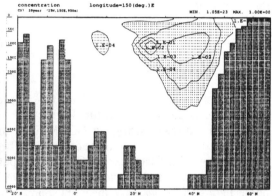

Fig. 3-2 CO_2 dispersion at $150°$ E section

(Injection over 5 years)

Fig. 4-2 CO_2 dispersion at $150°$ E section

(Injection over 20 years)

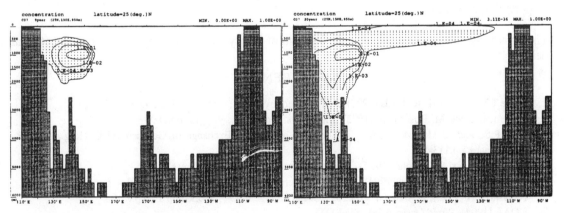

Fig. 3-3 CO_2 dispersion at $25°$ N section

(Injection over 5 years)

Fig. 4-3 CO_2 dispersion at $25°$ N section

(Injection over 20 years)

expressed in Eqs. (3), (4) and (5).

Figure 5 shows the vertical distribution of the partial pressure of CO_2 in the Pacific Ocean and Atlantic Ocean (Millero and Sohn, 1992). The natural background values of the pCO_2 is over 1,000 μ atm at the depth of 1,000m in the Pacific Ocean, so that the dilution 10^{-4} (i.e., $\Delta pCO_2 = 38$ μ atm) means small perturbation.

Figures 4-1, 4-2 and 4-3 show that CO_2 diffused over a limited area at the intermediate depth of the western North Pacific Ocean and the potential impact might be small, even in the case of the discharge over 20 years.

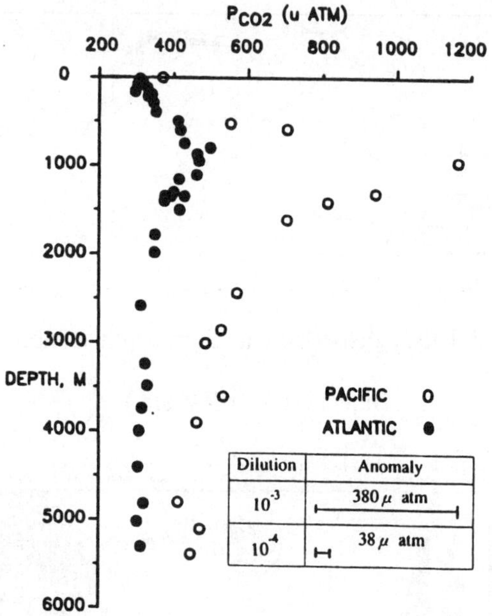

3. CONCLUSION

Fig.5 Profile of pCO_2 and anomaly due to injection

The dispersion of CO_2 injected at the intermediate depth of the North Pacific Ocean was calculated using an OGCM with the isopycnal mixing scheme, where the model performance was discussed in the comparison between the observed and computed profiles of the anthropogenic chemical tracer CFCs. The computed dispersion of CO_2 continuously injected over 20 years suggests that the potential impact might be limited in the small region around the injection point. In this study, we gave the point source in the model, which corresponds to the assumption of homogeneous mixing in the coarse OGCM cell. Particle tracing approach with the random walk process or nesting grid scheme combining with OGCM might be effective to study such a sub-grid scale dispersion of CO_2.

Further investigations should be needed to reveal the initial dilution of the injection pipe (injection method and behavior of the CO_2 plume) and diffusion just after the injection including the local dispersion mechanism.

The authors would like to thank Ms. F. Taguchi of Denryoku Computing Center for the technical help.

REFERENCES

1. Haine T.W.N and K.J. Richards (1995), Jour. of Geophys. Res., Vol.100, No. C6,pp.10727-10744
2. Hellerman, S and M. Rosenstein (1983), Jour. Phys. Oceanogr., Vol. 13, pp.1093-1104
3. Liss, P.S. and L. Merlivat (1986), in The Role of Air Sea Exchange in Geochemical Cycling ed. by P. Buet-menard, pp.113-128
4. Millero, F.J. and M.L. Sohn (1992), "Chemical Oceanography", CRC Press
5. Nakashiki, N. and T. Hikita (1994), Energy Convers. Mgmt, Vol.36, No.6-9, pp.453-456
6. Nakashiki, N.(1995), IEA/GHG, Proc. of " Ocean Storage of CO_2 Workshop 1", pp.71-98
7. NOAA (1994), World Ocean Atlas, CD-ROM
8. Redi,M.H. (1982), Jour. of Phys. Oceanogr., Vol. 12, pp.1154-1158
9. Tokieda, T and S. Watanabe and S. Tsunogai. (1996), Jour. of Oceanogr.,Vol.52, No.4, pp.475-490
10. Warner M.J. and Weiss (1985), Deep-sea Res., Vol. 32, No. 12, pp-1485-1497
11. Warner, M. J., J.L. Bullister, D.P. Wisegarver, R.H. Gammon and R.F. Weiss, "Basin -wide Distributions of Chlorofluocarbons CFC-11 and CFC-12 in the North Pacific: 1985-1989", (submitting in JGR)

Energy Convers. Mgmt Vol. 38, Suppl., pp. S361–S366, 1997
© 1997 Elsevier Science Ltd. All rights reserved
Printed in Great Britain
0196-8904/97 $17.00 + 0.00

Pergamon

PII: S0196-8904(96)00295-6

LES STUDY OF FLOW AROUND A CO_2-DROPLET PLUME IN THE OCEAN

F. THORKILDSEN and G. ALENDAL

Nansen Environmental and Remote Sensing Center
Edv. Griegsvei 3A, N-5037 Solheimsviken, NORWAY

ABSTRACT

Transport of liquid CO_2 to depths of 500-2000 m, and subsequent solution of CO_2 in the sea water, is one of several CO_2 mitigation options which have the potential to become both technically and economically feasible, and environmentally acceptable. A three dimensional Large Eddy Simulation(LES) code with conservation equations for salt and total carbond concentration is used to simulate spreading from a CO_2 dropet plume, in an ocean linearly stratified through a salinity gradient of 0.025 psu km^{-1}. Results from an integrated model for release of CO_2 droplets rising in the water column are used to prescribe the carbon source for the LES model. For a plume with release of 25 kg–CO_2 s^{-1} the CO_2 enriched water spreads out at a neutral depth in a layer with center at a depth about 300 meter below the release port ranging from an upper bound at the center of the plume down to 1600 meter. The pH value in this layer becomes low and will affect the marine life. © 1997 Elsevier Science Ltd

KEYWORDS

Ocean disposal of CO_2, CO_2 droplet plume, pH value fields, numerical simulation.

INTRODUCTION

A wide variety of methods for sequestration of CO_2 in the ocean have been proposed (Marchetti, 1977; Block *et al.*, 1992; Kondo *et al.*, 1995). One of them is release of liquid CO_2 at 500-2000 m depth (Herzog *et al.* (1991),Thorkildsen *et al.* (1996)). At these depths CO_2 is a liquid with less density than sea water (Ely *et al.*, 1989), generating a rising plume driven by the buoyancy of the CO_2 droplets.

The analysis of bubble and droplet plumes has been developed from that of ordinary buoyant plumes in combination with a conservation model for mass in each droplet (McDougall, 1978; Liro *et al.*, 1992; Asaeda and Imberger, 1993). In the case of CO_2 droplets, the mean droplet size is reduced due to diffusion of CO_2 into the ambient liquid.

Solution of CO_2 in (sea-)water leads to a density increase of the water (Haugan and Drange, 1992). This effect counteracts the lifting force from the buoyant droplets and eventually must lead to water being expelled or peeled off from the rising plume, see Fig. (1). The spread out phase of the peeled water will here be modelled with a Navier Stokes solver. The code resolves the small scale dynamics around the rising droplet plume and produces distribution fields for CO_2 concentration and hence the reduced pH value.

The model is forced with the flux from droplet plume model of Thorkildsen *et al.* (1996) caused by the peeling events, included as a line source of CO_2 and buoyancy in the center of the model domain. The ocean is initially linearly stratified through a salinity gradient.

MODEL DESCRIPTION AND SOLUTION METHODS

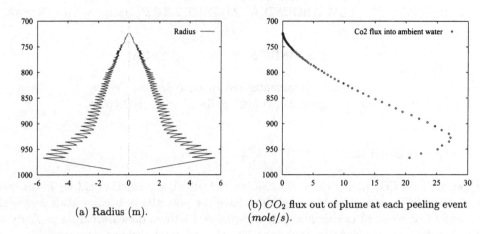

(a) Radius (m).

(b) CO_2 flux out of plume at each peeling event *(mole/s)*.

Fig. 1. A CO_2 droplet plume with a release of 25 kg CO_2 per second as droplets with an radius of 1 cm from an injection device placed at 1000 m depth(Thorkildsen *et al.*, 1996). The radius of the rising plume is less than 6 meter, All of the droplets are are dissolved into the plume water after an ascent of about 250 m.

The model solves the following nondimensionalised equations:

$$\frac{\partial}{\partial t}\mathbf{u} + \mathbf{u}\cdot\nabla\mathbf{u} = -\nabla p + (\mathbf{Ri}_C\theta + \mathbf{Ri}_S\psi)\mathbf{g} + \nabla\cdot\frac{1}{\mathbf{Re}_\tau}\tilde{S} \tag{1}$$

$$\nabla\cdot\mathbf{u} = 0 \tag{2}$$

$$\frac{\partial\theta}{\partial t} + \mathbf{u}\cdot\nabla\theta = \frac{1}{\mathbf{RePr}_C}\nabla^2\theta \tag{3}$$

$$\frac{\partial}{\partial t}\psi + \mathbf{u}\cdot\nabla\psi = \frac{1}{\mathbf{RePr}_S}\nabla^2\psi \tag{4}$$

where $\mathbf{u}$ is the velocity vector, p pressure, $\mathbf{g}$ is a directional vector for the gravity force, θ and ψ are respectively normalized CO_2 concentration and salinity, bouth to be defined later, and $\tilde{S}$ is the stress tensor

$$S_{ij} = \frac{1}{2}\left(\frac{\partial u_i}{\partial x_j} + \frac{\partial u_j}{\partial x_i}\right) \tag{5}$$

The equations have been made dimensionless by introducing a characteristic velocity U and length L, giving a characteristic time scale $L\,U^{-1}$. The CO_2 concentration C and the salinity S have been nondimensionalized against reference values C_0 and S_0, and characteristic differences ΔC and ΔS, according to

$$\theta = \frac{C - C_0}{\Delta C}, \quad \psi = \frac{S - S_0}{\Delta S} \tag{6}$$

These scalars are used as active scalars making it possible to increase the density for the CO_2 enriched water and to create a stratified environment. The Boussinesq approximation has been applied and a linear equation of state with expansion coefficients for CO_2 concentration, β_C and salinity, β_S has been used

$$\rho = \rho_0\left(1 + \beta_C(C - C_0) + \beta_S(S - S_0)\right). \tag{7}$$

The non-dimensional numbers in Eq. (1–3) are;

$$\mathbf{Re} = \frac{UL}{\nu_0}, \quad \mathbf{Ri}_C = \frac{g\beta_C \Delta CL}{U^2}, \quad \mathbf{Ri}_S = \frac{g\beta_S \Delta SL}{U^2}, \quad \mathbf{Pr}_C = \frac{\nu_0}{\kappa_C} \quad \mathbf{Pr}_S = \frac{\nu_0}{\kappa_S}$$

which are the Reynolds, the Richardson and the Prandtl numbers, respectively, ν_0 is the molecular viscosity and κ denotes the diffusion coefficients for the respective scalars.

In LES modelling all variables and equations are filtered giving rise to extra terms due to the nonlinearities. For the momentum equations, these terms are usually divided into a Leonard stress term and a dissipation term, the last one often assumed to be dependent on the resolved strain and an eddy viscosity. The Leonard stress can be of importance for the backscattering of energy from subgridscale eddies to the resolved scales. Antonopoulos-Domis (1981) however, claimed that for staggered grid better results can be obtained if the Leonard stress is neglected. For this reason, and for the sake of simplicity, this term is left out of the present calculations.

The eddy viscosity, ν, in the present code is based on a result from Renormalization Group Theory (Yakhot and Orszag, 1986) which involves solving the third order equation

$$\nu = \nu_0 \left[1 + H \left(\frac{a\Delta^4}{2(2\pi)^4 \nu_0^3} \nu \left(\frac{\partial v_i}{\partial x_j} + \frac{\partial v_j}{\partial x_i} \right)^2 - C \right) \right]^{1/3} \tag{8}$$

for ν. The fixed constants are $a = 0.12$, $C = 75$, and Δ is the filtering scale. The function H is defined

$$H(x) = \begin{cases} x, & \text{for } x > 0 \\ 0, & \text{for } x < 0 \end{cases} \tag{9}$$

If the strain is large, or in other words if

$$\frac{a\Delta^4}{2(2\pi)^4 \nu_0^3} \bar{\epsilon} >> C, \quad \bar{\epsilon} = \nu \left(\frac{\partial v_i}{\partial x_j} + \frac{\partial v_j}{\partial x_i} \right)^2 \tag{10}$$

Eq. (8) reduces to

$$\nu = c_s \Delta^2 \left| \frac{\partial v_i}{\partial x_j} + \frac{\partial v_j}{\partial x_i} \right| \tag{11}$$

with $c_s = \sqrt{a/2(2\pi)^4} = 0.0062$. The viscosity has then returned to a Smagorinsky type model with c_s close to 0.005 which according to Yakhot and Orszag (1986) is the value reported to give the best results for Smagorinsky type models.

The local turbulent Reynolds number appearing in Eq. (1) is given by

$$Re_\tau = \frac{UL}{\nu} \tag{12}$$

Yakhot and Orszag (1986) also supplies a turbulent dissipation coefficient for turbulent transport but the present code has shown to be slightly too dissipative when tested against lock-release experiments, so only molecular dissipation has been included in the scalar transport equation (Alendal, 1996).

The momentum equations are solved using a fractional timestep method (Gresho, 1991). First step solves the momentum equation neglecting the pressure gradient, the second step solves an elliptic equation to find the pressure which is used in the final step to adjust the velocity to be divergence free, Eq. (2) is implicitly solved in the process.

The elliptic pressure field problem is solved with black-red ordering SOR (Golub and Van Loan, 1989) updating the relaxation parameter for each half step to accelerate convergence (Press et al.,

1992). For the transport equations, Eq. (3), an upwind scheme for the advection has been used to avoid false extremal values that are likely to occur for ordinary central difference discretization. This introduced extra diffusivity into the code (Drange and Bleck, 1996).

RESULTS

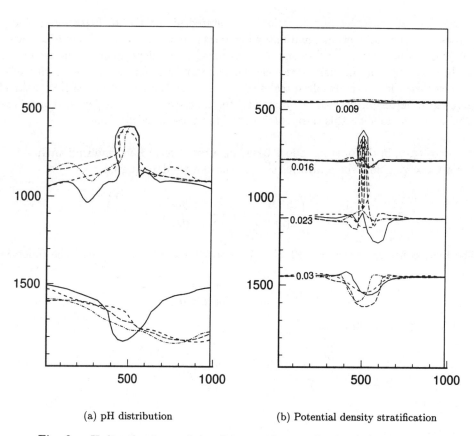

(a) pH distribution   (b) Potential density stratification

Fig. 2. pH distribution and density stratification for day 3 (solid line), day 6 (dashed), day 9 (dash-dotted) and day 12 (long dashed). The pH contours are at 7.75, about 0.25 pH values below the ambient value, the potential density contours are given as deviation from the value at the sea surface in kg m⁻³. The potential density stratification is only weakly altered by the sinking CO_2 enriched water.

The model is set up for simulation of the CO_2 distribution around a CO_2 droplet plume. The CO_2 flux from the plume is given as a source to Eq. (3), distributed in the vertical direction according to the results from an integrated droplet plume model (Thorkildsen et al., 1996), see Fig. (1). In the horizontal direction the source is put in one grid cell and placed in the center of the model domain. The model domain is discritized on a regular mesh with horizontal and vertical resolution of 30 and 60 meter, respectively. The Richardson number for CO_2 and salt are set to 9.81 and 1.6 respectively, and the Reynolds number is 10^8. The Prandtl numbers for CO_2 and salt in water are ~580 and ~700 respectively. Initially the model domain is linearly stratified through a salinity gradient of 0.025 psu km⁻¹. The vertical momentum equation is forced with the buoyancy of the droplets in the plume area. The horizontal boundaries of the model are open, and the bottom is closed with noslip conditions.

The simulation has been continued until a nearly stationary solution has been obtained. For this case after six days the CO_2 distribution only changes slightly with time, Fig. (2), and the total

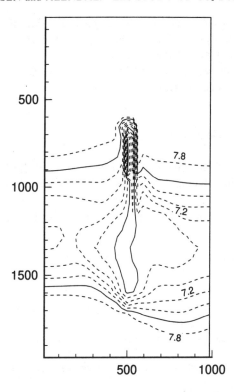

Fig. 3.   pH distribution after 12 days of CO release. The pH level is reduced
by a value of 0.2 all the way from 800 m, 200 m above the CO_2
release depth, nearly down to the bottom.

amount of CO_2 in the model domain is nearly constant.

When the simulation is started the plume water rises due to the buoyancy forcing, and spreads
out into the nearest area above and around the plume. This water sinks and the drag against the
rising plume reduces the rising velocity of the plume water. When the parcel has moved below the
plume area the plume accelerates again and the process repeats itself, i.e. a pulsing occurs. Parcels
of CO_2 rich water sinks rapidly down below the plume and continue to penetrate down below its
neutral depth. This can be seen in Fig. (2b) for day six, where a parcel with high density is seen
at about 1300 m depth. The period of this pulsing is approximate 2.5 hours.

The bulb on the lowermost isopycnal below the plume in Fig. (2b) shows that the dense sinking
plume water penetrates below its neutral depth before it starts to travel horizontaly out of the
region.

In Fig. (3), contours for a 0.2 reduction or more in pH are shown. The pH is reduced by at least a
value of 0.2 all the way from 150 m above the release depth of the CO_2 droplets and nearly down
to the bottom. When the pH level is reduced with this amount, or more the marine environment
will be affected (Magnesen and Wahl, 1993). But, the great thickness of the layer of CO_2 enriched
water may result in a fast dilution of this water outside the model domain.

CONCLUSIVE REMARKS

A Navier Stokes solver has been used to model sinking of CO_2 enriched water around a CO_2
droplet plume. The fine resolution of the model domain resolves the small scale dynamic around
the plume. The size of the model domain suffer under the enormous numerical demand of this

fine resolution. But the small scale results give a strong indication of the vertical distribution of the carbon, and can be used as input to for instance ocean general circulation models. The earth rotation and backgound crossflow are at this time omitted. Introducing this features will alter the CO_2 distribution around the release point, and is may be necessary before using this results for realistic CO_2 disposal studies. Further studies are in progress.

ACKNOWLEDGMENTS

The authors would like to acknowledge financial support from Norsk Hydro a.s. and Saga Petroleum a.s. We would like to tank Dr. Ryuichi Nagaosa at the National Institute for Resources and Environment in Japan for letting us using his Navier Stokes Solver. Dr. Helge Drange and Dr. Paul Samuel at NERSC have been of great help, and stimulated us in our work.

REFERENCES

Alendal, G. (1996), Implementation and testing of an RNG-based subgridscale eddy viscosity in a Navier Stokes solver, *J. Comp. Phys.*, Submitted.

Antonopoulos-Domis, M. (1981), Large-eddy simulation of a passive scalar in isotropic turbulence, *JFM*, **104**:55–79.

Asaeda, T. and Imberger, J. (1993), Structure of bubble plumes in linearly stratified environments, *J. Fluid Mech.*, **249**:35–57.

Block, K., Turkenburg, W. C., Hendriks, C. A., and Steinberg, M., editors (1992), *Proceedings of the first international conference on Carbon Dioxide removal*, Amsterdam.

Drange, H. and Bleck, R. (1996), Multi-dimensional forward-in-time and upstream-in-space based differencing for fluids, Monthly Weather Review, In print.

Ely, J., Haynes, W., and Bain, B. (1989), Isochoric (p, V_m, T) measurements on CO_2 and on $(0.982CO_2 + 0.018N_2)$ from 250 to 300 K at pressure to 35 MPa, *J. Chem. Thermodyn.*, **21**:879–894.

Golub, G. H. and Van Loan, C. F., editors (1989), *Matrix computations*, Number 3 in Johns Hopkins Series in the Mathematical Sciences, Johns Hopkins, London, second edition.

Gresho, P. M. (1991), Some current CFD issues relevant to the incompressible Navier-Stokes equations, *Comp. Methods in Appl. Mech. Eng.*, **87**:201–252.

Haugan, P. M. and Drange, H. (1992), Sequestration of CO_2 in the deep ocean by shallow injection., *Nature*, **357**:318–320.

Herzog, H., Golomb, D., and Zemba, S. (1991), Feasibility, Modeling and Economics of Sequestering Power Plant CO_2 Emissions in the Deep Ocean, *Environ. Prog.*, **10**(1):64–74.

Kondo, J., Inui, T., and Wasa, K., editors (1995), *Proceedings of the second international conference on carbon dioxide removal*, Kyoto, 24-27 October 1994. Published as Volume 36, No. 6-9 of the journal *Energy Convers. Mgmt.*

Liro, C. R., Adams, E. E., and Herzog, H. J. (1992), Modeling the release of CO_2 in the deep ocean, *Energy Conv. Mgmt.*, **33**(5-8):667–674.

Magnesen, T. and Wahl, T. (1993), Biological impact of deep sea disposal of carbon dioxide, Technical Report 77A, Bergen, Norway.

Marchetti, C. (1977), On geoengineering and the CO_2 problem, *Climatic Change*, **1**:59–68.

McDougall, T. J. (1978), Bubble plumes in stratified environments, *J. Fluid Mech.*, **85**:655–672.

Press, W. H., Teukolsky, S. A., Vetterling, W. T., and Flannery, B. P. (1992), *Numerical Recipies in FORTRAN*, Cambridge University press, Cambridge, second edition.

Thorkildsen, F., Alendal, G., and Haugan, P. M. (1996), Modeling of co_2 droplet plumes, *J. geophys. Res.*, **(Submitted)**.

Yakhot, V. and Orszag, S. A. (1986), Renormalization group analysis of turbulence. I. Basic theory, *J. Scient. Comp.*, **1**(1):3–51.

Pergamon

Energy Convers. Mgmt Vol. 38, Suppl., pp. S367–S372, 1997
© 1997 Elsevier Science Ltd. All rights reserved
Printed in Great Britain
0196-8904/97 $17.00 + 0.00

PII: S0196-8904(96)00296-8

PHOTOSYNTHETIC GREENHOUSE GAS MITIGATION BY OCEAN NOURISHMENT

IAN S. F. JONES

Ocean Technology Group,
University of Sydney, NSW 2006, Australia

and

D. OTAEGUI

Earth Ocean & Space Pty Ltd,
Australian Technology Park, Sydney, NSW 2016, Australia

ABSTRACT

The phytoplankton of the upper ocean remove carbon dioxide from the atmosphere by photosynthesis. Their detritus or that of their grazers falls into the deeper ocean taking carbon with it. The ocean uptake of carbon dioxide is limited by the availability of nitrogen in the upper waters over much of the global ocean. This paper examines the cost of providing nitrogen to the upper ocean from a pilot plant with a capacity to sequester 2,000,000 tonnes of carbon dioxide per year. The plant would provide reactive nitrogen at the edge of the continental shelf and monitor the enhanced phytoplankton growth by satellite. The costs compare very favourably with other strategies of carbon dioxide capture and direct placement in carbon sinks. This comes about because the capture mechanism exploits solar energy and the large surface area of the ocean. The sequestration is shown to be permanent and not dependent on the overturning time of the ocean.

© 1997 Elsevier Science Ltd

KEYWORDS

Climate change; ocean; nitrogen; phytoplankton; nourishment; Greenhouse gas; mitigation.

INTRODUCTION

Nations ratifying the United Nations Framework Convention on Climate Change, FCCC, have agreed to address the problem of potential rapid climate change induced in large part by agricultural practices and the burning of fossil fuels. The Intergovernmental Panel on Climate Change, IPCC, has suggested that carbon dioxide will increasingly be the Greenhouse gas of most concern (Houghton et al., 1992). Thus there is a need to demonstrate an efficient method of reduction of Greenhouse Gas concentration in the atmosphere.

While the current focus has been on capturing or reducing the carbon dioxide at its sources, an economical method of reducing the amount of carbon dioxide in the atmosphere is to increase the sinks of this gas. The ocean offers the most promising site for doing this on a scale commensurate with the predicted global problem. The ocean covers 70% of the planet and the deep ocean has an enormous capacity to accept additional carbon. Enhanced carbon dioxide uptake by the world's ocean by nitrogen nourishment of the central ocean gyres has been advocated by Jones (1996). This appears to be economically attractive when compared with other large scale mitigation strategies and has the advantage of contributing to the global food supply by enhancing the primary production of the world's oceans. The environmental impacts of such a scheme, if adopted generally, are discussed in Jones and Young (1997) who concluded that overall impacts are positive and that the negative factors are small.

Laboratory scale experiments and mesocosm tests have supported the general proposition that much of the ocean surface will support phytoplankton growth with the addition of nutrients. On a much larger scale Martin et al (1994) have demonstrated enhanced photosynthetic activity in one area in the Pacific that is deficient in iron. Generally however the limiting nutrient is nitrogen.

The next step is to carry out a pilot scale demonstration over a number of years. The size proposed is one that sequesters 2,000,000 tonnes of carbon dioxide per year. This paper is concerned with the concept design of such a plant using ammonia as the nutrient.

PGGM PROCESS DESIGN

The PGGM process involves the supply of nitrogen to the upper sunlit region of the ocean away from the shoreline and in regions of current that will carry the nutrient into the central ocean gyres. The phytoplankton that prosper in the path of this nitrogen enriched water are monitored from space and this provides the quantitative measure of the success of the system in sequestering carbon dioxide from the atmosphere.

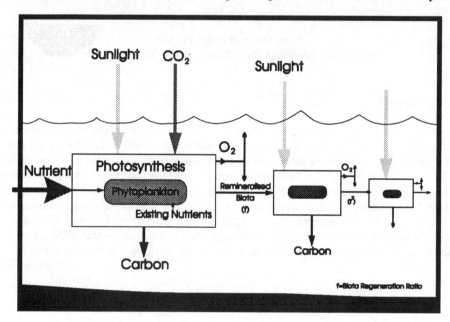

Fig. 1. The process design shows how the addition of nutrient increases carbon uptake of the ocean.

The process should be designed to provide the nutrient at a depth and an early dilution that ensures uptake by the standing stock of phytoplankton. Injection of the nutrient at too great a depth will place it out of reach of the phytoplankton, too shallow will increase the concentration of reactive nitrogen at the sea surface and some will be lost to the atmosphere. The initial uptake of nitrogen might be as low as 70% of that supplied. The Redfield ratio of carbon to nitrogen used is 6.6 to 1 for typical phytoplankton. The process then sequesters 20.7 units by weight of carbon dioxide for one unit of nitrogen taken up by the phytoplankton. However, the production of nitrogen may imply some generation of carbon dioxide. The overall sequestration efficiency for the pilot plant with specific operating conditions is given below.

After the initial uptake of the nitrogen by the phytoplankton, the phytoplankton are grazed and enter the food chain or die, to be partially remineralised (broken down into reactive nitrogen, mostly ammonia, and other constituents including carbon). The fraction of the new nutrient which is subsequently uptaken by new phytoplankton further down current from the point of initial release is designated (f) in Figure 1. It is not counted in the performance of the PGGM process, as the subsequent nitrogen uptake only consumes remineralised carbon, not carbon found in the atmosphere.

PGGM PLANT

The pilot plant to demonstrate the PGGM process comprises the five parts shown in Figure 2:

1. The nutrient source
2. The nutrient delivery system
3. The photosynthetic process
4. Monitoring the carbon uptake
5. Assessing the performance of the process.

The pilot plant has been designed to use reactive nitrogen in the form of ammonia. Ammonia is a suitable nutrient as it occurs naturally in the phytoplankton recycling and can be manufactured from natural gas. This nutrient is to be supplied at a rate as required for the various conditions encountered for the site. The first step in the pilot plant is to dissolve the ammonia in sea water to produce aqueous ammonia at a concentration of 300g/litre. An aqueous form allows the ammonia to be easily pumped and dispersed into the ocean.

The delivery system pumps the ammonia through a 150mm diameter pipe laid on the seabed. This diameter provides a good balance between pumping costs and weight of steel. The pipe extends to the edge of the continental shelf where depths begin to exceed 200m. This allows the ammonia to be dispersed over the deep ocean where the carbon will be permanently stored. A vertical riser attached to the pipe will release the aqueous ammonia at a depth of 50m allowing the ammonia to be dispersed into the sunlit mixed layer. Dispersion is enhanced by adding a diffuser to the vertical riser. It is anticipated that the diffuser will not pose any hazards to shipping. The details are summarised in the next section.

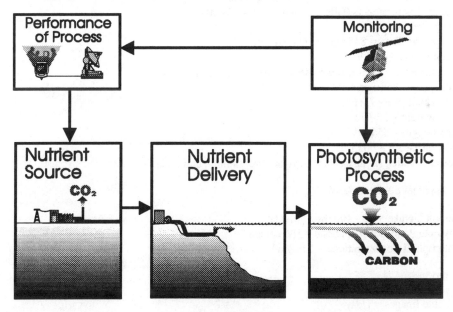

Fig. 2. The conceptual design of the pilot plant showing five parts

Various parameters are to be monitored at the riser to control and monitor the amount of aqueous ammonia released into the ocean. Appropriate flow rates of ammonia can be predetermined based on modelling of these parameters for the dispersion conditions likely to be encountered. Modelling is to take into account the desired amount of carbon uptake while minimising any adverse effects to the local environment. Maximum carbon uptake values of 100 mg C m^{-2} day^{-1} are to be set, lower than those found in nearshore regions.

Measuring the actual carbon uptake is achieved through regular retrieval of appropriate satellite images for the area. These images will measure the amount of chlorophyll at the surface of the ocean, and combined with estimates of the ocean mixed layer depth provide a measure of the primary production induced. In addition, regular surveying carried out by survey vessels will show the reliability of this form of estimation of carbon dioxide uptake for the PGGM pilot scheme.

To adequately assess the effectiveness of the PGGM process, it is anticipated the pilot plant will initially operate for three years with a 60% availability cycle. At this point a review will be carried out. If appropriate the plant can then be refurbished to operate for the economic life of the equipment.

PILOT PLANT SPECIFICATIONS

A summary of the capacity of various components of the pilot plant are given below.

1. Nutrient Source Ammonia Manufacture

   Ammonia Output = 190,000 tonnes/year
       (Natural Gas Requirements = 135,000 tonnes/year)
       (Water Requirements = 125,000 tonnes/year)
       (Electricity Requirements = 6 MW)

2. Nutrient Delivery System

   Dilution of Ammonia = 300g NH_3/kg water
   Sea Water Requirements = 630,000 tonnes/year
   Storage Requirements (Diluted Ammonia) = 15,000 tonnes
   Pump Pressure = 32 bar
   Pump Flow Rate = 845 kg/min (1130 l/min)
   Pump Power = 90 kW
   Pipe Material = Carbon Steel with anti-corrosion lining
   Pipe Inside Diameter = 150 mm
   Pipe Thickness = 7.2 mm
   Pipe Weight = 28t/km
   Pipe Length = 55 km
   Diffuser Depth = 50 m
   Instrumentation (at diffuser) = Temperature, salinity, ammonia flow, current

3. Photosynthetic Process Goal

   Carbon Uptake = 50 - 100 mg C/m^2/day
   Phytoplankton Concentrations = 100 cells/ml
   Chlorophyll Concentrations = 5 mg/m^3

4. Monitoring

   Satellite Images = 1/day
   Mixed layer depth = 12/day

5. Performance of Process

    Continuous Review and Verification
    Survey Vessel Expeditions = one per month

PILOT PLANT PERFORMANCE

Based on a pilot scheme with the capacity to sequester 2,000,000 tonnes of carbon dioxide per year, it is estimated 190,000 tonnes of ammonia per year is needed to increase the phytoplankton uptake of carbon. This calculation is based on the following assumptions:

* For a Redfield ratio of carbon to nitrogen taken as 6.6:1 by weight, the ratio of carbon dioxide to ammonia (NH_3) is 20:1.

* Allowing for losses in the system (approx. 30%) the carbon dioxide to ammonia ratio is reduced to 14:1.

* Assuming a low energy process is used to make the ammonia (34GJ/t) 260,000 tonnes of carbon dioxide is released from the ammonia plant annually if the energy and feed stock is natural gas. This reduces the net carbon dioxide to ammonia ratio to 12:1. For this ratio approximately 190,000 tonnes of ammonia is required to sequester 2,000,000 tonnes of carbon dioxide.

PILOT PLANT OPERATING COSTS

The cost of the pilot plant has been estimated on a life of three years. The capital costs and operating costs for 3 years are as in the following table.

Table 1.   Cost estimates for pilot plant and impact studies

Capital	$
pipe @ $1500 per tonne + lining	2.5M
riser	0.5M
installation	6.9M
barge establishment costs	4.0M
site office, pumps, instrumentation	0.7M
Expense	
ammonia @ 60% duty	34.2M
electricity	0.3M
labour	1.7M
satellite images	0.1M
Investigation	
surveys before and after plant installation	4.0M
Contingency	
9%	5.1M

As well as operating the pilot plant, we have assumed that some R&D would be carried to examine the achieved efficiency and to monitor the environmental impacts. The total cost of the pilot plant project is estimated at US$60M over the three years of operation.

While the pilot plant operating at 60% capacity would sequester 3.6 million tonnes of carbon dioxide for $51M, a full scale plant would be able to take advantages of the economy of scale. If a plant operator wished to obtain an internal rate of return of 8%, assuming that he was being paid under a government contract guaranteed for the 15 year life time of the plant, this scheme is able to sequester large amount of carbon dioxide at a favourable cost of order $7.5 per tonne of CO_2.

CONCLUSIONS

The process design for sequestering carbon breaks away from the approach of capture and disposal of carbon dioxide, an approach which Riemer (1996) of the IEA, classifies as able to be implemented using known but expensive, technology. The capture of carbon dioxide is an expensive process. This paper studies the design of a pilot plant which demonstrates Photosynthetic Greenhouse Gas Mitigation, an alternative process of direct sequestration in the ocean. It extends a concept of providing iron to some regions of the ocean, first advocated by Martin et al (1990) by examine the feasibility of providing the more generally limiting nutrient, nitrogen. This study shows it is feasible to carry out the PGGM process with known technology and proposes a trial which will enable the efficiency of ammonia uptake in the ocean to be determined and the environmental effects monitored on a realistic scale. There are uncertainties to do with the public acceptability of increasing the productivity of the ocean even though the increased supply of protein is needed to feed the world's increasing population.

ACKNOWLEDGEMENTS

Michael Gunaratnam and Helen Young of Earth Ocean & Space discussed many of the ideas in this paper with the authors. Sid French of Worley Limited provided valuable comments on the manuscript.

REFERENCES

Houghton, J. T., B. A. Callander and S. K. Varney (1992). *Climate Change 1992*. Cambridge University Press, Cambridge.

Jones, I. S. F. (1996). Enhanced carbon dioxide uptake by the world's ocean. *Energy Convers. & Mgmt*, 37, 1049-1052.

Jones, I. S. F. and H. E. Young (1997). Engineering a large sustainable world fishery. Currently under review.

Martin, J. H., S. E. Fitzwater and R. M. Gordon (1990) Iron deficiency limits phytoplankton growth in Antarctic waters. *Global Biogeochemical Cycles*, 4, 5-12.

Martin, J. H. et al (1994). Testing the iron hypothesis in ecosystem of the equatorial Pacific Ocean, *Nature*, 371, 123-124.

Riemer, P. W. F. (1996). Greenhouse gas mitigation technologies, an overview of the CO_2 capture and storage and the future activities of the IEA Greenhouse Gas R & D Programme. *Energy Convers. & Mgmt.*, 37, 665-671.

SECTION 5

CHEMICAL UTILIZATION

Pergamon

Energy Convers. Mgmt Vol. 38, Suppl., pp. S373–S378, 1997
© 1997 Elsevier Science Ltd. All rights reserved
Printed in Great Britain
0196-8904/97 $17.00 + 0.00

PII: S0196-8904(96)00297-X

CARBON DIOXIDE UTILISATION IN THE CHEMICAL INDUSTRY.

M. ARESTA and I. TOMMASI.

Centro METEA e Dipartimento di Chimica,
Università di Bari, Campus Universitario, 70126, Italy.

ABSTRACT

The amount of carbon dioxide available for industrial utilisation may expand to unprecedented levels if the recovery of carbon dioxide from energy plants flue gases will be implemented. The potential of each of the three possible uses (technological, chemical, and biological) is far from being clearly defined.
The chemical utilisation option, that has intrinsic thermodynamic and kinetic constraints, may rise controversial positions, depending on the criteria used for the analysis. The estimate of its real potential demands a thorough comparative analysis, using the Life Cycle Assessment methodology, of existing processes/products with the new ones based on CO_2, in order to establish whether, or not, the latter avoid carbon dioxide (either directly or indirectly) and their economics. The rejection/consideration assessment methodology will produce reliable results only if an exhaustive number of parameters is used. The analysis cannot be limited to practiced industrial processes, but must be extended to an exhaustive inventory of cases. © 1997 Elsevier Science Ltd

KEY WORDS

Carbon Dioxide Utilisation. Thermodynamics. Economics. Life Cycle Assessment.

INTRODUCTION

Recovery of carbon dioxide from energy-plant flue gases may contribute to control its accumulation in the atmosphere. Large amounts of carbon dioxide would be, thus, available either for disposal or for utilisation. This scenario has increased the interest to assess the extent to which the utilisation option (technological, biological and chemical), may be expanded. Such analysis is valuable for two main reasons:
* recycling carbon dioxide may potentially contribute to both avoiding carbon dioxide and saving primary resources.
* The utilisation may be economically more advantageous than disposal.
As there is a general positive attitude towards the chemical utilisation of carbon dioxide, as a "clean synthetic methodology", the assessment study seems to be very timely.
In this paper we present the current status of carbon dioxide utilisation in the chemical industry and discuss the complexity of the assessment methodology for the rejection/consideration of a given utilisation option finalized to the mitigation of carbon dioxide.

PRESENT AND FUTURE OF CARBON DIOXIDE CHEMICAL UTILISATION

Carbon dioxide, either recycled from industrial processes (reforming, fermentation, ammonia synthesis, water gas shift reaction, other sources) or extracted from natural wells, is presently used for two main industrial purposes:
(i) technological application.
(ii) Fixation into chemicals.
Case *(i)* includes the industrial uses that do not convert carbon dioxide into other chemicals:

*Enhanced oil recovery (EOR) *Extraction/reaction solvent *Additive for drinks *Antibacterial and antifungi agent *Refrigerators *Food packaging *Fire estinguishers *Soldering *Moulding *Anti-dust *Water treatment (although carbon dioxide is converted into HCO_3^-, this use is not considered a "conversion" technology).

In several Countries, until recently, depending on the availability, a large amount of the carbon dioxide used for these purposes has been extracted from natural wells, that may supply quite pure CO_2 (up to 99%): the use of recycled carbon dioxide would be recommended and appropriate. This consideration applies to EOR, in particular. If recycled carbon dioxide were used, part of it (ca. 50%) would remain in the oil reservoir, resulting in a "natural field" storage of carbon dioxide.

Case *(ii)* includes all possible conversions of carbon dioxide, namely:

a) Fixation of the entire molecule into organic products (synthesis of species in which the -COO- moiety is present: RCOOH, RCOOR', ROC(O)OR', RHCOOR', polycarbonates, polyurethans, etc. Urea, H_2NCONH_2, and its derivatives are also included in this list).

b) Fixation into inorganic carbonates (synthesis of Group 1- and 2-element carbonates, that already find industrial application (see below), and interaction of carbon dioxide with natural basic silicates, mimicking the "silicate weathering" process, that produces carbonates).

c) Reduction to other C1 molecules (HCOOH, CO, H_2CO, CH_3OH, CH_4) or fixation of a reduced form (synthesis of Cn-alcohols and hydrocarbons, formamides, etc).

This classification of the carbon dioxide chemical conversion (Aresta 1987, 1990, 1992, 1993), groups the reactions according to their energetics and other chemicals requirement. Noteworthy, cases *(a)* and *(b)* do not require hydrogen, while case *(c)* demands an external source of hydrogen for carbon dioxide conversion. This difference is of crucial importance, as the use of hydrogen may represent a limitation to the application of a process, standing the condition of scarce availability of hydrogen produced from water (or methane, as a second choice) at a low "carbon dioxide emission" rate (e.g., low energy process, use of solar energy).

The industrial utilisation of carbon dioxide is limited today to a very few processes:

* Synthesis of urea (ca 30 Mt y^{-1}).
* Synthesis of salycilic acid (ca 20 kt y^{-1}).
* Synthesis of Group 1 and 2-element inorganic carbonates, such as Na_2CO_3, K_2CO_3, $BaCO_3$ (a few tens Mt y^{-1}).
* Synthesis of polycarbonates from epoxides (only a few kt y^{-1}, at present).
* Additive in the synthesis of methanol (variable amounts, up to several Mt y^{-1}).

The first three processes, old one century or more, do not require any "catalyst". The last two, developed more recently, require a catalyst (either metal systems, or not).

The engineering of catalysts for carbon dioxide conversion was started after the first transition metal-carbon dioxide complex was discovered (Aresta 1975) and for long time remained a "scientific curiosity". Only recently it has reached the level of "interest for industrial application". However, the conversion of carbon dioxide is not an easy and straightforward reaction, due to its molecular properties, thermodynamics (carbon dioxide lays in a potential energy well) and kinetics. Despite the enormous potential (Nature uses carbon dioxide as source of carbon in many different ways), the development of new catalytic reactions was not too much encouraged in the recent past, as there was no urgent need to change the existing synthetic technologies, mostly based on cheap raw materials and intermediates. The concern of the environmental impact of some of these technologies, new regulations on carbon dioxide emission, and the expected availability of large masses of CO_2, may dramatically change the scenario.

The mitigation of carbon dioxide and the use of carbon dioxide for developing a "green chemistry" or "clean synthetic methodologies" are, indeed, complementary strategies. In fact, the reduction of the emission of carbon dioxide can be achieved in a direct (fixation of the molecule into chemicals) or indirect way (saving energy and reducing waste production). For this reason, the estimation of the utilisation potential may not be restricted to the existing industrial applications (only five processes!), nor limited by the exiguous number of catalytic reactions and photochemical processes recently developed at the level of "industrial application". As we have reported above, this approach is very young and there are still many unexplored areas. Therefore, the exclusion of possibilities should follow an exhaustive "inventory of cases": the contribution the chemical fixation can give to the utilisation option is the integral of all the "feasible" applications. Nature uses hundreds different systems (plants, algae, micro-organisms) to drive the "carbon cycle"!

The assessment study must consider the implementation and exploitation of options not only in the short-, but also in the medium-, and long-term, as it is actually done for other technologies, such as

the ocean disposal. Our attitude must be to evaluate if a tenfold expansion of the use of carbon dioxide, from the actual limit of ca. 100 Mt per year, is eventually possible. If we adopt a simplified approach we may possibly reach the conclusion that even 100 Mt is an impossible limit! The exact definition of the potential of the chemical utilisation technology, based on the inventory of feasible processes, must be completed with an economic study. The acceptance/rejection of new processes/products requires a comparative analysis with existing ones, through the "Life Cycle Assessment, LCA", that should consider:
- Thermodynamics, Kynetics, Energy content of reagents, Yield, Selectivity, Energy requirements for processing main- and side-products and solvents, Waste treatment.
This study should give the amount of: Recycled carbon dioxide, Primary resources saved, Avoided carbon dioxide per each process/product.
The reaction rate, the life-time of the product, and the total market demand or amount used per year, will allow to calculate the specific fraction of recycled carbon dioxide per year.
To complete the economic evaluation, the "added value" of the product has to be considered.
The useful datum is, indeed, the "cost of avoided carbon dioxide". Supposed that a process/product "avoids carbon dioxide", we have to decide if "that particular utilisation has to be preferred to disposal". In fact, if the "availability of carbon dioxide" is set as the "zero point", the choise between "disposal" and "utilisation" can be driven by the economic convenience: the disposal will always rise a "cost" (both in terms of "avoided carbon dioxide" and "economics of the process") derived from the energy used for compressing, pumping, housing carbon dioxide, while the utilisation might result to be a "profit" (due to reduced amount of solvents and reagents used, lower processing costs, lower amount of wastes, etc. proper of the carbon dioxide based technology).
The Scheme reported below summarizes a comparative analysis

SCHEME

DISPOSAL	UTILISATION
CO₂ really disposed per t housed	CO₂ fixed per t reacted
$1 - X$	$1 - Y + Z$

$$X = \Sigma (e_c, e_p, e_h, ...)$$

$$Y = \Sigma (e_r, e_{oc}, e_{pi}, e_{wt}, ...)$$

$$Z = \Sigma (e_{sc}, e_r, e_{oc}, e_{pi}, e_{wt}, ...)$$

as CO₂ equivalent   as CO₂ equivalent

Where: e=energy; c=compression; p=pumping; h=housing; r=reaction; oc=other chemicals involved into the reaction; sc=substituted chemicals; pi=product isolation; wt=waste treatment.

However, a simplified study, based on a limited number of parameters, choosen among those listed above, may produce unclear results, as we discuss below.
Major issues for the utilisation of carbon dioxide are the amount of fixed CO₂ and the life of the product. The two operators (amount and life) can operate simultaneously or indipendently. If we choose the former mode, we must take into consideration only those processes that fix large amounts (> 10 Mt/y) of anthropogenic CO₂ (ca. 20 Gt y^{-1}) in long-living products. Easily we reach the conclusion that the chemical utilisation could either be excluded *a priori* ("forget-it" position) as relevant to the mitigation option or, at best, only inorganic carbonates and polymers would be considered, that means to use a few tens Mt of carbon dioxide per year.
If we decide to adopt the latter mode, we may greatly enlarge the number of products to take into consideration: fuels, organic carbonates, carbamates, isocyanates, urea, etc. These species have a market of several Mt per year, but are short-lived.
As a further example, let us consider the energetics of reaction. For a preliminary consideration/ rejection of processes, thermodynamics is a useful tool. We may decide to consider, for example, only those processes characterized by a negative "free energy" change (Free Energy, more than Enthalpy, should be used, as all reactions that use carbon dioxide have a high entropic content). All reactions using dihydrogen would have an "accepted" mark, as they have a negative free energy change. But dihydrogen availability is a limiting factor, as discussed above. Should we consider, or not, the fuel synthesis from carbon dioxide as a process worth to implement?
Kinetics, yield and selectivity of a reaction may be choosen as consideration/rejection operators. In this case, too, a careful analysis is necessary. Reactions that at a glance are judged to be of low or no practical interest, can be mastered to application. We discuss below the case of the use of carbon dioxide in the synthesis of carbamates, an apparently "unuseful" reaction because of the by-

products, that has been driven towards high yield and excellent selectivity (Aresta and Quaranta 1992, 1993, 1996; McGhee et al 1995).

An n-dimension matrix (or n-variable function) should be used in the assessment study, and the result optimized by varying the weight of each parameter.

A point of great interest would be to attach to each studied option a list of "recommendations" or "must", for making the process of practical application.

RECENT PROGRESS IN CARBON DIOXIDE CONVERSION

In this paragraph a short review of recent achievements for a selected number of reactions that convert carbon dioxide will be presented, with a comment on their status and perspectives. The selection is not exhaustive of possibilities. In general, chemicals with a market of the order of one Mt per year are taken into account. No attempt is made at this stage to use a "yes/not", "consideration/rejection" operator.

Phosgene Substitution.

Phosgene ($COCl_2$) is largely used in the chemical industry (6-8 Mt y^{-1}) for the synthesis of urethans, polyurethans, isocyanates, carbonates, polycarbonates. Carbon dioxide is a good substitute in the reaction of synthesis of carbamic esters. (Aresta and Quaranta, 1996)

Synthesis of carbamic esters and isocyanates.

Phosgene based route

$$CH_4 + H_2O = CO + 3H_2 \qquad (1)$$
$$C + H_2O = CO + H_2 \qquad (2)$$
$$CO + Cl_2 = COCl_2 \qquad (3)$$
$$COCl_2 + ROH + Base = ROCOCl + BaseHCl \qquad (4)$$
$$ROCOCl + 2 R'R''NH = R'R''NC(O)OR + R'R''NH_2Cl \qquad (5)$$

Carbon Dioxide based route

$$CO_2 + 2R'R''NH = R'R''NCOOH_2NR'R'' \qquad (6)$$
$$R'R''NCOOH_2NR'R'' + RX = R'R''NCOOR + R'R''NH_2X \qquad (7)$$

Side reaction

$$R'R''NCOOH_2NR'R'' + RX = CO_2 + R'R''NH + RR'R''NHX \qquad (8)$$

The thermodynamics and kinetics of both reaction routes are quite favourable. The alkylation of the amine (eq. 8) is a side-reaction that until now prevented the practical application of the carbon dioxide process. Quite recent studies have shown that this reaction can be prevented (Aresta and Quaranta 1992, 1993, McGhee *et al.* 1995) and a new interest in this synthetic methodology for the synthesis of carbamic esters at the industrial level grew up .

By the way, the carbon dioxide route would allow to avoid chlorine (that is an important goal for the chemical industry). Alkylating agents other than halogenated species can be also employed.

Primary amine ($R'NH_2$) carbamates can be easily converted into isocyanates:

$$R'HNCOOR \text{ -------> } R'NCO + ROH. \qquad (9)$$

Single-, or multi-step procedures have been developed, characterized by high yield and selectivity (close to 100%) that deserve consideration for full exploitation.

Polycarbonates and polyurethanes can be prepared by this way. The co-polymerization of unsaturated amines with carbon dioxide in mild conditions is a new process (Tsuda 1995).

The market of these products is sizeable at several tens Mt per year.

Methanol

The synthesis of methanol from carbon dioxide, instead of carbon monoxide, implies the use of one extra mole of hydrogen:

$$CO + 2 H_2 = CH_3OH \qquad (10) \qquad\qquad CO_2 + 3 H_2 = CH_3OH + H_2O \qquad (11)$$

The yield and selectivity of the reaction based on carbon dioxide have been quite improved (Saito *et al.* 1995) and the reaction conditions are getting milder with new catalysts. A fact of interest would be the development of catalysts for the direct conversion of carbon dioxide into methanol, avoiding the preliminary conversion into carbon monoxide.

Methanol can be used for the synthesis of hydrocarbons (Inui *et al.* 1996, Lee *et al.* 1996). Hydrogen could be provided, in the future, by biological systems that produce dihydrogen from water (Miura 1996), or developing the methane reforming with carbon dioxide (Inui et al. 1996). The synthesis of methanol from biomasses (Steinberg 1996) is a process of great interest. The market of methanol, used as a chemical or energy product, may be of hundreds Mt per year.

Dimethylcarbonate and Homologues, Diphenylcarbonate.

These compounds find a large industrial application as monomers for polymers, solvents, additives for fuels, alkylating-(arylating) (supplanting alkyl halides) or acylating-(aroylating) agents (substituting phosgene and other reagents, to afford carbamic esters and isocyanates of industrial interest) (Aresta et al. 1995, Aresta and Quaranta, 1996).
The thermodynamics strongly depend on the nature of the group R.
$$2\,ROH \;+\; CO_2 \;=\; (RO)_2CO + H_2O \qquad\qquad (12)$$
The free energy change is negative when R=methyl or higher alkyl, but is positive when R=phenyl ($+19.6$ kcal mol^{-1}, with the reagents in their stardard state).
Despite the favourable thermodynamics, up to date the synthesis of dimethylcarbonate, DMC, and its higher homologues, has found an obstacle in the very low yield and selectivity (Ko et al. 1995, Wagner at al. 1995).
Diphenylcarbonate, that has a large industrial application, is prepared from DMC and phenol through a transesterification reaction. On the other hand, the unfavourable thermodynamics of the direct synthesis from phenol and carbon dioxide can be almost circumvented by combining two reactions, i.e. using a third species (olefin) that is converted into marketable products.
A new synthetic methodology produces cyclic carbonates (that are used as monomers for the synthesis of polymers) from olefins, carbon dioxide and dioxygen (Aresta et al. 1996).
$$RHC{=}CHR' \;+\; CO_2 \;+\; 0.5\,O_2 \;=\; RHC\text{-}CHR''OC(O)O \qquad (13)$$
This reaction is of great interest as the starting materials are easily accessible.
The potential market of carbonates is of tens Mt per year.

Urea

Urea, $(H_2N)_2CO$, is produced at a rate of a few tens Mt per year and finds a large use as agrochemical and intermediate in the chemical industry. It reacts with alcohols to afford carbamates and carbonates. The strict control of the reaction conditions and a catalyst are required in order to avoid the conversion into ammonia and the formation of the trimer of isocyanic acid (eq. 16).
$$(H_2N)_2CO + ROH \;=\; NH_3 \;+\; H_2NCOOR \qquad\qquad (14)$$
$$H_2NCOOR + ROH \;=\; NH_3 \;+\; (RO)_2CO \qquad\qquad (15)$$
$$H_2NCOOR \;=\; 1/3\,(HNCO)_3 + ROH \qquad\qquad (16)$$
The utilisation of urea as intermediate has not been fully explored yet; the chemistry of urea and that of carbonates can be combined to give an interesting network of reactions (Aresta and Quaranta, 1996).
The market is already of 30 Mt per year and could be expanded.

Polymers

These materials have a long-life and a large potential market. Up to date, only polycarbonates and polyurethanes have been synthesized from carbon dioxide, in low amount.
The synthesis of propylene poly-carbonate, a co-polymer of carbon dioxide and propyleneoxide, is an example of industrial exploitation.
New reactions have been developed that co-polymerize carbon dioxide with unsaturated amines under very mild conditions. (Tsuda, 1995) The potential of this reaction and the application of the new polymers are unknown (Saito et al. 1995).

Inorganic Carbonates

These are very long-lasting species. A few tens Mt per year of carbonates of Group 1 and 2-elements (Na, K, Ba, other) are currently synthesized often using natural calcium carbonate, as source of carbon dioxide, that is converted into calcium chloride. Instead, it would be of great interest to substitute recovered carbon dioxide to calcium carbonate.

Another interesting reaction is the fixation of carbon dioxide by some natural silicates, mimicking the natural weathering process, that affords silicates (Aresta, Quaranta and Tommasi 1992) and a carbonate. We report below a few examples of reactions that would be interesting to exploit.

$$2\,Mg_2SiO_4 + 2\,H_2O + CO_2 === H_4Mg_3Si_2O_9 + MgCO_3 \tag{17}$$
$$\quad olivine \qquad\qquad\qquad\qquad serpentine$$

$$3\,KAlSi_3O_8 + H_2O + CO_2 === KH_2Al_3SiO_{12} + K_2CO_3 + 6\,SiO_2 \tag{18}$$
$$\quad orthoclase \qquad\qquad\qquad muscovite$$

$$2\,KAlSi_3O_8 + 2\,H_2O + CO_2 === H_4Al_2Si_2O_9 + K_2CO_3 + 4\,SiO_2 \tag{19}$$
$$\quad orthoclase \qquad\qquad\qquad kaolin$$

$$3\,MgCa(SiO_3)_2 + 2\,H_2O + 3\,CO_2 === H_4Mg_3Si_2O_9 + 3\,CaCO_3 + 4\,SiO_2 \tag{20}$$
$$\quad diopside$$

The kynetics of these processes are very slow in Nature and should be modified for practical application. The investigation of these systems is giving promising results. (Kojima, 1996)

CONCLUSIONS

Carbon dioxide is used by the chemical industry at the rate of 100 Mt per year. This limit must be expanded for the utilisation would be a technology significant for the mitigation of carbon dioxide. The assessment of the real potential of this technology, requires an inventory of feasible processes and the defintion of the conditions for exploitation. A comprehensive analysis is necessary in order to avoid trivial errors in comparing the utilisation option with disposal.

REFERENCES

Aresta M, C.F. Nobile, V.G. Albano, E. Forni and M. Manassero (1975), J.C.S. Chem Comm., 636.

Aresta M. (1987) in "Carbon Dioxide as a Source of Carbon: Chemical and Biochemical Uses", M. Aresta and G. Forti eds., Reidel Publ., pp. 1-20.

Aresta M. (1990) In : "Enzymatic and Model Reactions for Carboxylation and Reduction Reactions", M. Aresta and J. V. Schloss eds., Elsevier Publ., pp. 1-25.

Aresta M, E. Quaranta and I. Tommasi (1992), In : Energy Convers. Mgmt, Elsevier Science Ltd., Vol. 33, No. 5-8, pp. 495-504.

Aresta M. and E. Quaranta (1992), Tetrahedron, 48, p. 1515.

Aresta M. (1993), In : Energy Conv. Mgmt, Elsevier Science Ltd., Vol. 34, No. 9-11, pp 745-752.

Aresta M. and E. Quaranta (1993), Ital. Pat. 1237208.

Aresta M., E. Quaranta and I. Tommasi (1994), New. J. Chem., 18, 133-142.

Aresta M., C. Berloco, E. Quaranta (1995), Tetrahedron, 51, pp. 8073 - 8088.

Aresta M., I. Tommasi, E. Quaranta, C. Fragale, J. Mascetti, M. Tranquille, F. Galan, M. Fouassier (1996), Inorg. Chem., 35, p. 4254 - 4260 and references therein.

Aresta M. and E. Quaranta (1996), Chem. Tech., in the press.

Inui T., H. Hara, T. Takeguchi, J.B. Kim, S. Iwamoto (1996), In : ICCDR-III, Cambridge (USA), 9-11 September 1996.

Ko K., F. Ogata, (Chem. Abstr. 1995), Jpn. Kokai Tokkyo Koko JP 0) 33715 and references therein.

Kojima T., A. Nagamine, N. Ueno, S. Uemina (1996), In : ICCDR-III, Cambridge (USA), 9-11 September 1996.

Lee K.-W., P.H. Choi, K.W. Jun, S.J. Lee and M.J. Choi (1996), In : ICCDR-III, Cambridge (USA), 9-11 September 1996.

McGhee W., C. Riley, K. Christ, Y. Pan, B. Parnas (1995), Org. Chem., 60, p. 2820.

Miura Y. (1996), In : ICCDR-III, Cambridge (USA), 9-11 September 1996.

Saito M., T. Fujitani, I. Takahara, T. Watanabe, M. Takeuchi, Y. Kanai, K. Moriya, T. Kakumoto (1995), In : Energy Convers. Mgmt., Elsevier Science Ltd., Vol. 36, p. 577.

Steinberg M. (1996), In : ICCDR -III, Cambridge (USA), 9-11 September 1996.

Tsuda T. (1995), Gazz. Chim. Ital., 125, p. 101.

Wagner A., W. Loeffer, B. Haas, (Chem. Abstr. 1995), Ger. DE4310 109 and references therein.

Pergamon

PII: S0196-8904(96)00298-1

Energy Convers. Mgmt Vol. 38, Suppl., pp. S379–S384, 1997
© 1997 Elsevier Science Ltd. All rights reserved
Printed in Great Britain
0196-8904/97 $17.00 + 0.00

CO$_2$ REMOVAL AND FIXATION
SOLAR HIGH TEMPERATURE SYNGAS GENERATION
FOR FUEL SYNTHESIS

T. Weimer[1], M.Specht[2], A. Bandi[2], K.Schaber[1], C.U. Maier[2]

[1] Institute of Technical Thermodynamics and Cryogenics, Karlsruhe University,
Richard Willstaetter Allee 2, 76128 Karlsruhe, Germany
[2] Center for Solar Energy and Hydrogen Research, Hessbruehlstr. 21c,
70565 Stuttgart, Germany

ABSTRACT-For the use of remote solar energy sources, the generation of liquid fuels is an interesting extension because an existing infrastructure for transportation and distribution is available. In this paper two processes for climate neutral methanol synthesis working with high temperature processes for syngas generation from CO$_2$ and water are described. They offer a better efficiency for the use of solar energy compared to alternative processes with higher electricity demand. The CO$_2$ enrichment from the atmosphere has a higher energy demand compared to the recovery from concentrated emissions, but the use of solar energy for the enrichment process leads to the decision to use atmospheric CO$_2$. © 1997 Elsevier Science Ltd

KEYWORDS- CO$_2$ fixation, syngas, high temperature solar processes, atmospheric CO$_2$

INTRODUCTION

The chemical storage of hydrogen as liquid carbon containing fuels is of interest as an alternative to liquefaction of hydrogen for long distant transportation of renewable energy. Using recycled CO$_2$ and renewable energy for the fuel generation, the resulting synfuels are climate-neutral energy carriers. An economic advantage of liquid fuels compared to liquefied hydrogen is the possibility of using the existing infrastructure for transportation and distribution. Basic products for the generation of syngas are CO$_2$ and water. Although the CO$_2$ enrichment from air has a higher energy demand compared to the CO$_2$ recovery from concentrated emissions solar-powered high temperature processes such as limestone decomposition lead to a efficient use of renewable solar energy for this application. Furthermore the transportation of CO$_2$ from concentrated emissions to the remote renewable energy source can be avoided (Weimer et al., 1996). The synfuel production consists of three processes (I) CO$_2$ enrichment, (II) hydrogen/syngas generation, and (III) fuel production, e.g. methanol or Fischer-Tropsch synthesis. Using the high temperature processes solid oxide electrolysis (SOE) and the reverse water gas shift (RWGS) reaction operating at similar temperature ranges of about 1200 to 1300 K as limestone decomposition a high overall solar efficiency for the fuel synthesis can be achieved.

CO$_2$ ENRICHMENT FROM THE ATMOSPHERE

The higher solar efficiency of thermal processes compared to the generation of electricity from solar energy is basis for the process design for the CO$_2$ enrichment from atmosphere. Because of the low CO$_2$ content in air (360 vpm) any compression of the processed air will lead to an exorbitant electricity demand. If the CO_2 is separated from air by chemical absorption with irreversible reaction, compression can be avoided. Columns with modern structurized packings have a low pressure drop and minimize the electricity demand. Potassium hydroxide solution is suitable for scrubbing.

$$2KOH + CO_2 \rightarrow K_2CO_3$$

The absorption of CO_2 from air in packed columns was investigated extensively. The experimental results are in agreement with a developed model for this process. Based on these results the electric energy demand for a fan and a pump is estimated to 50 kJ/mol CO_2 at a gas velocity of 1 m/s, a temperature of 293 K and a packing height of 3-5 meters. For the regeneration of the scrubbing solution the addition of quicklime to a part of the scrubbing solution leads to a precipitation of limestone.

$$CaO + H_2O + K_2CO_3 \rightarrow CaCO_3 \downarrow + 2KOH$$

For a continuous absorption and regeneration process a scrubbing solution containing carbonate and hydroxyl ions is required. First experiments in laboratory scale show that with solutions containing 0.8 mol/l OH^- and 0.25 mol/l CO_3^{2-} ions a 90 percent conversion of the quicklime to limestone can be achieved with the addition of less quicklime than the equation indicates. The energy demand for the precipitation process can be neglected compared to the other process steps for CO_2 enrichment. The energy demand for limestone drying will be discussed below. The last process step, the calcination of limestone is part of the solar high temperature processes.

SOLAR HIGH TEMPERATURE PROCESSES

Calcination of limestone

For the CO_2 enrichment from air calcination at high temperatures is a suitable process to release the CO_2 and for regeneration of quicklime from the precipitated limestone

$$CaCO_3 \xrightarrow{1200 \text{ K}} CaO + CO_2 \uparrow$$

Thermal decomposition of limestone is a well-known process with widespread industrial application. The reaction temperature is about 1200 K and the energy demand in the range of 210 kJ/mol CO_2 for thermal decomposition of dry and pure limestone (Wuhrer, 1958). If this value is taken as the basis for the calculation of the energy demand for the thermal decomposition of a mixture containing 90 mol percent limestone and 10 mol percent slaked lime the resulting thermal energy demand is lower than 250 kJ/mol CO_2. If the 40 percent lower thermal energy demand for the decomposition of slaked lime compared to the calcination of limestone (Bahrin, 1993), is taken into account with the same process efficiency as for calcination the resulting energy demand is 225 kJ/Mol CO_2 when CO_2 and quicklime leave the process at a temperature of 400 K. With reaction products at a temperature of 1200 K the energy demand is 300 kJ/mol CO_2. An additional energy demand has to be taken into account for the drying of the precipitated limestone. But the generated steam can be fed to the high temperature electrolysis process for hydrogen generation. Therefore this energy demand for drying is as part of the energy consumption of the hydrogen generation.

Hydrogen Generation with Solid Oxide Electrolysis (SOE)

High temperature electrochemical technology based on yttria-stabilized zirconia electrolytes can be applied for high temperature electrolysis of steam (and carbon dioxide, see below). This offers the best efficiency for electrolytic hydrogen production because it allows in part the substitution of electric energy by solar high temperature heat with higher solar efficiency. In a high temperature electrolysis unit a solid oxide electrolyte shows ionic conductivity due to the existence of oxygen ion site vacancies. If a DC voltage is applied to an anode/electrolyte/cathode assembly, H_2O is split to form O^{2-} ions. The oxygen ions, formed at the cathode, migrate through the electrolyte to the anode, where they are oxidized to generate oxygen. Due to the impermeability to gas, the cathodic reaction products and O_2 are separated by the ceramic ZrO_2 electrolyte. The operation temperature is about 1300 K. In our experiments the high temperature electrolysis set-up is a multicell tubular cell configuration. The high temperature electrolyte material is a yttria-stabilized zirconia (YSZ) with an yttria content of 8 mol%. The multicell stack consists of a number of tubes, connected by Ni. The cathode material is a Ni cermet (metallic Ni embedded in a YSZ matrix) and a $La_{1-x}SrMnO_3$ perowskite served as anode. The electrochemical set-up can be represented by H_2O, $Ni(YSZ)/(ZrO_2)_{0.92}(Y_2O_3)_{0.08}/La_{1-x}SrMnO_3$, air. The energy demand of the SOE is given below.

Retro Water Gas Shift Reaction

The water-gas shift reaction (WGS) and the retro water-gas shift reaction (RWGS) can be described by the equation:

$$CO + H_2O \rightarrow CO_2 + H_2$$

$$\Delta H = -41 \text{ kJ / mol } CO_2$$

The values of the equilibrium constant, K_p, expressed by

$$K_{p=} \frac{p_{CO_2} p_{H_2}}{p_{CO} p_{H_2O}}$$

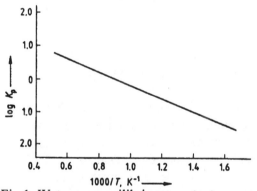

Fig.1. Water-gas equilibrium constant dependence on the temperature

depends on the temperature of the reacting gases. The equilibrium constant is not influenced by the total pressure, because the number of molecules on the right and on the left side of the equation is identical. The temperature dependence of the equilibrium constant (log K_p vs. 1/T) is shown in Fig. 1. As it can be deduced from Fig. 1, at increasing temperature the equilibrium is shifted towards CO, whereas at lower temperatures dominates CO_2.

The shift reaction (WGS) is mainly performed to regulate the proper ratio between CO and H_2 in synthesis gas or to purify feed gases of CO. The retro shift reaction (RWGS) can be used in the reduction of CO_2 to CO for the purpose of syngas production. At high temperatures higher than 1200 K, which are required for shifting the equilibrium toward CO production, no catalysts are needed. At 1300 K the conversion rate of CO_2 to CO will be about 80 % if the concentrations of CO_2 and H_2 have a ratio of 1:3 at the beginning.

COMBINATION OF PROCESSES

Syngas Generation with Solid Oxide Electrolysis

A new approach for synthesis gas production is the simultaneous reduction of H_2O and CO_2 using a solid oxide high temperature electrochemical cell. The simultaneous reduction of steam and CO_2 with a solid oxide electrolyte leads to a synthesis gas ($H_2/CO/CO_2$) which can be used for the subsequent generation of synthetic fuels. The thermodynamic data of CO_2 and H_2O splitting at 1300 K show that the CO_2 reduction to CO has a lower electric energy demand than the H_2O reduction to H_2 ($\Delta G_c < \Delta G_h$). However, for CO_2 splitting a higher total energy input is necessary ($\Delta H_c > \Delta H_h$). For an operation without an additional solar thermal energy input, higher current densities and potentials with a higher joulean heat formation have to be applied compared to allothermal operation. Fig.2 and 3 represent the dependence of ΔH, ΔG, and ΔQ on the temperature for water and CO_2 splitting. The following values are given for a cell operation temperature of 1300 K.

$$CO_2 \rightarrow CO + \frac{1}{2}O_2$$

$$\Delta G_C = 170 \, kJ \, / \, mol \, CO_2 \quad \Delta H_C = 282 \, kJ \, / \, mol$$

$$H_2O \rightarrow H_2 + \frac{1}{2}O_2$$

$$\Delta G_H = 176 \, kJ \, / \, mol \, H_2 \quad \Delta H_H = 250 \, kJ \, / \, mol$$

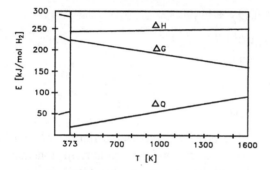

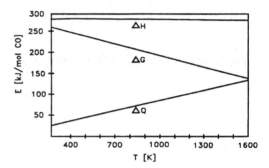

Fig.2. Dependence of ΔH, ΔG, and ΔQ for H_2O electrolysis on temperature

Fig.3. Dependence of ΔH, ΔG, and ΔQ for CO_2 reduction on temperature

The heat energy demand can be supplied in part by solar high temperature heat at a temperature level of about 100 K higher than the operating temperatur of the electrolysis unit. The stoichiometry of the generated syngas is dependent on the feed gas composition, the applied cell voltage, the operating temperature and the contact time. Due to the RWGS reaction, which is affected by these parameters, it is still unclear how much of the CO is produced by the direct and how much by the indirect reduction mechanism. In principle, the high temperature electrolysis of H_2O/CO_2 is applicable to syngas generation of proper composition for the subsequent fuel synthesis process. The generation of CO from CO_2 at 1300 K requires a 13 % higher energy demand (ΔH), than water electrolysis. The electric energy demand for the production of a synthesis gas of proper stoichiometry ($CO:H_2 = 1:2$) for methanol synthesis is only 6.3% higher. The methanol synthesis from syngas is exothermic ($\Delta H = -91 \, kJ/mol$) and the excess heat can be used for steam generation of water, which is fed to the solid oxide electrolyzer. This offers an about 90% waste heat utilization of the chemical reactor at a temperature level of about 500 K. For the syngas generation of CO_2 /steam in a process without heat supply an electric energy demand of 800 kJ_{el}/mol CH_3OH is required in the solid oxide electrolysis unit. This offers a 72% efficiency for energy storage in the form of the hydrogen carrier methanol with regard to the lower heating value of methanol (624 kJ/mol CH_3OH), when CO_2 is available for this process and an additional energy of 65 kJ/mol CH_3OH for syngas compression is taken into account.

A part of the electric energy demand of the high temperature solid oxide electrolysis system can be replaced by solar high temperature heat (allothermal operation). This heat transfer can be realized by feeding the electrolysis unit with CO_2/steam at a temperature level at about 1400 K. In this case the electric energy demand for syngas generation from high temperature steam and CO_2 is only 670 kJ/mol CH_3OH.

Calcination in a Hydrogen Feed

Another possible combination of solar high temperature processes is to combine the thermal decomposition of limestone with the RWGS reaction. For this purpose the generated hydrogen with a temperature of 1300 K is used as feed gas stream for the calcination reactor. The overall reaction is

$$CaCO_3 + 3H_2 \rightarrow CaO + H_2O + 2H_2 + CO$$

As mentioned above, only about 80 % of the released CO_2 will be converted to CO at 1300 K. After separate cooling of the quicklime and the gas in a heat exchanger with generation of high temperature steam for the SOE, a selective absorption of H_2O by the quicklime is possible at temperatures below 600 K. The energy demand for the syngas generation from water and limestone is in the range of

$$l_{el} = 630 \, kJ \, / \, mol \, CH_3OH$$

$$q = 490 \, kJ \, / \, mol \, CH_3OH$$

with a utilization of the waste heat from H_2O absorption and the methanol synthesis reaction for the evaporation of water and assuming a 90 % efficiency for thermal processes.

FUEL SYNTHESIS PROCESSES

Two different processes for methanol synthesis from water and atmospheric CO_2 with solar high temperature process units are represented in figures 4 and 5. The total electric energy demand per mol methanol in the process with SOE syngas generation consists of 50 kJ for absorption, 670 kJ for SOE and 65 kJ for syngas compression to 60 bar. The ideal thermal energy demand is 270 kJ for limestone decomposition, 25 kJ for superheating of steam and CO_2 and 110 kJ for the SOE. The energy demand for evaporation of water, the preheating of steam and the rectification is supplied by heat exchange. Assuming again a 90 % efficiency for thermal processes the total energy demand of this process for climate neutral methanol synthesis with SOE syngas generation is

$$l_{el} = 785 \, kJ \, / \, mol \, CH_3OH$$

$$q = 450 \, kJ \, / \, mol \, CH_3OH$$

For the process with combination of RWGS reaction and thermal decomposition of limestone the electric energy demand of the SOE is only 630 kJ. The thermal energy demand is given above. Therefore the total energy demand is here

$$l_{el} = 745 \, kJ \, / \, mol \, CH_3OH$$

$$q = 490 \, kJ \, / \, mol \, CH_3OH$$

Compared to the lower heating value of methanol (624 kJ/mol) both processes achieve an overall energetic efficiency of 50 %. The energy demand for the enrichment of CO_2 from atmosphere is about 20 % of the total energy demand. If solar efficiencies for the generation of electricity in the range of 0.2 and for heat generation of 0.7 are assumed the process with RWGS reaction during calcination has a slightly higher solar efficiency of 14 % compared to 13.6 % for SOE generation of syngas.

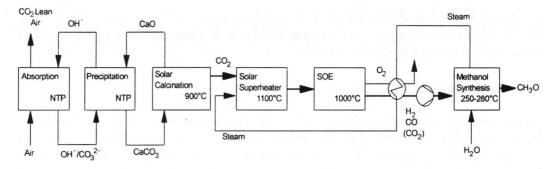

Fig.4. Process flowsheet for solar methanol production from atmospheric CO_2 via syngas generation in a combined steam/CO_2 solid oxide electrolyser

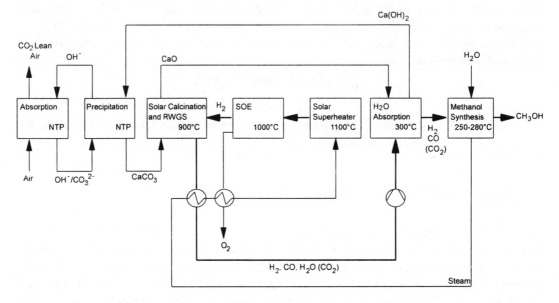

Fig.5. Process flowsheet for solar methanol production from atmospheric CO_2 via syngas generation in a combined calcination/retro water-gas shift reactor

CONCLUSIONS

The described processes are a promising alternative for the climate neutral synthesis of liquid fuels with renewable energy. Because of the lower solar efficiency in the range of 14 % compared to the generation of electricity with efficiencies about 20 % they are interesting alternatives only if there is no direct connection of the solar energy source to an electric grid possible. The use of atmospheric CO_2 will avoid the transportation of CO_2 to the distant located renewable energy source from concentrated emissions. A further increase in the energetic efficiency can be achieved by the integration of thermal processes for hydrogen generation. The combination of thermal hydrogen generation from steam and the RWGS reaction during limestone decomposition will lead to higher efficiencies for methanol synthesis compared to the electricity generation from solar energy.

REFERENCES

Bahrin,I., VCH, Weinheim, 1993
Weimer,T., K. Schaber, M. Specht and A. Bandi, *Energy Conv. Mgmt.*, 37, 1351 (1996)
Wuhrer, J., *Chem. Ing. Tech.* 30, 19 (1958).

Energy Convers. Mgmt Vol. 38, Suppl., pp. S385–S390, 1997
© 1997 Elsevier Science Ltd. All rights reserved
Printed in Great Britain
0196-8904/97 $17.00 + 0.00

Pergamon

PII: S0196-8904(96)00299-3

HIGHLY EFFECTIVE GASOLINE SYNTHESIS FROM CARBON DIOXIDE

T. INUI, H. HARA, T. TAKEGUCHI, K. ICHINO, J.B. KIM, S. IWAMOTO and S.B. PU

Department of Energy and Hydrocarbon Chemistry, Graduate School of Engineering, Kyoto University, Sakyo-ku, Kyoto 606-01, Japan.

ABSTRACT

Effective catalytic conversion of CO_2 into gasoline was investigated by one-pass operation using a serially connected flow-type reactor. Rapid CO_2-reforming of methane into syngas was conducted by a newly developed Rh-modified Ni-Ce$_2$O$_3$-Pt catalyst at a low furnace-temperature around 400~500°C. A high methane conversion was achieved by an in-situ heat supply due to catalytic combustion of added ethane or propane which is more combustible with higher combustion heat than methane. Thus obtained syngas or a CO_2-enriched syngas was then converted into methanol on a novel highly active methanol synthesis catalyst, composed of Pd-modified Cu-Zn-Cr-Ga-Al mixed oxides, and the methanol produced was introduced to the successive reactor packed with a metallosilicate catalyst having MFI structure, and totally converted into gasoline with a high space-time yield and selectivity.

© 1997 Elsevier Science Ltd

KEYWORDS

Gasoline synthesis; CO_2-rich syngas; methanol synthesis; syngas synthesis; methane reforming; CO_2-reforming; on-site heat supply; seriesly connected reactor; one-pass operation; metallosilicate.

INTRODUCTION

Effective catalytic conversion of CO_2 into valuable compounds such as methanol (Kanai *et al.*, 1994) and a high quality gasoline (Inui *et al.*, 1992) and other chemicals (Keim, 1987) is expected as one of the possible ways to mitigate CO_2. Since the product has high value, it has a potential to compensate a cost of hydrogen to be used as the reducing reagent. Nevertheless, the highly effective synthesis of hydrogen must be the most indispensable condition to realize the CO_2 hydrogenation process in practically. Therefore, in the present study, first of all, a rapid methane reforming involving CO_2-reforming into hydrogen and carbon monoxide was investigated by using a newly developed Ni-based composite catalyst at much lower reaction temperature than the conventional steam reforming catalysts used in practically (Inui *et al.*, 1995).

Thus obtained syngas or hydrogen-and CO_2-enriched syngas was then converted into methanol. Methanol to hydrocarbon conversion progresses on a solid acid catalyst via dimethylether as the intermediate product. This reaction generally involves the autocatalysis, and therefore, the concentration of light olefins formed at the first step by dehydration from dimethylether is the higher the better for the successive oligomerization of olefins to a gasoline range hydrocarbon (Inui, 1987a). This leads the necessity for methanol synthesis catalyst that the catalyst has to an activity to produce methanol as much as possible. When one uses a conventional Cu-Zn-oxide based methanol synthesis catalyst, which is prepared by the precipitation method, the catalytic activity is not so high that, in order to obtain a higher methanol concentration in the effluent gas, the reaction gas must be introduced to the reactor with a lower space velocity or a longer contact time. As a results one can obtain merely a low space-time yield of gasoline. That is the reason why the methanol synthesis catalyst must be improved very much to achieve the effective gasoline conversion by the one-pass operation condition. Thus, the methanol synthesis catalyst had been improved by our intrinsic preparation procedure, the uniform gelation method (Inui *et al.*, 1992), and furthermore, Pd and Ga were combined to modify the catalyst (Inui, 1996a).

Methanol to gasoline conversion was then carried out by using metallosilicate catalysts having a pentasil pore opening structure (MFI), because, when a typical shape-selective zeolite catalyst such as H-ZSM-5 is chosen as the catalyst for methanol conversion, the products would be limited mainly in gaseous paraffinic hydrocarbons owing to the intrinsic activity of H-ZSM-5 such as hydrogen shift and/or hydrogenation activities (Inui *et al.*, 1993a).

Finally, the requisite and rationality of the combination of different functions of catalysts to achieve gasoline synthesis from CO_2 will be summarized.

EXPERIMENTAL

Catalyst

A Ni-based four-component catalyst composed of 10 wt% Ni-5.6 wt% Ce_2O_3-1.1 wt% Pt-0.2 wt% Rh (Inui *et al.*, 1995) was prepared by the step-wise impregnation method. A ceramic fiber in a plate shape of 1 mm thickness, which has a voidness of 88%, was adopted as the catalyst support. Before loading the catalyst components, the support was washcoated with alumina layer by 17 wt%.

A methanol synthesis catalyst composed of 38.1 wt% Cu-29.4 wt% ZnO-1.6 wt% Cr_2O_3-17.8 wt% Ga_2O_3-13.1 wt% Al_2O_3 was prepared by the uniform gelation method (Inui *et al.*, 1992).   A concentrated mixed nitrate solution was put in a thin layer and it was contacted with an NH_3-H_2 vapor and totally transformed to a gel, followed by drying, thermal decomposition of ammonium nitrate and calcination in air.   Modification of 1 wt% Pd on the basis total catalyst weight was made by physical mixing with the four component catalyst mentioned above and a Pd supported on γ-alumina (Inui, 1996).   Property of hydrogen reduction for the preoxidized catalyst was measured by a temperature-programmed reduction (TPR) method using a Shimadzu Thermal Analyzer DT-40.

As the gasoline synthesis catalysts, an H-Fe-silicate having a Si/Fe atomic ratio of 400 or an H-Ga-silicate having a Si/Ga atomic ratio 400 or 200 was packed into the second stage reactor connected in series with the first reactor in which methanol synthesis catalyst was packed.   These metallosilciates were synthesized by the rapid crystallization method (Inui, 1989).

Reaction Method

An ordinary flow-type reactor was used under the atmospheric pressure for methane reforming reactions.   The catalyst of circular plate form of 10 mm in diameter and 1 mm in thickness was placed rectangularly in the transparent-quartz tubular reactor of 10 mm inner diameter.   For methanol synthesis, a pressurized flow-type reactor was used.   The reaction conditions are described in Fig. 4. For hydrocarbon synthesis, a seriesly connected two-stage series reactor was used.   The reaction conditions for 2nd reactor were controlled independently from 1st reactor.   All the products and unconverted reactants come out from the 1st reactor were directly introduced to the 2nd reactor.   The catalyst 10~24 mesh in size was packed into the reactor.

RESULTS AND DISCUSSION

Syngas Synthesis by Rapid Methane-Reforming

A reaction gas mixture of 10% CH_4-10% CO_2-80% N_2 was allowed to flow the reactor packed with the Ni-Ce_2O_3-Pt-Rh catalyst at a space velocity of 730,000 h^{-1} (based on the net catalyst volume involving catalyst substance and the support) or contact time of 4.9 m-sec with elevating the reaction temperature from 300 up to 700°C.

As shown in Fig. 1, ever at such a very short contact time, the reaction progressed stoichiometrically and produced an equivalent moles of H_2 and CO along the reaction equilibrium of the following equation, at the temperature range from 300 to 700°C.

$$CH_4 + CO_2 \rightarrow 2H_2 + 2CO + 259 \text{ kJ/mol (at 500°C)}$$

An extraordinary high space-time yields of H_2 and CO, 5,930 mol/l·h were achieved at 700°C at a methane conversion of 86.2%. From the comparison with other component catalysts, the high active of the four-component catalyst is attributed to the prominent effect of Rh as the porthole for hydrogen

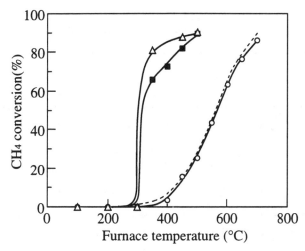

Fig. 1. Change in methane conversion by the combination of catalytic combustion of ethane or propane
Reaction gas composition:
○: 10% CH$_4$-10% CO$_2$/N$_2$
△: 10% CH$_4$-10% CO$_2$-5% C$_2$H$_6$-17.5% O$_2$/N$_2$
■: 10% CH$_4$-10% CO$_2$-3.3% C$_3$H$_8$-16.5% O$_2$/N$_2$
Dotted line: Equilibrium conversion of methane in the CO$_2$-reforming reaction
SV: 730,000 h^{-1} (CT: 4.93 m-sec)

Table 1. Reaction data of CO$_2$-reforming of methane with and without combination
of hydrocarbon combustion

Reaction	CH$_4$ fed	Furnace temp.	Conversion (%)		CH$_4$ STC	STY (mol/l·h)	
	(%)	(°C)	CH$_4$ fed	CO$_2$ fed	(mol/l·h)	H$_2$	CO
CO$_2$-reforming	10 [a]	700	86.2	88.2	2,756	5,814	6,046
	35 [a]	700	87.3	58.9	3,815	6,922	9,918
Combined with ethane combustion	10 [b]	500	95.0	13.8	2,945	6,645	4,869
	35 [b]	500	47.8	8.2	5,440	10,798	9,557
Combined with propane combustion	10 [c]	500	89.3	24.2	2,860	8,891	6,753
	35 [c]	500	57.3	58.9	6,263	14,457	12,581

a) other components were 10% CO$_2$ and N$_2$
b) other components were 10% CO$_2$ + 5% C$_2$H$_6$ + 17.5% O$_2$ and N$_2$
c) other components were 10% CO2 + 3.3% C$_3$H$_8$ + 16.5% O$_2$ and N$_2$

spillover to maintain more reduced state of catalyst during the reaction (Inui, 1993b). Since this catalyst also exhibited high activities for H$_2$O-reforming, H$_2$O-CO$_2$ co-reforming, and partial oxidation reforming, an appropriate composition of syngas could be obtained by regulation the feed gas composition (Inui *et al.*, 1995).

Furthermore, in order to achieve more rapid reforming, the catalytic combustion of more easily combustible hydrocarbons such as ethane or propane were added into the reaction gas with corresponding amount of oxygen to supply the combustion heat directly at the vicinity of catalytic sites. As shown in Fig. 1, the catalytic combustion of added hydrocarbons occurred at around 300°C and at that furnace temperature conversion of methane jumped up to 65~80% which is far beyond the reaction equilibrium. This indicates that the catalyst-bed temperature elevated ca. 300°C by the in-situ supply of heat and maintained above 600°C even at much lower furnace temperature.

The results are summarized in Table 1 numerically. As shown, by combining the catalytic combustion, extraordinary high space-time yields of hydrogen and carbon monoxide could be realized even at such a relatively lower furnace temperature (500°C).

Table 2. Performance of methanol synthesis from carbon oxides on the Pd-Ga-modified Cu-Zn oxide-based composite catalyst

Fed Gas	Catalyst	SV	Temp.	Conversion to (C-mol%)				STY (g/l·h)	
		(h⁻¹)	(°C)	MeOH	DME	HC	CO_x	MeOH	DME
CO_2-rich [a] syngas	Pd-non-modified	18,800	250	12.8	0.0	0.0	3.3	756	0
		18,800	270	21.3	0.0	0.0	3.3	1,261	
	Pd-modified	18,800	250	19.2	0.0	0.1	3.0	1,135	
		18,800	270	22.0	0.0	0.2	3.9	1,300	0
CO-rich [b] syngas	Pd-modified	18,800	270	45.8	18.8	0.5	1.4	3,690	500
		37,600	270	35.6	4.1	0.2	1.1	5,730	480

a) 22% CO_2-3% CO-75% H_2, b) 30% CO-3% CO_2-67% H_2, Reaction pressure: 80 atm.

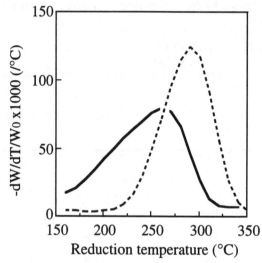

Fig. 2. TPR profiles for the Pd-modified and non-modified methanol synthesis catalysts. Solid line: Pd-modified; Broken line: Pd-non-modified.

Methanol Synthesis from Syngas and Mixture of CO₂ and Hydrogen

The methanol synthesis catalyst prepared by the uniform gelation method (MSCg) exhibited a 50% higher activity than conventional catalyst prepared by precipitation method (MSCp) (Inui, 1992). Modification of MSCg with Ga and Pd markedly enhanced the activity for both CO_2-rich and CO-rich syngases. The results are summarized in Table 2. In both hydrogenations of CO_2 and CO, extraordinarily high space-time yields of methanol could be obtained at considerably high conversion level near the equilibrium conversion even at a very high space velocity, 37,600 h⁻¹ or a considerably short contact time (7.66 sec) under 80 atm. The high activity is ascribed to the reason that the desired reduced state of the catalyst metal oxides, mainly copper oxide, for exhibiting the optimum catalytic performance could be controlled by the hydrogen spillover through the Pd metal parts and the inverse-spillover from the Ga parts (Inui *et al.*, 1996a).

In order to identify the opposite roles between Pd and Ga, the preoxidized catalyst with and without combination of Pd, and the effect of Ga_2O_3 in the catalysts were measured by means of TPR. The results are shown in Figs. 2 (Inui, 1996a) and 3 (Inui *et al.*, 1996b), respectively. As shown in Fig. 2, the combination of Pd decreased the reduction temperature very much due to the effect of hydrogen spillover, and the temperature, at which the reduction rate becomes the maximum, shifted from 293°C to 261°C, suggesting that the catalytic activity would be also enhanced at lower temperature. In fact, as can be seen from Table 2, space-time yield of methanol increased very much on the Pd-modified catalyst. The effect of Ga_2O_3 addition to the Cu-Zn-Cr-Al mixed oxides on the hydrogen reduction can be evaluated from Fig. 3. Obviously, the addition of Ga_2O_3 up to 10% retards the reduction indicating the function of inverse spillover like as observed in the paraffin aromatization reaction (Inui *et al.*, 1987a).

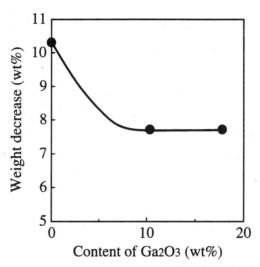

Fig. 3. Effect on Ga₂O₃ content on the decrease in the methanol synthesis catalyst weight measured by temperature-programmed reduction.

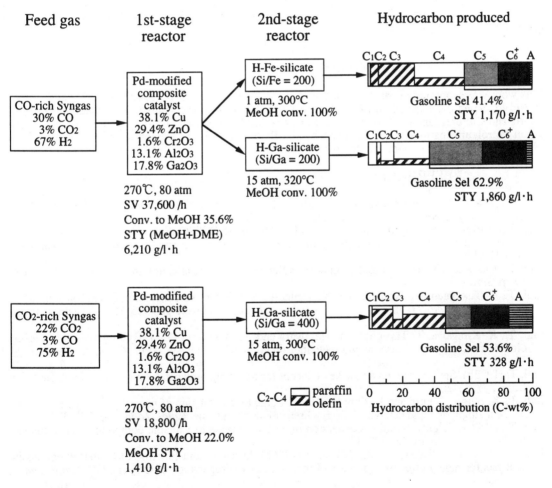

Fig. 4. Gasoline synthesis from a CO-rich syngas and a CO₂-rich syngas via methanol by using the two-stage series reactors packed with different functional catalysts.

Gasoline Synthesis from CO$_x$ via Methanol as the Intermediate Product

The results are summarized in Fig. 4 for a CO-rich syngas and a CO$_2$-rich syngas, respectively.

In case of CO-rich syngas, methanol formed in the 1st reactor was totally converted in the 2nd reactor packed with H-Fe-silicate at 300°C, 1 atm to a gasoline fraction with a selectivity 41.4%, and 1,170 g/l·h space-time yield/g of gasoline. On the other hand, when H-Ga-silicate was used as the catalyst for 2nd reactor and was operated at a pressurized condition (15 atm) and at 320°C, methanol converted totally and the gasoline formed was 62.9% in selectivity and 1,860 g/l·h.

From the CO$_2$-rich syngas, H-Ga-silicate having Si/Ga ratio 400, methanol introduced from 1st reactor converted totally and an aromatic poor gasoline was obtained with a selectivity 53.0%, and a STY of 328 g/l·h.

It is expected that since light olefins still remains in the product, these can be converted successively into gasoline fraction by recycliar feed or under somewhat longer contact time.

CONCLUSIONS

Carbon dioxide could be hydrogenated with hydrogen to methanol with a high space-time yield and its successive conversion on H-Fe-silicate or H-Ga-silicate made possible the gasoline synthesis from CO$_2$ by one-pass conversion. In stead of hydrogen, methane or natural gas could be used as the source of hydrogen through the rapid CO$_2$-reforming reaction with in-situ hydrocarbon oxidation producing an appropriate ratio of H$_2$ and CO.

ACKNOWLEDGMENT

This work was supported in part by Grand-in-Aid for Scientific Research (A)(2) 07405037 and Grand-in-Aid for Scientific Research on Priority Areas 07235215, from the Ministry of Education, Science, Sports and Culture, Japan. A part of this work also has been carried out as a research project of The Japan Petroleum Institute Commissioned by the Petroleum Energy Center with the subside of the Ministry of International Trade and Industry.

REFERENCES

Inui, T. (1987a). Selective conversion of light olefins to high octane-number gasoline on novel Fe-silicate catalysts. *React. Kinet. Catal. Lett.*, 35, 227-236.

Inui, T., Y. Makino, F. Okazumi, S. Nagano and A. Miyamoto (1987b). Selective aromatization of light paraffins on platinum-ion-exchanged gallium-silicate bifunctional catalysts. *Ind. Eng. Chem. Res.*, 26, 647-652.

Inui, T. (1989). Mechanism of rapid zeolite-crystallization and its application to catalyst synthesis. *ACS Symp. Series*, 398, 479-492.

Inui, T., T. Takeguchi, A. Kohama and K. Tanida (1992). Effective conversion of carbon dioxide to gasoline. In: *Proc. 1st Intern. Confer. Carbon Dioxide Removal*, pp. 513-520. Pergamon Press, Oxford.

Inui, T., K. Kitagawa, T. Takeguchi, T. Hagiwara and Y. Makino (1993a). Hydrogenation of carbon dioxide to C$_1$~C$_7$ hydrocarbons via methanol on composite catalysts. *Appl. Catal. A: General*, 94, 31-44.

Inui, T. (1993b). Spillover effect as the key concept for realizing rapid catalytic reaction. *Stud. Surf. Sci. Catal.*, 77, 17-26.

Inui, T., K. Saigo, Y. Fujii and K. Fujioka (1995). Catalytic combustion of natural gas as the role of on-site heat supply in rapid catalytic CO$_2$-H$_2$O reforming of methane. *Catal. Today*, 26, 295-302.

Inui, T., (1996a). Highly effective conversion of carbon dioxide to valuable compounds on composite catalysts. *Catal. Today*, 29, 329-337.

Inui, T., H. Hara, T. Takeguchi and J.B. Kim (1996b). Structure and function of Cu-based composite catalysts for highly effective synthesis of methanol by hydrogenation of CO$_2$ and CO. *Catal. Today*, in press.

Kanai, Y., T. Watanabe and M. Saito (1994). Catalytic conversion of carbon dioxide to methanol over palladium-promoted Cu/ZnO catalysts. In: *Carbon Dioxide Chemistry; Enviormental Issues* (J. Paul and C.M. Pradier ed.), pp. 102-109. The Royal Soc. of Chemistry.

Keim, W. (1987). Industrial uses of carbon dioxide. In: *Carbon Dioxide as a Source of Carbon* (M. Aresta and G. Forti ed.), *NATO ASI Series*, pp. 23-31.

Energy Convers. Mgmt Vol. 38, Suppl., pp. S391–S396, 1997
© 1997 Elsevier Science Ltd. All rights reserved
Printed in Great Britain
0196-8904/97 $17.00 + 0.00

Pergamon

PII: S0196-8904(96)00300-7

SEPARATE PRODUCTION OF HYDROGEN AND CARBON MONOXIDE BY CARBON DIOXIDE REFORMING REACTION OF METHANE

Osamu TAKAYASU, Fumihisa SATO, Kyoko OTA, Takamasa HITOMI,
Takafumi MIYAZAKI, Tsutomu OSAWA, and Ikuya MATSUURA

Faculty of Science, Toyama University, Toyama City, Toyama 930, Japan

ABSTRACT

First, modeling of the carbon dioxide reforming reaction (CO_2RR) mechanism was carried out by the chemical affinity values of the individual elementary reactions which would be involved in the overall reaction. The chemical affinity values were obtained from a composition of an outlet stream passed through a catalyst bed. The catalysts used were Ni, Ru, Rh, Pd and Pt supported on MgO. As a result, the decomposition of CH_4 to give carbonaceous deposits and H_2 was the slowest step in CO_2RR. Then, a separate production of H_2 and CO over Ni catalysts from an alternate stream of CH_4 and CO_2 was proved by the use of a flow system equipped with two reactors in parallel, two selector valves, and two gas reservoirs. As a result, such mixtures were obtained as H_2(74%)+CH_4(18%) and CO(71%)+CO_2(17%) when Ni/α-Al_2O_3 was used as a catalyst. © 1997 Elsevier Science Ltd

KEYWORDS

carbon dioxide reforming reaction, steam reforming reaction, chemical affinity, equilibrium, separate production of hydrogen, separate production of carbon monoxide, carbon dioxide, methane

I. A REACTION PATH INFERRED WITH THE AID OF CHEMICAL AFFINITY

Studies of the carbon dioxide reforming reaction (CO_2RR) were carried out with the intention of inferring its reaction path by the use of the chemical affinity values obtained from a composition of an outlet stream passed through a catalyst bed in a standard flow reactor. Needless to say, a chemical reaction terminates when its equilibrium is attained. In other words, if a composition of the outlet stream is close to an equilibrium of an elementary reaction, it is recognized that the elementary reaction is fast enough to terminate during the period that the stream passes through a catalyst bed. This means that individual reaction steps can be classified as fast or as slow depending upon whether the composition of the outlet stream is close to or far from their individual equilibriums. Here, chemical affinity (A) is a convenient function to be used to decide a reaction step being close to or far from its equilibrium quantitatively (Takayasu *et al.*, 1995). When the value of A of a given reaction equals zero, the reaction attains at equilibrium after the stream passed through a catalyst bed. That is, the reaction is fast enough to terminate during the period. When $A > 0$, the reaction is capable of taking place further. When $A < 0$, the reverse reaction is able to take place further. These reactions are so slow as not to terminate during the period.

The CO_2RR is a desirable system for the intention, because almost all equilibrium constants of the reaction steps involved in CO_2RR are available in literature. The reaction steps are shown as follows:

$$CH_4 + CO_2 \rightarrow 2CO + 2H_2 \qquad \Delta H_{R1} = 247 \text{ kJ} \qquad \text{(R1)}$$
$$CH_4 + H_2O \rightarrow CO + 3H_2 \qquad \Delta H_{R2} = 206 \text{ kJ} \qquad \text{(R2)}$$

These overall reactions would involve the reaction steps shown below.

$$CH_4 \rightarrow C + 2H_2 \qquad \text{(R3)}$$
$$C + CO_2 \rightarrow 2CO \qquad \text{(R4)}$$
$$C + H_2O \rightarrow CO + H_2 \qquad \text{(R5)}$$
$$H_2 + CO_2 \rightarrow CO + H_2O \qquad \text{(R6)}$$
$$CO + H_2O \rightarrow H_2 + CO_2 \qquad \text{(R7)}$$

It has been recognized that R3 is a rate determining step over Ni and Ru catalysts (Rostrup-Nielsen 1984).   It also has been believed that the reactions, R6 and R7, are at equilibrium over Ni and Ru catalysts.

EXPERIMENTAL

The catalysts used were MgO-supported Ni, Ru, Rh, Pd and Pt. The amount of Ni loaded was 10 wt.%, and those of the others were 1 wt.%. The MgO used was ultrafine single-crystal MgO (99.98% pure, 100 m^2g^{-1}, #100 furnished by Ube Industries Ltd.) Metals were introduced through the corresponding metal acetylacetonate (Takayasu *et al.*, 1990). CO_2RR tests were carried out using a conventional standard flow system. The reactor used was made of quartz, 8 mm in diameter. 0.1 g of the catalyst, except 0.2 g of Pd/MgO, was charged into the reactor. The feed stream was CH_4(15 mL/min)+CO_2(15)+He(70). Flow rates of individual gases were controlled by mass flow controllers (Stec Inc.). Reaction temperatures were decreased from 1173 K to 573 K at a 50 K interval, maintaining each temperature for more than 30 min. Each 2 mL gas of the outlet steam at each temperature was sampled and submitted to G.C. The amount of H_2O involved in the outlet stream was determined from material balance equations.

RESULTS

Figure 1 shows the results of CO_2RR tests. The reaction took place above 673 K in the presence of the catalysts. Inlet CH_4 and CO_2 are normalized to be 100 in the vertical axis.   [Result 1]: The activities

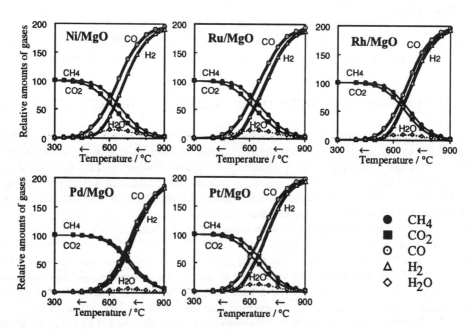

Fig. 1. Results of the carbon dioxide reforming reaction at decreasing temperatures.

over the different metal catalysts did not differ greatly. The amounts of CO were greater than those of H_2 possibly due to the reverse water gas shift reaction (R6) in all five catalysts. However, [Result 2]: the differences depended upon the metals in the sequence

$$Ni, Ru, Pt \gg Rh > Pd \qquad (1)$$

[Result 3]: Small amounts of H_2O were observed around 900 K in Fig. 1. The amounts were also expressed by the same sequence as Eq. 1.

DISCUSSION

Chemical affinity (A) is defined for a given reaction (RG) as

$$A_{RG} = -RT \ln (K_{obsRG} / K_{eqRG}) \qquad (2)$$

where R is gas constant, T is a reaction temperature in K, K_{eqRG} is an equilibrium constant of a given reaction at T K. And K_{obsRG} is a value obtained by the same equation as K_{eqRG} was obtained, but using partial pressures of gases involved in the outlet stream.

In addition, there are a few mathematical connections among the chemical affinities of the reactions R1-R7. First, since R1 is obtained by the summation of R3 and R4, the equilibrium constant of R1, K_{eqR1}, is expressed by the product of K_{eqR3} and K_{eqR4}.

$$K_{eqR1} = K_{eqR3} K_{eqR4} \qquad (3)$$

Similarly,

$$K_{obsR1} = K_{obsR3} K_{obsR4} \qquad (4)$$

Eqs 3 and 4 give Eq 5:

$$K_{obsR1}/K_{eqR1} = (K_{obsR3} / K_{eqR3}) (K_{obsR4} / K_{eqR4}) \qquad (5)$$

Therefore, the chemical affinity of R1, A_{R1}, should be equal to the summation of those of R3 and R4.

$$A_{R1} = A_{R3} + A_{R4} \qquad (6)$$

Similarly following equations are given:

$$A_{R2} = A_{R3} + A_{R5} \qquad (7)$$
$$A_{R1} = A_{R3} + A_{R5} + A_{R6} \qquad (8)$$
$$A_{R2} = A_{R3} + A_{R4} + A_{R7} \qquad (9)$$

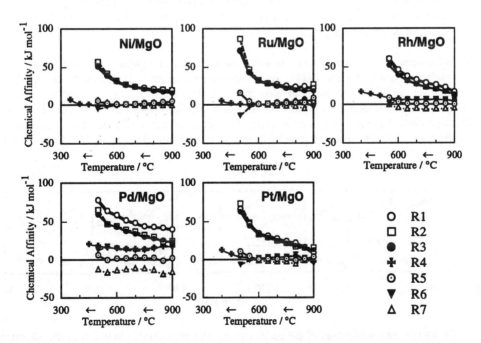

Fig. 2. Chemical affinity values for individual reactions obtained from the results shown in Fig. 1.

Chemical affinity values for R1-R7 were calculated by using the partial pressures of gases involved in an outlet stream that passed through a catalyst bed. Equilibrium constants used for R1-R2 and R6-R7 (Grau 1968) and R3 and R5 (Rostrup-Nielsen 1984) were quoted from literatures. However, the constants for R4, K_{eqR4}, at various temperatures were not available in literature, so they were obtained by Eq 3 using K_{eqR1} and K_{eqR3}. The values A calculated are shown in Fig. 2.

It is obvious in Fig. 2 that the relations shown by Eqs. 6-9 are realized. The values, A_{R6} and A_{R7}, take the same value but in an opposite sign, because R7 is the reverse reaction of R6. A reverse reaction takes the same value in chemical affinity but in an opposite sign. It has been believed that R6 and R7 are at equilibrium in CO_2RR. This is shown clearly by the chemical affinity values in Fig. 2. The decomposition reaction of CH_4, R3, is farthest from equilibrium in the reactions involved in CO_2RR. This means that R3 is a rate determining step. It is well recognized that the decomposition of CH_4 is a rate-determining step in the steam reforming reaction. (Rostrup-Nielsen 1984) The same result was obtained by the chemical affinity calculation in this paper. Also, it is recognized that the reaction path can be estimated reliably by the chemical affinity values which are obtained from an outlet stream passed through a catalyst bed.

It is shown in Fig. 2 that [Result 4] activities of the overall reactions, R1 and R2, are in the sequence which was obtained from the chemical affinity values, A_{R1} and A_{R2}.

$$Ni, Ru, Rh > Pd, Pt \qquad (10)$$

Also, it is clear in Fig. 2 that the reactions R4-R7 are attained at equilibrium over the catalysts of Ni/MgO, Ru/MgO and Pt/MgO, except Pd/MgO and Rh/MgO. [Result 5]: The reactions R4, R6 and R7 are a little bit far from equilibrium over Pd/MgO and Rh/MgO. [Result 6]: The reaction R4 is far from equilibrium over Pd/MgO and Rh/MgO even though R5 is close to equilibrium. [Result 7]: The activities of R4 and R6, i.e., A_{R4} and A_{R6}, depend upon the metal, and they give the sequence:

$$Ni, Ru, Pt > Rh >> Pd \qquad (11)$$

Though R3 is a rate determining step in CO_2RR, the reactions R4 and R5 that follow to R3 are at equilibrium over Ni/MgO, Ru/MgO, and Pt/MgO as is shown in Fig. 2. The equilibriums of R4 and R5 are inclined to shift toward the carbon-side and an amount of carbon would be retained on the metal surfaces especially at lower temperatures. The amount of carbon retained on the surfaces seems to differ greatly depending upon whether the reaction temperatures are brought from a higher temperature or from a lower temperature. The hysteresis curves were reported in the catalytic activity of CO_2RR for elevating temperatures and for decreasing temperatures.

The approach-to-equilibrium temperature is often used for the comparison of catalytic activities in industrial practice. (Rostrup-Nielsen 1984) A comparison of catalytic activities by the chemical affinity method which is originally proposed in this paper is considered to be superior to that by the approach-to-equilibrium temperature, because the chemical affinity values for fundamental reaction steps can be obtained easily from an outlet stream. Also, a reaction path can be discussed in detail through the chemical affinity values. The reaction quite far from equilibrium should not be involved in the overall

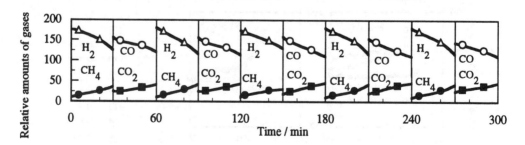

Fig. 3.   Change in the composition of the outlet stream over Ni/α-Al₂O₃ at 973 K in the alternate stream of CH_4 (5 mL/min) and CO_2(5).

reaction. Though the reaction R4 is reported to be a side reaction in the reforming reactions.[Rostrup-Nielsen 1984] We could not obtain such evidence in this chemical affinity calculation method. The reaction R4 would be a part of the main reactions.

The catalyst bed used was wide and thin in order to inhibit second reactions of the products. When a catalyst bed was narrow and thick (3 mm in diameter and 10 mm in length) in our experiment, we had a problem that values in chemical affinity which should be positive were obtained as negative.

II. SEPARATE PRODUCTION OF HYDROGEN AND CARBON MONOXIDE BY AN ALTERNATE STREAM OF METHANE AND CARBON DIOXIDE

It was concluded in Part I that CO_2RR of CH_4 consisted of the reactions, R3 and R4, over metal catalysts. Nickel was the most active catalyst for the rete determining step R3 in the catalysts used. H_2 and CO might be produced separately by passing an alternate stream of CH_4 and CO_2 separately over the catalyst. This separate production was examined over Ni catalysts by a flow system equipped with two catalyst beds in parallel, and two selector valves. The outlet streams from CH_4 and from CO_2 could be stored in two gas reservoirs separately.

EXPERIMENTAL AND RESULTS

The catalysts used were 10 wt.%-Ni/α-Al₂O₃, Ni/MgO, and Ni/SiO₂ prepared by the same method shown above. The α-Al₂O₃ used (99.98% pure, 55 m²g⁻¹) was prepared by calcining γ-Al₂O₃ at 1473 K for 3 h. The SiO₂ used (99.98% pure, 350 m²g⁻¹) was Zeolite Powder Ex-504 (Si/Al=215) furnished from Süd Chemie Ag. The apparatus used was a conventional flow system equipped with two reactors in parallel, two selector valves, and two gas reservoirs. The first selector switched the inlet streams, CH_4 and CO_2, to the two reactors alternatively. The second selector switched the outlet streams from the catalyst beds to the individual gas reservoirs corresponding to the individual inlet streams, CH_4 and CO_2. Therefore, the two selector valves were switched simultaneously. Figure 3 shows an example of the results over 3 g of Ni/α-Al₂O₃ at 1023 K. Inlet CH_4 and CO_2 are normalized to be 100 in the vertical axis. Switching the inlet stream from methane (5 mL/min) to carbon dioxide (5 mL/min) and vice versa was repeated alternately after every 30 min. This was the best condition for the 3 g catalyst around 1050 K. As a result, the gases of 450 mL in each reservoir were obtained after 8 hours. The compositions are shown in Table 1. It is clear that the compositions were dependent on the catalysts and that the best catalyst is Ni/α-Al₂O₃ of the three.

Table 1. Compositions of gases stored in the two reservoirs

Catalyst / 3 g	From methane stream CH_4:H_2:CO_2:CO / %	From carbon dioxide stream CH_4:H_2:CO_2:CO / %
Ni/α-Al₂O₃	18 : 74 : 2 : 6	3 : 9 : 17 : 71
Ni/MgO	16 : 74 : 4 : 6	2 : 12 : 27 : 59
Ni/SiO₂	57 : 36 : 6 : 1	9 : 6 : 46 : 39

The carbon deposits produced on the catalysts at 923 K in a CH_4 stream were identified as CH_x by elemental analysis (Yanaco MT-3). The results showed x=0 over Ni/α-Al₂O₃ and Ni/MgO, and x=0.02 over Ni/SiO₂. It is clear that the carbonaceous deposits scarcely involve hydrogen on those catalysts.

Observations by scanning electron microscopy were performed with a Nippon Denshi JEM-100 SX. As a result, both filamentous carbon and carbon having a mosslike morphology were observed. The amounts of filamentous carbon and also the total amounts were in this order: Ni/α-Al₂O₃ > Ni/MgO >> Ni/SiO₂.

DISCUSSION

It is shown in Fig. 3 and Table 1 that H_2 and CO can be produced separately by the alternate streams of CH_4 and CO_2 instead of their mixed stream. The H_2 and CO produced separately were contaminated by impurities of CO and CO_2 and of H_2 and CH_4, respectively, as shown in Table 1. The contamination was a result of the residues that remained in the connection tubes after switching the streams.

The conversions of both CH_4 and CO_2 decreased gradually during each run as shown in Fig. 3. The conversion of CH_4 would decrease with the increasing amount of carbon on a catalyst surface, because larger amount of carbon would interfere with the decomposition of CH_4. On the other hand, the conversion of CO_2 would also decrease rapidly with the decreasing amount of the carbon as the time proceeds. Therefore, the time interval of each run from switching to switching, and flow rates of individual streams are very important factors in connection with those conversions. When the interval is too short, the contamination by the impurities would increase. When the interval is too long, the conversions would decrease due to the larger amounts of the reactants passing through a catalyst bed without reacting.

Amounts of carbon deposited in a CH_4 stream and in a CO_2 stream were observed by a thermogravimetric analyzer (Shimadzu TGA-50). The amounts were dependent on the catalysts, and all carbon could not be removed in a CO_2 stream.   95% of carbon on Ni/α-Al$_2$O$_3$ was removed with CO_2 in spite of only 70% on Ni/MgO. Though it is believed that carbon deposited during CO_2RR interferes with the reaction, it may be removed in a high percentage by CO_2. It needs further investigation to know the relation between carbon deposited and its deactivation effect.

Matsukata and co-workers reported a separate production of H_2 and synthesis gas from CH_4 and CO_2 over various metal catalysts using a circulating fluidized bed. (Matsukata et al., 1995) Our method could produce both H_2 and CO separately instead of syngas.

The CO_2RR is an endothermic reaction. It requires a great amount of heat energy and a high temperature. If the heat energy is supplied by a fossil fuel, it is unavoidable to generate a great amount of CO_2. For this problem, the hydrogen produced by this separate production method should be burned as fuel. If so, this method should be adjusted to focus only on the production of CO as an industrial raw material. CO is more important than CO_2 itself as an industrial raw material. The heat of combustion of H_2 produced by R1 (ΔH_{R8}) is two times as great as the heat required for R1 (ΔH_{R1}).

$$2H_2 + O_2 \rightarrow 2H_2O \qquad\qquad \Delta H_{R8} = -483 \text{ kJ} \qquad\qquad (R8)$$

ACKNOWLEDGMENT

We are grateful that Mr. Hideto Hashiba and Mr. Masaichi Hashimoto (Nippon Shokubai Co., Ltd.) for the SEM and TEM measurements and the elemental analysis.

REFERENCES

Grau, G.G., (1968). Homogene Gasgleichgewichte. In: *Landolt-Beornstein Zahlenwerte und Funktionen* (J. Dans, J. Bartels, P.T. Bruggencate, A. Eucken, G. Joos, and W.A. Roth, Vol. II-5a, pp. 360-384. Springer-Verlag, Berlin.

Matsukata, M., T. Matsushita and K. Ueyama (1995). A circulating fluidized bed CH_4 reformer: Performance of supported Ni catalysts. *Energy and Fuels*, **9**, 822-828.

Rostrup-Nielsen, J. R. (1984). Catalytic steam reforming. In: *Catalysis Science and Technology* (J.R. Anderson and M. Boudart, ed.) Vol. 5, pp. 1-117. Springer-Verlag, Berlin.

Takayasu, O., T. Hata and I. Matsuura (1990). Particle size of nickel supported on crystalline magnesium oxide prepared in a solvent-free condition. *Chemistry Express*, **5**, 829-832.

Takayasu, O., F. Sato and I. Matsuura (1995). Reaction path in the carbon dioxide reforming reaction of methane. *Shokubai*, **37**, 482-485.

Pergamon

Energy Convers. Mgmt Vol. 38, Suppl., pp. S397–S402, 1997
© 1997 Elsevier Science Ltd. All rights reserved
Printed in Great Britain
0196-8904/97 $17.00 + 0.00

PII: S0196-8904(96)00301-9

CATALYTIC CONVERSION OF CARBON DIOXIDE INTO HYDROCARBONS OVER ZINC PROMOTED IRON CATALYSTS

Sang-Sung NAM*, Soo-Jae LEE, Ho KIM, Ki-Won JUN, Myuong-Jae CHOI
and Kyu-Wan LEE[†]

K R I C T., P.O. Box 107, Yusong, Taejon 305-600, Korea

ABSTRACT

The hydrogenation of carbon dioxide to hydrocarbons over iron catalysts was studied in a fixed bed reactor under pressure of 10 atm and temperature of 573 K. Iron catalysts promoted with V, Cr, Mn and Zn prepared by precipitation method were adopted in the present study. The catalysts were characterized by XRD, carbon dioxide chemisorption and ^{57}Mössbauer spectroscopy. The hydrocarbons were formed directly from carbon dioxide over iron catalysts. The iron promoted with Cr and Mn improved conversion of carbon dioxide and increased the selectivity of C_2 - C_4 alkenes. Whereas, the Zn promoted iron catalyst showed unusually very high selectivity for C_2 - C_4 alkenes. With varying Fe:Zn ratio, the smaller ratio of Zn increased the alkene selectivity. © 1997 Elsevier Science Ltd

KEYWORDS

Conversion of CO_2, Fischer -Tropsch reaction, Zn promoted iron catalysts, Iron carbides, Alkene selectivity

INTRODUCTION

Fixation and utilization of carbon dioxide as a useful carbon source might be one of the effective ways to reduce global warming that is caused from the increasing carbon dioxide concentration in the atmosphere. One of the promising goals in the carbon dioxide utilization is the synthesis of valuable chemicals by the hydrogenation of carbon dioxide (Choi et al., 1995; Park et al., 1995; Chanchlani et al., 1992; Inui et al., 1993; Fujiwara et al., 1995a). Although numerous efforts have been made to improve the selectivity to higher hydrocarbons or lower olefins in coal derived syngas reaction (Janardanarao et al., 1990; Dry, 1982), a few approaches have been reported on the hydrogenation of carbon dioxide to hydrocarbons (Lee et al., 1989; Fujiwara et al., 1995b; Lee et al., 1991; Choi et al., 1996). In this report, the hydrogenation of carbon dioxide to hydrocarbons over various iron catalysts were studied. V, Cr, Mn and Zn. promoted Iron catalysts were prepared by precipitation, reduced with hydrogen, have been

---

[†] To whom correspondence should be addressed.

used in the present study.   These catalysts enhanced conversion of carbon dioxide and Zn promoted iron catalyst showed a remarkable change in the distribution of hydrocarbons.

EXPERIMENTAL

The iron promoted with V, Cr, Mn and Zn catalysts were prepared by coprecipitation method with $Fe(NO_3)_3 9H_2O$ and corresponding nitrates as precursor of promoters.   Coprecipataion was carried out by mixing requisite amounts of metal nitrate solutions and then NH_4OH (10% ammonia by weight) was added slowly into an aqueous solution up to pH 7.   The atomic ratio of iron to promoted metal was 9 : 1.   For more study, Zn promoted Iron catalysts with different ratio (Zn/Fe atomic ratio : 0 - 1.0) were prepared by this method.   The coprecipitated material was initially filtered, washed with distilled water and dried at 383 K for 24 hrs.   The catalysts were calcined in air at 773 K for 24 hrs and reduced in situ of pure hydrogen at 723 K for 20 hrs.   The powder X-ray diffraction analysis was carried out by a Rigaku 21155D6 using CuKα radiation.   The ^{57}Mössbauer spectra of fresh and used catalysts were recorded at room temperature using the spectrophotometer equipped with a 308 channel pulsed-height analyzer.   The source used for this study was ^{57}Co in Rh matrix in the standard transmission geometry.   The spectra were taken over the range ± 10 mm/sec and were shown in Fig. 1.   The CO_2 chemisorption was conducted at 308 K ; first, the sample (1.0 g) was evacuated for 4 hrs at 573 K and reduced under flowing H_2 for 12 hrs before measuring the CO_2 chemisorption.   The hydrogenation of CO_2 was carried out in a 1/4 inch I.D. stainless-steel fixed-bed reactor with an effective length 30 cm.   About 0.5 g of catalyst (particle size: 21 - 35 mesh) was packed between quartz wool plugs in the reactor having on-line analytical provisions and was reduced in-situ at 773 K.   The reaction was conducted at 10 atm total pressure, 573 K temperature, $H_2/CO_2 = 3$, and 1900 ml/g.cat/hrs space velocity. The activity and selectivity data were collected when the catalyst attained steady activity that remained unchanged for 24 hrs.

RESULTS AND DISCUSSIONS

Characterization of Catalysts

The XRD patterns of fresh catalysts showed that α-Fe_2O_3 phase in the form of large particle in an amorphous matrix.

Because of similar ^{57}Mössbauer spectra of all prepared catalysts, we presented two typical cases (Fe-V and Fe-Zn). ^{57}Fe Mössbauer spectra of these fresh catalysts (Fig. 1-A1 and B1) showed isomer shift to 0.37 mm/sec with magnetically split of sextet ($H_f = 515$ KOe) and showed the broadening of the central quadrupole (QS = 0.64 mm/sec) which were corresponded to large Fe_2O_3 particles (180 Å) (Topsoe et al., 1980, Berry et al., 1985) and consistent with XRD results. As shown in Fig. 1-A2 and B2, ^{57}Mössbauer spectra of used catalyst showed superposition of patterns for iron carbides and high-spin Fe^{3+}, as has been observed in the case of iron Fischer-Tropsch catalysts (Galuszka et al., 1992). It also showed that the greater proportion of catalyst was covered by carbon containing species (Fe_xC_y [x,y : integer]: Isomer Shift = 0.22, 0.27, 0.44 mm/sec) according to the computer fitted spectra. These carbides increase the probability of chain growth of hydrocarbons.

Chemisorption experiments suggested that the amount of chemisorbed carbon dioxide on the prepared catalysts was increased in the order of V < Cr < Mn < Zn. Therefore, the relative proportion of chemisorbed hydrogen and carbon species is a very important parameter in the distribution of hydrocarbons. Further research will be carried out to investigate the nature of these structures.

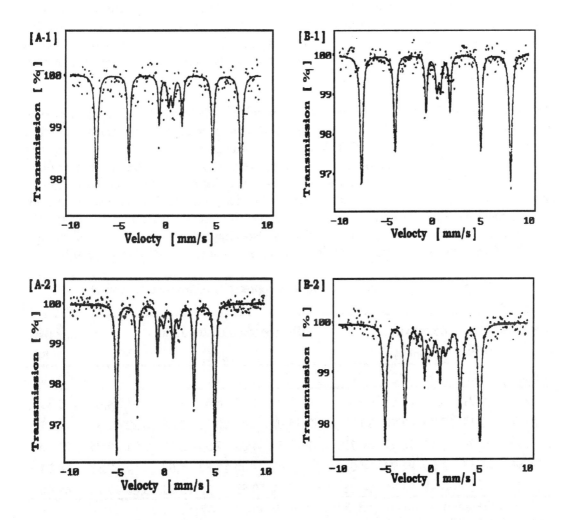

Fig. 1. Mössbauer spectra of the promoted iron catalysts at room temperature.
[A-1] fresh V-Fe, [A-2] used V-Fe , [B-1] fresh Zn-Fe, [B-2] used Zn-Fe
(The ratio of M : Fe is 10 : 90)

Carbon Dioxide Hydrogenation

Hydrogenation of carbon dioxide over iron promoted V, Cr, Mn and Zn catalysts have been carried out in a continuous fixed bed flow reactor.   These results were presented in Table 1. The hydrocarbons were synthesized in the Fischer-Tropsch process and a typical Anderson-Schults-Flory distribution was observed.   Compared with iron catalysts, the addition of 10 % of V to iron decreased the conversion of carbon dioxide and did not show significant advantages for the selectivity of C_2 - C_4 alkenes.   In the mean while, the addition of 10 % of Cr, Mn and Zn to iron catalysts increased the conversion of carbon dioxide and the selectivity of C_2 - C_4 alkenes.   During the course of hydrogenation of carbon dioxide, we believed that the acidic carbon dioxide favors the basic properties of catalyst surface in order to chemisorb, which is an essential step in the formation of hydrocarbons by the hydrogenation of carbon dioxide.   If more carbon containing iron species were formed during activation of fresh catalysts, the possibility of the chain growth of surface carbons will be increased.   And, the produced olefins are strongly adsorbed on the catalysts sites and may undergo further reaction (Kuipers et al., 1995; Satterfield, 1991).   Eventually, produced olefins are further hydrogenated and converted to paraffins.

Therefore, the addition of V to iron catalysts easily oxidized the catalyst surface and prevented the transformation of iron carbide species (Fig. 1-A2), this decreased the conversion of carbon dioxide.   However, the addition of Cr, Mn and Zn to iron generated more iron carbide species and created more basic sites on the catalyst surface, these sites increased the selectivity of C_2 - C_4 alkenes.   Moreover, Zn promoted iron catalyst greatly increased the selectivity of C_2 - C_4 alkenes, especially light olefins.   This suggests that Zn promoted iron catalyst favors for reverse water gas shift reaction and acts as a base which preferably adsorbs and activates the carbon dioxide.   The readsorption of olefins is also retarded by the presence of strongly adsorbed CO_2 onto the Zn surface and this reduces their secondary reactions into paraffins.

Table 1. CO_2 hydrogenation on various coprecipitated iron-metal catalysts[a]

Catalysts	CO₂ conv.	Selectivity (C mol %)		Hydrocarbon Distribution (C mol %)								Ol (%)[b] (Ol+Pa)
(Fe-M)	(%)	CO	HC	C_1	$C_2^=$	C_2	$C_3^=$	C_3	$C_4^=$	C_4	C_5>	C2-C4
Fe	16.21	36.39	63.61	49.67	0.24	19.26	1.17	16.61	1.15	7.51	4.41	5.55
Fe-V	11.17	27.30	72.70	39.72	1.57	15.02	-	22.05	5.88	8.22	7.54	14.13
Fe-Cr	25.70	21.53	78.47	64.75	1.01	15.54	4.05	7.39	2.10	2.91	2.25	21.70
Fe-Mn	23.15	8.06	91.94	38.06	1.24	17.42	8.00	13.82	5.09	8.84	7.53	26.34
Fe-Zn	26.54	4.35	95.65	24.26	6.95	7.11	19.58	4.92	13.93	5.60	17.29	69.64

[a]CO_2 hydrogenation at 1900 ml/g/h, 573 K, and 10 atm,
[b]Selectivity to olefins (C mol %)

Based on above result, the performance (activity, selectivity, and stability with time) of Zn promoted iron catalysts with varying Fe:Zn ratio were more investigated and results were presented in Table 2. The catalyst with the ratio of Fe:Zn = 90:10 showed the highest conversion of carbon dioxide and highest selectivity for olefin formation. When the ratio of Fe:Zn was less than 1, the conversion of carbon dioxide and C_2 - C_4 alkene selectivity was decreased.

Table 2. Effects of catalyst composition on the catalytic activity of Fe-Zn[a]

Catalysts	CO_2 conv.	Selectivity (Cmol%)		Hydrocarbon Distribution (C mol %)								$\frac{Ol. (\%)[b]}{(Ol+Pa)}$
(Fe:Zn)	(%)	CO	HC	C_1	$C_2^=$	C_2	$C_3^=$	C_3	$C_4^=$	C_4	$C_5>$	C2-C4
10 : 0	16.02	36.39	63.61	49.67	0.24	19.26	1.17	16.61	1.15	7.51	4.41	5.55
9 : 1	26.54	4.35	95.65	24.62	6.95	7.11	19.58	4.92	13.93	5.60	17.29	69.64
7 : 3	25.36	14.21	85.79	30.35	3.06	10.04	14.38	7.56	9.57	7.63	17.31	51.61
5 : 5	24.82	8.02	91.98	43.05	1.93	16.52	9.89	12.09	5.40	6.95	4.18	32.63
3 : 7	23.27	20.51	79.46	37.84	1.62	15.74	7.54	13.58	7.50	5.75	10.54	32.08
1 : 9	19.67	22.43	77.57	54.87	0.12	18.38	0.48	14.25	0.44	6.82	4.66	2.25
0 : 10	4.40	99.31	0.69	73.70	0	7.88	0	18.43	-	-	-	0

[a]CO_2 hydrogenation at 1900 ml/g/h, 573 K, and 10 atm,
[b]Selectivity to olefins (C mol %)

CONCLUSIONS

Cr , Mn and Zn Promoted iron catalysts increased the conversion of carbon dioxide during the course of reaction. Zn promoted iron catalyst showed very high selectivity for C_2 - C_4 alkenes. Decreasing the Fe:Zn ratio increased the basicity of Fe-Zn surface and increased the amount of carbon dioxide chemisorption. Therefore, it is a promising application to introduce Fischer-Tropsch reaction to our experimental systems and the product distribution can be controlled by these combinations.

REFERENCES

Berry, F.J., L. Lin, W.C. Yu, R. Tang, S. Zhang, Y.D. Bai, An in situ Mössbauer investigation of the influence of metal-support and metal-metal interactions on the activity and selectivity of iron-Ruthenium catalysts J. Chem. Soc. Faraday Trans. 1, 81, 2293, (1985)

Chanchlani, K.G., R.R. Hudgins and P.L. Silveston, Methanol synthesis form H_2, CO, and CO_2 over Cu/ZnO catalysts. J. Catal. 136, 59, (1992)

Choi, P.H., K.W. Jun, S.J. Lee, I.H. Park, M.J. Choi and K.W. Lee, Effect of potassium in iron/alumina catalysts for the synthesis of hydrocarbons. ICCDU.-3. USA. (1995)

Choi, P.H., K.W. Jun, S.J. Lee, M.J. Choi and K.W. Lee, Hydrogenation of carbon dioxide over alumina supported Fe-K catalysts. Catal. Letter, 40, 115, (1996)

Dry, M.E., Catalytic aspects of industrial Fischer-Tropsch synthesis. J. Mol. Cat. 17, 133, (1982)

Fujiwara, M., H. Ando, M. Matsumoto, Y. Matsumura, M. Tanaka, Y. Souma, Hydrogenation of carbon dioxide over Fe-ZnO/Zeolite composite catalysts. Chem. Letters. 839, (1995a)

Fujiwara, M., R. Kieffer, H. Ando and Y. Souma, Development of composite catalysts made of Cu-Zn-Cr oxide/zeolite for the hydrogenation of carbon dioxide. Appl. Catal. A 121, 113, (1995b)

Galuszka, J., T.Sano and J.A.Sawicki, Study of carbonaceous deposite on Fischer-Tropsch oxide-supported iron catalysts. J. Catal.136, 96, (1992)

Inui, T., K. Kitagawa, T. Takeguchi, T. Hagiwara and Y. Makino, Hydrogenation of carbon dioxide to C₁-C₇ hydrocarbons via methanol on composite catalysts Appl. Catal. A 94, 31, (1993)

Janardanarao, M., Direct catalytic conversion of synthesis gas to lower olefins. Ind. Eng. Chem. Res. 29, 1735, (1990)

Kuipers, E.W., I.H. Vinkenburg, and H. Dosterbeck, Chain Length dependence of α-olefin readsorption in Fischer-Tropsch synthesis. J. Catal. 152, 137, (1995)

Lee, M.D., J.F. Lee, C.S. Chang, Hydrogenation of carbon dioxide on unpromoted and potassium-promoted iron catalysts. Bull. Chem. Soc. Jpn. 62, 2756, (1989)

Lee, M.D., J.F. Lee, C.S. Chang and T.Y. Dong, Effects of addition of chromium, manganese, or molybdenum to iron catalysts for carbon dioxide hydrogenation. Appl. Catal. 72, 267, (1991)

Park, S.E., S.S. Nam, M.J. Choi and K.W. Lee, Catalytic reduction of carbon dioxide. The effects of catalysts and reductants. Energy Convers. Mgmt. 36, 573, (1995)

Satterfield, C.N., Heterogeneous Catalysis in Practice, 2nd ed, McGraw-Hill, 432, (1991)

Topsoe, H., J.A Dumesic, and S. Morapin, R.L. Cohen (1980). Application of Mössbauer spectroscopy Vol. 2, Academic Press, New York.

 **Pergamon**

Energy Convers. Mgmt Vol. 38, Suppl., pp. S403–S408, 1997
© 1997 Elsevier Science Ltd. All rights reserved
Printed in Great Britain
0196-8904/97 $17.00 + 0.00

PII: S0196-8904(96)00302-0

METHANOL SYNTHESIS FROM CO₂ AND H₂
OVER A Cu/ZnO-BASED MULTICOMPONENT CATALYST

Masahiro SAITO[1],*, **Masami TAKEUCHI**[2], **Taiki WATANABE**[2],
Jamil TOYIR[3], **Shengcheng LUO**[3], **Jingang WU**[3]

1) National Institute for Resources and Environment (NIRE), 16-3 Onogawa, Tsukuba-shi, Ibaraki 305, Japan
2) Research Institute for Innovative Technology for the Earth (RITE), 9-2 kizukawadai, kizu-cho, Soraku-gun, Kyoto 619-02, Japan
3) RITE (NEDO Industrial Technology Researcher), 16-3 Onogawa, Tsukuba-shi, Ibaraki 305, Japan

ABSTRACT

The optimum operation conditions for preparing a Cu/ZnO-based multicomponent catalyst (Cu/ZnO/ZrO₂/Al₂O₃/Ga₂O₃) by a coprecipitation method were established. The operation conditions for preparing the precipitate rarely affect the activity of the catalyst, whereas thoroughly washing the precipitate with distilled water to remove Na in the catalyst is greatly important for improving the activity of the catalyst. The Cu/ZnO-based multicomponent catalyst prepared under the optimum preparation conditions was highly active and stable for a long period in a continuous methanol synthesis operation at 523 K with a total pressure of 5 MPa. In the case of the methanol synthesis over the multicomponent catalyst using a recycle reactor employed as a model reactor for a practical methanol synthesis process, the conversion of CO₂ in a make-up gas to methanol was more than 99.9%, indicating that the methanol synthesis over the multicomponent catalyst is extremely selective. © 1997 Elsevier Science Ltd

KEYWORDS

CO₂ hydrogenation; methanol synthesis; Cu/ZnO-based multicomponent catalyst; catalyst stability; recycle reactor.

INTRODUCTION

The greenhouse effect of carbon dioxide has been recognized to be one of the most serious problems in the world, and a number of countermeasures have been proposed so far. Catalytic hydrogenation of CO₂ to produce various kinds of chemicals and fuels has received much attention as one of the most promising mitigation options. In particular, methanol synthesis by CO₂ hydrogenation has been considered to play an important role in the transportation of hydrogen energy produced from natural energy such as solar energy, hydropower and so on, as shown in Fig. 1 (Sano, 1993). According to some estimations (Kubo, 1995), an electric power of 300 MWh could be obtained from a methanol fired power plant in Japan, if methanol synthesized from CO₂ and H₂ produced by a electrolysis of water using an electric power of 1000 MWh is transported to Japan through the system shown in Fig. 1.

A practical methanol synthesis process greatly requires a high performance catalyst, which must be highly active and selective for methanol synthesis and also stable for a long period in a continuous operation. NIRE and RITE have been jointly implicated in the development of high performance methanol synthesis catalysts since 1990. The authors elucidated the role of metal oxides contained in Cu/ZnO-based ternary catalysts, and then developed Cu/ZnO-based multicomponent catalysts containing two or three metal oxides (Saito, *et al.*, 1995). The present

---

* To whom correspondence should be addressed.

study is devoted to optimize the operation conditions for preparing a Cu/ZnO-based multicomponent catalyst by investigating the effects of various preparation conditions on the activity of the catalyst. In the following step, the change in the activity has been examined for the multicomponent catalyst prepared under the optimum preparation conditions during a long-term (> 2000 h) methanol synthesis test at 523 K with a total pressure of 5 MPa using a conventional fixed bed flow reactor. Furthermore, we have investigated the methanol synthesis over the multicomponent catalyst using a recycle reactor serving as a model reactor for a practical methanol synthesis process.

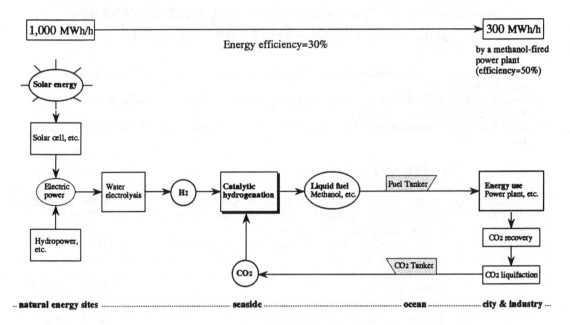

Fig. 1 Global energy network combined with catalytic hydrogenation of CO_2.

EXPERIMENTAL

A Cu/ZnO-based multicomponent catalyst ($Cu/ZnO/ZrO_2/Al_2O_3/Ga_2O_3$) was prepared by a coprecipitation method. A mixed aqueous solution of metal nitrates and an aqueous solution of Na_2CO_3 were added dropwise to distilled water. Subsequently, the precipitate was filtered out, washed with distilled water, dried in air at 383 K overnight, and calcined in air at 623 K for 2 h. The operation conditions for preparing the catalyst such as the temperature during coprecipitation, the concentration of an aqueous solution of mixed metal nitrate, the time of aging precipitate, the extent of washing precipitate with distilled water and so on were varied.

A conventional fixed bed flow reactor was used both for short-term and long-term methanol synthesis tests. A commercial catalyst ($Cu/ZnO/Al_2O_3$) for methanol synthesis from syngas was used for comparison. Furthermore, a recycle reactor equipped with a compressor for recycling unreacted gases was used for investigating practical methanol synthesis operations, as shown in Fig. 2. The catalyst fixed in a reactor was reduced in a gas mixture of H_2 (10%) and He (90%) at 573 K with a total pressure 5 MPa. The hydrogenation of CO_2 was then carried out at 523 K with a total pressure of 5 MPa in a fixed bed flow reactor by feeding a gas mixture of H_2 and CO_2 with a mole ratio of $H_2/CO_2=3$. The reaction products were analyzed by means of gas chromatographs directly connected to the reactor. The main products of CO_2 hydrogenation over Cu/ZnO-based catalysts were methanol, CO and water. Methane, dimethyl ether and methyl formate were also detected in the reaction products, but the selectivities for the by-products were less than 0.1%. The catalyst activity was measured 2h after supplying the feed gas to the reactor except for a long-term test and for a test using the recycle reactor.

The total copper surface area of the catalyst after the reaction (Cu_{total}) was determined by the technique of N_2O reactive frontal chromatography (RFC) after re-reducing the post-reaction catalyst with H_2 at 523 K (Chinchen, *et al.*, 1986). X-ray diffraction measurements were performed for analyzing the structure of the catalyst.

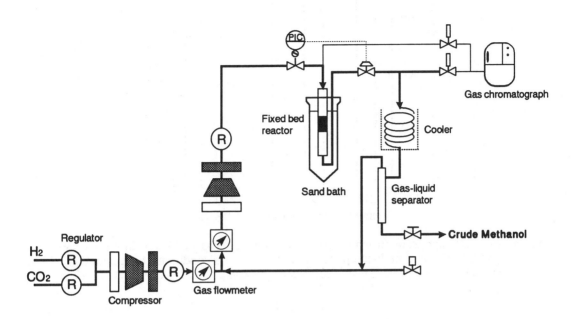

Fig. 2 Schematic diagram of a recycle reactor with a catalyst volume of 50 ml.

RESULTS AND DISCUSSION

Optimization of Operation Conditions for Preparing the Catalyst

The various operation conditions for preparing the precipitate except the temperature during coprecipitation had no significant effect on the activity of the multicomponent catalyst. This finding is very favorable for preparing a practical catalyst.

Figure 3 shows the methanol synthesis activity and the total Cu surface area of the catalyst as a function of coprecipitation temperature ranging from 273 K to 333 K. No significant difference among the activities of the catalysts prepared at temperatures between 273 K and 313 K has been observed, whereas the activity of the catalyst prepared at 333 K was slightly (only 7%) lower. XRD measurements showed that the crystallite size of the precipitate prepared at 333 K was slightly larger than those of the precipitates prepared at temperatures ranging from 273 K to 313 K. This finding suggests that the activity of the catalyst depends on the crystallite size of the precipitate.

In the following step, the effect of extent of washing the precipitate with distilled water to remove Na coming from Na_2CO_3 on the methanol synthesis activity of the catalyst was investigated. The amount of Na remained in the catalyst decreased with an increase in the number of washing the precipitate. The activity of the catalyst and the Cu surface area of the catalyst greatly increased with a decrease in the amount of Na in the catalyst, as shown in Fig. 4. The precipitates after different number of washings gave almost the same XRD patterns, whereas the XRD pattern for the catalyst calcined at 623 K became sharper with an increase in the amount of Na in the catalyst. This finding clearly indicates that the Na remained in the catalyst crystallizes the components of the catalyst, and thus decreases the surface area, the Cu surface area and the activity of the catalyst.

In summary, the operation conditions for preparing the precipitate rarely affect the activity of the catalyst, whereas thoroughly washing the precipitate with distilled water to remove Na in the catalyst is greatly important for improving the activity of the catalyst.

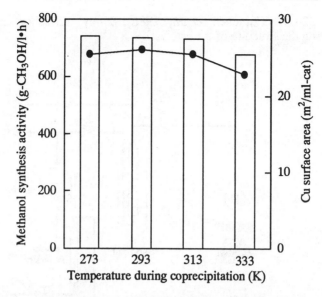

Fig. 3 Effect of temperature during coprecipitation on the activity (□) and Cu surface area (●) of a Cu/ZnO-based multicomponent catalyst. Reaction conditions: temperature =523 K, total pressure = 5 MPa, SV = 18,000, feed gas composition = CO_2 (25) / H_2 (75).

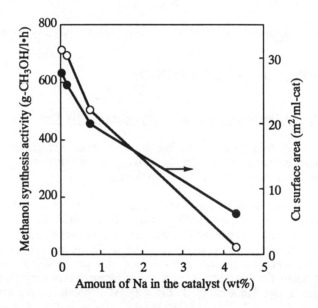

Fig. 4 Effect of amount of Na in the catalyst on the activity (O) and Cu surface area (●) of the multicomponent catalyst. Reaction conditions were the same as shown in Fig. 3.

Long-term Stability of the Catalyst

A methanol synthesis catalyst for a practical process is highly required to have a stable activity for a long period in a continuous operation. A long-term methanol synthesis test was performed at 523 K with a total pressure of 5 MPa by using a gas mixture of CO_2, CO and H_2, because unreacted gases and CO must be recycled to the reactor in a practical process. Figure 5 shows the change in the activity with time on stream of the multicomponent catalyst compared with a commercial catalyst used for methanol synthesis from syngas. The activity of the multicomponent catalyst decreased only by 17% in 1000 h during the test, but unchanged from 1000 h to 3400 h.

On the other hand, the activity of the commercial catalyst decreased by 20% in 1000 h, and still decreased to 75% of the initial activity in 2100 h. The Cu particle size of the multicomponent catalyst increased by 50% after the methanol synthesis for 3400 h, whereas that of the commercial catalyst increased by 75% in 2100 h. Therefore, one explanation for the deactivation could be related to the decrease in the Cu surface area of the catalyst. These findings described above clearly indicate that the multicomponent catalyst developed in the joint research is very stable for a long period in a continuous methanol synthesis operation.

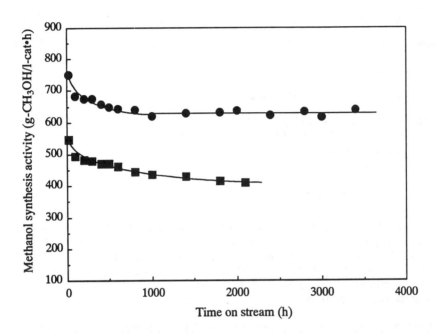

Fig. 5 Change in the activities of the multicomponent catalyst (Cu/ZnO/ZrO$_2$/Al$_2$O$_3$/Ga$_2$O$_3$, ●) and a commercial catalyst used for methanol synthesis from syngas(Cu/ZnO/Al$_2$O$_3$, ■) during a long term methanol synthesis. Reaction conditions: temperature =523 K, total pressure = 5 MPa, SV = 18,000, feed gas composition = CO$_2$ (25)/CO(3)/H$_2$ (75).

Methanol Synthesis using a Recycle Reactor

Practical methanol synthesis must be performed by using a reactor with recycling equipments for unreacted gases, because the conversion of CO_2 to methanol at reaction equilibrium is very low under ordinary reaction conditions, for example, 17% at 523 K and 5 MPa. Therefore, methanol synthesis using a recycle reactor was also investigated. The products of CO_2 hydrogenation at 473 K to 548 K with a total pressure of 5 MPa in the recycle reactor were methanol, CO, water, methane, ethane, dimethyl ether, methyl formate, ethanol, propanol and butanol, but the yields of the products other than methanol, CO and water were very small. The selectivity for methanol in the products except CO and water was more than 99.8%. The reaction products were cooled to 270 K in a gas-liquid separator connected to the reactor. Liquid products collected in the gas-liquid separator were taken out of the reactor, and unreacted gases as well as gaseous products such as CO, methane, ethane and dimethyl ether were recycled back to the reactor. The concentrations of methane, ethane and dimethyl ether in a recycled gas mixture remained constant after some initial period in CO_2 hydrogenation without purging unreacted gases. Table 1 shows the composition of liquid products except H$_2$O taken out of the recycle reactor. The purity of the crude methanol produced in the present work was 99.96 wt%, and higher than that of the crude methanol produced in a commercial methanol synthesis from syngas. The CO concentration of the feed gas to the reactor in the methanol synthesis from CO_2 and H_2 is much lower than that in the methanol synthesis from syngas. It is well known that CO is much more reactive with H_2 to produce higher alcohols and higher hydrocarbons than CO_2. Therefore, this finding suggests that the lower CO concentration in the feed gas to the reactor should result in the lower yield of by-products and thus the higher methanol purity.

Table 1 The composition of liquid products (except water) from a recycle reactor in this work[a], compared with that from a commercial plant for methanol synthesis from syngas[b]

Compound		Composition	
		This work	A commercial plant
Methanol	CH$_3$OH	**99.96 wt%**	**99.59 wt%**
Methyl formate	HCOOCH$_3$	330 ppm	700 ppm
Higher alcohols (C$_2$-C$_4$)	ROH	30 ppm	530 ppm
Hydrocarbons (C$_6$-C$_{10}$)	C$_n$H$_m$	-	50 ppm
Dimethyl ether	(CH$_3$)$_2$O	-	230 ppm

a) Reaction conditions : catalyst=Cu/ZnO/ZrO$_2$/Al$_2$O$_3$/Ga$_2$O$_3$, temperature=517K, total pressure=5 MPa, H$_2$/CO$_2$ ratio in the make-up gas=3.
b) cited from a booklet on ICI methanol synthesis catalyst.

CONCLUSIONS

(1) The optimum operation conditions for preparing a Cu/ZnO-based multicomponent catalyst (Cu/ZnO/ZrO$_2$/Al$_2$O$_3$/Ga$_2$O$_3$) by a coprecipitation method were determined. The operation conditions for preparing the precipitate except the temperature during coprecipitation had no significant effect on the catalyst activity. The temperature during coprecipitation should be less than 313 K. The removal of Na from the catalyst by washing the precipitate is greatly important for improving the catalyst activity.

(2) The multicomponent catalyst prepared under the optimum conditions is highly effective and stable for a practical methanol synthesis from CO$_2$ and H$_2$.

(3) The purity of crude methanol produced from CO$_2$ and H$_2$ in a recycle reactor is more than 99.9%, and higher than that of crude methanol produced from syngas in a present commercial plant.

ACKNOWLEDGMENT

The present work was partly supported by New Energy and Industrial Technology Development Organization (NEDO).

REFERENCES

Chinchen, G. C., K. C. Waugh and D. A. Whan (1986). The activity and state of the copper surface in methanol synthesis catalysts. *Appl. Catal.*, **25**, 101.
Kubo, Y. (1995). private communication.
Saito, M., T. Fujitani, I. Takahara, T. Watanabe, M. Takeuchi, Y. Kanai, K. Moriya, T. Kakumoto (1995). Development of Cu/ZnO-based multicomponent catalysts for methanol synthesis by CO$_2$ hydrogenation. *Energy Convers. Mgmt*, **36**, 577-580.
Sano H. (1993). New version of energy transportation system for CO$_2$ global recycling. *Proc. Int. Symp. on CO$_2$ Fixation & Efficient Utilization of Energy*, Tokyo, Japan, 117-122.

Energy Convers. Mgmt Vol. 38, Suppl., pp. S409–S414, 1997
© 1997 Elsevier Science Ltd. All rights reserved
Printed in Great Britain
0196-8904/97 $17.00 + 0.00

Pergamon

PII: S0196-8904(96)00303-2

AN ASSESSMENT PROCEDURE FOR CHEMICAL UTILISATION SCHEMES INTENDED TO REDUCE CO_2 EMISSIONS TO ATMOSPHERE

H. AUDUS[1] and H. OONK[2]

[1]IEA Greenhouse Gas R&D Programme, CRE, Stoke Orchard, Glos. GL52 4RZ, UK.
[2]TNO Milieu, Energie en Procesinnovatie, Postbus 342, 7300 AH, Apeldoorn, The Netherlands.

ABSTRACT

The concept of reducing emissions of CO_2 to the atmosphere by producing chemicals has been suggested by many people as a potential greenhouse gas mitigation option. The goal of such schemes is either: (i) fixation of CO_2 in a chemical compound for a significant time, or, (ii) reduction of emissions by replacing an existing process with an alternative releasing less CO_2. This paper describes a simple assessment procedure which can be used as a preliminary screen to test the validity of schemes for chemicals utilisation of CO_2. The procedure is presented as a template (or worksheet); two summaries of its use are presented. The template gives research workers a decision methodology which leads to the proposed scheme either being rejected or accepted as being of sufficient potential to merit more detailed investigation. © 1997 Elsevier Science Ltd

KEYWORDS

Carbon dioxide; Fixation; Chemical utilisation.

INTRODUCTION

Increasing concern about anthropogenic emissions of greenhouse gases and their potential effect on the climate have led research workers to investigate the possibility of storing CO_2 in the form of a chemical compound. The concept is referred to as 'fixation' because CO_2 is fixed in the chemical for a long period of time, and only enters the atmosphere when released by a chemical reaction. Fixation of CO_2 as a chemical is an alternative to greenhouse gas mitigation by storage schemes based on forest, or underground, or deep-ocean storage.

Emissions of CO_2 to atmosphere could also be reduced by:

- Implementation of alternative processes which produce the same product and have a lower net CO_2 emission than the established process.
- Replacement of an existing product with an alternative, the production of which releases less CO_2 to atmosphere.

These product and production based mitigation options are referred to as chemical 'off-set', the amount of off-set being the net change in CO_2 emitted per constant unit of output.

A comprehensive assessment of the above concepts to determine the extent of mitigation likely to be achieved by any particular scheme can be an extensive exercise. For example, it should include assessment of the energy requirements of the process and allow for the emissions arising from use of fossil fuels. There is therefore, a need for a screening procedure whereby potential chemicals and processing routes can undergo a preliminary evaluation. Such a procedure would enable researchers to assess quickly the suitability of a chemical utilisation scheme, using only simple information that can be readily obtained from standard texts.

Previous authors have described a number of prerequisites which must be fulfilled if a chemical utilisation option is to be valid as a fixation or off-set route (Aresta et al., 1992, Pechtl, 1991). These prerequisites, and a number of others, are discussed in this paper and converted to a set of criteria against which potential chemicals or chemical production routes can be tested for validity as CO$_2$ mitigation options.

A 'template' or 'work sheet' using these criteria is discussed, and its application to the assessment of 2 chemical utilisation schemes summarised. The chemicals assessed have been suggested as mitigation options by various authors; they are discussed in a survey of CO$_2$ utilisation options by (Ormerod, et al., 1995).

This paper does not discuss chemical utilisation options relying on the use of photosynthesis or any other imported 'carbon-free' energy. Carbon dioxide is assumed to be derived, however indirectly, from a fossil fuel source.

UTILISATION CRITERIA

The prerequisites for useful chemical fixation or off-set fall into two categories. Firstly is it scientifically robust ? Secondly is it effective ? These prerequisites and the ease of applying them as simple test criteria are discussed below.

Scientific prerequisites

Net emissions of CO$_2$. The proposed utilisation option has to reduce the net emission of CO$_2$ to atmosphere. Material and energy balances for a process route would provide the basis for an accurate assessment of the carbon balance. Such calculations require the process to be designed in some detail and complex calculations which are best carried out using a computer model. They are not practical as simple test criteria. Fortunately, there are two indicators which can be easily used to give guidance on net emissions of CO$_2$.

The first, applicable to fixation, is the carbon-to-hydrogen (C/H) ratio of the chemicals. Only processes which consume carbon can have products with a C/H ratio greater than the raw material (Pechtel, 1991). Consider the production of methanol from CO$_2$ and hydrogen:

$$CO_2 + 3H_2 \quad \Leftrightarrow \quad CH_3OH + H_2O$$

Each mole of CO$_2$ originates from a fossil fuel. The molar representation of C/H ratios for the major fossil fuels are: natural gas, CH$_4$; oil, CH$_{1.3}$; coal, CH$_{0.8}$. Methanol has the same molar C/H ratio as CH$_4$; it has not consumed carbon because the C/H ratio is no greater than any of the three raw materials. If the CO$_2$ was oil or coal derived substantial amounts of extra hydrogen must be provided which releases more CO$_2$.

(Oonk and Heslinga, 1995) have extended the logic of the Pechtel-criterion to a comparison of the heat of the proposed reaction with the heat of combustion of fossil fuels. They have shown that if a net fixation of CO$_2$ is to occur, the heat of reaction can be no more than 1.25 times the heat of combustion of the reference fossil fuel. These criteria give good agreement for chemicals which are predominantly carbon and hydrogen; they can give conflicting results with other chemicals. It is concluded that at least one of these criteria must be passed for net fixation of CO_2 to occur.

The second indicator, applicable to off-set, is based on a comparison of energy consumption in the proposed and alternative processes. See table 1.

Table 1. Process characteristics.

Characteristics of processes reducing net CO_2 emissions (c.f. the established alternative).
(a) A reduction in the number of processing steps. (A one-step process is generally more efficient than a three-step process.)
(b) Milder operating conditions. (Processes operating at ambient temperatures and pressures are generally more efficient than processes operating at high temperatures and pressures.)
(c) There are less discontinuities in the process conditions. (e.g. an operating sequence of, 1,2,and 8 bar pressure steps would be more efficient than, 8,1, and 2 bar steps.)
(d) Changes of phase are avoided. (Distillation, absorption, freezing, etc. require large amounts of energy.)
(e) There are improved possibilities for process integration. (i.e. the net efficiency can be increased.)

The adoption of an alternative processing route is likely to result in a net reduction in CO_2 emissions if one or more of the characteristics given in table.1 apply.

Chemical feasibility. Carbon dioxide is the end product of reactions that release energy and little 'driving force' is left. It can for example, react with water but in general CO_2 requires an input of energy to react. Consideration of free enthalpy (ΔG) can be used to give quick, approximate guidance on whether a reaction is promising. A negative or slightly positive free enthalpy value indicates that the equilibrium for the reaction favours the desired product. There is no definite point at which a reaction becomes untenable as an industrial process because the reaction can be forced in the desired direction, e.g. by applying high pressures; but the situation becomes increasingly unfavourable as the free enthalpy increases. (It is important to note that thermodynamics can not give any information about reaction rates. The application of a catalyst may alter the rate at which a given reaction proceeds it can not influence the equilibrium.) The free enthalpy of a reaction can frequently be found in the literature, if not a theoretical value can be calculated from information in standard texts.

Effectiveness

There are three prerequisite which determine the effectiveness of a chemical utilisation scheme as a means of reducing emissions of CO_2: lifetime, market size, and availability of co-reactants. These prerequisites can be specified as criteria such that failure to satisfy any one of the three indicates that a chemical utilisation scheme is not likely to be effective.

Lifetime. Chemical fixation has to store CO_2 for many years for it to be an alternative to other storage methods, such as in forests, underground, or in the ocean. Ideally, storage for centuries would be preferred but storage for several decades could be acceptable. The approximate lifetime of a chemical can be determined by considering its use. Table. 2 derived from (Aresta et al.,1992) gives an indicative lifetime for various types of chemical product. Assuming storage must be for decades at least, the range of possibilities is reduced to polymers and inorganic carbonates.

Table 2. Lifetime of chemicals.

Lifetime	Chemical product
centuries	inorganic carbonates
decades	polymers
years or less	agro-chemicals, pharmaceuticals, fuels

Market size. Given the anthropogenic emissions of CO_2 at present equate to 6GtC/year, any worthwhile storage option must total an appreciable percentage of this figure.   It can be argued that market size is not critical as a lot of small contributions could add up to be significant, but research effort should be focused on effective solutions. If we apply the criterion that an effective store must be equivalent to 0.01% of annual emissions (1/10 000th) this means that the market size for the chemical in question must be equivalent to about 2 million tonnes of CO_2/year. The 'market' is widely defined as meaning 'total capacity for'. Table. 3, again derived from (Aresta *et al*, 1992), presents indications of market size for various generic types of chemical.

Table 3. Market size for chemicals.

Chemical	Present size (tonnes of CO_2/year)	Future potential (tonnes of CO_2/year)
agro-chemicals	10 million	100 million
fuels (as a specific chemical)	10 million	>> 100 million
carboxylated products	10 000	100 000
polymers	10 000	100 million
new materials	-	100 million

The figure given in table. 3 are total market size, a view needs to be taken on what fraction of the total market could be captured by the chemical being assessed. It can be seen that based on market size the effective options are at present limited to agro-chemicals and a fuel substitute.

Co-reagents. The availability of co-reagents is a similar prerequisite to market size. Any co-reagents need to available as a bulk chemical or as raw materials. We apply the simple criterion that to be effective sufficient co-reagent is needed for processing the equivalent of 10 million tonnes CO_2/year.

THE TEMPLATE

A simplified version of the overall template is used to illustrate the principles, see tables 4&5; a more comprehensive template which will be in the form of one or two A4 sheets with guide notes on its use is still being finalised. Examples follow for one 'fixation' and one 'off-set' chemical.

Fixation of CO_2 as an inorganic carbonate

Does fixation of CO_2 in an inorganic carbonate contribute to a reduction of CO_2 emissions to atmosphere ? This reaction takes place in various minerals with a complex structure but the underlying reaction is: $CaO + CO_2 \Leftrightarrow CaCO_3$. Table 4. is the template summary.

Table 4. Template summary for fixation of CO_2 in an inorganic carbonate.

CRITERION	RESULT	REJECT POINT	CONCLUSION
Scientific:			
A) Net CO_2 emission			
C/H ratio (fixation)	n.a	C/H ratio decreases	-
Heat of reaction (fixation)	negative	1.25xHr > Hc	pass
Process characteristics (off-set)	n.a.	See table 1.	-
B) Chemical feasibility			
ΔG reactions	ΔG's negative	ΔG's distinctly +ve	pass
Effectiveness:			
Lifetime (fixation)	centuries	< 10 years	pass
Market size	huge, (can be stored)	< 2 million tonnes/yr	pass
Co-reagents	> 10 million Tonnes	< 10 million tonnes/yr	pass
		Overall conclusion :	**pass**

According to our test criteria the fixation of CO_2 in inorganic carbonates seems a feasible method of reducing emissions of CO_2 to atmosphere. The next step should be a more detailed examination of the likely economics, technology requirements including reaction rate, and overall environmental implications.

Production of methanol by an alternative to the established process

At present methanol is produced from synthesis gas (essentially a mixture of CO and H_2) which is in turn mostly produced by steam reforming of methane. An alternative route for the production of methanol may be to produce it from CO_2 and hydrogen. Does this change of processing routes contribute to a reduction in emissions of CO_2? We have for the established process the following equations:

$$CH_4 + H_2O \Leftrightarrow CO + 3H_2 \Leftrightarrow CH_3OH + H_2$$

The excess hydrogen is used as a supplementary fuel because the steam reforming reaction is highly endothermic. In the suggested alternative, CO_2 is imported and reacted with hydrogen. Hydrogen is not readily available and has to be produced as part of the process. The steam reforming of methane is the most effective method of producing hydrogen from a fossil fuel.

Hence, the reactions are:

$$CH_4 + H_2O \Leftrightarrow CO + 3H_2$$
$$CO_2 + 3H_2 \Leftrightarrow CH_3OH + H_2O$$
Net reaction: $CO_2 + CH_4 \Leftrightarrow CH_3OH + CO$

Note that there is no surplus of hydrogen to provide energy for the reaction. Note also, that if 'fixation' were being tested the reaction would fail because the C/H ratio has not increased. However, the proposition being tested is that the alternative route to methanol reduces emissions of CO_2, hence, applying the template produces the results given in table 5.

Table 5. Template summary for off-set by alternative production process for methanol

CRITERION	RESULT	REJECT POINT	CONCLUSION
Scientific:			
A) Net CO_2 emission			
C/H ratio (fixation)	n.a	C/H ratio decreases	-
Heat of reaction (fixation)	n.a	$1.25 \times Hr > Hc$	-
Process characteristics (off-set)	more steps	See table 1.	fail
B) Chemical feasibility			
$\Delta G_{reactions}$	ΔG's negative	ΔG's distinctly +ve	pass
Effectiveness:			
Lifetime (fixation)	n.a.	< 10 years	-
Market size	> 100 million tonnes	< 2 million tonnes/yr	pass
Co-reagents	> 100 million tonnes	< 10 million tonnes/yr	pass
		Overall conclusion :	**fail**

According to our test criteria the alternative route for the production of methanol is unlikely to reduce emissions of CO_2. This is because the overall process is more complex which is likely to lead to a lower overall efficiency and an increase in net emissions of CO_2. This conclusion has been confirmed by more detailed calculations which show both routes to produce emissions in the region of 0.5 kg CO_2/ kg of methanol. Hence, the change in process routes is unlikely to make a significant contribution to a reduction in CO_2 emissions to atmosphere.

ACKNOWLEDGEMENT

The work reported here was undertaken by Hans Oonk and commissioned by The IEA Greenhouse Gas R&D Programme. This paper is a shortened overview of the work by Harry Audus who is responsible for any errors.

REFERENCES

Aresta, M., Quaranta, E., Tommasi, I. (1992), Prospects for the utilisation of carbon dioxide. *Energy Conversion and Management.*, **33**, 495-504.

Oonk, H., Heslinga, D., (1995), Energetic aspects of carbon dioxide. Poster presented at IEA GHG conference: Greenhouse Gases - Mitigation Options, 22-25 August 1995, London.

Ormerod, W., Riemer, P.W.F., Smith, A., (1995), *Carbon dioxide Utilisation.* Pub., IEA Greenhouse Gas R&D Programme, ISBN 1 898373 17 5.

Pechtl, P.A., (1991), Zur Beurteilung von CO_2- Emissionsminderungsmaßnahmen aus kalorischen Kraftwerken. *Erdöl und Kohle.* , **44**, 159-165.

Energy Convers. Mgmt Vol. 38, Suppl., pp. S415–S422, 1997
© 1997 Elsevier Science Ltd. All rights reserved
Printed in Great Britain
0196-8904/97 $17.00 + 0.00

Pergamon

PII: S0196-8904(96)00304-4

GREENHOUSE GAS CHEMISTRY

A. BILL, A. WOKAUN

Paul Scherrer Institute, CH-5232 Villigen, Switzerland

and

B. ELIASSON, E. KILLER, U. KOGELSCHATZ

ABB Corporate Research Center, CH-5405 Baden, Switzerland

ABSTRACT

One of the major problems facing mankind is the global warming of the atmosphere due to man-made emissions of greenhouse gases. Mitigation of these greenhouse gas emissions to the atmosphere can be achieved by using direct control technologies (capture, disposal or chemical recycling). In this paper, we report on carbon dioxide and methane recycling with other chemicals, especially with hydrogen and oxygen, to produce liquid fuels such as. methanol. Methanol synthesis from pure CO_2 is investigated with various catalysts at moderate pressures (≤ 30 bar) and temperatures (≤ 300 °C). The catalysts show good methanol activity and selectivity. The remarkable finding is that the equilibrium yields per pass over the catalysts are already reached at temperatures above 250 °C. For a stoichiometric feed, at 225 °C, 20 bar and at a space velocity of 4500 h^{-1}, a best methanol yield equal to 7.2 % is reported. The conversion of CO_2 and CH_4 to methanol is also studied in a silent electrical discharge at pressures of 1 to 4 bar and temperatures close to room temperature. Methanol yields are given for mixtures of CO_2/H_2, CH_4/O_2 and also for CH_4/air mixtures. © 1997 Elsevier Science Ltd

KEYWORDS

Greenhouse gases; carbon dioxide; methane; methanol synthesis; catalyst; silent discharge; CO_2 hydrogenation; CH_4 oxidation.

INTRODUCTION

One of the major global environmental problems facing mankind is the global warming of the atmosphere. The ever-increasing consumption of fossil fuels since the industrial revolution - some 250 years ago - has caused a marked increase in the global concentration of carbon dioxide, which now amounts to 360 ppm (Fricke *et al.*, 1993). The amount of carbon dioxide released to the atmosphere by man is currently about 30 Gt per annum (8 Gt C every year, Eliasson, 1994a). This infrared adsorbing gas, together with other gases, such as methane, chlorofluorocarbons, nitrous oxide and ozone, threatens to cause a rise in the Earth's temperature, thus leading to an intensified greenhouse effect. Due to the global population growth and increase in living standards - especially in developing countries - the greenhouse gas emissions will undoubtedly increase during the next years.

One possible approach to mitigate the emissions of carbon dioxide and methane to the atmosphere would be to recycle the carbon in a chemical process to form useful products such as methanol or dimethyl ether, for instance. Methanol has the advantage that it is liquid under normal conditions. It can be stored and transported as easily as gasoline, and can be used in conventional combustion engines without requiring any major adjustments. Methanol has twice the energy density of liquid hydrogen. Methanol synthesis can thus be looked upon as a way of converting hydrogen into an energy carrier that can be more conveniently stored and transported.

If pure hydrogen from renewable sources (e.g. hydroelectric power) is available, the easiest method for converting it to methanol with CO_2 is to combine both gases in a thermal reactor at about 220 °C under moderate pressure (20 - 50 bar). Such experiments have been performed by us and are reported in this paper. This conversion requires no additional energy as the reaction is exothermic and the heat and energy needed originate from the hydrogen and the reaction itself. The function of carbon dioxide as a storage matrix for hydrogen is evident in this case. If no pure hydrogen is available but only hydrogen containing molecules then greenhouse gases can, for instance, be recycled in a silent discharge reactor with the aid of a non-equilibrium high pressure electrical discharge. It generates high energy electrons, which in turn dissociate the carbon dioxide or the hydrogen donor molecules and allow them to form new molecules. In this case we have to supply external energy, namely the electrical energy to run the discharge. But we have two decisive advantages compared to the thermal reactor. In principle, we can use any kind of hydrogen donor molecule, which can be dissociated, for instance water vapor or hydrogen sulfide. Maybe more important, we do not have to use a catalyst. Therefore we are not bound by the boundary condition that the reactions have to run at high temperature. Due to the nature of the chemical kinetics methanol formation prefers low temperatures. As we don't need a catalyst we can now run at moderate or low temperature in the discharge. The use of methane in a silent discharge is equivalent to supplying part of the hydrogen required from a fossil fuel source. Furthermore, as CH_4 represents a greenhouse gas itself, its activation in a silent discharge to produce methanol as a synthetic liquid fuel is of intrinsic interest.

RESULTS AND DISCUSSION

Experimental Setup

The experimental configurations of the packed bed reactor, as well as the silent discharge reactor, are described elsewhere (Bill *et al.*, 1996, Highfield *et al.*, 1996, Eliasson *et al.*, 1996).

To evaluate the hydrogenation catalysts, a stainless steel, conventional plug-flow, packed-bed reactor is used. It is located inside an oven. Experiments are performed with 4 - 6 g of catalyst under isothermal conditions ($T_{max.}$ = 350 °C) and at pressures below 30 bar. Reactants are supplied from pre-calibrated mass flow controllers and the pressure is controlled by an electronic back-pressure regulator. The product stream from the reactor is analyzed by gas chromatography.

The discharge reactor is mounted in a high pressure vessel of cylindrical shape, which can be run at pressures up to 10 bar and controllable temperatures up to 400 °C. A silent discharge (Eliasson *et al.*, 1996) is maintained in an annular gap of 1 mm width formed by an outer steel cylinder and an inner cylindrical quartz tube. The steel cylinder serves as the ground electrode and an alternating high voltage of 18 kHz frequency is applied to a gold coating on the inside of the quartz tube. The power is accurately measured and can be regulated between 50 W and 900 W by adjusting the amplitude of the applied voltage. Different feed gases can be mixed before entering the discharge reactor. The individual flow rates are controlled by mass flow controllers. A regulating valve at the exit of the reactor controls the pressure in the discharge. Most of the products are analyzed by gas chromatography. Oxygen concentration can be measured by a special oxygen monitor based on paramagnetism. The ozone concentration is measured by UV absorption at 254 nm.

Performed experiments in a packed-bed reactor

We focus on the methanol synthesis from H_2 and CO_2, carbon dioxide being used as the sole C source:

$$CO_2 + 3\,H_2 \Leftrightarrow CH_3OH\,(g) + H_2O\,(g) \tag{1}$$

Copper based catalysts from academic institutions as well as industrial catalysts (B) have been tested at 20 bar, at a space velocity of 4500 h^{-1} (residence time 0.8 s) and with a feed ratio $H_2/CO_2 = 3:1$.

CO_2 hydrogenation in a thermal reactor not only leads to the production of methanol but also to the formation of CO, H_2O and some CH_4. Figure 1 shows all the products measured as a function of temperature.

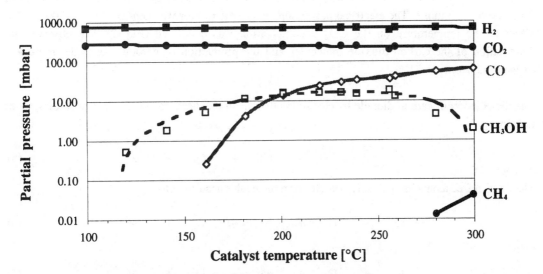

Fig. 1:   Mixture of hydrogen and carbon dioxide. Partial pressures of new molecules produced as well of the input gases a function of the temperature
(Catalyst B, $H_2/CO_2 = 3:1$, Pressure = 20 bar, Q_{tot} = 0.6 Nl.min^{-1}, S_V = 4500h^{-1})

Water was not measured directly but its amount can be estimated by the sum of CH_3OH and CO (mass balance calculations). The methanol yield increases with temperature up to about 220 °C. However it starts to decrease above 220 °C (as expected for an exothermic reaction), in spite of a continuous increase in the conversion of CO_2. The selectivity rapidly shifts in favor of carbon monoxide, which is produced by the reverse water gas shift reaction:

$$CO_2 + H_2 \Leftrightarrow CO + H_2O \text{ (g)} \tag{2}$$

This reaction (2) occurs in competition with reaction (1). Despite of a higher activation energy (E_A (2) $= 116.3 \pm 1.9$ KJ.mol^{-1} and E_A (1) $= 58.5 \pm 1.9$ KJ.mol^{-1} for catalyst B), the reverse water gas shift reaction is much faster than the methanol synthesis. At a temperature close to 280 °C, also methane is produced. According to all catalysts investigated, the optimum for the methanol yield is obtained in the temperature range between 220 and 240 °C. Best activity is obtained with catalyst B. It gives, for the selected experimental conditions, a methanol yield of 124.8 $g_{MeOH}.h^{-1}.g_{cat}^{-1}$.

Performed experiments with the silent discharge reactor

We report on different kinds of feed gases, namely: pure CO_2, pure CH_4, $H_2 + CO_2$ mixtures , CH_4 plus O_2 and CH_4 plus air.

These molecules can be activated in a discharge. For example, CO_2 and H_2 are excited according to:

$$e \quad + \quad CO_2 \quad \rightarrow CO_2^* \quad + \quad e \tag{3}$$

$$e \quad + \quad H_2 \quad \rightarrow \quad H_2^* \quad + \quad e \tag{4}$$

The molecules thus excited can then initiate a wide range of reactions.

In the first experiment we led pure carbon dioxide through the reactor at a pressure of 1 bar and at a flow rate of 1 Nl.min^{-1}. The electrical power supplied to the reactor was kept constant at 200 W. The effect of the temperature on the CO_2 decomposition was examined (Fig. 2). We observed the dissociation of the carbon dioxide molecules and the production of carbon monoxide, molecular oxygen and ozone.

The dissociation of carbon dioxide by electron collisions in the discharge takes place according to the reaction:

$$e \quad + \quad CO_2 \quad \rightarrow \quad CO \quad + \quad O \quad + \quad e \tag{5}$$

The oxygen free atoms immediately react to form molecules according to:

$$O \quad + \quad O \quad + \quad M \quad \rightarrow \quad O_2 \quad + \quad M \tag{6}$$

and:

$$O \quad + \quad O_2 \quad + \quad M \quad \rightarrow \quad O_3 \quad + \quad M \tag{7}$$

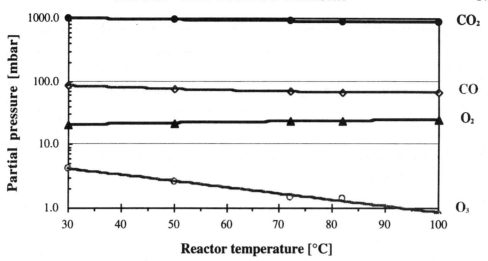

Fig. 2: Partial pressures of molecules produced in pure carbon dioxide as a function of temperature
(Electrical power = 200 W, Pressure = 1 bar and Q_{CO_2} = 1 Nl.min^{-1})

The ozone yield decreases with rising temperature. This behavior is well known and understood from studies on ozone generation in silent discharges (Eliasson *et al.*, 1987). It is interesting to note that CO is not oxidized by ozone.

In a second experiment, we led a mixture of H_2 and CO_2 (3:1) through the reactor at a pressure of 1 bar and at a total flow rate of 1 Nl.min^{-1}. The electrical power in the reactor was 400 W. The effect of the temperature was again examined and results are given in Fig. 3. We observe the formation of carbon monoxide, water, methane and methanol. No oxygen and ozone were detected in this case.

Carbon monoxide and water are the major products. They are produced in comparable amounts which slightly increase with temperature. In this case, CO_2 again dissociates according to the reaction (3). Oxygen atoms preferably recombine with hydrogen to form water molecules according to:

$$O \quad + \quad H_2 \quad + \quad M \quad \rightarrow \quad H_2O + \quad M \tag{8}$$

Methanol and methane are also produced in smaller amounts that depend on temperature. Methanol formation decreases with rising temperature, as expected for an exothermic reaction (($\Delta H^0)_{298}$ = -49.5 kJ.mol^{-1}) (Chase *et al.*, 1985). A maximum methanol yield of 0.2 % is obtained at the lowest investigated temperature (50 °C). Surprisingly enough, methane formation increases with rising temperatures although the reaction is also exothermic (($\Delta H^0)_{298}$ = -165.0 kJ.mol^{-1}) (Chase *et al.*, 1985). At 200 °C, a methane yield of 0.5 % is reached.

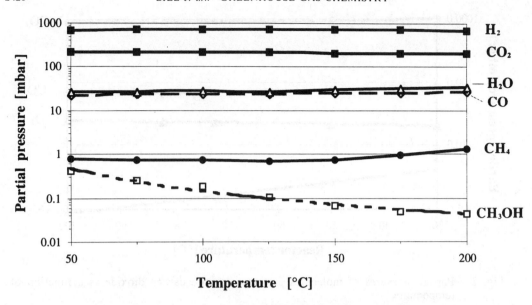

Fig. 3: Mixture of hydrogen and carbon dioxide. Partial pressures as a function of temperature (*Electrical power = 400 W, H_2/CO_2 = 3:1, Pressure = 1 bar and Q_{tot} = 1 Nl.min^{-1}*)

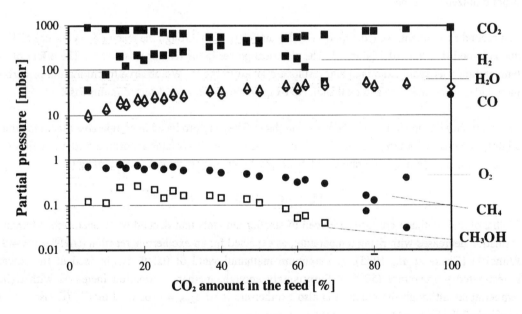

Fig. 4: Mixture of hydrogen and carbon dioxide. Partial pressures as a function of the CO_2 amount in the feed
Electrical power = 400 W, Pressure = 1 bar, Temp. = 80 °C and Q_{tot} = 1 Nl.min^{-1})

Figure 4 shows the partial pressures of all the products when varying the CO_2 amount in the feed gas. Again CO and H_2O are produced in comparable amounts. There is an increase when raising the CO_2 amount the feed. As an obvious check, we confirmed that, no water is formed when no hydrogen is present in the feed, and correspondingly, there is no CO detected when no CO_2 is fed in. CH_4 is produced again in smaller amounts. Its concentration decreases when the amount of CO_2 in the feed increases. For methanol, the performed experiment shows a maximum yield for a gas feed containing about 20 % CO_2. This result is close to theoretical predictions (Eliasson *et al.*, 1994b).

As already reported (Eliasson *et al.*, 1996), we also studied the decomposition of pure methane in the silent discharge reactor. The major products were H_2 and C_2H_6 and smaller concentrations of higher alkanes. Using a mixture of CH_4 and CO_2 (2:1) we found some methanol and much more CO and H_2. Starting from a mixture of oxygen in methane (1:19), we could also form methanol. Its concentration saturated at about 1.2 % after all oxygen was consumed. Similar results have been reported (Okazaki *et al.*, 1995) for a pulsed silent discharge reactor. In a final exploratory experiment we led methane and air mixtures of different mixing ratios through the reactor. The highest methanol concentrations of about 0.6% were obtained for 30% air in methane. These experiments show that it is possible to synthesize methanol from methane and air at ambient conditions (pressure of 1 bar and room temperature) in a silent discharge.

CONCLUSIONS

For a packed-bed reactor, it has been shown that under moderate pressure and temperature conditions, fair methanol activity can be obtained and that equilibrium values are already reached at 250 °C.
In a silent discharge reactor, it has been demonstrated that methanol can be synthesized from mixtures of methane with carbon dioxide, oxygen or air and mixtures of carbon dioxide with hydrogen. The species generated by the dissociation of the feed gases are energetic enough to initiate reactions at room temperature and atmospheric pressure that would normally require the use of catalysts and elevated temperatures and pressures. It is our goal to investigate and optimize these processes and find optimum mixing ratios, pressure and temperature conditions. Perhaps, the combined use of catalysts and electrical discharges may lead to the high yields necessary for these processes to become economic.

REFERENCES

Bill, A., B. Eliasson and E. Killer (1996). Hydrogenation of carbon dioxide in a packed bed reactor. *Proc. 11th World Hydrogen Energy Conf. (HYDROGEN '96)*, Stuttgart, Germany. Proc. Vol. II, pp. 1989-1995.

Chase, M. W., C. A. Davies, J. R. Downey, D. J. Frurip, R. A. Mc Donald and A. N. Syverud. (1985). *J. Phys. Chem. Ref. Data*, **14**, Suppl. 1. (JANAF TABLES).

Eliasson, B., M.Hirth and U. Kogelschatz. (1987). Ozone synthesis from oxygen in dielectric barrier discharges. *J. Phys. D: Appl. Phys.* **20**, pp. 1421-1437.

Eliasson, B. (1994a). CO_2 Chemistry: An Option for CO_2 Emission Control?. In: *Carbon Dioxide Chemistry: Environmental Issues*, (J. Paul and C.-M. Pradier Ed.), The Royal Society of Chemistry, Cambridge, pp. 5-15.

Eliasson, B., W. Egli and U. Kogelschatz. (1994b). Modeling of dielectric barrier discharge chemistry, *Pure & Appl. Chem.*, **66**, pp. 1275-1286.

Eliasson, B., U. Kogelschatz, E. Killer and A. Bill (1996). Hydrogenation of carbon dioxide and oxidation of methane in an electrical discharge. *Proc. 11th World Hydrogen Energy Conf (HYDROGEN '96),* Stuttgart, Germany. Proc. Vol. III, pp. 2449-2459.

Fricke, W. and M. Wallasch. (1993). Atmospheric CO_2 records from sites in the UBA air sampling network. In: *Trends 93: A compendium of Data on Global Change.* (Boden, T. A., D. P. Kaiser, R. J. Sepanski and F. W. Stoss Ed.), Carbon Dioxide Information Analysis Center, Oak Ridge National Laboratory, Oak Ridge, TN, USA. pp. 135-144.

Highfield, J.G., A, Bill, B. Eliasson, F. Geiger and E. Uenala. (1996). The role of simple alcohols in renewable hydrogen energy. *Proc. 11th Int. Symp. Alcohol Fuels (ISAF XI),* Sun City, South Africa.

Okazaki, K., T. Nozaki, Y. Uemitsu and K. Hijikata. (1995). Direct conversion from methane to methanol by a pulsed silent discharge plasma. *Proc. 12th International Symposium on Plasma Chemistry (ISPC-12),* Minneapolis, USA. Proc. Vol. II, pp. 581-586.

 Pergamon

Energy Convers. Mgmt Vol. 38, Suppl., pp. S423–S430, 1997
© 1997 Elsevier Science Ltd. All rights reserved
Printed in Great Britain
0196-8904/97 $17.00 + 0.00

PII: S0196-8904(96)00305-6

METHANOL AS AN AGENT FOR CO₂ MITIGATION

M. STEINBERG

Dept. Of Advanced Technology, Brookhaven National Laboratory
Upton, N.Y. 11973

ABSTRACT

The Carnol System consists of methanol production by CO_2 recovered from coal fired power plants and natural gas and the use of the methanol as an alternative automotive fuel. The Carnol Process produces hydrogen by the thermal decomposition of natural gas and reacting the hydrogen with CO_2 recovered from the power plant. The carbon produced can be stored or used as a materials commodity. A design and economic evaluation of the process is presented and compared to gasoline as an automotive fuel. An evaluation of the CO_2 emission reduction of the process and system is made and compared to other conventional methanol production processes including the use of biomass feedstock and methanol fuel cell vehicles. The CO_2 emission for the entire Carnol System using methanol in automotive IC engines can be reduced by 56% compared to the conventional system of coal fuel power plants and gasoline driven engines and by as much as 77% CO_2 emission reduction when methanol is used in fuel cells for automotive purposes. The Carnol System is shown to be an environmentally attractive and economically viable system connecting the power generation sector with the transportation sector which should warrant further development.

© 1997 Elsevier Science Ltd

KEYWORDS

CO_2 mitigation, power plant flue gas, methanol production, automotive fuel, fuel cells.

INTRODUCTION

Coal and natural gas are abundant fuels. Because of their physical and chemical properties, coal and natural gas are difficult to handle and utilize in mobile as well as stationary engines. The infrastructure is mainly geared to handle clean liquid fuels. In order to convert coal to liquid fuel, it is generally necessary to increase its H/C ratio either by increasing its hydrogen content or decreasing its carbon content. On the other hand, in order to convert natural gas to liquid fuels it becomes necessary to decrease its hydrogen content. Thus, by coprocessing the hydrogen-rich natural gas with hydrogen deficient coal, it should be possible to produce liquid fuels in an economically attractive manner. For environmental purposes of decreasing CO_2 greenhouse gas emissions, several approaches can be taken. The CO_2 emission from central power stations can be removed, recovered and disposed of in deep ocean (Cheng et al., 1984). Alternatively, carbon can be extracted from coal and natural gas and only the hydrogen-rich fractions can be utilized from both of these fuels to reduce CO_2 emissions while storing the carbon (Steinberg, 1989). Because of its physically properties, carbon is much easier to dispose of either by storage or used as a materials commodity than sequestering CO_2. A third alternative CO_2 mitigation method is to utilize the stack gas CO_2 from coal burning plants by reacting with hydrogen obtained from natural gas to produce methanol, which is a well-known liquid automotive fuel. In this paper, we describe and evaluate the Carnol Process (Steinberg, 1993), which connects the power generation sector with the transportation sector resulting in an overall CO_2 mitigation system.

THE CARNOL PROCESS

The Carnol Process is composed of three unit operations described in the following.

1.   Carbon dioxide is extracted from the stack gases of coal fired power plants using monoethanolamine (MEA) solvent in an absorption-stripping operation. The technology for this operation is well known in the chemical industry for CO_2 recovery and has recently been significantly improved upon for use in extracting CO_2 from power plant stack gases (Sudo et al., 1994). The power required to recover CO_2 from an integrated coal fired power plant to recover 90% of the CO_2 from flue gas can be reduced to about 10% of the capacity of the power plant. This energy requirement can be reduced to less than 1% when the CO_2 recovery operation is integrated with a methanol synthesis step described in step 3 below.

2.   The hydrogen required to react with CO_2 for producing methanol can be obtained from either of two methods involving natural gas. In the conventional method for producing hydrogen natural gas is reformed with steam; $CH_4 + 2H_2O = CO_2 + 4H_2$. This process produces CO_2 and, thus, CO_2 emission is increased. However, hydrogen can be produced without CO_2 emission by the non-conventional method of thermally decomposing methane to carbon and hydrogen; $CH_4 = C + 2H_2$. The energy requirement in conducting this process is less than that required by the above conventional process. A fluidized bed reactor has been used to thermally decompose methane and more recently we are attempting to improve reactor design by utilizing a molten metal bath reactor (Steinberg, 1996). The carbon is separated and either stored or can be sold on the market as a materials commodity, such as for strengthening rubber for tires. The temperatures required for this operation are 800°C or above and pressures of less than 10 atm.

3.   The third step in the process consists of reacting the hydrogen from Step 2 with the CO_2 from Step 1 in a conventional gas phase catalytic methanol synthesis reactor; $CO_2 + 3H_2 = CH_3OH + H_2O$. This is an exothermic reaction so that the heat produced in this step can be used to recover the CO_2 from the absorption/stripping operation described in Step 1, thus reducing the energy required to recover the CO_2 from the power plant to less than 1% of the power plant capacity. This combination has a decided advantage over the energy cost of derating the power plant when CO_2 is disposed of by pumping into the ocean in which case more than 20% of the power plant capacity is consumed. The gas phase methanol synthesis usually takes place at a temperature of 260°C and a pressure of 50 atm using a copper catalyst. The synthesis can also be conducted in the liquid phase by using a slurry zinc catalyst at lower temperature of 120°C and 30 atm of hydrogen pressure (Steinberg, 1993) in which case the CO_2 is connected directly without recovery.

CARNOL PROCESS DESIGN

A computer process simulation equilibrium model has been developed for the Carnol Process based on the flow sheet shown in Figure 1. A material and energy balance is shown in Table 1 selected from a number of computer runs. This run shows that 112.1 kg of methanol can be produced from 100 kg of natural gas (CH_4) and 171.1 kg CO_2 with a net emission of only 25.8 lbs CO_2/MMBTU of methanol energy including combustion of the methanol. This is an 85.7% reduction in CO_2 emission compared to the conventional emission from a steam reforming methanol plant which emits 182 lbs CO_2/MMBTU including the combustion of methanol. The power plant at the same time has a 90% reduction in CO_2 because only 10% of the CO_2 from the MEA solvent absorption plant remains unrecovered and emitted to the atmosphere.

METHANOL AS AN AUTOMOTIVE FUEL

The Carnol Process can be considered as a viable coal CO_2 mitigation technology because the resulting large production capacity of liquid methanol can be used in the large capacity automotive fuel market. Most processes which utilize CO_2 produce chemical products which tend to swamp the market and thus cannot be used. Methanol as an alternative automotive fuel has been used in internal combustion (IC) engines as a specialty racing car fuel for a long time. More recently, the EPA has shown that methanol can be used in IC engines with reduced CO and HC emissions and at efficiencies exceeding gasoline fuels by 30% (Steinberg, 1989). Compared to gasoline, the CO_2 emission from methanol in IC engines is 40% less. Methanol can also be used either directly or indirectly in fuel cells at several times higher efficiency than IC gasoline enginess for automotive use. A great advantage of methanol is that, as a liquid fuel it fits in well with the infrastructure of storage and distribution compared to compressed natural gas and gaseous or liquid hydrogen which are being considered as alternative transportation fuels.

It should also be pointed out that removal and ocean disposal of CO_2 is only feasible for large central power stations. For the dispersed domestic industrial and transportation power sectors, the Carnol Process provides the capability of CO_2 reduction by supplying liquid methanol fuel to these more diverse CO_2 emitting sources.

ECONOMICS OF CARNOL PROCESS

A preliminary economic analysis of the Carnol Process has been made based on the following assumptions:

1. 90% recovery of CO_2 from a nominal 900 MW(e) coal fired power plant.
2. Capital investment based on an equivalent 3 step conventional natural gas steam reforming methanol plant which amounts to $100,000/ton MeOH/day (Kordinak, 1994).
3. Production cost which includes 19% financing, 1% labor, 3% maintenance, and 2% process catalyst and miscellaneous cost adding up to a fixed charge and operating cost of 25% of the capital investment (IC) on an annual basis.
4. Natural gas varies between $2 and $3/MSCF.
5. Carbon storage is charged at $10/ton; Market value for carbon black for tires is as high as $1000/ton.
6. Methanol market price is $0.45/gal. but has varied historically from $0.45 to $1.30/gal. in the last few years because of the use in producing MTBE as a mandated gasoline additive.

At $18/bbl oil and 90% recovery as gasoline and $10/bbl for refining cost, gasoline costs $0.78/gal. and methanol being 30% more efficient than gasoline competes with gasoline at $0.57/gal. methanol.

Table 2 summarizes the economics of production cost factors and income factors for a range of cost conditions. In terms of reducing CO_2 cost from power plants, with $2/MSCF natural gas and a $0.55/gal. methanol income the CO_2 reduction cost is zero. At $3/MSCF natural gas and $0.45/gal. income from methanol, the CO_2 disposal cost is $47.70/ton CO_2, which is less than the maximum estimated for ocean disposal (IEA, 1993). More interesting, without any credit for CO_2 disposal from the power plant, methanol at $0.55/gal. can compete with gasoline at $0.76/gal. (~ $18/bbl oil) when natural gas is at $2/MSCF. Any income from carbon makes the economics look even better.

CO_2 EMISSION EVALUATION OF ENTIRE CARNOL SYSTEM

Although we can show 90% or more CO_2 emission reduction for the coal fired power plant, the other two parts of the system, methanol production and automotive emissions, have relatively less CO_2 emission reduction compared to conventional systems. Therefore, the entire Carnol System must be evaluated as shown in Figure 2.

Alternative methanol production processes are evaluated in Table 3. The yield of methanol per unit of methane feedstock is shown for 1) conventional process in two parts: A) steam reforming of natural gas process, and B) using CO_2 addition in conventional steam reforming process; 2) Carnol Process, in two parts: A) using methane combustion to decompose methane for hydrogen in methane decomposition reactor (MDR), and B) hydrogen combustion to decompose the methane in MDR; and 3) a steam gasification of biomass process. The Carnol Process with H_2 and the biomass process (solar energy) reduces CO_2 to zero emission compared to conventional, but with a loss of 35% and 47% methanol yield respectively. The Carnol Process when using methane combustion in the decomposer reduces CO_2 emission by 43% while the production yield is only reduced by 26% compared to conventional. The conventional process with CO_2 addition (1B) is interesting because there is a 32% increase in production, although the CO_2 emission is only reduced by 23%.

For purposes of clarification, of the above analysis, the overall stoichiometry for the Carnol Process is shown in the following together with the conventional processes for methanol production.

Carnol Process $CH_4 + 0.67 CO_2 = 0.67 CH_3OH + 0.67 H_2O + C$
Conventional Steam Reforming Methane $CH_4 + H_2O = CH_3OH + H_2$
Conventional Steam Reforming of Methane With CO_2 Addition:
$$CH_4 + 0.67 H_2O + 0.33 CO_2 = 1.33 CH_3 OH$$

It is noted that in the Carnol Process a maximum amount of CO_2 is utilized and an excess of carbon is produced. In the conventional process, no CO_2 is used and an excess of hydrogen is produced. With CO_2 addition to the conventional process, no excess of carbon or hydrogen is formed and methanol per unit natural gas is maximized.

Methanol can also be produced using biomass and since the net CO_2 emission is zero with CO_2 being converted to biomass by solar photosynthesis, the biomass process must also be included in the evaluation and the stoichiometry is as follows:

Biomass Steam Gasification Process for Methanol Synthesis:
$$CH_{1.4}O_{0.7} + 0.3H_2O = 0.5CH_3 OH + 0.5CO_2$$
photosynthesis $CO_2 + 0.7 H_2O = CH_{1.4} O_{0.7} + O_2$

The entire Carnol System is evaluated in Table 4 in terms of CO_2 emissions and compared to the alternative methanol processes and to the base line case of a conventional coal fired power plant and gasoline driven automotive IC engines. Methanol in a fuel cell automotive engine is also evaluated. All the cases are normalized to emissions from 1MMBTU of a coal-fired power plant which produces CO_2 for a Carnol methanol plant equivalent to 1.27 MMBTU of methanol for use in an automotive IC engine. The assumptions made are listed at the bottom of Table 4. The conclusions drawn from Table 4 are as follows:

1.   The use of conventional process methanol reduces CO_2 by 13% compared to the gasoline base case and is mainly due to the 30% improved efficiency of methanol in IC engines.

2.   By addition of CO_2 recovered from the coal fired power plant to the conventional methanol process, the CO_2 from the power plant is reduced by about 25% (161 lbs/MMBTU compared to 215 lb CO_2/MMBTU) and the CO_2 emissions for the entire system is reduced by 24%. It should be pointed out that the CO_2 can also be obtained from the flue gas of the reformer furnace of the methanol plant itself.

3.   The Carnol Process reduced the coal fired power plant CO_2 emission by 90% and the overall system emission is reduced by 56%.

4.   Since the use of biomass is a CO_2 neutral feedstock, there is no emission from the power plants because the production of biomass feedstock comes from an equivalent amount of CO_2 in the atmosphere generated from the coal-fired power plant. Thus, the only net emission comes only from burning methanol in the automotive IC engine and thus, the CO_2 emission for the entire system is reduced by 57%, only slightly more than the Carnol System. However, the main point is that at present the cost of supplying biomass feedstock is higher than that of natural gas as a feedstock for methanol production.

5.   Another future system involves the use of fuel cells in automotive vehicles. The efficiency of fuel cells is expected to be 2.5 times greater than gasoline driven engines (World Car Conference, 1996). Applying the Carnol Process to produce methanol for fuel cell engines reduces the CO_2 emission for the entire system by a maximum of 77%. Furthermore, because of the huge increase in efficiency, the capacity for driving fuel cell engines can be increased by 92% over that for the IC engines using the same 90% of the CO_2 from the coal burning power plant in the Carnol process.

CONCLUSIONS

The Carnol Process can reduce CO_2 emissions from coal fired power plant while producing methanol for automotive IC engines with virtually no derating of the power plant. With natural gas at $2/MSCF, the methanol cost appears to be competitive with gasoline for IC engines at $18/bbl oil. The CO_2 emission for the entire Carnol System is reduced by 56%. Compared to the conventional system, steam reformed natural gas with CO_2 addition from the power plant, reduces CO_2 emissions by only 13%, but can have a higher production capacity per unit natural gas than the Carnol Process. Biomass as a methanol feedstock can reduce CO_2 by 57%. The development of methanol fuel cell engines can reduce CO_2 emissions by 77% for the entire system with a large increase in production capacity. The use of methanol as an automotive fuel produced from coal fired power plant CO_2 stack gas and natural gas appears to be an environmentally attractive and economically viable system connecting the power generation sector with the transportation sector and, therefore, should warrant further development effort.

REFERENCES

1.   Cheng H. and M. Steinberg, "A System Study for the Removal, Recovery and Disposal of Carbon Dioxide from Fossil Fuel Power Plants in the U.S.," DOE/CH-0016-2 TRO16, U.S. Dept. Of Energy, Washington, D.C. (December 1984).
2.   IEA Greenhouse Gas R and D Programs, "Greenhouse Issues," No. 7, Cheltenham, U.K. (March 1993).
3.   Korchnak, J., John Brown Engineers Co., Houston, Texas, Private Communication (1994).
4.   Larson, E.D. and R.E. Katofsky, "Production of Hydrogen and Methanol via Biomass Gasification, in <u>Advance in Thermochemical Biomass Conversion</u>," Elsevier Applied Science, London, U.K. (1992).

5.   Motor Vehicle Emissions Laboratory (MVEL), "An Analysis of the Economic and Environmental Effects of Methanol as an Automotive Fuel," EPA Report No. 0730, Ann Arbor, Michigan (Sept. 1989).

6.   Steinberg, M., "Coal to Methanol to Gasoline by the Hydrocarb Process," BNL 43555, Brookhaven National Laboratory, Upton, N.Y. 11973 (August 1989).

7.   Steinberg, M., "The Carnol Process for Methanol Production and Utilization with Reduced CO_2 Emissions," BNL 60575, Brookhaven National Laboratory, Upton, N.Y. 11973 (October 1993).

8.   Steinberg, M. "The Carnol Process for CO_2 Mitigation from Power Plants and the Transportation Sector," EPA report NMRL-RTP-015. USEPA, Research Triangle Park, N.C. 27711 (1996).

9.   Suda, O., et.al., "Development of Fuel Gas Carbon Dioxide Recovery Technology," Chapter in Carbon Dioxide Chemistry Environment Issues, Eds., T. Paul and C.M. Pradies, pp. 222-35, the Royal Society of Chemistry, Hemovan, Sweden (1994) and T. Mimura, Kansai Electric Power Co., Osaka, Japan, Private Communication (June 1995).

10.   World Car Conference '96, Bourns College of Engineering, Center for Environmental Research and Technology, University of California, Riverside, CA (Jan. 21-24, 1996).

11.   Wyman, C.E., et.al., "Ethanol and Methanol from Cellulosic Biomass," pp. 866-923, in Renewable Energy, Eds., T.B. Johansson, et.al., Island Press, Washington, D.C. (1993).

Table 1
Carnol Process VI Design
Process Simulation - Mass and Energy Balances

UNIT (See Fig. 1)	CARNOL VI MDR and Liquid Phase MSR
MDR	
Pressure, atm	1
Temp. °C	900
CH_4 Feedstock, Kg	100
Preheat Temp. °C	837
CH_4 Fuel for MDR, Kg	6.6
CH_4 Conversion, %	96.3
Carbon Produced, Kg	72.1
Heat Load, Kcal	65,006
Purge Gas from Fuel, Kmol	2.1
MSR - Liquid Phase	
Pressure, atm	50
Temp. °C	120
Recycle Ratio	0.5
CO_2 Feedstock, Kg	171.1
CO_2 Conversion, %	90.3
Methanol Prod., Kg	112.1
Water Cond., Kg	63.2
Energy for Gas Compression to MSR	
Primary, Kcal	53,306
Secondary for Recycle, Kcal	694
Performance	
Ratio MeOH/CH_4, Kg/Kg	1.12
Carbon Efficiency, MeOH, %	56.0
Thermal Eff., MeOH, %	42.9
Thermal Eff., C+MeOH,%	85.7
CO_2 Emission Lbs/MMBTU	25.8
CO_2 Emission Kg/GJ	11.1

Table 2
ADVANCED CARNOL VI PRELIMINARY PROCESS ECONOMICS

Plant Size - To Process 90% Recovery of CO_2 from 900 MW(e) Nominal Coal Fired Power Plant
90% Plant Factor, CO_2 Rate = 611 T/Hr. = 4.82 x 10^6 Tons CO_2/Yr.
Feedstock: Natural Gas Rate = 2.82 x 10^6 T/Yr. = 407,000 MSCF/D
Carbon Production = 2.03 x 10^6 T/Yr.
Methanol Production = 3.16 x 10^6 T/Yr. = 69,300 Bbl/D
Plant Capital Investment (IC) = 9607 T/D x 10^5 = $961 x 10^6

Production Cost Factors							=	Income Factors					
0.25 IC	Natural Gas			C Storage		CO_2 Cost		C Income		MeOH Income		Cost for Reducing CO_2	
10^8/Yr	10^8/Yr	($/MSCF)	10^8/Yr	($/Ton)	10^8/Yr	($/Ton)	10^8/Yr	($/Ton)	10^8/Yr	($/Gal)	10^8/Yr	($/Ton)
2.40	2.67	(2)	0.20	(10)	0	(0)	0	(0)	5.27	(0.55)	0	(0)
2.40	4.00	(3)	0.20	(10)	0	(0)	0	(0)	5.27	(0.55)	-1.34	(-27.60)
2.40	2.67	(2)	0.20	(10)	0	(0)	0	(0)	5.27	(0.55)	0	(0)
2.40	4.00	(3)	0	(0)	0	(0)	1.13	(55.60)	5.27	(0.55)	0	(0)
2.40	2.67	(2)	0.20	(10)	0	(0)	0	(0)	4.30	(0.45)	-0.97	(-20.00)
2.40	2.67	(2)	0	(0)	0	(0)	0.77	(37.90)	4.30	(0.45)	0	(0)
2.40	4.00	(3)	0.20	(10)	0	(0)	0	(0)	4.30	(0.45)	-2.30	(-47.70)
2.40	4.00	(3)	0	(0)	0	(0)	2.10	(103.00)	4.30	(0.45)	(0)	(0)

Table 3
METHANOL PRODUCTION AND CO_2 EMISSION
PROCESS COMPARISON

PROCESS		PRODUCTION YIELD		CO_2 EMISSION [4]	
		Moles MeOH Mole Feedstock	% Reduction from Conventional	Lbs CO_2 MMBTU (MeOH)	% Reduction from Conventional
1A	Conventional Process Steam Reforming of CH_4	0.76 [1]	0%	44	0%
1B	Conventional Process with CO_2 Addition	1.00	(32%) [2]	34	23%
2A	Carnol Process Heating MDR with CH_4	0.56	26%	25	43%
2B	Carnol Process Heating MDR with H_2	0.50	35%	0	100%
3	Steam Gasification of Biomass	0.40 [3]	47%	43	100%

(1 Based on thermal efficiency of 64% (Wyman et al., 1993)   (3 Based on BCL process (Larson et al., 1992)
(2 This represents a 32% increase in yield vs conventional.   (4 CO_2 emission only from fuel production plant.

Table 4 CO_2 EMISSION COMPARISON FOR SYSTEMS CONSISTING OF
COAL FIRED POWER PLANT, FUEL PROCESS PLANT AND AUTOMOTIVE POWER PLANT
Basis: 1 MMBTU for coal fired 900 MW(e) power plant
1.27 MMBTU of liquid fuel for IC engine - other fuel efficiencies proportions energy up and down
CO_2 Emission units in Lbs CO_2/MMBTU (multiply by 0.43 for KG/GJ)

System Unit	Coal Fired Power Plant	Fuel Process Plant	IC Automotive Power Plant	Total System Emission	CO_2 Emission Reduction
Baseline Case: Coal Fired Power Plant and Gasoline Driven IC Engine	215	15	285	515	0%
Case 1A Coal Fired Power Plant With Conventional Steam Reformed Methanol Plant	215	56	175 [2]	448	13%
Case 1B Coal Fired Power Plant With CO_2 Addition to Conventional Methanol Plant	161 [4]	54	175	390	24%
Case 2 Coal Fired Power Plant with CARNOL Process Methanol Plant	21 [3]	32	175	·228	56%
Case 3 Coal Fired Power Plant with Biomass for Methanol Plant	0	43	175	219	57%
Case 4 Coal Fired Power Plant with CARNOL Methanol and Fuel Cell Automotive Power	11 [5]	17	Fuel Cell 89 [3]	117	77%

1) 90% recovery of CO_2 from coal fired plant,
2) Methanol is 30% more efficient than gasoline in IC engines,
3) Fuel cell at 2.5 times more efficient than conventional gasoline IC engine.
4) Only 25% recovery of CO_2 from coal plant is necessary for supply CO_2 to conventional methanol plant.
5) Only 52% emissions of coal plant CO_2 is assigned to Carnol for fuel cells

Figure 1

Carnol VI Process for CO$_2$ Mitigation Technology
Combining CO$_2$ Recovery From Power Plants With Liquid Metal Methane Decomposition and Liquid Phase Methanol Synthesis

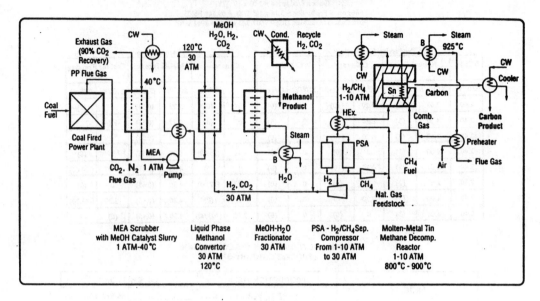

Figure 2

CARNOL System Configuration For CO$_2$ Emission Mitigation

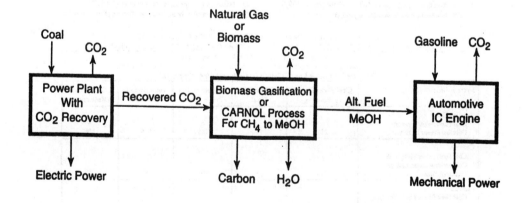

Energy Convers. Mgmt Vol. 38, Suppl., pp. S431–S436, 1997
© 1997 Elsevier Science Ltd. All rights reserved
Printed in Great Britain
0196-8904/97 $17.00 + 0.00

Pergamon

PII: S0196-8904(96)00306-8

HYDROGEN AND ELECTRICITY FROM DECARBONISED FOSSIL FUELS

Olav Kaarstad[1] and Harry Audus[2]

[1] Statoil R&D Centre, Postuttak, N-7004 Trondheim, Norway,
tel: +47 73 58 4266, fax: +47 73 58 4576, e-mail: okaa@fou.tr.statoil.no
[2] IEA Greenhouse Gas R&D Programme, CRE, Stoke Orchard, Glos. GL52 4RZ, UK,
tel: +44 1242 680753, fax: +44 1242 680758, e-mail: harry@ieagreen.demon.co.uk

ABSTRACT

Hydrogen and electricity may become the favoured twin energy carriers in a possible "greenhouse driven" future due to their lack of CO_2-emissions at the point of use. It is shown that decarbonising fossil fuels to hydrogen with CO_2 being stored in deep geological formations represents the least expensive way of producing CO_2-free hydrogen. For fossil fuels there seems to be a natural progression from fuel switching (coal to oil to natural gas) for low to moderate emission reduction pressure. For increasing pressure for reduction of emissions CO_2-free (decarbonised) electricity from coal will be introduced followed by decarbonised electricity from natural gas and decarbonised hydrogen from natural gas.

© 1997 Elsevier Science Ltd

KEYWORDS
Hydrogen, decarbonisation, CO_2, electricity, energy

THE QUESTION

The main question being addressed by this paper is: *Under what circumstances will hydrogen become one of the major energy carriers?* In order to attempt an answer to this question it is necessary to look at the energy markets, the technologies, electricity as an energy carrier and also at the driving forces that may make tomorrow's energy system different from what we can observe today.

THE HYDROGEN BACKGROUND

The science fiction writer Jules Verne envisioned in his novel *The Mysterious Island* (1874) a world powered by hydrogen. For a while it also looked as if his vision would at least in part be realised when "town gas", a mixture of typically 50% hydrogen and 50% carbon monoxide made from coal, lit gas street lights and provided light, heat and hot water in homes in hundreds of towns in the then industrialising world. Delivery of town gas is rare today. In its place we often find natural gas linked by huge pipelines from often distant gas reservoirs. The use of pure hydrogen is still quite common in the industrialised societies, but mostly in industrial plants and other areas outside public awareness. About 1200 kilometre's of hydrogen pipelines exist mainly in USA, Germany, the Netherlands and Great Britain. Hydrogen is a very important molecule with an enormous breadth and extent of industrial application and use. Its primary use is as a reactant in chemical processes, but it is also used as a fuel in space applications. Other uses of hydrogen as a fuel are minuscule for simple cost reasons.

The notion of a hydrogen economy was popularised in the 1970s partly by the success of the hydrogen fuelled space program and partly by Marchetti and DeBeni in Europe and by Gregory in the U.S. Their assertion was that hydrogen could be a major energy carrier that was compatible with, and complementary to electricity. The popular view at that time was of an increasingly electric future with abundant low cost electricity coming first from nuclear reactors, later from breeder reactors and eventually from fusion reactors. Electricity could not, however, be stored in quantity or be used for many modes of transportation and for some applications in industry. The hydrogen vision was therefore to cover the storage and non-electric end uses by making hydrogen from electrolysis or, in the case of high temperature gas reactors, as thermochemically split water. This nuclear-based vision collapsed when problems started to appear for the nuclear industry worldwide.

After the collapse of the nuclear hydrogen vision, the 1980s and 1990s have seen the growth of a new vision for a electricity and hydrogen fuelled future based on renewable energy sources. Almost without exception all national R&D programmes dealing with hydrogen energy are based on the assumption that hydrogen has to come from either hydropower, solar PV electricity, wind power, biomass and so forth. The problem with this vision, by now realised also by many of those promoting hydrogen energy, is that transition to hydrogen from renewables will result in very high energy costs. A great deal of soul searching seems therefore to be going on in the hydrogen energy communities with respect to the viability of renewable based hydrogen as an energy carrier.

HYDROGEN TODAY

Today the least expensive way of making hydrogen is through steam reforming of natural gas. The resulting cost of hydrogen is typically about 5-6 US$/GJ when made from natural gas at a price of 3 US $/GJ. About 95% of all hydrogen used in the world goes to produce ammonia (50%), petroleum processing (37%) and methanol (8%). Most of the hydrogen used for making these three products is both produced and consumed within the plant fences. The remaining 5% of the hydrogen use is distributed over a range of metallurgical applications, the hydrogenation of fat and edible oil and so forth.

As illustrated above the conversion of natural gas to hydrogen almost doubles the price per unit of energy (from 3 to 5-6 $/GJ). It follows from this that hydrogen has to show one or more substantial advantage over natural gas if it is to compete in the energy market. One such advantage would be the lack of CO_2 and other greenhouse gas emissions if hydrogen is produced in a CO_2-free manner. Another advantage, depending on the sector of end use, can sometimes be that the technology for using hydrogen offers higher efficiency (i.e. fuel cells) or other favourable effects (e.g. low weight for space fuel application). At the present state of technologies for utilising hydrogen it seems that the technology factor alone cannot justify the change to hydrogen as an energy carrier. Going from fossil fuels to hydrogen, it would seem, needs to be motivated almost solely by the lack of CO_2-emissions. This in turn would at first sight seem to favour hydrogen produced by either nuclear energy or renewables since CO_2 is inevitably produced when burning fossil fuels. We will show, however, that there is a CO_2-free option for producing hydrogen that is less expensive than nuclear or any form of renewable energy.

DECARBONISING NATURAL GAS AND COAL TO HYDROGEN

Carbon released to atmosphere from fossil fuels was about 6 GtC in 1990. Approximately 2,3 GtC were emitted by the energy sector mainly during conversion to electricity. The majority of carbon emissions i.e. 3,7 GtC were emitted at the point of end use. There are many alternatives for reducing end use emissions amongst which are increases in efficiency and use of less carbon intensive fuels. For instance, natural gas could be substituted for petroleum fuels in many transport applications. Another approach is to decrease the use of primary energy and increase the use of clean energy carriers.

In 1995 the IEA Greenhouse Gas R&D Programme and Statoil jointly commissioned the Foster Wheeler company, one of the worlds major designers and builders of large hydrogen plants, to conduct a study in which the cost of hydrogen production by conventional means from coal and natural gas was derived (Foster Wheeler/IEA, 1996). The cost of hydrogen from state-of-the-art plant that release CO_2 to atmosphere was compared with processing schemes which includes capture, compression and storage of the associated CO_2 in deep geological formations (a depleted natural gas field) at a distance of 70 km from the hydrogen plant. The production of hydrogen involves separation of CO_2; relatively minor process changes can be made to recover CO_2 instead of venting it to atmosphere. All the studied plants produced 278 000 Nm³/h (in 3 streams of 94 000 Nm³/h) 99,9% purity hydrogen at 60 bar at battery limits. The plant was assumed to be located in the Netherlands and be self-sufficient in utilities including electricity. Energy input costs were taken as a central price of 3 US $/GJ for natural gas and 2 US$/GJ for coal, reflecting in broad terms the market price of these two primary energy sources. The centre price of hydroelectricity for the electrolyser plant (the renewable comparison) was arbitrarily set at 3,6 US cents/kWh, equivalent to 10 $/GJ$_{el}$. The economic results of the Foster Wheeler study are illustrated in fig. 1 for coal and natural gas. The investment- and operating cost of generation of hydrogen from electricity (electrolysis) is taken from a recent study by Norsk Hydro Electrolysers AS and Electricité de France (Norsk Hydro/EdF, 1996).

WHAT IS THE CO_2 STORAGE CAPACITY?

Carbon dioxide may be stored either in deep geological formations or in the deep ocean. For the latter, fundamental research is in progress and the technological and environmental feasibilities are as yet not established. The underground storage of CO_2 on the other hand, is fairly well proven through about 50-60 enhanced oil recovery (EOR) projects injecting CO_2 into oil reservoirs in the USA and Canada.
Statoil will, by the end of 1996, start injecting about 1 million tonnes of CO_2 per year into a deep saline aquifer in the Sleipner offshore area in the North Sea. This will be the first time that CO2 will be injected purely for climate change reasons.
In early 1996 a 2,5 year study under the European Union Joule II R&D Programme concluded that "Underground disposal is a perfectly feasible method of disposing of very large quantities of carbon dioxide such as are produced by fossil fuel fired power plants and most other point sources of CO2" (EU, 1996). The same Joule II study tentatively estimates that there is space available underground in the European Union and Norway (mostly under the North Sea) to store approximately 800 Gt of CO_2. This is about a quarter of the present atmospheric CO_2 content. No global estimate based on studies of similar detail is as yet available, but there is no reason to suppose that geology of Europe is atypical for the world as a whole.

COMPARING THE ECONOMICS OF DECARBONISATION

The key results are shown in figure 1 and 2. For the central natural gas case the cost of hydrogen is increased by a surprisingly low 25% going from a standard steam reforming process to steam reforming plus CO_2-capture, transport and storage underground. For the coal case the equivalent penalty for capture and storage of CO_2 is a 28% cost increase in the produced hydrogen. Similar processes for making hydrogen based on decarbonised oil have not yet been studied, but it may be assumed that the resulting price of hydrogen will lie between that of natural gas and coal.

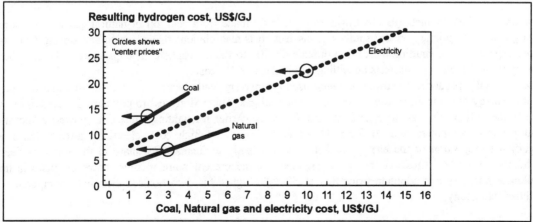

Figure 1 Cost of Decarbonised Hydrogen (10% DCF)

Figure 1 shows the cost of producing decarbonised hydrogen from coal, natural gas and by electrolysis based on hydropower (or any other renewable or nuclear form of electricity). At the centre prices of 2 and 3 US$/GJ for coal and natural gas respectively and 10 $/GJ (3,6 cents/kWh$_e$) for hydropower it is seen that hydrogen from natural gas is least expensive, hydrogen from coal is nearly twice as costly and hydrogen from hydropower is more than 3 times as costly as from natural gas. The cost of hydropower has to go down to about 5 $/GJ (1,8 cents/kWh$_e$) to compete with hydrogen from coal.

The cost of avoiding carbon emissions in the production of hydrogen from fossil fuels is in the range of 15-20 US$/ton CO_2 (50-70 $/ton carbon) with coal in the lower and natural gas in the upper end of this range.

Even in a possible future greenhouse driven world, hydrogen would only be one among many energy carriers. Figure 2 attempts to shed some light on the relative competitiveness of the various carbon containing and non-carbon energy carriers. The production of fossil fuel based electricity with CO_2-removal and underground storage is included in order to illustrate the possible future role of decarbonised electricity alongside hydrogen and other energy carriers. The decarbonisation of fossil fuels to electricity has, unlike decarbonised hydrogen, been studied for a number of years within the IEA Greenhouse Gas R&D Programme and elsewhere. In general terms it is expected that decarbonising electricity made from natural gas or coal will increase the production cost by between 50 and 75%.

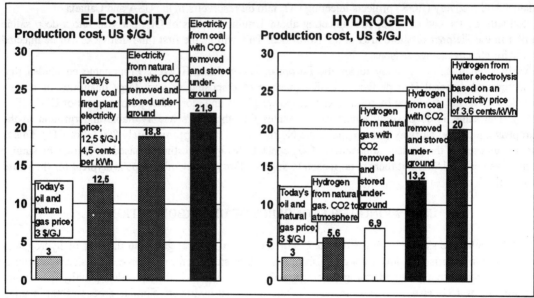

Figure 2 The Economics of Decarbonisation to Electricity and Hydrogen
(at today's fuel prices and 10% DCF)

It is obvious that products (like hydrogen or electricity) made from coal, oil or natural gas will be more costly than these primary energy sources themselves. It is also obvious that catching and storing CO_2 is more expensive than emitting the carbon dioxide directly to the atmosphere. The question is; Under what conditions will the energy markets be willing to pay these added costs?

Conventionally produced electricity is being used extensively while being 4-5 times more expensive per unit of energy than the direct use of coal, oil or natural gas. This willingness to pay more for electricity is due to the technological advantages of using this energy carrier for lighting houses, for powering motors and personal computers and so forth. Given that, despite its high cost, electricity already takes a significant proportion of the energy market, it seems likely to increase its share in the event of deep reductions in carbon emissions being investigated. At the present point in time, however, there is no willingness to pay a production price that is 50-75% higher for decarbonised electricity compared to ordinary electricity.

WHEN AND WHY WILL HYDROGEN BECOME A MAJOR ENERGY CARRIER?

Conventionally produced hydrogen gas (with CO_2 emitted to atmosphere) costs about twice that of natural gas or oil and about 3 times more than coal. At present only the space industry seems to be willing to pay for the high cost of hydrogen energy. Making an analogy with electricity, we may infer that hydrogen has not got the required technological advantages in end use to pay for the added costs. This lack of superior end use technology for hydrogen compared to for instance natural gas seems to be a greater barrier to introducing hydrogen energy than the cost of decarbonisation.

Let us for a moment imagine a possible future greenhouse driven world in which costly CO_2 counter measures like tradable permits, voluntary agreements, joint implementation, regulations or CO_2-taxes are in place. We may catch a glimpse of the energy sector of this greenhouse driven world in figure 3.

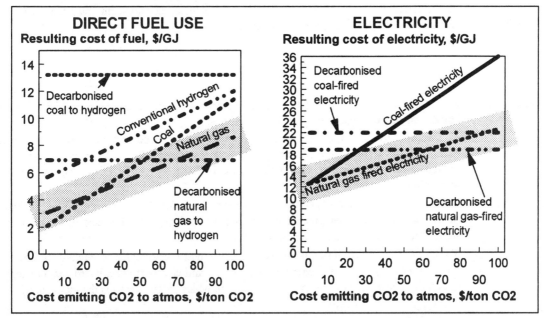

Figure 3 Fuel and electricity cost when the cost of emitting CO_2 to atmosphere is varied
Shaded area indicates the most likely areas of interest for increasing CO_2 mitigation cost. (Oil, which is not included in figure 3 for clarity reasons, will generally be located between coal and natural gas)

Looking first at *electricity production* in the right side of fig. 3, the shaded area indicates that with increasing CO_2-cost we will see a dominance of fuel switching from coal to oil to natural gas up to a CO_2-cost in the order of 40-60 \$/ton. There may, however, be full or partial decarbonisation of coal or oil fired electricity generation already in the CO_2-cost range of 30+ \$/ton by accepting lower coal and oil producer prices (or developing new technology) in order to compete with natural gas. In the 60-100 \$/ton CO_2 cost range we may expect to see decarbonised coal-, oil- and natural gas fired electricity generation along with a reduction in ordinary natural gas fired electricity generation.

When looking at *hydrogen production* in the left side of fig. 3, the shaded area indicates that decarbonised hydrogen will be of little general interest in the energy market until the CO_2-cost reaches the 60 or 70 \$/ton CO_2 mark. Only then can we expect to see hydrogen used on a large scale for heating purposes, for steam raising in industry and in the transportation sector. The hydrogen would be produced from natural gas. Sectors like aviation, requiring liquefied hydrogen, will have the additional considerable penalty of liquefying. Aviation will therefore be one of the last sectors to hang on to oil in a greenhouse driven world.

What, then is the answer to the question asked on the first page of this paper? We believe that it can be formulated as follows;

In energy market terms hydrogen, decarbonised or not, is a very expensive energy carrier which at present lacks the technological advantages in end use that can compensate for the high cost.

Breakthrough is needed in at least one important end use technology (e.g. 65-70% efficiency in converting hydrogen to electricity in fuel cells for house complexes/hotels/shopping malls) plus a substantial reduction in hydrogen production cost even from fossil fuels. These technological factors combined with a high and sustained pressure to reduce CO_2-emissions may propell hydrogen out of the current energy market backwater.

Decarbonised electricity is today a better proposition than hydrogen since a 50% increase in the electricity production price, resulting in about 25% increase in the end use price, is a less formidable barrier.

Both decarbonised hydrogen and electricity would, however, be needed if deep cuts were to be made in the total CO_2 emissions.

Factors like an expected increased price difference between coal, oil and natural gas due to greenhouse-driven changes in demand, differing regional availability and cost of the fuels, the availability of CO_2 storage sites, different transportation costs and technological development over the next few decades will undoubtedly modify the above reflections as time goes by.

NEXT STEPS?

On a short to medium timescale it is our belief that promoting the concept of decarbonisation of fossil fuels to electricity and hydrogen can have the following positive effects;

- To inform energy and climate change actors within the private sector and in governments that *there is a backstop technology with a known price tag* that under pressing circumstances can slice a substantial share off anthropogenic CO_2-emissions to the atmosphere

- This will act as a *pricing benchmark* to strive to reach for those promoting other methods of reducing emissions of greenhouse gases (e.g. nuclear, solar PV electricity, biomass, wind, efficient energy use, fusion energy etc.).

- To motivate coal, oil and natural gas companies and governments to *engage in R&D* in order to move the cost of decarbonised hydrogen and electricity downwards as an energy option for a possible greenhouse driven future in which fossil energy is the main culprit

REFERENCES

EU (1996). *The Underground Disposal of Carbon Dioxide.* Joule II project no. CT92-0031. British Geological Survey, Keyworth, Nottingham, UK

Foster Wheeler/IEA (1996). *Decarbonisation of fossil fuels.* Report Number PH2/2, March 1996 of the IEA Greenhouse Gas R&D Programme

Norsk Hydro/EdF (1996). *Analysis and Optimisation of Equipment Cost to Minimise Operation and Investment for a 300 MW Electrolysis Plant.* Hydrogen Energy Progress XI, Proceedings of the 11th World Hydrogen Energy Conference, Stuttgart, Germany, 23-28 June 1996

Pergamon

Energy Convers. Mgmt Vol. 38, Suppl., pp. S437–S442, 1997
© 1997 Elsevier Science Ltd. All rights reserved
Printed in Great Britain
0196-8904/97 $17.00 + 0.00

PII: S0196-8904(96)00307-X

LIQUID PHASE METHANOL SYNTHESIS CATALYST

H.MABUSE[1], K.HAGIHARA[1], T.WATANABE[1], and M.SAITO[2]

1) Research Institute of Innovative Technology for the Earth(RITE)
   9-2, Kizugawadai, Kizu-cho, Soraku-gun, Kyoto, 619-02, JAPAN
2) National Institute for Resources and Environment(NIRE)
   16-3, Onogawa, Tsukuba-shi, Ibaraki-ken, 305, JAPAN

ABSTRACT

This work focuses on the investigation of the stability of catalyst activity in the liquid phase methanol synthesis process. The effects of various kinds of metfhods to inhibit the deactivation of catalyst have been experimentally examined. The activity of catalyst was stabilized without lowering of activity by a suitable hydrothermal treatment, although the activity of the untreated catalyst decreased gradually with time. The additive and coating of hydrophobic material were effective for slowing down the crystallite size growth and inhibition of deactivation of catalyst. It was considered that hydrophobic material inhibited the adsorption of produced water onto the hydrophilic catalyst. Hydrophobic treatment of catalyst was found to be effective for liquid phase methanol synthesis from CO_2 and H_2 using hydrophobic solvent. © 1997 Elsevier Science Ltd

KEYWORDS

methanol; liquid phase; catalyst; hydrophobic

INTRODUCTION

Methanol synthesis from CO_2 and H_2 has recently much attention as one of promising processes to convert CO_2 into chemicals. Gas phase methanol synthesis process should recycle a large quantity of unconverted gas and furthermore the single pass conversion is limited by large heat release in the reaction. A liquid phase methanol synthesis in solvent has received considerable attention, since a temperature control is much easier in the liquid phase than in the gas phase. Several types of reactors have been proposed such as the liquid entrained reactor (Lee et al.,1992) and the Trickle bed reactor (Akgerman et al.,1993). Lee et al., (1989). have concluded that produced water is one of the most strongly suspected species promoting the crystallite size growth.

As a new methanol synthesis process, basic study of liquid phase method has been made. This synthesis method is characterized by the continuous water phase recovery of the methanol generated. The effects of various parameters such as temperature, pressure, solvent recycling velocities were investigated in previous work (Hagihara et al 1995). It was found that the crystallite size of copper and zinc oxide changed very much before and after reaction in liquid phase methanol synthesis comparing with that in gas phase. In this work, the performance of the treated catalyst in the catalytic reaction for liquid phase methanol synthesis was investigated and compared with that of untreated catalyst.

EXPERIMENTAL

Catalyst

Various Cu/ZnO-based catalysts containing three to four metal oxide components were prepared by a coprecipitation method. A mixture aqueous solution of metal nitrates and an aqueous solution of Na_2CO_3 were added dropwise to distilled water. Ssubsequently, the precipitate was filtered out, washed with distilled water, dried in air at 393K overnight, calcined in air at 623K for 2hr. Finally the power of calcined catalyst was overloaded under a pressure of 20MPa and crushed to the size of 1~2mm.

Hydrothermal treatment. The reduced catalyst was heated at 523K in presence of water and solvent mixture.

Addition of high purity graphite and hydrophobic silica. The powder of calcined catalyst was mixed physically with high purity graphite and hydrophobic silica(two types). The mixture was overloaded under a pressure of 20MPa and crushed to the size of 1~2mm.

Coating of methyl hydrogen silicone oil. The powder of calcined catalyst was impregnated with solution of methyl hydrogen silicone oil (replaced a part of methyl group in dimethyl silicone oil by hydrogen) in isopropylalcohol and was dried. The treated catalyst was heated in a furnace for a reaction with oxygen in order to polymerize methyl hydrogen silicone oil and was overloaded under a pressure of 20MPa and crushed to the size of 1~2mm.

Apparatus and procedures

The effect of treatment was evaluated for the variation of crystallite size of Cu under heating (in presence of mixture of water and solvent) and compared with catalyst lifetimes using untreated catalyst.

Accelerated test The treated catalysts were evaluated in the following method. The stainless steel autoclave was equipped with a stirrer and it had an internal volume of $100cm^3$. The reduced catalyst, solvent(n-dodecane) and water were charged to the autoclave and heated at 623K. The concentration of water corresponds to about ten times of that in the practical reactor. After heating, the crystallite size of Cu was determined by means of X-ray diffraction measurements.

Catalyst lifetime. The hydrogenation reaction of carbon dioxide was conducted using a $220\ cm^3$ autoclave under a pressurized condition. The continuous reactor consisted of a draft tube, a catalytic basket type impeller, a gas feed section and a liquid-liquid separator. The catalyst was charged to the catalyst basket type impeller. The stirring rate was 500 rpm, which was sufficient to stir the reaction system so that higher stirring rates did no increase the reaction rate. As a model solvent, n-dodecane ($n\text{-}C_{12}H_{26}$) $170cm^3$ was charged to the reactor. The mixture of gases ($H_2/CO_2=3/1$) was admitted into the reactor by operating a forward pressure regulator, and its flow was metered by a mass flow meter. A mixture of products was introduced to the liquid-liquid separator by a high pressure pump. Then, the liquid phase was separated into upper phase (solvent phase) and lower phase (aqueous methanol phase), according to the difference of specific gravity. After separation, the solvent phase was continuously recycled to the reactor. The separation temperature was the room temperature. The aqueous phase was sampled for analysis. The methanol and produced water were analyzed by a gas chromatography. (Hagihara et al 1995)

RESULTS AND DISCUSSION

Hydrothermal treatment.

The catalyst used in this study was $Cu/ZnO/ZrO_2 = 40/30/30$. Fig. 1 shows the variation in the activities of the hydrothermal treated catalysts.   It shows the variation in the activity of the thermal treated (723K 1hr in H_2/N_2 atmosphere) and untreated catalysts for comparison. The activity of the hydrothermal treated catalyst was stable, although the activity of the untreated catalyst decreased gradually with time. The variation in the activity of the thermal treated catalyst was low as compared with the untreated catalyst. The activity of the excessive treated catalyst was very low, although it was stable.   Fig. 2 shows the X-ray diffraction pattern of the hydrothermal and thermal treated catalyst before and after reaction.   There was a difference between two catalysts in crystallite sizes of ZnO, although the crystallite sizes of Cu were not very much change before and after reaction. It seems that the stability of activity is closely connected with the variation of crystallite size of Cu and ZnO. It was found that the crystallite size growth was inhibited by the hydorothermal treatment. The moderate hydrothermal treatment contributes to the stability of activity without the decrease of the activity remarkably.

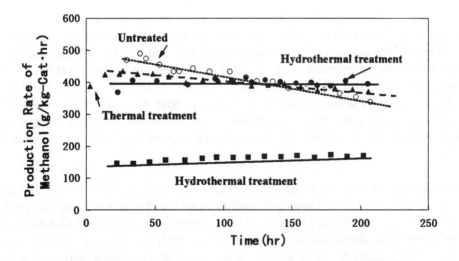

Fig. 1 Variation in the activities of treated and untreated catalyst with time
   (●) (■) Hydrothermal treatment : treated time 50,100hrs
   (▲)   Thermal treatment
   (O)   Untreated

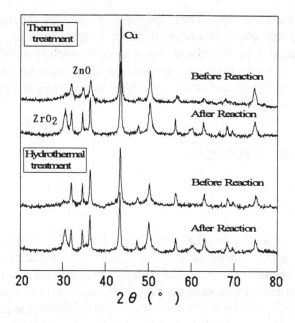

Fig. 2 X-ray diffraction patterns of treated catalyst before and after reaction

Addition and coating of material.

The catalyst used in this study was $Cu/ZnO/ZrO_2/Al_2O_3$ = 40/30/25/5. The accelerated test was performed in order to evaluate the catalysts in a brief period. The contact angle between catalyst and water was measured in order to evaluate the hydrophobicity of catalysts. The catalyst lifetime tests were carried on for the typical catalysts.

Accelerated test. The results in Table 1 show the contact angles between the catalysts and water (the character of the wetting) before the accelerated test, and the crystallite sizes of Cu in various treated catalysts and untreated catalyst after the accelerated test. The crystallite sizes of treated catalysts were lower than the untreated one. The coating of methyl hydrogen silicone oil was most effective among another treatment. It is obvious from these results that the sintering of Cu is retarded by the addition of high purity graphite and hydrophobic silica and the coating of methyl hydrogen silicone oil. It seems that the hydrophobicity of materials are connected with the sintering.

Catalyst lifetime. It was found that the treated catalysts were effective for the inhibition of sintering. Hence, a long-term methanol synthesis test was performed using the catalysts mixed physically with high purity graphite (25wt%) and coated methyl hydrogen silicone oil (5wt%). It was performed using the non additive catalyst for comparison. Fig. 3 shows the variation in the activities of these catalysts. The activity of the non additive catalyst decreased gradually with time. The activity of the catalyst mixed physically with the high purity graphite was stable, although it was low due to the decrease of the active component. On the other hand, the activity of the catalyst coated methyl hydrogen silicone oil was increased gradually with time and was stable at higher level after 200hrs.

Table 1. Difference of crystallite size of Cu

Additive	Component (wt%)	Contact angle (degree)	Crystallite size of Cu (nm)
High purity Graphite	25	(*1)	22.0
High purity Graphite	35	(*1)	20.6
Hydrophobic silica A	10	30	22.4
Hydrophobic silica B	10	60	20.0
Methy hydrogen silicon oil	5	85	12.6
Methy hydrogen silicon oil	10	150	9.2
Non additive		(*2)	26.1

(*1) : The measurement was impossible, because water slowly percolated.
(*2) : The measurement was impossible, because water rapidly percolated.

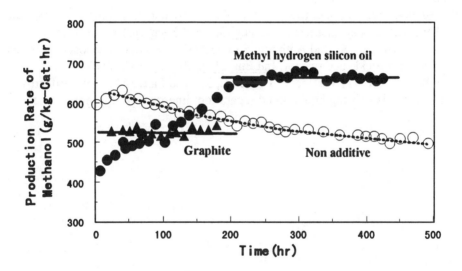

Fig. 3 The variation in the activity of catalysts
  (▲) the catalyst mixed physically with high purity graphite
  (●) the catalyst coated methyl hydrogen silicone oil
  (○) the non additive catalyst

CONCLUSION

The performance of the treated catalyst in the catalytic reaction for liquid phase methanol synthesis was investigated and compared with that of untreated catalyst. It was found that the moderate hydrothermal treatment contributed to the stability of activity without the decrease of activity remarkably. It seems that the stability of activity is closely connected with the variation of crystallite size of Cu and ZnO. The additive of high purity graphite and hydrophobic silica, and the coating of methyl hydrogen silicone oil were effective for slowing down the crystallite size growth.   The activity of the catalyst mixed physically with high purity graphite was stable, although it was low due to the decrease of active component.   The activity of the catalyst coated methyl hydrogen silicone oil was increased gradually with time and was stable at higher level after 200hrs. It seems that the results obtained are connected with the hydrophobicity of materials. It was found that the hydrophobic treatment of catalyst is effective in the liquid phase methanol synthesis by CO_2 hydrogenation.

ACKNOWLEDGMENT

This work is partly supported by New Energy and Industrial Technology Development Organization.

REFERENCES

Lee, S., Parameswaran, V.R., Wender, I., and Kulik, C.J. (1989). The roles of carbon dioxide in methanol synthesis. Fuel Sci. and Tech. Int'l,7,899-918

Lee, S.,Vijayaraghavan, P., and Kulik, C.J. (1992). Modeling of a liquid entrained reactor for liquid phase methanol synthesis process. Fuel Sci. and Tech. Int'l,10,1501-1521

Akgerman, A.,Tjandran, S., and Anthony, R.G. (1993) Low H2/CO rariosynthesis gas conversion to methanol in a trickle bed reactor. Ind.Eng.Chem.Res.,32,2602-2607

Hagihara, K., Mabuse, H., Watanabe, T., Kawai, M., and Saito, M.,(1995) Effective liquid-phase methanol synthesis utilizing liquid-liquid separation. Energy Convers.Mgmt.,36,581-584

Energy Convers. Mgmt Vol. 38, Suppl., pp. S443–S448, 1997
© 1997 Elsevier Science Ltd. All rights reserved
Printed in Great Britain
0196-8904/97 $17.00 + 0.00

Pergamon

PII: S0196-8904(96)00308-1

CARBON RECYCLING SYSTEM THROUGH METHANATION OF CO_2 IN FLUE GAS IN LNG POWER PLANT

T. YOSHIDA, M. TSUJI, Y. TAMAURA, T. HURUE*, T. HAYASHIDA*, K. OGAWA*

Research Center for Carbon Recycling & Utilization, Tokyo Institute of Technology, Ookayama,
Meguro-ku, Tokyo 152, Japan

* Kyushu Electric Power Co. Inc, Shiobaru, Minami-ku,
Fukuoka 815, Japan

ABSTRACT

A bench scale test at ambient pressure has been performed for carbon recycling system using a ferrite process in LNG power plant. The chemical reaction consists of two steps; (i) adsorptive separation and decomposition of CO_2 to carbon on oxygen-deficient Ni ferrite and (ii) methanation of deposited carbon. CO_2 decomposition and methanation reactivities were much improved when the honeycomb support was used instead of a packed-bed-type reactor. The overall methanation rate of CO_2 in flue gas from LNG power plant could achieve $1,000 \, Nm^3 \, h^{-1}$ when $12,000 \, kg$ Ni ferrite loaded on a honeycomb support was used. The present methanation and carbon recycling system could be extended to other CO_2 sources such as IGCC power plant and depleted natural gas plant. © 1997 Elsevier Science Ltd

KEYWORDS

methanation; CO_2 decomposition; ferrite; carbon recycling system; power plant; LNG boiler.

INTRODUCTION

Current global carbon recycling systems are causing a considerable buildup of CO_2 in the atmosphere. It may be due to the large difference between burning rate of fossil fuels and fixation rate of CO_2 by vegetation and forestation. CO_2 buildup in the atmosphere is considered to cause the global warming and subsequent changes in the climate. A fossil fuel combustion power plant is one of the major CO_2 emission sources, amounting 25% of CO_2 emissions in Japan, and mitigation has been the current subject. To achieve the goal, the current oil energy system will be forced to shift to the hydrogen energy system from carbon-free renewable sources, i.e., solar hydrogen. However, one needs to keep utilizing much more efficiently fossil fuels in current carbon-based infrastructure till the full-scale hydrogen age be realized. A long term will be necessary to overcome some significant barriers for introducing H_2 gas into the primary energy market: carbon-free H_2 production method, storage, transportation, and H_2 turbine in large scale. In a transient age between the current oil-based era and hydrogen energy era, a solar H_2 gas will be insufficiently supplied to the primary energy market. The CO_2/CH_4 conversion system by ferrite process may serve an option of using such H_2 gas in power plant. The proposed ferrite process allows CO_2 separation from flue gas, subsequent decomposition to carbon, and methanation reactions by alternative flow of H_2 and flue gas in a single column at the same temperature. Hence, pure CH_4 gas can be theoretically synthesized from CO_2 with low concentration in flue gas and H_2 gas with the minimum process energy loss, while conventional catalytic processes need an additional separation process of CH_4 gas formed. This paper describes the bench scale test of CO_2/CH_4 conversion system by use Ni ferrite process having compatibility with natural gas-fired power plant.

EXPERIMENTAL METHODS

Synthesis of Ni ferrite

Ni ferrite with the Ni/Fe mole ratio 0.15 was used throughout the experiment.   This was synthesized by air-oxidizing Fe(II)/Ni(II) mixed hydroxide suspension for 6 h at pH 9 and at 65°C according to the previous reports (Kiyama *et al.*, 1973, Tamaura 1979).   The procedure was briefly described as follows. Requisite quantities of $FeSO_4 \cdot 7H_2O$ (312g) and $NiSO_4 \cdot 6H_2O$ (44.2g) were dissolved in oxygen- and CO_2-free distilled water ($4 \, dm^3$) prepared by passing N_2 gas through them for a few hours.   The pH of the solution was adjusted to 9 by adding $13 \, mol \, dm^{-3}$ NaOH solution to form hydroxide suspension.   The formed mixed precipitate was oxidized by air/N_2 mixed gas bubbling for 6 h at 65°C while kept at the same pH by adding $2 \, mol \, dm^{-3}$ NaOH solution using an autotitrator.   The product was collected by decantation, washed with distilled water and acetone, successively, and then heated to remove adsorbed water and acetone in N_2 gas stream at 300°C for 3 h.

Methanation Equipment

Ni ferrite was activated by flowing hydrogen gas at a flow rate of $9.0 \, dm^3 \, h^{-1}$ for 4.0 h in a home-made stainless steel housing at 330°C and CO_2 was allowed to decompose by passing feed gas (CO_2: 20%, N_2: 80%) through the activated ferrite.   Generated gases were determined using a gas analyzer (Shimadzu; CGT-10). The carbon deposited on the surface of ferrite was determined by elemental analyzer (Perkin Elmer 2400) .   Ferrite powder was used in packed bed (65mm$\phi \times$ 150mm long) or in a honeycomb support made of cordierite (Fig.1).   Table 1 shows the characteristics of the honeycomb support.

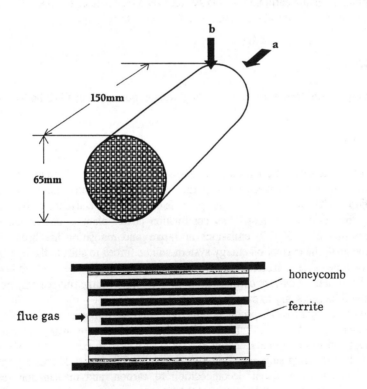

Fig. 1.   Outer appearance (top) and cross sectional view (bottom)
of a honeycomb support

Table 1. Specification of honeycomb support

Material	cordierite
Chemical composition	$2MgO \cdot 2Al_2O_3 \cdot 5SiO_2$
Size	65 mmϕ ×150mm
Thermal expansion	$1 \times 10^{-6} K^{-1}$
Specific heat	$0.8 J g^{-1} K^{-1}$
Thermal conductivity	0.0025 cal sec cm^{-1} K^{-1}
Maximum temperature	1400 °C
Pore volume	0.2 cm^3 g^{-1}
Porosity	35 %
Crushing strength(a)	>85 kg cm^{-2}
Crushing strength(b)	>11 kg cm^{-2}

(a): longitudinal direction, (b): normal direction to longitudinal direction

RESULTS AND DISCUSSION

Methanation Process

Conversion process of CO_2 to CH_4 on ferrite is composed of the following three steps;
(i) activation step of Ni ferrite by H_2 at flue gas temperature,

$$Ni_xFe_{3-x}O_4 \;+\; \delta H_2 \;\rightarrow\; Ni_xFe_{3-x}O_{4-\delta} \;+\; \delta H_2O, \tag{1}$$

(ii) decomposition step of CO_2 in flue gas into carbon on the reduced Ni ferrite (Kodama *et al.*, 1995, Kato *et al.*, 1994),

$$Ni_xFe_{3-x}O_{4-\delta} \;+\; (\delta/2)CO_2 \;\rightarrow\; Ni_xFe_{3-x}O_4C_{\delta/2}, \tag{2}$$

and (iii) methanation of deposited nanocarbon with H_2 (Tsuji *et al.*, 1994, Yoshida *et al.*, 1993),

$$Ni_xFe_{3-x}O_4C_{\delta/2} \;+\; 2\delta H_2 \rightarrow Ni_xFe_{3-x}O_{4-\delta} \;+\; (\delta/2)CH_4 \;+\; \delta H_2O. \tag{3}$$

The activation step (i) is endothermic and a portion of its process energy is supplied by thermal energy of the flue gas and partially by the exothermic methanation process (iii).   In the process (ii), CO_2 in the flue gas is decomposed to carbon and separated from the main constituent N_2 gas by H_2 activated Ni ferrite.

The chemical composition of synthesized Ni ferrite was $Ni^{2+}_{0.39}Fe^{2+}_{0.59}Fe^{3+}_{2.01}O_{4.00}$.   XRD pattern of the product showed a single phase of the spinel-type structure and the lattice constant a_0 was determined to be 0.8375 nm.   Fig. 2 shows CO_2 decomposition performance of Ni ferrite loaded in honeycomb support and in packed bed.   The feed gas was passed at a flow rate of 0.6 dm^3 h^{-1}.   The efficiency of CO_2 decomposition was much improved when the honeycomb support was used, compared to Ni ferrite powder packed in reactor.   The amount of deposited carbon on Ni ferrites after CO_2 decomposition was 0.22 g (conversion CO_2 of to C: 77%) in the latter and 0.49g (conversion of CO_2 to C: 35%) in the former.   In the latter case which the powder was directly packed in, hydrogen gas excessively reduced the ferrite near the inlet part and formed less reactive metallic iron (α-Fe) on Ni ferrite.   The CO_2 decomposition efficiency has been improved by about two times compared to the case where the powder was directly filled in. CO_2 decomposition and methanation were alternatively repeated on a bench scale (Fig.3). CO_2 and CO were below the detection limit.   The amount of methane was 0.3dm^3 at one cycle of process and conversion of CO_2 to CH_4 was 100%.   The CO_2 decomposition and methanation rates in bench scale test are shown in Table 2.   Methanation rate of deposited carbon was much lower in magnetite and was not studied further.

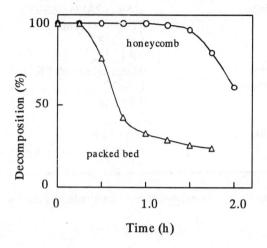

Fig. 2. CO_2 decomposition performances of packed bed (Ni ferrite 0.3kg)
and honeycomb support (Ni ferrite 0.15kg) at 350°C.
Activation time by H_2 : 4h at flow rate of H_2 = 30dm^3 h^{-1}, 0.1MPa,
flow rate of feed gas= 0.6dm^3 h^{-1} CO_2 (feed; CO_2:20%, N_2:80%), 0.1MPa

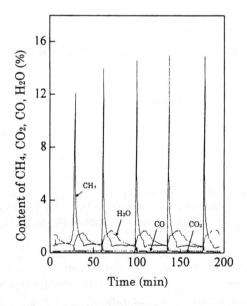

Fig. 3.   Bench scale test of CO_2 decomposition and successive methanation
reaction with honeycomb supported Ni ferrite at 350°C.
Activation time by H_2 4 h at flow rate of H_2 = 30dm^3 h^{-1}, 0.1 MPa,
cycling time: CO_2 decomposition : 0.17 h, at flow rate of feed gas
= 1.8 dm^3 h^{-1} CO_2 (feed; CO_2:20%, N_2:80%), 0.1MPa, Methanation
by H_2 : 0.17 h, flow rate of H_2 = 9.0 dm^3 h^{-1}

Table 2. Effect of specimen form on rates of CO_2 decomposition to carbon and methanation by the ferrite process on a bench scale at 350°C

Ferrite*	Form	CO_2 decomposition rate** $(dm^3 h^{-1})$	Methanation rate*** $(dm^3 h^{-1})$
Fe_3O_4	packed bed	0.18	<0.1
$Ni_{0.39}Fe_{2.6}O_4$	packed bed	0.22	0.44
$Ni_{0.39}Fe_{2.6}O_4$	honeycomb support	1.8	1.8

*   : 0.2 kg
** :activation time by H_2: 4 h at a flow rate of H_2=9.0 dm^3 h^{-1}, 0.1MPa
flow rate of feed gas=1.8 $dm^3 h^{-1}CO_2$ (feed gas; CO_2:20%, N_2:80%, 0.1MPa)
***:flow rate of H_2 =9.0 $dm^3 h^{-1}$ (H_2:100%, 0.1MPa)

Process Design

Based on the data obtained in bench-scale test, the amount of ferrite necessary for the pilot plant processing flue gas of 1000m^3 h^{-1} is estimated to be 99,000 kg in the case of packed bed and 12,000 kg in the case of honeycomb type (Table 3).   The honeycomb support reactor dimension is 4.5 m wide × 4.5 m deep × 3.0 m high.   In the proposed recycling system, H_2O in flue gas is removed prior to treating flue gas in Ni ferrite reactor and the gas volume is decreased to 748 Nm^3/h.   The target constituents CO_2, NO_x and O_2 are completely decomposed into each element. N_2 gas is discharged as exhaust.   The CO_2 decomposition rate reaches 94 Nm^3 h^{-1}.   The deposited carbon is methanated in a rate of 50 kg h^{-1}.   409 Nm^3 of H_2 is used for methanation and keeps Ni ferrite oxygen-deficient.   However, the product gas still contains 20% of unreacted H_2 gas on a volume basis.   The content could be decreased by improving the utilization efficiency of Ni ferrite reactor.   This CO_2/CH_4 conversion system has an advantage that relatively low concentration of CO_2 in flue gas can be directly converted to methane with minor change in the conventional facility.

Table 3. Quantity of ferrite required for scaling up ferrite process to pilot plant scale with flue gas volume 1000$m^3 h^{-1}$

Ferrite	Form	Quantity required (kg)	(m^3)
$Ni_{0.39}Fe_{2.6}O_4$	packed bed	99,000	16
$Ni_{0.39}Fe_{2.6}O_4$	honeycomb support	12,000	30

Fig. 4 is a typical design that we have obtained in a feasibility study on a pilot scale processing 1,000 m^3 h^{-1} of LNG boiler flue gas volume 1,400,000m^3 h^{-1}. H_2O in flue gas from an LNG boiler is removed by using a conventional adsorption method.   Activated 12,000 kg Ni ferrite on a honeycomb support can also decompose NO_x in flue gas to form N_2 which is discharged to the atmospheric environment as an exhaust, while CO_2 is decomposed to carbon in nanoscale on the Ni ferrite.   Then the deposited carbon is hydrogenated to form CH_4 which is circulated in a different roop to obtain as pure gas as possible and then recycled to the LNG boiler. Chemical processes of CO_2 separation from flue gas, decomposition, methanation and simultaneous NO_x removal can be implemented in the same reactor. The conventional catalytic NO_x decomposition unit can be eliminated in the proposed system.

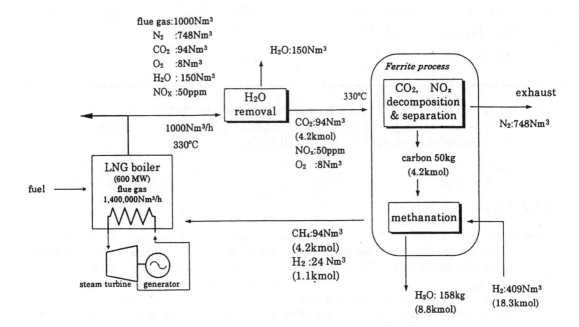

Fig. 4.   Schematic of CO_2/CH_4 conversion system by the ferrite process on a pilot scale.

CONCLUSIONS

A feasibility study for carbon recycling system has been carried out on a scale of 1000 $m^3 h^{-1}$ of flue gas. The system is based on methanation of CO_2 in flue gas from LNG power plant in a Ni ferrite reactor supported in cordierite support. The proposed system is straightforward for scaling up by additional Ni ferrite reactor. This is still open to be studied for improvement of the Ni ferrite reactor size and performance. The conventional catalytic NO_x decomposition unit can be eliminated because NO_x is decomposed along with CO_2 in H_2 activated Ni ferrite.

REFERENCES

Kato, H., T. Kodama, M. Tsuji and Y. Tamaura (1994). Decomposition of carbon dioxide to carbon by hydrogen-reduced Ni(II)-bearing ferrite. *J. Mater. Sci.*, **29**, 5689-5692.

Kiyama, M. (1973). Condition for the Formation of Fe_3O_4 by the air oxidation of $Fe(OH)_2$ suspension. *Bull. Chem. Soc. Japan*, **47**, 1646-1650.

Kodama, T., T. Sano, T. Yamamoto, M. Tsuji and Y. Tamaura (1995). CO_2 decomposition to carbon with ferrite derived metallic phase at 300°C. *Carbon*, **33**, 1443-1447.

Tamaura, Y. (1979). The Fe_3O_4-Formation by the Ferrite process. *Water Research*, **13**, 21-31.

Tsuji, M., H. Kato, T. Kodama, S. G. Chang, N. Hasegawa and Y. Tamaura (1994). Methanation of CO_2 on H_2-reduced Ni(II)- or Co(II)-bearing ferrites at 200°C. *J. Mater. Sci.*, **29**, 6227-6230.

Yoshida, T., K. Nishizawa, M. Tabata, A. Hiroshi, T. Kodama, M. Tsuji and Y. Tamaura (1993) Methanation of CO_2 with H_2-reduced magnetite. *J. Mater. Sci.*, **28**, 1220-1226.

Energy Convers. Mgmt Vol. 38, Suppl., pp. S449–S454, 1997
© 1997 Elsevier Science Ltd. All rights reserved
Printed in Great Britain
0196-8904/97 $17.00 + 0.00

Pergamon

PII: S0196-8904(96)00309-3

CATALYTIC CONVERSION OF CARBON DIOXIDE USING PHASE TRANSFER CATALYSTS

D. W. PARK, J. Y. MOON, J. G. YANG AND J. K. LEE

Department of Chemical Engineering, Pusan National University, Pusan, 609-735, KOREA

ABSTRACT

In the present study, synthesis of 5-membered cyclic carbonates from CO_2 and epoxides, such as glycidyl methacrylate(GMA) and diglycidyl 1,2-cyclohexane dicarboxylate(DCD), were investigated in view of the characteristics of phase transfer catalysts, reaction mechanisms and kinetics. Quaternary salts showed a good conversion of epoxide at 1atm of CO_2 pressure. Among the salts tested, the one having a larger alkyl group and a more nucleophilic counter anion exhibited a better catalytic activity. Kinetic studies in a semi-batch reactor, through which a slow stream of CO_2 is continuously passed, showed that the reaction rate was pseudo-first order with respect to the concentration of epoxide. In a batch autoclave reactor with high CO_2 pressure, however, the reaction rate showed second order kinetics. The reaction was also carried out with an insoluble phase transfer catalyst, quaternary ammonium chloride(QCl) anchored to metal oxide, to facilitate the recovery of catalyst. © 1997 Elsevier Science Ltd

KEYWORDS

Carbon dioxide ; catalytic conversion ; kinetics ; phase transfer catalyst ; synthesis of cyclic carbonate

INTRODUCTION

The chemistry of carbon dioxide has received much attention, since it is considered to be responsible for the green house effect and yet the most abundant and the least expensive carbon source. Various chemical and biological methods to fix and utilize CO_2 are under study. Among them, the reactions with oxiranes or some polymers containing pendant groups are of special interest since they lead to the production of cyclic carbonates(Rokicki *et al.*, 1984). These cyclic carbonates can be used as aprotic polar solvents and as sources for reactive polymer synthesis(Inoue *et al.*, 1982).

It is reported by Kihara and Endo(1992) that (2-oxo-1,3-dioxolan-4-yl)methyl methacrylate(DOMA) can be prepared by fixation of CO_2 into glycidyl methacrylate in the presence of metal halide or phase transfer catalysts. But detailed kinetic studies have not been undertaken. Recently, the authors (Kihara and Endo, 1993) discovered the synthesis of poly(hydroxyurethane)s from bis(cyclic carbonates) and diamine. The polyurethane-bearing hydroxy group is difficult to be synthesized by polyaddition of polyol and diisocyanate.

In this paper, the synthesis of DOMA was studied to understand the effect of catalyst structure. A detailed kinetic study was carried out for this reaction. The synthesis of di(2'-oxo-1',3'-dioxolan-4'-yl) methyl-1,2-cyclohexane dicarboxylate (DOCD) from CO_2 and diglycidyl 1,2-cyclohexane dicarboxylate(DCD) are also studied in view of kinetics and performance of metal oxide supported quaternary ammonium chloride catalyst. In the case of a poly(hydroxyurethane)($\overline{Mn}$ =9100) from DOCD and hexamethylene diamine for example, the amount of CO_2 reduction may be estimated to be 1056 g of CO_2.

EXPERIMENTAL

Glycidyl methacrylate(GMA), DCD and reaction solvent were used after distillation on CaH_2. Quaternary ammonium salts, such as tricaprylylmethylammonium chloride(Aliquat 336), tetrabutylammonium chloride(TBAC), tetrabutylammonium bromide(TBAB), tetrabutylammonium iodide(TBAI), tetraoctylammonium chloride(TOAC) and tetrapropylammonium chloride(TPAC) were all reagent grade and used as purchased without purification.

The synthesis of cyclic carbonate was carried out under both atmospheric and high pressures of CO_2. For the former case, 1 mmol of catalyst was introduced to a 150 mL four-neck pyrex reactor containing the mixture of 30 mmol of GMA(or 50 mmol of DCD) and 50 mL of solvent, and the solution was heated up to a desired temperature (60-100°C). Reaction was started by stirring the solution under a slow stream of CO_2(10 mL/min), and continued for 6 hours. Periodically, a small portion of reaction mixture was taken out and analyzed by gas chromatograph(HP5890A) or HPLC(Waters5061). The identification of cyclic carbonate was performed by FT-IR(Mattson Polaris), ^{1}H-NMR and ^{13}C-NMR(Jeol PMX-60 SI, TMS as an internal standard).

For the synthesis of DOMA under high CO_2 pressure, 120 mL of diglyme solution containing both 60 mmol of GMA and 6 mmol of catalyst was introduced to an autoclave(Parr 4841), and the reactor was pressurized to a desired level by a CO_2 cylinder. The variation of reactor pressure during the experiment was measured by a pressure transducer. The final reaction mixture was analyzed by the same method as described above.

RESULTS AND DISCUSSION

Reaction of GMA with CO_2

The synthesis of DOMA was carried out by the reaction of GMA with carbon dioxide in toluene in the presence of various quaternary ammonium salts. The formation of the five-membered cyclic carbonate was characterized by an IR absorption peak at $1800 cm^{-1}$. In a semi-batch reactor at a constant flow of CO_2, only the concentration of GMA varies since the concentration of dissolved CO_2 in toluene can be assumed constant. If the rate constant is k', the consumption rate of GMA is expressed as:

$$-r_{GMA} = -d[GMA]/dt = k'[GMA] \qquad (1)$$

Integration of eq. (1)gives eq.(2)

$$\ln\{[GMA]_0/[GMA]\} = k't \qquad (2)$$

where $[GMA]_0$ is the initial concentration of GMA.

Experimental results in a semi-batch reactor are shown in Fig. 1. Since the plots of $\ln\{[GMA]_0/[GMA]\}$ vs. time give good straight lines, the reaction can be considered as first-order with respect to [GMA]. From the slope, the order of the catalytic activity is determined as TOAC>THAC>TBAC. Bulky quaternary salts, having longer distances between cations and anions, are generally known to exhibit higher activity in activating anions(Starks *et al.*, 1994). This explains why they are more effective in nucleophilic attack of the anion to the oxirane ring of GMA.

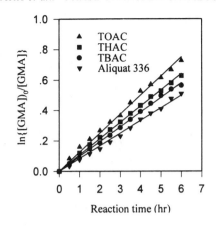

Fig. 1. Effects of cations of quaternary ammonium salts on the ln{[GMA]$_0$/[GMA]}.

Table 1. Pseudo first-order rate constant(k') with various catalysts

Catalysts	k'(1/hr)	Catalysts	k'(1/hr)
TOAC	0.125	TBAI	0.155
THAC	0.107	TBAB	0.123
Aliquat 336	0.083	TBAC	0.097

When the different halide ions were used for the quaternary ammonium salts, the first-order rate constant decreased in the order of I$^-$ > Br$^-$ > Cl$^-$ as shown in Table 1. This is contrary to our previous results on the reaction of glycidyl vinyl ether and CO$_2$ in N-methyl pyrollidinone(Park *et al.*, 1996). The solvation of the anion is an important factor for the liquid phase reaction. In a protic solvent, like toluene in this study, stronger solvation may be expected with a hard anion like Cl$^-$ than with a soft anion like I$^-$. Therefore the order of nucleophilicity will decrease in the order of I$^-$ > Br$^-$ > Cl$^-$ which is consistent with the order of the reactivity of quaternary ammonuim halides. According to the mechanism proposed by Kihara *et al.*(1993), nucleophilic attack of anion to epoxide ring is known as the limiting step of the reaction.

The effect of temperature on the reaction rate was studied at 60, 80, 90 and 100°C with Aliquat 336 catalyst(Table 2). The rate constant increased with temperature and the activation energy was estimated as 8.48 kcal/mol from Arrhenius plot.

Table 2. Pseudo first-order rate constant(k') at 60, 80, 90 and 100 °C

Temp. (°C)	k'(1/hr)	Temp. (°C)	k'(1/hr)
60	0.04	90	0.11
80	0.08	100	0.17

When the addition reaction between GMA and CO$_2$ is carried out in a batch reactor, both [GMA] and [CO$_2$] vary with time. If we assume that the reaction rate is dependent on both reactants and follows second-order kinetics, the reaction rate of GMA can be expressed as :

$$-r_{GMA} = -d[GMA]/dt = k[GMA][CO_2] \qquad (3)$$

Because the reaction proceeds stoichiometrically, the concentration of GMA is :

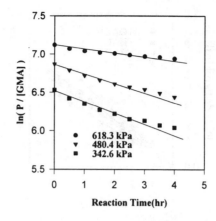

Fig. 2. Linear plot of $\ln\{P/[GMA]\}$ vs. time.

$$[GMA] = \{[GMA]_0 - [CO_2]_0 + [CO_2]\} \tag{4}$$

where $[CO_2]_0$ is the initial concentration of CO_2. Then, eq. (3) becomes :

$$-r_{GMA} = -d[GMA]/dt = -d[CO_2]/dt = k[CO_2]\{[GMA]_0 - [CO_2]_0 + [CO_2]\} \tag{5}$$

Eq. (5) is rearranged to eq. (6) :

$$\left\{\frac{1}{[CO_2]} - \frac{1}{[CO_2] + [GMA]_0 - [CO_2]_0}\right\}d[CO_2] = k\{[CO_2]_0 - [GMA]_0\}dt \tag{6}$$

By integrating eq. (6) and substituting eq. (4), we obtain :

$$\ln\{[CO_2]/[GMA]\} = k\{[CO_2]_0 - [GMA]_0\}t + \ln\{[CO_2]_0/[GMA]_0\} \tag{7}$$

If $[CO_2]$ is proportional to its pressure in the gas phase(P), according to the Henry's law, eq.(7) can be arranged to eq.(9) :

$$[CO_2] = H'P \tag{8}$$

$$\ln\{P/[GMA]\} = k\{[CO_2]_0 - [GMA]_0\}t + \ln\{P_0/[GMA]_0\} \tag{9}$$

Here, P_0 is initial pressure of CO_2 and H' is the Henry's constant. If CO_2 is the only component in the gaseous phase, we can easily estimate the extent of reaction by measuring P with time. Fig. 2 shows the plots of $\ln\{P/[GMA]\}$ vs. time at different P_0 of 308, 446 and 611 kPa at 80°C with Aliquat 336 catalyst. Since each plot shows a good linear relationship, we can get $k\{[CO_2]_0 - [GMA]_0\}$ and $\ln\{P_0/[GMA]_0\}$ from the slope and intercept respectively. Replacing $[CO_2]_0$ in the slope by $H'P_0$, eq. (11) can be obtained :

$$k\{[CO_2]_0 - [GMA]_0\} = k\{H'P_0 - [GMA]_0\} \tag{10}$$

$$P_0 = \frac{k\{[CO_2]_0 - [GMA]_0\}}{kH'} + \frac{[GMA]_0}{H'} \tag{11}$$

Since the plot of P_0 vs. $k\{[CO_2]_0 - [GMA]_0\}$ in Fig. 3 gives another linear relationship, the reaction rate could be considered as second order. This is consistent to the mechanism of some other cyclic carbonate syntheses proposed in earlier reports(Rokicki *et al.*, 1984; Kihara *et al.*, 1993). From the slope and inter-

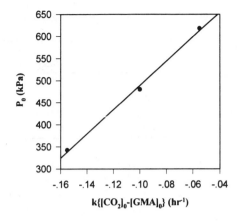

Fig. 3. Linear plot of P_0 vs. $k\{[CO_2]_0-[GMA]_0\}$.

cept corresponding to $1/(kH')$ and $1/H'$, the second-order rate constant k and Henry's constant H' at 80°C are estimated to be 0.031 L/mol·hr and 6.6×10^{-4} mol/L·kPa respectively. This method is specially advantageous considering that sampling from a pressurized reactor and subsequent analysis of the reaction mixture are very difficult. The H' value of CO_2 in NMP is not known in the literature, but that of CO_2 in water at 80°C is 1.3×10^{-4} mol/L·kPa(Wilhelm *et al.*, 1977).

Reaction of diglycidyl-1,2-cyclohexane dicarboxylate(DCD) with CO_2

The reaction of DCD and CO_2 leads to the formation of a bis(cyclic carbonate) of diglycidyl 1,2-cyclohexane dicarboxylate. Experiments with different types of cation structure and halide anion of the quaternary ammonium salt catalysts in DMF solvent are performed. The same trends of increasing reactivity was observed with the increase of cation size from TBAC to TOAC. But the catalytic activity increases in the order of $I^- < Br^- < Cl^-$ as shown in Fig. 4, which is contrary to the result of DOMA synthesis using toluene as solvent, as previously described in the solvation effect.

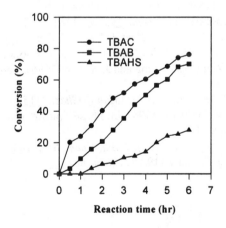

Fig. 4. Effects of counter anions of quaternary ammonium salts on the conversion of DCD.

Table 3. Effects of supported catalyst on the conversion of DCD

Support	Surface area	%N	Amount of catalyst (mmol/g-support)	Conversion (%)
γ-Al$_2$O$_3$	155	1	071	7
α-Al$_2$O$_3$	4	0	0	0
TiO$_2$ (ana.)	70.8	0.02	0.014	25.2
SiO$_2$	82	0.03	0.021	39.3

In order to facilitate the recovery and the reuse of phase transfer catalyst, metal oxide-supported quaternary ammonium salt catalysts are prepared according to the reported method(Tundo *et al.*, 1982). The quantity of active sites was estimated by measuring the amount of nitrogen with an elemental analyzer(Carlo Erba E. A. 1108). Table 3 shows the attached amount of catalyst and conversion of DCD at 100°C. The supported catalysts prepared with high surface area metal oxide exhibited higher conversion. QCl/γ-Al$_2$O$_3$ catalyst maintained its reactivity even after 5 more times use and recovery.

CONCLUSION

1) Quaternary ammonium salts showed a good catalytic activity in addition reactions of GMA and DCD with CO$_2$ even at atmospheric pressure.

2) Among the quaternary ammonium salts tested, the ones with higher alkyl chain length showed a increased catalytic activity and, in case of counter anions, catalytic activity was increased as in the order of I$^-$>Br$^-$>Cl$^-$ in the reaction of GMA. But in the reaction of DCD, the order of catalytic activity with the counter anions was vice versa.

3)In a high pressure batch reactor, the reaction rate was first order with respect to both [GMA] and [CO$_2$]. A method to evaluate Henry's constant and rate constant is also proposed.

4)Immobilization of quaternary ammonium salts on metal oxide supports was also performed in this work and QCl/γ-Al$_2$O$_3$ showed the highest catalytic activity. Metal oxide supported catalyst can be recovered and reused for at least five runs.

REFERENCES

Inoue, S., N. Yamazaki and F. Higashi (1982). In: Organic and Bioorganic Chemistry of Carbon Dioxide (S. Inoue, ed.), Chap. 4, pp. 167-182. Kodansha Ltd., Tokyo.

Kihara, N and T. Endo (1992). *Makromol. Chem.*, 193, pp. 1481-1492.

Kihara, N and T. Endo (1993). *J. Polym. Sci., Part A. Polym. Chem.*, 31, pp. 2765-2773.

Kihara, N., N. Hara and T. Endo (1993). *J. Org. Chem.*, 58, pp. 6198-6202.

Park, D. W., J. Y. Moon, J. G. Yang, S. M. Jung and J. K. Lee (1996). *React. Kinet. & Cat. Lett*, presented for the publication.

Rokicki, G., W. Kuran and B. P. Marcinak (1984). *Manat. Chem.*, 115, pp. 205-214.

Starks, C. M., C. L. Liotta and M. Halpern (1994). In: Phase Transfer Catalysis, Chap. 1, pp. 11-12. Chapman & Hall, New York.

Tundo, P., P. Venturello and E. Angelett (1982). *J. Am. Chem. Soc.*, 104, pp. 6547-6551.

Wilhelm, E., R. Battino and R. J. Wilcock (1977). *Chem. Rev.*, 77, pp. 219.

Pergamon

Energy Convers. Mgmt Vol. 38, Suppl., pp. S455–S460, 1997
© 1997 Elsevier Science Ltd. All rights reserved
Printed in Great Britain
0196-8904/97 $17.00 + 0.00

PII: S0196-8904(96)00310-X

Technology Assessment of Alternative Fuels by CO₂ Fixation Use in Passenger Cars

Seiji Matsumoto,   Atsushi Inaba
National Institute for Resources and Environment(NIRE),
16-3, Onogawa, Tsukuba, Ibaraki, 305, Japan

Yukio Yanagisawa
Research Institute of Innovative Technology for the Earth(RITE),
Kizu, Kyoto 619-02 Japan,
/ Harvard School of Public Health

ABSTRACT

Alternative fuel vehicles were investigated as a possible mitigation measure to reduce CO_2 emissions. Fuel economies of several alternative cars were compared, and integrated CO_2 emissions from fuel mining to consumption in Japan were calculated. The alternative fuel vehicles we investigated were methanol, compressed natural gas, electric, hydrogen, hybrid (internal combustion engine and motor), and fuel cell vehicles.
Our calculations showed that a combined approach discharged the least amount of CO_2 when compared to individual alternative fuel vehicles studied in this paper. The combined approach was to use methanol and electricity coming from a coal fired power plant. Methanol was produced from CO_2 collected from the flue gas of the coal fired power plant and hydrogen originated from a non-carbon natural energy source. The MeOH-powered vehicle emitted 28.5 kg- CO_2 per 100 km driven. Electricity generated at the coal fired power station could be supplied to the electric vehicles. The CO_2 emission per 100 km driven was reduced to 16.7 kg by the combination of the methanol and electric vehicles.
CO_2 recycling for methanol production will be one of the CO_2 mitigation strategies in the transportation sector, if hydrogen can be produced plentifully from renewable energy sources. The methanol vehicle is also advantageous when considering available infrastructures. © 1997 Elsevier Science Ltd

KEYWORDS

CO_2, Alternative fuel, vehicle, Methanol, CO_2 recycling

INTRODUCTION

Carbon dioxide (CO_2) emission from the transportation sector accounts for approximately 20% of the Japanese total CO_2 emissions, and its growth rate is faster than other sectors. More than 90% of CO2 emission in the transportation sector comes from vehicles mostly using petroleum as an energy source. There are several mitigation measures to reduce CO_2 emissions from the transportation sector, such as improving fuel efficiency and systematic control of traffic flows. In this paper, we focused on the use of alternative fuel vehicles as a mitigation measure to reduce the CO_2 emission. Methanol, compressed natural gas, electricity, and hydrogen were selected as alternative energy sources for driving vehicles. We carried out technological assessments on these vehicles from viewpoints of energy efficiency and CO_2 emission potential.

METHODS

A vehicle needs specific forms of energy converted from primary sources such as oil, natural gas, coal and renewable energy sources including solar, wind and hydraulic power. Since CO_2 is discharged from each step of the energy conversion, a full fuel cycle analysis, covering the mining of the primary energy source, transportation, conversion and consumption by the vehicle, was carried out to assess an integrated CO_2 emission for each alternative fuel. We included CO_2 emissions not only from mining and conversion, but also transportation processes from producing countries to Japan, because Japan imports almost all primary energy sources.

Standardizing driving modes is important for comparing fuel economy, because mileage or fuel economy depends on the driving modes of the vehicle. In Japan, the fuel economy is indicated by several ways such as 60 km/h constant driving mode, 10-mode representing driving conditions in an urban area, 11-mode simulating early morning driving situations beginning from cold engine start, and 10-15 mode including high speed drives on free ways. We selected 10-mode as the basis for the fuel economy comparison because it represents typical driving patterns in urban areas and because of the availability of the mileage data.

RESULTS

FUEL ECONOMY

Methanol Vehicles. Fuel economy data for the Corolla-FFV (TOYOTA) are available for the firing both gasoline and methanol. Fuel consumption using high-octane gasoline is 13.7 km/l(compression ratio, Rc=11), while the mixture of 85% methanol and 15% of gasoline is consumed at the levels of 7.88 km/l-M85. This value is equivalent to 13.8km/lGE (km traveled per one liter gasoline equivalent). Fuel economy of the Corolla-1600, which is a base car of the FFV, is 13.2km/l using normal gasoline. This mileage is lower than that of the FFV because of the lower compression ratio. Since the octane value of methanol is higher than that of gasoline and methanol can be burned at leaner conditions than gasoline, we expected improvement of thermal efficiency of the methanol vehicle.

Compressed Natural Gas (CNG) Vehicles. Fuel consumption of a CNG vehicle is reportedly 13.8 km/lGE (Wagon, Rc=9.2). The disadvantages of the CNG vehicle are the larger aerodynamic friction due to a larger fuel tank and energy loss due to gas compression. These reduce the fuel economy by 17%. The octane value of methane is 130, higher than those of methanol and gasoline. Improvement in thermal efficiency is expected by increasing the compression ratio. For example, when Rc is raised from 8.2 to 13.2, thermal efficiency is reported to be improved by 22.2%. Lean burning of CNG has a similar effect to that in the methanol engine, where a 5% improvement of fuel consumption was found.

Electric Vehicles (EV). The EV model used for comparison purposes was the Roadster (MAZDA), for which gasoline consumption is 11.2 km/l. Energy consumption of Roadster-EV is 43.3 km/lGE in 10-mode and 69.8km/l at the 60km/h constant mode. The 10-mode efficiency is nearly 4 times greater than that of the gasoline vehicle.

These results were then compared to those of the IMPACT (GM), an original EV car designed to reduce body weight and aerodynamic friction as much as possible. The fuel consumption of the IMPACT is approximately half of the Roadster EV. The technologies applied to the IMPACT to enhance energy efficiencies can be used in the other alternative fuel vehicles. In these calculations, we excluded energy loss due to electricity generation and battery charging.

Hydrogen Vehicles (HV). Since vehicles with an internal combustion hydrogen engine are still at an experimental stage, we could not get fuel consumption data of HV at standardized running modes such as 10-mode. Therefore, fuel consumption of HR-X2 (MAZDA) in 10-mode was estimated to be about 10km/l based on the data taken in constant speed running mode.

Hybrid Vehicles. Hybrid vehicles can be made by various combinations of power generation units such as an internal combustion engine and motor. DASH21 (Daihatsu) is a light duty car equipped with a 660ml gasoline engine, electric generator and motor. Its fuel consumption rate at 10-mode is 29 km/lGE and 32 km/lGE at 10-15 mode. The advantage of the hybrid vehicle is its flexibility in engine use. For example, the main engine is not always required to meet full loads by expecting assistance from a sub-power unit.

Fuel Cell Vehicles (FCV). A fuel cell (FC) is one of the direct power generation systems converting chemical energy to electric energy without significant heat generation. Higher thermal efficiency is theoretically expected than internal combustion engines. Among several types of FC under development, we think that FC equipped with Proton Exchange Membranes (PEM) is the most promising for passenger cars, since ① the power per unit weight of FC is large, ② response time is quick, and ③ operation temperature is relatively low. If we operate FC using carbon-based fuels such as alcohol and hydrocarbon, CO must be carefully removed to avoid catalyst poisoning in the PEM. A prototype FCV made by MAZDA showed the fuel consumption rate of 44 km/lGE. However, details of testing conditions were not known.

INTEGRATED CO2 EMISSIONS

For the full fuel cycle CO_2 emission analysis for alternative fuel vehicles, CO2 emissions per 100 km driven by 1500 cc or 1600 cc alternative fuel passenger cars were calculated for fuel consumption under the 10-mode driving condition. CO_2 emissions for fuel production corresponding to 100 km driven were added. CO_2 emissions due to mining and transportation from the mining country to Japan must also be added to CO_2 emissions for driving and fuel production. Table B summarizes the results of our calculations.

Gasoline-Fueled Vehicles. CO_2 emission of a gasoline-fueled Otto-cycle engine car was calculated as follows. Gasoline consumption is 7.3 l/100km under the 10-mode driving condition, shown in Table A. Since a calorific value of gasoline, carbon mass ratio and specific gravity are 43.5MJ/kg, 0.85 and 0.75, gasoline consumption and CO_2 emission per 100 km driven became 5.47 kg and 17.1 kg- CO_2 respectively $\{7.3 \times 0.75 \times 43.5 \times (1/43.5) \times 0.85 \times (44/12)=17.1(\text{kg- }CO_2/100\text{km})\}$. The integrated CO_2 emission, 20.8 kg- CO_2, was obtained by adding 3.66 kg- CO_2 emitted from the refinery assuming CO_2 emissions of 0.67 kg/kg gasoline produced from crude oil.

Methanol Vehicles (MV). Fuel consumption of an M85 (85% methanol and 15% gasoline) fueled Otto-cycle engine car under the 10-mode driving condition is 7.25 lGE/100km, as shown in Table A. If the energy consumption of M100 (100% methanol) methanol fueled car was equal to the M85 engine, 11.83 kg of MeOH was required to drive for 100km, and CO_2 emission was 16.3 kg using the calorific value of methanol, 20MJ/kg $\{7.25 \times 0.75 \times 43.5 \times (1/20) \times (12/32) \times (44/12)=16.3(\text{kg- }CO_2/100\text{km})\}$.

The integrated CO_2 emission became 23.7 kg/100 km by adding CO_2 emissions of 0.23 kg/kg MeOH produced from the conversion of LNG and 0.395 kg/kg MeOH produced corresponding to the mining of natural gas.

CNG Vehicles. Fuel consumption for a CNG car is 7.27 lGE/100km, as shown in Table A. As caloric value of natural gas is 45.1 MJ/kg, 5.25 kg of methane is consumed to drive 100 km, corresponding to 14.5 kg- CO_2 emitted $\{7.27 \times 0.75 \times 43.5 \times (1/45.1) \times (12/16) \times (44/12)=14.5(\text{kg- }CO_2/100\text{km})\}$.

If 0.608 kg- CO_2 emission per kg methane is assumed for mining and transportation, 17.7 kg of the integrated CO_2 would be emitted per 100km driven.

EV. Electricity consumption of Roadster (MAZDA) on the 10-mode driving condition was calculated to be 2.31 lGE/100km, as shown in Table A. Assuming 30 MJ/kg as the calorific value for coal, 35% thermal efficiency for a coal fired power plant, and a carbon mass ratio for coal of 0.7, coal consumption at the power station and CO_2 emissions per 100 km driven was calculated to be 7.18 kg and 18.4 kg-CO2, respectively $\{2.31 \times 0.75 \times 43.5 \times (1/30) \times (1/0.35) \times 0.7 \times (44/12)=18.4(\text{kg- }CO_2/100\text{km})\}$.

If 30% energy loss for transmission and charging is assumed, CO_2 emitted per 100 km driven is 23.6 kg $\{18.4 \div 0.7=26.3 \text{ (kg- }CO_2/100\text{km})\}$. If 0.131 kg CO_2 emitted per kg coal is assumed for mining and transportation, 1.34 kg of CO_2 is emitted from mining and transportation of 10.3 kg-coal, which was calculated by 7.18 kg-coal divided by 0.7. The integrated CO_2 emission was 25.0 kg- CO_2/100km.

HV. Energy consumption by HV is 6.21 l-GE/100km at constant speed running mode and the calorific value of hydrogen is 121 MJ/kg, leading to hydrogen consumption per 100km driven of 1.674 kg-H2. Although hydrogen vehicles do not emit CO_2 in the exhaust, CO_2 emissions for electricity generation from coal and for electrolysis must be counted. Here, 48.8kWh of energy consumption per kg hydrogen produced by electrolysis was assumed $\{6.21 \times 0.75 \times 43.5 \times (1/121) \times 48.8 \times (3600/1000) \times (1/0.35) \times (1/30) \times 0.7 \times (44/12)=71.9(kg\text{-}CO_2/100km)\}$. As 28.0 kg of coal is required in this system for 100 km driven, CO_2 emissions for mining and transportation are 3.65 kg, as shown in Table B. Assuming HV consumed hydrogen at a rate of 10 l-GE/100km under the 10-mode driving condition, CO_2 emissions were calculated to be 115.8 kg- CO_2 $\{71.9 \times (10/6.21)=115.8(kg\text{-}CO2/100km)\}$. When hydrogen is produced from methane by shift reaction using water, 17.87kg- CO_2 is emitted to produce 1 kg of hydrogen.   CO2 emissions then become 29.8 kg- CO_2 $\{1.674 \times 17.87=29.8(kg\text{-}CO_2/100km)\}$.

<u>MV and EV based on a CO_2 recycling system.</u>   Methanol vehicles need 11.83kg-MeOH to travel 100km, as was calculated in a previous section. When methanol is produced from the recycled CO_2 separated from flue gas of a coal-fired power station, energy for CO_2 separation of 0.256 kWh/kg- CO_2 was needed for a membrane technology of 60% separation efficiency. A 0.35 kWh/kg-MeOH is required to produce methanol under 95% of CO_2 conversion. The total energy to produce methanol from flue gas was determined to be 0.721 kWh/kg-MeOH.   During these CO_2 transformation steps, 43% of the separated CO_2 from the flue gas was exhausted to the atmosphere and remaining 57% was converted to methanol.

The amount of coal consumed by the power plant to produce 11.83 kg methanol for 100 km driven is 11.12 kg $\{11.83 \times (12/32) \times (1/0.7) \times (1/0.57)=11.12 (kg\text{-}coal/100km)\}$.

Although 32.4 kWh of electricity was generated by the plant $\{11.12 \times 30 \times 0.35 \times (1000/3600)=32.4(kWh)\}$, 8.53 kWh was consumed to produce 11.83 kg of methanol $\{0.721 \times 11.83=8.53(kWh)\}$. CO_2 was discharged from the power plant, CO_2 transformation steps and MeOH vehicle operation at the rates of 28.5 $\{11.12 \times 0.7 \times (44/12)=28.5(kg\text{-}CO2)\}$, 12.3 $\{11.12 \times 0.7 \times (44/12) \times (0.43)= 12.3(kg\text{-} CO_2)\}$ and 16.2 $\{11.12 \times 0.7 \times (44/12) \times (0.57)=16.2(kg\text{-} CO_2)\}$ per 100km driven.

EV can run 79.9 km $\{23.9 \times (1-0.3) \times \{2.31 \times 0.75 \times 43.5 \times (1000/3600) \times 100=79.9km\}$ using 23.9 kWh $\{32.4-8.53=23.9\}$ under 30% electricity loss due to transmission and charging.   This yields 179.9 km total driving distance by both the EV and methanol vehicle. As EV does not exhaust CO_2 during the driving, only 15.8 kg of CO_2 is emitted per 100 km driven $\{28.5 \div (179.9/100)=15.8(kg\text{-} CO_2/100km)\}$ if no CO_2 emission for hydrogen production can be assumed.

If hydrogen would be produced from methane using shift reaction, 2.7 kg of CO_2 is emitted per 1 kg of methanol production. This process adds 31.9 kg of CO_2 to the 28.5kg- CO_2 emitted per 100 km driving of MeOH vehicle. CO_2 emissions per 100 km driven for MeOH vehicle and EV are then increased to 33.6 kg $\{6 \times 11.83 \div 32=2.22(kg\text{-}H2), (31.9+28.5) \div (179.9/100)=33.6(kg\text{-}CO_2/100km)\}$.   Hydrogen must be supplied by a non-carbon renewable source.

CONCLUSIONS

We compared performances of alternative fuel vehicles and their resulting CO_2 emissions from the full fuel cycle aspects. CO_2 emissions for CNG vehicles were the lowest among the alternative fuel vehicles investigated, even if fuel treatment was taken into account. A vehicle using MeOH produced from natural gas emits 23.7 kg of CO_2 per 100 km driven.   If MeOH is produced from CO_2 separated from flue gas of the coal-fired power plant and hydrogen is produced by non-carbon natural energy process, a MeOH vehicle emits 28.5kg- CO_2 per 100 km driven. Since electricity generated at the coal-fired power plant can be used to power EV, CO_2 emissions can be reduced to 16.7 kg per the 100 km driven for this case which powers both EV and MeOH vehicles.   Although a non-carbon natural energy process must be used to produce hydrogen for this system, the result of this calculation shows the possibility of CO_2 reduction in the transportation sector by recycling CO_2 from flue gas to methanol.

ACKNOWLEDGEMENT

This study was supported by NEDO (New Energy and Industrial Technology Development Organization).

REFERENCES

Arakawa, M., et al., MAZDA Technical Paper No.12, p.97,(1994)
Araki,S., et al., Proc. of 11th Energy-System & Economics Conf. JAPAN, p.319,(1995)
Ariyoshi,M.,'Hybrid Vehicle', Proc. of EV Forum '95 JAPAN, p.254,(1995)
Dept. of Global Environ., EA. JAPAN ed., Global Warming Measure Technol. Handbook No.4, p.130, (1992), Daiichi-Hoki
Fujinaka,M., 'Electric Vehicles', Sangyo-Kogai, Vol.29, No.3, p.283(1993)
Hasegawa,Y., et al., 'Fuel Cell Vehicle', MAZDA Technical Review No.12, p.103, (1994)
IAE, IAE-C-8922(1990)
MAZDA Co., Brochure of Hydrogen Vehicle, (1994)
SAE of Japan ed., Index Tables of Japanese Automobiles 1994,(1994), SAE of Japan
Tsukasaki, et al., 'Collora-FFV', TOYOTA Technical Review, Vol.42, No.1, p.93, (1992)
Takahashi,T., ''Vivio-EV', Proc. of 3rd Transport. & Disribut. Div. JSME, p.83, (1994)
Watanabe, T., 'Natural Gas Vehicles' Sangyo-Kogai, Vol.29, No.3, p.283, (1993)
Yoshino, M., Proc. of EV Forum'95 JAPAN, p.246, (1995)

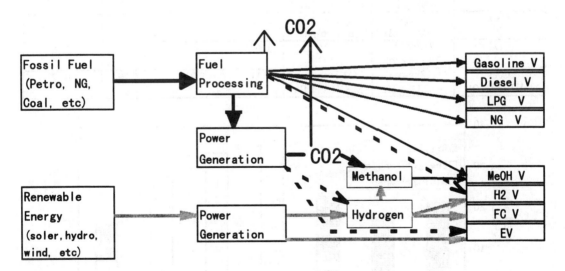

Fig. 1   Energy Flow of Vehicles

Table A Comparison of Vehicles

	Driving mode	energy consump.	Mileage of gasoline eq. (km/lGE)	Energy Consump. per 100km drive (lGE/100km)	Driving Range per 1MJ (km/MJ)
Toyota Corolla-1600	10-mode	13.7km/l	13.7	7.3	0.42
Toyota Corolla-FFV	ibid.	7.88km/l	13.8	7.25	0.424
WAGON-type 1500cc(CNG)	ibid.	180km/ 11.2Nm^3	13.8	7.27	0.422
GM-IMPACT	50km/h	200km/13.3kWh	136	0.732	4.18
Subaru-VIVIO - EV	10-mode	75km/17.5kWh	38.9	2.57	1.19
	40km/h	141km/kWh	73	1.37	2.24
Mazda Roadster(EV)	10-mode	-	43.3	2.31	1.32
	10-15mode	-	48.1	2.08	1.47
	60km/h	-	69.8	1.43	2.14
Mazda-HR-X2(H2)	60km/h	230km/3.84kg	16.1	6.21	0.495
Daihatsu DASH21(hybrid)	10-mode	-	29	3.13	0.98
	10-15mode	-	32	3.45	0.889
Mazda FCV		40km/2.8Nm^3	44	2.27	1.35

Table B Energy Consumption and CO2 Emission per 100km Driven

Resources	Energy Consumption	Vehicle Type	Mining +Transp.	Driving	Energy Transform	TOTAL (kg-C02
			(kg-CO2/100km)			/100km)
Crude Oil	5.47kg-gasoline	Gasoline V.	0.33	17.1	3.66	21.1
Natural Gas	5.25kg-NG	NG Vehicle	3.1	14.5		17.7
Natural Gas	11.83kg-MeOH (7.68kg-NG)	MeOH Vehicle	4.67	16.3	2.72	23.7
Natural Gas	1.674kg-H2 (10.87kg-NG)	Hydrogen V (NG transform)	6.61	-	29.9	36.5
Coal	1.674kg-H2 (28kg-Coal)	Hydrogen V (Electrolysis)	3.668	-	71.9	75.6
Coal	21.0kWh (10.3kg-coal)	EV	1.348	-	26.3	27.6
Coal +H2(CO2=0)	6.18kg-coal +1.233kg-H2	CO2 Transform MeOH V + EV	0.86	9.01	6.79	16.7

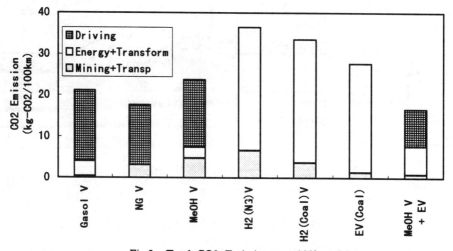

Fig.2   Total CO2 Emission per 100km drive

Energy Convers. Mgmt Vol. 38, Suppl., pp. S461–S466, 1997
Pergamon

PII: S0196-8904(96)00311-1
0196-8904/97 $17.00 + 0.00

ABSORPTION AND FIXATION OF CARBON DIOXIDE BY ROCK WEATHERING

T. KOJIMA, A. NAGAMINE, N. UENO and S. UEMIYA

Department of Industrial Chemistry, Seikei University,
Musashino, Tokyo 180, Japan

ABSTRACT

The weathering of alkaline rocks, such as alkaline or alkaline earth silicate is thought to have played a great role in the historical reduction in the atmospheric CO_2 of this planet. To enhance the process artificially, we should increase the surface area of the rocks. However, some additional pulverization energy is necessary, which leads to the additional CO_2 emission. In the present paper, first, we reviewed the possibilities of the utilization of the reaction as a countermeasure against the CO_2 problem from the view points of resources and global carbon circulation. Second, we report the experimental results on weathering kinetics conducted for various kinds of silicates. Lastly, the amounts of pulverization energy of wollastonite and olivine sand were evaluated using industrial data of their pulverization. It was concluded that the CO_2 absorption by rock weathering is one of the most promising measures for CO_2 problem.

© 1997 Elsevier Science Ltd

KEY WORDS

Carbon Dioxide, Rock Weathering, Silicate, Carbonate, Dissolution Kinetics, Pulverization Energy

INTRODUCTION

The weathering of alkaline rocks, such as alkaline or alkaline earth silicate is thought to have played a great role in the historical reduction in the atmospheric CO_2 of this planet. The initial CO_2 pressure on the planet is thought to have been several tens atm, but recently, it is less than several hundreds ppm. It is usually believed that the overall chemistry of the CO_2 absorption by wollastonite, a typical silicate, is given by the following equation

$$CaSiO_3 + CO_2 \rightarrow CaCO_3 + SiO_2 \qquad (1)$$

and the produced $CaCO_3$ is stored as limestone like a coral reef. Thus, the coral reef has so far been believed as a carbon stock. This is superficially true. However, the intrinsic reaction of the coral reef formation is expressed as

$$Ca^{2+} + 2HCO_3^- \rightarrow CaCO_3 + CO_2 + H_2O \qquad (2)$$

which means that in this process, a part of the carbon dioxide is again released to the atmosphere (Kojima, 1994) from the ocean surface. Instead of the reaction (2), we should focus attention on the reaction of rock weathering,

$$CaSiO_3 + 2CO_2 + H_2O \rightarrow Ca^{2+} + 2HCO_3^- + SiO_2 \qquad (3)$$

which occurs on land. Thus two moles of CO_2 are absorbed. They produce two moles of bicarbonate ions that are transported with one mole of calcium ion to the ocean, but one mole of CO_2 is again released with the formation of one mole of calcium carbonate (exactly, 0.6 mole of carbon dioxide is released with one mole of calcium carbonate deposition under the present ocean surface alkaline condition, Kojima, 1994). Thus the overall CO_2 fixation by eq. (1) is one mole (or a little more) per one mole of silicate.

In the present paper, first, we review the possibilities of the utilization of the reaction as a countermeasure against the CO_2 problem from the view points of resources and global carbon circulation. Second, we report the experimental results on the weathering kinetics conducted for various kinds of silicates. Lastly, the amounts of pulverization energy of silicates are evaluated using their industrial operation data.

PRELIMINARY EVALUATION

<u>Resources</u>. First of all, the amount of the silicate resources is discussed. The composition of the earth crust is shown in Table 1. It is clearly observed that alkaline silicates are abundant in the crust.

Table 1 Composition of Crust in vol% (Nesbitt and Young, 1984)

Minerals and representative Composition	Crust	Cont. Crust	Exposed Crust
Quarts, SiO_2	18	23.2	20.3
Plagioclase, $NaAlSi_3O_8$ or $CaAl_2Si_2O_8$	42	39.9	34.9
Glass, SiO_2	–	0.0	12.5
Potassium feldspar, $KAlSi_3O_8$	22	12.9	11.3
Olivine, $(Mg,Fe)_2SiO_4$	1.5	0.2	0.2
Pyroxene, Pyroxenoid, $(Ca,Mg,Fe)SiO_3$	4	1.4	1.2
Amphibole, $(Ca,Mg,Fe)_7Si_8O_{22}(OH)_2$	5	13.7	12.0
Phyllosilicate, $K_2Al_6Si_6O_{20}(OH)_4$	4	8.7	7.6
Magnetite, Fe_3O_4	2	1.6	1.4

<u>Order of Magnitude Estimation of Rate</u>. To utilize the weathering process artificially, the most important factor is its rate. To enhance it, we should increase the surface area of the rocks, however, some additional pulverization energy is necessary, which leads to the additional CO_2 emission. To evaluate the necessary degree of pulverization, we roughly evaluated its rate from the view point of global change in atmospheric CO_2 concentration.

It is well known that the carbon dioxide concentration was drastically reduced in the order of 1 Gy. Now it is required that the present CO_2 concentration is reduced in the order of 100y. Here, one layer of the silicate rocks with diameter of 100m was assumed to suffer the natural weathering process, though definite value of the equivalent diameter is unknown. When we use these rocks to absorb CO_2 from the atmosphere, the rocks should be pulverized into the order of diameter of

(100m) (100y)$\diagup$(1Gy)=10μm under the assumption of linear rate to the concentration of CO_2. If we capture 10-20% CO_2 in flue gas from power station, and select most suitable rocks, the rate is increased by three orders of magnitude because of higher concentration of CO_2 in the flue gas. Therefore if 0.1 % of the exposed silicate resources is mined, pulverized to 10 μm or larger, and used for CO_2 absorption from flue gas, the increase in atmospheric CO_2 is prevented for the order of 100 y. The rate will be larger in the CO_2 saturated water than under the natural, usually dry condition.

Possibility of Increase of Net CO2 Fixation Using Deep Ocean and Utilization of Weathering of CaCO3. Though net fixation of carbon dioxide is possible even by eq. (1), i.e., combination of eqs. (3) and (2), the carbon dioxide fixation becomes more efficient by preventing the process of eq. (2). When we dispose the product from eq. (3) into the deep ocean, eq. (2) does not occur under the chemical equilibrium condition. Furthermore, the acidification of deep sea water, which is worrysome in case of direct injection of liquefied CO_2, does not occur. (Even if eq. (2) occurred in the deep ocean, the produced CO_2 would be dissolved into the oceanic water by its high pressure and would not be released into the atmosphere for the order of thousand years of ocean water circulation.)

The products from eq. (3) are expected to contain fine SiO_2 powder. This causes the increase in density of the produced slurry, which enhance its sedimentation rate into the deep ocean, without any additional energy input. The unreacted fine minerals also contribute to increase the apparent density.

Taking the above possibility into account, the weathering of calcium carbonate,

$$CaCO_3 + CO_2 + H_2O \rightarrow Ca^{2+} + 2HCO_3^-$$ (4)

the reverse reaction of eq. (2), is also thought to contribute to the net CO_2 fixation. In the present study, calcium carbonate was therefore also chosen as a sample.

EXPERIMENTAL EVALUATION OF RATE

Next, we conducted the experimental rate evaluation for various kind of silicates; wollastonite, olivine sand, potassium and sodium feldspars (orthoclase, albite and nephelite with cyanite), talc and calcium carbonate.

Preliminary Experiments on Direct Carbonation at High Temperature. In a recent publication (Lackner *et al.*, 1995), it was pointed out that CO_2 could be absorbed by eq. (1) using same kinds of minerals as in the present study, directly at high temperature. They indicated that eq. (1) progresses at the temperature less than 554 K for wollastonite and

$$1\diagup 2 \cdot Mg_2SiO_4 + CO_2 \rightarrow MgCO_3 + 1\diagup 2 \cdot SiO_2$$ (5)

eq. (5) under 515 K for olivine at 1 atm, from the thermodynamic consideration of $\triangle G<0$. Therefore, we tried to examine the possibilities of the reaction by using TGA and to compare with our results shown below, because they did not conduct any experimental demonstration. However our TGA results indicated no apparent weight change for several hours under the temperature range shown above while the following aqueous absorption rate by eq. (3) was obviously higher than the high temperature reaction of eq. (1). Therefore, in the present paper, we only focused on the aqueous absorption.

Experimental. The expected chemical reaction schemes for the various kinds of rocks are shown in Table 2. The dissolution kinetics of minerals suspended in water (mostly 0.1 wt% of solid) was measured for 0 - 600 hr at 25℃. The 500 cm^3 of deionized water presaturated with pure CO_2 was employed and sufficient amount of CO_2 (2 cm^3/s) was continuously introduced with mechanical stirring at 600 rpm. The experimental apparatus is shown in Fig. 1. The filtered aqueous solution was acidified and then analyzed for the main metal element of the mineral by ICP-AES (Kyoto Koken, UOP-1 MKII).

In comparison, almost the same experiments were conducted under various conditions as follows, by keeping other factors same as the standard conditions; diameters for Sicatec, Olivine sand, and calcium carbonate K1; stirring speed; N_2 instead of CO_2; 1 wt% or 0.01 wt% instead of 0.1 wt%.

Table 2 Assumed Chemical Reaction Scheme

wollastonite	$CaSiO_3 + 2CO_2 + H_2O \rightarrow Ca^{2+} + 2HCO_3^- + SiO_2$	$(Ca:CO_2=1:2)$
calcium carbonate	$CaCO_3 + CO_2 + H_2O \rightarrow Ca^{2+} + 2HCO_3^-$	$(Ca:CO_2=1:1)$
olivine	$Mg_2SiO_4 + 4CO_2 + 2H_2O \rightarrow 2Mg^{2+} + 4HCO_3^- + SiO_2$	$(Mg:CO_2=1:2)$
orthoclase	$2KAlSi_3O_8 + 2CO_2 + 3H_2O$	
	$\rightarrow 2K^{2+} + 2HCO_3^- + 4SiO_2 + Al_2Si_2O_5(OH)_4$	$(K:CO_2=1:1)$
albite	$2NaAlSi_3O_8 + 2CO_2 + 3H_2O$	
	$\rightarrow 2Na^+ + 2HCO_3^- + 4SiO_2 + Al_2Si_2O_5(OH)_4$	$(Na:CO_2=1:1)$
talk	$Mg_3Si_4O_{10}(OH)_2 + 6CO_2 + 2H_2O$	
	$\rightarrow 3Mg^{2+} + 6HCO_3^- + 4SiO_2$	$(Mg:CO_2=1:2)$

Table 3 Experimental results

	metal wt % by EPMA	rep. d. SEM, μm	S. area SEM m^2/g	rate, mmol/m^2h metal,	CO_2
wollastonite ($CaSiO_3$)	CaO:48.3*				
LF-60 (China)	45.2	75.9	.0205	0.73	1.46
Sicatec (China) powder	37.0	44.2	.0355	0.52	1.04
olivine (chrysolite, Mg_2SiO_4)	MgO:57.3*				
Olivine sand, PAN	36.4	40.8	.0549	0.11	0.22
calcium carbonate ($CaCO_3$)	CaO:100.0*	(except CO_2)			
(Japan, Fukushima) K1	92.3	168	.0135	0.45	0.45
heavy calcium carbonate	97.2	1.72	1.35	0.37	0.37
orthoclase ($KAlSi_3O_8$)	K_2O:16.9*				
orthoclase (Australia)	12.5	11.8	0.207	.021	.021
indian orthoclase	14.5	18.8	0.132	.038	.038
nephelite ($NaAlSiO_4$)	Na_2O:19.8*				
with cyanite (Al_2SiO_5)	8.8**	9.67	0.267	.024	.024
albite ($NaAlSi_3O_8$)	Na_2O:11.8*				
indian soda feldspar	13.6	11.6	0.203	.034	.034
talk ($Mg_3Si_4O_{10}(OH)_2$)	MgO:31.9*				
99 talk (China)	34.1	18.3	0.298	.025	.050

*theoretical, **10.3% of K_2O is also included.

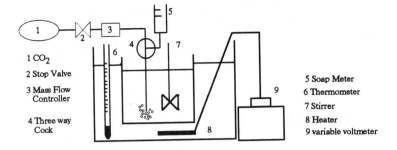

1 CO₂
2 Stop Valve
3 Mass Flow
  Controller

4 Three way
  Cock

5 Soap Meter
6 Thermometer
7 Stirrer
8 Heater
9 variable voltmeter

Fig. 1 Experimental apparatus.

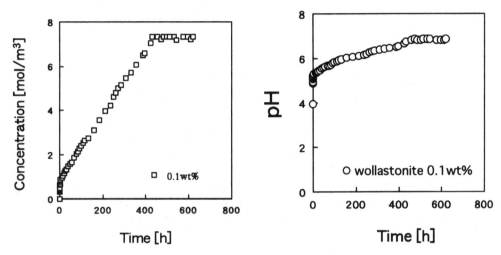

Fig. 2 Time variation of Ca²⁺ concentration.   Fig. 3 Time variation of pH.

Characterization of Samples.   The characterization of samples was done by SEM with
EPMA.   From SEM photographs, mean representative diameter and surface area were
determined.   The surface area was determined regarding particles as pillars for
wollastonites.   Their chemical composition was also determined by EPMA.   The
results are shown in Table 3.

Results.   The time variation curves of calcium ion concentration and pH are shown
in Figs. 2 and 3 respectively, for wollastonite (0.1 wt%) as examples with CO₂
introduction, while no obvious increase in the metal ion concentration was found
when nitrogen was introduced instead of CO₂.   After an initial rapid increase in
the metal concentration, it linearly increased with time and then attained a
plateau value as reported previously (Nagamine *et al.*, 1994).   The initial rapid
increases in concentration and pH were thought to be caused by rapid ion exchange
between surface metal ions and aqueous protons.   The plateau concentration was found
to be determined by the dissolution equilibrium.   From the linear dissolution
stage, the intrinsic dissolution rate of the mineral was determined.   The rate was
found not to be affected by the stirring speed or gas flow rate under the conditions
employed.   The rate was also found to be proportional to the surface
area of the sample, irrespective of the wt. ratio of sample to water, or its average
diameter.   Thus the concentration attained to the plateau value more rapidly (within
several tens hr) for the case of 1 wt%.   The dissolution rate per unit area, of

particles once utilized for the experiment was almost same as that of fresh
particles. It was concluded that the dissolution kinetics was controlled at the
mineral surface area. The calculated linear rates per unit surface area are shown
in Table 3. The dissolution rates of wollastonite and calcium carbonate were
found to be faster than others.

PULVERIZATION ENERGIES

Lastly, the pulverization energies of wollastonite and olivine sand were evaluated
using industrial data of their pulverization. The required amounts of electricity
of the pulverizers for 6t/h of wollastonite from 200 mm to 75μm are 55 kW for a
single crusher, 100 kW for a hammer crusher, and 100 kW for a roller mill, in
series. The CO_2 emission from pulverizers is given as 18.7 kg-CO_2/t-wollastonite
using average CO_2 emission from power plants, 0.44 kg-CO_2/kWh, while the CO_2
uptake by wollastonite is theoretically 758 kg-CO_2/t from eq. (3). Thus the the
CO_2 emission from pulverizers was 2.47 % of the fixed amount of CO_2 by rock
weathering. When the wollastonite is pulverized to 10μm, the pulverizatio energy
will be increased to 6.76 %, according to the Bond's theory. On the other hand,
when more efficient pulverization process is employed, its energy can be saved.
In case of olivin sand, pulverization energy to 41μm for a 30,000 t scale plant
using 10 crushers or mills, was given as 32 kWh/t and 14.1 kg-CO_2/t while CO_2 uptake
is 1251 kg-CO_2/t from table 2.   Thus the the CO_2 emission from pulverizers was 1.13
% of the fixed amount of CO_2 by rock weathering for 41μm and 2.28 % for 10μm
though its rate is 15% of that of wollastonite.

CONCLUSION

It was concluded that the CO_2 absorption by rock weathering is one of the most
promissing measures for CO_2 problem from the present experimental and energy
evaluation. Wollastonite was found to be most reactive among the minerals tested
in the present study.

REFERENCES

Kojima, T. (1994). *CO_2 Problem*, Agune, Tokyo (will be published in English, Gordon
   and Breach, 1996)
Lackner, K. S., Wendt, C. H., Butt, D. P., Joyce Jr., E. L. and Sharp, D. H. (1995).
   Carbon dioxide disposal in carbonate minerals. *Energy*, **20**, 1153-1170.
Nagamine, A., Kojima, T. and Uemiya, S. (1994). Absorption of CO_2 by rock
   weathering, in Japanese, presented in Tsukuba meeting of Soc. Chem. Eng. Jpn.
   (will be submitted for publication in *Kagaku Kougaku Ronbunshu*)
Nesbitt H. W. and Young G. M. (1984). Prediction of some weathering trends of
   plutonic and volcanic rocks based on thermodynamic and kinetic considerations.
   Geochim. et Cosmochim. Acta, **48**, 1523-1534.

SECTION 6

BIOLOGICAL UTILIZATION

 **Pergamon**

Energy Convers. Mgmt Vol. 38, Suppl., pp. S467–S473, 1997
© 1997 Elsevier Science Ltd. All rights reserved
Printed in Great Britain
0196-8904/97 $17.00 + 0.00

PII: S0196-8904(96)00312-3

BIOLOGICAL FOSSIL CO₂ MITIGATION

Evan Hughes[1] and John R. Benemann[2]

[1]Electric Power Research Institute, Palo Alto, California, and
[2]Consultant, Walnut Creek, California

ABSTRACT

Over ten times more CO_2 is fixed by plants into biomass, and annually released by decomposers and food chains, than is emitted to the atmosphere due to the burning of fossil fuels. Human activity is already directly and indirectly affecting almost half of the terrestrial biological C cycle. Management of even a small fraction of the biological C cycle would make a major contribution to mitigation of this greenhouse gas.

Electric power generation is responsible for roughly one third of fossil CO_2 emissions. Direct CO_2 mitigation processes are those that reduce fossil CO_2 emissions from specific power plants. Direct biological CO_2 mitigation processes include the cultivation of microalgae on flue-gas or captured CO_2, and the cofiring of wood with fossil fuels. Indirect biological processes, such as growing trees for C storage or for fueling dedicated biomass power plants, recapture CO_2 that already has entered the atmosphere. Indirect and direct CO_2 mitigation processes have the same overall effect in reducing global warming potential.

Reducing global CO_2 emissions from forest destruction and unsustainable agricultural and land use practices is one of the most cost-effective, and environmentally beneficial actions that can be taken now to arrest global climate change. Another near-term option is to enhance the substitution of fossil fuels with biofuels. Biofuels are a major source of fuel for the poorer half of mankind. Globally, biofuels could replace a substantial fraction of current fossil fuel usage.

Cofiring biomass wastes and residues with coal is one of the lowest-cost, nearest-term options for reducing fossil CO_2 emissions at existing power plants. Long-term demonstrations of biomass cofiring are required at full-scale coal-fired power plants to document efficiencies, ash characteristics, biomass preparation and feeding, and other technical issues. Biomass fuel resources for cofiring can be expanded in the near-term through greater recovery of wastes and residues in forestry and agriculture, and in the mid-term through systems that produce biomass specifically for use as fuels (energy crops). © 1997 Elsevier Science Ltd

KEYWORDS: Cofiring, Biomass, CO₂ Mitigation, Biofuels, Carbon Cycle.

INTRODUCTION

CO_2 is responsible for well over half the total warming potential of all greenhouse gases. Globally about 20 billion tons of fossil CO_2 are emitted each year from the burning of fossil fuels, and another 2 to 8 billion tons are released through human-mediated oxidation of the biosphere: intensive agriculture, deforestation and other unsustainable practices. Cumulatively, net biospheric CO_2 emissions due to human activities over the past century have rivaled emissions from fossil CO_2 sources. Indeed, human activities are currently impacting a large fraction, approaching half, of the total annual terrestrial primary biological productivity of our planet, estimated at 500 billion tons of CO_2 fixed annually (Vitousek, 1994).

The magnitude of the biological C cycle, together with the already large human influence on it, suggest it as a major lever for the management and mitigation of rising atmospheric CO_2 levels. Indeed, already, net carbon storage in northern hemisphere forests accounts for perhaps half, or more, of the "missing carbon" - the almost half of anthropogenic CO_2 emissions not appearing in the atmosphere. Preservation and management of the biosphere is urgently required to assure that such C sinks will continue to function and even expand in the future, and to reduce CO_2 emissions from destructive forestry and agricultural practices. In addition, the production of biomass fuels, which provide about 15% of primary energy sources used by humanity (Scurlock and Hall, 1990), must be expanded, and made more efficient and sustainable.

Of course, harnessing the biosphere for the purposes of mitigating the potential effects of global warming and producing fuels is not without cost or controversy. Although the overwhelming scientific consensus, represented by the reports of the Intergovernmental Panel on Climate Change (IPCC, 1996), is that global warming is a very real possibility, some, in particular economists who heavily discount the future, argue for only a minimal response in the near-term. However, the precautionary principle and the goals of economic and environmental sustainability require that at least some actions be initiated now to reduce the rate of atmospheric changes. There are many actions and technologies which can reduce a large fraction of the anthropogenic CO_2 and other greenhouse gas emissions at relatively low cost (Rubin et al., 1993). And a concerted R&D effort would further reduce the cost of such technologies, even of the more expensive ones, such as mitigation of fossil CO_2 emitted by power plants.

THE U.S. ELECTRIC UTILITIES AND CLIMATE CHANGE

The U.S. electric industry, with about 500,000 MW(e) of generating capacity, is responsible for the emission of over 1.8 billion metric tons (Gt) of fossil CO_2, over a third of total U.S. emissions. Coal-fired power plants are responsible for about 85% of CO_2 emissions by the electric industry, making these the obvious, though not necessarily easiest, targets for CO_2 emission reductions. Thus R,D&D of CO_2 reduction technologies by coal-fired power plants deserves special attention and funding by industry and government.

The U.S. electric industry's commitment to reducing greenhouse gas emissions is evidenced by over 600 utilities joining the voluntary U.S. DOE "Climate Challenge" program, pledging to reduce the equivalent of 44 million tons of fossil CO_2 emissions (Kane, in these Proceedings). This despite the current deregulation of this industry, that is creating a competitive business environment inherently unfavorable to such voluntary programs. Clearly, incentives and actions that encourage continued and increased participation and flexibility towards the goals of greenhouse gas reductions by utilities would be helpful.

Any program to reduce greenhouse gas emissions must take a long-term view, allowing for turnover of capital equipment. This is particularly true for the utility industry, facing huge asset devaluations of large power plants, resulting from shifts in power generation economics due to the deregulation process. It is, however, this long-term nature of capital investment turnover, that also requires concerted and immediate action in the research, development, demonstration, and implementation of greenhouse gas mitigation technologies for the utility sector. EPRI has been instrumental in bringing government and industry together in this endeavor, aiding the electric utilities in actions to reduce fossil CO_2 emission and by taking a leadership position in developing the economic models and evaluation methodologies for CO_2 mitigation technologies (Rhudy, in these Proceedings).

EPRI has studied and developed biomass energy for almost two decades (Benemann, 1978), from wood combustion and gasification systems, to energy plantations and even plants grown on or in seawater (halophytes and algae). Recently, EPRI's biomass power program has been active in the development of biomass conversion and utilization, with projects in dedicated biomass power systems (Lamarre, 1994), energy crops (Hughes and Wiltsee, 1995) and the cofiring of biomass with coal (Tillman et al., 1994). This paper provides a brief introduction to biological CO_2 mitigation processes, emphasizing direct processes able to utilize or reduce emissions by specific power plants, such as microalgae ponds (see Benemann, in these Proceedings) and, in particular, the cofiring of coal with biomass (see also Hunt et al., and Benjamin et al., in these Proceedings).

BIOLOGICAL CO₂ MITIGATION PROCESSES

Direct CO_2 mitigation processes reduce fossil CO_2 emissions by specific power plants. Indirect processes reduce fossil CO_2 emissions independently from any particular power plant, but could be linked to them through offset agreements (emissions allowances trading agreements). Indirect mitigation processes have the same overall effect as direct processes, due to the global fungibility of greenhouse gases, and can be remote in space and even time from the fossil CO_2 emission sources. Indeed, there can be even a trade-off between different greenhouse gases (e.g. CO_2 and methane).

Due to favorable climatic and economic conditions, many tropical developing countries (not currently required to reduce greenhouse gas emissions under the Rio Convention), offer many relatively large and low-cost opportunities for indirect biological CO_2 mitigation, through forestry, agricultural, and biofuel projects. Financed by governmental and private organizations in developed countries, such "Joint Implementation" projects could have significant potential for abating global greenhouse gas emissions. However, many technical and political issues remain to be resolved, including allocations of benefits, long-term viability and verifiability, and transactional costs. In any event, major reductions in greenhouse gas emissions could also be achieved in developed countries using biofuels technologies and biological CO_2 sequestration.

CO_2 sequestration activities include prevention of deforestation, increasing forest productivities, reforestation, afforestation, and improved agricultural and range management, all directed towards maintaining and increasing C storage both above and below ground.   Of course, these biological CO_2 mitigation actions would become limiting over the next century, as maximal forest protection and standing biomass levels are approached, slowing above ground C storage. Still, very large and long-term positive effects would be realized, not only in reducing the potential for climate change but also in preserving ecosystems and biodiversity, and in increased productivity of agricultural and forestry systems.

Such actions are urgently needed to reverse the accelerating rate of environmental degradation - the oxidation of the biosphere, resulting from unsustainable human economic activities. And these goals can be accomplished with reasonable and cost-effective changes to forestry, agricultural and economic practices. Such activities also provide large near-term opportunities for emissions trading both within the U.S. and internationally allowing a globally efficient response to the threat of global warming. The U.S. electric utilities have been at the forefront of developing such options for greenhouse gas mitigations (Hinsman et al., in these Proceedings).

Replacement of fossil fuels with biofuels provides the other major biological opportunity for CO_2 mitigation. A consensus estimate by a panel of experts (Sampson et al. 1993), estimated that before the end of the next century about 4 to 16 Gt per year of CO_2, 20% to 80% of current fossil CO_2 emissions, could be mitigated world-wide through replacement of fossil fuels with biofuels. This wide range in estimates reflects uncertainties in the development of efficient biomass production and conversion technologies, the environmental consequences of large-scale biomass fuels production, the economic constraints on global greenhouse gas mitigation, and the fertilizing effect of increasing atmospheric CO_2 concentrations. Achieving the higher range of this estimate will require a global effort to manage a significant fraction of the biosphere, a major undertaking in view of the even greater needs for food and fiber, preservation of natural ecosystem biodiversity, and living space for a projected 10 billion population, hopefully with a higher median standard of living than today. Although some may object to such visions, humankind has little choice but to strive towards the planetary management of both biological and abiotic resources.

BIOMASS POWER

Biological and thermochemical processes for conversion of raw biomass - wood, straw, manures, sludges, etc., - to alternative liquid and gaseous biofuels - ethanol, biodiesel, methane, hydrogen - have in some cases already been commercialized and in general merit continuing R&D (Benemann, 1996). However, the largest potential for biomass as an energy source is as a solid fuel for power generation, in direct combustion and possibly, through gasification (Wiltsee et al., 1993).

In the U.S., about 1% of the power generation capacity, about 5,000 MWe was until recently provided by wood combustion (Turnbull, 1994). These are mainly small scale plants (typically < 20 MWe) attached to the forestry industries, although a few are larger stand-alone facilities (up to 50 MWe). These small-scale wood-fired power plants are not as efficient or economical to operate as larger fossil fuel power plants. Much of this wood-fired capacity was developed in the past fifteen years, because of preferential power purchase contracts designed to promote renewable power. The phase-out of these subsidies and the ongoing deregulation and restructuring of the electric power industry, has already resulted in the reduction of generation capacity by about 30% in the past five years. Recent developments in California, where the first comprehensive law deregulating the electric utility industry was recently enacted, do not augur well for renewable electricity generally and biomass power, as the new law provides inadequate incentives to allow these renewable energy facilities to survive in a deregulated environment.

If the benefits of CO_2 mitigation were valued even modestly, it would provide a significant stimulus to the biomass power industry, as well as other alternative energy sources. And CO_2 mitigation is only one of the benefits of such systems, which include local and regional air pollution reduction (through lower SOx emissions), and local economic development. Indeed, the use of waste biomass resources that would otherwise be landfilled, resulting would carry additional greenhouse gas mitigation credits, as landfilling results in the generation of substantial amounts of methane gas, a much more potent greenhouse gas than CO_2, (Augenstein et al., 1994). Greenhouse gas mitigation is one of the environmental issues that needs to be addressed at both the national and international levels as part of the ongoing global deregulation of the electric power industry.

Many of the limitations of biomass power can be ameliorated, if not solved, by cofiring biomass with coal. Cofiring allows for the flexible use of biomass fuels and for efficiencies and economics of scale, compared to stand-alone units, as discussed next.

BIOMASS   COFIRING

Burning wood, either in a dedicated wood-fired power plant or co-fired with coal, also results in CO_2 emissions, of course. However, the CO_2 released from biomass does not contribute to greenhouse gases, to the extent that the biomass is a renewable resource, which is generally the case. Thus, a distinction must be made between fossil and biomass CO_2 emissions. CO_2 fixed by plants temporally reduces atmospheric CO_2 levels, it is re-emitted when the plant biomass decomposes. Permanent reductions take place only when and where a fossil fuel is replaced with biomass, as when wood is co-fired with coal. Thus greenhouse gas mitigation takes place at the boiler, not where CO_2 is fixed into the biomass. Thus, the CO_2 reduction credit should be applied at the power plant, not to the cultivation of biomass. Although in a global analysis it makes no difference how and where CO_2 reductions are achieved, this is an important accounting issue, and justifies the classification of cofiring as a direct mitigation technology, equivalent to "end-of-pipe" physical capture and disposal schemes, addressed elsewhere in these Proceedings.

Direct mitigation technologies are inherently attractive as they avoid many of the problems of indirect ("offset") processes, such as accountability, verifiability, transactional costs, and even equity issues when offsets are carried out in economically disadvantaged areas (e.g. less developed countries). However, direct processes are generally much more expensive than most indirect methods. Cofiring is the exception, being the lowest cost direct CO_2 mitigation technology available to reduce fossil CO_2 emissions at existing power plants.

Cofiring of biomass wastes and residues with coal has, of course, been studied for many years (McGowin and Hughes, 1992), including the cofiring of "refuse derived fuels" (RDF), in the late 1970's by EPRI, electric utilities, and DOE. More recently cofiring has emphasized wood-based fuels in a variety of different types of coal-fired power plants, as wood fuels are much cleaner and more uniform than the RDF derived from municipal wastes. Over the past two decades, cofiring tests and commercial applications have ranged from 50 MW to just under 500 MWe. Co-firing of wood wastes with coal at low levels (only about 1% by heat rate) is currently carried out commercially at a cyclone unit of Northern States Power in Minnesota, and a pulverized-coal (PC) unit of the Southern Company in Georgia. A long-term test of cofiring at low heat rates (less than 3% biomass fuel) in a PC boiler was performed by Santee Cooper Electric Cooperative in South Carolina, and shorter terms tests were carried out by the Southern Company and TVA. Southern Company achieved a cofiring level above 4% heat in tests at Georgia Power's Plant Hammond (Boylen, 1993).

Recent studies sponsored by EPRI, and co-funded by TVA and DOE, led to conceptual designs and cost estimates for wood cofiring options (Foster Wheeler Environmental Corp., 1993). These studies included issues of fuel preparation and feeding, thereby extending the range and supply of acceptable fuels, minimizing cofiring costs, establishing optimal fuel mixtures, and optimizing performance parameters generally. Results from tests conducted by EPRI and TVA since 1993, have included sampling and analysis of over 50 sources of wood fuels, cold flow and storage tests of wood wastes mixed with coal, and full-scale combustion tests at TVA power plants, including a set of ten 4-hour test runs in a 250 MWe cyclone boiler at the Allen plant in Memphis, Tennessee. The maximum test was 9% by heat value (20% by mass). Full-scale tests of cofiring at low levels (<3% by heat value) at two TVA pulverized coal plants, Colbert and Kingston, were also performed (Tillman et al., 1994).

These recent EPRI sponsored cofiring projects and supporting research, have led to the development of a database concerning biofuel properties and to information and test results on safety, boiler performance during cofiring, combustion mechanisms, blended biofuel/coal storage and transport, pulverizer issues, fuel chemistries associated with blending, ash quality and related issues. The database is now being used to evaluate issues associated with blending biofuels with coal, such as ash behavior related to slagging and fouling.

Despite uncertainties, results from the EPRI/TVA collaboration suggest that for cyclone burners and small-size biomass fuel, capital costs for the fuel preparation and feeding systems are only $100 - 200 /kWe. Even for pulverized coal boilers, requiring more feed preparation and potential boiler modifications, the cost of CO$_2$ mitigation by cofiring is substantially lower than any other end-of-pipe CO$_2$ mitigation. Indeed, with the right combination of power plant design, waste resource availability, and tipping fee credit, cofiring can be economically viable today, even without a CO$_2$ credit.

These cofiring projects have resolved some technical issues, demonstrated that biofuels can be used with coal in existing large utility boilers, and identified potential benefits for both the utility and its customers. However, due to marginal economics, complications of dealing with dual fuels, risks, uncertainties, and some technical issues that remain to be resolved (often on a case-specific basis), cofiring has yet to be adopted to a significant extent by the electric power industry.

The EPRI cofiring program has also defined issues that remain at least in part unresolved:

1. Effect of cofiring on ash properties and fly ash sales.
2. Influence of cofiring on boiler slagging and fouling problems.
3. Impacts of fuel blending on pulverizer performance.
4. Maximum percentage of cofiring as a function of materials-handling and combustion technologies;
5. Characteristics of biofuels required to achieve defined goals, such as avoiding derate and achieving dispatch and reduced NOx emissions;
6. Ability to meeting customer-service, environmental and cost goals.

Overcoming these technical, economic, and institutional barriers is the goal of a recently initiated DOE-PETC/EPRI cooperative R&D program, discussed next.

THE DOE-PETC/EPRI COOPERATIVE BIOMASS-COAL COFIRING R&D PROGRAM

Increasing interest in developing CO$_2$ emission control technologies for existing coal-fired power plants, has made cofiring a major focus of a recently initiated collaborative program on CO$_2$ mitigation between EPRI and the U.S. Department of Energy, Pittsburgh Energy Technology Center (DOE-PETC). The DOE-PETC/EPRI joint program on biomass cofiring began in 1995 as a co-sponsored project at the General Public Utilities (GPU) Shawville generating station (Hunt et. al., These Proceedings). The DOE-PETC/EPRI Cooperative Agreement, signed in September 1996, will co-fund a number of projects to develop CO$_2$ mitigation technologies, emphasizing new and ongoing cofiring projects carried out in collaboration with electric utility industry partners. Activities currently planned include:

TVA.	The EPRI-TVA studies, including full-scale cofiring tests, were discussed above. Resource assessments demonstrated the potential for widespread cofiring at TVA plants. Further full-scale tests are planned with TVA under this Agreement.
NYSEG.	EPRI is co-sponsoring a test program with New York State Electric and Gas Company, to evaluate size reduction and drying equipment for wood fuels and co-firing in a tangentially-fired 100 MWe PC boiler, using a separate wood fuel feeding system. Mid-level (10 - 15 % heat) feed rates will be tested, having already reached 10% levels in previous tests.
GPU.	EPRI and General Public Utilities (GPU), will help co-fund a mid-level cofiring test in a wall-fired PC unit at the Seward Plant, near Johnstown Town, Pennsylvania. This test will also use a separate feed for the wood.
SCS.	Southern Company Services (SCS) has carried out short test-runs at a tangentially-fired PC boiler in Savannah, Georgia, which indicated that separate wood feeding of up to 40% (heat) is possible. A longer-term pre-commercial test, including some further tests with natural gas overfire, would allow exploration of the upper limits of solid fuel cofiring.
MGE.	EPRI is co-funding with Madison Gas & Electric (MGE) a test program being conducted by the University of Wisconsin at the MGE Blount Street Station in Madison. There an existing retrofit to burn refuse-derived fuel and shredded paper waste in a wall-fired PC unit will be used to conduct the first U.S. test of cofiring switchgrass, a proposed energy crop, with coal in a full-size utility boiler.
Duke Power.	EPRI, Duke Power Co., and the National Plastics Council are co-sponsoring laboratory, engineering analysis and, as a next step, a full-scale test, of cofiring plastic manufacturing wastes in a PC boiler. Wood cofiring will be considered later. This service area has a large biomass resource potential.

NIPSCO.   EPRI is completing a study co-funded by Northern Indiana Public Service Company to evaluate fuel supply and power plant operations for cofiring wood in a full-size cyclone boiler, specifically to reduce CO_2 emissions.

The above are only the current projects being considered by this DOE-PETC / EPRI Cooperative Agreement. To support and direct these efforts and to convert the test results into predictions and assessments for other fuels and power plants, EPRI will use the data from these tests in existing and developing models and databases:

COFIRE1   - a cost and performance spreadsheet for cofiring cases;
ALTFUELS   - an alternative fuels database;
SOAPP   - a state-of-the-art power plant model to be adjusted for cofiring; and
CQIM   - a coal quality model for predicting efficiency, emissions, and fouling/slagging.

The need for the above, and additional, cofiring tests at actual operating power plants, is the large variability encountered in the boiler-specific requirements for biomass preparation, quality, and feeding. These tests will provide the basis for rapid commercialization of this technology, wherever and whenever waste or other low-cost biomass fuels are available. Biomass gasification directly coupled to existing boilers could extend the co-firing concept to coal-fired power plants where solid biomass fuels can not be readily used. This could become a future R&D topic within this program.

CONCLUSIONS

Biological systems offer a range of options for CO_2 mitigation, from forestry protection and enhancement to substituting fossil fuels with biofuels. One of the options available today, with relatively low cost and significant potential for implimentation, is cofiring of biomass in coal-fired power plants.
The major limitations to biomass cofiring are often not preparation and feeding of biomass fuels, important as they are, but the availability of low-cost wood fuels. It is the availability of low-cost waste biomass fuels and the opportunities for low capital cost retrofit, that make cofiring a near-term option for reducing fossil CO_2 emissions by existing power plants. In the future the biomass resource suitable for cofiring can be greatly expanded by use of intensively produced wood fuels. EPRI has recently assessed the potential economics of biomass fuels produced by such energy plantations (Hughes and Wiltsee, 1995). The conclusion were that such dedicated energy farms can be considered in many regions where surplus agricultural land is available. This will, of course, require mid- to long-term research and development to achieve the projected productivities and performance objectives on which the cost estimates are based. At present, only projects that have a substantial subsidy or derive most of their income from fiber production, with biofuels as co-products, are economically viable.

One argument against cofiring, and biomass power generally, is that biomass could be used for the production of liquid transportation fuels, much more valuable than solid fuels. However, the technology for conversion of lignocellulosic biomass to, for example, ethanol, is still undeveloped and very expensive. More importantly, if CO_2 mitigation is the goal, then replacement of coal with biomass would have a large advantage over alternative biofuels.

In conclusion, the cost of CO_2 mitigation by the cofiring option is substantially lower than any other end-of-pipe CO_2 mitigation or most alternative fuels technologies. Although cofiring would reduce CO_2 emissions by the electric industry by only a few percent, this is a near-term technology that can make important contributions to the goal of CO_2 mitigation. Biological CO_2 mitigation processes generally can greatly reduce the threat of global warming. They require a major, world-wide, R,D&D and implementation effort.

REFERENCES

Augenstein, D.C., J.R. Benemann, and E. Hughes (1994). Electricity from Biogas. In Proceedings of the 2nd Biomass Conference of the Americas, National Renewable Energy Laboratory, Golden, Colorado.

Benemann, J.R. (1978). Biofuels: A Survey. Electric Power Research Institute, Palo Alto, California, EPRI ER-746-SR.

Benemann, J.R. (1996). The future of renewable biological energy conversion systems. In Miyamoto, K., (ed.), Biological Energy Conversion, in press.

Boylan, D.M. (1993). Southern Company Tests of Wood/Coal Cofiring in Pulverized Coal Units. In Proceedings: Strategic Benefits of Biomass and Waste Fuels. Electric Power Research Institute, Palo Alto, California, TR-103146 pp. 4-33 - 4-43 .

Hughes, E.E., and G.A. Wiltsee Jr. (1995). Comparative Evaluation of Fuel Costs from Energy Crops. In Proceedings of the 2nd Biomass Conference of the Americas. National Renewable Energy Laboratory, Golden, Colorado, pp. 1291 - 104 (1995).

IPCC (Intergovernmental Panel on Climate Change) (1996). Climate Change 1995 (three volumes). Cambridge Press, Cambridge.

Lamarre, L. (1994). Electricity from Whole Trees, EPRI Journal, 19, 1 - 4 (1994).

McGowin, C.R., and E.E. Hughes (1992). In Kah, M.R. (ed.), Clean Energy from Waste and Coal, American Chemical Society, Symp. Ser. No. 515, pp. 14 - 24.

Rubin, E.S., R.N. Cooper, R.A Frosch, T.H. Lee, G. Marland, A.H. Rosenfeld, D.D. Stine (1992). Realistic Mitigation Options for Global Warming. Science, 257, 148-149, 261.

Sampson, N.R., L.L. Wright, J.K. Winjum, J.D. Kinsman, J.R. Benemann, E. Kursten, and J.M.O. Scurlock (1993). Biomass Management and Energy. Water, Air, and Soil Pollution, 70, 129 - 141.

Scurlock, J.M.O., and D.O. Hall (1990). Energy from Biomass. Biomass, 21: 75 - 86. Foster Wheeler Environmental Corp. (1995). Status and Prospects of Wood Cofiring. Electric Power Research Institute, Palo Alto, California.

Tillman, D.A, E.E. Hughes, E. Stephens, and B. Gold (1994). Cofiring Biofuels in Coal-fired Boilers: case Studies and Assessments. Bioenergy 94, pp. 147 - 160.

Turnbull, J.H. (1993). Use of Biomass in Electric Power Generation: The California Experience. Biomass and Bioenergy 4.

Vitousek, P.M. (1994). Beyond Global Warming: Ecology and Global Change. Ecology 75: 1861 - 1876.

Wiltsee, G.A., C.R. McGowin, and E.E. Hughes (1993). Biomass Combustion Technologies for Power Generation. In Proc. 1st Biomass Conf. of the Americas. National Renewable Energy Laboratory, Golden, Colorado, pp. 347 - 367.

Pergamon

Energy Convers. Mgmt Vol. 38, Suppl., pp. S475–S479, 1997
© 1997 Elsevier Science Ltd. All rights reserved
Printed in Great Britain
0196-8904/97 $17.00 + 0.00

PII: S0196-8904(96)00313-5

CO_2 MITIGATION WITH MICROALGAE SYSTEMS

John R. Benemann
343 Caravelle Drive, Walnut Creek, CA 94598, U.S.A.

SUMMARY

Microalgae present one of the few technologies for the capture and utilization of CO_2 emitted by power plants. These microscopic plants would be grown in large open ponds, into which power plant flue gas or pure CO_2 (captured from power plants) is sparged, and, after harvesting, the biomass would be converted to a fossil fuel replacement, preferably a high value liquid fuel such as biodiesel. The requirements for large areas of land, favorable climate, and ample water supplies will restrict the potential of this technology. Also, even with rather favorable technical assumptions, the currently projected costs of microalgae-fuels are high, similar to most power plant CO_2 capture and disposal options. However, if the technology of microalgae could achieve very high productivities, equivalent to 10% solar energy conversion, and if projected low-cost cultivation, harvesting and processing techniques could be developed, microalgae technology could become a low-cost CO_2 mitigation option, particularly if prices for fossil fuels increase in the future. In the nearer-term microalgae CO_2 utilization can be integrated with wastewater treatment and reclamation, providing an early application of this technology. Long-term basic and applied R&D are required to develop this technology, as one of the many options that may be required in the future to help preserve our planetary atmosphere and biosphere. © 1997 Elsevier Science Ltd

KEYWORDS: Microalgae, CO_2 Mitigation, Greenhouse Gases, Global Warming.

INTRODUCTION

Reducing the build-up of atmospheric CO_2, the major driving force in projected global warming, can be accomplished by three conceptually different methods:
1. reducing the use of fossil fuels;
2. removing CO_2 from the atmosphere; and
3. capturing and sequestering or utilizing the CO_2 emitted by fossil fuel combustion before it enters the atmosphere.

The first can be accomplished, for examples, by increasing the efficiency of power generation, decreasing demand through more effective energy utilization, or by substituting fossil fuels with non-CO_2 generating energy sources (nuclear, biofuels, other renewables). The second option, removing CO_2 from the atmosphere, is currently only achievable with higher plants, e.g. trees, to build-up the carbon content of the biosphere, mainly in standing forests and soils. These "indirect" mitigation options are the focus of most of the current technological development and CO_2 mitigation activities. However, it is likely, even certain, that these will not be enough to stabilize atmospheric CO_2 levels sufficiently to avoid a future greenhouse world.

Thus, the third option, the capture sequestration or utilization of CO_2 from power plants and other fossil fuel combustion systems, the focus of This Conference, will also need to be considered in the future. However, most such "direct" CO_2 mitigation technologies are quite expensive, compared to the indirect options. One such technology is the use of microalgae cultivation systems that use power plant flue gases or captured CO_2 to produce fuels.

MICROALGAE FOR CO₂ MITIGATION

Microalgae, like other direct CO_2 mitigation technologies typically are projected to have costs exceeding \$50, even \$100, per ton of CO_2 emissions avoided (Benemann, 1993; Kadam, in These Proceedings), an order of magnitude higher than most of the indirect approaches. Furthermore, even these economic projections for microalgae systems are based on many highly favorable extrapolations and assumptions, that are well beyond current technological capabilities. Thus, such cost projections define future R&D goals and needs, rather than current reality. And, perhaps most important, microalgae systems require land, water, and climate resources that are seldom found in juxtaposition, or near a power plant. These factors constrain the likely potential CO_2 reductions by microalgae systems in the U.S. to a few million, at most a few tens of millions, of tons of CO_2, a small percentage of present power plant fossil CO_2 emissions.

Thus, the argument can be made that investments into microalgae R&D, even if successful, are unlikely to make a significant contribution to solving the CO_2 problem, and, therefore, should not be pursued. Certainly, R&D expenditures must be prioritized, and the development of technologies that are not likely to make a significant dent in CO_2 emissions should not be supported. The issue is what is "significant" in this context. With about 5 billion tons of fossil CO_2 emitted by the U.S. annually, and over twenty billion tons world-wide, and even more projected in the future, technologies that promise reductions of thousands, even hundreds of thousands, of tons annually, such as most schemes to fix CO_2 into chemicals, can, indeed, be considered irrelevant. And compared to higher plant biomass systems, projected by many to be able to reduce global atmospheric CO_2 concentrations by billions of tons annually, microalgae systems present, indeed, a rather minor opportunity.

However, biomass systems are not a single technology. Rather, they comprise scores of different plant species and types, a large diversity of agronomic system, climates, soils, and biomass conversion pro-cesses, each will require both generic and specific research, development, and demonstration (R,D&D), at multiple sites over many years, even decades. Indeed, no single plant biomass system or technology is likely to have an overwhelming impact. Only in aggre-gate can biological systems make a major contribution. problem. Thus R&D investments into specific biological technologies are justified, even if individually limited to a few million to tens of millions of tons of CO_2 reduction annually.

There are several additional arguments in favor of microalgae for CO_2 mitigation that justify at least a modest effort in this area:

1. The use and recycling of CO_2 is inherently preferable to the disposal options.
2. Direct CO_2 mitigation processes are inherently preferable to indirect ones.
3. As fossil fuel prices increase, microalgae CO_2 mitigation costs decrease, unlike other methods for CO_2 capture and sequestration.
4. Microalgae systems R&D is essentially generic - easily translated to new sites.
5. Microalgae R&D can have "spin-offs" from wastewater treatment to chemicals.
6. Microalgae R&D can be easily extrapolated from small scales to larger systems.
7. Microalgae R&D is inherently faster than other biomass systems, due to the very rapid growth rates of these microscopic plants.

To amplify on some of the points: Microalgae, growing in water, have fewer and more predictable process variables (sunlight, temperature) than higher plant systems (which also must deal with soil and precipitation), allowing easier extrapolation from one site, even climatic condition, to others. Thus, fewer site-specific studies are required for microalgae than, for example, tree farming. Also, microalgae grow much faster than higher plants, with generation times in hours to days, vs. years to decades for trees. This would allow the R&D to be completed in years, not decades, and/or to tackle more challenging problems. And, although, certainly, site requirements for locating microalgae processes will be much more restrictive than for higher plants, the higher productivities projected for microalgae systems overcome at least some of the limitations of higher plant systems - principally the need for very large land areas. Indeed, microalgae systems can use land, such as hard-pan and high clay soils, and water resources, such as waste or brackish waters, not suitable for conventional agriculture or forestry, minimizing the competition with food and fiber production.

This defense of microalgae R&D for CO_2 mitigation is not only required to counter the arguments made by detractors of this technology, but, perhaps most critical, to neutralize the overblown rhetoric and unsupportable claims by many of its sometimes overenthusiastic proponents. Micro-algae will not likely be a major future fuel source, either in the U.S. or globally, despite some claims to the contrary. Nor can the development of highly complex and clearly exorbitantly expensive photobioreactors be justified in the context of CO_2 mitigation and fuel production.

Here I present the conclusions of a recent, updated, conceptual economic analysis of microalgae power plant CO_2 utilization and conversion to high value liquid fuels (Benemann and Oswald, 1996), and discuss a potential pathway for the future development of this technology. First, however, I address one of the critical issues in microalgae production: the achievable productivities of microalgal systems.

MICROALGAE PRODUCTIVITIES

One of the claims for microalgae cultures is that they have the potential for achieving very high productivities. However, such projections are often based on selective extrapolations of limited data from short-term and small-scale pond operations, even of laboratory experiments. Notwithstanding some assertions to the contrary, microalgae have no particular or inherent claims to unusually high productivities. Microalgae contain the same photosynthetic machinery present in higher plants. Indeed, the estimated (in the absence of reliable published data) that the productivity of microalgae in commercial operations and from wastewater treatment ponds is about 20 to 30 mt/ha/yr, not remarkable when compared to irrigated crops under similar climatic conditions. However, microalgae cultures exhibit some features that argue for potentially higher productivities:

- continuous production avoids the establishment periods of conventional plants;
- ability to provide optimal nutrient levels (e.g., CO_2, N, P, etc.) at all times;
- absence of non-photosynthetic supporting structures (roots, stems, fruits, etc.).
- ability to adjust harvest rates to keep culture densities at optimal levels at all times; and
- control over cell composition (for high oil contents, for example), without decreasing productivities - maximizing potential fuel production and CO_2 utilization.

Still, these arguments are not strong - and maximal productivities for microalgae are unlikely to exceed the maximal levels observed with higher plants, such as sugar cane, grown under optimal conditions of sunlight, temperature, water supply and fertilizers - about 100 mt/ha/yr of dry bio-mass, corresponding to about 3.5 - 4% total (7 - 8% visible) solar light energy) conversion efficiency. Even achieving such levels will require considerable R&D. For example, techniques must be developed to maintain unialgal cultures of specific, desirable algal strains, preventing invasion by other algae, zooplankton grazers and biotic infections generally. Another issue is how to minimize the respiration by algal cells at the high oxygen tensions present in mass culture ponds.

However, in the laboratory, photosynthetic visible light conversion efficiencies (light energy transformed into biomass higher heating value) are routinely measured at 20 - 24 %. But this is only achieved at low light intensities: at high light levels the photosynthetic apparatus is unable to process all the photons absorbed by the chlorophyll and other pigments. This results in the algal cells wasting from 50% to 80% of the incident light, and is one the major reasons that algal culture productivities at the high light intensities typical of sunlight are much lower than in laboratory experiments.

This problem has been recognized for over forty years 1950's (Vandevar Bush, 1953), and a solution proposed nearly as long ago: reduce the so-called antenna pigments of the photosynthetic apparatus, to prevent this wastage of light energy. Although this approach is yet to be demonstrated in the laboratory, let alone in practice, it provides a clear experimental approach to achieving the high solar conversion efficiencies required for microalgal CO$_2$ fixation and fuel production systems. The alternative approach, also recognized for many decades, is to use devices such as optical fibers, to disperse light evenly throughout an algal culture. This can be demonstrated in the laboratory to increase culture productivities, but is impractical in any scale-up.

MICROALGAE PRODUCTION SYSTEMS

Microalgal biomass, like other plant biomass, is potentially suitable for conversion to liquid (gasoline, biodiesel, ethanol) and gaseous (methane, and hydrogen) fuels. However, even laboratory data on such conversions is limited and scale-up has not been attempted. Furthermore, the production of the algal biomass for energy production has not been attempted beyond a few small-scale outdoor studies, of limited duration, with the largest study using two 0.1 ha ponds operated for a two seasons in New Mexico (Weissman and Tillett, 1990, unpublished). Thus, any discussion and analysis of microalgae technology for purposes of CO$_2$ mitigation and fuel production must be based on extrapolations from other systems, in particular the commercial production of microalgae for human foods and, perhaps most applicable, microalgae wastewater treatment ponds.

The standard commercial technology for microalgae biomass production uses raceway-type, paddle wheel mixed ponds. Such systems are flexible, relatively easily scalable, and of low cost. Three commercial plants based on this design, ranging from about 5 to 15 ha in size, are operating in the U.S., producing high-value food supplements (beta-carotene, Spirulina biomass). No alternative design is apparent, despite scores of photobioreactor concepts proposed. For CO$_2$ mitigation and energy production raceway ponds would be used, exceptthat individual growth ponds would be over 5 ha in size (vs. 0.5 ha now), and they would be unlined (or use a simple clay liner, not expensive plastic liners as used in commercial systems). Harvesting would also need to be of low cost, which restricts consideration to simple settling or straining processes. This could be followed by a centrifugation or chemical flocculation step, to increase the density of the biomass to allow extraction of the oils or fermentation of carbohydrates to ethanol. The residues would be fermented to methane, to recover as much fuel values as possible, and water, nutrients, CO$_2$, and wastes recycled.

Initial proposals of this concept of microalgae for CO$_2$ utilization and conversion to fuels dates back over forty years and it has been developed experimentally and conceptually by Oswald and Golueke (1960). They conceptualized large earthen growth ponds located near the power plant, with the biomass converted to methane by anaerobic digestion. Most of the water and residues (containing the N, P and other nutrient elements) were to be recycled to the ponds, along with any CO$_2$ from the conversion process. Although this initial assessment was superficial, more detailed development of this concept carried out in the 1970's and early 1980's, in response to the energy crisis (reviewed in Benemann and Oswald, 1996), also concluded that, with favorable assumptions, such systems could potentially provide a competitive fuel source. Since the 1980's work in this field has emphasized the production of high value liquid fuels, in particular biodiesel, the methyl or ethyl esters of vegetable oils.

A recently updated analysis (Benemann and Oswald, 1996) of several alternative cases (direct use of flue gas and captured CO$_2$, at productivities equivalent to 5% and 10% solar conversion efficiencies, and crude oil prices of $25 and $35/barrel) concluded that CO$_2$ mitigation costs would be $100/mt CO$_2$ for the least favorable case and less than $10/mt for the most favorable. The major conclusion was that no insurmountable "show-stoppers" were apparent, but that considerable long-term R&D, in particular in the areas of biomass productivity and culture stability will be required to achieve such cost-goals.

CONCLUSIONS

Despite over forty years of R&D, with some $50 million in support by the U.S. Department of Energy over the past two decades, microalgae technologies for CO$_2$ fixation and fuel production have barely advanced beyond the conceptual stage. An even larger, but more recent, Japanese program (Usui and Ikenouchi, in These Proceedings) has concentrated on closed photobioreactors. However, despite fifty years of development, closed photobioreactors have not yet been even applied in the commercial production of high value (> \$100/kg biomass) microalgae products. And the current commercial microalgae production systems, using open ponds designs, produce only a few hundred tons of biomass annually, at costs approaching \$10,000/mt. The goals of microalgae production for CO_2 capture and fuel production require almost one order of magnitude higher productivities (mt/ha/yr), two orders of magnitude lower production costs (\$/mt), and three orders of magnitude higher production levels (mt/yr) than current systems - daunting goals indeed. Coupled with the climatic restrictions, land requirements, and limited numbers of sites where such systems could be realistically established in the U.S., the prospects for developing microalgae-based CO_2 mitigation processes might appear doubtful. However, past R&D programs did not, in most part, address central issues of this technology, resulting in limited progress. And resource and technology limitations are apparent for most of other CO_2 capture and disposal/ utilization options. Indeed, even a dozen such facilities, comprising a few thousand hectares, would result in over a million tons of CO_2 mitigated, a significant impact.

Finally, and perhaps most important, some nearer-term approaches to microalgae CO_2 utilization/ recycling are possible by combining such systems with wastewater treatment. Indeed, the earlier work in this field clearly identified the potential of such approaches both conceptually (Oswald and Golueke, 1960) and experimentally (Benemann et al., 1980). Wastewater treatment provides both resources (water, nutrients) and economic incentives for such systems, allowing smaller-scale and more economic systems for CO_2 utilization-providing a pathway for the development and eventual commercialization of this technology.

REFERENCES

Benemann, J.R., "Utilization of Carbon Dioxide from Fossil Fuel-Burning Power Plants with Biological Systems," Energy Conserv. Mgmt., 34: 999 - 1004 (1993).

Benemann, J.R., and W.J. Oswald, Systems and Economic Analysis of Microalgae Ponds for Conversion of CO_2 to Biomass. Final Report to the Pittsburgh Energy Technology Center. U.S. Dept. of Energy (March 1996).

Benemann, J.R., in Algal Biotechnology, in R.C. Cresswell, T.A.V. Rees, and N. Shah, eds., Longman, London pp. 317 (1990).

Benemann, J.R., B.L. Koopman, J.C. Weissman, D.E. Eisenberg and R.P. Goebel, in G. Shelef, and C.J. Soeder, eds., Algal Biomass, Elsevier, Amster., p. 457 (1980).

Bush, V., quoted in J. Burlew, Algae Culture: From Laboratory to Pilot Plant, Carnegie Institute, Washington D.C. (1953).

Oswald, W.J., and C.G. Golueke, "Solar Energy Conversion with Microalgae Systems," Adv. Appl. Microbiol., 11: 223 (1960).

Pergamon

Energy Convers. Mgmt Vol. 38, Suppl., pp. S481–S486, 1997
© 1997 Elsevier Science Ltd. All rights reserved
Printed in Great Britain
0196-8904/97 $17.00 + 0.00

PII: S0196-8904(96)00314-7

A NEW MARINE MICROALGA CULTIVATION IN A TUBULAR BIOREACTOR AND ITS UTILIZATION AS AN ADDITIVE FOR PAPER SURFACE IMPROVEMENTS

K. HON-NAMI, A. HIRANO, S. KUNITO, Y. TSUYUKI,*
T. KINOSHITA,** and Y. OGUSHI**

Tokyo Electric Power Co., Energy and Environment R&D Center,
4-1 Egasaki-cho, Tsurumi-ku, Yokohama 230, Japan
*Mitsubishi Heavy Industries, Ltd., Chemical Plant Engineering & Construction Center,
3-3-1 Minatomirai, Nishi-ku, Yokohama 220, Japan
** Mitsubishi Heavy Industries, Ltd., Hiroshima R&D Center,
4-6-22 Kan-on-shin-machi, Nishi-ku, Hiroshima 733, Japan

ABSTRACT

A microalga isolated from Sagami Bay in Japan was cultivated in a tubular reactor and its productivity was obtained. The alga productivity indicated the maximum value at the initial algae concentrations of nearby 0.6g/ ℓ and was influenced by the flow rate. It was compared with those simulated based on basic research. Relatively good correlations between the actual productivities and the simulated ones were found and a flow rate dependency on productivities was confirmed. Utilization studies of the alga as a pulp additive indicated its effectiveness for anti-printthrough, smoothness and tolerance for deterioration, and the alga could therefore be a possible pulp substitute.

© 1997 Elsevier Science Ltd

KEYWORDS

microalga, tubular reactor, productivity, utilization, pulp substitute,

SCREENING OF MICROALGAE

With the intent of procuring high-productivity microalgae for the purpose of carbon dioxide fixation, collections were made in the sea water along the coasts of Sagami Bay and the Seto Inland Sea, and 8 types of microalgae were obtained using the survival screening method. *Tetraselmis* sp. (Tt-1; Sagami Bay collected stock) was selected based on a productivity rate of Nannochroris sp. (NANNO-2) using a 20 liter cultivation flask and on cellular size. Table 1 shows the cultivation data. Tt-1 had a productivity of 10-15 g/m² · day, a cellular grain size of 8-10 μ m, superior stability, and no adhesive tendency.

Table 1. Evaluation Results from a Small-batch Cultivation Experiment

Strain	Cell size (μm)	Stability	Adhesiveness ; Cohesiveness	Production rate[1] (g/m$^2 \cdot$day)
Ts-1	2~3 ϕ	Good	None	7~11
Ts-2	3~5 ϕ	No good	Some	—
Ts-3	1~2 ϕ	Passable	None	3~5
Ts-4	2~3 ϕ	Good	None	5~10
Ts-5	2~3 ϕ	Good	None	5~10
Tc-1	6~12 ϕ	Good	None	5~13
Tc-2	6~10 ϕ	Passable	None	5~8
Tt-1	8~10 ϕ	Good	None	10~15
NANNO-2 [2]	2~3 ϕ	Passable	None	5~7

*1 Overall value under conditions of 10,000lux of artificial light (light/dark=14/10 hr).
*2 For comparative reference purposes.

INDOOR CULTIVATION

The production rate of Tt-1 was measured using a tubular reactor ($70^{ID} \times 8000^{l}$, 50 ℓ , exposure area 0.91m^2) made of transparent acryl resin and installed in an artificially lighted room (air-conditioned at 25℃, 30,000 lux using metal haloid lamp). The productivity of Tt-1 in semi batch cultivation was affected by flow rate and initial algae concentration. Algae productivity had a maximum value at an initial algae concentration of 0.5-0.7 g/ ℓ , and a fast flow rate gave a high productivity. The production rate obtained in this experiment was compared to a simulated value.

PRODUCTIVITY SIMULATION

The algae production rate of a tubular reactor was calculated using the methods given below. The peculiar property values of photosynthesis, photosynthesis constant (K_p), light dependency constant (ϕ), oxygen consumption rate (ρ) and absorptivity coefficient (ε) were measured and are listed at Table 2.

Table 2. Photosynthetic Values for Tt-1

	ε 1/m ·1/kg·m^3	Measured Temp. (℃)	K_p mℓ/kg·h	ϕ lux	ρ mℓ/kg·h	α —
Tt-1	154	25	2.84×10^5	7.54×10^3	7.17×10^3	39.61
		15	1.38×10^5	2.67×10^3	—	—

K_p, ϕ and ρ were given by the following temperature functions because of their high temperature dependency.

$$K_p = 1.46 \times 10^4 \ (T + 273.15) - 4.069 \times 10^6 \qquad (1)$$

$$\phi = 487 (T + 273.15) - 1.377 \times 10^5 \qquad (2)$$

$$\rho = 1267 \times 2^{\frac{T}{10}} \qquad (3)$$

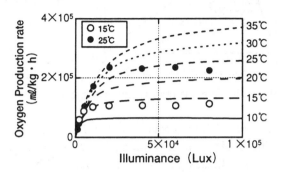

Fig.1 Photosynthesis curve at each temperature

Figure 1 shows the correlation between the measured values for photosynthesis oxygen production rate (k_p) and the same values calculated by equation (4). Since the measured and calculated values showed good correlation under 15℃ and 25℃, they were taken as the function for the temperature range of 10-30℃ as well.

$$k_p = K_p \left(\frac{I}{I + \phi} - \frac{I}{\alpha} \right) \qquad (4)$$

Where

$$\alpha = K_p / \rho \qquad (5)$$

The overall oxygen production rate of algae (q) and the algae production rate (P) based on the oxygen production rate (q) were expressed by equation (6) and equation (8), respectively (Nishikawa *et al.*, 1992).

$$q = \frac{K_p}{\varepsilon} \left\{ ln \left(\frac{I_0 + \phi}{I_H + \phi} \right) - \left(\frac{I_0}{I_H} \right) \right\} \qquad (6)$$

Where

$$I_H = I_0 e^{-CH\varepsilon} \qquad (7)$$

$$P = \gamma \sum_{t_1}^{t_2} q \qquad (8)$$

Further, all algae in a tubular reactor respirate day and night, and therefore the oxygen consumption rate of algae (q') was expressed by equation (9).

$$q' = C \rho (\pi H / 4) \qquad (9)$$

The overall production rate (P') including the respiration loss was finally expressed by equation (10). In the low concentration range of ($<0.6g/ \ell$), the simulated production rate of algae was far off the experimental values, and for this reason the maximum division of this algae was two hold in a day.

$$P' = \gamma \sum_{t_1}^{t_2} q - \gamma' \sum_{t_{2'}}^{t_{1'}} q' \qquad (10)$$

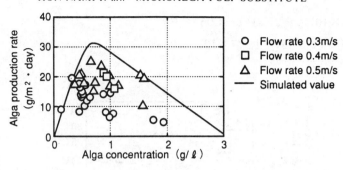

Fig.2 Correlation of initial aigae concertration and algae production rate

The simulation results under indoor cultivation conditions are as shown in Fig.2. P' had a maximum production value at near the initial algae concentration of 0.7g/ ℓ . The key in Fig.2 shows flow rate, although a higher flow rate indicates that production rate will have a high value. Because there was no factor for flow rate for the simulation, the oxygen production rate (q) was hypothesized to be affected by such rate. With the addition of a compensating factor f(u) in accordance with flow rate in case this equipment is used, parameter fitting shown in equation (11) was carried out.

$$q = \frac{K_p}{\varepsilon} \left\{ ln \left(\frac{I_0 + \phi}{I_H + \phi} \right) - \left(\frac{I_0}{I_H} \right) \right\} f(u) \tag{11}$$

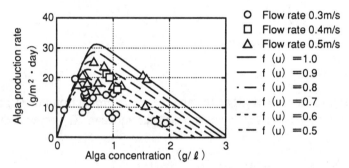

Fig.3 Correlation of initial algae concentration and algae production rate

Figure 3 shows the simulation results with the experimental values under a flow rate of 0.3, 0.4, and 0.5 m/s and f(u)=0.5-1.0. Then, with a constant having f(u), a simulation with relatively good correlation could be obtained.

ALGAE UTILIZATION STUDIES

An effective use of algae is as a paper feedstock and/or its substitute. Since micro-fibers in wood pulp can greatly improve surface features related to paper strength and suitability for printing, the effectiveness of adding microalgae to wood pulp was studied.

In this research, a hand-made paper making machine was used, and paper was made by mixing micro algae into the paper feedstock for newsprint. A 500 ppm of a polyacrylamide was added to the pulp to improve its productivity. The quality of the paper made with microalgae additives was evaluated for density, gas permeability, smoothness, ink absorptivity and tensile index. Further, the paper deterioration characteristics over time were also examined.

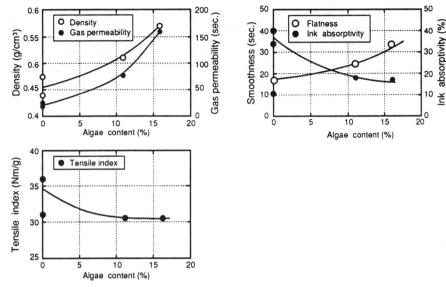

Fig.4 Quality of paper with microalgae additive

The paper quality as regards the content of algae is as given in Fig.4. As the content of algae is increased, density increased, and it was confirmed that the algae filled up the voids of the paper. Furthermore, the effectiveness of the microalgae as additives was confirmed when it was seen that the quality of gas permeability, smoothness, and ink absorptivity, which supports void characteristics in paper, were all improved. On the other hand, the tensile index showed a slight decrease with an increase of the algae content, although it was in the range satisfying standards. It may be suggested that this phenomenon was caused by the chlorophyll and fats in microalgae blocking the couplings between fibers. The characteristics of the algae are as shown in Table 3.

Table 3. The Characteristic of Algae

Stain	Chlorophyll & Fats	Rolysaccharides & Protein	Starch	Cell wall Constituent
Tt-1	19.3%	55.9%	4.5%	20.2%

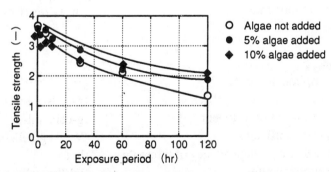

Fig.5 Tensile strength against exposure period

The tensile strength in the durability tests is shown in Fig.5. The strength of paper with microalgae indicated a low value at the initial stage as compared with the original paper, but it became equal after 60 hours, and showed a high value after an even longer period. This would seem to indicate that the algae as additive lessened the strength deterioration, raising the possibility of its use as a pulp substitute.

CONCLUSION

The microalga, *Tetraselmis* sp. strain Tt-1, was cultured in a tubular reactor and measured for its productivity. The alga productivity indicated the maximum value at the initial algae concentrations of 0.5-0.7g/ ℓ and was influenced by the flow rate. These were well corresponding to algae production simulation. Tt-1 as pulp additive indicated improvement effects for anti-printthrough, smoothness and tolerance for deterioration, and Tt-1 could therefore be a possible pulp substitute.

REFERENCE

Nishikawa, N., Hon-nami, K., Hirano, A., Ikuta, Y., Hukuda, Y., Negoro, M., Kaneko, M., and Hada M. (1992). Reduction of Carbon Dioxide Emission from Flue Gas with Microalgae Cultivation. *Energy Convers. Mgmt*, <u>33</u>, 553-560.

SYMBOLS

C	:	Microalgae harvesting concentration	(kg-cells/m^3)
$f(u)$	:	Flow rate adjustment factor	$(-)$
H	:	Culture depth	(m)
I	:	Light intensity in the culture	(lux.)
I_H	:	Light intensity in the culture bottom	(lux.)
I_0	:	Light intensity on the culture surface	(lux.)
k_p	:	Photosynthesis oxygen production rate	(ml-O$_2$/kg-cells·h)
K_p	:	Photosynthesis constant	(ml-O$_2$/kg-cells·h)
P	:	Algae production rate	(g/m^2·d)
P'	:	Overall algae production rate	(g/m^2·d)
q	:	Oxygen production rate	(ml-O$_2$/m^2·h)
q'	:	Oxygen consumption rate	(ml-O$_2$/m^2·h)
t_1	:	Sunrise time	$(-)$
t_2	:	Sunset time	$(-)$
T	:	Temperature	(℃)
α	:	Microalgae respiration constant	$(-)$
γ	:	Microalgae weight increase coefficient	(g-cells/ml-O$_2$)
γ'	:	Microalgae weight loss coefficient	(g-cells/ml-O$_2$)
ε	:	Light absorption coefficient	(l/m·(kg-cells/m^3))
ϕ	:	Constant related to light dependency of microalga	(lux.)
ρ	:	Oxygen consumption rate	(ml/(kg-cells·h))

Energy Convers. Mgmt Vol. 38, Suppl., pp. S487–S492, 1997
© 1997 Elsevier Science Ltd. All rights reserved
Printed in Great Britain
0196-8904/97 $17.00 + 0.00

Pergamon

PII: S0196-8904(96)00315-9

The Biological CO$_2$ Fixation and Utilization Project by RITE(1)
— Highly-effective Photobioreactor System —

Naoto Usui, Masahiro Ikenouchi

Research Institute of Innovative Technology for the Earth （RITE）
2−8−11 Nishishinbashi Minatoku Tokyo 105, JAPAN

ABSTRACT

This paper reports on the development of highly efficient photobioreactor system for the fixation of CO$_2$ from LNG thermal power plant. Nine types of photobioreactor has been developed as part of a study for developing a technology for biological CO$_2$ fixation. Using new types of photobioreactor, *Chlorella* sp. UK-001 cultured under a cycle of 10 hours irradiation and 14 hours of darkness, attaining a CO$_2$ fixation rate almost ten times that of tree. We also developed the sunlight collection and transmission system with high efficiency. In the process development of biomass to useful substances, the extraction of liquid fuel from *Botryococcus* sp. and the applicability of algae as a filler such as for plastics are also described.

© 1997 Elsevier Science Ltd

KEYWORDS

Biological CO$_2$ Fixation; Photobioreactor; Utilization; *Chlorella* sp.; *Botryococcus* sp.

INTRODUCTION

One of the major causes of global warming is the increase of CO$_2$ in the atmosphere. Reduction of CO$_2$ emissions caused by the consumption of fossil fuel is an urgent and crucial challenge which may affect the very survival of the human race. Biological fixation of CO$_2$ by making use of our abundant solar energy is one interesting area being studied as a possible solution to global warming.

In 1990, we launched a ten-year research program aimed at developing an ability to fix CO$_2$ ten times more efficiently than trees by focusing on microalgae with a high photosynthetic potential.
This research was entrusted to the Research Institute of Innovative Technology for the Earth (RITE) by the New Energy and Industrial Technology Development Organization (NEDO) under the supervision of Japan's Ministry of International Trade and Industry (MITI). Some 150 researchers from 16 private companies are involved with this research, working in their companies' research laboratories.

The research program is divided into two parts. Theme 1 deals with "Research and Development on Photosynthetic Bacteria and Microalgae as Highly Efficient Photosynthetic Microorganisms" and theme 2 relates to "Research and Development of High-density, Large-volume Culture System for Microalgae and Utilization of Biomass". The conceptual diagram of our proposed "Biological CO$_2$ Fixation and Utilization System" is shown in Figure 1.

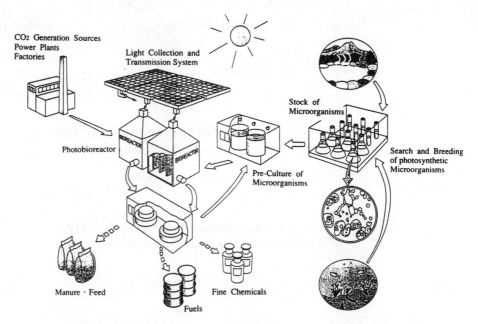

Fig.1 Biological CO₂ Fixation and Utilization System

The CO_2 contained in flue gas emitted by thermal power plants burning fossil fuel is treated as raw material in a system. The CO_2 is fixed within a photobioreactor by photosynthesis using microalgae. The sunlight needed for photosynthesis to take place is collected by a sunlight collection and transmission system for delivery to the photobioreactor. Once the algae have fixed CO_2, they are put to effective use in the form of fuel, fodder, fertilizer, or building materials.

The group working on theme 2 are improvement of CO_2 fixation rate by using a photobioreactor, research to improve the efficiency of the sunlight-collection device, work on peripheral technology such as solid-liquid separation, and a review on the effective use of biomass. We are also implementing a case study in which it is assumed that photosynthetic microalgae will be used to fix the CO_2 in flue gas from a fossil-fuel power plant. We are also engaged in evaluating the material balance, energy balance, and economic efficiency of a total biological CO_2 fixation and utilization system operating.

This paper focuses on one of the studies that was carried out under theme 2, in which highly efficient photobioreactors were developed. These photobioreactors were evaluated using a typical strain of *Chlorella* sp. UK-001 (an algae screened from theme 1 research).

PHOTOBIOREACTOR SYSTEM FOR BIOLOGICAL CO₂ FIXATION

High-density, high-volume culturing system

The first part of the study aimed at developing a high-density, high-volume culturing system with nine types of reactor, including the direct sunlight utilizing reactor. The culturing characteristics and equipment requirements for both *Chlorella vulgaris* and *Chlorococcum littorale* in each type of reactor were reviewed. Based on the results of this review, three types of photobioreactor — the flat-plate type, internally luminous stirrer type, and fountain type (Fig. 2) —were selected for further work.

The flat-plate type photobioreactor delivers light to the culture solution through optical fibers and plate-shaped illuminators mounted inside the reactor. This is considered a standard design in our study, for a large volume is available for a specific radiation area and can easily be scaled up.

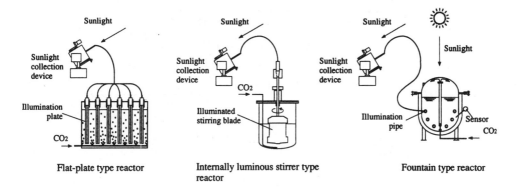

Fig.2 Configuration of photobioreactor under Development

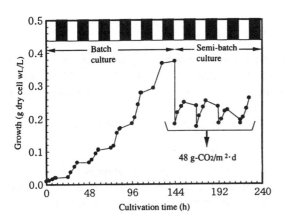

Fig. 3 Growth Curve of Chlorella UK-001 with 10 Hr-light and 14 hr-dark Cycle

Table 1. Culture conditions

Item	Conditions
Algae	Chlorella sp. UK-001
CO_2 concentration in supply gas	10%
Culture temperature	30°C
Working volume (V)	0.0015 m³
Specific radiation area (A/V)	31.2 m⁴
Light source	Artificial light (halogen lamps)
Lighting conditions	10 h light, 14 h dark

The internally luminous stirrer type photobioreactor incorporates agitation blades with light irradiation function; it features simple construction and good agitation efficiency.

The fountain type photobioreactor takes in solar radiation directly through the top, and a liquid film forms on the upper walls within the reactor. Algae are distributed evenly inside the reactor with ease, and the gas-liquid contact efficiency is high. It also incorporates illumination pipes.

Figure 3 shows *Chlorella* sp. UK-001 cultured for 10 days in a flat-plate type photobioreactor, while Table 1 lists the culturing conditions used. After six cycles of 10 hours irradiation and 14 hours of darkness for batch culturing , the algae increased to a concentration of almost 0.4 g/L. Semi-batch culturing was then carried out: observations of a range of cell concentrations were made to maximize the daily growth rate during these initial six days. An average CO_2 fixation rate of 130 g-CO_2/m³ per day was attained, indicating good reactor efficiency. This value corresponds to about 48 g-CO_2/m² per day in terms of sunlight collecting area on a sunny day.

Based on these cultivation results, we are now proceeding to construct a 200L experimental plant based on a flat-plate type photobioreactor. The aim of this experimental system is to further improve CO_2 fixation rate and obtain more engineering data.

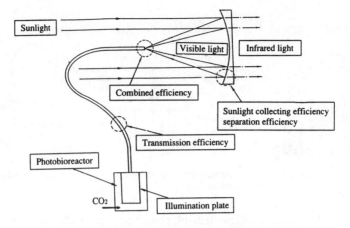

Fig. 4   Efficiency of Sunlight Collection System

Sunlight collection and transmission system

The sunlight collection and transmission system is an integral part of the photobioreactor system. Sunlight, or more specifically the visible light needed for photosynthesis, is collected with a concave mirror. The infrared (IR) radiation passes through the mirror and the visible light is then guided to the reactor through optical fibers. There are plans to use the filtered IR radiation as a power source. The sunlight collection device automatically follows the sun under the guidance of a built-in sensor.

The three factors shown in Fig. 4 are related to the efficiency of this system.
Overall efficiency improvements of over 55% were achieved by implementing various improvements, including the development of a more sophisticated protective cover for the system and selecting an appropriate optical fiber diameter, as well as optimizing the ratio of fiber core area to bundle size.

The illumination plates to irradiate visible light in the reactor were also studied. A latice of strips between two frosted acrylic plates developed for the purpose was found to be effective, and capable of distributing light evenly inside the reactor.

UTILIZATION OF THE PRODUCTS OF PHOTOSYNTHESIS

Until 1994, this study focused on the development of techniques for efficiently extracting and refining such substances as hydrocarbons, lipids, carbohydrates, β-carotene, and eicosapentaenoic acid(EPA) from the cultured photosynthesizing microorganisms, and on the development of analysis and evaluation technique for these products.

 A system used to fix the CO_2 in exhaust gases emitted from a thermal power plant can be expected to produce an enormous amount of algae. It is desired to find useful means of harnessing these algae. We therefore turned our attention to fuel, manure, fodder, and building material applications (Table 2), since these may use large amounts of algae.

Recovery and use of carbohydrates

Botryococcus sp. is a unique algae that fixes CO_2 through photosynthesis and, as a result, produces a hydrocarbon. The amount of hydrocarbon produced often reaches 30% of the total weight of each cell.

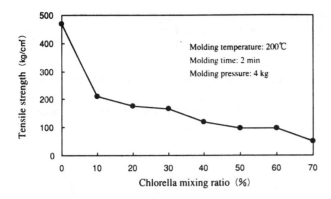

Fig. 5   Effect of Chlorella and PVC Mixing Ratio
on Strength of Molding

Table 2   Applications Considered Effective

Item	Method and Application
Fuel	Extraction of carbohydrate from *Botryococcus* sp.
	Direct liquefaction using coal liquefaction technology
Manure	Compost
Animal feed	Fodder or feed for domestic animals or fish cultivation
Building materials	Plastic filler
	Concrete additives for high efficiency concrete
Biodegradable plastic	
Physiologically active material	Reformation of *Botryococcus* sp.-generated carbohydrates

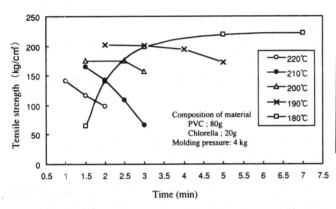

Fig. 6   Effect of Molding Temperature and Molding
Time on Strength of Molding

Table 3   Comparison between Extract and Class-C Heavy Oil

Item	Extract	Class-C oil
Concentration (g/cm³)	0.86	0.95
Calorific value (cal/g)	10.400	10.500
Nitrogen content (%)	0.068	0.18
Sulfur content (%)	0.005	0.95

This hydrocarbon is excreted from the cells, but since it is stored between the cells it is not into the culture solution. A study of extraction methods, however, confirmed that successful extraction is easily achieved by using ethyl acetate as a solvent. A comparison of the extracted oil with class-C heavy oil, which has similar characteristics, revealed that the extracted oil has almost the same calorific value, while it is more environmentally friendly fuel since it contains less nitrogen and sulfur. See Table 3.

Application to building materials

We studied the use of CO₂-fixing algae as fillers, such as in plastics. The idea was to shape the products of photosynthesis in molds, creating a chemically stable form with the potential for long-term storage, and to make an effective use of the molded shapes as building materials.

Chlorella vulgaris and polyvinyl chloride (PVC) were mixed and formed into molds after heating and compression. The strength characteristics of these moldings were studied to determine the most appropriate molding conditions.

Judging from the attained tensile strength (Figs. 5 and 6), a mix of 10% to 30% was found to yield relatively stable moldings. It was also confirmed that the optimum range of heat treatment is 180 to 190℃.

These molding experiments produced glossy, green plate-shaped moldings from *Chlorella vulgaris* and PVC. These plates are sufficiently strong for use as building materials. If the material were spread into a thin film and applied to wooden boards, it would make good decorative finish.

Another possible application currently under study is the use of polysaccharides produced via photosynthesis as an additive in concrete to improve fluidity, since building materials with the polysaccharides produced by the algae content have excellent pre-hardening fluidity. Such additives would eliminate the need for concrete compaction.

CONCLUSION

A sunlight-collecting bioreactor has been developed as part of a study for developing a technology for biological CO_2 fixation . Using this reactor, *Chlorella* sp. UK-001 was cultured under a cycle of 10 hours irradiation and 14 hours of darkness, attaining a CO_2 fixation rate almost ten times that of tree.

The sunlight collection and transmission system is an integral part of the photobioreactor system. Overall efficiency improvements of over 55% were achieved by implementing various improvements.

In the development of useful substance production system from biomass, an environmentally friendly liquid fuel with high calorific value was successfully extracted from *Botryococcus* sp. , one type of algae that fixes CO_2 by photosynthesis and produces a hydrocarbon in the process. Algae obtained by fixing CO_2 have also been successfully used as a filler, such as in plastics, and formed into green, glossy molded shapes.

An experimental 200L plant based on the flat-plate type photobioreactor is now being constructed. This plant will be used to work on further improvements in the CO_2 fixation rate.

This technique for CO_2 fixing is being studied from a total-system viewpoint. The idea is to develop a system that encompasses everything from the supply of exhaust gas to production of useful products at a million-kW LNG thermal power plant. In this light, the power requirements, CO_2 balance, economic efficiency, and other considerations are now being investigated.

REFERENCES

N.Murakami, M.Ikenouchi (1996.7). Carbon dioxide Fixation and Utilization by Microalgae Photosynthesis *The japan Institute of Energy,5th Annual Meeting* , 281-284

S.Hirata,M.Hayashitani,M.Taya,S.Tone(1996).Carbon Dioxide Fixation in Batch Culture of *Chlorella* sp. Using a Photobioreactor with a Sunlight-Collection Device.*Journal of Fermentation and Bioengineering*, 81,470-472

Research Institute of Innovative Technology for the Earth — Introduction to Biology-Related Projects — .*Science & Technology in japan*, 55,8-10

Zhong-Fu Li,M.Yamasita,H.Kabeya,T.Otsuki,J.Hosokawa(1996).Application of Microalgae as a Filling of Plastics. *Reports of The Shikoku National Industrial Research Institute,* 28,1-5

ACKNOWLEDGMENTS

These studies are being carried out under the sponsorship of NEDO as part of "The Project of Biological CO_2 fixation and Utilization."

Energy Convers. Mgmt Vol. 38, Suppl., pp. S493–S497, 1997
© 1997 Elsevier Science Ltd. All rights reserved
Printed in Great Britain
0196-8904/97 $17.00 + 0.00

Pergamon

PII: S0196-8904(96)00316-0

THE BIOLOGICAL CO2 FIXATION AND UTILIZATION PROJECT BY RITE (2)
- Screening and Breeding of Microalgae with High Capability in Fixing CO2 -

Masakazu Murakami and Masahiro Ikenouchi

Research Institute of Innovative Technology for the Earth (RITE)
2-8-11, Nishi-Shimbashi, Minato-ku, Tokyo 105, JAPAN

ABSTRACT

In order to mitigate the potential problems associated with increasing atmospheric CO_2 levels, RITE has started the biological CO_2 fixation and utilization project in 1990 using photosynthetic microorganisms, such as microalgae and photosynthetic bacteria. Photosynthetic microorganisms utilize flue gas CO_2 as a carbon source and solar energy as the energy source to produce biomass which can be used for useful materials.

Extensive screening has been conducted to obtain photosynthetic microorganisms from nature with high capability in fixing CO_2 and/or with the ability to produce useful materials. More than ten strains of microalgae with high capability in fixing CO_2 and/or with the ability to produce useful materials have been obtained. Two green algal strains obtained by the screening, *Chlorella* sp. UK001 and *Chlorococcum littorale*, showed the high CO_2 fixation rates exceeding the initial target value of $1 gCO_2/l/day$ at 24h illumination. *Botryococcus braunii* SI-30 was obtained as the alga producing hydrocarbons which can be used as recyclable fuel resource. This alga contains more than 15% of its dry weight as hydrocarbons and shows relatively high growth rate than other *Botryococcus braunii* strains.

Efforts are now under way to increase their CO_2 fixation rates by optimizing the culture conditions and to develop applications for the products. © 1997 Elsevier Science Ltd

KEYWORDS

Photosynthetic microorganism; microalgae; photosynthesis; carbon dioxide; optimization of culture condition; genetic engineering

INTRODUCTION

The Biological CO_2 Fixation and Utilization Project was started in 1990 by Research Institute of Innovative Technology for the Earth (RITE) under a commission from the New Energy and Industrial Technology Development Organization (NEDO), which is subsidized by the Ministry of International Trade and Industry (MITI).

The project aims at developing a new system for removing CO_2 from industrial exhaust gases by using aquatic photosynthetic microorganisms, such as microalgae and photosynthetic bacteria. The photosynthetic microorganisms were selected because of their following distinctive features; 1) capability to assimilate CO_2 into carbohydrates and other useful substances such as lipids and proteins by using solar energy, 2) higher rates of CO_2 fixation rates than land plants, and 3) better suitability for incorporating the CO_2 removing system into industrial processes than other photosynthetic systems using higher plants. In addition, many microalgal strains are able to grow well in saline water and are able to tolerate widely fluctuating temperatures, high CO_2 concentrations and high light intensities.

The project consists of two major research groups; the first group (named Theme 1) focuses on the biological research on the photosynthetic microorganisms and the second group (named Theme 2) develops the CO_2 fixation and utilization system by using the organisms obtained by Theme 1. The activities of Theme 2 are described elsewhere in this volume.

The detailed research purposes of Theme 1 are as follows; 1) screening of the photosynthetic microorganisms with high capability in fixing CO_2 from nature, 2) screening of the photosynthetic microorganisms which produce useful materials, such as hydrocarbons and polysaccharides, 3) establishment of optimal culture conditions for the microorganisms obtained in terms of fixing CO_2 and/or producing useful materials, and 4) development of biotechnology for manipulating the photosynthetic microorganisms by using the tools of molecular and cellular biology to genetically engineer the organisms for altered photosynthetic activities or altered capability in producing useful materials.

In this paper, the summarized results obtained by several research teams joining the Theme 1 are presented.

SCREENING AND BREEDING OF MICROORGANISMS

Screening of Photosynthetic Microorganisms with High Capability in Fixing CO_2

To obtain the superior photosynthetic microorganisms, microalgae were collected from hot springs, lakes and oceans. The algal samples were then placed in culture medium to allow their growth under the experimental conditions. Single cell colony was isolated by picking up the isolated colony grown on the agar plate. Isolated algal cells were then characterized for their CO_2 fixation rates by measuring the conversion rate of $H^{14}CO_3^-$ to organic compounds or by measuring O_2 evolution rate. The isolated cells were then examined their ability to grow under high temperature and/or high CO_2 concentration, because the cells will be cultivated in the presence of flue gas from power plants.

More than ten strains of microalgae which show high capability in fixing CO_2 were obtained (Kurano et al., 1995, Sakai et al., 1995, Michiki, 1995). They are classified as green algae, cyanobacteria and others. Most of them are green algae such as *Chlorococcum* species and *Chlorella* species, probably because the decreased pH in the culture medium caused by high concentration of CO_2 was favorable for green algae.

Two green algal strains obtained by the screening, *Chlorococcum littorale* and *Chlorella* sp. UK001, have been further examined for the applicability to large scale culture in the photobioreactors developed in this project.

In the experimental small scale cultures, the CO_2 fixation rates of both algae exceeded the initial target value of 1gCO_2/l/day at 24h illumination. The initial target value was then raised to 1gCO_2/l/day at 10h illumination so as to simulate the actual daylight period. The efforts are now under way to attain the newly revised target value and to develop ways of utilization.

Screening of the Photosynthetic Microorganisms which Produce Useful Materials

The large amount of algal biomass resulted from CO_2 fixation should be utilized as resources for fuel, feed, building materials or other chemical stocks. Therefore, the microalgae which produce useful substances such as hydrocarbons, lipids, proteins, or polysaccharides have been screened. *Botryococcus braunii* is well known as hydrocarbon producing microalga and can be used as recyclable fuel resource, however, its growth rate has been thought to be low (Iwamoto, 1986). Therefore, the screening of *Botryococcus* sp. which shows the relatively higher growth rate, i.e., the higher rate in CO_2 fixation, has been attempted in this project. The algal samples were collected from lakes and the microalgae which secreted oily substances outside the cells were isolated. Isolated cells were examined for the ability to produce hydrocarbons and the growth characteristics were determined. Among the isolated algal clones, *Botryococcus braunii* SI-30 was selected as the candidate for mass culture, because of its high CO_2 fixation rate and hydrocarbon content(Murakami,N. et al., 1996) . This alga contains more than 15% of its dry weight as hydrocarbons and its CO_2 fixation rate exceeds the target value of 1gCO_2/l/day at 24h illumination under the small scale culture (Fig. 1). The optimum growth temperatures are between 25 - 30℃ (Fig.2).

Optimization of Culture Conditions for Microalgae in CO_2 Fixation

As a result of the intensive screening of microalgae in terms of high CO_2 fixation rates and the capability to produce useful substances, a large varieties of microalgal samples were obtained. Although the early stage screening revealed the desirable characters of the algae, such as high CO_2 fixation rate, their ability should be improved prior to large-scale cultivation. Therefore, the algae considered to be the candidates for mass cultures have been subjected to the optimization of culture conditions.

Since *Synechocystis aquatilis* SI-2 isolated from the ocean showed the high specific growth rate suggesting its high CO_2 fixation rate, its culture conditions for fixing CO_2 were optimized in 5L photobioreactor. Parameters examined were temperature, pH, CO_2 concentration, and NaCl concentration. Under the optimized conditions, the specific growth rate increased to 0.23/h from 0.19/h and the CO_2 fixation rate shows more than twofold increase, reaching 1.5gCO_2/l/day (Fig.3).

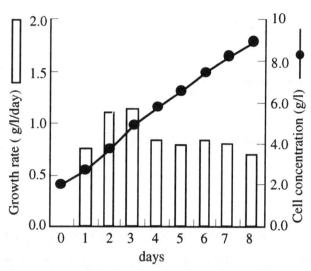

Fig. 1. Growth of *Botryococcus* sp. SI-30

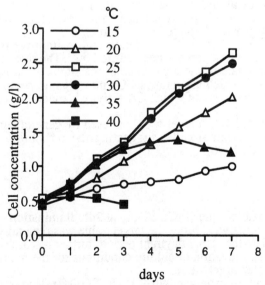

Fig. 2. Effects of temperatures on the
growth of *Botryococcus* sp. SI-30

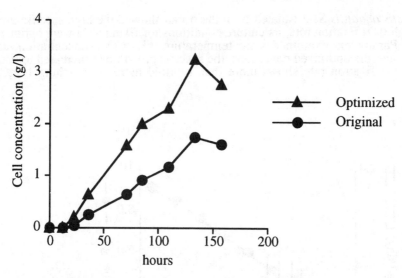

Fig. 3. Effect of optimization on the growth of *S. aquatilis*

Development of Biotechnology for Manipulating the Photosynthetic Microorganisms

The recent progress in molecular and cellular biotechnology enables us to manipulate the genes of microorganisms to alter the metabolic pathways or to give them the ability to produce exogenous products. Although biotechnology for bacteria, plants and mammalian cells has developed rapidly, biotechnology for photosynthetic microorganisms lags considerably behind, probably because of the lack of a driving force to develop technology in the commercial production of algae. Potentiality of biotechnology to improve the ability in CO_2 fixation, productivity of useful materials or tolerance to environmental factors provides an incentive to promote the research on development in this area. Since the physiological and biochemical information on the microorganisms is essential for the development of biotechnology, basic research on physiology and biochemistry has been conducted along with development of applied biotechnology.

In order to assess the rate limiting factors in CO_2 fixation and to improve the photosynthesis, research on carbonic anhydrase (CA) function has been carried out. The role of CA is to catalyze the formation of CO_2 from bicarbonate in the algal cells at the site of ribulose 1,5-biphosphate carboxylase/oxygenase (RuBisCO), which assimilates CO_2 into organic compound. The cDNA coding for CA of *Porphyridium purpureum*, of which CA activity is much higher than other algae, has been cloned and characterized.

Synechococcus sp. PCC7942, of which CA level was increased by genetic engineering, showed higher rate of CO_2 fixation than original cells, suggesting the possibility to improve algal photosynthesis (Murakami,M. *et al.*, 1996).

CONCLUSIONS

The initial target value for CO_2 fixation rate ($1gCO_2/l/day$ at 24h illumination) was achieved as a result of intensive screening of microalgae. Since the target value was then raised to $1gCO_2/l/day$ at 10h illumination so as to simulate the actual daylight period, further research is needed to attain the newly revised target value. Optimization of culture conditions for the microalgae obtained is most important in order to attain the target value.

In addition to screening of the microalgae with high CO_2 fixation rates, screening of microorganisms which produce hydrocarbons, polysaccharides or fine chemicals such as β-carotene are in progress. These organisms will serve not only as the next candidate for mass culture in a bioreactor, but as the sources of genes for producing useful materials by genetic engineering in future.

REFERENCES

Iwamoto,H.(1986). Production of hydrocarbons by microalgae (in *Japanese*), *Bio Science and Industry*, **44**, 1160-1167.

Kurano,N., H.Ikemoto, H.Miyashita, I.Hasegawa, H.Hata and S.Miyachi (1995). Fixation and utilization of carbon dioxide by microalgal photosynthesis. *Energy Convers. Mgmt.*, **36**, 689-692.

Michiki,H. (1995). Biological CO_2 fixation and utilization project. *Energy Convers. Mgmt.*, **36**, 701-705.

Murakami,M., N.Yamaguchi, H.Murakami, T.Nishide, T.Muranaka, F.Yamada and Y.Takimoto (1996). Over-expression of carbonic anhydrase and its localization in carboxysome in cyanobacteria *Synechococcus* sp. PCC7942. *Abstract Paper at the annual meeting of the Society for Fermentation and Bioengineering Japan*. No. 1059.

Murakami,N., A.Sakugawa, K.Nagahama, K.Suzuki, G.Maesato and K.Hisazuka (1996). Isolation and cultivation of *Botryococcus braunii* SI-30. *Abstract Paper at the annual meeting of Japan Society for Bioscience, Biotechnology, and Agrochemistry*. No. 20a7.

Sakai,N., Y.Sakamoto, N.Kishimoto, M.Chihara and I.Karube (1995). *Chlorella* strains from hot springs tolerant to high temperature and high CO_2. *Energy Convers. Mgmt.*, **36**, 693-696.

ACKNOWLEDGEMENT

This work was supported by the New Energy and Industrial Technology Development Organization (NEDO).

Pergamon

Energy Convers. Mgmt Vol. 38, Suppl., pp. S499–S503, 1997
© 1997 Elsevier Science Ltd. All rights reserved
Printed in Great Britain
0196-8904/97 $17.00 + 0.00

PII: S0196-8904(96)00317-2

DEVELOPMENT OF A PHOTOBIOREACTOR INCORPORATING
Chlorella sp. FOR REMOVAL OF CO_2 IN STACK GAS

YOSHITOMO WATANABE and HIROSHI SAIKI

Biotechnology Department, Abiko Research Laboratory,
Central Research Institute of Electric Power Industry, Abiko1646,Abiko city,
Chiba pref., 277,Japan

ABSTRACT

We developed a new design photobioreactor incorporating *Chlorella* sp. for removal of CO_2 in stack gas. Photosynthetic conversion of CO_2 into *Chlorella* biomass was investigated in a photobioreactor, which we termed a cone-shaped helical tubular photobioreactor. The laboratory scale photobioreactor (0.48 m high $\times$ 0.57 m top diameter) was set up with a 0.255 m^2 installation area. The photostage was made from transparent polyvinyl chloride (PVC) tubing (1.6 cm internal diameter with 2 cm wall thickness and 27 m in length). The inner surface of the cone-shaped photostage (0.50 m^2) was illuminated with a metal halide lamp, the energy input into the photostage[photosynthetically active radiation (PAR, 400-700 nm)] was 2127 KJ day^{-1} (12 h day / 12 h night) . The maximum daily photosynthetic efficiency was 5.67% (PAR) under an air-lift operation at a flow rate of 0.3 litre min^{-1} 10 % CO_2 enriched air. Maximum increase of *Chlorella* biomass was 21.5 g dry biomass m^{-2} (installation area) day^{-1} or 0.68 g dry biomass litre $medium^{-1}$ day^{-1}. Also, a helical tubular photobioreactor for outdoor culture was constructed with a 1.1 m^2 installation area (1.2 m top diameter) and photosynthetic productivity was investigated in July,1996. © 1997 Elsevier Science Ltd

KEYWORDS

Global warming;CO_2; CO_2 fixation ; microalgae; *Chlorella* ; photobioreactor ; photosynthesis

INTRODUCTION

Microalgae such as *Chlorella* have high photosynthetic capability; microalgae would be applicable to the direct conversion means of CO_2 in the stack gas into valuable microalgal biomass. The utilization of the microalgal biomass so produced as food or animal feed etc. help ameliorate global warming(Watanabe *et al.*,1992).

In order to establish the CO_2 conversion technology using microalgae we isolated and screened *Chlorella* sp. strain HA-1 ,which was resistant to high concentrations of CO_2 and NOx. The resulting strain was expected to be applicable to treatment of stack gas from thermal power stations(Watanabe *et al*,1992, Yanagi *et al*,1995.).Furthermore we investigated the photosynthetic performance of a new design photobioreactor;the helical tubular photobioreactor(HTP)(Watanabe *et al*,1995 , Watanabe and Hall,1996). The cone-shaped tubular photobioreactor have several advantages over a conventional microalgal culture system and may be especially suitable when the sun's angle is high and direct solar irradiance is overhead.

In this study, we examine (1) photosynthetic performance of a laboratory scale cone-shaped helical tubular photobioreactor(HTP) incorporating *Chlorella* sp. strain HA-1, (2) outdoor culture experiment of a larger scale cone-shaped helical tubular photobioreactor, for the development of the microalgal CO_2 fixation technologies for removal of CO_2 in stack gas from thermal power stations.

MATERIAL AND METHODS

Photobioreactor

A schematic diagram of the laboratory scale cone-shaped HTP system is shown in Fig. 1. The photobioreactor parts comprise; a cone-shaped helical photostage made of transparent polyvinyl chloride(PVC) tubing (27 m in length and 1.6 cm internal diameter with 0.2 cm wall thickness) ; a helical heat exchanger set in a water bath ; degasser ; metal halide lamp(400W) as a light source; air pump, CO_2 gas cylinder and gas flow meter for CO_2 enriched air supply. The photobioreactor was set up with a installation area of 0.255 m² and light absorbing surface(inside surface of cone-shape) was 0.50 m². The outer surface of the photostage was covered with aluminum foil to prevent the absorption of outside light when determining the photosynthetic efficiency.

The inside surface of the cone-shaped photostage was illuminated and the photosynthetic photon flux density(PPFD) at the surface of the photostage was measured using a quantum meter and the input of photosynthetically active radiation(PAR) into the reactor was calculated by using the conversion factor[1].

A schematic diagram of the larger scale cone-shaped helical HTP systems for outdoor culture experiment is shown in Fig. 2. Total length of transparent PVC tubing (1.6 cm internal diameter with 0.2 cm wall thickness) of photostage was 110 m and PVC tubing was divided into two parts at the intermediate level so as to allow an air-lift operation. A helical heat exchanger was set in a water bath and it was also divided into two pieces so as to allow culture medium circulation. The photobioreactor was set up with a installation area of 1.1 m² and the light absorbing surface was 2.2 m². The outer surface of the photostage was covered with aluminum foil and white cloth to prevent the absorption of outside light. The photosynthetic photon flux density at the horizontal level surface was measured automatically at 5 minute intervals and the input of photosynthetically active radiation into the photobioreactor was calculated .

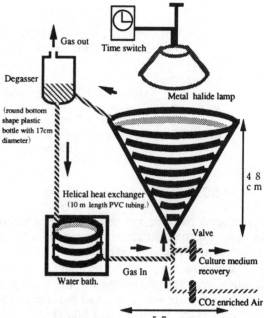

Fig. 1. A schematic diagram of the laboratory scale cone - shaped helical tubular photobioreactor system. Cone-shaped helical photostage is made of 27 m length transparent PVC tubing (1.6 cm inner diameter with 0.2 cm thickness; total outer diameter is 2.0 cm). The photobioreactor is constructed with a 0.255 m² installation area. Photostage is illuminated using metal halide lamp (400W) equipped with a time switch. Air/ CO2 enriched air is prepared by mixing air supplied by an air pump and CO2 from a CO2 cylinder.

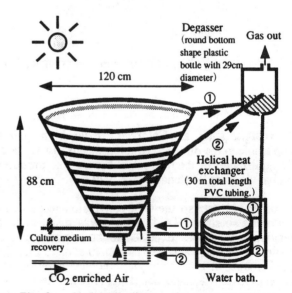

Fig. 2. A schematic diagram of a cone - shaped helical tubular photobioreactor with 1.1 m² installation area for outdoor culture experiment.

Organisms, Culture Medium, Culture Operation and Biomass Assay

Chlorella sp. strain HA-1(Watanabe *et al.*,1992, Yanagi *et al.*,1995) was used for culture experiments. The culture medium(M4N medium) had the following composition: (amounts in mg/litre): KNO_3, 5000; $MgSO_4 \cdot 7H_2O$, 2500; KH_2PO_4, 1250; $FeSO_4 \cdot 7H_2O$, 3;

H_3BO_3, 2.86; $MnSO_4 \cdot 7H_2O$, 2.5; $ZnSO_4 \cdot 7H_2O$, 0.222; $CuSO_4 \cdot 5H_2O$, 0.079; Na_2MoO_4, 0.021. The initial pH value was 6.0.

Batch culture was investigated using the laboratory scale photobioreactor to determine the photosynthetic performance. An air/CO_2(10% CO_2) mixture(CO_2 enriched air) or air/CO_2/NO (10% CO_2 and 100 ppm NO) mixture was injected at a flow rate of 0.3 litre min^{-1}. Optical density at 750 nm was measured and biomass concentrarion was calculated using a standard curve between biomass concentration(dry weight) and optical density. Also batch culture was investigated using photobioreactor for outdoor culture experiment to determine the productivity under field conditions. An air/CO_2 (10% CO_2)mixture was injected at a flow rate of 1.2 litre min^{-1}.

RESULTS AND DISCUSSION

Photosynthetic performance in the laboratory scale photobioreactor

Photosynthetic performance during batch culture is shown in Fig.3. The results showed a higher growth rate up to the 5th day when the biomass concentration reached approximately 2 or 2.5 g per litre. The evaluation of photosynthesis is summarized in Table 1. The photosynthetic performance data indicated that the maximum daily growth rate was 5.48 g per reactor (0.60 g dry biomass per litre) corresponding to a photosynthetic efficiency(P.E.) of 5.67 %(PAR). The productivity of *Chlorella* sp.strain HA-1 of this photobioreactor was 21.5 g dry biomass per m^2 (installation area) per day.

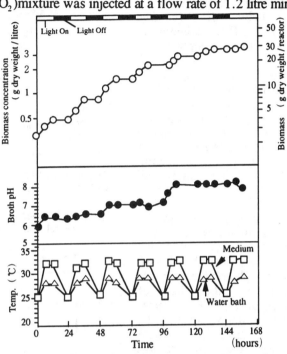

Fig.3. Changes in dry weight biomass of *Chlorella* sp.strain HA-1, culture medium pH, water bath and culture medium temperature in the laboratory scale cone-shaped helical tubular photobioreactor

Table 1   Maximum growth yield and photosynthetic efficiency of *Chlorella* sp. strain HA-1 in the laboratory scale cone-shaped helical tubular photobioreactor [a].

Maximum daily growth. (g dry biomass/ reactor / day)	Carbon content of biomass (%)	Enthalpy of dry biomass[b] (kJ/ g D.W.)	Maximum daily energy recovery (kJ / reactor / day)	Input of photosynthetically active radiation(PAR) into reactor (kJ / reactor / day)	Photosynthetic efficiency (%) (PAR)
5.48	46.1	22.0	120.6	2127	5.67

[a] Calculated from Results of Fig.2. [b] Converted from carbon contents (11.4 Kcal ≒ calorific value for 1 g-C of microalgal biomass) (Platt and Irwin,1973)

Fig. 4 shows the carbon and NO balance in the laboratory scale photobioreactor. The light energy input was estimated as the average solar irradiance of Japan from March to October. Assuming that P.E.was 5.67 %(PAR), CO_2 utilization efficiency was 21.9% and NO removal rate was 85%(detailed data not shown). These results indicate that this photobioreactor showed good photosynthetic performance.

Outdoor culture experiment

Results of the outdoor culture experiment, started on 4 th July, 1996, using the photobioreactor described in Fig.2 is shown in Fig. 5. The results show a steady growth during batch culture. The maximum growth was observed from 68 hr.(3 rd day night) to 85 hr(4 th day moon) . The growth rate from the night of 6 th July to 1 PM of 7 th July was 8.81 g dry weight /

photobioreactor and it corresponded to a P.E. of 3.04 %(PAR).

Several researchers reported the microalgal productivities using several culture systems as follows;productivities of the panel type reactors in Summer in Italy ranged from 18 to 24 g m^{-2} day^{-1} (Tredici and Materassi,1992) ; productivity of *Spirulina platensis* SP-G cultured in outdoor ponds equipped with a paddle wheel was up to 21 g m^{-2} day^{-1} in Israel (Vonshak and Guy ,1992). daily gross productivities in open pond outdoor cultures with the marine chlorophyte *Tetraselmis suecica* were 15 to 20 g m^{-2} day^{-1} under full sunlight conditions in Hawaii (Laws and Berning, 1991).

The results achieved in the outdoor culture experiment in this study show that the productivity of the helical tubular photobioreactor is similar to the that of conventional culture systems. However, improvement of the photobioreactor would increase the productivity. The photobioreactor for outdoor culture experiment in this study had a distortion or slackening of tubing, therefore it had many gas pockets and light passed through the tubing. This phenomenon was a significant reason for the insufficient photosynthetic performance. We are improving the photobioreactor, by investigating the double layer tubular photobioreactor so as to absorb the solar irradiance effectively or by optimizing the photobioreactor size to allow good circulation of the culture medium(preventing a gas pocket accumulation), as well as accumulating photosynthetic performance data under outdoor field condition using the present photobioreactor system.

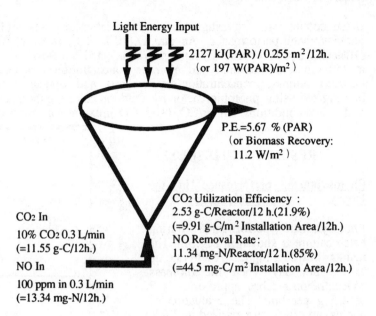

Fig. 4. Carbon and NO balances in the laboratory scale cone- shaped helical tubular photobioreactor.

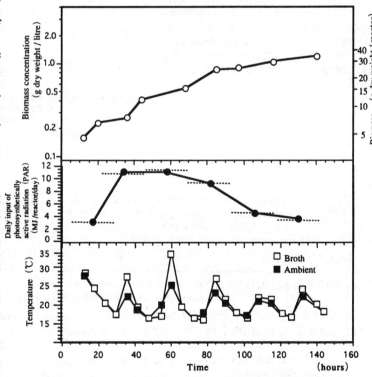

Fig.5 Changes in dry weight biomass of *Chlorella* sp.strain HA-1, daily photosynthetically active radiation (PAR) input, ambient and culture broth temperature in the cone-shape helical tubular bioreactor outdoor culture experiment. Broken lines indicate daily input of PAR in term of PPFD measurement. Day of start was 4th July,1996.

REFERENCES

Hall,D.O.and J.M.O.Scurlock (1993). Biomass production and data. In: *Photosynthesis and production in a changing environment* (D.O.Hall , J.M.O. Scurlock , H.R.Bolhar-Nordenkampf , R.C.Leegood and S.P.Long,ed.), pp.425-444.Chapman & Hall,London.

Laws, E.A. and J.L. Berning (1991). A study of the energetics and economics of microalgal mass culture with the marine chlorophyte *Tetraselmis suecica* : Implications for use of power plant stack gases. *Biotechnol. Bioeng.*, 37, 936-947.

Platt,T. and B.Irwin(1973).Coloric content of phytoplankton. *Limnol.Oceanog.*, 18, 306~310.

Tredici, M.R.and R. Materassi (1992) .From open ponds to vertical alveolar panels: the Italian experience in the development of reactors for the mass cultivation of phototrophic microorganisms. *J.Appl.Phycol.*, 4, 221-231

Vonshak ,A. and R. Guy (1992). Photoadaptation, photoinhibition and productivity in the blue-green alga,*Spirulina platensis* grown outdoors. *Plant,Cell and Environ*.,15,613-616.

Watanabe,Y.,Ohmura,N. and H.Saiki(1992). Isolation and determination of cultural characteristics of microalgae which function under CO_2 enriched atmosphere. *Energy Convers.Mgmt.*, 33, 545-552.

Watanabe,Y., de la Noüe,J. and D.O.Hall(1995). Photosynthetic performance of a helical tubular photobioreactor incorporating the cyanobacterium *Spirulina platensis. Biotechnol. Bioeng.*, 47, 261-269.

Watanabe,Y. and D.O.Hall(1996).Photosynthetic production of the filamentous cyanobacterium *Spirulina platensis* in a cone-shaped helical tubular photobioreactor, *Appl.Microbiol.Biotechnol.*,44, 693-698.

Yanagi, M.,Watanabe,Y.and H.Saiki(1995). CO_2 fixation by *Chlorella* sp. HA-1 and its utilization.*Energy Convers.Mgmt.*, 36,713-716.

Pergamon

Energy Convers. Mgmt Vol. 38, Suppl., pp. S505–S510, 1997
© 1997 Elsevier Science Ltd. All rights reserved
Printed in Great Britain
0196-8904/97 $17.00 + 0.00

PII: S0196-8904(96)00318-4

POWER PLANT FLUE GAS AS A SOURCE OF CO₂ FOR MICROALGAE CULTIVATION: ECONOMIC IMPACT OF DIFFERENT PROCESS OPTIONS

KIRAN L. KADAM

Biotechnology Center for Fuels and Chemicals
National Renewable Energy Laboratory,
1617 Cole Boulevard, Golden, CO 80401, USA

ABSTRACT

As CO_2 plays a central role in the economics of microalgae cultivation, an accurate estimate of its cost is essential. Toward this end, an economic model was developed for CO_2 recovery from power-plant flue gas and its delivery to microalgae ponds. A design basis was devised for recovering CO_2 from flue gas emitted by a typical 500 MW power plant located in the Southwestern United States. For the standard process, which included monoethanolamine (MEA) extraction, compression, dehydration, and transportation to the ponds, a delivered CO_2 cost of $40/t was estimated. The model was also used to evaluate the efficacy of directly using the flue gas, however, this option was found to be more expensive. The economics of microalgae cultivation using power-plant flue gas can be evaluated by integrating this model for CO_2 recovery with a previously developed model for microalgae cultivation. The model predictions for a long-term process are: a lipid cost of $1.4/gal (unextracted) and a mitigation cost of $30/t CO_2 (CO_2 avoided basis). These costs are economically attractive and demonstrate the promise of microalgal technology.   © 1997 Elsevier Science Ltd

KEYWORDS

microalgae cultivation, power plant, flue gas, lipid production, CO_2 mitigation, monoethanolamine (MEA) process, economic analysis

INTRODUCTION

The United States (US) generates about 4.8 Gt (G ≡ 10^9) CO_2, per year amounting to about 22% of the worldwide anthropogenic emissions (Herzog *et al.*, 1994). Electrical power plants are responsible for more than one-third of the US emissions, and for about 7% of the world's CO_2 emissions from energy use. Physical capture of CO_2 from fossil-fuel power-plants, which are stationary, concentrated sources of CO_2, has been regarded as a possible mitigation option since Marchetti (1977) first suggested disposing the captured CO_2 in deep ocean. Several investigators have since investigated a variety of options for CO_2 capture from power plants and its subsequent disposal or use (Herzog *et al.*, 1993). Thus, CO_2 capture is a common step to most of the remediation options, and getting a precise handle on its cost is essential.

Because CO_2 cost plays a pivotal role in the economics of microalgae cultivation, developing an economic model for its capture from flue gas is important to rigorously derive a CO_2 resource cost to be charged to the process. The model reported addresses CO_2 extraction using a typical monoethanolamine (MEA) process and delivery to the ponds; this allows us to precisely determine the delivered CO_2 cost to the process. The previously derived economic model for microalgae cultivation (Kadam and Sheehan, 1996) and this model for CO_2 capture from flue gas can be integrated to develop realistic costs for lipid production and CO_2 mitigation, and to gain an insight into the overall process.

MEA PROCESS

To date, the MEA process has been the technology of choice in the recovery of CO_2 from low-pressure flue gases (Maddox, 1977; Kohl and Riesenfeld, 1979). A detailed flow diagram for the MEA process is shown in Figure 1 (adapted from Maddox, 1977; Anada *et al.*, 1982). The process flow scheme changes little, irrespective of the aqueous amine solution used as the absorptive agent. The primary pieces of equipment are the absorber and stripper columns, together with the associated piping, heat exchange, and separation equipment. The MEA process was chosen for CO_2 capture and was modeled as described below.

CO_2 COST ESTIMATION

Design Basis

The key design and calculated performance parameters are given in Table 1. A 500-MW capacity was assumed for the standard recovery process. The power plant was assumed to be located in the Southwestern US. Based on typical Southwestern plants, flue gas emission and CO_2 concentration of 2.25 million SCFD/MW (standard cubic feet per day/megawatt) and 14% by vol., respectively, were assumed. The SO_2 and NO_x levels are defined as those in the 1990 CAAA (US Clean Air Act Amendments), i.e., 200 and 150 ppm, respectively. Based on these conditions, the total CO_2 recovered was 2,830,300 t/yr for a CO_2 recovery of 88.6%. A lower plant capacity of 50 MW, which had a correspondingly lower CO_2 production capacity, was also analyzed.

The standard process includes MEA extraction, compression to 1500 psi, dehydration, and transportation to the ponds, and delivers a gas that is ~100% CO_2. Dehydration of the gas to water concentrations of ≤ 50 ppm by volume is necessary to prevent corrosion of the pipeline by carbonic acid. In order to assess whether the flue gas could be directly utilized, a case was devised that omitted the extraction step and only included compression, dehydration, and transportation to the ponds. As this direct pumping option does not further process the flue gas, it delivers the gas to the ponds at a CO_2 concentration of only 14%.

A transportation distance of 100 km was assumed. The pond system would likely not be adjacent to the power plant, and a 100-km (ca. 60-mile) distance is a judicious choice from the standpoint of reasonable transportation costs and increased land availability.

Basis for Capital and Operating Costs

Table 2 lists the major equipment as shown in Figure 1, and Table 3 shows items that contribute to the operating costs. As the direct pumping option does not include MEA extraction, Table 2 does not apply. However, because this option has to handle a sevenfold higher volume of the gas for a given CO_2 input to the algae ponds, the size of compressors and pipeline increase, and electricity consumption rises substantially. A higher gas-injection capacity at the ponds is also needed; however, this was not factored into the calculations.

The Questimate® software (ICARUS Corp., 11300 Rockville Pike, Rockville, Maryland) was used for estimating capital costs. For major pieces of equipment such as compressors and blowers, vendor quotations were solicited, and they compared reasonably well with the software estimates. For consistency, all model calculations were done using capital costs derived from Questimate.®

Table 1. Design Basis and Plant Performance: 500 MW Plant, Standard MEA Process

Parameter	Value
Input to the Model	
Power plant type	Coal-fired
Location	SW US
Rated capacity, MW	500
Coal consumption, t/d	5,556
FGD (Flue Gas Desulfurization)	Yes
Atmospheric pressure, psia	12.2
Flue gas emission, MM SCFD/MW	2.25
Flue gas flow rate, SCFM sm^3/min	781,250 22,503
CO_2 concentration, vol. %	14.0
Compression pressure, psig	1,500
Pipeline distance, km	100
Calculated Performance Parameters	
Total CO_2 recovered, t/yr	2,830,311
CO_2 recovery, %	88.6
Equivalent microalgae facility size for algal productivity of 45 g/m^{2}d, ha	13,862

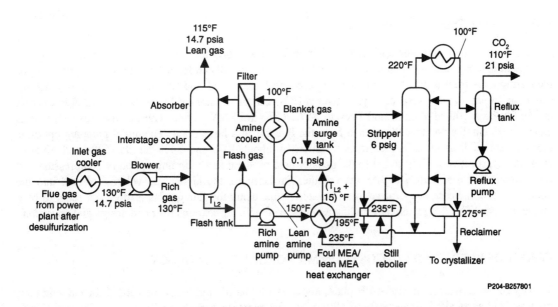

Fig. 1. Detailed flow diagram of the MEA process.

Table 2. Costs Centers: MEA Extraction

Capital Costs			Operating Costs
Absorber column	Lean amine cooler	Rich amine pump	MEA makeup
Stripper column	Inlet gas cooler	Lean amine pump	Na_2CO_3, anhydrous
Stripper reboiler	Interstage cooler	Reflux pump	Cooling tower water
Stripper reclaimer	Byproduct gas blower	Crystallizer	Coal for steam gen.
Stripper condenser	Reflux tank	Boiler	Electricity
Amine-amine heat exchanger	Surge tank	Cooling tower	Labor
Flash tank	Amine filter		

Table 3. Cost Centers: Compression, Dehydration and Transportation

Capital Costs		Operating Costs
Inlet Gas Cooler	Compressor 2 inter-cooler	Cooling tower water
Compressor 1: high capacity, centrifugal	Dryer	Electricity
Compressor 2: high pressure, reciprocal	Pipeline	Labor
Compressor 1 inter-cooler		

Pipeline costs were estimated by updating the data of Hare *et al.* (1978) to 1995 by using an average of *Chemical Engineering* and Marshall-Swift indices; these indices were used since the pipeline cost index is no longer published by *Oil & Gas Journal*. Another approach for estimating pipeline costs is to use *Oil & Gas Journal*'s annual Pipeline Economics Report (True, 1994). Both methods gave similar numbers for 1995 pipeline costs; the *Oil & Gas Journal* costs were used in model calculations. Coal costs were obtained from the following US Department of Energy's publications: "Coal Industry Annual 1993," and "Cost and Quality of Fuels for Electric Utility Plants 1993" (Energy Information Administration, 1994a; 1994b). Materials costs were obtained from *Chemical Marketing Reporter*.

Comparison of Processing Options

The MEA extraction, compression and dehydration, and transportation sections were analyzed for capital and operating costs, and an annualized cost in $/t CO_2 was calculated for each case. Cost summaries for recovering CO_2 from a 500 MW plant using the MEA extraction and the direct pumping options are shown in Tables 4 and 5, respectively. Delivered CO_2 cost for the standard process with MEA extraction is $40.5/t for a 500-MW plant and rises to $57.1/t for a 50-MW plant. The 50-MW case represents either a very small plant or a slip stream from the flue gas of a large plant. The direct pumping option is more expensive than the MEA extraction option by about 40% and 54% for the 500- and 50-MW capacities, respectively. This is due to the penalty paid by the direct pumping option in compression and transportation of a sevenfold higher gas volume. Furthermore, the higher gas volume injected into the ponds can also increase evaporative loss of water. Thus, using the flue gas 'as is' does not seem to be a viable strategy. This is a significant finding and has profound implications for pond design and operation.

EFFECT OF PREDICTED CO_2 COSTS ON MICROAGAL TECHNOLOGY

To trap all the CO_2 from the 500-MW plant, about 14,000 ha of land would be needed for the long-term process that has an algal productivity of 45 $g/m^2/d$. The microalgae model of Kadam and Sheehan (1996) for a 1000-ha pond system was used for studying the effect of the CO_2 resource costs on overall process economics. Scale-up of the microalgal system is assumed to be on a modular basis size as ponds larger than 20 ha are considered to be unwieldy, if not impossible. A large-scale microalgal facility would

Table 4. Cost Summary for CO_2 Recovery from 500 MW Plant: Standard Process with MEA Extraction

	Capital Costs, $	Operating Costs, $	Annualized Costs, $/t CO_2	Annualized Costs, % of Total Cost
MEA extraction	112,785,600	60,155,900	28.72	70.9
Compression and drying	52,359,500	13,564,500	8.48	20.9
Transportation	44,508,300	445,100	3.30	8.2
Total	209,653,400	74,165,500	40.50	100.0

Table 5. Cost Summary for CO_2 Recovery from 500 MW Plant: Direct Pumping of Flue Gas

	Capital Costs, $	Operating Costs, $	Annualized Costs, $/t CO_2	Annualized Costs, % of Total Cost
Compression and drying	287,928,100	91,423,800	46.64	81.5
Transportation	142,564,700	1,425,600	10.58	18.5
Total	430,492,800	92,849,400	57.21	100.0

Table 6. Economic Performance of Microalgal Processes with Different Maturities

	Mid-Term Process	Long-Term Process
Cell concentration in pond , g/L	1.0	1.2
Algal lipid content, % wt	45	50
Residence time, d	5.5	4
Operating season, d/yr	275	300
Algal productivity, g/m^2/d	27.3	45
Photosynthetic efficiency, %	3.6	6.1
Algae cost, $/t	282.5	209.5
Lipid cost, $/bbl, $/gal (unextracted)	87.7 / 2.09	58.6 / 1.40
CO_2 cost, % of annualized cost	26.6	37.9
CO_2 mitigation cost[1], $/t CO2		
as is basis	63.8	20.0
CO_2 avoided basis[2]	95.7	30.0

[1]Based on credit at the following rate: lipid = $240/t, protein = $120/t, carbohydrate = $120/t.
[2]CO_2 avoided basis cost = 1.5 x (as is basis cost). Based on data of Herzog (1994).

consist of several 20-ha ponds, and hence, the economies of scale would not apply in a significant way. Furthermore, these estimates are conceptual in nature. Therefore, the above model can be assumed to reasonably predict the process economics of a 14,000-ha system. If anything, the model predictions would be on the conservative side.

Table 6 shows a comparison of two cases that can be envisioned: a mid-term and a long-term process that exemplify the potential of this technology. Microalgal technology can be looked upon as an avenue for producing a lipid feedstock for biodiesel production; alternatively, it can be an option for CO_2 mitigation. The model was used to predict process performance from both of these angles. Using a CO_2 resource cost of $40.5/t, the mid-term process yields a CO_2 mitigation cost of $96/t ($CO_2$ avoided basis), which falls in the range of cost estimates reported by Herzog (1994) for the capture and ocean-disposal option. The lipid cost for the mid-term process is also comparable to the current crude soybean oil prices of $2.25 – $3.25/gal; soybean oil is currently the preferred feedstock used in biodiesel production. The long-term process yields a lipid cost of $1.4/gal (unextracted) and a CO_2 mitigation cost of $30/t ($CO_2$ avoided basis), both of the costs being highly competitive. Thus, the microalgal technology is a promising technology for both lipid feedstock production and CO_2 mitigation.

CONCLUSIONS

The economic model developed allows us to rigorously derive CO_2 recovery costs and make decisions regarding processing options. Pumping the flue gas directly to the ponds does not seem to be a viable alternative as it is 40% more expensive than the MEA extraction option. This finding is important since it has a significant effect on pond design and operation. Integrating the models for microalgae cultivation and CO_2 capture from flue gas allows us to predict overall process performance. The predicted lipid and CO_2 mitigation costs are economically viable, given that the defined performance targets are met and the long-term process comes to fruition. Thus, microalgal cultivation is a potentially useful technology.

ACKNOWLEDGMENTS

This work was supported by the following projects: Biological Trapping of Carbon Dioxide, funded by DOE/PETC (Dept. of Energy, Pittsburgh Energy Technology Center, Pittsburgh, Pennsylvania; and Bioutilization of Coal Combustion Gases, funded by DOE/FE (Dept. of Energy, Fossil Energy section, Germantown, Maryland).

REFERENCES

Anada, H., D. King, A. Seskus, M. Fraser, J. Sears and R. Watts. (1982). *Feasibility and Economics of By-product CO_2 Supply for Enhanced Oil Recovery*, Final Report: DOE/MC/08333-3 (DE82004814).

Energy Information Administration. (1994a). US DOE (Dept. of Energy) publication: *Cost and Quality of Fuels for Electric Utility Plants 1993*, Washington, DC.

Energy Information Administration. (1994b). US DOE (Dept. of Energy) publication: *Coal Industry Annual 1993*, Washington, DC.

Hare, M., H. Perlich, R. Robinson, M. Shah and F. Zimmerman (1978). *Sources and Recovery of Carbon Dioxide for Enhanced Oil Recovery*, Final Report: FE-2515-24.

Herzog, H., E. Drake and J. Tester (1993). *A Research Needs Assessment for the Capture, Utilization and Disposal of Carbon Dioxide from Fossil Fuel-Fired Power Plants*, Report for DOE (Dept. of Energy) Grant No. DE-FG02-92ER30194.A000.

Herzog, H. (1994). CO_2 mitigation strategies: How realistic is the capture and sequestration option? Paper no. 94-RA113.02, *Proc. 87th Air & Waste Manage. Assoc. Ann. Mtg. & Exhibi.*, Cincinnati, Ohio.

Kadam, K.L. and J.J. Sheehan. (1996). Microalgal technology for remediation of CO_2 from power plant flue gas: a technoeconomic perspective. *World Resource Review*, 8(4), 493-504.

Kohl, A.L. and F.C. Riesenfeld (1979). *Gas Purification*, 3rd ed., Gulf Publishing Co., Houston.

Maddox, R. N. (1977). *Gas and Liquid Sweetening*, John M. Campbell (Campbell Petroleum Series), Norman, Oklahoma.

Marchetti, C. (1977). On geoengineering and the CO_2 problem. *Climatic Change*, 1, 59-68.

True, W.R. (1994). Pipeline economics report. *Oil & Gas J.*, 92(47), 41-58.

Energy Convers. Mgmt Vol. 38, Suppl., pp. S511–S516, 1997
© 1997 Elsevier Science Ltd. All rights reserved
Printed in Great Britain
0196-8904/97 $17.00 + 0.00

PII: S0196-8904(96)00319-6

OPTIMIZATION OF THE FERTILIZATION BY NUTRIENTS OF THE OCEAN, TAKING FIXATION BY PHYTOPLANKTON INTO ACCOUNT

KUNIO HORIUCHI, TOSHINORI KOJIMA
Dept. of Ind. Chem., Seikei Univ., Musashino Tokyo 180 Japan
and
ATSUSHI INABA
National Institute for Resources and Environment, Onogawa Tsukuba 305 Japan

ABSTRACT

Fertilization of the ocean with nutrients such as nitrate and phosphate is one of the countermeasures against the global warming problem. In the present paper, we evaluated the effective ratio of the nutrients sprinkling into the ocean to the nutrients assimilation by phytoplankton. When the nutrients equivalent to the 1/500 of the annual accumulated amount of CO_2 in the atmosphere were sprinkled within the circle of 7.1 - 100 km radius by a ship, less than 0.01 % of the nutrient was found to be transferred to the deep ocean without assimilation. The rest remained in surface ocean, most of which was found to be incorporated in the phytoplankton within one year. We also evaluated the energy balance of transportation of the fertilizers by ship to that taken up by the fertilization in ocean. The energy evaluation indicated that the amount of CO_2 produced from the transportation process was about 0.23 % of the amount that is expected to be taken up into the ocean from the atmosphere. Furthermore, we estimated the energy balance of fertilization with activated sludge instead of fertilizer. It was found that fertilization by activated sludge containing a great deal of water needed a large amount of energy for the transportation by ship. © 1997 Elsevier Science Ltd

KEYWORDS

Environment; Global warming; Carbon dioxide; Ocean; Fertilization

INTRODUCTION

The rise in atmospheric CO_2 concentration causes global warming which is one of most important global environmental issues. The ocean plays an important role in the global carbon cycle and the amount of carbon included in the ocean is several tens times as much as that in the atmosphere. Furthermore the inorganic carbon concentration in the deep ocean is not in equilibrium with the atmospheric CO_2 partial pressure. Thus the ocean is expected to be a large sink of the CO_2 in the atmosphere from flue gas emitted by the thermal power generation and so on.

The fertilization of the ocean with nutrients such as nitrate and phosphate has been considered as one of the measures using ocean. The surface ocean, excluding the upwelling area, is known to be poor in the nutrients which are rapidly consumed by phytoplankton. It suggests that the fertilization of the ocean can offset the continuing increase in the atmospheric CO_2 concentration. The technique using fertilizer or activated sludge containing nutrients is expected to promote the propagation of phytoplankton and to assimilate inorganic carbon such as HCO_3^- and CO_3^{2-}. Consequently this situation leads to the decrease in the surface-ocean partial pressure of CO_2, drawing down CO_2 from the atmosphere.

In the previous paper (Horiuchi et al., 1995), we calculated the desired amount of nutrients to offset the increase in the atmospheric CO_2 concentration, based on the PNC ratio (Redfield ratio) of the phytoplankton in the ocean. It was found using the unsteady state model modifying the conventional two box model, in which the circulation of carbon and nutrients in ocean and the sedimentation of carbonaceous particle into the deep ocean were taken into account that atmospheric CO_2 could be absorbed quickly, within a few years, by the fertilization. From the estimation of energy requirements it was indicated that the ratios of amount of CO_2 emitted from the production and from the transportation of fertilizer to that absorbed into the ocean were about 9 % and negligible small, respectively. However, the assumption in the two box model differed from the practical fertilization of ocean with nutrients because all nutrients sprinkled in the surface ocean were assumed to be consumed rapidly by the phytoplankton. The practical way of fertilization of ocean is considered to be that high concentration of nutrients are sprinkled on a limited area of surface ocean in order to minimize the CO_2 emission from the energy requirement for the transportation of the nutrients by ship. The nutrients thus sprinkled diffuse in the surface of sea with assimilation by the phytoplankton.

In the present study, in order to optimize the fertilization by high concentrations of nutrients to a limited area of surface ocean, we first developed a two box model with horizontal diffusion of the organic carbon and the nutrient containing the nitrogen considering the nutrient assimilation by the phytoplankton, the sedimentation of carbonaceous particle into the deep ocean and the exchange between the surface and the deep oceans. Secondly, considering the fraction of the nutrients not consumed by the phytoplankton, we estimated that the ratio of CO_2 emitted from the transportation by ship to that taken into the ocean from the atmosphere. Finally, we also estimated the energy balance of fertilization with activated sludge instead of fertilizer.

ASSUMPTIONS ON FERTILIZATION TO OCEAN AND MODEL

Outline of Fertilization to Ocean

In this paper, it was assumed that high concentration of nutrients containing nitrogen and phosphorus are supplied into the a limited area of ocean on the ratio of 1:15:120 of P:N:C, the Redfield ratio of phytoplankton. Only nitrogen was considered in our estimations of assimilation rate and diffusion because the amount of nitrogen observed in the surface ocean is 0.94 time as much as that of nitrogen required in the photosynthesis on the basis of phosphorus (Handa, 1992; Redfield *et al.*, 1963). As far as the ratio of nitrogen and phosphorus sprinkled to the ocean was to be kept on 1:15, it can be assumed that the behavior of phosphorus is easily predicted from the results of nitrogen, because the behavior of phosphorus is the same as that of nitrogen. The nutrients sprinkled to the surface ocean will be assimilated by the phytoplankton or transfer into the deep ocean due to the exchange between surface and deep ocean. The organic carbon produced by phytoplankton in the surface ocean transfers into the deep ocean due to the exchange between surface and deep ocean, and by the sedimentation of particle matter.

It is known that the annual increment of carbon in the atmospheric CO_2 is 3.5×10^{15} $g \cdot y^{-1}$ in 1985. The whole surface area (Sur) is 3.60×10^8 km^2 (National Astronomical Observatory, 1992) and the average depth of the ocean mixed layer (Ls) was assumed to be 75 m (Bolin *et al.*, 1981). It was assumed that the nutrients equivalent to the 1/500 of the annual accumulated amount of CO_2 in the atmosphere was sprinkled within the circle of 7.1 - 100 km radius in the surface ocean. Thus, the nitrogen concentration in the surface ocean ranged from 6.6×10^{-3} $mol \cdot m^{-3}$ to 3.3×10^{-2} $mol \cdot m^{-3}$ under the assumption that the nutrients diffused instantaneously in the limited area.

Model Description

This model is composed of the two boxes of horizontal diffusively mixed layers representing the surface ocean and the deep ocean which were divided. The depth of surface ocean (Ls) and deep ocean (Ld) were parameterized with the same values of conventional two box model (Bolin *et al.*, 1981). It was assumed that the carbon and nitrogen in surface and deep ocean are kept at the steady state without fertilization. The initial carbon and nitrogen concentrations were determined to be suitable from the observational values (Broecker, 1981; Sugimura and Suzuki, 1988). The transfer of carbon and nitrogen between surface ocean and deep ocean was assumed to occur the mutual exchange and the sedimentation of carbonaceous particles from the surface ocean into the deep ocean. The values of the exchange rates of carbon and nitrogen between the surface and the deep ocean (kmd) were determined by the observation results of the distribution of $^{14}C/^{12}C$ ratio in the reservoirs (Bolin *et al.*, 1981). The sedimentation rate of carbonaceous particles from the surface ocean into the deep ocean (Usp) was determined to be 0.69 $m \cdot y^{-1}$ by the steady state mass balance in the two box model taking the biological uptake into account using the ratio of particle to dissolved organic matter of 0.1 (Inaba *et al.*, 1990). The emission rates of nitrogen and inorganic carbon from the organic matter in the deep ocean (Kn, Kc) were also given by the initial steady state concentrations of the organic carbon and the nitrogen. The values of diffusion coefficients in the surface and deep ocean (Ds, Dd) were determined by the observation results of the distribution of radioactive decay of the elements in each reservoirs (Broecker, 1981). These diffusion coefficients were overall values considering various effects, e.g., current and wind. Table 1 shows the initial concentrations of organic carbon and nitrogen in both reservoirs and the parameters at the steady state without fertilization.

Table 1. Initial conditions for model

Ccs_0=0.330 [mol•m^{-3}] Cns_0=7.07x10^{-6} [mol•m^{-3}] Ccd_0=0.0550 [mol•m^{-3}] Cnd_0=3.61x10^{-2} [mol•m^{-3}]

Ls=75 [m] Ld=3125 [m] Ds=1.00x10^{10} [m^2•y^{-1}] Dd=1.00x10^4 [m^2•y^{-1}]

Vx=2.36x10^{-2} [y^{-1}] Hm=1.13x10^{-4} [mol•m^{-3}]

kmd={Ls•Vx•Cns$_0$/(Cns$_0$+Hm)}/(Cns$_0$-Cnd$_0$) [mol•m^{-2}•y^{-1}]

Usp= {8•Ls•Vx•Cns$_0$/(Cns$_0$+Hm)+kmd•(Ccs$_0$-Ccd$_0$)}/Ccs$_0$ [mol•m^{-2}•y^{-1}]

Kc=Usp•Ccs$_0$-kmd•(Ccs$_0$-Ccd$_0$) [mol•m^{-2}•y^{-1}] Kn=kmd•(Cnd$_0$-Cns$_0$) [mol•m^{-2}•y^{-1}]

C, concentration: L, Depth: D, diffusion: Vx, maximum specific growth rate: Hm, saturation constant: Usp, sedimentation rate: kmd, exchange rate between the surface and the deep ocean: K, emission rates: Subscripts; c, organic carbon: n, nitrogen: s, surface ocean: d, deep ocean: 0, steady state

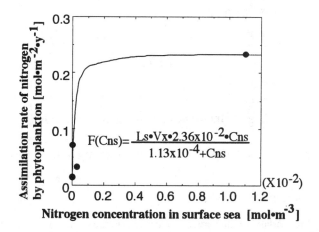

Fig. 1 Determination of constants of nitrogen assimilation rate by phytoplankton

The nitrogen assimilation rate by phytoplankton (F(Cns)) was given by Monod type equation, which is shown in (1).

$$F(Cns)=Ls•Vx•Cns/(Cns+Hm) \tag{1}$$

The constants in Eq. (1) were determined by the observed values in the surface ocean excluding the upwelling area (Platt and Horrison, 1985, Honda, 1992) and in the upwelling area (Smith *et al.*, 1991). The results are shown in Fig. 1. The saturation constant (Hm) and maximum specific growth rate of phytoplankton (Vx) were given as 2.36x10^{-2} y^{-1} and 1.13x10^{-4} mol•m^{-3} respectively. The partial differential equations of the organic carbon and the nitrogen balance, used in our model, are shown as follows.

$$Ls•∂Ccs/∂t=Ls•Ds•(∂^2Ccs/∂r^2+1/r•∂Ccs/∂r)+8•F(Cns)-Usp•Ccs-kmd•(Ccs-Ccd) \tag{2}$$
$$Ls•∂Cns/∂t=Ls•Ds•(∂^2Cns/∂r^2+1/r•∂Cns/∂r)-F(Cns)-kmd•(Cns-Cnd) \tag{3}$$
$$Ld•∂Ccd/∂t=Ld•Dd•(∂^2Ccd/∂r^2+1/r•∂Ccd/∂r)+kmd•(Ccs-Ccd)+Usp•Ccs-Kc \tag{4}$$
$$Ld•∂Cnd/∂t=Ld•Dd•(∂^2Cnd/∂r^2+1/r•∂Cnd/∂r)+kmd•(Cns-Cnd)+Kn \tag{5}$$

CALCULATION RESULTS

Organic carbon and nitrogen transfer from the surface to the deep ocean

The transferred amounts of the organic carbon and the nitrogen from the surface ocean to the deep ocean under our assumption on the fertilization to the ocean were shown in Fig. 2. The transferred amounts of organic carbon due to the exchange of the organic carbon, the sedimentation of the organic carbon were CKmd and CUsp respectively, and NKmd is that of nitrogen due to the exchange. In Fig. 2, NKmd is multiplied by 8 according to Redfield ratio to compare directly with CKmd and CUsp. The calculation results showed that CUsp is negligible because the fraction of carbon included in organic particle to the total organic carbon is only 1/10 in the surface ocean and the rate of sedimentation is smaller than that of the exchange. Furthermore, it was found that the total transferred amounts of nitrogen from the surface ocean to deep ocean was only 0.01 % of the sprinkled amount to ocean for the first year, two third of which was incorporated in the organic matter. The rest of 99.99 % remained in the surface ocean.

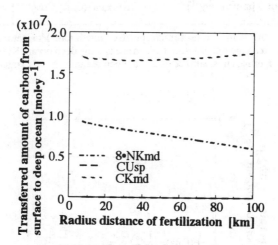

Fig. 2   Transferred amount of organic carbon and nitrogen from surface ocean to deep ocean

Carbon and nitrogen concentrations in the surface ocean

The time variations of the carbon and nitrogen concentrations in the surface ocean for 10 km radius sprinkling were shown in Fig. 3. The calculation results showed that the profile of nitrate concentration in the surface ocean returned to the initial condition before fertilization, while that of organic carbon concentration was not back to the initial value, within one year. This indicates that most of the nitrogen in the surface ocean is incorporated in the organic matter within one year. The high concentration area of the organic carbon became wider than that of the nitrogen because nitrate sprinkled to the ocean was quickly diffused and assimilated by the phytoplankton in the surface ocean.

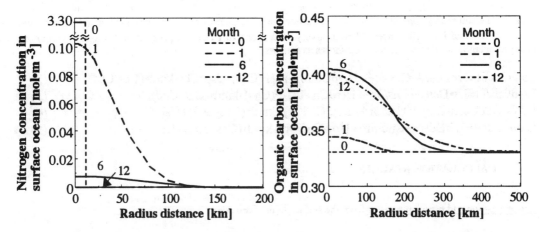

Fig. 3   Horizontal concentration profiles for nitrogen and organic carbon in surface ocean

ENERGY BALANCE OF FERTILIZATION TO OCEAN WITH FERTILIZER

From the viewpoint of the energy balance, the ratio of CO_2 emitted from the transportation by ship to that taken into the ocean from the atmosphere was calculated in order to judge the impact of ocean fertilization on the CO_2 problem. The amount of fertilizer equivalent to 0.2 % of annual increase of the atmospheric CO_2 was assumed to be transported to 500 km offshore by ship and sprinkled in a limited area. Then, the nitrogen and phosphorus of 7.75×10^{10} g-N and 5.17×10^9 g-P were required respectively . Other assumptions on the production and the transportation of the N and P fertilizers were same as those in the previous paper (Horiuchi *et al.*, 1995). The amounts of fertilizers of urea involving 46 wt% of nitrogen and fused phosphate involving 20 wt% of P_2O_5 were 2.35×10^{12} g and 4.00×10^{11} g respectively. The fraction of emission from the production of the fertilizer was 9 % of the annual increase of the atmospheric CO_2 as shown in the previous paper (Horiuchi *et al.*, 1995). The effect of the radius of the fertilization on the fraction of emission from the transportation to absorption was shown in Fig. 4. Even for 100 km sprinkling, the CO_2 emission from the required sprinkling energy was less than 0.23 % of the absorbed CO_2. It was found that the fertilizer should be transferred and sprinkled as widely as possible.

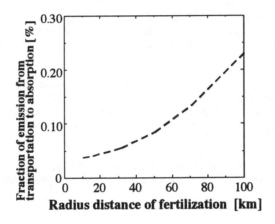

Fig.   4 Fraction of CO_2 emission from transportation of fertilizer by ship

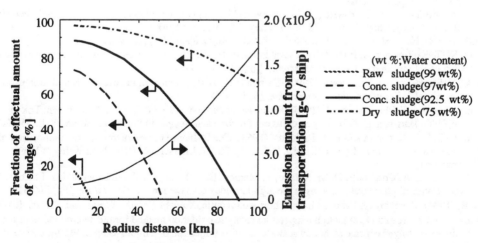

Fig. 5 Amount and fraction of CO_2 emission from transportation
by ship of 4.8×10^{10} g of sludge

ENERGY BALANCE OF FERTILIZATION TO OCEAN WITH ACTIVATED SLUDGE

The activated sludge is expected to be used as a nutrient source instead of the fertilizer because it does not need any energy for its production. The CO_2 emission from energy required for its transportation was calculated to compare with the above case of fertilizer . The activated sludge (water content 99 wt%) involves 5.8 wt% of nitrogen and 4.0 wt% of phosphorus on the dry base (Sudoh, 1984). The concentrated sludges involving 97 and 92.5 wt% of water and the dry sludge involving 75 wt% of water were also evaluated. The assumption of the transportation of each sludge by ship was same as the above. Each sludge of 4.80×10^{10} g was assumed to be transported by one ship to 500 km offshore and sprinkled in the limited area.   The effects of the radius on the amount of CO_2 emission from the transportation and the fraction of emission from the transportation to the absorption were shown in Fig. 5. This figure shows that the fertilization of the activated sludge containing a great deal of water needed large energy for the transportation by ship and should have an optimum area to be sprinkled to the ocean when we consider the possibility of occurrence of red tide by over fertilization.

CONCLUSION

We developed a modified two box model taking two dimensional horizontal diffusion of the organic carbon and nutrients, and the nitrate fixation kinetics by the phytoplankton into account additionally to our previous model. In the present evaluation, 7.3×10^{10} mol of nitrate, equivalent to 0.2 % of annual increase of atmospheric CO_2, was assumed to be transported and sprinkled in a limited area. The radius of the sprinkling area ranged from 10-100 km. It was found that the nitrate concentration in the surface ocean was returned to the initial condition before fertilization within one year.

The fraction of the nitrate not consumed by the phytoplankton or transferred from the surface ocean to the deep ocean as organic matter but transferred as inorganic nitrate was less than 0.01 % of the sprinkled amount to ocean within 1 year. Even for 100 km sprinkling, the CO_2 emission from the required sprinkling energy was less than 0.23 % of the absorbed CO_2. It was found that the fertilizer should be transferred and sprinkled as widely as possible.

We also estimated the energy balance of fertilization with the activated sludge instead of fertilizer. It was found that the fertilization by activated sludge containing a great deal of water needed a large amount of energy for the transportation by ship.

REFERENCES

Broecker, W.S. (1981). *Chemical oceanography*. (Translated by N. Niizuma, Kaiyo Kagaku Nyumon). Tokyo
    Daigaku Suppankai, Tokyo
Bolin, B., A. Bjorkstorm, C. D. Keeling, R. Bacastow and U. Siegenthaler (1981). In: *Carbon Cycle Modeling*
    (Bolin, B. ed.), Chap. 1, pp 1-28. John Wiley & Sons, New York
Handa, N. (1992). Hasseishita nisankatansoha dokoheyuku. *Kagaku Asahi*, July, 19-22.
Horiuchi, K., T. Kojima and A. Inaba (1995). Evaluation of fertilization of nutrients to the ocean as a measure for
    CO_2 problem, *Energy Convers. & Mgmt.*, 36 (6-9), 915-918.
Inaba, A., Y. Shindo and H. Komiyama (1990). Carbon cycle at steady state in the ocean - Simulation using a two
    box model -, *Kagaku Kogaku Ronbunshu*, 16, 5, 1120-1124
National Astronomical Observatory ed. (1995). *Rika Nenpyo (dai 68 satsu 1995)*. pp. 676 Maruzen, Tokyo
Platt, T and W. G. Harrison. (1985). Biogenic fluxes of carbon and oxygen in the ocean, *Nature, 318*, 55-57.
Redfield, A.C., B.H. Ketchum and F. A. Richards. (1963). The Influence of organisms on the composition of sea
    water. In: *In The Sea*: (M.N. Hill, ed.), Vol. 2, pp.26-77, Interscience Pub., John Wiley & Sons,
    New York
Smith, W.O. Jr., L. A. Codispoti, D. M. Nelson, T. Manley, E. J. Buskey, H. J. Niebauer and G. F. Cota (1991).
    Importance of phaeosystis blooms in the high-latitude ocean carbon   cycle, *Nature, 352*, 514-516.
Sudoh, R. (1984). Korekara no Seikatsu haisui shori wo megutte, *Kagaku Kogaku Symposium Series*, 4, 1-10.
Sugimura, Y and Y. Suzuki (1988). High temperature catalytic oxidation method for the determination of non-
    volatile dissolved organic carbon in seawater by direct injection of a liquid sample, *Marine Chemistry*, 24,
    105-131

Energy Convers. Mgmt Vol. 38, Suppl., pp. S517–S521, 1997
© 1997 Elsevier Science Ltd. All rights reserved
Printed in Great Britain
0196-8904/97 $17.00 + 0.00

Pergamon

PII: S0196-8904(96)00320-2

CALCIUM CHLORIDE AS A BIOMIMETIC INTERMEDIATE FOR THE MINERALIZATION OF CARBONATE IONS OF WATER AS CALCIUM CARBONATE IN GELATINOUS MATRICES OF CHITOSAN AND CHITIN

Shigehiro HIRANO, Koichi YAMAMOTO, and Hiroshi INUI

Chitin/Chitosan R&D Room, 445-Sakuradani, Tottori 680, Japan, and
Department of Agricultural Biochemistry and Biotechnology, Tottori
University, Tottori 680, Japan

ABSTRACT

A novel biomimetic method is described for the mineralization of CO_3^{2-} ions of water as $CaCO_3$ in each gelatinous matrix of chitin and chitosan using $CaCl_2$ as a water-soluble CO_3^{2-} ions. Each gelatinous $CaCl_2$-composite (40-87% $CaCl_2$) of chitin and chitosan was prepared by two methods: 1) the entrapping of $CaCl_2$ into each gelatinous granule of chitosan and chitin, and 2) the soaking of each gelatinous granule or square slice of chitosan and chitin in aqueous $CaCl_2$ solutions. The gelatinous composite was soaked in aqueous 1M K_2CO_3 solution at room temperature to give the corresponding white $CaCO_3$-composite (17-30% $CaCO_3$). In these products, CO_3^{2-} ions were detected by FT-IR absorptions at 1425, 876 and 712 cm^{-1}.

KEYWORDS

$CaCO_3$-chitosan composite; $CaCO_3$-chitin composite; biomimetic mineralization: crab shells.

INTRODUCTION

From their external to internal layer, crab shells are composed of 1) the exocuticle layer (chitin, proteins and pigments), 2) the endocuticle layer (chitin, $CaCO_3$ and proteins), and 3) the membrane layer (chitin) (Jungreis, 1979). In the ecdysis stage of crabs as shown in Fig. 1, calcium ions of the solid $CaCO_3$ in the endocuticle layer are solubilized and stored in crab body fluids, and re-utilized for the mineralization of carbonate ions of water as $CaCO_3$ to form a new endocuticle layer (Yano, 1972, 1974, 1977). As a biomimetic exocuticle layer, some sponge sheets of chitin-protein composite materials and N-succinyl chitosan-collagen composite materials (Yasutomi et al., 1995) have been reported. As a biomimetic endocuticle layer, $CaCO_3$-chitosan composite materials have been prepared by dipping chitosan in a supersaturated $CaCO_3$ solution for the crystal growth of $CaCO_3$ on chitosan (Zhang and Gonsalves, 1995), and beads and sheets of $CaCO_3$-chitin composite materials have also been prepared from a mixture solution of alkaline chitin and $CaCO_3$ (Hirano, et al., 1995; S. Hirano et al., 1996). As a biomimetic membrane layer, chitin membranes and sponge sheets were prepared by air-drying of thin layers of N-acetylchitosan gel (Hirano, 1978).

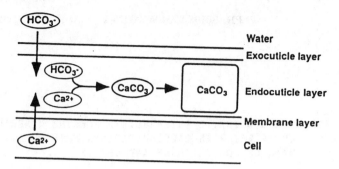

Fig. 1.   A model for the mineralization of $CO_3{}^{2-}$
ions of water as $CaCO_3$ in crab shells

The mineralization of $CO_3{}^{2-}$ ions of water as metal salts is important
to prevent the global green house effect. The present work reports a novel
method for a biomimetic mineralization of carbonate ions as $CaCO_3$ in each
gelatinous matrix of chitin and chitosan using $CaCl_2$ as a water-soluble
Ca^{2+} ion source.

EXPERIMENTAL

Materials

Crab shell chitosan (C/N = 6.1) was a product of Katakurachikkarin Company,
Tokyo, and N-acetylchitosan gel was prepared by N-acetylation of chitosan
in aqueous 2% acetic acid-methanol (Hirano, 1977).

Methods

FT-IR spectra were recorded on a Jasco FTIR 5300 spectrometer (Jasco
Company, Tokyo).   Chlorine ions were quantitatively analyzed by titrating
with 0.01 N $AgNO_3$ in 10 % K_2CrO_4 solution. Elemental analyses were
performed at the Micro-elemental Analysis Center of Kyoto University,
Kyoto.

Entrapping of $CaCl_2$ into Each Gelatinous Granule of Chitosan and Chitin

1) Chitosan (0.16 g) was dissolved in aqueous 2% acetic acid (15 ml), and
solid $CaCl_2$ (1.1 g, 10 mol eq./GlcN) was mixed to give a viscous solution.
The solution was added dropwise into acetone (200 ml) to give gelatinous
wet granules (diameter _ca._ 50 mm), which were collected by centrifugation
and dried over P_2O_5 _in vacuo_ to give hygroscopic gelatinous dry granule
(_ca._ 1.0 mm)(1.2 g yield).

2) N-Acetylchitosan (0.2 g) was dissolved in $CaCl_2$ $2H_2O$-saturated methanol
(20 ml) by stirring at room temperature for several days (Shirai _et al._,
1995).   The solution was added dropwise into acetone (200 ml), and the
mixture was kept at room temperature overnight to give gelatinous wet
granules (diameter _ca._ 50 mm) of $CaCl_2$-chitin composite materials.

Soaking of Each Gelatinous wet Granule or Square Slice of Chitosan and Chitin into Aqueous 10% $CaCl_2$ Solution

Each gelatinous wet granules (diameter ca. 50 mm) or square slices (ca. 2x2 cm with 0.3 mm thickness) of chitosan gel (Hirano et al., 1990) and N-acetylchitosan gel (Hirano et al., 1977) were soaked into aqueous 10% $CaCl_2$ solution at room temperature overnight to give $CaCl_2$-chitosan and $CaCl_2$-chitin composite materials (40% $CaCl_2$).

Mineralization of CO_3^{2-} Ions of Water as Water-Insoluble $CaCO_3$ in Each Gelatinous Matrix of Chitosan and Chitin

The granules or square slices of each $CaCl_2$-chitosan and $CaCl_2$-chitin composite materials were soaked in aqueous 1M K_2CO_3 solution, and the suspension was stirred occasionally at room temperature for 5 h to give the corresponding $CaCO_3$-composite granule and free $CaCO_3$. The composite granules or square slices only were collected, and washed thoroughly with distilled water. The products were dried to give the corresponding $CaCO_3$-composite product (17 to 30 % $CaCO_3$ as estimated from the C/N ratio of the elemental analyses), which was negative for the chlorine test with $AgNO_3$. The free $CaCO_3$ precipitate was discarded. $CaCO_3$-chitosan composite materials: FT-IR (KBr) 1599 (NH_2), 1426, 876 and 712 cm^{-1} (CO_3^{2-}). $CaCO_3$-chitin composite materials: FT-IR (KBr) 1655 and 1559 (C=O and NH of NAc), 1431, 878 and 712 cm^{-1} (CO_3^{2-}).

RESULTS AND DISCUSSION

Liberation of Chloride Ions from the $CaCl_2$-composite Gelatinous Granules of Chitosan into Distilled Water and Aqueous 1M K_2CO_3 Solution

The $CaCl_2$-composite gelatinous granules of chitosan were hygroscopic, and their $CaCl_2$ content was from 40% to 87%. Calcium chloride was liberated from the gelatinous granule into distilled water at room temperature for 10 min, and was liberated into aqueous 1M K_2CO_3 solution for 40-60 min (Fig. 2). The delay of $CaCl_2$-liberation rate from the gelatinous matrices is due to the formation of water-insoluble $CaCO_3$ not only on the granule surface but also in the granule matrices, resulting in the prevention of the $CaCl_2$-liberation from the gelatinous matrices.

Mineralization of CO_3^{2-} Ions of Water as Water-insoluble $CaCO_3$ in Each Gelatinous Matrix of Chitosan and Chitin

Each composite material of $CaCl_2$-chitosan and $CaCl_2$-chitosan prepared above was soaked in aqueous 1M K_2CO_3 solution, and CO_3^{2-} ions of water were mineralized as $CaCO_3$ not only in the gelatinous matrices but also on the matrix surface.

Each calcificated gelatinous granule or slice of chitosan and chitin was obtained as a white product and kept almost the original shapes even after drying because of solid $CaCO_3$ immobilized in the matrices, and $CaCO_3$ was detected not only on the matrix surface but also within the matrices as examined by IR absorptions at 1425, 876 and 712 cm^{-1} in the FT-IR spectra (Fig. 3). However, the noncalcificated one shrunk into small transparent granules or thin slices. No chlorine was detected in both the products. This fact indicates that CO_3^{2-} ions of water are calcificated as $CaCO_3$ both on the matrix surface and within the matrices, resulting in the slow

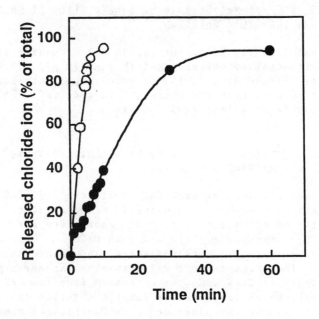

Fig. 2. Time course for the liberation of chloride ions into distilled water (o) and into aqueous 1 M K_2CO_3 solution (o) from one granule (diameter <u>ca</u>. 50 mm) of $CaCl_2$-chitosan composite materials. Each value is the average of three experiments.

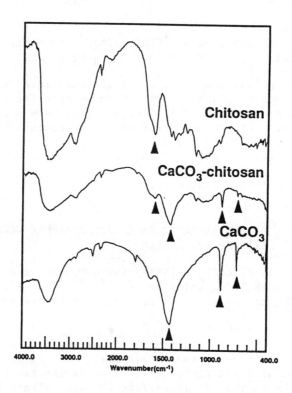

Fig. 3. FT-IR Spectrum of $CaCO_3$-chitosan composite materials in reference to those of chitosan and $CaCO_3$.

liberation of $CaCl_2$ in aqueous K_2CO_3 solution. However, the morphological structure of the matrices have almost no order, and differs distinctly from an orderly cylindrical structure of the endocuticle layer of crab shells (Hirano et al., 1995).

CONCLUSION

The present work proves a biomimetic in vitro mineralization of CO_3^{2-} ions of water in the gelatinous chitin and chitosan matrices. The future task is to develop a novel method for the mineralization of CO_3^{2-} ions of water as solid $CaCO_3$ only within the gelatinous matrices.

REFERENCES

Hirano, S., R. Yamaguchi and N. Matsuda (1977). Architecture of chitin gels as examined by scanning electron microscopy, Biopolymers, 16, 1987-1992.

Hirano, S. (1978). A facile method for the preparation of novel membranes from N-acyl- and N-arylidene-chitosan gels, Agric. Biol. Chem., 42, 1981-1982.

Hirano, S.,R. Yamaguchi, N. Fukui and K. Horiuchi (1990). A chitosan oxalate gel: its conversion to an N-acetylchitosan gel via a chitosan gel, Carbohydr. Res., 201, 145-149.

Hirano, S., H. Inui and K. Yamamoto (1995). The mineralization of CO_3^{2-} ions in crab shells, and their mimetic composite materials, Energy Convers. Magmt., 36, 783-786.

Hirano, S., K. Yamamoto, H. Inui, M. Ji and M. Zhang (1996). The mineralization of CO_3^{2-} ions on chitin- and chitosan-metal composite granules in water, Macromol. Symp., 105, 149-154.

Jungreis, A. M. (1979). Physiology of molting in insects, Advan. Insect. Physiol., 14, 109-183.

Shirai, A., K. Takahashi, R. Rujravanit,N. Nishi, and S. Tokura (1995). Regeneration of chitin using a new solvent system, Chitin and Chitosan, The Versatile Environmental Friendly Modern Materials, eds. M. B. Zakaria, W. M. W. Muda, and M. P. Abdullah, Penerbit UKM, Bangi, pp. 53-60.

Yano, I. (1972). Histochemical study on the exocuticle with respect to its calcification and associated epidermal cells in a shore crab, Nippon Suisan Gakkaishi, 38, 733-739.

Yano, I. (1974). Incorporation of glucose-1-^{14}C into chitin of the exoskeleton of a shore crab with special reference to the period of endocuticle formation. Nippon Suisan Gakkaishi, 40, 783-787.

Yano, I. (1977). Structure and morphology of the outer integument of some crustaceans, Kagaku to Seibutsu, 15, 328-336.

Yasutomi, Y., Nakakita,N.,Shioya, N. and Y. Kuroyanagi (1993). Development of biosynthetic wound dressing with drug delivery capability composed of polyurethane membrane and biological sponge material: a clinical applications. Nessho, 19, 102-216.

Zhang, S. and K. E. Gonsalves (1995). Synthesis of $CaCO_3$-chitosan composites via biomimetic processing. J. Appl. Polym. Sci., 56, 687-695.

Energy Convers. Mgmt Vol. 38, Suppl., pp. S523–S528, 1997
© 1997 Elsevier Science Ltd. All rights reserved
Printed in Great Britain
0196-8904/97 $17.00 + 0.00

Pergamon

PII: S0196-8904(96)00321-4

CYANOBACTERIAL BIOCONVERSION OF CARBON DIOXIDE FOR FUEL PRODUCTIONS

KAZUHISA OHTAGUCHI, SUSUMU KAJIWARA,
DANI MUSTAQIM AND NAHOKO TAKAHASHI

Department of Chemical Engineering, Tokyo Institute of Technology,
2-12-1 O-okayama Meguro-ku, Tokyo, Japan

ABSTRACT

Laboratory-scale systems for bioconversion of CO_2 to biofuels, H_2 and ethanol, have been developed in an attempt to present data for baseline evaluation of a resource recycling technology utilizing biomass of cyanobacteria. The procedure employs cultivation in the light of photolithotroph *Synechococcus leopoliensis*, onset, by shift into anaerobic dark, of hydrogenase-mediated H_2 formation reaction, and conversion of biomass waste into ethanol. Cyclic turnover of carbon by harnessing photosynthesis was successfully demonstrated by the experiment, but the research for a large-scale fuel production is still in its preliminary stage. © 1997 Elsevier Science Ltd

KEYWORDS

cyanobacteria; bioconversion; CO_2 fixation; photosynthesis; biofuel; H_2 production; ethanol production; *Synechococcus leopoliensis*; photoautotroph; open pond

INTRODUCTION

The subject matter of this paper concerns the bioconversion of CO_2 for fuel production. One of the key determinant in bringing any of the CO_2-removal systems into actual use is probably the capital requirements, since CO_2, which naturally presents in the atmosphere, has long been known as an innocent gas and disastrous consequences (sea-level rise and climatic change) of CO_2 accumulation are the events yet arisen but only foretold by specialists. Furthermore, the problems on disposal of large volume of captured CO_2 is still unsolved. To make CO_2-removal process be both technically and economically more feasible, the development of well-engineered systems, that convert CO_2 into useful products, will be advocated.

Of most interested phenomenon for this purpose is the photosynthetic bioconversion of CO_2 to biomass as a renewable resource supporting combustible fuels. In a case of dry crop residue, it has been estimated that the energy value of ash-free biomass reaches about 17.7 MJ/kg (Wereko-Brobby, *et al*, 1996). Practical among such has been the production of fuel ethanol from sugar cane, which is in operation on a large scale (annual production: more than 15,000,000m³) in Brazil. In the case of current ethanol production, the attention of biomass feedstock has been mostly directed to agricultural products, *viz.*, sugar cane, molasses, cassava, corn, wood and agricultural residues, which incorporate atmospheric CO_2 in the farms and woodlands. This approach suggests another opportunities of the technology for biomass, which will be used to remove CO_2 at the emission source. Until very recently little attention has been given to the CO_2-removal biotechnologies, however, much work on the biofuel production from cultured photosynthetic organisms has been performed beginning with the classical work in 1951 of Arthur D.Little, Inc.(Fisher, 1961), in which 405,000m² plant for the production at 236kg/h of the algae *Chlorella pyrenoidosa* was designed. The estimated power

Table 1 Classification of CO_2 emission sources.

emission source	Demand [kw]	CO_2 evolution rate [t-CO_2 /h]
<LARGE>		
Steam-electric plant LNG	600,000	272
Coal	1,000,000	900
<MIDDLE>		
Power station of 68,000t/y		
polystyrene plant	3,500	1.4
CO_2-production for EOR,Tex.		45.8
CO_2-production for soda ash		
production, Calif.		27
Food processing, Okla.		8.3
<SMALL>		
Vehicles		0.004

production from it was no more than 600 kilowatts. The need for the reduction of operating cost was emphasized.

Alternatives for biofuels may involve H_2 which presents many favorable opportunities as a fossil substitute. The first reported reaction was photohydrogen production of green algae *Scenedesmus obliquus*w which, in the dark, evolved H_2 by the anaerobic degradation of a reserve substance(Gaffron, *et al.*, 1942). Later procaryotic H_2 formation was also reported. In glutamic and lactic acids, the photosynthetic bacterium *Rhodospirillum rubrum* evolved H_2 at a specific rate of 0.803mmol/(g·h) (Zurrer *et al.*, 1979). Heterocystous cyanobacerium *Anabaena cylindrica* was used for the demonstration of simultaneous production of H_2 and O_2. Outdoor experiments to study solar energy conversion with this strain resulted in the specific H_2 evolution rate of 0.244mmol/(g·h) which corresponded to the H_2 evolution rate of 0.175mmol/(dm^3·h) (Hallenbeck *et al.*, 1978). For application purposes, H_2 production was investigated with the filamentous, non-heterocystous marine cyanobacterium *Oscillatoria* sp. strain Miami BG7 of which H_2 production rate reached 0.364mmol/(g·h) (Phlips *et al.*, 1983). Cyanobacterium, *Spirulina platensis* M-185, was described to evolve H_2 in the dark and under anaerobic or microaerobic condition with the rate of 0.0473mmol/(g·h)(Asada, *et al.*, 1986). An extensive study on cyanobacterial H_2 metabolism under a microaerobic condition has shown the capability of H_2 production by strains *Synechococcus* PCC6803, 602, 6301(*Anacystis nidulans*), 6307 (*Coccochloris peniocystis*), *Microcystis* PCC7820, *Gloeobacter* PCC7421, *Synechocystis* PCC6308 (*Aphanocapsa* and *Gloeocapsa alpicola*), 6714(*Aphanocapsa* sp.), and *Apanocapsa mentana*(Howarth *et al.*, 1985). The condition for H_2 production has been variously reported, however, these observations seem to stimulate further experimental investigations for biofuel productions.

Our study on the fixation and utilization of CO_2 through the cultivation of cyanobacterium *Synechococcus leopoliensis*(*Anacystis nidulans* IAM M-6) was initiated as a baseline evaluation of above technologies, since this is one of the fastest-growing unicellular photolithotroph which requires radiant energy, using water as electron donor, and exhibits the simplest nutritional requirements of any known organisms. Laboratory-grown cells of *S.leopoliensis* stored a significant amount of absorbed light energy in the reaction products including starch. Preliminary experiment showed that the cell suspension, which was shaded from fluorescent lamps of the photosynthetic photon flux density(*PPFD*) at 0.341mol/(m^2·h) for 120hours, resulted in the conversion efficiency of about 4.6g/MJ. The present study was therefore designed to find a resource recycling technology using the produced biomass of *S.leopoliensis*. The maximum use of biomass has been argued with multiple biofuel production schemes. The basic components of the proposed multiple reaction systems include:

1. Photobioreactor for bioconversion of CO_2 into biomass of *S.leopoliensis*.
2. Primary reactor for biofuel production, where hydrogenase in the biomass of *S.leopoliensis* performs the chemical reduction of hydrogen ions to form hydrogen(H_2).
3. At the end of the systems, secondary reactor for biofuel production, where saccharide in HCl-treated biomass extract is converted to ethanol by the culture of *Saccharomyces sake*.
On the basis of the experience in the laboratory environment, we evaluated above technologies.

PROBLEM DEFINITION

It is fundamental to a proper understanding of CO_2-emission-control problem that the discharge rate varies considerably according to sources (Table 1). The MEA process would show appreciable economics for the

600,000 kw power station over other technological processes (Yokoyama *et al.*, 1995). On the other side, biomass approach, which has a prospect of concomitant synthesis of biofuels(Ohtaguchi *et al.*, 1991), is seemingly suited for smaller scale emission sources. Although CO_2 problem is issued from global environment, we are of the opinion that integration of all the possible technologies furnishes a key for the solution of the problem, as implied by the maxim "Think globally and *act locally*". Of these this study concerns with the middle class of emission source, with an example of 10t/h CO_2 evolution plant. One leading event, which set the world's goal to freeze CO_2 emission, was the UN-sponsored Conference on Environment and Development, held in Brazil in 1992. It aimed at freezing CO_2 emissions by the end of the nineties to restrain them at the 1990 levels. To achieve this, we aim to cut in CO_2 emissions of 25 per cent for above plant and to estimate the size of 2.5 t/h CO_2 -removal and bioconversion processes.

EXPERIMENTAL

Incorporation of CO2 into Biomass

For CO_2-removal purposes, cyanobacterium *Synechococcus leopoliensis* (also known as *Anacystis nidulans* IAM M6 and *Synechococcus* PCC6301) was grown at 40°C on modified Detmer medium(Watanabe, 1960), which was aerated continuously by 6% CO_2 in air in the external-loop airlift bioreactor of the effective volume of 1.2dm³ at the superficial gas velocity of the riser part of 12m/h. The vessel was made of Pyrex glass tubes. The bioreactor was illuminated, at all hours of the day and night, from one side with light from a bank of fluorescent lamps(FPL 27 EX-N, Hitachi Illumination Co. Ltd., Japan). The value for *PPFD* of the incident light was fixed throughout at 0.684mol/(m²·h).

Biomass-Mediated H2 Production

The cyanobacterial H_2 production was performed in the dark with DEC-N solution as described before (Ohtaguchi *et al.*, 1995). Attempts in this work were made to investigate the effects of temperature on H_2 production. The pH of the culture was controlled at 7.5. Starch was extracted from sonication-treated biomass at 60°C with the mixture of DMSO and 8 mol/dm³ HCl (DMSO:HCl=7:2). Assay was carried out using F-kit starch (Boehringer Mannheim Co.).

Bioconversion of Biomass to Ethanol

Bioconversion of extracted-glucose to ethanol in the culture of *Saccharomyces sake* IFO 2347 was performed under anaerobic condition according to our earlier methods (Dani Mustaqim *et al.*, 1996).

RESULTS AND DISCUSSION

CO2 fixation   Figure 1 shows the time course of the concentration of *S.leopoliensis*(X), the photosynthetic photon flux density of the transmitted light(*PPFD*), the mass of starch per unit cell mass(ρ_S), and pH. The growth curve presents the specific growth rate of biomass($\mu = (dX/dt)/X$) of 0.0617h⁻¹. The curve of *PPFD* versus time shows that the time-averaged interception efficiency, which is defined as the ratio of intercepted to incident radiation, is 0.776, and that time-averaged conversion efficiency, which is defined as the ratio of the increase of biomass to intercepted radiation, is 6.24 g/MJ, which is higher than that of our previous observation by 36%. It is also found to be higher than that of higher plant *Zea mays* which was no more than 5 g/MJ under solar radiation through the crop-growing season(Farage *et al.*, 1987).

Table 2 tabulates the amounts of elements in dry biomass of *S.leopoliensis* from which its cell formula of $CH_{1.78}N_{0.170}O_{0.565}$ is determined. From this, one C-mole of biomass, which indicates the amount of biomass containing one gram-atom carbon, is calculated to be 25.2. Since *S.leopoliensis* is a photoautotroph, which requires only CO_2 to supply their carbon needs with relying on light energy, overall stoichiometry

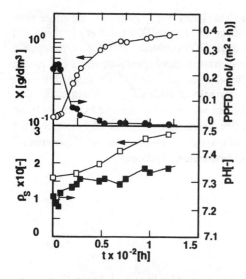

Fig.1   Time courses of reaction parameters in a
photosynthetic experiment using *S.leopoliensis*.

Table 2 The elemental composition of *S.leopoliensis*

Element	Percentage of dry mass	Element	Percentage of dry mass
Carbon	47.32	Hydrogen	7.07
Oxygen	35.64	Sulfur	0.56
Nitrogen	9.41	Ash	0.0

Table 3 Effect of temperature on H_2 production

Temperature [°C]	H_2 production rate [mmol/(dm³·h)]	$\rho_S X$ at the end of cultivation [g/dm³]
25	0.0208	1.56
30	0.127	0.306
35	0.130	0.684
45	0.0115	1.47

during doubling is modeled as follows:

$$CH_{1.78}N_{0.170}O_{0.565} + CO_2 + \text{other substrates} \rightarrow 2CH_{1.78}N_{0.170}O_{0.565} + \text{other products} \qquad (1).$$

Thus the CO_2 uptake rate(CUR) of *S.leopoliensis* is evaluated by the relation that

$$CUR = (44/25.2)dX/dt \qquad (2).$$

Following this, CUR of the order of 0.0354 g/(dm³·h) is calculated from the data shown in Fig.1.

If above estimate is applicable to large size bioreactors in which *S.leopoliensis* utilizes solar energy for 9 hours per day(a year-round averaged photoperiod) with the relative efficiency of photosynthesis at 40% , the volume of cyanobacterial culture required for the 25% removal of CO_2, discharged unceasingly with the emission rate of 10t/h, can be calculated as follows:

> volume of the CO_2-removal bioreactor
> = 0.25x 10 t/h x 24h/d / {0.0354x10⁻³t/(m³·h) x 0.4 x 9h/d}
> = 471,000m³.

Application of open-ponds or an artificial lake is likely one of the most favorable options for this reaction. To have a rough estimate, we postulate that *S.leopoliensis* is cultivated in outdoor ponds with a diurnal light(9h)-dark(15h) cycle of sunlight illumination, of which solar flux value in daytime is approximated to 1,830 kJ/(m²·h) and that the mixing of culture is perfect by the continuous sparging of 6% CO_2 in air. To make the discussion simple, assume that during the daytime the culture is in a steady state under the chemostat operation while that during the dark period it is out of chemostat operation and cell concentration remains constant. Although this dynamical system is in essence in a unsteady state, it could be regarded as in a semi-steady state(Van Liere *et al.*, 1979). In order to avoid photodamage situation(such as photobleaching) at high light irradiance of sunlight, it is further assumed that the open pond has a dark volume, under the sunny surface layer, in which *S.leopoliensis* is not exposed to the full solar flux and that the ratio of the volume in the dark to the volume in the light is 1: 9 considering the effect of intermittence of flashing light illumination. Above experiment results in time-averaged phosynthetic biomass production rate of 0.0203g/(dm³·h), so that, if the flashing-light- intermittence effect appears, then the illuminated area to receive radiant energy for supporting the daytime cell growth of the 40% efficiency is estimated as:

> surface area of the CO_2-removal bioreactor
> = 0.0203 kg/(m³·h) x 0.4 x 471,000m³/(6.24x10⁻³ kg/MJ)/{1,830x10⁻³MJ/(m²·h)}
> = 335,000m²,

and overall liquid height as:

liquid height
$$= 471,000m^3/(335,000m^2)$$
$$= 1.41m.$$

The liquid height of sunny layer is evaluated to be 0.14m($= 1.41m \times 1 /(1+9)$), which is about of the same order of magnitude as the widely accepted depth of open ponds. To fulfil the effect of intermittence of flashing light illumination, more work needs to be done on a method of suitable liquid circulation in vertical direction.

When chemostat operation is made at the dilution rate of 0.0247h^{-1}, which is about 40% of above μ value, the effluent flow rate of the process is calculated to be 11,600m^3/h. Biomass concentration of this stream is likely of the order of 0.4g/dm^3, and hence biomass production would reach the rate of 4.65t/h. It should be further noted that these systems also act as an prime supplier of oxygen for atmosphere. If phtosynthetic quotient, which depends on the cellular composition and physiological state, is assumed to be 1.0, a rough estimate of oxygen evolution in daytime reaches 1,820 kg-mol/h($= 0.25 \times 10,000$ kg/h $\times 32/44$).

H2 Production

Table 3 is a tabulation of experimental data of H$_2$ production, in which *S. leopoliensis*-biomass of 6.22g/dm^3 was used. The optimal temperature is found to be 35°C. At this condition H$_2$ production rate reached 0.130mmol/(dm^3•h), which is of the same order of the values reported for *A.cylindrica*.

If the biomass-rich-effluent from above large-scale CO_2-removal process is allowed to settle in an appropriate separation process, for increasing the biomass concentration, and if the concentrated biomass is collected every 3 days, then the amount of biomass for each treatment is estimated as 126t($= 4.65$t/h $\times$ 3d $\times$ 9h/d), so that the use in the dark of anaerobic closed-systems having the effective volume of 20,300m^3 ($=126$t/(6.22 $\times 10^{-3}$ t/m^3)) is needed for the efficient production of H$_2$. Underground reservoir seems to be a prototype of this reactor. Attainable level of the H$_2$ production rate of above model plant is thus estimated as 2.64 kg-mol/h ($= 0.130$ mol/ (m^3•h) $\times$ 20,300m^3). Ongoing H$_2$ production in chemical processing industry is based on the reforming of natural gas, propane, butane and naphtha, which are non-renewable resources. For instance, CAR(Combined Autothermal Reforming) process, installed in Chemko s.p. Strazse Co. in 1991, produces H$_2$ at the rate of 13,000Nm^3/h(Mashiko *et al.*,1993), which is of the order of 580 kg-mol/h. Thus above CO_2-utilization technology is found to be used as a 0.46% substitute for hydrocarbon-reforming process, if necessary. Only one installation of above systems does not seem to stand up to current international economic criteria, however, they would play an important role if there is an increase in the number of such cyanobacterial CO_2 -removal processes.

Of special interest is the presence of unreacted biomass-starch in the effluent from H$_2$-production-process. If reaction is performed ar 35°C and if biomass is renewed every 72 hours, it discharges the starch, in the form of biomass, at the concentration of 0.94 g/dm^3. The amount of starch in that effluent reaches 19.1 t ($= 0.94$ kg/m^3 $\times$ 20,300m^3).

Ethanol Production

In our previous work, ethanol production from *S.leopoliensis* -biomass was performed using aerobic culture of *S.sake*. The present experiment was designed to find out fermentation characteristics of *S.sake* in anaerobic condition. The result showed that the yeast converted the raw glucose, extracted from biomass of *S.leopoliensis*, to ethanol and yeast-biomass with the molar yield factor, which is defined as moles ethanol formed/moles glucose consumed, of 1.73. This figure suggests the possibility of the production of ethanol from effluent of H$_2$ process. The possible amount of ethanol is estimated as 184 kg-mol($= 1.73 \times 19,100$ kg/180). The energy value of ethanol is described as 26.8 MJ/kg, and thus above ethanol is found to have the energy of 227,000MJ.

CONCLUSION

Bioconversion of CO_2 to fuels, H_2 and ethanol, was challenged by investigating the feasibility of a model system in which we concerned the use of biomass of photolithoautotroph *Synechococcus leopoliensis* . From these studies it would appear that cyanobacerial biomass is of great significance in formulating technology of ecological perspective. On the other hand, process variables were roughly figured. More data will be needed before the engineering implications of these estimates become clear.

This paper is dedicated to the memory of Dr.K.Koide, late emeritus professor of Department of Chemical Engineering, Tokyo Institute of Technology. The authors are indebted to Prof.A.G.Fredrickson, Department of Chemical Engineering and Materials Science, University of Minnesota, for his helpful suggestions.This work was supported in part by Grants-in Aids for Scientific Research(No.05650793, 06650907, 07650951) of the Ministry of Education, Science and Culture, Japan.

REFERENCES

Asada,Y. and S.Kawamura(1986). Screening for cyanobacteria that evolve molecular hydrogen under dark and anaerobic conditions. *J.Ferment.Technol.*, 64(6),553-556.

Dani Mustaqim and K.Ohtaguchi(1996). A synthesis of bioreactions for the production of ethanol from CO_2. *Energy.* in press.

Farage,P.K. and S.P.Long(1987). Damage to maize photosynthesis in the field during periods when chilling is combined with high photon fluxes. Progress in photosynthesis research, vol.4, 139-142, Martinus Nijhoff, Dordrecht.

Fisher,A.W.Jr.(1961). Economic aspects of algae as a potential fuel. In: *Solar Energy Research.* pp.185-189. University of Wisconsin Press, Madison, Wisconsin.

Gaffron,H. and J.Rubin(1942). Fermentative and photochemical production of hydrogen in algae'"*J.Gen. Physiol.*, 26,209.

Hallenbeck P.C.,L.V.Kochian, J.C.Weissman and J.R. Benemann(1978). Solar energy conversion with hydrogen-producing cultures of the blue-green alga, *Anabaena cylindrica. Biotechnol.Bioeng.Sym.*, No.8,283-297.

Howarth,D.C. and G.A.Codd(1985). The uptake and production of molecular hydrogen by unicellular cyanobacteria. *J of Gen.Microbiol.*, 131,1561-1569.

Mashiko,Y. and T.Sasaki(1993). *Petrotech*, 11(16),1042-1043.

Ohtaguchi,K. and T.Kobayashi(1991). Re-evaluation of biotechnologies. *Sci. & Technol. in Japan*, 11(3),21-24

Ohtaguchi,K., A.Ohta, N.Takahashi, M.Ogawa and K.Koide(1995). Kinetics of hydrogen production by the photolithotroph *Synechococcus leopoliensis. IONICS*, 21, 69-72.

Phlips,E.J. and A.Mitsui(1983). Role of light intensity and temperature in the regulation of hydrogen photoproduction by the marine Cyanobacterium *Oscillatoria* sp. Strain Miami BG7. *Appl. Environ. Microbiol.*,45(4),1212-1220.

Van Liere,L., L.R.Mur, C.E.Gibson and M.Herdman(1979). Growth and physiology of Oscillatoria *agardhii* Gomont cultivated in continuous culture with a light-dark cycle. *Arch. Microbiol.*, 123, 315-318.

Watanabe,A.(1960). *J of Gen.Microbial.*, 6,283.

Wereko-Brobby,C.Y. and E.B.Hagen(1996). *Biomass conversion and technology* •John Wiley & Sons, New York.

Yokoyama,T. and S.Kudou(1995). Evaluation of chemical absorption process for CO_2 removal from the flue gases of LNG-fired power plant. *CRIEPI Report - Komae Research Lab. Rep.*,No.T94057.

Zurrer,H. and R.Bachofen(1979). Hydrogen production by the photosynthetic bacterium *Rhodospirillium rubrum. Appl. Environ. Microbiol.*, 37(5),789-793.

Energy Convers. Mgmt Vol. 38, Suppl., pp. S529–S532, 1997
© 1997 Elsevier Science Ltd. All rights reserved
Printed in Great Britain
0196-8904/97 $17.00 + 0.00

PII: S0196-8904(96)00322-6

Pergamon

DESIGN OF THE BIOREACTOR FOR CARBON DIOXIDE FIXATION BY *SYNECHOCOCCUS* PCC7942

SUSUMU KAJIWARA, HIDENAO YAMADA, NARUMASA OHKUNI, and
KAZUHISA OHTAGUCHI

Department of Chemical Engineering, Tokyo Institute of Technology,
2-12-1 O-okayama Meguro-ku, Tokyo, Japan

ABSTRACT

The growth characteristics of *Synechococcus* PCC7942 for carbon dioxide (CO_2) fixation was studied. When the culture was illuminated at 8 klx and sparged with 5% CO_2 (volume/volume), high CO_2 uptake rate (*CUR*) of this strain appeared. In the batch reactor under above condition, the highest values for both the specific growth rate and final cell mass concentraion were achieved by the pH controled cultivation at pH 6.8. This condition also led to the highest *CUR* on the second day of the cultivation. Therefore, based on these observations, the additional cultivation of *Synechococcus* PCC7942 for keeping the highest CUR value was performed in chemostat (1.1 dm³). This experiment achieved the result that the cell suspension of *Synechococcus* PCC7942 per 1 dm³ of culture in chemostat took in 3.6 g of CO_2 for six days.

© 1997 Elsevier Science Ltd

KEYWORDS

carbon dioxide; CO_2 uptake rate; *Synechococcus* PCC7942; optimal condition; chemostat.

INTRODUCTION

Excess CO_2 accumulation in the atmosphere during the latter half of this century has become one of the environmental concerns on the global scale, from which in the recent years various CO_2-removal techniques has been developed in the world. Of these, photosynthetic organisms, particularly microalga, are used in the studies utilizing solar energy. Only cyanobacteria are procaryotic microalga and have a high photosynthetic capability. Also, these procaryotes perform oxgenic photosynthesis like the chloroplasts in plant cells and their photosynthetic mechanism are studied widely(Bryant *ed.*, 1994). Therefore, cyanobacteria are good biomaterials for ditailed studies of CO_2 fixation techniques. *Synechococcus* PCC7942, that is unicellular cyanobacterium, provide a higher growth rate under light and is capable of growing in the culture aerated with a CO_2-enriched air containing 5% CO_2 (Marcus *et al.*, 1986). Moreover, since the transformation system of *Synechococcus* PCC7942 has been already established (Kuhlemeier *et al.*, 1981, Golden *et al.*, 1984), this strain is a useful host for genetic manipulation. Many reports about the forign gene expression in *Synechococcus* PCC7942 were described previously (Friedberg *et al.*, 1986, Van der Plas *et al.*, 1989, Gruber *et al.*, 1990). Hence, the CO2 fixation technique utilizing *Synechococcus* PCC7942 have the potential to convert high concentration of CO_2 in industrial boiler flue gas to renewable resources. In this paper, we recognize the growth characterstics of 'host' *Synechococcus* PCC7942 and design the bioreacter of this strain for CO_2 fixation.

EXPERIMENTAL

Strain and medium

Synechococcus PCC7942 was obtained from the Algal Culture Collection at the Institute of Molecular and Cellular Biology, University of Tokyo. All experiments were done with this strain. The medium used for the routine maintenance and the cultivation of *Synechococcus* PCC7942 was BG11 medium consisting of 1.5 g of NaNO$_3$, 0.075 g of MgSO$_4$·7H$_2$O, 0.036 g of CaCl$_2$·2H$_2$O, 0.04 g of K$_2$HPO$_4$, 0.02 g of Na$_2$CO$_3$, 6 mg of citric acid, 6 mg of Ferric ammonium citrate, 1 mg of EDTA and 1 ml of A-5 trace metal solution per dm^3 buffered with 10 mM HEPES to pH 8.0. The A-5 trace metal solution contained 2.86 g of H$_3$BO$_3$, 1.81 g of MnCl$_2$·4H$_2$O, 0.222 g of ZnSO$_4$·7H$_2$O, 0.39 g of Na$_2$MoO$_4$·2H$_2$O, 0.079 g of CuSO$_4$·5H$_2$O and 0.0492 g of Co(NO$_3$)$_2$·6H$_2$O per dm^3.

Cultivation and analysis

Synechococcus PCC7942 was subcultured in oblong flat flasks (37 mm thickness, 115 mm width, and 380 mm vertical length) containing 0.5 dm^3 of BG11 medium at 30°C for 7-10 days under light illumination of 5 klx. Batch cultivations of *Synechococcus* PCC7942 were made using a external loop airlift column (1.1 dm3)(Yamada *et al.*, 1996). The reactor provided sufficient liquid circulation by sparging CO$_2$-enriched air from a gas sparger. The pH during the cultivations were adjusted by the addition of appropriate volume of HCl solution once per day. In chemostat, above airlift column was used as the reactor and supplied with sterile feed using a peristaltic pump. The feed rate (F) was set of 0.0417 (=0.0379x1.1) dm3/h. The growth medium was withdawn from the reactor by the same rate as the feed. The temperature of the cultivations was maintained at 30°C. Cell growth was evaluated by measuring the optical density at 600 nm. The total amount of fixed CO$_2$ was calculated from the final dry cell mass concentration which was obtained from elementary analysis of *Synechococcus* PCC7942 cells.

RESULTS AND DISCUSSION

The growth condition in batch reactor

In the first step to the optimal growth condition of *Synechococcus* PCC7942, we tried to do three series of experiments in batch reactor. Parameters investigated in these runs were the gas flow rate, CO$_2$ concentration of inlet gas, and the intensity of incident light, whereas the cultivation time and the initial cell mass concentration were fixed of 150 hours and 0.0864 g/dm^3, respectively. The cultivations, in which gas flow rates were varied from 12 dm^3/h to 60 dm^3/h, but CO$_2$ concentration of inlet gas was fixed at 5% and light intensity at 5.4 klx, showed that the gas flow rate within the ranges of above did not affect the growth of *Synechococcus* PCC7942. At 12 dm^3/h of gas flow rate and 5.4 klx of light intensity, the cells of *Synechococcus* PCC7942 were cultivated with the CO$_2$ concentration of inlet gas ranging from 5% to 30%. The highest specific growth rate was appeared at 5% CO$_2$ concentration. On the orther hand, the specific growth rate at CO$_2$ concentration between 10% and 30% was higher than that at 0.03% CO$_2$ (air) and the final cell mass concentration was not affected by CO$_2$ concentration of inlet gas except 0.03% (Fig. 1). This finding suggests that the bioreactor using *Synechococcus* PCC7942 is capable of applying directly CO$_2$ removal techniques for industrial gaseous waste streams such as flue gas which contains CO$_2$ at a level of about 15%. In the series of experiments under illumination with different light intensities between 3 klx and 10 klx, 3 klx of light intensity slightly affect the specific growth rate and the final cell mass concentration of *Synechococcus* PCC7942 (Fig. 2). This result represented that it was necessary for the culture of this strain to achieve high specific growth rate and high cell mass concentration to illuminate of greater than 5.4 klx. Since *Synechococcus* PCC7942 metabolized only CO$_2$ as a carbon source, the amount of CO$_2$ uptake in the biomass was estimated from the amount of this cell mass. Therefore, data presented here indicated that the great amount of CO$_2$ fixation was achieved at 5% CO$_2$ concentration of inlet gas and 8 klx of light intensity.

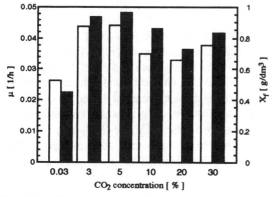

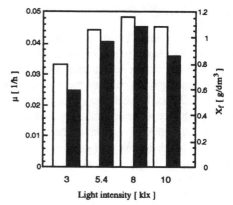

Fig. 1. Effect of CO$_2$ concentration of inlet gas on the specific growth rate (□) and the final cell mass concentration (■) of *Synechococcus* PCC7942.

Fig. 2. Effect of light intensity on the specific growth rate (□) and the cell mass concentration (■) of *Synechococcus* PCC7942.

The control of pH in batch reactor

To evaluate the effect of pH on the CO$_2$-removal characteristics of *Synechococcus* PCC7942, we cultivated the cells of *Synechococcus* PCC7942 under the control of pH ranging from 5.4 to 8.0. The orther parameters were fixed to the optimal condition obtained above. The time courses for these runs were shown in Fig. 3. At pH 5.4, little growth of *Synechococcus* PCC7942 appeared. On the orther side, the cells cultivated at pH 7.4 and 8.0 were capable of growing for the early two days, but then stopped growing. In the latter case, especially, cell color changed from blue-green to yellow at the third day and then these cells died. This phenomenon has been known as chlorosis or bleaching and results from the changes of environmental conditions such as the deprivation of essential nutrient (Collier et al., 1992). Because of the biolysis of *Synechococcus* PCC7942, the data of the cell mass concentrations (X) at pH 8.0 were absent after the second day. The specific growth rate and the final cell mass concentration of the cultivation at pH 6.8, which were 0.0551 1/h and 1.12 g/dm^3 respectively, were the highest in all pH-controlled runs. The elementary analysis of *Synechococcus* PCC7942 showed that carbon atoms contained in 1 g dried cell mass was 0.465 g and this value was constant among all growth phases. Therefore, the CO2 uptake rate (CUR) of *Synechococcus* PCC7942 was obtained by the following equation;

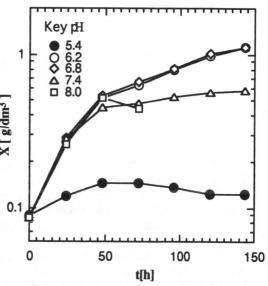

Fig. 3. Time courses of cultivation of *Synechococcus* PCC7942 variables in batch reactor with pH control. X : cell concentration, t : time.

$$CUR \, (g/dm^3h) = \left(\frac{dX}{dt}\right) \cdot 0.465 \cdot \frac{44}{12} \qquad (1)$$

Using this equation, the *CUR* of the cells at pH 6.8 were calculated. Hence, it is found out that the highest *CUR* (0.025 g/dm^{3}h) is achieved when cell is cultivated at the cell mass concentration of 0.286 g/dm^3.

The chemostat cultivation

Keeping the highest *CUR* of *Synechococcus* PCC7942, the total amount of CO_2 uptake of this strain ought to become the highest value. As the *CUR* is defined by the cell mass concentration (Eq. (1)), it is necessary to hold the cell mass concentration at the appropriate value for the continuance of the *CUR* at the highest level. We examined to cultivate the cells of *Synechococcus* PCC7942 in chemostat under the above optimal growth condition. The cell mass concentration was about 0.3 g/dm^3 and kept constant during the cultivation. The time courses of the cell mass concentration and the *CUR* were shown in Fig. 4, which indicates that the *CUR* values were kept constant at about 0.025 g/dm^3h, that was the highest *CUR*, for 150 hours. This result means that an chemostat cultivation of *Synechococcus* PCC7942 in 4000 m³ scale takes in 1 t/h of CO_2, which suggested that it was possible to design the bioreactor for CO_2 fixation of boiler flue gas from 1 t/h of emission sources.

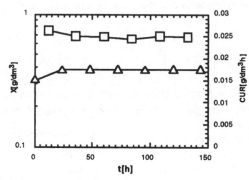

Fig. 4. Changes in cell concentration and CO_2 uptake rate of *Synechococcus* PCC7942 in chemostat
X: cell concentration, CUR: CO_2 uptake rate (= gram of CO_2 fixed by cells in 1 dm³ culture per hour), t: time

REFERENCES

Bryant, D.A. (1994). *The Molecular Biology of Cyanobacteria.* Kluwer Academic Publishers, Netherlands.
Collier, J.L. and Grossman, A.R. (1992) Chlorosis induced by nutrient deprivation in Synechococcus sp. strain PCC 7942: not all bleaching is the same. *J. Bacteriol.* **174**, 4718-4726.
Friedberg, D. and Seijffers, J. (1986) Controlled gene expression utilising Lambda phage regulatory signals in a cyanobacterium host. *Mol. Gen. Genet.* **203**, 505-510.
Golden, S.S. and Sherman, L.A. (1984) Optimal conditions for genetic transformation of the cyanobacterium *Anacystis nidulans* R2. *J. Bacteriol.* **158**, 36-42.
Gruber, M.T., Glick, B.R., and Thompson, J.E. (1990) Cloned manganese superoxide dismutase reduces oxidative stress in *Escherichia coli* and *Anacystis nidulans* . *Proc. Natl. Acad. Sci. USA* **87**, 2608-2612.
Kuhlemeier, C.J., Borrias, W.E., Van den Hondel, C.A.M.J.J., and Van Arkel, G.A. (1981) Vectors for cloning in cyanobacteria: construction and characterization of two recombinant plasmids capable of transformation to *Escherichia coli* K-12 and *Anacystis nidulans* R2. *Mol. Gen. Genet.* **184**, 249-254.
Marcus, Y., Schwarz, R., Friedberg, D., and Kaplan, A. (1986) High CO_2 requiring mutant of *Anacystis nidulans* R2. *Plant Physiol.* **82**, 610-612.
Van der Plas, J., Bovy, A., Kruyt, F., de Vrieze, G., Dassen, E., Klein, B., and Weisbeek, P. (1989) The gene for the precursor of plastocyanin from the cyanobacterium *Anabaena* sp. PCC7937: isolation, sequence and regulation. *Mol. Microbiol.* **3**, 275-284.
Yamada, H., Ohkuni, N., Kajiwara, S., and Ohtaguchi, K. (1996) CO_2-removal characterstics of *Anacystis nidulans* R2 in airlift bioreactors. *Energy* in press

 **Pergamon**

PII: S0196-8904(96)00323-8

Energy Convers. Mgmt Vol. 38, Suppl., pp. S533–S537, 1997
© 1997 Elsevier Science Ltd. All rights reserved
Printed in Great Britain
0196-8904/97 $17.00 + 0.00

STABLY SUSTAINED HYDROGEN PRODUCTION BY BIOPHOTOLYSIS IN NATURAL DAY/NIGHT CYCLE

YOSHIHARU MIURA[1], TORU AKANO[1,2], KIYOMI FUKATSU[1,2], HITOSHI MIYASAKA[1,2],
TADASHI MIZOGUCHI[3],
KIYOHITO YAGI[3], ISAMU MAEDA[3],
YOSHIAKI IKUTA[4], HIROYO MATSUMOTO[4],

KANSAI ELECTRIC POWER CO.[1], TECHNICAL RESEARCH CENTER,
ENVIRONMENTAL RESEARCH CENTER,
11-20 NAKOJI, 3-CHOME, AMAGASAKI, HYOGO 661, JAPAN
RITE AMAGASAKI 2ND AND NANKOH LABORATORIES[2],
OSAKA UNIVERSITY[3], MITSUBISHI HEAVY INDUSTRIES LTD[4],

ABSTRACT

Our photobiological hydrogen production system was successfully scaled up to a pilot plant scale in the photosynthetic starch accumulation and the fermentative production of organic compounds from starch by *Chlamydomonas* MGA 161 and the hydrogen photoproduction by *Rhodovulum sulfidophilum* W-1S from organic compounds produced by *Chlamydomonas* MGA 161 in the natural day/night cycle. The conversion yield of organic compounds from starch of *Chlamydomonas* MGA 161 was 80 to 100 % of the theoretical yield, but the molar yield of hydrogen photoproduction of *Rv. sulfidophilum* W-1S was approximately 40 %. One main cause for the low hydrogen yield of *Rv. sulfidophilum* W-1S was a competition between the accumulation of intracellular Poly (3-hydroxybutyric acid), PHB, and the hydrogen production under light anaerobic conditions. The PHB accumulation was repressed and the hydrogen photoproduction was promoted by enhancing the nitrogenase activity of *Rv. sulfidophilum* W-1S. © 1997 Elsevier Science Ltd

KEYWORDS

Hydrogen; green alga; photosynthetic bacteria; biophotolysis; PHB accumulation; nitrogenase; photosynthetic production of H_2

INTRODUCTION

The use of hydrogen produced by biophotolysis based on microalgal photosynthetic fixation of carbon dioxide is one promising procedure for the reduction of carbon dioxide emission which is a main cause of greenhouse warming. This has induced a number of researchers to study its feasibility (Miura, 1995). However, the key enzyme of the photosynthetic hydrogen production, hydrogenase or nitrogenase is inhibited or inactivated by oxygen produced photosynthetically. Focusing attention on the induction of hydrogenase and hydrogen evolution of green algae under dark anaerobic conditions (Gaffron *et al.*, 1942), we have proposed a stable biophotolysis system which consists of three main steps ; (i) photosynthetic starch accumulation in microalgae in the day ; (ii) algal fermentation to produce organic compounds and hydrogen in the night ; (iii) further conversion of the organic compounds to produce hydrogen by photosynthetic bacteria in the day (Miura *et al.*, 1982). In this system, hydrogen production is temporally separated from oxygen production, and therefore the hydrogen evolution system is able to function free of oxygen inhibition. Photosynthesis can actively occur during each day time. These advantages make the hydrogen production activity sustainable. In addition, pure hydrogen is easily obtained because separation of hydrogen from the oxygen gas is not required.

In this work, the scaling up of our biophotolysis system was investigated, based on the basic researches of photosynthetic hydrogen production in a combination of a marine green alga *Chlamydomonas* MGA 161 and a marine photosynthetic bacterium *Rhodovulum sulfidophilum* W-1S in bench-scale experiments (Miura *et al.*, 1992). The molar yield of hydrogen was discussed. The relationship between the accumulation of intracellular polymer and the hydrogen evolution in *Rv. sulfidophilum* W-1S was investigated.

MATERIALS AND METHOD

The microorganisms used in this work were a marine green alga *Chlamydomonas* MGA 161 and a marine photosynthetic bacterium *Rhodovulum sulfidophilum* W-1S which were isolated and purified in our previous works (Miura *et al.*, 1986, 1992). Those strains were cultivated on a modified Okamoto medium (Miura *et al.*, 1986). The cultivation and hydrogen evolution of those strains were carried out as described previously (Miura *et al.*, 1986, 1992).

The three steps of our biophotolysis system were scaled up to scales shown in Fig. 1. The cultivation of *Chlamydomonas* MGA 161 was done in a raceway type of cultivator of 400L in the day. The fermentative production of organic compounds and hydrogen by *Chlamydomonas* MGA 161 were carried out at 30° C under dark anaerobic conditions in a fermentation tank of 135L. The production of hydrogen from organic compounds by *Rhodovulum sulfidophilum* W-1S was done at 30° C under anaerobic condition with argon in the day in three kinds of photobioreactors *i.e.* raceway, tubular and parallel plate types of reactors of 200L, respectively.

The intracellular polymers were analyzed by 200MHz ^{1}H-NMR, using tetramethylsilane as a standard compound at a concentration of 1 %(v/v) in a chloroform D, $CDCl_3$. The ^{1}H-NMR spectrum was recorded on a Varian VXR-200 apparatus. The 200MHz ^{1}H-NMR was taken from a $CDCl_3$ solution of polymer using a 60° pulse, 3.904 s acquisition time, 3000.3 Hz spectral width, 23×10^3 data points, and 32 accumulations.

The content of a intracellular polymer Poly (3-hydroxybutyric acid), PHB, was analyzed by the following procedure ; (i) pigments were extracted from the lyophilized cells by acetone for 5 h at 70° C ; (ii) the extracts were carefully removed from the cell debris and the remaining cell debris was again suspended with chloroform and the extraction of PHB was carried out at 70° C ; (iii) after extraction for 6 h, the cell debris was removed by passing the suspension through a cellulose filter, and the chloroform solution was concentrated using a rotary evaporator ; (iv) the precipitated polymer was separated and the weight of total PHB content was measured.

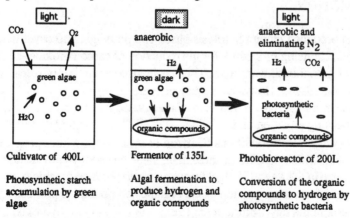

Fig. 1 Process of photobiological hydrogen production by a combination of a marine green alga *Chlamydomonas* MGA161 and a marine photosynthetic bacterium *Rv. sulfidophilum* W-1S .

The activity of nitrogenase was obtained by measuring the reduction rate of acetylene. The assay was done in L-shaped test tubes (15ml) with 4 ml of cell suspension containing 50 mM sodium succinate and 50 μg/ml chloramphenicol. The gas phase in the L-tube was argon containing 10% acetylene. Ethylene formed during 15 min of incubation in the light (75 μmol photon / m^2 / s) at 30° C was measured by gas chromatograph (GC-12A, Shimadzu).

RESULTS AND DISCUSSION

In the scaling up of our biophotolysis system, the stable photosynthetic accumulation of starch and fermentative production of organic compounds from the starch were achieved by *Chlamydomonas* MGA 161 in natural day/night cycle as shown in Fig. 2. The conversion yield of acetic acid, ethanol and glycerol from starch by this alga was 80 to 100 % of the theoretical conversion yield. The hydrogen production by this alga under dark anaerobic conditions was very low.

F : Dark and anaerobic fermentation in 135L tank.

G : Algal cell growth and starch synthesis in 400L raceway type culture vessel.

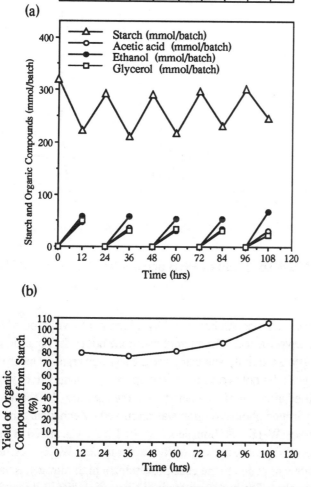

Fig. 2   (a) Photosynthetic starch accumulation in 400L cultivator and fermentative production of organic compounds (acetic acid, ethanol and glycerol) in 135L fermentation tank by *Chlamydomonas* MGA161 in a day/night cylce. (b) Conversion yield of organic compounds (acetic acid, ethanol and glycerol) from starch.

Hydrogen photoproduction rates of *Rv. sulfidophilum* W-1S in three different kinds of photobioreactors were compared and the rate of hydrogen production in a parallel plate type of reactor was higher than those in a tubular and a raceway types of reactors.   Therefore, the parallel plate type of reactor was used for the hydrogen photoevolution by *Rv. sulfidophilum* W-1S.   The stable hydrogen photoevolution by *Rv. sulfidophilum* W-1S was repeatedly observed in the day time of natural day/night cycle, but the molar yield of hydrogen was approximately 40 %.   The cause for the low yield of hydrogen of this photosynthetic bacterium was investigated.

Under light anaerobic conditions which were the same with those for the hydrogen photoevolution, *Rv. sulfidophilum* W-1S accumulated a considerable amount of polymers.   The nature of the intracellular polymers were analyzed by 200MHz ^{1}H-NMR.   The three typical signals which fully correspond to the signals of the standard Poly(3-hydroxybutyric acid), PHB, appeared as shown in Fig. 3.   This suggests that the intracellular polymer is PHB.

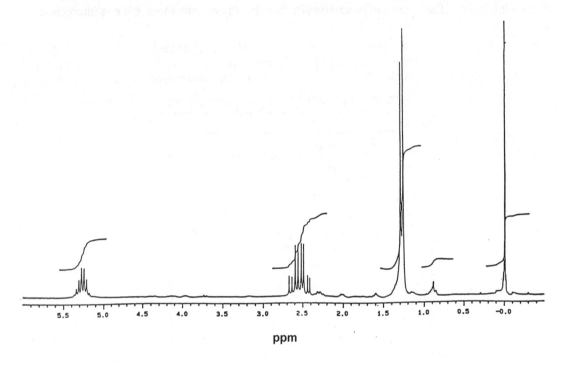

Fig. 3   200MHz ^{1}H-NMR spectrum of PHB accumulated by *Rv. sulfidophilum* W-1S.

The relationship between PHB accumulation and hydrogen photoevolution of *Rv. sulfidophilum* W-1S was investigated.   An attempt was done under the conditions having the cells with no initial nitrogenase activity and using acetate as a substrate.   Under no initial nitrogenase activity, the cells did not evolve hydrogen up to 6 h of incubation, whereas the PHB accumulation was increased after 3 h of incubation.   On the contrary, the accumulation of PHB was repressed and the hydrogen photoevolution was promoted by enhancing the nitrogenase activity of *Rv. sulfidophilum* W-1S.   Sodium succinate of 1 mM was added at the beginning of incubation to enhance the nitrogenase activity of *Rv. sulfidophilum* W-1S .   After 27 h of incubation, sodium acetate was added to the cells possessing a high nitrogenase activity of 1087 nmol C_2H_4/ h /mg dry weight.   The hydrogen evolution was high after 1 h and 3 h from the addition of acetate, whereas the increase of PHB accumulation was very low as shown in Table 1.

Table 1. Hydrogen photoevolution and PHB accumulation of *Rv. sulfidophilum* W-1S having a high nitrogenase activity of 1087 nmol C_2H_4/h/mg dry weight.

Incubation time (h) after acetate addition	H_2 evolution (µmol/200 ml culture)	PHB content (% of cell dry weight)
0	0	47.9
1	37.5	46.8
3	63.1	48.4

PHB is widespread intracellular storage compound in procaryotic organisms (Anderson and Dawes, 1990). The polymer is accumulated when carbon and energy sources are in excess, but growth is limited by the lack of a nitrogen, sulphur or phosphorous source. Both hydrogen evolution and PHB accumulation could be considered to be electron sinks in *Rv. sulfidophilum* W-1S. The competition for the consumption of reducing equivalents would appear to exist between PHB accumulation and hydrogen evolution under light anaerobic conditions in the photosynthetic bacterium *Rv. sulfidophilum* W-1S. The nitrogenase activity of *Rv. sulfidophilum* W-1S should be enhanced in the preculture to repress the PHB accumulation and promote the hydrogen production in our biophotolysis system.

ACKNOWLEDGEMENT

This research is conducted as a part of the Industrial Technology Development Promotion Program of Research Institute of Innovative Technology for the Earth (RITE) for the global environmental problems, supported by the Ministry of International Trade and Industry.

RERERENNCES

Anderson, A. J.and E. A. Dawes (1990). Occurrence, metabolism, metabolic role, and industrial uses of bacterial polyhydroxy alkanoates. *Microbiol. Rev.*, **54**, 450-472

Gaffron , H. and J. Rubin (1942). Fermentative and photochemical production of hydrogen in algae. *J. Gen. Physiol.*, **26**, 219-240.

Miura, Y. , K. Yagi, M. Shoga and K. Miyamoto (1982). Hydrogen production by a green alga, *Chlamydomonas reinhardtii*, in an alternating light/dark cycle. *Biotech. Bioeng.*, **24**, 1555-1563.

Miura, Y., S. Ohta, M. Mano and K. Miyamoto (1986). Isolation and characterization of a unicellular marine green alga exhibiting high activity in dark hydrogen production. *Agric. Biol. Chem.*, **50**, 2837-2844

Miura, Y., C. Saitoh, S. Matsuoka and K. Miyamoto (1992). Stably sustained hydrogen production with high molar yield through a combination of a marine green alga and a photosynthetic bacterium. *Biosci. Biotech. Biochem.*, **56**, 751-754 .

Miura, Y. (1995). Hydrogen production by biophotolysis based on microalgal photosynthesis. *Process Biochem.*, **30**, 1-7.

Energy Convers. Mgmt Vol. 38, Suppl., pp. S539–S544, 1997
© 1997 Elsevier Science Ltd. All rights reserved
Printed in Great Britain
0196-8904/97 $17.00 + 0.00

Pergamon

PII: S0196-8904(96)00324-X

THE SYNTHESIS OF ALTERNATIVES FOR THE BIOCONVERSION OF WASTE-MONOETHANOLAMINE FROM LARGE-SCALE CO₂-REMOVAL PROCESSES

KAZUHISA OHTAGUCHI[1] AND TAKAHISA YOKOYAMA[2]

[1]Department of Chemical Engineerig, Tokyo Institute of Technology,
2-12-1 Ookayama, Meguro-Ku, Tokyo 152, Japan
[2]Komae Research Laboratory, Central Research Institute of Electric Power Industry,
2-11-1 Iwamoto Kita, Komae-Shi, Tokyo 201, Japan

ABSTRACT

The alternatives for bioconversion of monoethanolamine(MEA), which would appear in large quantities in industrial effluent of CO_2-removal process of power companies, have been proposed by investigating the ability of some microoganisms to deaminate MEA. An evaluation of biotechnology, which include productions from MEA of acetic acid and acetaldehyde with *Escherichia coli*, of formic and acetic acids with *Clostridium formicoaceticum*, confirms and extends our earlier remarks on availability of ecotehnology for solving above problem. © 1997 Elsevier Science Ltd

KEYWORDS

carbon dioxide; MEA process; bioconversion; waste treatment; biodegradation; acetic acid; *Escherichia coli*; *Clostridium formicoaceticum*

INTRODUCTION

Of most available CO_2-removal process for the massive CO_2, discharged into boiler flue gas of large fossil-fueled power plants, is the "MEA (monoethanolamine, 2-aminoethanol) process " (Bottoms, 1933) in which 2-aminoethanol is used as an acid-gas absorbent. The technical term represents the combination of an acid-gas absorber with a MEA regenerator. The regenerated solution of MEA is continuously recycled back to the top of the absorber, while a portion of it is periodically distilled in a reclaimer to remove insoluble materials(including amine degradation products) which promote foaming of the circulating solution and thence result in high amine losses(Pauley *et al.*, 1988). When the amine degradation becomes the most concern, the contaminated solution of MEA is replaced with new solution. To date, no large-scale schemes for treating the industrial effluent of the MEA process have come close to recognizing the massive discharge of it.

In order to treat large quantities of MEA in the effluent we have been pursuing studies involving the bioconversion approach to generate useful products. Initial attempt was made to create and solve a primitive problem, associated with bioprocessing, using the laboratory culture of *Escherichia coli* K12(Ohtaguchi *et al.*,1994). The nitrogen requirements were fulfilled by the microbial breakdown of MEA to ammonium ion and acetaldehyde. Carbons, produced in this reaction, were converted to acetic acid and biomass. The continued research is under way to create alternatives, since the cleavage of MEA is known to occur also with several other organisms. Our present report confirms and extends the earlier remarks on availability of bioconversion technology with other alternatives.

PROBLEM DEFINITION

Figure 1 shows the basic flow scheme for the MEA process of which principal reactions of MEA with CO_2 are represented as follows:

$$2\ HOCH_2CH_2NH_2 + H_2O + CO_2 \quad \rightarrow \quad (HOCH_2CH_2NH_3^+)_2CO_3^{2-} \tag{1}$$
monoethanolamine

$$(HOCH_2CH_2NH_3^+)_2CO_3^{2-} \quad \rightarrow \quad 2HOCH_2CH_2NH_2 + H_2O + CO_2 \tag{2}.$$

The temperatures, stated in the references for existing processes, are 38-49°C at the absorber (Eq.(1)), and 93-127°C at the regenerator (Eq.(2)). Of crucial to operating cost of existing MEA process is the purchase of MEA to offset MEA-loss, mainly caused by the irreversible reactions. MEA is thermally stable even at the normal regenerator temperature, however, it is shown to be vulnerable to oxidative degradations (Kohl and Riesenfeld, 1960). One such degradation product, N-(2-hydroxyethyl)-ethylenediamine, has been shown to be unregenerated and present in lean-MEA. About 20% of carbon oxysulfide reacts irreversibly with MEA to form degradation products such as oxazolidone-2. In cases where carbon disulfide is an impurity of flue gas, it irreversibly reacts with MEA to form N,N,N',N'-tetra(2-hydroxyethyl)-thiocarbamide, which causes considerable MEA-loss. The formation of irreversible products significantly contributes to the duties of reclaimer and filters. Normal reclaimer in LNG-fired plants treats about 5% of circulating solution, which is effected by semicontinuous steam distillation, followed by withdrawal of a portion of degradated solution and feeding of fresh solution. Such reclaimer-operation which is experienced in chemical processing industries by others occurs once per 3-5months. When the accumulation of contaminants becomes a grave operational emergency, the daily operation of MEA process is followed by restart after operational shutdown, which further generates a large amount of wastewater that must be disposed of safely and economically. The bulk of the compositions of it, although there appear degradation products, are water and MEA (usually, 10-20%), among which MEA presents a potential risk for causing severe injuries to the eyes and skin and its threshold level in vapor concentration is set at $6mg/m^3$ in guidelines from the public health aspects(Patty, 1963).

Table 1 tabulates design and operating data, which were assumed in the most recent paper concerning a CO_2-removal MEA plant in 600,000 kilowatts of generating capacity by the thermal combustion of LNG, which emit 272 tons of CO_2 gas per hour(Yokoyama et al., 1995). Design feed gas flow rate to this plant was so voluminous as $1,263,000Nm^3/h$ that four trains, each of which involves 2 water-wash processes (5.538m ID x 15m height), 2 absorbers (4.7m ID x 44.2m height) and 2 regenerators(4.66m ID x 26.7 m height), were considered. From these data, the amount of circulation rate of MEA in the systems is evaluated as 1,306 t/h, that is, 6,540t-20% MEA/h. If reclaimer runs for 30 days per year, it would be approximated that the annual discharge of MEA waste at the reclaimer bottom reaches a mass of the order of 10,000t, that is, 1 t/h. This figure is conservative for the evaluation of total production of the plant effluents since the discharge at above operational shutdown is not accounted. The gas treatment of existing MEA process in chemical processing industries, such as Gas/Spec FT1, Kerr-McGee CO_2 recovery, Econamine FG, has been confined to CO_2-removal of the order of no more than 50t/h, which is effected by MEA of circulation rate of no more than 200 t/h. The scale-up ratio from these processes to the 600,000 kw-power-station scale is extremely large at approximately 1:10, and hence the consideration in advance of waste disposal of MEA is found to be crucial in the design of large-scale plant.

Destruction by fire is likely regarded as economically infeasible due to high content of water in the effluent which requires considerable energy for its evaporation. Land treatment systems might be capable of decomposing a small amount of MEA, however, they reluctant to a large amount of effluents for the above risk. On the other hand, bioreactor treatment, which in general shows appreciable operating cost over other treatments, seems to be suited to play paramount roles for this purpose.

The microbial decomposition of MEA, which is a compound assembled from four of the most predominant elements of organisms and appears in eucaryotic membranes as phosphatidylethanolamine, has already been exhibited by biochemists(Blackwell et al., 1978). A model for the biological treatment of MEA, using *Escherichia coli* K12, was formulated in our previous work, which described biotransformation of MEA to

Table 1 Design and operating data of MEA process in 600,000 kilowatts of LNG-fired generating capacity.

<SCOPE>	
Gas rate	$1,263,000 \ Nm^3 / h$
CO_2-removal	245 t/h
% of CO_2-removal	90 %
Product gas	99.0% CO_2
<MEA>	
Conc.	20 wt %
CO_2 loading(rich)	$0.4 mol\text{-}CO_2/mol\text{-}MEA$
Circulation rate per path	1338.3 kg-mol/h
Number of path	16
<REGENERATOR>	
Reboiler pressure	400,000 Pa
Heat duty	$1,200 \ kcal/kg\text{-}CO_2$

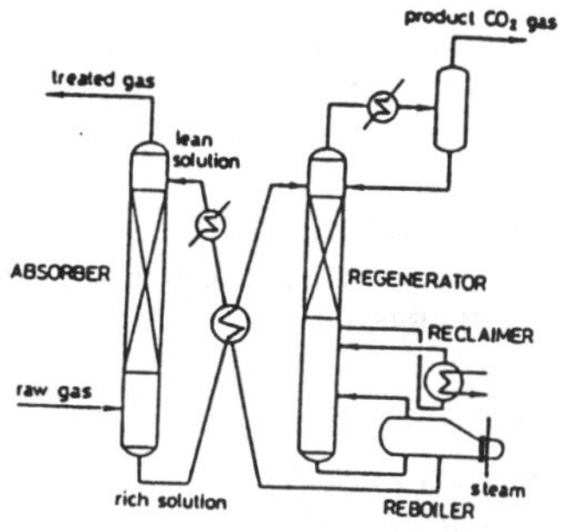

Fig. 1   MEA process for CO_2-removal

ammonium ion for biomass formulation, acetaldehyde and acetic acid by the following reactions:

$$H_3N^+CH_2CH_2OH \rightarrow NH_4^+ + CH_3CHO \quad (3),$$
ethanolamine ammonia-lyase

$$CH_3CHO + CoA + NAD^+ \rightarrow CH_3COCoA + NADH + H^+ \quad (4),$$
aldehyde dehydrogenase

$$CH_3COCoA + Pi \rightarrow CH_3CO_2PO_3^{2-} + CoA + 2H^+ \quad (5),$$
phosphate acetyltransferase

$$CH_3CO_2PO_3^{2-} + 2H^+ \ ADP \rightarrow CH_3COOH + ATP \quad (6).$$
acetate kinase

METABOLIC ALTERNATIVES

Essential to proper treatment in bioreactors is transformation of MEA to renewable resources under the controlled condition. In order to compare the *E.coli* - mediated-biotransformation of MEA with other biological substitutes, metabolic alternatives were created (**Table 2**) with reference to the previous works(Blackwell *et al.*, 1976, Atkinson *et al.*, 1980, Krieg *et al.*, 1984, Sneath *et al.*, 1986, Moller *et al.*, 1986). The presence of the cobamide-dependant enzyme ethanolamine ammonialyase was also described for *Klebsiella pneumoniae*, *Salmonella tryphimurium*, *Clostridium* strains, of which reaction is defined by Eq.(3). In contrast the coenzyme B_{12}-independent deamination of MEA, in which ethanolamine *O*-phosphate appears as the intermediate, was known to occur in *Erwina* species, *Morganella morganii*, most *Pseudomonas* strains, *Flavobacterium phenanum*, *Achromobater* strains. Acetaldehyde formation of these organisms from MEA is shown by the following equations(Jones *et al.*, 1973):

$$H_3N^+CH_2CH_2OH + ATP \rightarrow H_3N^+CH_2CH_2OPO_3^{2-} + ADP \quad (7),$$
ATP-amino alcohol phosphotransferase

$$H_3N^+CH_2CH_2OPO_3^{2-} + OH^- \rightarrow CH_3CHO + NH_4^+ + PO_4^{3-} \quad (8).$$
amino alcohol *O*-phophate phospholyase

The mechanism by which MEA is assimilated by yeast *Saccharomyces cerevisiae* mutant auxotrophic for MEA(*CHO* 1) differs from that of procaryotic cells(Atkinson *et al.*, 1980). In this mutant, MEA is incorporated into cells to confer the situation where phosphatidylserine is not available as a precursor.

The first screening was performed to eliminate the alternatives which are inevitable of the production of toxins. The survivors from this selection involve followings:
(1) production of livestock feed SCP, amino acids and acetic acid in the aerobic culture of *E.coli* K12,
(2) production of amino acids and fuel ethanol in the aerobic culture of exogenous *ALDH* (alcohol dehydrogenase)bearing *E.coli* C600/pLOI 297,

Table 2 Metabolic alternatives for MEA degradation

Strain	Major products and toxicity
<Enterobacteriaceae>	
E.coli K12	acetic acid, acetaldehyde, amino acids, biomass; nontoxic
E.coli NCIMB 8114	acetic acid; pathogenic
K.pneumoniae NCIMB418,8267	acetic acid; pathogenic
E.ananas NCPPB441, *E.carotovora* NCPPB1280	acetaldehyde; pathogenic to plants
E.milletiae NCPPB955	ethanol; pathogenic to plants
M.morganii	acetaldehyde, NH_4^+; pathogenic
S.typhimurium	pathogenic
<Pseudomonadaceae>	
P.fluorescence NCIMB9046, *P.aureofaciens* NCIMB9030, *P.aureofaciens* NCIMB9030	nontoxic
P.pseudoalcaligenes NCIMB9946, *P.flava* DSM619, *P. psedoflava* DSM1034	PHB; nontoxic
<Clostridium>	
C.absonum DSM599, *C.barkeri* NCIMB 10623, *C. cadaveris* NCIMB 10676, *C.propinicum* NCIMB10656, *C. scatologenes* NCIMB8855, *C.subterminale* NCIMB 9384	H_2; culture supernatant are non-toxic to mice
C.difficile NCIMB10666, *C.histolyticum* NCIMB503, *C.limosum* NCIMB 10638, *C.perfringens* DSM1801	toxic
C.formicoaceticum DSM92	acetic and formic acids; nontoxic
C.hastiforme DSM1786	H_2S;culture supernatant are nontoxic to mice
C.lituseburense NCIMB 10637	butyric,acetic and iso-valeric acids; nontoxic
<Yeast>	
S.cerevisiae MC13, KA101	biomass; nontoxic

(3) production of biodegradable polymer; poly-ß-hydroxybutyric acid(PHB) in the aerobic culture of *Pseudomonas pseudoalcaligenes*, and
(4) production of acetic acid and formic acid in the anaerobic culture of *Clostridium formicoaceticum*.

DESIGN SCOPE

Experimental

To illustrate the scope for above alternatives, *Escherichia coli* K12(IFO 3301) and *Clostridium formicoaceticum* ATCC 27076 were selected as test organisms and batch cultivations for each of them were performed in our laboratory. *E.coli* was grown aerobically at 37°C on the synthetic medium containing carbon source glycerol($20g/dm^3$) and nitrogen source MEA·HCl($10g/dm^3$), according to the method described before (Ohtaguchi *et al.*, 1995). *C.formicoaceticum* ATCC 27076 was grown anaerobically in N_2 /CO_2 at 37°C on the synthetic medium containing(in g/dm^3) K_2HPO_4 (3.5), $NaHCO_3$ (5), $MgSO_4·7H_2O$ (0.2), $CaCl_2·2H_2O$(0.01), fructose(20), MEA· HCl(4.2), yeast extract(0.2), and (in mg/dm^3) EDTA(5), $FeSO_4·7H_2O$(2), $ZnSO_4·7H_2O$(0.1), H_3BO_3(0.3), $MnCl_2·4H_2O$ (0.03), $CoCl_2·6H_2O$ (0.2), $CuCl_2· 2H_2O$ (0.01), $NiCl_2·6H_2O$(0.02), $Na_2MoO_4·2H_2O$ (0.03), thiamine·HCl(14), nicotinic acid(7), pantothenic acid(3.5), pyridoxine (7), vitamin B_{12} (0.28), p-aminobenzoic acid (0.7), biotin (1.4), resazurin (1) and cystein·HCl (300). The pH of the culture for both strains was controlled at 7.5 with the addition of 1 mol/dm^3 KCl, guided by On/Off controller. Analytical methods were described previously.

Results and Discussion

Table 3 presents a summary of experimental results, from which the effective volume of bioreactor for the treatment of a 1 t/h waste MEA using *E.coli* is calculated as:
    1 t/h / (0.261x 10^{-3} t/($m^3·h$)) = 3,830m^3,

Table 3 Characteristics of biodegradation of MEA

	E.coli K12	C.formicoaceticum
Conversion	1.0	0.776
Reaction rate[g/(dm³•h)]		
MEA degradation	0.261	0.0851
Biomass production	0.269	0.0736
Acetic acid production	0.394	1.01
Formic acid production	nd.	0.391

and that using *C.formicoaceticum* as:

$1\ t/h\ /\ (0.0851x\ 10^{-3}\ t/(m^3•h)) = 11,800m^3$.

There is some references on large-scale bioreactors such as SCP(single cell protein) manufacturing process, developed by ICI Imperial Chemical Industries, Ltd), which has an effective volume of $1,560m^3(7m$ ID x 60m height, total volume:$2,300m^3$). In view of this, only the first case seems to be survived for practical purposes. If this ICI standard size $2,300m^3$ bioreactor designed by ICI is specified, then construction of 2 bioreactors is found to be effective for effluent treatment of LNG-fired power plant. The amount of acetic acid that may be recovered from these 2 units is calculated as:

$0.394\ x10^{-3}\ t/(m^3•h)\ x\ 3,830m^3 = 1.51\ t/h$,

and that of biomass as:

$0.269\ x10^{-3}\ t/(m^3•h)\ x\ 3,830m^3 = 1.03t/h$.

The bioreactor of *C. formicoaceticum* was eliminated from our candidates. Nevertheless there is a possibility that this reactor may be applied to MEA process of aforecited chemical processing industries. In our previous studies (Suzuki *et al.*, 1993), we have demonstrated the anaerobic incorporation of CO_2 into the carbon of acetic acid by the culture of homoacetogen *Acetobacterium woodii*, in the presence of H_2 and CO_2. It is especially noteworthy that *C. formicoaceticum* synthesized acetic acid in the absence of H_2, and hence this technology also appears promising for engineering purposes.

E.coli K12 is of most genetically-characterized organisms in which recombinant DNA techniques are available. We have transformed a K12-mutant strain C600 with plasmid pLOI297, which was constructed by others with *Zymomonas mobilis* genes encoding pyruvate decarboxylase and alcohol dehydrogenase II(Ingram *et al.*, 1988). The uses of such mutants, which are capable of ethanol production, and of PHB synthesizing bacteria may open new possibility of MEA treatment technology.

CONCLUSION

The volumetric scale-up ratio of the MEA process from chemical-processing-industry to steam-electric power company scale is extreamly large, and hence, in the scale-up-planning stage, environmental consideration of waste-MEA appeared to be more significant. This study was an extention of our investigations of bioconversion of waste-MEA to renewable products. Of the alternatives studied, the high possibility of indroduction of *Escherichia coli* was confirmed. This aspect will receive further attension.

This work was supported in part by Grants-in Aids for Scientific Research(No.03650763, 04650842) of the Ministry of Education, Science and Culture, Japan.

REFERENCES

Arnold,D.S., D.A.Barrett and R.H.Isom(1982). CO_2 can be produced from flue gas. *Oil & Gas J.*,81, Nov.22,130-136.

Atkinson,K., S.Fogel and S.A.Henry(1980). Yeast mutant defective in phosphatidylserine synthesis," *J. of Biol.Chem.*, 255 (14), 6653-6661.

Blackwell,C.M., F.A.Scarlett and J.M.Turner(1976). Ethanolamine catabolism in bacteria, including *Escherichia coli. Biochem. Soc. Trans.*, 4, 495-497

Blackwell,C.M. and J.H.Turner(1978). Microbial metabolism of amino alcohols. *Biochem.J.*, 176,751-757.

Bottoms,R.R.(1933). U.S. Pat. Re. 18,958, Sept.26.

Ingram,L.O. and T.Conway(1988). Expression of different levels of ethanologenic enzymes from *Zymomonas mobilis* in recombinant strains of *Escherichia coli. Appl.Environ.Microbiol.*,54(2),397-404.

Jones,A., A.Faulkner and J.Turner(1973). Microbial metabolism of amino alcohols.*Biochem.J.*,134,959-968.

Kohl,A.L. and F.C.Riesenfeld(1960). In: *Gas purification*. McGraw-Hill, New York, USA(1960).

Krieg,N.R. and J.G.Holt(1984). In: *Bergey's Manual of Systematic Bacteriology*, vol.1, Williams & Wilkins, Baltimore

Sneath,P.H.A., N.S.Mair, M.E.Sharpe and J.G.Holt(1986). In: *Bergey's Manual of Systematic Bacteriology*, vol.2, Williams & Wilkins, Baltimore.

Moller, B., H.Hippe and G.Gottschalk(1986). Degradation of various amine compounds by mesophilic clostridia. Arch. Microbiol., 145, 85-90

Ohtaguchi,K., K.Koide and T.Yokoyama(1995). An ecotechnology integrated MEA process for $CO2$-removal. *Energy Convers.Mgmt.*, 36,401-404.

Patty,F.A.(1963). In: *Industrial hygine and toxicology*. vol.2, Interscience, New York, USA.

Pauley,C.R. and B.A.Perlmutter(1988). Texas plant solves foam problems with modified MEA systems. *Oil & Gas J.*, 87, Feb29, 67-70.

Suzuki,T., T.Matsuo, K.Ohtaguchi and K.Koide(1993). Continuous production of acetic acid from CO_2 in repeated-batch cultures using flocculated cells of *Acetobacterium woodii. J.Chem.Eng., Japan*, 26(5), 459-462.

Yokoyama,T. and S.Kudou(1995). Evaluation of chemical absorption process for CO_2 removal from the flue gases of LNG-fired power plant. *CRIEPI Report :Komae Research Lab. Rep.*, No.T94057.

Energy Convers. Mgmt Vol. 38, Suppl., pp. S545–S549, 1997
© 1997 Elsevier Science Ltd. All rights reserved
Printed in Great Britain
0196-8904/97 $17.00 + 0.00

Pergamon

PII: S0196-8904(96)00325-1

Biomass Development and Waste Wood Co-Firing

Wallace Benjamin, Professional Engineer
New York State Electric & Gas, Binghamton, New York (USA)

Abstract

Waste wood fuels have been co-fired with coal for a number of years. Many utilities and industries have experimented in the development of this alternative fuel for a host of reasons. Utilities are facing major changes in the future and industries are continuously looking for ways to help the bottom line. Waste wood fuel can lower production costs and provide fuel flexibility at a coal-fired power plant. Wood waste from saw mills, wood products manufacturing, land clearing, furniture companies, pallets, particle board, plywood, railroad ties and utility poles all can serve as supplemental fuels. Reduced sulfur dioxide (SO_2) and carbon dioxide (CO_2) emissions will be realized with the offset of the non - renewable coal.

New York State Electric & Gas (NYSEG) has experience in mixing waste wood with coal for co-firing in stoker plants and Pulverized coal units. We are also developing a separate wood feed system to directly blow prepared wood fuel into a pulverized coal boiler. This approach has the potential of increasing the percentage of waste wood that can be burned with coal in a pulverized coal boiler above a rule of thumb 5 % limit when wood fuel is blended with coal.

The objective of our paper is to provide some history of waste wood co-firing, identify some pitfalls and address emissions impacts and greenhouse gas reductions from co-firing.
© 1997 Elsevier Science Ltd

Key Words

Biomass, cofiring, Dedicated Feedstock Supply System (DFSS), pulverizer, retrofitting

Introduction

With the effects of retail wheeling, open access, regulatory changes, and uncertainty in the electric industry, it is imperative that NYSEG respond by reducing operating and fuel costs while maintaining an increased focus on reducing emissions. A current NYSEG project is evaluating the adaptability of a typical pulverized coal (PC) plant built in the 1950s to use renewable wood fuels with coal. The goal is to reduce air emissions while maintaining good operational procedures and cost controls.

NYSEG has been researching the viability of cofiring renewable wood fuel with coal in an existing pulverized coal plant. The tests were conducted at NYSEG's Greenidge Station Plant in Dresden, New York, on the western shore of Seneca Lake. This plant's first generating unit came on line in 1938, with additional units added in 1942, 1950, and 1953. Since then, the first two units have been retired and Unit 3 is currently on reserve standby.

Realizing the importance of producing electricity in an efficient and environmentally acceptable manner, the co-fire option is a near-term alternative that can recover energy from waste wood instead of allowing the wood to be landfilled. With cofiring, a utility can continue near-normal operation and provide economic growth to the neighboring agriculture and lumber industries.

The current wood fuels for co-firing are residues from a variety of wood processing industries. The development of Dedicated Feedstock Supply System (DFSS) using willows as a new cash crop may bolster a weak farm economy in the Northeast and generate a renewable fuel that can be an asset to a utility as a predictable, uniform, and sustainable energy source.

NYSEG Program

NYSEG is in the second phase of developing resources and systems for cofiring biomass with coal. In the first phase, stoker boilers were fired with biomass (typically wood waste products). Encouraged by positive results at the older stokers, NYSEG decided to develop the process for its pulverized coal boilers beginning with Greenidge Station.

The test burns at Greenidge Station demonstrated that a parallel fuel feed system can effectively provide wood products to a PC unit. Emission results were promising but inconclusive. Additional testing, for longer durations, at varied loads and with different wood products needs to be conducted to clarify and establish relationships between the percent wood fired at varying moisture contents. Loads need to be varied to develop continuous emission monitor emission data that can be compared to coal-only data.

Economic analysis indicates that it will be beneficial to further refine the equipment and systems. Refinements may include chipping and drying equipment, plus installation of fuel storage and feed systems with permanent boiler penetration. NYSEG will attempt to identify the problems associated with cofiring by direct injection, compared to cofiring a biomass/coal mixture through the existing fuel handling system. Specifically, an examination will be made of fuel size criteria and the system modifications necessary for minimal impacts on coal-fired operation.

Boiler/Generating Unit Description and Modifications

Greenidge Station is well suited to burning wood fuel as a pilot project. This plant is not scheduled for the addition of a scrubber and is NYSEG's third most efficient unit. Greenidge Station is a 108-MW pulverized coal unit equipped with a General Electric turbine generator and a 665,000-lb Combustion Engineering tangentially-fired boiler. Greenidge Station is located in the center of New York, surrounded by farms, forests, vineyards, and orchards. It is located close to the New York State Thruway and adjacent to Route 14, with good truck access. At 10 percent cofiring, 10.8 MW of renewable energy could be produced.

The test program has used the active 108-MW reheat Unit 4. The tangentially fired Combustion Engineering boiler can produce steam at 1005°F at 1465 psi. The plant size, upgraded electronic boiler controls, and continuous emissions monitoring systems make this unit ideally suited for research and testing.

Fuel Delivery System Description

The mechanical/pneumatic test system that was selected and installed has a capacity of 7.5 tons per hour of alternate wood fuel input. The test system allows for direct firing of wood fuel in the boiler without the need to blend the fuel with coal prior to combustion. This flexibility allows for testing small quantities of alternate wood fuels with minor impacts on operations and maintenance at the plant.

Designs for retrofitting the current coal burning plant to co-fire biomass are based on the boiler receiving, via a separate fuel feed system, wood products reduced to a top size of 1/4 inch for burning in suspension. This method was selected because other utility experience with mixing wood with coal and then reducing in the pulverizers showed that, at 5 percent wood by weight or above, the mills lose efficiency, which impacts the coal sizing[1]. There are additional advantages to using a separate fuel feeding system. One is the ability to quickly change the blend of alternative fuel being fired to match various operational needs. This system can be designed to feed wood at much higher rates than could be fed through a coal pulverizer.

Fuels Used and Pre-Combustion Preparation

The technology for size reduction and drying is available today. Size reduction methods and technology must be developed from an economic standpoint, as well as for the practical applications used by either fuel vendors or owners of pulverized coal combustors. Processing and drying techniques need further economic evaluation, research and development.

Fuel can be obtained with a 2-in. nominal size from whole tree chippers and wood processors. Grinders do not normally produce a product that has good flow characteristics. The wood fibers are sticky, stringy, and elongated when produced from a grinding operation. The fuel product needs to be processed by equipment that produces a chip. A stoker or cyclone boiler unit can burn 2 in. chips with limited or no modifications. To burn a 2-inch chip in a pulverized coal unit is not possible without the addition of a grate system.

Our research indicates that a wood fuel product of less than 1/4 in. diameter is desired. The primary particle size distribution should be between 1/8 in. and 1/16 in.

The fuel source used for this project was obtained from the sawmill industry. The sawdust from the sawmill was delivered on walking floor trucks and ranged in quantity from 15 to 23 tons. The first test load was delivered in a dump trailer and dumped on the parking lot adjacent to the fuel feed hopper. The walking floor trucks were unloaded directly to the fuel feed hopper.

There is some competition for sawdust because dairy farms use this material for bedding. Pellet fuel manufacturers also use sawdust for fuel for pellet wood stoves. However, large lumber producers and furniture manufacturers have expressed the desire for a steady long-term outlet for their byproducts. They are also motivated by a desire to find an environmentally sound method of waste disposal.

---

[1]The results of these tests are documented in an EPRI report "Wood Fuel Co-Firing at TVA Power Plant," Palo Alto, CA, June 1993.

Emissions Projections

Fuel use estimates from 1990 to 1995 indicates that NYSEG has burned of 60,600 tons of wood with coal at three of its generating stations. This has offset 27,900 tons of non renewable coal and reduced CO_2 emissions by 71,600 tons and SO_2 emissions by 1100 tons. NYSEG has been burning Tire Derived Fuel (TDF) with coal since 1991 also offsetting 21,400 ton of coal from the fuel stream.

Projected estimates of burning 40,000 tons of wood (6500 btu's per lb.) product per year, NYSEG will eliminate the combustion of 20,000 tons of coal (13,000 Btu's per lb.) at the 108 mw Greenidge Station. This could also eliminate 600 tons of SO_2 emissions, and possibly decrease NO_x emissions by 120 tons and replace 20,000 tons of fossil CO_2 with biomass CO_2 (a renewable resource). If 100 similar wood processing and fuel feed systems are in use by the year 2010, then 2,000,000 tons of coal would be saved, SO_2 emissions would be reduced by 60,000 and 42,9000 tons of fossil CO_2 would be saved by renewable biomass CO_2.

Viability of Co-Firing

The retrofit of an existing pulverized coal unit to cofire up to 15 % will cost approximately $2M - $5M, or considerably less than a new wood-firing facility. The need for new generation in the Northeastern U.S. is very questionable at best. Co-firing can help key account customers dispose of waste products. Operational impacts can be held to a minimum. A separate fuel feed system allows for fuel switching to respond to load requirements.

Conclusions

The test burns at Greenidge Station demonstrated that a separate fuel feed system can effectively feed wood products to a pulverized coal unit. A 10 percent boiler Btu input was fed at a continuous rate for approximately 2 hours with no noticeable negative impacts on boiler operation. The fuel feed rate produced about 9 MW of the 89 MW system output. The sawdust used had a moisture content of approximately 10.4 percent. Other testing was conducted using green sawdust, having a moisture content of approximately 38 percent. With this material, an output of about 4-5 MW was obtained of the 94 MW system output. The duration of this test was 4 hours.

Retrofitting an existing plant would not require any new points of emissions, and the required equipment installations would be at an existing industrial site. A new wood-fired facility which would require the development of a new site and a new stack with air emissions and water discharge permits. Other environmental considerations are noise, traffic, and visual impacts.

Emission results were promising. NO_x reductions were observed when feeding the higher moisture content wood, but not the dry wood. This needs further evaluation to determine the cause. Additional testing, for longer durations, at varied generating loads and with different wood products need to be conducted to clarify and establish relationships between the wood fired (wet vs. dry) and the emissions. Generating loads need to be varied to develop continuous emission monitor (CEM) emission data that can be compared to coal-only data. EPA factor calculations need to be established for wood co-firing.

The economic analysis indicates that it may be beneficial to pursue further refinements to the equipment and systems. These refinements may include chipping and drying equipment, plus installation of fuel storage and feed systems with permanent boiler penetration. The project team concluded that the next step is to further evaluate the generation and emission impacts with woods of varied moisture contents and at varied Btu input rates. This data then can be used to determine if a drying system would be a cost effective option.

Energy Convers. Mgmt Vol. 38, Suppl., pp. S551–S556, 1997
© 1997 Elsevier Science Ltd. All rights reserved
Printed in Great Britain
0196-8904/97 $17.00 + 0.00

 **Pergamon**

PII: S0196-8904(96)00326-3

The Shawville Coal/Biomass Cofiring Test:

A Coal/Power Industry Cooperative Test of Direct Fossil-Fuel CO$_2$ Mitigation

Earl F. Hunt [1], David E. Prinzing [2], Joseph J. Battista [3], and Evan Hughes [4]

ABSTRACT

Under the sponsorship of the Pennsylvania Electric Company (Penelec), and with the support of EPRI, the U.S. Department of Energy, and the Pennsylvania Department of Conservation and Natural Resources, Penelec and Foster Wheeler Environmental Corporation conducted low percentage (3% by weight) wood cofiring tests at Units 2 and 3 of the Shawville Generating Station in November 1995. Unit 2 is a 138 MW$_e$ (gross) wall-fired pulverized coal boiler equipped with ball and race mills, table feeders, and low-NO$_x$ burners. Unit 3 is a 190 MW$_e$ (gross) tangentially-fired pulverized coal boiler equipped with bowl mills, paddle feeders, and low-NO$_x$ burners. There is no spare capacity in the pulverizing systems of either unit when operating under full load conditions.

This project was unique in a number of respects, expanding the knowledge and cofire experience base achieved in other previous projects (ref. 1). First, the project tested the use of blended biofuels in boilers equipped with low NO$_x$ burners. Additionally, three types of biofuel were tested: (1) mill waste sawdust, (2) utility right-of-way trimmings, and (3) harvested hybrid poplar. Biofuels were processed off-site, blended at a nearby coal blending yard, and then trucked to the unmodified Shawville Station on a "just-in-time" schedule basis. The off-site fuel preparation and blending operations provided an adequate supply of blended biofuel and coal to allow the project to meet its overall test objectives, and also provided much useful new data and information needed to help establish commercial processing facilities.

Plant operating technical objectives were to determine the impacts of using low percentage biofuel blends on boiler capacity, efficiency, stability, temperature, and air/solid waste emissions. Significant boiler capacity limitations were experienced on both units when cofiring 3 percent biomass blends. The 138 MW$_e$ boiler lost 8-10 MW$_e$ of capacity due to feeder limitations, and the 190 MW$_e$ boiler lost 15 MW$_e$ of capacity due to significant reductions in mill outlet temperatures. For both units, the 3 weight percent biofuel blends behaved like wet coal.

Significant new information was developed with respect to both fuel processing and plant performance issues. This paper describes the project and the test results and explores the CO$_2$ mitigation impacts of firing blends of coal and biomass that are prepared off-site. © 1997 Elsevier Science Ltd

Keywords: wood cofiring, pulverized coal boiler, biomass power

[1] Earl F. Hunt. Foster Wheeler Environmental Corporation. 326 Hampshire Road, Sinking Spring, PA 19608. Phone: 610-777-1675 Fax: 610-777-1076 Email: 103440.2144@compuserve.com.
[2] David E. Prinzing, PE. Foster Wheeler Environmental Corporation. 2525 Natomas Park Drive, Suite 250, Sacramento, CA 95833. Phone: 916-921-2525 Fax: 916-921-5124 Email: 102503.3537@compuserve.com
[3] Joseph J. Battista. GPU Generation Corporation. 1001 Broad Street, Johnstown, PA 15907. Phone: 814-533-8234 Fax: 814-533-8315
[4] Evan Hughes. Electric Power Research Institute. 3412 Hillview Ave., P.O. Box 10412, Palo Alto, CA 94303. Phone: 415-855-2179 Fax: 415-855-8501

DESCRIPTION OF SHAWVILLE UNITS 2 AND 3

The biomass cofiring test program involved two of the four units at Shawville, Units 2 and 3. Unit 2 is a 138 MW$_e$ (gross) Babcock and Wilcox (B&W) wall-fired pulverized coal boiler that typically generates 900,000 lb/hr of 1020 deg F 1850 psig superheated steam at full load. It typically generates 950 deg F, 390 psig reheated steam. The unit is equipped with four rows of four low-NO$_x$ burners situated on the front wall of the boiler. There are four B&W Ball and Race mills (with associated table feeders), each supplying pulverized coal to four burners. The pulverized fuel is above specification; over 80 percent passes 200 mesh. The coal used at this unit has a very high Hardgrove Grindability Index (HGI) of 80 to 90. There is no spare capacity in the mills. If one mill is out of service, boiler capacity is reduced by approximately 27 MW$_e$.

Shawville Unit 3 is a 190 MW$_e$ (gross) Combustion Engineering (CE) tangentially-fired, twin-furnace pulverized coal boiler that typically generates 1,200,000 lb/hr of 1020 deg F, 2450 psig superheated steam at full load. It typically generates 980 deg F, 475 psig reheated steam. The unit is equipped with four rows of eight low-NO$_x$ burners situated at the corners of the twin furnaces. There are four CE Raymond Bowl mills (with associated paddle feeders), each supplying pulverized coal to a complete elevation of eight burners. The pulverized fuel is above specification; over 80 percent passes 200 mesh. The coal used at this unit has a very high Hardgrove Grindability Index (HGI) of 80 to 90. There is no spare capacity in the mills. If one mill is out of service, boiler capacity is reduced by approximately 45 MW$_e$. If two mills are out of service, oil is used to maintain capacity in the unit.

FUEL CHARACTERIZATION

Four different fuels were used for the biomass cofiring test program: a reference coal and three biofuels. The reference (baseline) coal was a composite of 20 coals, bid against a specification produced by the Penelec fuels department.   The three biofuels included mill waste sawdust, utility right-of-way tree trimmings (ROW), and hybrid poplar, a short rotation woody crop (SRWC). The biofuels were processed prior to blending with coal by grinding (if necessary) in a tub grinder and screening with a trommel screen to ensure a particle size of less than ¼ inch. The right-of-way trimmings and short rotation woody crops, with longer, more stringy fibers, proved to be more difficult to handle during fuel preparation and blending operations than the sawdust. In fact, only a small amount of the hybrid poplar were fired in a single test blend because of an inability to successfully handle the fuel during blending operations. Diesel fuel consumed for processing expressed as a ratio of MMBTU diesel consumed to MMBTU wood heat produced in the power plant ranged widely from 2.5 percent for sawdust where only screening was required to about 25 percent for ROW and 75 percent for SRWC (grinding and multiple screening required). It is believed that these values can be improved substantially in an optimally designed commercial process facility.

The fuels were fully characterized, including a proximate analysis, ultimate analysis, higher heating value, and ash elemental analysis. The right-of-way trimmings, which had been recently buried and recovered, had unusually high levels of ash for that kind of material (15 percent, as-received), indicating the presence of significant amounts of dirt. The composition of the ash from the right-of-way trimmings and short rotation woody crops also reflected the dirt in those fuels. The sawdust ash composition was more typical of clean wood fuels, showing significant calcium and potassium concentrations. The moisture contents of the biofuels were typical of green wood, and ranged from 39 to 45 percent by weight.

Wood is much more difficult to grind in a ball mill than coal, and consequently has much lower Hardgrove Grindability Index (HGI) values. The HGI is typically used when sizing pulverizers, and is a measure of the difficulty that pulverizers have in reducing the fuel to an acceptable particle size distribution. The HGI is influenced significantly by both the moisture and the fibrous nature of the wood fuel. Consequently, the wood was found to lower the HGI of a 3 percent wood fuel blend by approximately 6 points compared to the reference coal alone, which had a HGI of 89. This is a very soft coal; the reference HGI for design purposes used by B&W is 55. Shawville's very high HGI contributes to high capacities in the pulverizers.

BIOMASS COFIRING TESTS

A total of 20 tests were conducted, 10 on Unit 2 and 10 on Unit 3 in early November 1995. The test variables were load condition (full or minimum), amount of biomass vs. coal, and type of biomass in the 3 percent biofuel blend. With a couple of exceptions, there were two repetitions of each test for each unit. Each test consisted of 3 to 4 hours of steady state conditions. The first four tests on each unit were with 100% coal for baseline reference purposes at full and light load. The remaining six tests on each unit were with 97% coal and 3% biomass of various types as indicated in tables in the following sections.

BIOMASS COFIRING IMPACTS

For the low-percentage biomass cofiring strategy used in this test, the blended fuel proceeded through the coal pulverizing system in the same manner as coal alone. Since the first noticeable impacts of wood cofiring were expected to appear in the coal pulverizing system, the pulverizers and feeders were carefully monitored during the tests. Specifically, the feeder speeds, mill amps, and mill outlet temperatures were recorded and analyzed for each pulverizer. Also, pulverized coal samples were extracted at the mill outlet to assess any impacts on mill fineness. Other technical objectives associated with the cofiring tests focused on boiler stability, achieving capacity, boiler efficiency (and related parameters), airborne emissions, and solid waste impacts. The only significant impacts of wood cofiring encountered were related to the coal pulverizing systems and the ability of the boilers to achieve normal full capacity. These impacts are discussed below.

Shawville Unit 2: Wall-Fire PC Boiler, Ball and Race Mills, and Table Feeders

Unit 2 mill fineness results were unaffected by 3 percent wood cofiring. The weight percent passing 200 mesh was essentially the same for each fuel, at approximately 80 percent. Although the Hardgrove Grindability Index of the fuel blends was lowered relative to the reference coal, blending three percent wood with coal had no significant impact on the ability of the ball and race mills to meet fuel particle size requirements.

Mill amps for the four ball and race mills were carefully monitored as a function of fuel heat input rate during the tests. There was a consistent, although small, increase in the total amperage required to pulverize the fuel as biofuels were blended with the coal. For the coal/sawdust blend, there was an increase of approximately 7 amps total for a given heat input rate, while for the coal/ROW blend, there was an increase of approximately 5 amps total for a given heat input rate. This correlates to about a 5 percent increase in mill amps at full load for the coal/sawdust blend when compared to the baseline coal, and about a 4 percent increase for the coal/ROW blend over the baseline coal at full load.

The Unit 2 average mill outlet temperatures were largely unaffected by wood cofiring. With the exception of the two full load ROW blend tests, the desired mill outlet temperature of 160 deg F was achieved in each case. Although the mill outlet temperatures were all above the minimum 150 deg F required, the operator did experience some difficulty in maintaining the desired temperature on right-of-way material blends under full load conditions.

The most significant impact of wood cofiring on Unit 2 operations was the feeder limitations experienced with the biofuel blends, and the consequent reductions in maximum achievable boiler capacity. Unit 2 is equipped with table feeders, which use centrifugal force to feed the fuel to the ball and race mills. It is possible that there was a build-up of material on the table feeders, or that there was some slippage which contributed to the feeder limitations experienced. Table 1 presents the average feeder speeds (in percent of maximum capacity) recorded during the tests, along with the boiler capacities achieved, expressed as gross load (MW), main steam flow (kpph), and average heat input to the boiler (10^6Btu/hr). The feeder speeds presented are averages for the four mills. Note that in all cases, the feeder speeds increased significantly as wood was introduced. Under full load conditions, the unit became feeder limited on wood blends and the maximum achievable capacity of the boiler was reduced. As can be seen in the table, the full load baseline coal tests were not performed at 100 percent of maximum capacity, whereas for the biofuel/coal blend tests, the feeders were performed at or near 100 percent of maximum capacity. When the feeders are operating at 100 percent of maximum capacity on coal alone, the unit can usually achieve gross loads of 138 to 139 megawatts in the winter. Wet coals have a similar impact on feeder speeds.

Table 1. Unit 2 Average Feeder Speeds and Boiler Capacities Achieved.

Fuel Blend	Test No.	Gross Load (MW)	Steam Flow (kpph)	Heat Input (10^6Btu/hr)	Feeder Speed (%)
Full Load Tests:					
Baseline Coal (100%)	1	132.6	895.7	1,500	72.0
Baseline Coal (100%)	2	133.1	899.9	1,610	78.6
Coal + 3% Sawdust	5	129.0	877.3	1,430	88.3
Coal + 3% Sawdust	6	128.8	880.6	1,370	93.9
Coal + 3% ROW	9	130.0	874.1	1,460	101.6
Coal + 3% ROW	10	128.7	858.0	1,390	105.1
Minimum Load Tests:					
Baseline Coal (100%)	3	91.3	624.0	1,080	57.6
Baseline Coal (100%)	4	90.9	625.2	1,090	57.9
Coal + 3% Sawdust	7	87.7	630.4	1,020	70.8
Coal + 3% Sawdust	8	88.4	628.8	970	71.6

A correlation of the average feeder speed for all four mills (in percent of maximum capacity), and the gross load achieved shows that under all load conditions, dramatic increases in feeder speeds were experienced when cofiring 3 weight percent wood with coal. Under full load conditions, these increases in feeder speeds translated to decreases in the maximum achievable capacity of the boiler. For the baseline coal, an average gross load of 133 megawatts was achieved at 75 percent of maximum feeder capacity. If the feeders were operated at 100 percent of maximum capacity on the baseline coal under similar conditions, a gross load of 138 or 139 megawatts would be expected. This compares with a gross load of 129 megawatts achieved at an average feeder speed of 91 percent on the coal/sawdust blend and a gross load of 129 megawatts achieved at an average feeder speed of 103 percent on the coal/ROW blend. The average feeder speed includes all four feeders. In some cases, individual feeders were operated as high as 119 percent of maximum capacity. These data support the conclusion that between 8 and 10 megawatts of capacity were lost as a consequence of feeder limitations associated with the 3 weight percent biofuel cofiring relative to dry reference coal firing.

Shawville Unit 3: Tangentially-Fired PC Boiler, Bowl Mills, and Paddle Feeders

The Unit 3 mill fineness results, like the Unit 2 results, appeared to be unaffected by 3 percent wood cofiring. The weight percent passing 200 mesh was essentially the same for each fuel, at approximately 82 percent. Although the Hardgrove Grindability Index of the fuel blends was lowered relative to the reference coal, blending three percent wood with coal had no significant impact on the ability of the bowl mills to meet fuel particle size requirements.

The impact of wood cofiring on mill amperage was not as clear for Unit 3 as it was with Unit 2. There was a small, though inconsistent, increase in mill amps associated with wood cofiring at 3 percent by mass.

Although Unit 2 experienced significant feeder limitations, the feeder speeds encountered on Unit 3 were largely unaffected by wood cofiring. Unit 3 is equipped with paddle feeders, which use gravity as a driving force to supply fuel to the bowl mills. Although the paddle feeders ran at slightly higher speeds on biofuel blends, the speed increase did not have any impact on the ability of the boiler to achieve normal capacity.

Although mill fineness, mill amps, and feeder speeds were largely unaffected by three percent wood cofiring on Unit 3, the average mill outlet temperatures were significantly affected. The biofuel blends behaved very much like wet coal. The unit experienced a reduction in the maximum achievable capacity with coal/wood blends as a consequence of the need to maintain adequate mill outlet temperatures. Table 2 presents the average mill outlet temperatures experienced for the full load tests, along with the gross load achieved and the measured average fuel moisture content. During typical operation, the mill outlet temperature set-point is 154 deg F, as was the case for the baseline coal tests. This set-point was retained for the coal/sawdust tests to show the impact in loss of capacity (3 MW_e). However, when wetter coals are encountered, it is standard practice to lower the temperature set-point to as low 150 or 151 deg F in order to maintain a higher capacity. This was done during the coal/ROW/sawdust tests to show the maximum achievable capacity on that material (175 megawatts, gross). When firing a dry baseline coal with mill outlet temperatures maintained at 151 deg F, a maximum gross capacity of 190 MW_e would be expected for Unit 3. This corresponds to a boiler capacity decrease of 15 MW_e when cofiring the coal/ROW/sawdust blend.

Table 2. Unit 3 Mill Outlet Temperatures.

Fuel Blend	Gross Load (MW)	Fuel Moisture (%)	Mill Outlet Temp. (deg F)
Baseline Coal (100%)	180.0	7.3	154
Coal + 3% Sawdust	177.2	8.0	154
Coal + 2% ROW + 1% Sawdust	175.0	9.1	151

Following the full-load coal/sawdust tests and the full-load coal/sawdust/ROW test, experiments with temperature set-points and maximum achievable capacity were performed for the two fuel blends. As the mill outlet temperature set-point was varied, the unit was allowed to settle for approximately ten minutes, and the corresponding gross load was recorded. The results of these experiments show a negative linear relationship between gross plant output in MW and mill outlet temperature and that the slope of this relationship is the same for the two biomass fuel blends tested (3% sawdust and 2% ROW/1% Sawdust). While the negative slope of each function was essentially the same, the intercept for the line corresponding to the coal/sawdust blend was about 3 megawatts higher than the line corresponding to the coal/ROW/sawdust blend. A similar correlation could be developed for the reference coal, which would likely be about 4 megawatts higher than the coal/sawdust line.

The coal/sawdust blend, if it had been fired with a mill outlet temperature of 150 deg F, would likely have been able to match the 180 megawatts gross capacity achieved with the baseline coal at 154 deg F. The normal, desired mill outlet temperature for the bowl mills on Unit 3 is 154 or 155 deg F. The minimum allowed mill outlet temperature is 150 deg F. This limitation was established to avoid problems with plugging in the fuel supply piping to the burners, dropping below acid dew points, corrosion problems, etc. These problems have been experienced in the past when mill outlet temperatures were allowed to drop as low as 140 deg F.

Since the pulverizer functions effectively as a fuel dryer, an increase in the moisture content of the fuel will require additional heat input to maintain the desired mill outlet temperature. The control system will open the hot air dampers associated with the pulverizer to allow more air into the pulverizer, supplying the additional heat required. However, this additional air flow will increase the pressure in the pulverizer. A slightly negative pressure, typically -1.0 inches of water, is required to keep the pulverizer from "puffing" pulverized fuel and to keep the pulverized fuel out of the bearings. If the pressure increases too high in the pulverizer and the mill outlet temperature is still too low, then the only recourse is to decrease the fuel flow rate through the pulverizer, which corresponds to a reduction in the maximum achievable capacity of the boiler.

CO_2 MITIGATION ASPECTS

Wood fuel used to displace typical eastern bituminous coal will result in a plant net CO_2 emission savings of about 2.5 pound of CO_2 per MMBTU displaced or about 1.1 pound of CO_2 net savings for each pound of green wood burned. This savings will be reduced by CO_2 emissions increases associated with processing the wood. The net savings may also be increased substantially depending on the disposal method used for the wood in the absence of the project. For example, if the biomass fuel employed is being diverted from a waste stream that would otherwise be landfilled (such as the sawdust and right-of-way materials used in this project), the effective greenhouse gas savings of a project may be a multiple of the direct CO_2 offset due to the avoidance of methane production from anaerobic decomposition. Assuming methane's greenhouse gas impact to be 10 times that of CO_2 (estimates of over 20 are sometimes used), it may be possible to offset 10% or more of a coal plant's CO_2 impact by using biomass fuel diverted from landfills if such biomass supplies only 1% of the plant's heat requirements, the approximate percentage of heat supplied by biomass in this project.

The economics of using biomass to generate CO_2 credits are potentially attractive where a fuel cost savings is being achieved (i.e., the cost of wood delivered to the boiler is less than the cost of coal), in which case the net cost per ton of CO_2 offset is zero or negative. However, economically successful biomass cofiring in a pulverized coal boiler depends to a great extent on whether biomass fuel is fed through the pulverizers (vs. directly fed to the boilers) and if fed through the pulverizers, on the amount of pulverizer mill capacity available.

CONCLUSIONS

The wood cofiring test at the Shawville Generating Station was successful in that significant information was gathered on the impacts of low percentage wood cofiring on coal pulverizing systems with little or no spare capacity. Three percent wood cofiring produced significant negative impacts in the pulverizing systems, leading to significant boiler capacity reductions in both a wall-fired PC boiler and a tangentially-fired PC boiler. For both units, the biofuel blends behaved very much like wet coal. Consequently, an alternative method of wood cofiring is recommended for these units, where the wood fuel would be kept separate from the coal. A separate injection system would bypass the coal pulverizing system and avoid the negative impacts experienced during the testing.

ACKNOWLEDGEMENTS

Funding for this work was provided by the Pennsylvania Electric Company, the Electric Power Research Institute, the Pittsburgh Energy Technology Center of the U.S. Department of Energy, and the Pennsylvania Department of Conservation and Natural Resources (formerly the Pennsylvania State Energy Office).

Reference

(1) Hughes, E. and D. Tillman. 1996. Biomass Cofiring: Status and Prospects 1996. Proc. Engineering Foundation Conference on Cofiring Biofuels in Utility and Industrial Boilers. Snowbird, UT. April 28 - May 3.

Pergamon

Energy Convers. Mgmt Vol. 38, Suppl., pp. S557–S562, 1997
© 1997 Elsevier Science Ltd. All rights reserved
Printed in Great Britain
0196-8904/97 $17.00 + 0.00

PII: S0196-8904(96)00327-5

POTENTIAL OF COFIRING WITH BIOMASS IN ITALY

M. ARESTA, I. TOMMASI and M. GALATOLA

METEA Research Center,
Via Puglie, 93 - 74100 - Taranto - Italy
Department of Chemistry University of Bari
Campus Universitario, Via Orabona, 4 - 70126 - Bari - Italy

ABSTRACT

Biomass is considered a potential fuel and a renewable source for the future. In Italy, the utilization of biomass nowadays is addressed, above all, towards thermal energy production. In the near future, however, it is predictable a higher differentiation in order to use biomass with the more suitable technology. In this paper we review the utilization of residual biomasses. © 1997 Elsevier Science Ltd

KEYWORDS

Biomass, renewable sources, cofiring, italian energetic consumption.

INTRODUCTION

After the coal-age and the oil-age the world is setting off on a polyenergetic scenario where several sources and technologies will be utilized to meet the increasingly energetic demand. In this scenario an important role will be played by new and renewable energetic sources like wind-, solar-, geothermal -hydroelectric-, photovoltaic energy and biomass.
The development in the utilization of these renewable energetic sources has several strategic consequences:
* It makes it easier to achieve a better position in the energy market by increasing the utilization of national resources.
* The employment of renewable energetic sources could contribute to reduce the environmental impacts connected to energy production and waste disposal.
In 1990 the Italian internal energy consumption amounted to 166 Mtpe with a 2.8% increase compared to the previous year. In Italy the utilization of coal as an energetic source was quite low (15 Mtpe, corresponding to less than 10% of the total energy demand). The oil consumption amounted to 94 Mtpe (+2.9%) while the natural gas consumption was 37 Mtpe (+8.2%); in the same period there was a reduction in the internal production of fossil fuels (-4%). As a consequence of the increase in the net import of energetic sources (+6.4%), in 1990 Italy imported 82% of the global energy demand. In this paper we will discuss the importance of the thermal utilization of residual biomass in the Italian context, also exploring the combined biologic plus thermal biomass utlization in conjuction with fossil cofiring opportunities.

BIOMASS AS ENERGY SOURCE

Among the several renewable resources, biomass is looked at with great attention because it is mainly made up of carbon and can be, then, either directly used as fuel in order to produce energy (in the form of heat and/or electricity) or turned to gaseous or fluid fuels after appropriated treatments.
In this paper we shall deal only with residual biomass that includes several organic products and by-products such as agricultural products (e.g. cereals, tomatoes, wheat, sugar beet, potatoes, etc.), agricultural by-products (straws,

trimming residues), wood, municipal solid waste, wastewater sludges, manure residues. Biomass grown for energy purposes will not be considered. Furthermore, we will restrict our comments to the use of biomass for thermal energy production.

Energy production by means of biomass is an ecologically worthy activity for two fundamental reasons:

- Because of its chemical composition, most of it is included in the natural cycle of carbon, whose ultimate destiny is to be turned to CO_2. From this point of view, it is not important whether such transformation occurs biologically or by means of thermochemical processes. CO_2 produced through thermal treatments, then, does not represent an additional quantity of this greenhouse gas, contrary to what happens when a fossil fuel is burned (coal, oil, natural gas).

- Fossil fuels are not, generally, "ready to use", but need to undergo a series of treatments (from extraction to refining). All these operations require energy, that, in some way, contributes to environmental pollution. Biomass employment as fuel could, then, contribute to the reduction of these additional flows.

As it is shown in Table 1, biomass provides worldwide about 70% of the total energy produced by means of renewable resources. In the future, it is reasonable to expect a reduction of its importance as a consequence of the technological progress in the utilization of other energy sources; anyway, biomass should still contribute for more than 50%.

Table 1. World consumption of renewable sources in 1980, 2000 and long-term potential expressed in Mtpe (source: Worldwatch Institute).

SOURCE	1980	2000	Long-term potential
SOLAR ENERGY	9.6	242 - 374	1080 - 3192
Direct	2.4	84 - 168	480 - 720
Residential collectors	2.4	40.8	120 - 192
Industrial collectors	2.4	69.6	240 - 480
Other	2.4	48-96	240 - 720 (1)
BIOMASS	1060	1514 - 1574	3835 - 4195
Wood	840	1152	2400 +
Waste	156	168	-
Manure	48	48	-
Biogas (small digesters)	2.4	55.2	115.2
Biogas (large digesters)	2.4	4.8	120 (1)
MSW	7.2	36	360 (1)
Methanol from wood	2.4	36 - 73.68	480 - 720 (1)
Energy cultures	2.4	14.4 - 36	360 - 480 (1)
HYDROELECTRIC ENERGY	460.8	912 - 1152	2160 +
WIND ENERGY	2.4	24 - 48	240 (1)
PHOTOVOLTAIC ENERGY	2.4	2.4 - 9.6	480 (1)
GEOTHERMAL ENERGY	7.2	24 - 72	240 - 480 (1)
TOTAL	1543	2718 - 3230	8035 - 10747 (1)

(1) Advanced technologies may increase the long-term potential.

In Italy the 1988 PEN (National Energetic Plan) estimated for biomass a contribution to the national supply of 1 Mtpe in 1990 and 2.5 Mtpe in 2000. However, these figures underestimate the real contribution of biomass as energy source. They refer only to the portion of wood as fuel that enters the commercial energy market.

The real contribution of biomass to the national energy production has been estimated through a study financed by ENI-AGIP spa, in cooperation with ENEA, CNR, Confindustria and Universities. The results of this study calculate a biomass contribution higher than 4 Mtpe in 1990 and an expected contribution of 10 Mtpe in 2000. Table 2 presents part of the ENI study results. In 1989 the Italian energy production from biomass was 3.8 Mtpe (about 2% of the total energy consumption) and represented about 15% of the total potential energy from biomass (25.9 Mtpe). All data and references related to this study are collected in a Data Base at SOGESTA - Urbino.

The analysis of these data also suggests that it would be better to subdivide biomass into two main fractions:

* Organic waste (manure, wastewater sludges, fermentables, food industry residues), characterized by high humidity content (>45%) and C/N ratio lower than 30.

* Combustible wastes (wood residues, cellulose residues, paper, plastic, textiles) characterized by humidity lower than 45% and with a C/N ratio higher than 30.

As will be explained later, it is necessary to recognize the differences between these fractions in order to choose the most suitable treatment technology thereby achieving higher process efficiencies and lower treatment costs.

Table 2. Production and utilization of biomass (sources: ENI, CNR, ENEA, 1989).

SOURCE	DRY MATTER (Mt/y)	ENERGY CONTENT (Mtpe)	POTENTIAL ENERGY (% of the total)	ENERGY USED (Mtpe)
WOOD RESIDUES	13.9	5.5	21.2	3.6
Forest and trees management	9.8	3.9	15.1	2.4
Wood industry	1.1	0.4	1.5	0.2
Food industries	3.0	1.2	4.6	1.0
ORGANIC WASTES	24.7	9.2	35.5	
Manure	16.5	5.8	22.4	very low
Water treatment plants	1.0	0.4	1.5	very low
Food industry	3.9	2.0	7.7	very low
MSW	3.3	1.0	3.9	very low
CELLULOSE	19.5	7.8	30.1	
Straw	13.5	5.4	20.8	very low
Mais residues	6.0	2.4	9.3	very low
SOLID WASTES	9.1	3.4	13.1	
MSW (paper, plastic)	4.6	0.7	2.7	0.1
Industrial organic wastes (textiles)	4.5	2.7	10.4	0.1
TOTAL	67.2	25.9	100	3.8

There is not really an established market for the energy produced by means of biomass but rather a self-consumption system. The main biomass consumers in Italy are:
i)   Families living in rural areas.
ii) Food industries.
iii) Wood industries.
In Italy, according to a study of the National Institute for Rural Sociology there are about 2,127,000 families utilizing wood and other biomasses to produce energy. The average conspumtion is about 5.7 t /y per family, while a good estimate for the energy production is in the range of 2-2.5 Mtpe/y. As far as food industry is concerned, the main producer/consumer of biomass are distilleries with an average annual production of 3.0 Mt of by-products (above all grape-seeds and grape peels). The other main industries of this compartment are oil mills (0.5 Mt/y), rice-fields (0.2 Mt/y) and other transformation industries (0.3 Mt/y). Therefore the total biomass production of food industry amounts to about 4 Mt/y; of this quantity more than 80% (more or less 1 Mtpe) is utilized for energy production. The wood industry category has an annual biomass production of about 0.75 Mt, equivalent to about 0.2 Mtpe. Half of this by-product quantity is burned directly by the same industries while the other half is sold as fuel.

BIOMASS AVAILABILITY

The characteristics of the biomass utilization circuit are:
• Short-distance distribution.
• Self-production.
These characteristics depend on the low economic value and large volume that are typical of biomass. As a consequence, from the economic and energetic point of view, it is more reliable to create a system where the producer and the utilizer coincide, or at least, the distance between them is not beyond few kilometers (the cost of transportation should be kept below 3$/t). The development of such a "local market" is not easy to achieve because it is necessary to identify which type of biomass is more convenient to utilize in each situation, which is the right technology to transform it and which are the potential markets for the products and by-products of the process. There exists a geographical differentiation among products and by-products produced by the various italian regional areas (see Table 3). It is then clear the necessity to create homogenous markets for the utilization of biomass in Italy. In order to achieve this result it is necessary:
− To assess product and by-product availability and corresponding farmed areas.
− To determine the biomass main qualitative characteristics.
− To identify the various alternative utilizations.
− To specify the availability period (annual, seasonal).
The time aspect is particularly important because while municipal solid waste and wastewater sludges are available all the year along, most agricultural biomasses are seasonal and so it is necessary to identify exactly the available biomass quantity and the possibility to store it, in order to identify the more suitable transformation technology.

Table 3. Product and by-product mean availability for the main crops and corresponding farmed areas.

SOURCE	SURFACE (E3 ha)	SPECIFIC PRODUCTION (t/ha)	TOTAL PRODUCTION (E3 t)	REGIONAL AREAS (> 70% Total production)
WHEAT	3220	6.5 - 9.8	10500 - 15700	North and South Italy
CEREALS	527	3.6 - 5.4	1800 - 2700	Central and South Italy
RICE	185	9.0 - 11.0	1660 - 2000	North Italy
ZEA MAYS	940	15.0 - 17.5	14100 - 16300	North Italy
SUGAR BEET	270	56.0 - 70.0	14300 - 18400	North Italy
POTATOES	155	15.0 - 30.0	2200 - 4500	Central and South Italy
LEGUME	224	2.5 - 5.0	550 - 1110	South Italy
TOMATOES	111	39.0 - 52.0	4300 - 5700	South Italy
GRAPEVINE	1250	9.1 - 19.5	11600 - 23600	North and South Italy
OLIVE	1030	2.1 - 3.5	2100 - 3500	Central and South Italy
FRUIT PLANTS	174	16.5 - 27.5	2860 - 4730	North and Central Italy
CITRUS FRUITS	156	11.0 - 22.0	1650 - 3300	South Italy

The analysis of these data indicates that in Italy it should be advisable to promote small or even domestic plants, planned just to sutisfy the local needs. The wide geographical distribution of the various sources and their seasonal availability, indeed, makes it difficult to guarantee a biomass supply in adequate quantity and at a reasonable price. On the other hand, we must underline that in this way the investment and process costs will be higher because of unavoidable scale economies. These effects are presented in Table 4 (Caserta, 1990) showing the costs related to the production of 1 kW_t from wood-cellulose combustion utilizing plant of different size.

Table 4. Cost of 1 kW_t from wood-cellulose.

PLANT POWER	INVESTMENT (Lit / kW_t)	YIELD	FUEL COST (Lit / kg)	kW_t COST (Lit / kW_t)
DOMESTIC (75 kW_t)	50 000	0.70	50	76
SMALL (350 kW_t)	50 000	0.75	50	56
MEDIUM (1.2 MW_t)	50 000	0.80	50	30
LARGE (12 MW_t)	60 000	0.85	30	18

Some of the many agriculture biomasses available in Italy are used right now (wheat straw, zea mays, cereals straw, rice straw and sugar beet), however, a large quantity, as it is pointed out in Table 5, are still unutilized, presenting a lost opportunity. Most of these biomasses are burnt in fields, with unclear effects on soil and physical environment.

Table 5. Available biomass amount vs actual utilization.

SOURCE	HUMIDITY (H_2O) %	C/N	AVAILABLE AMOUNT t/ha	AVAILABLE AMOUNT dry (t/ha)	TOTAL AMOUNT (E3 t)	TOTAL AMOUNT dry (t/ha)	USED %
STRAW (WHEAT)	10-20	118-130	1.0-2.0	1.3-1.7	2500-3500	2200-2800	80-90
STRAW (other CEREALS)	10-20	60-65	1.6-2.4	1.4-1.9	800-1200	720-960	5-10
STRAW (RICE)	20-30	61-65	4.5-5.5	3.6-3.8	830-1000	660-700	5-10
ZEA MAYS	45-65	60-80	9.0-10.5	3.7-4.0	8500-9800	3400-3700	50-60
SUGAR BEET	75-85	14-15	16.0-20.0	3.0-4.0	4300-5400	810-1000	5-10
POTATOES	55-65	22-26	5.0-10.0	2.2-3.5	700-1500	340-540	-
STRAW (LEGUMES)	10-20	28-30	1.5-3.0	1.3-2.4	330-670	290-530	-
TOMATOES	80-90	23-24	9.0-12.0	1.2-1.8	1000-1300	130-190	-
GRAPEVINE	40-50	47-64	2.1-4.5	1.3-2.2	2600-5600	1600-2700	-
OLIVE TREES	30-40	67-68	0.6-1.0	0.4-0.6	600-1000	400-600	-

BIOMASS PROPERTIES AND ENERGY PRODUCTION

It is quite important to consider the chemical and physical properties of biomass before deciding which technologies to adopt for its conversion to energy. Two of the more important biomass characteristics, as already said, are the humidity content and the C/N ratio (see Table 5). These properties are fundamental to decide if it is more suitable for a thermal or a biologic treatment process.

The high water content of some biomasses such as wastewater sludges, tomatoes, potatoes, sugar beet and VGF (i.e. the fermentable fraction of municipal solid waste), makes thermal treatment of these substrates economically disadvantageous if not impossible. On the other hand, artificial drying of these biomasses would imply a too high energy consumption. The treatment of these biomasses cannot take place very far from the area where they have been produced because of two reasons:

• Their high water content, and, therefore, their high volume, increase the transport costs.

• These substrate are quite immediately attacked by microorganisms.

The more suitable treatment for these substrates is anaerobic fermentation. This process occurs in strictly anaerobic conditions and is typical both of constrained anaerobic bacteria and of bacteria that can work either in anaerobic or aerobic conditions (facultative anaerobic). Aim of the process is the organic substances degradation and methane and carbon dioxide production.

Thermal processes (incineration, gasification and pyrolysis) are usually utilized with biomass having a C/N ratio higher than 30 and a humidity content lower than 45%. These characteristics are typical of cellulosic products (see Table 5) such as straws, olive trees, grapevine, fruit plants, citrus fruits and of RDF (Refuse Derived Fuel, the combustible fraction of municipal solid waste, i.e. paper, plastic, textiles, wood); RDF can be obtained either by means of a mechanical separation process applied down-stream to the collection system or with an up-stream separation carried out by citizens (in this case it is more appropriate to speak about PDF, that is Packaging Derived Fuel).

When we speak about applying thermal treatments to biomass it is necessary to assess whether it is better to choose the treatment technology according to the available biomass or to fit the available biomasses to the already existing technologies. In the first case it is necessary to consider the chemical and physical characteristics of the biomass source and that the system would be conditioned by the quality of the feeding material. Actually, the performance of any combustor is strictly connected to the homogeneity of the substrate fed in; having biomass sources very different humidity, density and other physical properties, it is not possible to assure the homogeneity requested without resorting to pre-treatment steps. The less homogeneous the substrate fed into, the lower the thermal efficiency and the higher the environmental impact of the entire process. Besides this, we have to remember the problems connected to the constant refuelling of a plant fed only with biomasses.

The problems connected to the utilization of thermal treatments applied to biomass can be solved if, instead of using the biomass as single fuel, we apply a cofiring process. Cofiring means simultaneous combustion of two or more fuels in a single reactor and under the same reaction conditions. Cofiring dry biomass (e.g. RDF) and coal, for example, is more efficient than single biomass combustion and its plant and working costs are lower. The higher efficiency depends basically on the lower humidity content and the lower combustion air excess employed in cofiring. Thermal efficiency of fuels is a function of the boiler and turbine efficiencies and of the energy consumption within the plant. Of course utilizing biomass as co-fuel it is necessary to adapt its physical characteristics to those of coal. It is often necessary to operate a size reduction and a densification step in order to reduce the density difference between coal and biomass. This contributes to:

− The improvement of the process efficiency.

− The decrease of CO concentrations in the flue gases.

− The reduction of the formation of "cold spots" in the reactor.

These treatments also lead to an increase in process costs, an increase that is partially paid back by the reduction in coal consumption. The best combustors for biomass cofiring, considering the characteristics of the biomass we have dealt with, seem to be the FBC (Fluidized Bed Combustors) and the EBC (Entrained Bed Combustors); in the latter case it is necessary to proceed to a physical reduction of the biomass to a size lower than 0.6 cm. It should be noticed that, in any case, the use of biomass as auxiliary fuel reduces combustor thermal efficiency (i.e. the fraction of energy in the fuel converted to steam). As a matter of fact, the efficiency of a combustor (meant as the energy fraction contained in the fuel and that is changed into steam) progressively decreases with the increase of biomass contribution. From a 89% efficiency of an only coal combustion process we reach a 78% efficiency with a biomass burning combustion (McGowing and Hughes, 1993).

A possible alternative to coal and biomass cofiring is primary biomass combustion with natural gas as the auxiliary fuel. Some experiments carried out in this field (Beshai et al., 1994) have shown encouraging results. The use of natural gas as auxiliary fuel, indeed, seems to imply the following advantages:

− The biomass pre-treatment step is avoided.

− Reduction in CO emissions (72-89%), probably due to the fact that the employment of natural gas allows the achievement of higher temperatures in the lower zone of the furnace.

− Reduction in CO_2 emissions as a consequence of the lower C/H ratio of natural gas.

− Reduction of dioxins (PCDD) and furans (PCDF) concentrations (up to 90% for a 15% calorific contribution of natural gas on the total heat required). The achievable reduction depends, basically, on the natural gas burner locations. If the main aim is to counterbalance biomass heating value variations (due to compositional or load changes), then the burners must be located in the lower region of the furnace (above the grate); if, on the contrary, the main purpose is the organic micropollutants destruction, then it would be more useful to locate the burners in the higher region of the furnace (post-combustion zone). PCDD and PCDF production, indeed, seems to be strongly influenced by the quantity of unburnt material coming out of the furnace;

– Increase of the boiler efficiency (about 5%) due to the lower air excess required and the lower fuel humidity.

The considerable advantages induced by this technology need to be further checked, especially with regard to the economic implications brought by the use of natural gas. An interesting development could be carried out employing, instead of natural gas, the biogas produced by anaerobic fermentation applied to wet biomasses. This could represent an interesting example of closed industrial cycle. Obviously, the employment of biogas, instead of natural gas, would also imply some disadvantages as:

◊ Higher gas consumption for each ton of biomass (due to biogas lower heating value).

◊ Necessity to apply a biogas purification process in order to avoid the presence of corrosive compounds (H_2S and HCl).

◊ Problems connected to the gas storage.

CONCLUSIONS

At the moment in Italy the energy produced by means of residual biomass covers about 2% of the national need but it could be possible to double this contribution utilizing a more rational approach to the problem and identifying for each biomass the suitable treatment technology. The advantages are concrete both from the economic point of view and for the reduction of the environmental impacts connected to the production, transport and utilization of fossil fuels. Biomass cofiring is becoming one of the more interesting available solutions.

REFERENCES

Caserta G. (1990), La voce Republicana, 19 settembre 1990.

Caserta G. (1990), in "Prospettive e limiti nell'uso energetico delle biomasse", energie alternative HTE, anno 12 numero 68 novembre-dicembre 1990, pp 380-384.

McGowing C.R. and Hughes E.E., (1993), "Efficient and Economical Energy Recovery from Waste by Cofiring with Coal", in Clean Energy from Waste and Coal, American Chemical Society.

Beshai R.Z., Santal A., Charles C.H., Melick T.A. and Sommer T.A., (1994), "Natural gas cofiring in a Refuse Derived Fuel incinerator: results of a field evaluation", in Proceedings of National Waste Processing Conference, Boston, MA (USA), 5-6 June 1994.

Energy Convers. Mgmt Vol. 38, Suppl., pp. S563–S568, 1997
© 1997 Elsevier Science Ltd. All rights reserved
Printed in Great Britain
0196-8904/97 $17.00 + 0.00

Pergamon

PII: S0196-8904(96)00328-7

UTILITY FOREST CARBON MANAGEMENT PROGRAM/ UTILITREE CARBON COMPANY

John Kinsman, Edison Electric Institute, 701 Pennsylvania Avenue, N.W.,
Washington, D.C., 20004, U.S.A.

Gary Kaster, American Electric Power Company, 59 West Main Street,
McConnelsville, Ohio, 43756, U.S.A.

Eric Kuhn, Cinergy Corp., P.O. Box 960, Cincinnati, Ohio, 45201, U.S.A.

Ron McIntyre, Detroit Edison Company, 2000 2nd Avenue, Detroit, Michigan, 48226, U.S.A.

ABSTRACT

The U.S. electric utility industry has a long history of involvement with traditional forest management and tree-planting programs. Utilities have recently initiated numerous forestry projects specifically to conserve energy and to offset CO_2 emissions. Many electric utilities have included forestry activities in their agreements with DOE under a voluntary national program called the Climate Challenge. The Utility Forest Carbon Management Program (UFCMP) is a voluntary initiative developed by the Edison Electric Institute, with support from 55 electric utilities, to expand utility industry efforts to manage CO_2 via forestry projects, both domestic and international. With slightly over $2.4 million committed, 40 of these companies have established a non-profit corporation called the UtiliTree Carbon Company to sponsor five projects. The projects in the final pool represent a diverse mix of rural tree planting, forest preservation, forest management and research efforts at both domestic (eastern and western U.S.) and international sites. This paper overviews these programs, which were designed to advance the state of knowledge regarding options for managing greenhouse gases via forestry; establish low-cost forestry options to manage greenhouse gases; and promote environmental stewardship by the utility industry. © 1997 Elsevier Science Ltd

KEYWORDS

Carbon, carbon dioxide, forest, electric utilities, management, United States

INTRODUCTION

Human activities related to energy production and land use are increasing the atmospheric concentration of greenhouse gases such as carbon dioxide (CO_2), which in turn may change the energy flux of the Earth/ atmosphere, possibly causing global warming and other changes in climate. The impacts of greenhouse gas emissions are very uncertain, however -- the rate, magnitude and regional characteristics of human-induced climate change are difficult to predict and require additional research. Regardless, because of the potential consequences of climate change, policies and programs are being developed to adapt to or mitigate greenhouse gases and climate change.

Many options exist for managing greenhouse gas emissions and sinks: increasing the efficiency of energy supply and use, including use of environmentally beneficial electrotechnologies; increased use of renewable and nuclear energy systems; fuel switching from coal and oil to natural gas; capturing and using methane from coal mines and landfills; and increased motor vehicle fuel economy. Adaptation (e.g., planning for sea-level rise, or planting different crops) would lessen impacts.

Another option is to sequester CO_2 in "sinks" such as plant biomass. Trees are referred to as "carbon sinks" because they take CO_2 out of the air and sequester it in plant tissue. About one-half of a tree is carbon. Carbon can be managed through many different types of forestry activities, including: forest preservation and management projects to maintain carbon sequestered by reducing deforestation and harvest impacts; forest management to enhance existing carbon sinks; creation of new carbon sinks by planting on pasture, agricultural land or degraded forest sites; storing carbon in wood products; and energy conservation through shading buildings and homes. Carbon can be sequestered in halophytes (salt-tolerant plants), organic matter in soil, in oceanic seaweed, or in microalgae in the ocean. Biomass can be used as a substitute for fossil fuel to produce energy.

The technical potential for forest carbon management is great, able to counteract a meaningful portion of the 3 Pg (1 Pg = 1 billion tonnes) carbon annual addition to the atmosphere. In addition, vigorous efforts to control land degradation in these areas could result in a net sequestration of up to one Pg carbon per year. Carbon offsets, properly documented and monitored, should be a major component of an international strategy to respond to greenhouse gas concerns.

The subject of this article is the management of carbon in trees from the electric utility industry's perspective. The Climate Challenge Program, Utility Forest Carbon Management Program, and UtiliTree Carbon Company will be described.

ELECTRIC UTILITIES AND FOREST MANAGEMENT

The electric utility industry has a long history of involvement with traditional forest management and tree-planting programs, through preserving forest lands for both recreational use and wildlife habitat, tree maintenance around power lines, education of homeowners on tree placement around power lines, and commercial forestry on electric utility-owned lands. In association with events such as Earth Day and Arbor Day, many utilities supply seedlings for employees, children and others to plant. The electric utility industry owns a large amount of land in order to house and surround its current and future generation, transmission and distribution facilities.

Utilities have also recently initiated numerous forestry projects specifically to conserve energy and to offset CO_2 emissions (Kinsman and Trexler, 1993, 1995; Dixon et al., 1993). A dozen or more electric utility companies are involved in urban forestry energy conservation programs such as American Forests' Global ReLeaf and the DOE/American Forests' Cool Communities. Some U.S.

electric utility companies -- such as the New England Electric System, PacifiCorp, American Electric Power Company, Wisconsin Electric Power Company, Cinergy Corp., Detroit Edison Company and Southern Company -- have initiated forestry efforts targeted at managing carbon. Some utilities are using biomass as a fuel to produce electricity.

Some specific reasons for utilities to participate in forest carbon management include:

- There is a large technical potential for forest carbon management -- a project can offset millions of tons of carbon emissions.
- Forestry options to manage carbon are cost effective in many cases -- e.g., a few dollars per ton of carbon offset. Forest carbon management opportunities can be among the most economical ways to address CO_2 emissions (Sedjo et al., 1995).
- Forestry carbon management adds flexibility, thus expanding the repertoire of options.
- Forestry projects yield positive public relations -- using forestry to manage CO_2 is well received by the public and environmental groups.
- Forestry efforts have positive secondary environmental and social benefits -- e.g., restoration of degraded lands and protection of biodiversity.
- International projects demonstrate the effectiveness of joint implementation activities with other nations, which is a critical tool for economically addressing greenhouse gas issues.

THE CLIMATE CHALLENGE

The Climate Challenge Program is a joint, voluntary effort of the electric utility industry and the U.S. Department of Energy (DOE) to reduce, avoid, limit or sequester (hereinafter referred to as "reduce") greenhouse gases.

There are numerous incentives for voluntary actions by the electric utility industry to reduce greenhouse gases. First, the U.S. is a signatory to the Framework Convention on Climate Change, and the President has pledged action to limit greenhouse gases. Second, a "do-nothing-until-it's-proven-beyond-everyone's-doubt" approach would diminish the industry's influence in helping to determine policy responses. Utilities must act to pro-actively retain the operational flexibility to manage greenhouse gases via the most cost-effective methods. Third, electric utilities have unique contributions to make, possessing special competence in providing cost-effective customer service and in achieving environmental excellence through technical innovation, such as energy-efficient electrotechnologies; increasing supply-side efficiencies related to clean coal technologies, nuclear energy, natural gas and renewable energy technologies; and demand-side management.

In the Climate Challenge, individual (or groups of small) utilities enter into a Participation Agreement with DOE, specifying one or more of six types of commitments (such as: reduce greenhouse gas emissions to the electric utility's 1990 baseline level by the year 2000, or undertake specific projects or actions, or make specific expenditures on projects or actions, to reduce greenhouse gas emissions) and specific actions that it has taken or intends to take. The electric utility will report annually on activities and achievements, in a clear and understandable manner that is consistent with the guidelines adopted pursuant to subsection 1605(b) of the Energy Policy Act.

DOE reports that as of May 1996 there have been 114 agreements signed by almost 600 utilities. The 114 agreements represent 61% of 1990 electric generation and electric utility CO_2 emissions. DOE estimates that the Climate Challenge will reduce, avoid or sequester over 160 million metric tons of CO_2 equivalent in the year 2000. The estimate is conservative and, according to DOE, the electric utility industry is the number one U.S. industry in pledging greenhouse gas reductions.

In July, the Energy Information Administration (1996) reported that Energy Policy Act reporting for 1994 " was dominated by electric utilities, which accounted for 96 of the 108 reporters." Almost 90 percent of the utility projects were undertaken under the Climate Challenge Program. Utilities reported over 70 forestry projects.

UTILITY FOREST CARBON MANAGEMENT PROGRAM (UFCMP)

This program is an initiative developed by the Edison Electric Institute, with support from 55 electric utilities, to expand electric utility industry efforts to manage CO_2 via forestry projects, both domestic and international. The goals of the program are to:

- Advance the state of knowledge regarding options for managing greenhouse gases via forestry.
- Establish low-cost forestry options to manage greenhouse gases.
- Implement projects to manage greenhouse gases.
- Promote environmental stewardship by the electric utility industry, including helping to demonstrate that a voluntary approach to environmental protection can work.

The UFCMP developed criteria and a process to review proposed projects and, subsequently, a request for proposals was issued to hundreds of individuals and organizations in February 1995. Thirty-two proposals were received in March 1995 and reviewed by the UFCMP committees, an outside consultant and, to a limited extent, by the UFCMP Advisory Council (representatives from nine non-electric utility organizations -- American Forests, Resources for the Future, Trees Forever, Society of American Foresters, Smithsonian Tropical Research Institute, U.S. Country Studies Program, DOE, Oak Ridge National Laboratory, and USDA Forest Service). Proposed projects were located in the U.S., Central America, South America and Asia.

Technical criteria addressed estimation of full life cycle carbon sequestration benefits (taking into account "leakage" and the fate of harvested biomass), project greenhouse gas calculations, monitoring, contingency plans, and non-greenhouse gas impacts, as well as project developer qualifications and experience. The cost-effectiveness of the project in terms of $ per ton CO_2 managed was a key project criterion.

Projects were ranked and a "pool" of five projects emerged as the final product for which sponsorship was sought. Subsequently, most UFCMP members joined together to form UtiliTree Carbon Company and provide cooperative funding.

UTILITREE CARBON COMPANY

A new non-profit corporation called the UtiliTree Carbon Company was established by 40 utilities to sponsor the projects identified by the UFCMP. The five projects in the final pool represent a diverse mix of efforts at both domestic (eastern and western U.S.) and international sites. Projects involve rural tree planting, forest preservation and forest management and are located in Louisiana, California, Oregon, Belize and Malaysia. The UtiliTree Carbon Company has committed slightly over $2.4 million to fund these projects. Carbon dioxide will be managed at a cost of under $1 per ton, including administrative expenses. Over 2 million tons of CO_2 benefit will result from the five projects over their lifetimes. Participants will share on a pro rata basis reporting of CO_2 benefits into the voluntary Energy Policy Act section 1605(b) data base. Brief descriptions of the five projects are provided below.

Bottomland Hardwood Forest Restoration in the Mississippi River Valley: A Carbon Sequestration Opportunity

This project will investigate the feasibility of using bottomland hardwood forest restoration on marginal farmland in the Mississippi River Valley as a means of sequestrating atmospheric carbon. The project, conducted by the School of Forestry at Louisiana Tech University, will also seek to improve the methods of reestablishing such forests. The 80 acre study site in Catahoula Parish, Louisiana is owned by the Louisiana Department of Wildlife and Fisheries. The restored forest will be part of an adjacent state wildlife refuge. Anticipated CO_2 benefits are 47,000 tons over 70 years.

The Rio Bravo Carbon Sequestration Project

The Rio Bravo Carbon Sequestration Pilot Project is a partnership between Programme for Belize, The Nature Conservancy, Wisconsin Electric Power Co., Cinergy Corp., Detroit Edison Company, PacifiCorp, and UtiliTree Carbon Company. The project consists of two components. Component A includes the purchases of a 14,400-acre parcel of endangered forest land that will link two forested Rio Bravo properties owned by Programme for Belize in the northwestern corner of Belize. Component B establishes a sustainable forestry management program at the Rio Bravo Conservation and Management Area that will increase the total pool of sequestered carbon in the 120,000-acre area of Rio Bravo. The partners established the following as the primary objective of Rio Bravo project: explore and demonstrate incentive-based opportunities for private voluntary international cooperation in greenhouse gas mitigation, conservation and sustainable development. Anticipated CO_2 benefits attributable to UtiliTree Carbon Company's contribution equal about 1,060,000 tons over 40 years.

Reduced Impact Logging of Natural Forests in Sabah, Malaysia

A successful forest based greenhouse gas emission offset project in Sabah, Malaysia, is being expanded. The Reduced Impact Logging project involves implementation of techniques to reduce the release of sequestered CO_2 associated with uncontrolled logging of natural tropical forests. New England Power Company (NEP) developed the project. Rakyar Berjaya Sdn. Bhd. (RBJ) of Malaysia, will implement reduced impact logging on 2,500 acres within its 2.4 million acre timber concession. The anticipated greenhouse gas benefit is 379,000 tons CO_2 over the 40 year project life. Non-greenhouse gas benefits are substantial. The reduced damage to the forest and forest soils represent a giant step toward sustainable tropical forest management.

Maximizing Carbon Storage through Forest Stewardship

The Pacific Forest Trust (PFT) has created a new incentive for forest stewardship through utilizing the tremendous carbon storage capability of the Pacific Northwest forest to increase carbon sequestration through restoring older growth forests to the private landscape. In a pilot project on approximately 500 acres of prime redwood land in California, the PFT is working with private forest landowners to significantly increase short and long-term CO_2 stores through forest stewardship. By ensuring that this land will be managed for older age cohorts of trees, the project will increase both total volumes of timber harvested and CO_2 stores. The forest land owners will be compensated for their costs of forest management changes to maximize carbon storage. Through a perpetual conservation easement, the project also protects this land, including critical habitat for salmon and other threatened species, as forest forever. This project is expected to produce a CO_2 benefit of 242,000 tons over 100 years.

Western Oregon Carbon Sequestration Project

The Western Oregon Carbon Sequestration Project will sequester carbon by planting trees on 900 acres of unforested non-industrial private timberland in western Oregon that otherwise would not be replanted. Participating landowners will agree to grow the trees for at least 65 years before harvest and to engage in sustainable forest management practices to ensure maximum growth per acre. Any harvested timber will likely be used for long-term purposes such as construction, continuing sequestration of much of the carbon. The project will have numerous additional benefits including expanding wildlife habitat, improved water quality and increased forest industry employment. The project will be implemented by Trexler and Associates in cooperation with Oregon Woods, Inc. Anticipated CO_2 benefits are in the range of 564,000 - 747,000 tons over about 65 years.

SUMMARY AND CONCLUSIONS

Electric utilities are interested in all technically and economically feasible alternatives for managing greenhouse gases emissions. Utilities are being pro-active, through the Climate Challenge and other programs, to ensure operational flexibility to achieve greenhouse gas reductions using the most cost-effective methods. The electric utility industry recognizes the technical and economic potential for forest carbon management. Electric utility companies are supporting a broad range of activities on their own lands and at other sides in the U.S. and abroad. Carbon offsets, properly documented and monitored, should be a major component of any such program to respond to greenhouse gas concerns.

ACKNOWLEDGMENTS

The authors appreciate the advice and support of the Utility Forest Carbon Management Program's Steering, Policy, Technical and Advisory committees; the UtiliTree Carbon Company Board of Directors; and many other individuals associated with these programs.

REFERENCES

Dixon R.K., Andrasko K.J., Sussman F.G., Lavinson M.A., Trexler M.C. and Vinson T.S. (1993). Forest sector carbon offset projects: Near-term opportunities to mitigate greenhouse gas emissions. *Water, Air and Soil Poll.* 70, 561.

Energy Information Administration (1996). Voluntary Reporting of Greenhouse Gases, 1995. DOE/EIA-0608(95). U.S. Department of Energy, Washington, D.C.

Kinsman J.D. and Trexler M.C. (1993). Terrestrial carbon management and electric utilities. *Water, Air, and Soil Poll.* 70, 545.

Kinsman J.D. and Trexler M.C. (1995). Into the Wood. *Electric Perspectives.* March-April 1995.

Sedjo, R.A., Wisniewski, J., Sample, A.V. and J.D. Kinsman. (1995). The economics of managing carbon via forestry: Assessment of existing studies. *Environmental and Resource Economics* 6, 139.

Pergamon

Energy Convers. Mgmt Vol. 38, Suppl., pp. S569–S573, 1997
© 1997 Elsevier Science Ltd. All rights reserved
Printed in Great Britain
0196-8904/97 $17.00 + 0.00

PII: S0196-8904(96)00329-9

POTENTIAL LAND AREA FOR REFORESTATION AND CARBON DIOXIDE MITIGATION EFFECT THROUGH BIOMASS ENERGY CONVERSION

SHIN-YA YOKOYAMA

National Institute for Resources and Environment
16-3, Onogawa, Tsukuba, Ibaraki 305, Japan

ABSTRACT

Reforestation has a great potential to fix carbon dioxide (CO_2) in the atmosphere through photosynthesis of biomass and biomass energy conversion. The magnitude of CO_2 fixation depends on the potential land area . We have proposed a potential area of about 350 Mha for reforestation in tropical and semi-tropical areas. If such land is used for energy plantation aimed at substitutive energy production by power generation, the CO_2 mitigation effect can be evaluated on the assumption of short rotation energy plantation(6-year rotation), plantation area (1Mha), biomass productivity (10.5 dry ton/ha/y), energy efficiency of power generation(0.27), etc. Assuming these factors, the net carbon dioxide mitigation effect is 1.4 billion t-C on 340 Mha. © 1997 Elsevier Science Ltd

KEYWORDS

Biomass, plantation, reforestation, carbon dioxide fixation, energy conversion, power generation

POTENTIAL LAND AREA FOR REFORESTATION

A total amount of the CO_2 fixation by planted trees highly depends on the practical planting areas, and it is no exaggeration to say that the evaluation of the biological CO_2 fixation is controlled by the estimation of this value of area. Most past assessment of practically planted areas were roughly estimated by an approximation of already cut forest areas in the past, but recent estimates of the area of lands available for planting have become much more practical. There are two ways for the estimations of practical planting areas. One is the so-called "top-down method" that depends on the multiplication of factors based on the Food and Agriculture Organization (FAO) data, and another is the so-called "bottom-up method" that integrates practical planted areas published by federal and local governments. It is needless to say that the bottom-up method is superior, but in fact it is impossible to clarify the potential practical planting areas in the world by this method. Examples of estimated practical planting areas to date are shown in Table 1.

Table 1. Reported Potential Land Area for Reforestation

	Potential Land Area (Mha)
Grainger(1988)	758
Myers(1989)	300
Houghton(1990)	865
Winjum et al.(1992)	600–1200
Alpert et al.(1992)	952
Bekkering(1992)	385–553
Nakicenovic et al.(1993)	265

Because the practical planting areas are highly related to the cost of reforestation, it is possible to plant trees at desert areas when no limitation on the cost is allowable. Then, if plantations are established at sites where no large scale irrigation or soil improvement are necessary, with ranges of $1,000 - 2,000/ha for planting in the developing countries, grasslands and a part of shrub lands where irrigation is necessary become infeasible.

Table 2 Estimated Potential Land Area for Reforestation

	Potential Land Area(Mha)
Central Africa	16.2
Madagascar	7.2
Mexico	80.4
Argentine	16.5
Bolivia	8.6
Brazil	105.9
Chile	8.6
Colombia	9.0
Peru	16.8
Uruguay	0.1
Venezuela	23.5
Sub total	292.8
Turkey	11.3
Vietnam	13.7
Thailand	1.7
Myanmer	20.7
Sub total	47.4
Total	340.2

Bekkering (1992) presented an idea that surplus lands should be   used for reforestation in the tropical zone, with sufficient rainfall and with good conditions for the food supply. On the basis of this idea, the land becomes about 300 Mha for sites proposed for reforestation, as shown in Table 2. Eleven countries with fairly good stable conditions in food supplies are candidates in South America and Central Africa, though several of them are listed as the food importing countries. If four countries of fairly good food supplies in the Southeast Asian region are added, the total area becomes 340 Mha.

THE GENERATION OF ELECTRICITY BY BIOMASS COMBUSTION

The mitigation of CO_2 is estimated here when electricity is generated by combustion of biomass such as fast growing trees, for example, *Eucalyptus* , it is important to note that the net accumulation of CO_2 in the atmosphere can be maintained if the amount of biomass which is used for energy combustion is balanced with that of biomass planted. If forests are managed and utilized for energy in a sustainable manner, CO_2 emission from combustion of fossil fuels can be avoided. Two cases are discussed for the electricity generation
from biomass combustion.

Case 1 : *Eucalyptus* plantation of 20 km in diameter, with a 6- year rotation and 5230 ha/y harvested

Parameters:

net production of *Eucalyptus*	10.5 dry ton/ha
harvested area	5230 ha
calorific value of *Eucalyptus*	20 GJ/dry ton
efficiency of power generation	0.22 (biomass combustion)
efficiency of power generation	0.33 (coal combustion)
carbon emission from coal combustion	0.027 t-C/GJ

The net production of *Eucalyptus* (10.5 dry ton/ha/y) seems reasonable because that of *Eucalyptus* has been reported to be more than 40 wet ton/ha/y in Brazil. In this case, the carbon content in wood is about 45 - 50%. The calorific value of *Eucalyptus* is 20GJ/ dry weight (about 4800 kcal) and the efficiency of direct combustion is 22%, because a small scale power generation plant is adopted in this case. At this time, no consideration is given to energy consumption for the construction of the power plant.

The mitigation of CO_2 is calculated as follows:
10.5(dry ton/ha/y) x 31400 (ha) x 20 (GJ/dry ton) x 0.22/0.33 x 0.0247 (t-C/GJ)
= 108.6 kt-C/y

As for the energy used in cultivation of *Eucalyptus*, energy is necessary for the preparation of land, planting, fertilization, management and thinning. Furthermore, the energy needed for cutting and transportation must be considered when harvesting. These are assumed to be:

energy needed for tree growth   11.94 GJ/ha/y
energy needed for tree harvesting   4.1 GJ/ha/y
carbon emission from gasoline   0.0208 t-C/GJ

Taking into consideration the energy needed for the growth and harvest of trees, it is calculated to be 3.5kt-C/y. The above-mentioned value is based on the data reported by Larson (1995). The energy input of fossil fuel for tree growth and harvest is converted to the energy of gasoline. Thus, the true CO$_2$ mitigation effect becomes 108.6 - 3.5 = 105.1 kt-C/y. In other words, 105,100 t of carbon can be reduced on this area each year.

Case 2 : *Eucalyptus* plantation of 1 Mha, with a 6-year rotation and 0.167 Mha/y harvested

Parameters:

net production of *Eucalyptus*   10.5 dry ton/ha
harvested area   0.167 Mha
calorific value of *Eucalyptus*   20 GJ/dry ton
efficiency of power generation   0.27 (biomass combustion)
efficiency of power generation   0.33 (coal combustion)
carbon emission from coal combustion   0.027 t-C/GJ

In this case, the efficiency of power generation with biomass is raised to 0.27 from 0.22 due to the scale effect of power plant.

The mitigation of CO$_2$ is calculated as follows:
10.5(dry ton/ha/y) x 1.0 (Mha) x 20 (GJ/dry ton) x 0.27/0.33 x 0.0247 (t-C/GJ)
= 4.24 Mt-C/y

Energy needed for cultivation and harvest is as follows:
energy needed for tree growth   4.1 GJ/ha/y
energy needed for tree harvesting   13.76 GJ/ha/y
carbon emission from gasoline   0.0208 t-C/GJ

In this case, the energy for harvesting is larger than for Case 1 due to the longer distance for tree transportation. The true CO$_2$ mitigation effect for 1 Mha becomes 4.24-0.12=4.12 Mt-C/y. Therefore, the net CO$_2$ mitigation effect is 1.4 billion t-C on 340 Mha.

SUMMARY

Biomass energy is understood as a key technology against global warming. According to the Intergovernmental Panel on Climate Change (IPCC), the land for biomass plantation is evaluated about 600 Mha or more. However, it is considered that the available land area is smaller than expected in the scenario if looking at the real situation of the developing countries in the tropical region where forest has been cut down due to urbanization, industrialization, and conversion to farming. In this article, the available land is estimated to be about 340 Mha taking into account the necessity of irrigation and soil improvement in addition to the situation of food supply of each country.

As for the individual species of wood and energy conversion technology of biomass, fast growing *Eucalyptus* is chosen and the generation of electricity by direct combustion of biomass is considered. In this analysis, about 4 million tons of carbon can be reduced on 1Mha(100km x 100km) land. Thus, 1.4 billion tons of carbon can be reduced on 340 Mha if sustainable management of biomass energy plantation is carried out for power generation from biomass.

REFERENCES

Alpert, S. B., D. F. Spencer, . and G. Hidy (1992). Biospheric options for mitigating atmospheric carbon dioxide levels. Energy Conserv. Mgmt., 33, 729-736.

Bekkering, T. D. (1992). Using tropical forests to fix atmospheric carbon: The potential in theory and practice. Ambio, 21, 414-419

Graigner, A. (1988). Estimating areas of degraded tropical lands requiring replacement of forest cover. The International Tree Crops Journal, 5, 31-61.

Houghton, R. A. (1990). The future role of tropical forests in affecting the carbon dioxide concentration of the atmosphere. Ambio, 19, 204-209.

Larson, E. D., C. I. Marrison and R. H. Williams (1995). CO$_2$ mitigation potential of biomass energy plantations in developing regions. Private communications.

Myers, N. (1989). The greenhouse effect: A tropical forestry response. Biomass, 18, 73-75.

Nakicenovic, N., A. Grubler, A. Inaba, S. Messner, S. Nilsson, L. Y. Nishimura, H-H. Rogner, A. Schafer, L. Schrattenholzer, M. Strubegger, J. Swisher, D. Victor, and D. Wilson (1993). Long-term strategies for mitigating global warming, Energy, 18, 401-609.

Winjum, J. K., R. K. Dixon and P. E. Schroeder (1992). Estimating the global potential of forest and agroforest management practices to sequester carbon. Water, Air and Soil Pollution, 64, 312-227.

SECTION 7

ADDITIONAL TOPICS

Energy Convers. Mgmt Vol. 38, Suppl., pp. S575–S580, 1997
© 1997 Elsevier Science Ltd. All rights reserved
Printed in Great Britain
0196-8904/97 $17.00 + 0.00

Pergamon

PII: S0196-8904(96)00330-5

OXYGEN-BLOWN GASIFICATION COMBINED CYCLE: CARBON DIOXIDE RECOVERY, TRANSPORT, AND DISPOSAL

R.D. DOCTOR, J.C. MOLBURG, and P.R. THIMMAPURAM

Argonne National Laboratory, 9700 South Cass Avenue,
Argonne, Illinois 60439

ABSTRACT

This project emphasizes CO_2-capture technologies combined with integrated gasification combined-cycle (IGCC) power systems, CO_2 transportation, and options for the long-term sequestration of CO_2. The intent is to quantify the CO_2 budget, or an "equivalent CO_2" budget, associated with each of the individual energy-cycle steps, in addition to process design capital and operating costs. The base case is a 458-MW (gross generation) IGCC system that uses an oxygen-blown Kellogg-Rust-Westinghouse (KRW) agglomerating fluidized-bed gasifier, bituminous coal feed, and low-pressure glycol sulfur removal, followed by Claus/SCOT treatment, to produce a saleable product. Mining, feed preparation, and conversion result in a net electric power production for the entire energy cycle of 411 MW, with a CO_2 release rate of 0.801 kg/kWhe. For comparison, in two cases, the gasifier output was taken through water-gas shift and then to low-pressure glycol H_2S recovery, followed by either low-pressure glycol or membrane CO_2 recovery and then by a combustion turbine being fed a high-hydrogen-content fuel. Two additional cases employed chilled methanol for H_2S recovery and a fuel cell as the topping cycle, with no shift stages. From the IGCC plant, a 500-km pipeline takes the CO_2 to geological sequestering. For the optimal CO_2 recovery case, the net electric power production was reduced by 37.6 MW from the base case, with a CO_2 release rate of 0.277 kg/kWhe (when makeup power was considered). In a comparison of air-blown and oxygen-blown CO_2-release base cases, the cost of electricity for the air-blown IGCC was 56.86 mills/kWh, while the cost for oxygen-blown IGCC was 58.29 mills/kWh. For the optimal cases employing glycol CO_2 recovery, there was no clear advantage; the cost for air-blown IGCC was 95.48 mills/kWh, and the cost for the oxygen-blown IGCC was slightly lower, at 94.55 mills/kWh. © 1997 Elsevier Science Ltd

KEYWORDS

Gasification; combined cycle; IGCC; life cycle; supercritical CO_2; pipeline.

OVERVIEW OF ENERGY CYCLE FOR INTEGRATED COMBINED-CYCLE BASE CASE

The energy system definition for this study extends from the coal mine to the final geological repository for the CO_2. The location of the IGCC plant is specified as the midwestern United States; in the studies conducted (Doctor et al., 1994; 1996), it is assumed that the plant is 160 km by rail from the mine. Details of the IGCC portion of the system are taken from Gallaspy (1990a), who describes an electric power station using an O_2-blown KRW gasifier, while a follow-up report (Gallaspy 1990b) describes a plant using an air-blown KRW gasifier with in-bed sulfur removal. In each case studied, the CO_2 recovery technologies have been integrated into plant design as much as possible to limit efficiency losses. For each part of the energy system, CO_2 emissions have been either computed directly from process stream compositions or calculated from energy consumption on the basis of a "CO_2 equivalence" of 1 kilogram of CO_2 per kilowatt-hour (electric) (kg/kWhe). In this way, a total CO_2 budget for the system can be derived and compared with the total CO_2 budget for other options, thereby taking into account effects outside the immediate plant boundary.

All seven cases presented here have been adjusted to be on a consistent basis of 4,110 tons/d (stream day) of Illinois No. 6 coal from the Old Ben No. 26 mine. This bituminous, 2.5%-sulfur coal contains 9.7% ash. The underground mine is associated with a coal preparation plant. The assumption is that the IGCC power plant is 160 km from the mine and the coal is shipped by rail on a unit train. The impact of coal mining and shipment on the energy budget is 2.41 MW of power use and 2,879 kg/h of CO_2 emissions. Limestone is used for in-bed sulfur capture in the two air-blown gasifier cases.

The coal preparation system for the O_2-blown IGCC plant includes equipment for unloading the coal from the unit train, passing it through magnetic separators, and then send it to a hammermill. From there, the coal is conveyed to storage silos, from which it is recovered in a fluidized stream for use in the gasifier. The coal is not dried for the O_2-blown cases. The impact of coal preparation on the energy budget is 0.85 MW of power use, with no CO_2 emissions (these will be combined with the overall emissions from the IGCC plant). Drying the coal was not considered for this case.

In contrast, the coal preparation system for the air-blown IGCC plant includes drying by the hot (760°C) flue gas from the IGCC sulfator process. This drying results in CO_2 being emitted from the energy cycle that is not reclaimed and presents a possible opportunity for further reduction. Energy use for coal and limestone preparation is 3.49 MW.

Gasifier Island

The O_2-blown base case employs an air-separation plant producing 1,900 t/d of 95% oxygen. The KRW process is an O_2-blown, dry-ash, agglomerating, fluidized-bed process. Three parallel gasifier trains operating at 3,100 kP/a and 1,010°C are included in the design. Following gasification, cyclones recover 95% of the fines; gas cooling and high-efficiency particulate removal follow. For the base case, glycol H_2S recovery provides a feed to a conventional Claus tail-gas cleanup system. Hence, the significant differences between the O_2-blown and air-blown cases are that the O_2-blown cases cool the product gas for sulfur cleanup and produce a sulfur product for the market, while the air-blown cases employ hot-gas cleanup and produce a landfill product. The impact of the gasifier island operation on the energy budget is 36.82 MW of power use and 6,153 kg/h of CO_2 emissions for the O_2-blown base case.

The air-blown base case uses in-bed sulfur removal. Spent limestone and ash from the gasifier are oxidized in an external sulfator before disposal. The sulfator flue gas is taken to the coal preparation operation for drying coal and not integrated into the later CO_2 recovery operation. The hot-gas cleanup system for particulate matter consists of a cyclone followed by a ceramic-candle-type filter. Solids collected are sent to the external sulfator before disposal. Inlet gas temperatures are maintained at approximately 280°C. Supplemental hot-gas desulfurization is accomplished in a fixed-bed zinc-ferrite system. Off-gas from the regeneration of this polishing step is recycled to the gasifier for in-bed sulfur capture. The impact of the gasifier island operation on the energy budget is 20.12 MW of power use and 137 kg/h of CO_2 emissions for the air-blown base case.

Power Island

Both the O_2-blown and air-blown base cases employ a turbine topping cycle and a steam bottoming cycle based on two heavy-duty GE MS701F industrial gas turbines with a 680°C firing temperature. The impact on the energy budget of the power island operation is 7.02 MW of power use for the O_2-blown base case and 10.58 MW of power use for the air-blown base case. For the O_2-blown base case, gross power generation is 458.20 MW, with a net generation of 413.50 MW; for the air-blown base case, gross power generation is 479.63 MW, with a net generation of 445.44 MW.

INTEGRATED GASIFICATION COMBINED CYCLE WITH CO_2 RECOVERY

Several changes were made to the base-case IGCC plant to incorporate CO_2 recovery. For the turbine topping-cycle studies (Cases 1 and 2), these changes entailed processing the cleaned fuel gas through a "shift" reaction to convert the CO to CO_2, recovering the CO_2, and then combusting the low-CO_2 fuel gas in a modified turbine/steam cycle to produce electricity. Gas cleaning and sulfator performance were considered to be unaffected by these changes. In contrast, the fuel cell topping-cycle studies (Cases 3 and 4) required a highly cleaned gasifier without use of the water-gas shift reaction to be used by the fuel cells. A block diagram of the O_2-blown IGCC system with CO_2 recovery appears in Fig. 1.

The fuel gas from the KRW process is high in CO. Conversion of the CO to CO_2 in the combustion process would result in substantial dilution of the resulting CO_2 with nitrogen from the combustion air and with water from the combustion reaction. If the CO_2 is removed before combustion, a substantial savings in the cost of the CO_2 recovery system is possible because of reduced vessel size and solvent flow rate. The CO in the fuel gas must first be converted to CO_2 by the shift reaction, so that the resulting CO_2 can then be recovered, leaving a hydrogen-rich fuel for use in the gas turbine. The shift reaction is commonly

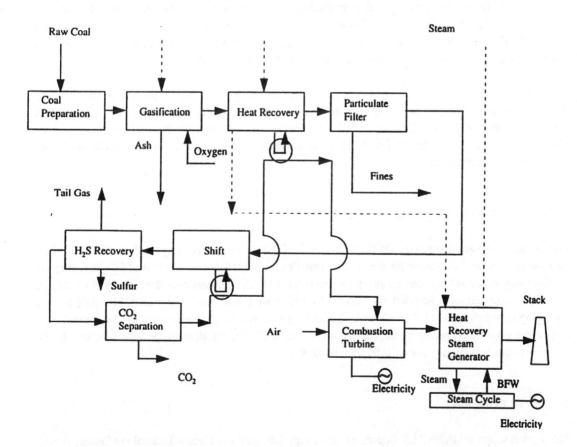

Fig. 1 Block Diagram of the Base-Case Oxygen-Blown KRW IGCC System Modified for CO_2 Recovery (Doctor et al. 1996)

accomplished in a catalyst-packed tubular reactor that uses a relatively low-cost iron-oxide catalyst. High CO_2 recovery is best achieved by staged reactors that allow for cooling between stages; a two-stage system configured to achieve 95% conversion of CO to CO_2 was found to be optimal.

Commercial CO_2-removal technologies all involve cooling or refrigerating the gas stream, with an attendant loss of thermal efficiency. To minimize the loss, the heat removed during cooling must be recovered and integrated into the system. Several options for this integration were evaluated, including steam generation alone, fuel-gas preheating with supplemental steam generation, and fuel-gas saturation and preheating. In the last case, moisture condensed from the fuel gas before CO_2 recovery is injected into the clean fuel-gas stream as it is heated by recovered heat following CO_2 removal. This option allows additional heat to be absorbed before combustion and increases the mass flow rate through the gas turbine. The balance of the thermal energy is used in the heat recovery steam generator for feedwater heating and steam generation.

For the optimal O_2-blown CO_2 recovery case (Case 1), the net electric power production was reduced by 37.6 MW from the base case, with a 0.277-kg/kWhe CO_2 release rate (when makeup power was considered). The low-pressure glycol system, which does not require compression of the synthesis gas before absorption, appears to be the best system of those studied.

PIPELINE TRANSPORT AND SEQUESTERING OF CO_2

Once the CO_2 has been recovered from the fuel-gas stream, its transportation, utilization, and/or disposal is assumed to be at supercritical pipeline pressures, so that it can be directly received in a geological repository. Costs for pipeline construction and use vary greatly by region within the United States. The recovered CO_2 represents more than three million normal cubic meters per day of gas volume. It is assumed that the transport and sequestering process releases approximately 2% of the recovered CO_2.

Levelized costs have been prepared, taking into account that the power required for compression will rise throughout the life cycle of these sequestering reservoirs. The first reservoirs to be used will, in fact, be capable of accepting all IGCC CO_2 gas for a 30-year period without requiring any additional compression costs for operation. The pipeline transport and sequestering process represents approximately 26 mills/kWh for the CO_2-recovery cases.

ENERGY CONSUMPTION AND ECONOMICS

Data on energy consumption and CO_2 emissions for the O_2-blown base case are provided in Table 1. These data can be compared with those for the optimal case that employs low-pressure glycol CO_2 recovery and a turbine topping cycle (i.e., Case 1), also provided in Table 1. A comparison of the costs of electricity for the CO_2-release base cases revealed that the cost for air-blown IGCC was 58.29 mills/kWh, while the cost for the O_2-blown case was 56.86 mills/kWh (Table 2). There was no clear advantage for the optimal cases employing glycol CO_2 recovery; the cost for air-blown IGCC was 95.48 mills/kWh, and the cost for the O_2-blown case was slightly lower, at 94.55 mills/kWh.

ACKNOWLEDGMENT

This work is supported by the U.S. Department of Energy, Morgantown Energy Technology Center, through contract W-31-109-Eng-38 (contract manager, Dr. Richard A. Johnson).

Table 1. Energy Consumption and CO_2 Emissions (Doctor et al. 1996)

Base Case - KRW O2-blown IGCC

Mining and Transport	Electricity MW	CO2 release kg/h
Raw Coal in Mine	-2.36	2,356
Coal Rail Transport	-0.05	523
Subtotal	-2.41	2,879
IGCC Power Plant		
Coal Preparation	-0.85	0
Gasifier Island	-36.82	6,153
Power Island	-7.02	320,387
Subtotal	-44.70	326,540
Power - Gas Turbine	298.80	
Power - Steam Turbine	159.40	
GROSS Power	458.20	
NET Power	413.50	
Pipeline/Sequester	0.00	0
Energy Cycle Power Use	-47.11	
NET Energy Cycle	411.09	329,419
CO2 emission rate/net cycle	0.801	kg CO2/kWh
Power use/CO2 in reservoir	N/A	kWh/kg CO2

Case #1 KRW O2-blown IGCC - Shift; Glycol H2S/CO2; Gas Turbine

Mining and Transport	Electricity MW	CO2 release kg/h
Raw Coal in Mine	-2.36	2,356
Coal Rail Transport	-0.05	523
Subtotal	-2.41	2,879
IGCC Power Plant		
Coal Preparation	-0.85	0
Gasifier Island	-36.82	6,153
Power Island	-7.02	320,387
Glycol Circulation	-5.80	-260,055
Glycol Refrigeration	-4.50	
Power Recovery Turbines	3.40	
CO2 Compression (to 2100psi)	-17.30	
Subtotal	-68.90	66,485
Power - Gas Turbine	284.80	
Power - Steam Turbine	161.60	
GROSS Power	446.40	
NET Power	377.50	
Pipeline/Sequester		
Pipeline CO2		260,055
Pipeline booster stations	-1.64	1,637
Geological reservoir (2% loss)	0.00	-254,854
Subtotal	-1.64	6,839
Energy Cycle Power Use	-72.95	
NET Energy Cycle	373.45	76,202
Derating from O2-Base Case	37.64	
Make-up Power	37.64	37,637
TOTAL	411.09	113,840
CO2 emission rate/net cycle	0.277	kg CO2/kWh
Power use/CO2 in reservoir	0.148	kWh/kg CO2

REFERENCES

Doctor, R.D., J.C. Molburg, and P.R. Thimmapuram, 1996, *KRW Oxygen-Blown Gasification Combined Cycle: Carbon Dioxide Recovery, Transport, and Disposal*, ANL-ESD-34, Argonne National Laboratory, Argonne, Ill.

Doctor, R.D., J.C. Molburg, P.R. Thimmapuram, G.F. Berry, and C.D. Livengood, 1994, *Gasification Combined Cycle:Carbon Dioxide Recovery, Transport, and Disposal*, ANL/ESD-24, Argonne National Laboratory, Argonne, Ill.

Gallaspy, D.T., T.W. Johnson, and R.E. Sears, 1990a, *Southern Company Services' Study of a KRW-Based GCC Power Plant*, EPRI GS-6876, Electric Power Research Institute, Palo Alto, Calif.

Gallaspy, D.T., 1990b, *Assessment of Coal Gasification/Hot Gas Cleanup Based Advanced Gas Turbine Systems: Final Report*, DOE/MC/26019.3004 (DE91002084), prepared by Southern Company Services, Inc., Birmingham, Ala., et al., for U.S. Department of Energy, Morgantown Energy Technology Center, Morgantown, W. Va.

Table 2. Summary of Comparative Costs of IGCC Systems (Doctor et al. 1996)

Case	Unit	BASE	BASE	Case #1	Case #2	Case#3	Case #4	Case #4 ESD-24/Glycol
Gasifier Oxidant		Oxygen	Air	Oxygen	Oxygen	Oxygen	Oxygen	Air
H_2S Recovery		Glycol	In-Bed/ZnTi	Glycol	Glycol	Methanol	Methanol	In-Bed/ZnTi
CO_2 Recovery		none	none	Glycol	Membrane	Glycol	Membrane	Glycol
Topping Cycle		Turbine	Turbine	Turbine	Turbine	Fuel Cell	Fuel Cell	Turbine
Bottoming Cycle		Steam	Steam	Steam	Steam	Steam	Steam	Steam
Component								
Base Plant Capital	$/kW	$1,332	$1,253	$1,485	$1,703	$2,560	$2,746	$1,487
CO_2 Control Capital	$/kW	$0	$0	$202	$602	$145	$905	$246
Total Plant Capital	$/kW	$1,332	$1,253	$1,687	$2,305	$2,705	$3,651	$1,733
Power Plant Annual Cost	$K	$137,253	$144,212	$203,238	$242,336	$249,786	$287,547	$204,288
Power Cost								
Base Plant Power Cost	mills/kWh	58.29	56.86	70.64	101.62	102.45	132.19	71.46
Pipeline Cost	mills/kWh	0	0	23.91	27.35	26.53	28.76	24.02
Net Power Cost	mills/kWh	58.29	56.86	94.55	128.97	128.98	160.95	95.48
Coal Energy Input	10^6 Btu/h	3839	3839	3839	3839	3839	3839	3839
Gross Power Output	MW	458.20	479.63	446.40	417.60	418.50	413.20	460.88
In Plant Power Use	MW	44.70	34.19	68.90	87.60	78.39	99.40	85.11
Net Plant Output	MW	413.50	445.44	377.50	330.00	340.11	313.80	375.77
Net Heat Rate	Btu/kWh	9284	8618	10170	11633	11288	12234	10216
Thermal Efficiency - HHV	%	36.78%	39.62%	33.58%	29.35%	30.25%	27.91%	33.42%
Out of Plant Power Use	MW	2.41	4.18	4.05	3.87	4.05	4.12	4.47
Net Energy Cycle Power	MW	411.09	441.26	373.45	326.13	336.06	309.68	371.30
Net Energy Cycle Heat Rate	Btu/kWh	9339	8700	10280	11771	11424	12397	10339
Thermal Efficiency - HHV	%	36.56%	39.25%	33.21%	29.01%	29.89%	27.54%	33.02%
Net Energy Cycle Power	MW	411.09	441.26	373.45	326.13	336.06	309.68	371.30
Net Replacement [Added] Power	MW	0.00	(30.17)	37.64	84.96	75.03	101.41	39.79
Net Grid Power	MW	411.09	411.09	411.09	411.09	411.09	411.09	411.09

Pergamon

Energy Convers. Mgmt Vol. 38, Suppl., pp. S581–S588, 1997
© 1997 Elsevier Science Ltd. All rights reserved
Printed in Great Britain
0196-8904/97 $17.00 + 0.00

PII: S0196-8904(96)00331-7

THE SEQUESTERING OF CARBON DIOXIDE, EMBODIED ENERGY AND COMMON TARGETS FOR THE MITIGATION OF GREENHOUSE GAS EMISSIONS

John H. Walsh
19 Lambton Avenue, Ottawa, Ontario, Canada K1M 0Z6.

ABSTRACT

A simple three-level spreadsheet has been extended to assess the emissions of carbon dioxide from pairs of countries cooperating in Common Targets and Activities Implemented Jointly. Using this methodology, actual data for 1995 may be compared conveniently with projected data for cooperative projects in the same year to illustrate both the changes in emissions of this greenhouse gas and in the consumption of primary energy. The technique was applied to a major energy-intensive industry of importance in international trade—steelmaking—in a scenario involving the production of ten million additional tonnes per year in each of Australia, Brazil, and Canada in cooperation with Japan, the largest steelmaker in 1995. If carbon dioxide can be captured and sequestered in each of the primary production countries, a major expansion of the industry is possible through the use of pooled sequestering facilities. The direct reduction of iron ore in natural gas-based processes has significant advantages when emissions of this gas must be controlled.

© 1997 Elsevier Science Ltd

KEYWORDS

Activities implemented jointly; carbon dioxide; common targets; embodied energy; energy-intensive industries; iron and steel.

INTRODUCTION

This paper provides a method to (1) monitor and (2) test scenarios for collaborative projects undertaken in pairs of countries to reduce carbon dioxide emissions through Common Targets or Activities Implemented Jointly. A spreadsheet technique is extended to cover the case of such agreements negotiated within the terms of the Conferences of the Parties (COP) to the Framework Convention on Climate Change (FCCC) agreed at the Earth Summit sponsored by the United Nations in Rio de Janeiro in June of 1992. The conditions pertaining to such cooperative agreements (including reporting requirements) were raised at the COP 1 Meeting held in Berlin March 28 - April 7 of 1995 and were discussed further at COP 2 held in Geneva July 8 - 19, 1996. The objective of such international joint activities is to encourage and enhance the adoption of technologies and procedures that lead to lower emissions of greenhouse gases on a world basis at least cost. This approach to coping with the greenhouse gas problem may be particularly important in developing countries, many of which, such as Brazil, China, and India, are experiencing large and consistent percentage increases in emissions at present. Despite these increases, many of the latter group of nations are reluctant to undertake such measures since their per capita emissions generally remain low as compared to the developed member countries of the OECD (Walsh, 1995).

Though the main objective of Common Targets and Activities Implemented Jointly is to provide flexibility in meeting mitigation objectives with the aim of minimizing the cost of these measures, the negotiation of such co-operative international projects, especially when developing counties are involved, presents an opportunity to encourage the transfer of technology and to stimulate international investment. Successful projects of this kind may lead to greater economic growth in the developing nations and thus may prove attractive on this account.

To date, such proposed collaborative activities have focused on more conventional approaches such as enhancing the efficiency with which energy is used (conservation measures), converted (such as increasing the efficiency with which electricity is generated from fossil fuels), and distributed (by installing improved transmission lines and pipelines), but including as well additional supplies of energy from benign sources from a greenhouse gas point of view. Off-set measures involving 'sinks' may also be considered to enhance the growth or protection of fast-growing forests in the hot countries of the developing world. In this paper, the possibilities for Common Targets which involve energy embodied in non-energy but energy-intensive commodities prominent in international trade are examined using steel and its semi-processed intermediate products as an example. The possible pooling of the carbon dioxide captured during the course of industrial production with that captured from the coal-based generation of electricity is also assessed.

NECESSARY CONDITIONS FOR COMMON TARGETS

Common Target Agreements (CTAs) and Activities Implemented Jointly (JI) are of interest when differing marginal costs of reduction in emissions of carbon dioxide (or their equivalent from other greenhouse gases) may be exploited in two or more countries. In some geographically large countries such as Canada and Russia, it is even possible some form of CTA/JIs might exist among regions (Loulou and Waaub, 1992). There are four objectives that should be met for CTA/JI activities to be successful: (1) Environmental Goal - the total emissions of greenhouse gases taken together should decline (Nevertheless, it is possible for the emissions from one of the partners to increase as a result of CTA/JI projects); (2) Economic Goal - the Gross Domestic Product (GDP) in both countries should increase as a consequence of the project; (3) Social Goal - the ratio of the total joint greenhouse gas emissions to the joint GDP should decrease with time; and (4) Personal Goal - the per capita GDP should also increase in both cooperating countries.

Two-country collaborations can be represented by a 45-degree graph as in Fig.1. Three countries—Australia, Brazil, and Canada—are each assumed to be collaborating with Japan, a large importer of energy. The joint CO_2 emissions for the three sets of partners are shown for 1995. After a new common project by any of the three other nations with Japan, a parallel line should lie lower on the graph than the original base-line position though, as noted above, emissions in any of the latter three nations could actually increase as a result of this additional activity. Any such increase, however, must be less than the decrease in the other partner.

EMBODIED ENERGY

Canadian non-energy exports are generally more energy-intensive than the country's corresponding non-energy imports and thus it is likely there is, in effect, a net embodiment of greenhouse gas emissions in the country's overall trade position. The first direct estimate of such a relationship to carbon dioxide emissions was published for the case of Brazil (Schaeffer and de Sá, 1996). The net embodied CO_2 was estimated at 11.4% of Brazil's total emissions in 1990 (some 8.3 million tonnes expressed as carbon) and this share was expected to rise with time. Such estimates are normally made by employing input-output tables which, unfortunately, are not usually available for some years after the relevant date (Hetherington, 1996). Despite the difficulties and uncertainties in making this calculation, this question is likely to be of increasing importance in the FCCC negotiations since many developing countries are energy rich and so tend to export energy-intensive commodities often as partly processed goods. In a two-country trading pattern, the emissions resulting from embodied energy can be tracked in graphs of the type in Fig.1 by assuming a closed system since both the combined and the individual country emissions can be represented together.

The specific example chosen for energy embodiment in traded non-energy materials focuses on the steel industry which accounts for some 4-6% of total world carbon dioxide emissions depending upon the methodology employed in the estimation. Though a mature industry, some 749.6 million tonnes of crude steel were produced in the principal steelmaking countries in 1995, an increase of 2.7% on the year (Anon.,

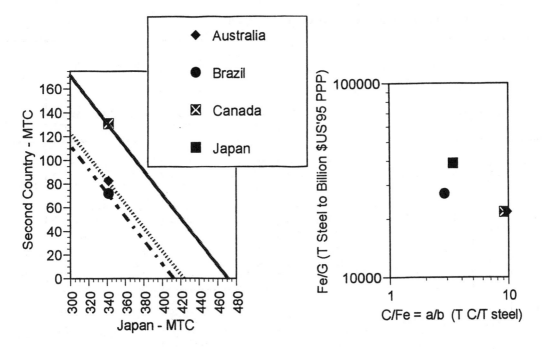

Fig.1. Paired-nation 45-degree chart for 1995.   Fig.2. Energy Map for steel in four countries, 1995.

1996). The industry has always based a large fraction of its production on re-circulated scrap: the industry obtains an average of about 45% of its production from this source of iron in Canada. Of the sources of virgin iron in 1995—iron produced from iron ore—some 526.1 million tonnes were obtained from the traditional blast furnace and 29.9 million tonnes from the newer direct reduced iron processes. Though the latter sources accounted for only 5.4% of the virgin iron supply, production of directly reduced iron increased 11.8% that year as compared to 2.3% for iron from the blast furnace. Moreover, directly reduced iron is becoming an increasing presence in international trade. For example, several facilities in Venezuela reduce local iron ore with natural gas to produce a metallic product much of which is exported to steelmaking countries. A special situation exists in Trinidad, where (mainly) Venezuelan iron ore is reduced to the metallic state using local natural gas which is then re-exported to steelmakers elsewhere. A new facility has entered production there at a reported capacity of 300,000 tonnes per year in which iron ore is first reduced and then converted to the carbide form: not only is energy embodied in the metallic product but some additional fuel in the form of carbon is incorporated as well to make the carbide. Three nations are thus involved when such processing facilities are located in Trinidad: the supplier of the iron ore; the host of the reduction stage; and the location of the final steelmaker. The trend to greater flows of energy-intensive commodities is likely to continue. Argentina, Brazil, Canada, India, Iran, Libya, Mexico, Peru, Quatar, Saudi Arabia, Trinidad, and Venezuela were among the countries with appreciable production of directly reduced iron in 1995.

Australia, Brazil, and Canada are considered here paired individually with Japan with the objective of providing additional steelmaking capacity. (It would also have been possible to examine the possibilities for the replacement of existing capacity reaching the end of its service life.) Japan was chosen because it was the largest steelmaker at 101.7 million tonnes of crude steel in 1995. Australia, Brazil, and Canada produced 8.5, 25.1 and 14.4 million tonnes of crude steel respectively in that year. Each of these countries are also exporters of iron ore but quite different local energy supply conditions apply. Both Australia and Canada are exporters of coal, including the coking grades, to Japan and other countries. Japan imports essentially all its iron ore as well as its requirements for fossil fuels. The relative importance of the steel industry to the four economies may be represented conveniently in a modified 'Energy Map' as in Fig.2 (Walsh, 1994). In this form of representation, the ratio of the total CO_2 emitted (designated C and expressed in million tonnes carbon) to the Gross Domestic Product (G in Purchasing Power Parity terms of $ Billion U.S.1995) may be decomposed into the product of the ratio of the steel production, Fe (in million tonnes), to G multiplied by C/Fe. Thus $C/G \equiv Fe/G * C/Fe$. The value of the abscissa, C/Fe, can be shown to be equal to the ratio of a/b

Table 1. Cooperation between Australia and Japan

1955 ACTUAL		Sources - M Tonnes			Total C	Prim. Energy	JOINT SCENARIO Million Tonnes C			Sources - M Tonnes			Prim. Energy
Fuel	Country	Electrical	Mobile	Stationary	M.T. C	PJ	C Consumed	F or S	C Emitted	Electrical	Mobile	Stationary	PJ
Oil	Jpn	46.32	79.51	96.88	222.71	11191	222.71	-	222.71	46.32	79.51	96.88	11191
	Aus	0.59	20.56	8.18	29.33	1474	29.33	-	29.33	0.59	20.56	8.18	1474
	SSubT	46.91	100.07	105.06	252.04	12665	252.04	-	252.04	46.91	100.07	105.06	12665
Nat. Gas	Jpn	16.08	0.80	14.90	31.78	2303	31.78	-	31.78	16.08	0.80	14.90	2303
	Aus	2.05	0.09	8.26	10.40	754	11.40	0.60	10.80	2.05	0.09	8.66	826
	SSubT	18.13	0.89	23.16	42.18	3057	43.18	0.60	42.58	18.13	0.89	23.56	3129
Coal	Jpn	27.47	-	59.20	86.67	3596	86.67	-	86.67	27.47	-	59.20	3596
	Aus	35.47	0.08	6.73	42.28	1754	45.33	7.86	37.47	29.54	-	7.93	1881
	SSubT	62.94	0.08	65.93	128.95	5350	132.00	7.86	124.14	57.01	-	67.13	5477
Total	Jpn	89.87	80.31	170.98	341.16	17090	341.16	-	341.16	89.87	80.31	170.98	17090
	Aus	38.11	20.73	23.17	82.01	3982	86.06	8.46	77.60	32.18	20.65	24.77	4181
	Sub T	127.98	101.04	194.15	423.17	21072	427.22	8.46	418.76	122.05	100.96	195.75	21271
	Jpn	Primary Electricity PJ (1 kWH = 3.6 MJ)				1349	Primary Electricity (1 kWH = 3.6 MJ)						1361
	Aus					54							54
	Sub T					1403							1415
	Jpn	Total Primary Energy PJ				18439	Total Primary Energy PJ						18451
	Aus					4036							4235
	Total					22475							22686

where a is the CO_2 emissions (in terms of million tonnes of C) per tonne of crude steel produced and b is the ratio of the emissions of CO_2 from the steel industry to the total emissions of this gas from all anthropogenic sources in a given year. The absolute values of these parameters depend upon the methodology chosen to attribute emissions to the industry concerned: for the steel industry in Canada in 1993, a was estimated to be 0.450 and b was 0.0514 excluding related emissions from the transportation sector.

THE CARBON DIOXIDE EMISSIONS SPREADSHEET

The CO_2 spreadsheet was introduced to provide a means of matching the emissions from the deep natural divisions in the economy—from electrical, mobile and stationary sources—to primary energy data (Walsh, 1996). Using this technique, it is possible to distribute emissions of this gas by fuel and by category of source. Total emissions by fuel were calculated by applying standard factors to data obtained from the *BP Statistical Review of World Energy*. In the absence of a recent detailed breakdown of this data by source, the National Energy Data profiles published by the World Energy Council in 1995 were used to derive coefficients for the four countries to permit the distribution of the emissions among the three sources specified. For Australia and Brazil, the data published by the WEC was for 1993 and for Canada and Japan, 1992. In effect, the structure of the energy economy was assumed not to have changed significantly between those years and 1995. (The known Canadian position in 1994 was found in reasonable agreement with that predicted from the 1992 WEC data.) Following this procedure, it was possible to complete the 'actual' left-hand side of Tables 1 (Australia and Japan), 2 (Brazil and Japan) and 3 (Canada and Japan) using 1995 energy consumption data. With the left-hand side complete, proposed co-operative projects may now be incorporated into 'scenarios' set out on the right-hand side of the tables. A separate column, designated F or S, is provided to account for carbon specifically off-set in forests (no allocation to F in this paper) and for CO_2 captured and sequestered (S). Using this procedure, the actual data for 1995 may be manipulated to

Table 2. Cooperation between Brazil and Japan

1995 ACTUAL		Sources - M Tonnes			Total C	Prim. Energy	JOINT SCENARIO Million Tonnes C			Sources - M Tonnes			Prim. Energy
Fuel	Country	Electrical	Mobile	Stationary	M.T. C	PJ	C Consumed	F or S	C Emitted	Electrical	Mobile	Stationary	PJ
Oil	Jpn	46.32	79.51	96.88	222.71	11191	222.71	-	222.71	46.32	79.51	96.88	11191
	Bra	1.05	31.80	25.39	58.24	2927	58.24	-	58.24	1.05	31.80	25.39	2927
	SSubT	47.37	111.31	122.27	280.95	14118	280.95	-	280.95	47.37	111.31	122.27	14118
Nat. Gas	Jpn	16.08	0.80	14.90	31.78	2303	31.78	-	31.78	16.08	0.80	14.90	2303
	Bra	-	-	2.54	2.54	184	3.54	0.60	2.94	-	-	2.94	257
	SSubT	16.08	0.80	17.44	34.32	2487	35.32	0.60	34.72	16.08	0.80	17.84	2560
Coal	Jpn	27.47	-	59.20	86.67	3596	86.67	-	86.67	27.47	-	59.20	3596
	Bra	1.32	-	9.27	10.59	439	12.55	0.81	11.74	1.10	-	10.64	521
	SSubT	28.79	-	68.47	97.26	4035	99.22	0.81	98.41	28.57	-	69.84	4117
Total	Jpn	89.87	80.31	170.98	341.16	17090	341.16	-	341.16	89.87	80.31	170.98	17090
	Bra	2.37	31.80	37.20	71.37	3550	74.33	1.41	72.92	2.15	31.80	38.97	3705
	SubT	92.24	112.11	208.18	412.53	20640	415.49	1.41	414.08	92.02	112.11	209.95	20795
	Jpn	Primary Electricity PJ (1 kWH = 3.6 MJ)				1349	Primary Electricity PJ (1 kWH = 3.6 MJ)						1361
	Bra					943							943
	SubT					2292							2304
	Jpn	Total Primary Energy PJ				18439	Total Primary Energy PJ						18451
	Bra					4493							4648
	Total					22932							23099

investigate the changes in emissions in the two-country groupings for joint projects undertaken that year. The change in primary energy consumption by fuel can also be estimated in this way.

THE COMMON TARGETS CO-OPERATION SCENARIO

The scenario selected to illustrate the spreadsheet method involves adding ten million tonnes of steelmaking capacity jointly between each of the three pairs of cooperating nations. The object is to compare the relative changes in joint emissions that would occur if the same expansion in capacity were undertaken in each of the three partnerships. Though this additional capacity appears large and would in fact take several years to install, it is about one-half the increase in world production of 19.8 million tonnes of steel in 1995. (Growth in the early months of 1996 has, however, been less robust.) Two process routes are considered. The first involves the classical blast furnace and oxygen steelmaking combination. In this case, it is assumed that semi-finished steel slabs would be exported to Japan for rolling and finishing. Processing of these slabs would require additional electrical energy for the rolling and other operations in Japan which, in the future however, will come increasingly from nuclear sources in that country. For this reason, no additional carbon emissions are attributed to this expansion in Japan as a first order approximation. Existing facilities in the three supplying countries could be expanded to supply this steel, or, in the case of Canada, a new west coast plant might be considered based upon supplies of coking coal from mines already exporting to Japan.

The second process route assumes additional natural gas-fueled direct reduction facilities will be used. In this case, the iron ore is converted to metallic form in each of the second countries and converted into steel in electric furnaces in Japan. Again the additional electrical requirement in Japan is assumed to come from

Table 3. Cooperation between Canada and Japan

1995 ACTUAL		Sources - M Tonnes			Total C	Prim. Energy	JOINT SCENARIO Million Tonnes C			Sources - M Tonnes			Prim. Energy
Fuel	Country	Electri-cal	Mobile	Station-ary	M.T. C	PJ	C Con-sumed	F or S	C Emitted	Electri-cal	Mobile	Station-ary	PJ
Oil	Jpn	46.32	79.51	96.88	222.71	11191	222.71	-	222.71	46.32	79.51	96.88	11191
	Can	3.53	37.82	25.31	66.66	3350	66.66	-	66.66	3.53	37.82	25.31	3350
	SSubT	49.85	117.33	122.19	289.37	14541	289.37	-	289.37	49.85	117.33	122.19	14541
Nat. Gas	Jpn	16.08	0.80	14.90	31.78	2303	31.78	-	31.78	16.08	0.80	14.90	2303
	Can	1.70	0.38	36.52	38.60	2797	39.60	0.60	39.00	1.70	0.38	36.92	2870
	SSubT	17.78	1.18	51.42	70.38	5100	71.38	0.60	70.78	17.78	1.18	51.82	5173
Coal	Jpn	27.47	-	59.20	86.67	3596	86.67	-	86.67	27.47	-	59.20	3596
	Can	20.81	-	4.11	24.92	1034	27.20	4.73	22.47	17.33	-	5.14	1129
	SSubT	48.28	-	63.31	111.59	4630	113.87	4.73	109.14	44.80	-	64.34	4725
Total	Jpn	89.87	80.31	170.98	341.16	17090	341.16	-	341.16	89.87	80.31	170.98	17090
	Can	26.04	38.20	65.94	130.18	7181	133.46	5.33	128.13	22.56	38.20	67.37	7349
	SubT	115.91	118.51	236.92	471.34	24271	474.62	5.33	469.29	112.43	118.51	238.35	24439
	Jpn	Primary Electricity PJ (1 kWH = 3.6 MJ)				1349	Primary Electricity PJ (1 kWH = 3.6 MJ)						1361
	Can					1547							1547
	SubT					2896							2908
	Jpn	Total Primary Energy PJ				18439	Total Primary Energy PJ						18451
	Can					8728							8896
	Total					27167							27347

the current round of expansion of nuclear capacity. The reduction processes in the supplying countries could be located anywhere convenient to the supply of high-grade ore or concentrate, natural gas, and shipping facilities. On the northwestern coast of Australia, iron ore and natural gas occur in reasonable proximity to each other. In Brazil, such plants could be fueled from the new planned trunk gas pipelines. In Canada, such facilities might be considered in Newfoundland where, based upon current exploration activity, natural gas may be discovered in sufficient quantity to support such a process yet still be insufficient to justify the construction of expensive pipeline facilities off the island ('trapped gas'). The reduction facility would thus provide an outlet for gas which would otherwise be restricted to serving only other minor local markets.

Five million tonnes of the additional steelmaking capacity required are assumed to come from expansions of the conventional blast furnace/oxygen steelmaking combination and the other five million tonnes from installations of the direct reduction facilities in each of the three countries. The emissions from this expansion as estimated below are then recorded on the right-hand 'scenario' side of Tables 1, 2 and 3.

THE CAPTURE AND SEQUESTERING OF CO_2 FROM STEELMAKING FACILITIES

The Conventional Process

There are only limited opportunities to capture CO_2 in conventional iron- and steelmaking. To have the least disruption to the established process, the sole source of the captured CO_2 is assumed to be that separated from blast furnace gas. Known techniques can be applied to this low-pressure gas and a conservative extraction of 60% of the (about) 21% by volume of carbon dioxide is assumed from this gas stream. A mid-range coal-injected practice is assumed (coke rate 380 kg and coal injection of 135 kg/tonne of hot metal [iron]) and, together with the contribution arising from fluxing agents, a total of 477 kg of carbon is released in the blast furnace gas per tonne of iron produced in each of the countries. With these assumptions, 28% of the total carbon charged to the furnace is captured as CO_2. Different blast furnace iron-to-scrap ratios are applied in the three countries as considered appropriate to each: for Australia - 70/30; for Brazil - 80/20; and for Canada - 60/40. The quantity of CO_2 captured (vented in brackets) for the three countries respectively for five million tonnes of crude steel production is, in terms of million tonnes C (MTC): Australia - 0.469 (1.200) MTC; Brazil - 0.536 (1.372) MTC; and Canada - 0.402 (1.029) MTC. Since a process step to remove CO_2 from the blast furnace gas is assumed installed, the possibility arises of re-cycling the remaining carbon monoxide-rich gas back to the blast furnace through the tuyères. This case is not considered here in this first estimate because the recirculated energy in this gas would have to be compensated by the combustion of other fossil fuels elsewhere in the plant where there is little chance carbon dioxide capture technologies could be applied: the savings in net emissions from this practice would not be great, if any.

Direct Reduction Processes

Most natural gas-based direct reduction processes require about 14 GJ per tonne of iron reduced. Carbon dioxide is already separated as standard practice in some of these processes in the course of re-generating the reducing gas. At present, this separated CO_2 is vented to the atmosphere. It is assumed here that, for the additional reduction capacity of five million tonnes specified for the calculation, 60% of the carbon associated with the natural gas can be captured for sequestering. For the production of five million tonnes of iron, the natural gas consumed contains almost exactly one million tonnes of carbon so that 0.60 million tonnes of the carbon contained may be captured for sequestering with the other 0.40 million tonnes released ultimately to the atmosphere. These values are applicable to all locations in the three countries under consideration. It is evident that this class of process has a major advantage when carbon dioxide emissions must be controlled for two reasons. First, because the fuel with the least carbon intensity—natural gas— is consumed and second, because some of these processes include a convenient means for separating this gas.

THE LINKAGE OF IRON REDUCTION PROCESSES WITH SEQUESTERING OPERATIONS

It is now clear that the cost of sequestering captured CO_2 in either aquifers or in deep oceans is both high and scale-dependent. Only when the CO_2 is used for such purposes as the enhanced recovery of oil, perhaps of interest in oil-producing regions such as Alberta in Canada, can these costs and scale considerations be off-set appreciably. In consequence, in the general case, it is unlikely that CO_2 separated from steelmaking or other industrial processes for that matter, such as in the production of hydrogen from natural gas, would be sequestered in stand-alone operations although application to the enhanced recovery of oil may prove an exception. However, should large-scale utility-based capture and sequestering operations be installed, the gas captured from the smaller industrial-scale activities could be transported by pipeline for sequestering at the larger sites operated by the utility to the mutual benefit of both parties. In the future, it is foreseeable that new industrial facilities requiring control of emissions of CO_2 may be built adjacent to the utility site. There may well be favoured sites for such industrial complexes at tidewater locations around the world.

It is assumed here that twenty percent of the emissions from the coal-based utility generation of electricity in each of the three collaborating countries are captured and that the steelmaking and direct reduction plants can make use of the necessary facilities for sequestering. The objective is to illustrate the advantages that arise when such processes are installed by the large utilities since, without this assumption, it is unlikely any capture and sequestering facilities could be considered in the industrial sector. The data on the right or 'scenario' side of Tables 1- 3 show the quantity of CO_2 sequestered (designated S in MT carbon) on the coal line as the 20% captured from the total of the coal-based generation by the utilities in the course of operation at 17% loss in efficiency combined with that from the steel plants producing five million tonnes per year by the conventional blast furnace-based process. The CO_2 captured from the five million tonnes produced in natural gas-based reduction processes is shown as S on the natural gas line in the three tables and has the same value for each country. In Brazil, hydroelectric power is important and in Canada both hydropower and nuclear energy are available at some locations. In Japan, the extra electrical requirement of 12.33 PJ (3425 GWH) is assumed to be supplied from the expansion of nuclear generation.

RESULTS AND CONCLUSIONS

The emissions spreadsheet method has been applied to the case of pairs of nations collaborating to reduce carbon dioxide emissions such as in Common Target Agreements or Activities Implemented Jointly negotiated under the terms of the Framework Convention on Climate Change. The case chosen to illustrate the technique was a significant expansion of steelmaking capacity (ten million tonnes per year) evaluated in each of Australia, Brazil, and Canada jointly with Japan. Provided the carbon dioxide captured in the course of the steel industry processes can be pooled with that captured from utilities generating electricity from coal, emissions fell in the case of the collaboration with Australia and Canada but rose slightly in the case of Brazil (highlighted cells in Tables 1, 2 and 3). The increase in Table 2 does not mean that such collaboration is not feasible with Brazil: the results reported here merely indicate that the particular scenario selected for study was not suited to conditions in that country.

The newer natural-gas based processes for the direct reduction of iron ore have a significant advantage over traditional blast furnace-based facilities when emissions must be reduced especially when the electricity required for the final steelmaking step can be supplied from nuclear or hydraulic sources in the country of final production. This is true with or without the capture of CO_2 in the course of the reduction process.

In the cases studied here, the capture of carbon dioxide from the coal-based generation of electricity in the primary production countries is indirectly related to the generation of electricity from nuclear reactors in the country of final production by the pooled sequestering of this gas captured during the initial ironmaking step because power from nuclear generation is used in the final stages of production. It is apparent that the capture and sequestering of carbon dioxide by utilities, when linked in this manner to the disposal of this gas captured from industrial processes, offers a way to significantly increase the output of energy-intensive products when emissions must be controlled. It is thus important to find locations where the pooling of captured emissions is feasible for joint sequestering operations. In the future it is possible that energy-intensive industries would be attracted to such sites. It is a matter of policy whether such efforts would qualify under the current negotiations in the FCCC but at the very least, such possibilities are a contribution to the principle of sustainable development.

REFERENCES

Anon (1996), *Iron and Steelmaker*, **23** (13), 2-8, (July).

Hetherington, Robert (1996), An Input-Output Analysis of Carbon Dioxide Emissions for the United Kingdom, IEA Greenhouse Gas R&D Conference, London, August 22-25, 1995. *Energy Convers.Mgmt*, **37** (6-8), 979-984.

Loulou, Richard and Jean-Philippe Waaub (1992), CO_2 Control with Cooperation in Quebec and Ontario: A MARKAL Perspective, *Energy Studies Review*, **4** (3), 278-296.

Schaeffer, Roberto, and Andre Leal de Sá (1996). The Embodiment of Carbon Associated with Brazilian Imports and Exports, IEA Greenhouse Gas R&D Conference, London. August 22-25, 1995. *Energy Convers.Mgmt*, **37** (6-8), 955-960.

Walsh, J.H. (1994), Application of Energy Maps to the Study of Carbon Dioxide Emissions from Energy-Intensive Industrial Sectors, *Proceedings of the Fourth Workshop of the International Energy Agency Energy Technology Systems Analysis Programme*, Banff, Alberta, September 2-8.

Walsh, J.H. (1995), 1994 Carbon Dioxide Fact Sheet, *Energy Studies Review*, **7** (1), 77-79.

Walsh, J.H. (1996), A Simple Spreadsheet for the Assessment of Options for the Mitigation of World Carbon Dioxide Emissions, IEA Greenhouse Gas R&D Conference, London, August 22-25, 1995. *Energy Convers.Mgmt*, **37** (6-8), 709-716.

Pergamon

Energy Convers. Mgmt Vol. 38, Suppl., pp. S589–S594, 1997
© 1997 Elsevier Science Ltd. All rights reserved
Printed in Great Britain
0196-8904/97 $17.00 + 0.00

PII: S0196-8904(97)00001-0

ESTIMATING THE EMBODIED CARBON EMISSIONS
FROM THE MATERIAL CONTENT

Kazuhiko Nishimura[†], Hiroki Hondo and Yohji Uchiyama

Socio-Economic Research Center,
Central Research Institute of Electric Power Industry,
1-6-1 Ohtemachi, Chiyodaku, Tokyo 100 Japan

ABSTRACT

The embodied carbon emissions includes the direct and indirect carbon emitted for the entire production process. These values are needed to estimate impacts on carbon emissions due to changes in consumption patterns. We have developed a model of the economic system to account for every production process. This model illustrates interrelations among production processes by incorporating sectors that produce multiple products. At each multiple product-producing sector, inputs differ in accordance with the material content of the output product. The model is consistent with available input-output coefficients and the physical law of material flow in each process. By exploring our model, we derived *carbon embodiment functions* that evaluate the embodied carbon emissions of an arbitrary product in terms of material content. Empirical analysis has been fulfilled using the available 405-sector input-output table of Japan. © 1997 Elsevier Science Ltd

KEYWORDS

Input-Output Table; Material Balance; Process Systems; Carbon Dioxide.

INTRODUCTION

Changes in consumption decisions can cause a major impact on total carbon emissions. In order to estimate this impact, it is important to add all of the carbon emitted at each step of the production process (i.e. embodied carbon) for every product that is consumed by households. We use the term *consumer goods* to represent this kind of product. Since consumption decisions are made among substitutable consumer goods, it is important to be able to estimate the embodied carbon of all consumer goods that substitute for each other.

Since the early 1970s, input-output (I/O) analysis and other related methods have been used to estimate the energy consumed or carbon emitted directly and indirectly (Bullard *et al.*, 1975). Although the I/O table describes the entire economic system, production processes are aggregated into sectors. In the actual I/O table, consumer goods-producing processes are often aggregated. Therefore, this concept does not provide a useful tool for estimating the embodied carbon of consumer goods.

Methods for incorporating process data into the I/O framework have been proposed (Bullard *et al.*, 1975, 1978) in order to overcome this aggregation problem and were applied in assessing new technologies (Uchiyama and Yamamoto, 1991, Hondo and Uchiyama, 1995). However, these procedures are generally very time consuming and are not suitable for estimating the embodied carbon in many different consumer goods. Moreover, these hybrid methodologies tend to underestimate the

---

† Author for correspondence.

embodied carbon emission due to the inevitable settings of system boundaries in actual cases. Hence, there is a need for an alternative methodology to estimate the embodied carbon emissions for consumer goods.

In order to estimate the embodied carbon emissions of arbitrary consumer goods without setting a system boundary, it is necessary to construct a model of the economic system in terms of interrelations between individual production processes. Accordingly, we have introduced a model for the economic system in which each sector of the model produces multiple products by adopting different production processes. This implies that relations between inputs and outputs in each sector change depending on the product produced as shown in Fig. 1. Since any product is identifiable in terms of its material content, inputs in each sector are determined by the material content of the product produced.

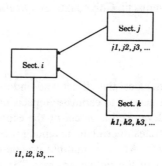

Fig. 1.   Input change at sector *i* for the process model.

It is possible to obtain the material contents of *average products* made within sectors by applying mass conservation to the I/O framework. We regard the relations between inputs and outputs in each sector in terms of material content of average products as one of the states that the model acquires. Consequently, we modeled the relations between inputs and outputs in each multiple-product-producing sector by extrapolating this state.

We have represented the economic system by two submodels.   The first submodel is called the *sector model* and assumes that each sector produces a unique product. By utilizing this model, material contents of average products are obtained. The second model is called the *process model* and relaxes the assumption of a unique product by incorporating the result obtained for the sector model. By utilizing the process model, the carbon-embodiment function which measures the embodied carbon emissions of arbitrary products in terms of material contents is obtained.

THE MODELS

Our models represent the material flow within the economic system. In order to do so, we have characterized the sectors of the I/O table by physical aspects. Thus we define the following sets: $N =$ set of all the sectors in the I/O table, $M =$ set of material-producing sectors (e.g., iron, copper, plastic resin, etc.), $K =$ set of intermediate product-producing sectors (e.g., bolts, nuts, bulbs, etc.), $R =$ set of material-recycling sectors, $F =$ set of final product-producing sectors (e.g., automobiles, refrigerators, television sets, etc.), $S =$ set of remaining sectors, where $N = M \cup K \cup R \cup F \cup S$. Note that there is no intersection among the sets M, K, R, F, S.

At sectors of K and F, inputs from sectors M, K and R are introduced into the products that these sectors produce. We call this type of input a *physical input*. On the other hand, inputs from sectors of F and S are used as tools or consumed rather than introduced into the product physically. We call this type of input a *non-physical input*. Note that the material-producing sectors we have chosen for the models are described in Table 1.

Sector Model

Since the purpose of the sector model is to derive the material content of the average product, material conservation is applied to the intermediate and final product-producing sectors in the I/O table, i.e.

$$\bar{c}_i^m = \sum_{j \in M \cup K \cup R} \bar{c}_j^m a_{ji} - \bar{g}_i^m , \qquad i \in K \cup F, m \in M, \tag{1}$$

where $\bar{c}_i^m$ = m-material content of a unit average product i, a_{ji} = I/O coefficient of sector i for sector j, $\bar{g}_i^m$ = average disposal coefficient expressed in terms of yen of disposed material m to produce a unit of average product i. This material balance is illustrated in Fig. 2.

Table 1. List of material-producing sectors.

Sector	Product
M1	Wood
M2	Fiber
M3	Synthetic rubber
M4	Thermo-setting resin
M5	Thermoplastics resin
M6	High functionality resin
M7	Other resin
M8	Glass
M9	Hot rolled steel
M10	Cast and forged steel
M11	Copper
M12	Lead
M13	Zinc
M14	Aluminum
M15	Other non-ferrous metals
M16	Other materials

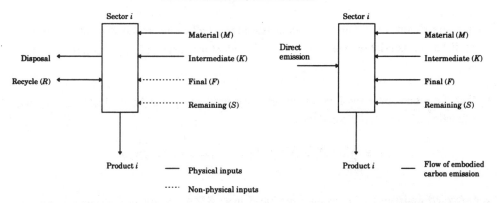

Fig. 2.   Physical and non-physical inputs and outputs at sector i.

Fig. 3. Balance of embodied carbon emission at Sector i.

In view of the definition of $\bar{c}_i^m$, the following statements are evident:

$$\bar{c}_i^m = \begin{cases} 1 & \text{when } i = m, \\ 0 & \text{otherwise,} \end{cases} \qquad i \in M, m \in M ; \tag{2}$$

$$\bar{c}_i^m = \begin{cases} \dfrac{\text{price of } m}{\text{price of } i} & \text{when } i \text{ recycles material } m, \\ 0 & \text{otherwise,} \end{cases} \qquad i \in R, m \in M . \tag{3}$$

Since the average disposal coefficient $\bar{g}_i^m$ is given for all sectors, Eq. (1) may be solved for $\bar{c}_i^m$ with

$i \in K \cup F$ in Eqs. (2) and (3). The values of $\bar{c}_i^m$ for selected F sectors, were calculated (as in Nishimura *et al.* 1996) by using the 405-sector 1990 I/O table and the disposal-coefficient matrix[†] of Japan.

Embodied carbon balance

We show that embodied carbon emissions conserve for each industry. The embodied (carbon) emissions for each sector can be expressed as follows:

$$b = e[I - A]^{-1}$$

where e = row vector of embodied emissions (direct and indirect emission due to unit output of each sector), b = row vector of direct emissions due to unit output of each sector, A = I/O matrix. Concerning

$$[I - A]^{-1} = I + A + A^2 + \cdots = [I - A]^{-1} A + I,$$

we have

$$b = e[I - A]^{-1} A + e = bA + e,$$

which implies that embodied emissions for each sector can be expressed in terms of embodied emissions of all the inputs bA plus the direct emission e. This is illustrated in Fig. 3.

Process Model

The materials balance in the intermediate and final sectors of the process model is

$$c_i^m x_i = \sum_{j \in M \cup K \cup R} c_j^m y_{ji} - g_i^m x_i, \qquad i \in K \cup F, \, m \in M, \tag{4}$$

where c_i^m = m-material content of a unit product produced in i, g_i^m = disposal coefficient expressed in terms of yen of disposed material m to produce a unit of product produced in i, x_i = output of i, y_{ji} = input from j to i.

We assume that the embodied emissions of an arbitrary product r_i can be evaluated as a linear homogeneous function of its material content as follows:

$$r_i = \sum_{m \in M} \delta_i^m z_i^m = \sum_{m \in M} \delta_i^m c_i^m x_i, \qquad i \in M \cup K \cup R \cup F, \tag{5}$$

where r_i = embodied emissions of a product produced in i, δ_i^m = embodied emissions in unit m-material that constitutes a product produced in i, z_i^m = m-material content of product produced in i. Equation (5) implies that

$$\delta_i^m = \partial r_i / \partial z_i^m = constant, \qquad i = M \cup K \cup R \cup F, \, m \in M. \tag{6}$$

Since the embodied emissions are balanced for each production process as shown above, the balance equation for the embodied emissions in each sector are

$$\sum_{m \in M} \delta_i^m c_i^m x_i = \sum_{m \in M} \sum_{j \in M \cup K \cup R} \delta_j^m c_j^m y_{ji} + \sum_{j \in F \cup S} b_j y_{ji} + \varepsilon_i \qquad i \in K \cup F, \tag{7}$$

[†] Disposal coefficients for each sector were estimated according to the survey on industrial residuals and treatments of Japan.

where ε_i = direct emission in i. Note that the embodied emission of disposed material is taken as zero.

Theoretically, changes in consumption patterns of consumer goods including adoption of new production processes can cause price changes and technological shifts in the economy. However, since consumer goods are not production factors for any production process and assuming all production processes have constant returns to scale production functions, it is possible to deduce that the structure of the economy is stable against changes in consumption decisions. Thus, the embodied emissions of products (apart from the consumer goods involved) will remain constant. Accordingly, it is assumed that

$$b_j, \; e_i = constant. \quad j \in F \cup S, \; i \in K \cup F \tag{8}$$

Inputs, subject to outputs, are written as follows in the process model when including the sector model:

$$c_j^m y_{ji} = \frac{\bar{c}_j^m a_{ji}}{\bar{c}_i^m} c_i^m x_i, \qquad m \in M, \, i \in K \cup F, \, j \in M \cup K \cup R, \tag{9}$$

$$y_{ji} = \frac{\sum_{m \in M} c_i^m x_i / p_m}{\sum_{m \in M} \bar{c}_i^m / p_m} a_{ji}, \qquad m \in M, \, i \in K \cup F, \, j \in F \cup S, \tag{10}$$

where p_m = price of material m. Equation (9) implies that the material contents of physical inputs vary in proportion to the material contents of the output for each material m. Equation (10) implies that non-physical inputs vary in proportion to the physical quantity of the output. For empirical analysis, we used weight as a physical quantity. In addition, direct emission is assumed to be the function of material content of the product produced such that

$$\varepsilon_i = e_i \frac{\sum_{m \in M} c_i^m x_i / p_m}{\sum_{m \in M} \bar{c}_i^m / p_m}, \quad m \in M, \, i \in K \cup F, \, j \in F \cup S. \tag{11}$$

These assumptions are consistent with the sector model, which is implicitly included as a normal point of the state that the process model can acquire. According to Eqs. (9) and (10), Eq. (7) is modified as follows:

$$\sum_{m \in M} \delta_i^m c_i^m x_i = \sum_{m \in M} \sum_{j \in M \cup K \cup R} \delta_j^m \frac{\bar{c}_j^m a_{ji}}{\bar{c}_i} c_i^m x_i + \sum_{j \in F \cup S} b_j \frac{\sum_{m \in M} c_i^m x_i / p_m}{\sum_{m \in M} \bar{c}_i^m / p_m} a_{ji} + e_i \frac{\sum_{m \in M} c_i^m x_i / p_m}{\sum_{m \in M} \bar{c}_i^m / p_m}, \; i \in K \cup F. \tag{12}$$

Taking the derivative of Eq. (12) with respect to material content $c_i^m x_i$ and using Eqs. (6) and (8) yields

$$\delta_i^m = \sum_{j \in M \cup K \cup R} \delta_j^m \frac{\bar{c}_j^m}{\bar{c}_i} a_{ji} + \sum_{j \in F \cup S} b_j a_{ji} \frac{1/p_m}{\sum_{m \in M} \bar{c}_i^m / p_m} + e_i \frac{1/p_m}{\sum_{m \in M} \bar{c}_i^m / p_m}, \; i \in K \cup F, \, m \in M. \tag{13}$$

Concerning that $\delta_j^m = b_m$ for $j = m \in M$, $\delta_j^m = 0$ for $j \neq m \in M$ and $\delta_j^m = 0$ for $j \in R$, equations (13) are solved for δ_i^m, $i \in K \cup F$, $m \in M$. Thus, the embodied energy of an arbitrary product is expressed as a function of material content as in Eq. (5).

For consumer goods that are products made in the final product-producing sectors, the coefficients for the carbon-embodiment functions δ_i^m for the selected F sectors, calculated by using the 405-sector I/O table and the carbon emission data for each sector developed by (Hondo *et al*., 1996), are shown in Table 2.

Table 2. Coefficients of carbon-embodiment functions for selected sectors (kg-carbon/kg-material).

Sectors/Materials	M1	M2	M3	M4	M5	M6	M7	M8	M9	M10	M11	M12	M13	M14	M15	M16
Wooden furniture and accessories	0.4	5.9	2.8	1.1	2.1	2.9	3.0	0.5	0.9	1.2	1.1	2.5	1.3	0.7	9.8	1.0
Metallic furniture and accessories	0.6	6.9	3.7	2.5	2.2	3.1	3.3	0.4	0.9	1.7	1.0	1.6	1.4	0.7	9.9	0.9
Radio and Television sets	14.3	6.3	3.2	3.1	2.2	3.2	3.2	2.5	1.6	3.2	2.8	2.8	3.0	2.8	13.3	2.9
VCR	20.0	5.8	4.1	5.0	3.2	4.1	3.8	3.5	2.4	3.2	3.0	3.8	3.6	2.5	12.8	3.9
Electric computing equipment	13.3	9.0	6.9	7.5	6.3	6.2	7.4	7.0	5.6	6.7	6.3	7.6	7.3	4.9	17.4	6.6
Applied electronic equipment	12.1	9.7	7.8	8.7	7.7	8.7	8.5	5.7	5.3	5.9	6.6	7.5	6.9	6.2	16.7	5.8
Electric lighting fixtures and equipment	5.2	5.6	2.7	2.7	2.2	3.2	3.0	2.1	1.5	2.8	1.7	2.3	1.8	1.3	11.5	1.8
Passenger cars	5.5	12.8	3.7	2.7	2.2	2.9	2.9	0.4	1.5	2.1	1.7	2.1	2.0	1.5	10.6	2.3
Trucks, buses and other cars	4.3	15.2	3.6	2.8	2.2	2.9	2.8	0.4	1.4	2.0	1.7	2.1	2.0	1.4	10.4	2.3
Two wheel motor vehicle	5.6	10.3	3.7	2.8	2.3	3.0	2.9	2.6	1.6	2.3	1.8	2.4	2.3	1.7	11.5	2.4
Steel ships	0.4	7.1	3.2	1.3	2.2	4.3	4.0	0.6	0.8	1.4	1.2	2.0	1.5	0.8	10.8	1.7
Other ships	0.5	12.0	4.8	1.3	1.2	2.6	2.4	1.4	1.0	1.4	1.5	2.3	1.7	0.9	10.9	2.2
Aircraft	22.0	14.2	6.8	6.6	6.1	7.2	6.9	1.6	3.8	6.5	5.5	4.9	6.1	4.5	13.6	5.3
Bicycles	38.1	39.8	6.2	7.4	7.4	4.9	9.8	66.8	4.3	7.6	3.6	3.6	4.0	3.2	12.2	4.6

CONCLUSION

We have presented mass-based economic models that are consistent with the fundamental principles of mass conservation, as well as with I/O analysis. We have derived the carbon-embodiment functions that estimate the embodied carbon emissions for an arbitrary consumer good since the model infers the production process in terms of its material content. The 405-sector I/O table of Japan has been used for model construction, and the coefficients for the carbon-embodiment functions were calculated. The simple procedure for estimating the embodied carbon emission using this function is useful in the analysis of consumer goods for which many substitutable goods exist such as automobiles, furniture and personal computers. Our methodology replaces the conventional hybrid methodology of I/O and process analysis for which process data is usually not complete. Also, further investigations including import and export treatments and recycle of materials must be accomplished for more decisive analysis.

REFERENCES

Bullard, C.W. and R.A. Herendeen (1975). Energy impact of consumption decisions. *Proceedings of the IEEE*, **63**, 484-493.

Bullard, C. W., P.S. Penner, and D.A. Pilati (1978). Net energy analysis: Handbook for combining process and input-output analysis. *Resources and Energy*, **1**, 267-313.

Hondo, H., K. Nishimura and Y. Uchiyama (1996). Energy requirements and CO2 emissions in the production of goods and services: Application of and input-output table to life cycle analysis. Economic Research Center, Rep. No. Y95013, Central Institute for Electric Power Industry, Japan.

Hondo, H. and Y. Uchiyama (1995). Effect of heat-resistant steel on CO2 emission from coal-fired power plants," *International Ecomaterials Conference*, Xi'an, China.

Management Coordination Agency Government of Japan (1995). *1990 Input-Output Tables*. Tokyo.

Ministry of Welfare (1990). *Survey on Industrial Residuals and Treatments*. Tokyo.

Nishimura, K., H. Hondo and Y. Uchiyama (1996). Derivation of energy-embodiment functions to estimate the embodied energy from the material content. *Energy*, **21**, 1247-1256.

Uchiyama, Y. and H. Yamamoto (1991). Greenhouse effect analysis of power generation plants. Economic Research Center, Rep. No. Y91005, Central Institute for Electric Power Industry, Japan.

Pergamon

Energy Convers. Mgmt Vol. 38, Suppl., pp. S595–S600, 1997
© 1997 Elsevier Science Ltd. All rights reserved
Printed in Great Britain
0196-8904/97 $17.00 + 0.00

PII: S0196-8904(97)00002-2

THE COSTS AND BENEFITS OF MITIGATION:
A FULL-FUEL-CYCLE EXAMINATION OF TECHNOLOGIES
FOR REDUCING GREENHOUSE GAS EMISSIONS

H. AUDUS and P. FREUND

IEA Greenhouse Gas R&D Programme, CRE Group Ltd, Cheltenham, Glos. GL52 4RZ, U.K.

ABSTRACT

Comparison of options for reducing greenhouse gas emissions must take account of many different matters in a consistent manner. A method of full-fuel-cycle analysis has been developed with this aim in mind. In this method, technical options are compared using a measure made up from the "private costs" of owning and operating a power station together with the "external costs" representing its environmental impact. The method is tested on three types of power generation plant, each of which incorporates CO_2 capture and sequestration technology - these include natural gas-fired and coal-fired power plant, as well as CO_2 storage in a disused gas field, in the deep ocean or in an off-set forest.

Emissions arising from all stages of the fuel cycle are estimated - from extraction of fossil fuel through to dispatch of power to the grid. All types of emission and their impact at local, regional and global level are assessed and valuation of these impacts is attempted. Previous work in this field has concentrated on local impacts, such as occupational health, and regional impacts, such as "acid rain". This study is one of the first to include a comprehensive assessment of global warming impacts, how these will vary with time as well as geographically, including allowance for any potential benefits. From this, the environmental external costs of the fuel cycles are determined. © 1997 Elsevier Science Ltd

The methodology developed for these assessment is described and key results are illustrated.

KEYWORDS

Carbon dioxide; capture and storage; full fuel cycle analysis; cost-benefit analysis

INTRODUCTION

The long time-scales involved in climate change are but one of the aspects which makes decision-making about new mitigation technology especially difficult. It may help the decision-maker to have access to information on the full range of mitigation options, expressed in similar terms and with adequate calibration of uncertainty. Such options include improvements in energy efficiency and use of low emission energy sources, as well as capture and storage of carbon dioxide, which is the subject of this conference.

Through the work of the IEA Greenhouse Gas programme and others, it has been shown that capture and storage would allow continued use of fossil fuels, even under circumstances requiring drastic reduction in emissions of greenhouse gases. The cost and performance of these technologies have been evaluated, using data on the actual (or private) costs of building and operating such plant. These assessments show that capture and storage of carbon dioxide would add at least 40% to the cost of electricity sent-out, and quite possibly more. However, these assessments do not take account quantitatively of the environmental impact of the plant. In order to fill this gap, the IEA GHG Programme commissioned a major study to examine the life-cycle emissions of power generation incorporating capture and storage of CO_2, to find out whether the original conclusions based on private costs would be significantly affected by such considerations. This has also allowed us to consider the wider implications of such assessments, which are discussed below, after a brief review of the study, the method used and the results. Further details are given elsewhere (IEA GHG 1995).

METHODOLOGY

The aim of the study was to find out whether the conclusions about CO_2 capture and storage based on evaluation of private costs were significantly altered by consideration of all of the aspects of the cycle of fuel production, electricity generation and emission control. The methodology developed for this work is based on life-cycle analysis. In a life-cycle analysis the cycle must be defined by clear boundaries and by a specific social and temporal context. The art lies in drawing boundaries around the system so that significant areas are included but insignificant or irrelevant areas are excluded. For example, the distribution and use of electricity was excluded from these studies as being common to all the fuel cycles to be evaluated. In contrast, different boundaries would have been needed for a comparison between centralised and distributed power-generation schemes.

Assessment of the benefits to society of a CO_2 mitigation option are complicated by the lack of accepted procedures to compare the benefits of mitigation with the cost of control. A common unit of measurement is required, and the only tangible universal measure is money. The expression of benefits and damages in monetary terms is the basis of cost-benefit analysis. This is a means of applying a consistent set of values to areas of the problem where no formal basis of cost exists, e.g. in areas such as visual intrusion or pollution which are outside the market system. Thus, the methodology described here is a combination of life-cycle and cost-benefit analysis.

Any industrial activity will have an impact on people not directly involved in it and these 'externalities' or 'external costs' are not incurred by the producer unless internalised by, for instance, the imposition of emission regulations. Knowledge of the external costs of fuel cycles is starting to be used as an aid to decision making (EEC, 1995), but previous work has concentrated on damages due to local and regional impacts. Our study extends this knowledge, e.g. by an assessment of global damages arising from the emission of greenhouse gases (GHG).

There are two fundamental inputs to the methodology: (i) the fuel cycle to be assessed, and (ii) the 'context' within which the fuel cycle operates.

(i) Each fuel cycle is defined as all stages of fuel abstraction, treatment and transport, power generation, and disposal of by-products, including captured CO_2. New plant with advanced technology is used.

(ii) The context within which the fuel cycle operates is defined by the dates of operation and location of the plant, and assumptions about the state of the world in respect of population, energy demand, national incomes, etc. These assumptions reflect the fact that many aspects of society are continuously evolving irrespective of climate change issues. The plant was located in The Netherlands, a site typical of the OECD, operating for 25 years from 2005. Two widely divergent scenarios, based on IPCC scenarios (IPCC, 1992) were used to test the sensitivity of the results.

PRIVATE COSTS

The fuel cycles used advanced technology which could be adopted by 2005 and included advanced emission controls. One natural-gas and two bituminous-coal-based cycles were examined. The carbon storage options considered were: the deep ocean; a disused natural gas field; and, for comparison, an 'offset' forest in which CO_2 is sequestered. Each storage option could be applied to any of the power generation options.

Fuel cost projections were derived from other sources (IEA, 1994). For the year 2005, it was assumed the cost of coal would be 2US$/GJ and natural gas, 4.5US$/GJ. Because of the extensive emission controls, the investment in energy recovery, the high fuel price, and the complexity of the stations, the private costs reported in Table 1 are considerably higher than would be expected for a conventional power station today. More details have been presented elsewhere. (IEA GHG, 1995, Summerfield et al., 1994)

Table 1: Full cycle emissions, efficiencies, and private costs.

FUEL CYCLE :	COAL (steam cycle)	COAL (steam cycle)	COAL (IGCC)	NATURAL GAS	NATURAL GAS
Carbon capture?	No	yes	yes	no	yes
Storage of CO_2	Atmosphere	Forest	Ocean	Atmosphere	Gas field
Full cycle CO_2 emission (gCO_2/kWh)	775	0	135	410	75
Full cycle efficiency (% LHV)	43	42	32	53	44
Cost of electricity (c/kWh)	5.7	7.7	8.6	4.6	6.6

The private costs were produced by an engineering contractor and are quoted as being accurate to +/- 25%; greater accuracy could be produced by a fully engineered design but this is not appropriate given the uncertainties of the overall assessment.

EMISSIONS AND THEIR IMPACT

For all of the fuel cycles, the emissions to atmosphere have the greatest impact; other emissions, such as to water, are smaller. Even with CO_2 capture, the main source of emissions is the power station exhaust gas, because it is not practicable to capture all of the CO_2. As a result, local and regional impacts are low.

A global climate model was used to calculate the effect of global GHG emissions on climate change. Then the emissions of GHGs from each fuel cycle were expressed as a fraction of global emissions and the contribution of each fuel cycle to climate change was calculated. The calculation was carried out in a series of 5 year steps over a 100 year period. The two scenarios used were IPCC scenarios IS92a and IS92d (IPCC, 1992). These scenarios represent different possible views of the future. The view based on IS92d is one of heightened environmental awareness and 'intervention' in which the capture and storage of CO_2 would be widely practised. The second view, IS92a, which can be described as a 'non-intervention' case, does not include major reductions in GHG emissions.

IMPACT VALUATION

Impact valuation is controversial. Although assessment of the impact of emissions is a matter of science and is, in principle, reproducible, valuation of these impacts is a matter of human judgement which will vary according to the beliefs and principles of the people involved.

Our primary interest is in global impacts, i.e. climate change and its consequences, but local and regional impacts were also assessed because damages arising from them are part of the overall cost-benefit equation. Local impacts tend to be small (for fossil-fuel-based cycles) because they tend to be tightly controlled and few people are affected. Regional impacts are significant and mainly associated with atmospheric releases, e.g. of acid gases. In this study, the values of local and regional impacts were found to be in the region of 0.1 - 0.25 cents/kWh (details have been reported elsewhere eg IEAGHG, 1995).

Climate change affects a large variety of receptors in many different ways. For example, in this study we examined impact on sectors such as wet lands and dry lands, agriculture, heating and cooling, biodiversity, etc. All impacts thought to be significant and open to calculation were assessed. For instance, we considered lives lost and economic damage due to increased magnitude and frequency of climate hazards. However, no allowance was made for secondary effects, such as social stress, as they are so many and so diffuse. The rate at which these changes occur is also important.

The valuation of damages was conducted using values conventionally applied in OECD countries and consistent with the scenario in use. For example, as the IS92d scenario implies active intervention on a world-wide scale to protect the environment, it would be consistent to place a high value on human life and environmental damage. Some of the issues about valuation we have recognised in this study are:

(i)   Equity: lack of action to combat climate change will impose inequitable costs on future generations and on regions where damages occur. Preliminary information has been generated on the regional distribution of impacts; damages have been calculated over time (from 2005 to 2100); "winners" and "losers" in different cost sectors have been identified.

(ii)   Discount rate: this has a major impact on the present worth of future damage. These studies have used a range of discount rates centred on 1.5% for environmental external costs and 10% for private costs. Therefore, the private and environmental costs are not expressed on the same basis.

(iii) Many analysts are uncomfortable with assumptions that lead to the statistical value of human life being higher for the wealthy than for the poor; others point to the fact that the rich spend more on their health than do the poor. Our study uses a value based on US$3 million/ life.

(iv)   Valuations of biological diversity tend to focus on species preservation, but there are wider issues of quantity, distribution, and shared habitat. A relatively low valuation is derived if based on potential usefulness of species; a high valuation is derived if based on willingness to pay for a highly regarded species. Our study centred round the upper end of such valuations (Barbier, et al, 1994).

CLIMATE CHANGE DAMAGES

Typically, global damages due to climate change are suggested (Pearce, 1995) as being equivalent to 1 to 2 % of Gross World Product (GWP), based on an instantaneous change in the atmospheric concentration of CO_2 to twice pre-industrial levels. In this study, damages due to climate change were calculated by comparing fuel cycles in a world with climate change against a world in which climate change does not occur. The central estimate for damages due to GHG emissions equates to $23/tC for the scenario in which CO_2 mitigation is widely adopted (IS92d), and $65/tC for the scenario (IS92a) in which the world continues to develop without a widespread reduction in the emission of GHGs. The global damage due to climate change was estimated to be US$18 000 billion in scenario IS92d ($\equiv$ 0.2% GWP), and US$77 000 billion in scenario IS92a ($\equiv$ 0.7% GWP).

The external cost due to GHG emissions by one fuel cycle was expressed as an average for the electricity produced over the life of the plant. For natural-gas cycles without CO_2 capture and storage, the results lay in the range 0.3 - 0.8 cents/kWh and, for coal-based cycles, 0.5 - 1.4 cents/kWh. For coal and natural-gas based cycles in which CO_2 was captured and stored, the external costs due to GHG emissions were reduced to the region of 0.1 - 0.3 cents/kWh (all these costs are at zero discount rate).

DECISION MAKING

The primary result of this study is that introduction of external costs would not change conclusions about capture and storage of CO_2 reached using private costs alone. This is not surprising given the large investment in capture and storage in the case studied, as this reduces the environmental impact to a minimal level. This gives us some confidence that the method is correct, at least qualitatively. As a result we can be sure that if we use private costs for selection of capture and storage technology there is little danger of serious error.

Looking further into the potential uses of this method, we can recognise that any decision on capture and storage is just one of a series beginning with a much broader question such as:
•   Is action to reduce greenhouse gas emissions justified?

Which leads to another question:
•   If action is justified, how much reduction in emissions is required?

More specific question follow, such as:
•   What combinations of mitigation options achieve the required reduction at minimum cost?

And, for each of these options, it might be asked:
•   Which particular technologies (e.g. for capture or storage) are most suitable?

All of these questions could, in principle, be addressed using the cost-benefit approach but information of sufficient quality is not yet available to do this with any confidence, certainly not for the first or the second question. Since we have now shown how the 4th question can be handled, it would be worth considering whether the third one might be tackled in the same way.

What would be needed to answer that question? We would need to know the cost of emission reduction (e.g. $/tC) as well as the value of the damages avoided by use of each technology. For this we would need to establish a baseline for comparison - say the situation where mitigation technology is not employed - and decide which scenario is most appropriate, but this is not as straightforward as we had expected.

For example, let us consider the case of a plant which did not use capture and storage (or an equivalent mitigation technique) in an interventionist scenario such as IS92d. This plant would gain an advantage over its competitors through lower costs but the environmental damage it would cause would be reduced, because the rest of the world had lowered its emissions. This plant can be recognised as being a "free-rider", taking advantage of the low external costs achieved by others, with the consequence that it would have little to gain from adopting mitigation itself.

Another example reinforces the point - consider the non-intervention scenario and a plant which has expensive mitigation technology, fitted before the rest of the world does so. This plant would incur extra private costs that its competitors do not have to bear. At the same time, the damages arising from it would be relatively high because every other plant is continuing to emit greenhouse gases, with consequent global damage. Thus the "environmental pioneer" is doubly penalised. Once again, using a scenario inappropriate for the particular degree of mitigation would result in unrepresentative results. These conclusions are reproduced in Figure 1, where we have put the 2 scenarios together with 2 possible levels of mitigation action - low or high.

Figure 1 Scenario - Mitigation Matrix

| | Scenario | |
Mitigation	Non-intervention	Intervention
Low	"No-regrets"	"Free-rider"
High	"Environmental pioneer"	Fossil fuels with capture and storage

Cases more representative of the 2 scenarios are: "no-regrets" under the non-intervention scenario and "capture and storage of CO_2" under the intervention scenario. In contrast, the "free-rider" and the "pioneer", whilst realistic of what would happen in practice, are not representative of the 2 scenarios considered. This suggests that if we assess the damages from a plant with mitigation by using a base-line set in the same scenario, we will obtain results appropriate to a pioneer or a free-rider but not to a typical situation. A pragmatic way around this difficulty might be to compare the results from 2 different scenarios in order to understand the effects of large-scale mitigation. However, this is contrary to normal practice in such assessments. For more realistic results, it would be better to model the changing impact of mitigation technologies through changing the model of the world (i.e. the scenario) as more plant with mitigation are built. Nevertheless, comparing the results under 2 different scenarios may be a more a practical route to improving our understanding of the benefits of mitigation. We are continuing to investigate this.

These conclusions are borne out by calculations we have recently completed in a limited extension of the original full fuel cycle study. This examined the sensitivity of the results to the scenario chosen and tested how the results change with degree of mitigation, all other factors being kept the same.

CONCLUSION

Life-cycle analysis with cost-benefit evaluation provides a basis for comparison of fossil-fuel based mitigation options. It could also become the basis for comparison of alternative approaches to mitigation, such as use of non-fossil sources of electricity, but further work is needed to understand how best to apply it.

ACKNOWLEDGEMENTS

The authors would like to acknowledge the help and advice given by Tom Downing, Nick Eyre and others in improving our understanding of these issues. This work has been carried out as part of the work of the IEA Greenhouse Gas R&D Programme but the views expressed are those of the authors and do not necessarily represent the views of the Programme, its members or the IEA.

REFERENCES

Barbier, E.B., Burgess, J.C., Folke, C. (1994), *Paradise Lost ? The Ecological Economics of Biodiversity.* Earthscan, London.

European Commission, DGX-II, (1995) *ExternE Externalities of Energy,* EUR 16520 EN, ISBN 92-27-5210-0, Europ, Luxembourg.

IEA, (1994), *World Energy Outlook,* ISBN 92 64 14074 3, IEA/OECD, Paris.

IEA Greenhouse Gas R&D Programme, (1995), *Global Warming Damage and the Benefits of Mitigation,* ISBN 1 89837303 5, IEAGHG, Cheltenham UK.

IPCC, (1992), *1992 IPCC Supplement Scientific Assessment of Climate Change,* Cambridge University Press, UK.

Pearce, D. (1995). *Blueprint 4 Capturing global environmental value* Chap.2, pp. 22-23. Earthscan, London.

Summerfield, I.R., Goldthorpe, S.H., Sheikh, K.A., Williams, N., Ball, P. (1994), The Full Fuel Cycle of CO_2 Capture and Disposal, *Energy Convers. Mgmt.,* **36**, No. 6-9, pp 849-852, 1995.

 **Pergamon**

Energy Convers. Mgmt Vol. 38, Suppl., pp. S601–S606, 1997
© 1997 Elsevier Science Ltd. All rights reserved
Printed in Great Britain
0196-8904/97 $17.00 + 0.00

PII: S0196-8904(97)00003-4

Full Fuel Cycle Emission Analysis for Electric Power Generation Options and Its Application in a Market-Based Economy

Don Macdonald, John Donner, and Andrei Nikiforuk

Alberta Department of Energy, Alberta Government
5th Flr, North Petroleum Plaza, 9945 - 108 Street,
Edmonton, Alberta, Canada, T5K 2G6
(the views contained herein are those of the authors and should not be construed as policy of the Alberta Government)

ABSTRACT

The view of full fuel cycle methodologies as a tool for government authorities to internalize the cost of environmental externalities related to power generation may be somewhat out of step with: the global move to more competitive, market-based economies; management of air emissions by broad-based cross-sectoral actions (such as voluntary programs) that harness market forces; and restructuring of electricity industries going on around the world.

This paper briefly describes a full fuel cycle study and how it was used in a jurisdiction that strongly supports economic development through free, and open market mechanisms and is supportive of market-based and stakeholder consultative principles in developing government environmental policy. Other useful applications of the full fuel cycle methodology are described, which do not involve externality adders. The province of Alberta, Canada is used as a microcosm for this kind of application.

© 1997 Elsevier Science Ltd

KEYWORDS

Full fuel cycle; electricity restructuring; market-based economies, externalities, greenhouse gases.

BACKGROUND: ALBERTA'S BUSINESS STRATEGY

The Alberta Government has made a strong commitment to make its own operations more business like and at the same time to get government out of private sector business. Alberta's business strategy consists of four pillars: 1) balanced provincial government budgets are the law, 2) government business plans ensure that government priorities match Albertans' priorities, 3) measuring results will improve Alberta's programs, and 4) Debt Retirement Act requires deliberate pay down of Alberta's debts. This strategy also involves a commitment to free enterprise and economic development; a competitive tax environment and strong infrastructure to attract investment and encourage job growth by the private sector; and less regulation to reduce costs for businesses, individuals and municipalities (overall Alberta's 1995 gross domestic product - real, 1986 dollars - was Cdn.$76.4B and the unemployment rate was 7.8%).

A set of performance criteria has been established that includes measuring environmental and economic performance. Indications are that Alberta's overall approach is working, as over the past five years Alberta has had the fastest growing economy in Canada with an average real rate of growth of 3.3% per annum.

Alberta is a resource-based province with abundant fossil fuel resources. Since the 1950's, abundant sub-bituminous coal resources allowed for the development of a 7,700 MW fleet of low cost, primarily, coal-fired generating stations in the province (88.8% coal, 7.7% natural gas, 3.3% small hydro, 0.1% wind and fuel oil). Electricity rates to industrial customers in Alberta have been some of the lowest in North America (Alberta range - 3.1 to 3.2 cents/kWh - North America range - 2.95 to 8.5 cents/kWh), thereby providing a strong economic reason for petrochemical and other industrial sectors to establish themselves in the province.

FULL FUEL CYCLE STUDIES

Some jurisdictions have used full fuel cycle (FFC) analysis as a first step toward determining damage costs associated with the monetary value of environmental externalities (Oak Ridge National Laboratory & Resources for the Future. 1994., Audus, 1996, Holland, *et al.*, 1996).   From a policy perspective, regulators in those jurisdictions have considered using this information to determine and select which power generation technologies would have the least cost to society - both conventional costs and environmental externalities.   However, FFC has uses independent of the sometimes controversial assessment of externalities.

A study in Alberta, Canada (FFCC, 1995) examined full fuel cycle emissions from electric power generation, based on a sampling of 13 different power generation technologies during the year 1990. The Alberta study did not attempt to calculate ecosystem or human health impacts or monetize environmental externalities associated with these emissions, nor did government intend to use the results for generation planning. The Alberta full fuel cycle study is based on a 1990 "snapshot" of eight existing stations and for five "future" stations (i.e., using projections of emissions data from commercially available technology that could be used in Alberta; stations in upper-case caps, Fig. 1. do not exist in Alberta). The full fuel cycle includes "upstream emissions" (exploration, gathering and processing, and transportation of the fuel - SO_x, NO_x, VOC's - Volatile Organic Compounds, CO, CO_2 and CH_4), as well as "burner tip" emissions from combustion of fuels to produce the electricity. The analysis focused on the full fuel cycle rather than attempting a complete life cycle analysis. Studies from Saskatchewan, Canada

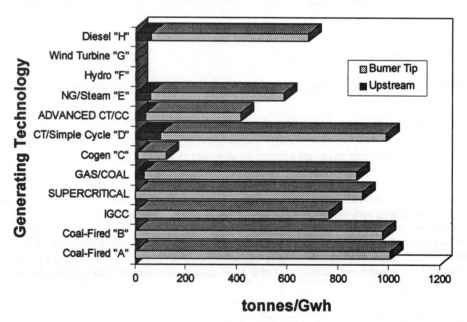

Fig. 1. Full Fuel Cycle CO2 Emissions

(SECDA, 1994) indicate that when all construction, manufacturing and related life-cycle emissions are totalled for various technologies, they are small in quantity or only account for <1% of the total emissions.

No upstream or burner/blade-tip emissions of any of the six primary pollutants could be identified for either the wind turbine station or the small hydro station cases examined. These technologies have the overall lowest full fuel cycle emissions profile. All fossil fuel-based technologies have total full fuel cycle CO_2 emissions in the 124 – 1006 t/GWh range. Of these, the existing natural gas-fired cogeneration unit had the lowest total full fuel cycle rate, while the coal-fired ("A") station had the highest. Similarly, most CO_2 emissions for fossil fuels generators come from the burner tip (>90%), as opposed to the upstream portion of the cycle (Fig. 1).

The Alberta Government has used FFC analysis results to demonstrate to potential buyers of Canadian natural gas that environmental regulation of natural gas development in the province is rigorous and that natural gas when combined with advanced combustion turbines can help them achieve emission reduction objectives for common pollutants and greenhouse gases - even on a full cycle basis. This application of FFC analysis by government supports private sector efforts to market Alberta natural gas, but does not interfere with the market itself.

Power generating companies considering ISO 14000 certification, may also need to demonstrate their services on a life-cycle basis as part of ISO certification. ISO 14040 calls for a measurement system to assess the environmental attributes and effects of products, processes, and services throughout their lifetime and examines all relevant inputs and outputs from raw material acquisition to final disposition (Begley, 1996). Having ISO 14000 certification - with a life cycle analysis - could provide a significant marketing advantage for companies in a deregulated, competitive market for electricity.

ELECTRIC INDUSTRY RESTRUCTURING

Industries and governments throughout the world are looking for new ways to remain competitive in an increasingly competitive global market. In Australia, New Zealand and the United Kingdom, the changes to the electricity sector involve breaking up government-owned monopolies. In the United States, excessively high electricity rates in some areas and large rate differences among utilities are driving the introduction of competitive forces, particularly in the generation sector (ADOE, 1996).

The provincial government initiated a restructuring of the electricity industry in Alberta in response to perceived inequities in the existing system and pressures from large industrial customers in the province to receive the most competitive price possible for electricity and to be allowed to make their own contracts for power. The Electric Utilities Act establishes the broad framework for Alberta's electric industry in the future. It was developed in consultation with stakeholders representing utilities, consumer groups, independent generators and environmental groups. The key purposes of the Act are: 1) to establish an efficient market for generation based on fair and open competition -- this is made possible because: a) vertical integration no longer serves the same purpose as it once did; and, b) small-scale generation technologies have become cost-effective -- 2) to ensure that the benefits and costs associated with existing, regulated plants continue to be shared equitably by current and future customers throughout the province, 3) to ensure that investment in new generation is guided by competitive market forces, 4) where regulation is still necessary, to minimize its cost and provide incentives for efficiency.

Unique elements of this new structure include: open competition for generation, an open-access power pool and system access for all generators and importers on a nondiscriminatory basis. The new structure also removes restrictions on who may import or export power in Alberta. Transmission and distribution are still natural monopolies, and continue to be regulated by the Alberta Energy and Utilities Board (EUB).

ENVIRONMENTAL POLICY

The environmental regulation of energy related emissions of common pollutants in Alberta is the responsibility of Alberta Environmental Protection (AEP) and the Alberta Energy and Utilities Board. Greenhouse gases are not regulated.   Environmental compliance by industry around the world is becoming more costly and in the U.S. there is a call for more "cost-benefit" analysis in reforming regulatory systems.

Future government response to air emission issues in Alberta is guided by a joint industry, environmental non-governmental, government stakeholder group - the Clean Air Strategic Alliance (CASA). This kind of stakeholder involvement and information sharing regarding environmental policy development is an element that most western countries are embracing (Portier, 1996).   A recent CASA stakeholder group that examined . . . "flexibility in electric power generation options that contribute to clean air objectives" . . . concluded that *air quality issues in the electric power sector should be addressed by broad-based, cross-industry sectoral actions that harness market forces, as opposed to sector specific actions.*

Canadian response to climate change is guided by the National Action Program on Climate Change (NAPCC, 1995). Two main components of the program, are the Voluntary Challenge and Registry Program (VCR) and the Canadian Joint Implementation Initiative.   These programs seek broad-based participation from all industrial sectors and provinces in Canada, and abroad, to address the climate change issue. The NAPCC is also committed to flexibility in response and does not focus on specific provinces or industrial sectors to deliver the Canadian stabilization objective.

DISCUSSION

In the 1980's, about 25 U.S. jurisdictions adopted specific public policy objectives and implemented electric industry-specific programs to account for environmental externalities through various mechanisms (EIA, 1995). At the time, most of these jurisdictions regulated their utilities and planned for new generation through Integrated Resource Planning (IRP) mechanisms. Accounting for environmental externalities in the IRP process introduced more renewable energy technologies into the generation mix by increasing the relative cost of technologies having high externality charges (usually fossil-fuel-based ones). In Australia, Diesendorf (1996) suggests that accounting for externalities in the restructured electricity industry would significantly change the mix of technologies away from fossil-based towards renewables and lead to greenhouse gas reductions.   In a regulated industry not undergoing competition, such an approach to environmental policy simply requires regulated utilities to pass these costs on to customers. However, as restructuring advances in the U.S., those jurisdictions seeking to maintain their existing environmental public policy objectives through externalities, are experiencing increasing difficulties. As interjurisdictional sales in electricity become more commonplace through competition, "interjurisdictional" market distortions arise causing some states to drop or modify their externality adders (Fang and Galen, 1996).   Electric industry restructuring efforts in all jurisdictions involve billions of dollars in capital and accounting for externalities exclusively in this industry puts a significant portion of this capital at risk.

Considering electricity restructuring, climate change policy directions, stakeholder advice and economic development issues, we believe that it is prudent to approach climate change externalities from a global basis. In the context of electricity restructuring in Alberta and in trying to be consistent with NAPCC, it does not make sense to apply climate change externalities, in the form of an adder to electricity, to only the electric power sector in Alberta. Should Alberta choose to do this, it would: 1) work against some of the NAPCC principles (shared responsibility among all sectors and provinces, competitiveness, flexibility and international cooperation); 2) be an unrealistic option for a resource rich jurisdiction like Alberta that derives over $2B annually from fossil fuel royalties and land sales, 3) cause a depressed economic growth in Alberta, 4) not significantly address global externalities or could in fact make matters worse if this resulted in some users simply moving to a jurisdiction that has a less energy efficient power generation system, 5) would work against global economic efficiencies in dealing with climate change, and 6) cause market distortions in electricity and competing industries

Market distortions can be intra or inter jurisdictional. In the first (intra jurisdictional), environmental externalities applied only to fossil-based electrical generation will almost certainly raise their operational costs relative to other energy sources. Within a given jurisdiction, if electricity is competing with other energy producers for the same customer (e.g. electricity and natural gas suppliers competing for home heating), incurring an environmental externality charge would then put electricity at a competitive disadvantage. In the second case (interjurisdictional), electricity restructuring invariably opens jurisdictional boundaries to competition from other regions. If environmental externalities (as adders) apply to one jurisdiction but not in competing regions, market distortions will arise because electricity from similar fossil-fuel resources will be relatively more expensive in one jurisdiction than in the others. The high cost of electricity within a jurisdiction would also discourage secondary industry from establishing (or staying within) such a region.

Other Applications of FFC

An important use of a FFC analysis that is facilitated by electric industry restructuring is in the marketing of "greenpower", or other differentiated electricity products, based on consumer preferences. In Sacramento, California residential customers voluntarily pay a premium for greenpower from a photovoltaic system that adds about 300-400 KW/year, based on about 100 new participants/year (Fang and Galen, 1996). A FFC analysis can be an important marketing tool in signing up customers for such a product by demonstrating that there are no, or few, upstream or hidden emissions associated with the generation technology being marketed.

Utilities with major capital invested in fossil-fired generation, particularly coal, may use FFC analysis to better understand their long-term risk to capital investments; particularly in the context of climate change and the Intergovernmental Panel on Climate Change's recent Second Assessment Report (IPCC, 1996) indicating ". . . a discernable human impact on the climate system." The long term risk to fossil-based capital because of climate change may continue to rise if the outcome of COP-3 of the Framework Convention on Climate Change is a "legally binding instrument or protocol" (FCCC, 1996) to achieve global reductions in greenhouse gases.

Also, in the context of climate change, many utilities in North America have made commitments to voluntary programs and many have committed to stabilization. To do this, most are carrying a portfolio of mitigation and offset (domestic and joint implementation) measures to achieve this goal. Some companies are also exploring and positioning themselves for a time when greenhouse gas mitigation or offsets actions may have some real financial value in the form of credits or corporate ownership of reductions achieved. In this context, a FFC analysis can be a powerful tool to evaluate corporate greenhouse gas assets or liabilities associated with domestic operations or to evaluate liabilities associated with an offset or a Joint Implementation project a company may be thinking of investing in.

Government policy analysts can also use the FFC methodology to monitor and better understand greenhouse and common pollutant shifting between industrial sectors in response to climate change initiatives or regulatory standards put in place.

CONCLUSION

For open, market-oriented economies undergoing electricity restructuring, and adopting stakeholder involvement in policy decisions, coupled with broad, global, market-oriented approaches to environmental policy, there are many new applications for the FFC methodology. These approaches support competition and do not interfere with or distort markets.

Jurisdictions around the world facing a similar policy framework as Alberta, will find it very difficult to utilize the full fuel cycle methodology to apply environmental externality adders to their electric power sectors, as a key measure to fulfilling their climate change objectives.

ACKNOWLEDGEMENTS

Constructive reviews of this paper were gratefully received from Mr. Bob Mitchell and Dr. Cynthia Carlson (Alberta Department of Energy) and two anonymous reviewers.

REFERENCES

ADOE, (1996). "Moving to Competition: A guide to Alberta's new electric industry structure", *Alberta Department of Energy*, Publication #E/T 7, 19p.

Audus, H. (1996). IEA Greenhouse Gas R&D Programme: Full Fuel Cycle Studies, *Energy Conserv. Mgmt*, **37**, Nos. 6-8, pp. 837-842.

Begley, R. (1996). ISO 14000 A Step Toward Industry Self-Regulation. *Environmental Science and Technology*. **30**, No. 7.

Diesendorf, M. (1996). How can a "competitive" market for electricity be made compatible with the reduction of greenhouse gas emissions?, *Ecological Economics*, **17**, pp. 33-48.

EIA,(1995). Electricity Generation and Environmental Externalities: Case Studies, *Energy Information Administration, U.S. DOE.*, 98p.

Fang, J.M. and Galen, P.S. (1996). Considering Environmental Issues in Electric Industry Restructuring: A Comparative Analysis of the Experience in California, New York and Wisconsin, *National Renewable Energy Laboratory*, U.S. DOE, 81p.

FFCC, (Full Fuel Cycle Consortium. 1995). Full Fuel Cycle Emission Analysis for Existing and Future Electric Power Generation Options in Alberta, Canada. *Alberta Department of Energy*, 62p.

FCCC, (1996). Review of the Implementation of the Convention and of Decisions of the First Session of the Conference of the Parties - Ministerial Declaration, *Framework Convention on Climate Change - 2nd Conference of the Parties*, July 18, 1996. 4p

Holland, M.R., Eyre.N.J., and T.E. Downing (1996). Assessment of the Costs of Global Warming on a Full-Fuel Cycle Basis. *Energy Conserv. Mgmt*, **37**, Nos. 6-8, pp. 819-824.

IEA, (1994). Full Fuel Cycle Study on Power Generation Schemes Incorporating the Capture and Disposal of Carbon Dioxide, Volumes 1-5. *IEA Greenhouse Gas R&D Programme*, Cheltenham, Glos, United Kingdom.

IPCC, (1996). IPCC Second Assessment Report of Intergovernmental Panel on Climate Change Framework Convention on Climate Change, *Cambridge University Press*, Volumes 1-3.

NAPCC, (1995). Canada's National Action Program on Climate Change, *Government of Canada*, 42p.

Oak Ridge National Laboratory & Resources for the Future, 1994. Estimating Externalities of Coal (Hydro, Nuclear) Fuel Cycles, *U.S. Department of Energy*. Report Nos. 3, 6 & 8.

Portier, M. (1996). Integrating Environment and Economy. *The OECD Observer*, No. 198, February/March 1996

SECDA, (1994). Levelized Cost and Full Fuel-Cycle Environmental Impacts of Saskatchewan's Electric Supply Options. *Saskatchewan Energy Conservation and Development Authority* Publication No. T800-94-P-004.

Pergamon

PII: S0196-8904(97)00004-6

Energy Convers. Mgmt Vol. 38, Suppl., pp. S607–S614, 1997
© 1997 Elsevier Science Ltd. All rights reserved
Printed in Great Britain
0196-8904/97 $17.00 + 0.00

Study on GHG Control Scenarios by Life Cycle Analysis
--- World energy outlook until 2100 ---

Keishiro Ito*, Yohji Uchiyama**, Toshihide Takeshita***and Hisashi Hayashibe*

* Institute for Policy Sciences, 2-4-8 Nagata-cho, Chiyoda-ku, Tokyo 100 JAPAN
** Central Research Institute of Electric Power Industry, 1-6-1 Ohtemachi, Chiyoda-ku, Tokyo 100 JAPAN
*** Technova Inc., 2-2-2 Uchisaiwai-cho, Chiyoda-ku, Tokyo 100, JAPAN

ABSTRACT

World energy supply up to the year 2100 has been analysed from a life cycle standpoint of the full fuel cycle. Five scenarios are set up and these scenarios are analysed and compared with regard to excess energy requirements, carbon emissions, health risks and investment costs for entire power generation systems. The current trend (fossil fuel-intensive) scenario of these scenarios is attractive from an investment standpoint, but deleterious froman environmental and risk standpoint. The CO2 removal and carbon recycle scenarios must get over the hard barriers of reducing excess energy use and investment costs through future technological innovations. Renewable energy technologies are environmentally appropriate and the renewable-intensive scenario could be introduced with acceptable cost burdens in long-term projections. The nuclear-intensive scenario is attractive from its economic, environmental and risk aspects, but must make efforts to acquire public and political acceptability for further worldwide adoption. © 1997 Elsevier Science Ltd

KEYWORDS

Life cycle analysis; primary energy mix; CO2 mitigation; health risk; investment cost.

INTRODUCTION

World energy demand is rapidly increasing with high growths of population and economies in the developing countries, especially in Asia. It is estimated that the energy demand will be two to three times in 2050 and four to five times in 2100 as much as the present demand. Under consideration for measures which reduce CO2 emissions from fossil fuels are the development of technologies for energy conservation, CO2 removal and C-recycle and the conversion to non-fossil energy such as nuclear and renewable energy. However, the more desirable options among them have not yet been evaluated. Therefore, they must be reviewed in detail to clarify their effects and possibilities. We have used the World Energy LCA Model which was developed for analysing and evaluating global energy supply scenarios from the life cycle standpoint of the full fuel cycle. Five global energy supply scenarios including CO2 removal and C-recycle systems consuming a lot of excess energy in construction and operation were evaluated considering resource availability, CO2 emissions and the magnitude of health risks and investment costs associated with power plants. As a result, acceptable scenarios are clarified in order to propose a desirable policy for mitigating future global CO2 emissions .

METHOD AND ASSUMPTIONS

World Energy LCA Model

In assessing the potential impact of low CO2-emitting energy supply scenarios for the long term up to 2100, it is not possible to identify and select a future energy supply scenario using a least cost analysis. This is because the relative cost of scenarios depends on resource constraints and future technological opportunities that are imperfectly known and, generally, a social and political selection is decided by considering various factors such as sustainability, environmental impacts and risks of the full fuel cycle.   Therefore, we have developed the World Energy LCA Model which is a simulation model for analysing and evaluating energy demand and supply, resource availability, CO2 emissions and the health risks and investment costs of power generating systems.   In this model, we integrated a method of analysing global energy supply scenarios with a life cycle approach based on a bottom-up process chain methodology, calculating energy and material requirements for all stages of power generating systems, that is, resource recovery, conversion, transportation, facility manufacturing, plant construction, plant operation and maintenance, plant dismantling and final disposal of wastes.   We have also constructed this model in order to support thought-experiments exploring desirable, but not optimized, future global energy supply scenarios. Therfore, this model will be helpful for introducing appropriate social and political criteria into the selection process.   The outline of the model is shown in Fig. 1.

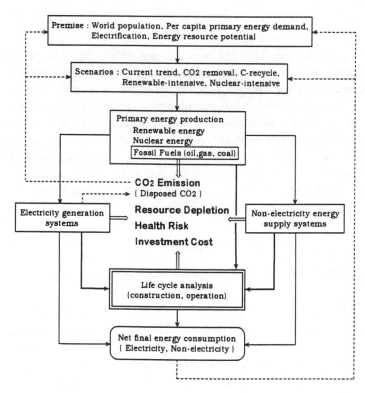

Fig. 1. Outline of the world energy L C A model

Population and Primary Energy Demand

World population to the year 2100 is followed using the middle case of UN projection, that is, 10 billion people in 2050 and 11 billion in 2100.   Nearly 90% of the increase will take place in developing countries.   The global primary energy demand reaches 35 ~ 45 Gtoe in the year 2100 (WEC and IIASA, 1995).   We, in the following study, adopted 27.8 Gtoe for the total demand which is the low demand case under the modest estimate of economic and technological development, considering improvement in energy efficiency of 20% (OECD, 1995).   It is because of getting the projection on the posibilities of the artificial selection of future energy sources.   In this projection, Non-OECD demand per capita doubles, approaching 60 % of the OECD one of 3.9 toe/Y/capita in 2100.

Resource Availability

In analysing world energy demand and supply, we have adopted the recoverable resource potential of fossil fuels and uranium, that is, 820Gtoe of oil, 870Gtoe of natural gas, 3,400Gtoe of coal and 260

Gtoe of uranium (15,550Gtoe in FBR) (WEC and IIASA, 1995). Fossil fuels among these resource potentials have been categorized by different production costs with the IFP classification (Bourrelier et al., 1992). On the other hand, for renewable energies such as hydro power, biomass, wind power, solar energy, geothermal and ocean, we have referred the WEC projections of 13Gtoe/Y in the year 2100 which are estimated from a geographical and economic point of view.

Excess Energy Requirement and Investment Cost

The power generating systems used in this study are commercially available and are expected to be installed in the future. The energy used by a power generating system is either directly used to generate electricity or indirectly used in the life cycle chain of processes associated with the power system. In our study, the indirect energy of entire power generating systems shown in Table 1 have been included in the model (Uchiyama, 1995).

Table 1. Indirect energy requirements and installed costs for power plants

		Plant Efficiency	Indirect Energy Requirements of Full Fuel Cycle				Energy Pay-Back Time	Assumed Plant Costs 1995 US$
			Manu. & Conct.		Ope. & Main.			
			Primary Energy	Electricity	Primary Energy	Electricity		
		% : HHV	MTOE / GW	GWH / GW	MTOE / TWH	GWH / TWH	YEAR	$ / KW
Oil Fired		39.3	0.092	105	0.0084	4.94	0.09	1,200
Natural Gas	Direct	39.3	0.082	106	0.0375	0.48	0.09	1,400
Fired	Combined	51.0	0.078	99	0.0294	0.50	0.08	1,600
	Direct	39.3	0.144	173	0.0099	8.82	0.14	1,600
Coal Fired	Combined	45.5	0.137	166	0.0087	7.78	0.13	1,900
Nuclear	LWR	33.0	0.118	136	0.0008	5.99	0.11	2,000
Power	FBR	40.0	0.207	262	0.0005	1.90	0.16	2,200
Hydro Power		98.0	0.152	917	0.0001	0.10	0.41	2,100
	Roof-Top	---	0.164	914	0.0009	5.22	1.27	2,500
Photovoltaic	C-Recycle	---	0.434	1,692	0.0033	10.84	2.98	1,500
Wind Turbine		---	0.149	384	0.0011	3.60	0.55	1,000
Geothermal		---	0.132	386	0.0035	8.92	0.20	1,900
Ocean(Therrmal, Tidal)		---	1.540	6,640	0.0059	25.80	4.38	6,400
Biomass Fired	LHV 45.0		0.137	166	0.0087	7.78	0.20	1,900

In particular, CO_2 removal and C-recycle systems require a huge excess of energy which includes the direct and indirect energy needed for recovery, liquefaction and transportation of CO_2 and for photovoltaic generation, hydrogen generation and methanol conversion. In the CO_2 removal systems, the excess energy requirements reaches 0.78 Gcal/T-CO2 for ACC plants (gas) and 0.84 for IGCC plants (coal) and in the C-recycle systems, 2.29 Gcal/T-CO2 for ACC plants (gas) and 2.46 for IGCC plants (coal). These excess energy requirements are shown in Table 2 and Fig.2 (NEDO, 1996).

Table 2. Excess energy requirements of CO_2 removal and C-recycle systems

	Plant (1,000MW)	Annual Capacity Factor %	Annual Fuel Demand Tcal / Y	Annual Recoverd CO2 MT / Y	Excess Energy		Annual Produced MeOH Tcal / Y
					Manu. & Const. Tcal / Y	Ope. & Maint. Tcal / Y	
CO2 Removal System *1)	ACC:gas	70	10,388	1.93	11	1,489	---
	IGCC:coal	70	11,664	3.98	23	3,312	---
C-Recycle System *2,3)	ACC:gas	70	10,388	1.93	2,088	2,327	7,215
	IGCC:coal	70	11,664	3.98	4,639	5,150	14,861

*1: CO2 Recovry, Liquefaction, Tanker Transportation(one way:2,000km) and Base for disposal on the Ocean.
*2: Energy Ratio of System (Produced Energy(MeOH) / Excess Energy) = 1.63 for ACC (gas), 1.52 for IGCC(coal)
*3: CO2 Recovery, Liquefaction and Storage, Tanker Transportation(one way:7,200km), Methanol Conversion by
    Hydrogen produced using PV Facilities, and storage and trannsportation of Methanol.
        PV Capacity required: 6.9 GW for ACC 1 GW (gas), 14.1 GW for IGCC 1GW (coal)

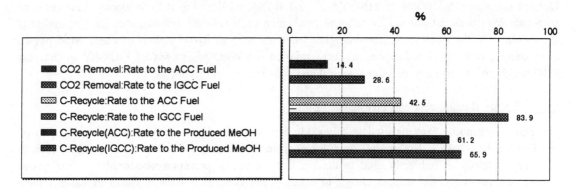

Fig. 2. Rates of the excess energy requirements of CO2 removal and C-recycle systems

Annual investment costs for the power plants in 2050 and 2100 are shown in 1995US$, but not discounted, using the values assumed in Table 1. Regarding the future installed costs for photovoltaic facilities, we assumed to be $2,500/kW for the roof-top use and $1,500/kW for the C-recycle use when cost reductions by the learning effect are considered. We also assumed lower capital costs due to mass production for other technologies: $1,000/kW for wind machine, $100/kWh for batteries, $0.34 billion/MT-CO2/Y for a CO2 removal system and $7.5 billion/MT-CO2/Y for a C-recycle system(NEDO, 1993 and 1996).

Health Risk of Power Generating System

Although there is significant progress recently in estimating the risks associated with power generating systems, the results still show a wide range of dispersion induced by differences in local conditions, economic situation, etc. Generally, risks are categorized into three types: ecological and environmental risks, health and safety risks to humans and social and economic risks. We restrict ourselves here to risks of fatal deaths, immediate and delayed, for both occupational workers and the public.

The magnitude of a risk is based on a GW-year of operation of each power generating system for an entire fuel cycle. The fatality data of different generating systems in our study are based on the values estimated by A. F. Fritsche et al. which applied worldwide data on the risks to the risks estimation in Switzerland and Germany (Fritsche, 1988). Because of the long period of projection, immediate and delayed effects are added to those risks both for occupational workers and the public. The fatality rate shows the highest value of 2.3 to 11.3 /GW-year for a coal fired generating system and the lowest value of 0.1 to 1.0 /GW-year for a nuclear generating system. The later values do not include any Chernobyl accident case because of the difference on safety concepts between ther former Soviet Union and West European/American types of reactors.

Regarding coal and oil fired systems, 50 to 80 % of the risks are accrued to the public by air pollution from fuel combustion, while the major risks for the nuclear systems are induced by occupational hazards during plant construction and operation. For renewable energy, the low intensity of the energy density requires much land and a lot of materials which makes the risks not very low.

RESULTS

Primary Energy and Electricity Supply

The main results of the simulations in 2050 and 2100 are summarized in Table 3 and 4. In the Current Trend Scenario, primary energy demand reaches 27.8 Gtoe in 2100 which is 3.1 times as large as the present demand in 1990. The CO2 Removal and C-Recycle Scenarios make the demands increase to 34.2 and 32.1 Gtoe, respectively, values which are larger by 23 % and 15 % than that of the Current Trend Scenario because of compensation for the deficit of final energy supply induced by constructing and operating the CO2 removal and methanol production facilities.

Table 3. A summary of primary energy and electricity supply for five scenarios in 2050 and 2100

			Current Trend	%	CO2-Removal caseA-Em	%	Renewable-Intensive	%	C-Recycle	%	Nuclear-Intensive	%
	Primary Energy Demand (GTOE/Y)	1990	8.8		8.8		8.8		8.8		8.8	
		2050	22.5		26.5		22.5		23.5		22.5	
		2100	27.8		34.1		27.8		32.1		27.8	
Electrification Ratio in 2100		%	60.0		67.4		60.0		65.3		60.0	
Primary Energy Mix in 2100	Electricity Use (GTOE/Y)	Fossil Fuel	11.2	67	11.2	67	3.3	20	3.3	20	3.3	20
		Renewable	3.0	18	3.0	18	8.3	50	8.3	50	5.0	30
		Nuclear	2.5	15	2.5	15	5.0	30	5.0	30	8.3	50
		Compensation Fuel *)	---		6.3	Coal	---		4.3	MeOH	---	
	Non-Electricity Use (GTOE/Y)	Fossil Fuel	9.3	84	9.3	84	2.2	20	3.2	28	2.2	20
		MeOH **)	---		---		---		2.4	22	---	
		Renewable	1.8	16	1.8	16	5.6	50	5.6	50	3.3	30
		Nuclear	---		---		3.3	30	---		5.6	50
Cumulative Fossil Fuel Consumption (GTOE)	Ratio 1990-2100		1.00		1.19		0.64		0.67		0.64	
			1,733		2,066		1,107		1,154		1,108	
Electricity Supply (T.P.) (PWH/Y)		1990	11.8		11.8		11.8		11.8		11.8	
		2050	51.2		70.3		51.5		57.1		51.0	
		2100	79.7		109.8		78.6		102.7		78.0	
Capacity of Power Plants (TW)		1990	2.5		2.5		2.5		2.5		2.5	
		2050	11.0		15.0		13.5		14.6		12.2	
		2100	17.2		23.5		23.0		28.0		20.2	
PV Capacity for MeOH Production		2100	---		---		---		66.8		---	

*) Fuel for compensating the deficit of net final energy supply. MeOH of C-Recycle: refer **)
**)Fuel produced by synthesizing recovered CO2 with hydrogen gas generated by electricity from photovoltaic facilities.

Both the Current Trend and CO2 Removal Scenarios highly depend on fossil fuel supplies. Coal supplies more than 50 % of the primary energy, while the share of coal decreases by 7 % in 2100 for other scenarios. Oil consumption for electricity is estimated at zero in the year 2100 for all scenarios. Natural gas is used as the primary fuel for electricity instead of oil. Renewable energy supplies 50 % of primary energy for the Renewable-Intensive Scenario and 17 % for the Current Trend Scenario. In order to reduce global CO2 emissions below the 1990 level, it is necessary to construct nuclear energy systems which supply 30 % of the primary energy. The share of nuclear energy for the Nuclear-Intensive Scenario is 50 % of the primary energy and only 9 % for the Current Trend Scenario.

The electrification rate of primary energy demand in the world tends to grow over a long period of the 21st century. In particular, economic growth in developing countries and the progress of global urbanization increases the proportion of electricity in the world energy demand. In our study, the electrification rate is assumed to reach 47 % in 2050 and 60 % in 2100. However, the rates increase further in both the CO2 Removal and C-Recycle Scenarios because of the additional generating plants required for compensating the energy supply deficit induced in these two scenarios. Because of the additional generating plants, the total amount of power generation capacity in both scenarios increases in 2100 by 37 % and 29 %, respectively, over that of the Current Trend Scenario.

CO2 Mitigation

The current trend scenario assumes a high dependence on fossil fuel supplies going from 85 % of the total energy supply in 1990 to 74 % in 2100. Global CO2 emissions for this scenario in 2100 are estimated to be 21 Gton-C/y which is 3.4 times larger than the 1990 level. Cumulative CO2 emissions become 1,587 Gton-C by 2100. They fall to 72 % with the emergent Case A of the CO2 Removal Scenario (see Table 4). In this Case A scenario, CO2 recovery systems are installed into all fossil fired generating plants in the world under the current trend scenario up to 2050. In this Scenario, 60 Gton-CO2 corresponding to 59 % of total emissions in 2100 must be disposed in the deep sea. Even if such a huge amount of CO2 gas could be recovered from electricity generating plants, annual CO2 emissions in 2100 can not reduce to the 1990 level and would remain 85 % larger than the level (see Table 4 and Fig. 3(a)). In order to reduce moreover global CO2 emissions in 2100 under the CO2 Removal Scenario, it is necessary to reduce fossil fuels consumption in the non-electricity use of the world energy supply. But, under this scenario, even if the fossil fuel share was reduced from 84 % of the present share to 50 % by converting the fossil fuel to renewable energy sources, the annual CO2 emissions in 2100 would remain 17 % larger than the 1990 level (see Fig. 3(a)-Case B).

Table 4. A summary of major characteristics for five scenarios in 2050 and 2100

		Current Trend	%	CO2-Removal caseA-Em	%	Renewable-Intensive	%	C-Recycle	%	Nuclear-Intensive	%
Ratio of Indirect Energy on	1990	7.3		7.3		7.3		7.3		7.3	
Electricity Supply *)	2050	6.9		30.5		7.5		19.8		7.0	
(%)	2100	5.8		31.6		5.6		26.8		4.8	
	1990	6.1	100	6.1	100	6.1	100	6.1	100	6.1	100
Annual CO2 Emissions	2050	15.9	258	10.5	172	9.6	157	10.1	165	9.7	158
(GT-C/Y)	2100	20.8	339	11.4	185	4.2	69	5.3	86	4.2	69
Cumulative Emissions(GT-C)	1990-2100	1,587		1,143		906		957		907	
CO2 Disposal (GT-CO2/Y)	2100	---		60.2		---		---		---	
Health Risk(Fatality) on	1990	5.3		5.3		5.3		5.3		5.3	
Electricity Supply **)	2050	23.2		39.2		13.0		15.0		12.3	
(1000 deaths / Y)	2100	42.9		68.2		15.6		20.3		12.2	
	1990	219		219		221		221		219	
Capital Cost for Electricity Supply	2050	823		1,518		1,048		2,714		904	
(Billion $/Y :1995US$, PPP)	2100	1,236		2,405		1,628		6,668		1,399	
Ratio to C. T. Value	2100	1.00		1.95		1.32		5.39		1.13	

*) Include additional energy demand associated with introducing CO2 Removal and C-recycle systems.
**)Based on the values estimated by A. F. Fritsche.

In order to reduce global CO2 emissions below the 1990 level up to 2100, it is necessary to use more renewable energy and nuclear energy for the world energy supply. In our CO2 mitigation study, three different scenarios were built to make global CO2 emissions reduce below the 1990 level up to 2100. Fig. 3(b) shows the CO2 mitigation curves of the C-Recycle, the Renewable-Intensive and the Nuclear-Intensive Scenarios which would realize the objective. Among these scenarios, the Renewable-Intensive and the Nuclear-Intensive Scenarios would be able to make the deepest reductions, i.e. up to 31 % greater than the 1990 level by 2100.

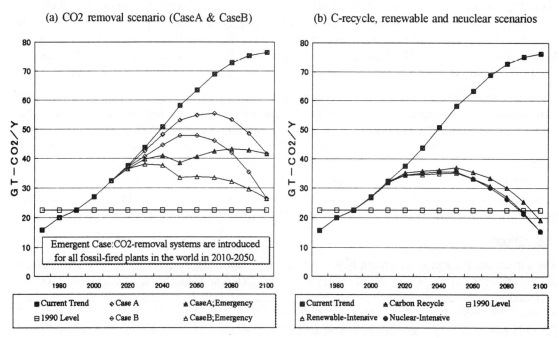

(a) CO2 removal scenario (CaseA & CaseB)   (b) C-recycle, renewable and neuclear scenarios

Fig. 3. Projections of global CO2 emissions for five scenarios

Life Cycle Analysis of Excess Energy, Investment Cost and Health Risk

A life cycle analysis can estimate the indirect energy used in the full energy chain associated with constructing and operating entire facilities including CO2 recovery and disposal systems. It was found in our study that indirect energy use is relatively low, approximately 5 ~ 6% in 2100 for the Current Trend, the Renewable and the Nuclear Scenarios (see Table 4). However, the CO2 removal system consumes a large amount of energy during plant operation, especially in the processes for recovering and liquefying CO_2 gas. In order to compensate for the deficit of net final energy supplies induced by the installation of CO_2 removal systems, additional generating plants are required. If the supply deficit is compensated by coal fired generating plants, a larger number of plants must be constructed. The additional energy required by the additional coal fired plants is estimated at 6.3 Gtoe in 2100 corresponding to 28 % of the total electricity supply. The total excess energy is estimated at 31.6 % of the primary energy demand for power generation, which includes the indirect energy demand induced by the additional installation of the CO_2 removal systems.

The primary energy mix for power generation of the C-Recycle Scenario is similar to the Renewable -Intensive Scenario. However, if the carbon recycle systems are installed to remove CO_2 emissions from all fossil fired generating plants in the world, these systems require additional methanol fired plants of 5 TW and additional PV facilities of 43TW for compensating for the deficit of the net final energy supply. Namely, the carbon recycle scenario generates only waste energy.

The annual investment costs for constructing power generating systems are estimated at 1,240 ∼ 1,630 billion \$ (1995US\$: P.P.P.) for the Current, the Renewable and the Nuclear Scenarios in the year 2100 (see Table 4). Both the CO_2 Removal and the .C-Recycle Scenarios accelerate the consumption of energy resources and require many additional facilities for CO_2 capture, storage, transportation, disposal, PV generation, hydrogen production, methanol conversion etc. As a result, the high burden of investment costs for constructing power generating systems is inevitable in these scenarios; total investment costs in the year 2100 would reach twice for the CO_2 Removal Scenario and five times for the C-Recycle Scenario as high as the costs of the Current Scenario.

Table 4. also shows the risk estimation of fatalities for the electricity supply systems of each scenario. Because of impacts on the public from air pollution caused by emissions from coal fired plants and occupational risks from coal mining accidents, the total risks for the fossil fuel intensive scenarios show 43,000 fatalities for the Current Trend Scenario and 68,000 fatalities for the CO_2 Removal Scenario. The C-Recycle, Renewable and Nuclear scenarios' risks range from 12,000 to 20,000 fatalities. In the latter two scenarios, the occupational hazards during construction and operation are twice as large as public fatalities. We may need to mention here that there are large uncertainties in future risk levels because of possible progress in air pollution control technologies, safety control technologies and medical care technologies.

POLICY IMPLICATIONS

The major policy implications based on these results are summarized as follows:

Current Trend and CO2 Removal Scenarios

The current trend scenario highly depending on fossil fuels is the most attractive option among the investigated five scenarios from the economic standpoint, but not desirable in environmental and risk standpoints. The CO_2 removal scenario would be one of the promising options to mitigate the global CO_2 emissions under the necessity of urgent measures to reduce the emissions up to the middle of the 21st century. Therefore, once the necessity of the urgent measures is clarified, the removal systems including the storages in aquifers and depleted oil and gas fields might be installed as one of several mitigating options in conjunction with energy conservation, afforestation and conversion to non-fossil energy.

However, this scenario is less acceptable as one of the CO_2 mitigating options from a long term point of view. It is because this scenario requires a huge amount of excess energy to remove CO_2 gas from flue gas of fossil fuel-fired plants and induces a high burden of investment costs, huge wastes of resources and high health risks for power plants. Moreover, in this scenario, it is difficult to reduce the global CO_2 emissions to the 1990 level. Therfore, in long term perspectives, it will be desirable for decision makers to choose other options such as Renewable and Nuclear Intensive Scenarios, from the standpoint of deep reduction of CO_2 emissions, economy, resource conservation and health risks.

Renewable-Intensive and C-Recycle Scenarios

The Renewable-Intensive Scenario accompanying with appropriate use of nuclear energy could achieve its goal to reduce the global CO_2 emissions below the 1990 level up to 2100 under the acceptable burden of the investment costs. However, regarding the implementation of this Scenario, it is necessary to carry out appropriate policies with respect to economic measures such as environmental taxes and tradeable emission quotas, international cooperation to support developing countries and land use for biomass and food production. The energy ratios of the carbon recycle systems will be 1.5 to 1.6 under the future technologies (see Table 2) and extremely depends on solar energy. In consequence, C-Recycle Scenario induces extraordinarily high burden of investment costs and a huge waste of energy. Therefore, regarding this scenario, it is appropriate to continue only basic research of the technologies because this scenario is indispensable to achieve the drastic innovations of technologies for getting over the hard varriers of improving the energy ratios of the systems and reducing the huge investment costs.

Nuclear-Intensive Scenario

The Nuclear-Intensive Scenario can achieve its goal to reduce the global CO_2 emissions below the 1990 level up to 2100 as well as the Renewable-Intensive Scenario. Moreover, this scenario has the great advantage of the lower cost burden which is necessary for the economical growth of the developing countries, sustainable energy supply for the forseeable future and prevention of air pollutions and acid rains, etc. Therefore, this scenario is considered the most promising CO_2 mitigating option in our study. However, in order to realize this scenario, it is necessary to strengthen international co-operation on the nuclear safety, non-proliferation and radioactive waste disposal, develop the reactors internationally standardized which are inherently safe and economic and acheive technology innovations on the nuclear fuel cycle which are able to reduce the generations of radioactive wastes.

CONCLUSIONS

It is inevitable that the world energy demand, especially in the developing countries, will continue to depend mainly on fossil fuels to the middle of the 21st century from the economical standpoint. In order to mitigate CO_2 emissions in the period, it is important to promote the improvements of energy efficiency in both demand and supply sides, the conversion to economical non-fossil energy sources and the suppress of deforestation. Moreover, it is necessary to make up and strengthen the international framework which promotes the international co-operation for the development and diffusion of the nuclear reactors based on the new concepts and technological innovations and cost reductions on the renewable energy use. However, once the necessity of urgent measures to reduce the global CO_2 emissions is clarified, the CO_2 removal systems might be installed as a transitional measure for mitigating CO_2 emissions to the middle of the 21st century.

In conclusion, global energy policies must be based on long term perspectives focused on changing world primary energy supply from fossil fuels to non-fossil energy supplied by nuclear and renewable, if the total amount of global CO_2 emitted is to be reduced in order to avoid the global warming.

REFERENCES

Bourrelier, P.H., X.B. Tour and J.J. Lacour (1992), Energy in the Long Term, Energy Policy March, 192-207.
Fritsche, A. F. (1988), Gesundheitsrisken von Energiersorgungs systemen, Verlag TUV Rheinland
NEDO 1993), Research on the Feasibility of CO_2 Global Recycle Systems using Renewable Energy, 167-201, NEDO, Tokyo.
NEDO (1996), Research for making the Implementing Program of the Earth Revival Plan, 82-105, NEDO, Tokyo.
OECD (1995). Global Warming-Economic Dimensions and Policy Responses-, 24, OECD, Paris.
Uchiyama, Y. (1995), Life Cycle Analysis of Power Generation Systems, CRIEPI, Tokyo.
WEC and IIASA (1995). Global Energy Perspectives to 2050 and Beyond, 4 and 26-40, WEC, London.

Energy Convers. Mgmt Vol. 38, Suppl., pp. S615–S620, 1997

Pergamon

PII: S0196-8904(97)00005-8

0196-8904/97 $17.00 + 0.00

EVALUATION OF CO_2 PAYBACK TIME OF POWER PLANTS BY LCA

Kiyotaka TAHARA, Toshinori KOJIMA and Atsushi INABA*

Dept. of Ind. Chem., Seikei Univ., Musashino Tokyo 180 Japan
*National Institute for Resources and Environment, Tsukuba Ibaraki 305 Japan

ABSTRACT

CO_2 emissions from construction of various power plants were calculated by the LCA (Life Cycle Assessment) methodology. The LCI (Life Cycle Inventory) was calculated by "NIRE-LCA", LCA software developed at the National Institute for Resources and Environment using a bottom-up approach. CO_2 payback times of renewable energy electric power plants (hydroelectric, OTEC and PV) were calculated vs. conventional fossil fuel-fired power plants (coal, oil and LNG). The evaluated payback times were much shorter than the typical operational lifetimes of the respective renewable energy electric power plants. © 1997 Elsevier Science Ltd

KEYWORDS

LCA; power plant; CO_2; payback time.

INTRODUCTION

Global warming, caused mainly by the increase of atmospheric CO_2 concentration, is one of the most important environmental issues. CO_2 emissions due to the use of electrical products account for close to 30% of total CO_2 emissions in Japan (Agency of Natural Resources and Energy, 1995). It is essential to develop highly-efficient energy utilization processes and substitute energy sources as a countermeasure against CO_2 problem, in addition to the presently commercial renewable energy power plants i.e., geothermal and large scale hydraulic power plants.

In this paper, CO_2 payback times were calculated for typical future renewable energy electric power plants (small scale hydroelectric, OTEC, and photovoltaic) compared to commercial fossil fuel-fired electric power plants (coal, oil, and LNG) in order to estimate CO_2 reduction potential of renewable energy electric power plants comparing to commercial fossil fuel-fired electric power plants. Air pollutant emissions resulting from plant construction and the production of plant construction materials were calculated for each plant. The amounts of materials for construction of each power plant were taken from previous papers (Uchiyama *et al.*,1991, Resource council,1983, Tahara *et al.*,1993, Inaba *et al.*,1995b). The resulting LCI (Life Cycle Inventory) was calculated by "NIRE-LCA", LCA (Life Cycle Assessment) software developed at the National Institute for Resources and Environment using a bottom-up approach (Kobayashi *et al.*,1994, Kobayashi *et al.*,1995).

POWER PLANT INPUT AND OUTPUT ENERGY

Annual Electric Supply and Plant Size

In this paper, we took PV (photovoltaic cell power plant) and OTEC (ocean thermal energy conversion) as alternative energy resources, and coal (coal-fired power plant), oil (oil-fired power plant), LNG (LNG-fired power plant) and hydroelectric for commercial energy resources. Annual electric supply and size of each plant are shown in Table 1.

Table 1. Annual electricity supply and size of plants.

Plant type	Annual electricity supply [kWh/year]	Plant size [MW]	Reference
coal-fired	6.084×10^9	1000	Uchiyama, 1991
oil-fired	6.169×10^9	1000	Uchiyama, 1991
LNG-fired	6.340×10^9	1000	Uchiyama, 1991
hydroelectric	3.932×10^7	10	Resource council, 1983
OTEC (2.5MW)	8.760×10^6	2.5	Resource council, 1983
OTEC (100MW)	5.696×10^8	100	Tahara, 1993
PV (U)	1.248×10^6	1	Uchiyama, 1991
PV (I)	1.183×10^7	10	Inaba, 1995b
PV (J)	8.640×10^6	10	Inaba, 1995b

The amounts of construction materials for each power plant were taken from previous papers. The PV(I) (Japanese-made solar cells installed in a power station constructed in Indonesia) (Inaba *et al.*,1995b) and PV(J) (Japanese-made solar cells installed in an analogous power station constructed in Japan) (Inaba *et al.*,1995b) were taken from our estimated data calculated by the same method.

Input Energy and Material for Construction

The amounts of construction materials used for fossil fuel-fired electric power plants are shown in Table 2.   The efficiency and operation rate of fossil fuel-fired electric power plants were 0.39 and 75% respectively (Uchiyama *et al.*,1991).

Table 2. Amount of construction material for fossil fuel-fired electric power plants.

Power plants		coal-fired	oil-fired	LNG-fired
steel	[t]	62,200	51,130	51,130
aluminium	[t]	624	230	230
concrete	[t]	178,320	71,270	71,270
Annual fuel consumption				
amount of fuel	[t/year]	2,336,000	1,448,800	1,114,500
Amount of energy for construction				
electricity	[GWh]	12,700	8,900	8,900
oil	[t]	709	282	282
coal	[t]	14,339	11,983	11,984

Natural energy electric power plants need only small amount of fuel for operation compared with fossil fuel-fired electric power plants, so renewable energy electric power plants require detailed data for materials for construction. Amounts of materials for renewable electric energy power plant construction are shown in Table 3. Material requirement for cases PV(I) and PV(J) are shown in our previous paper (Inaba *et al*.,1995b). Epoxy, insulation and insulation oil were not taken into account due to their small amounts and lack of detailed information.

Table 3. Amount of construction material for renewable energy electric power plants.

		hydroelectric	OTEC (2.5MW)	OTEC (100MW)	PV (U)
steel	[t]	1,097.4	1,665.1	14,606.8	1,035.0
cement	[t]	7,900.0	3,750.0	37,500.0	500.0
stainless steel	[t]	7.3	6.0	179.0	-
silicon steel	[t]	30.9	-	-	-
copper	[t]	19.0	12.2	269.7	108.0
aluminium	[t]	1.0	-	-	40.0
titanium steel	[t]	-	560.0	4,144.0	-
chrome steel	[t]	-	1.2	13.0	-
FRP	[t]	-	1,281.0	14,216.0	-
CFC	[t]	-	200.0	-	-
ammonia	[t]	-	-	2,000.0	-
silicone	[t]	-	-	-	50.0
glass	[t]	-	-	-	240.0
paper	[t]	0.2	-	-	-
PE	[t]	3.7	-	-	-
PVC	[t]	4.0	-	-	-
insulator	[t]	4.2	-	-	92.0
insulation oil	[t]	8.1	-	-	-
epoxy	[t]	0.05	-	-	-
Amount of energy for construction					
electricity	[GWh]	7.43	2.98	24.90	1.70
oil	[t]	1,646.21	1,490.78	5,679.61	97.09
coal	[t]	6.45	198.00	1,496.77	29.03

CALCULATION OF CO_2 EMISSIONS

The CO_2 emissions from production of materials were calculated by "NIRE-LCA". Most of data for LCI (Life Cycle Inventory) were taken from our previous paper (Inaba *et al*.,1995a), and the others data were taken from related companies, such as an iron making company.

CO_2 emissions not only from the combustion of fossil fuels, but also from the production of fossil fuels, electricity and other materials were calculated using the life-cycle assessment emissions inventory methodology. The system boundary for imported materials in this study was set at the ports of the producing countries, meaning that the CO_2 emissions in those foreign countries were excluded from this calculation while those from the transportation were include. The data for mining of raw materials including fossil fuels in the exporting countries were also excluded from this calculation. Japan now imports about 100 % of aluminum ingots and about 65 % of the naphtha used as industrial materials, whose production data were excluded. CO_2 emissions from liquefaction of natural gas in the producing countries was regarded as a part of transportation and was taken into account as emissions for transport was include. The transportation from the Japanese port to production facilities was also ignored. CO_2 emissions from the production of materials are shown in Table 4.

RESULT AND DISCUSSION

CO$_2$ Emissions from Power Plant Construction

The calculated total CO$_2$ emissions from the construction of each power plant were divided by the annual net generated electricity because the power plant generating capacity of each plant differed. These results are shown in Table 5. Oil- and LNG-fired power plants have the same input energy and materials but CO$_2$ emitted per annual net generated electricity of the LNG-fired power plant is smaller than the oil-fired power plant because of the higher efficiency of the LNG-fired power plant. CO$_2$ emitted per annual net generated electricity of PV(I) is smaller than PV(J), as PV(I) has a larger annual net electricity generation.

CO$_2$ Emissions for Power Plant Operation

Fossil fuel-fired power plants emit CO$_2$ from fossil fuel combustion for electricity production (coal : 5.57 $\times 10^9$, oil : 4.66 $\times 10^9$, LNG : 3.57 $\times 10^9$ kg-CO2/year). CO$_2$ emissions from operation of renewable energy electric power plants were assumed to be 1%/year for

Table 4. CO$_2$ emissions from materials and energy.

Items	NIRE [kg-CO$_2$/kg]
iron and steel	1.180
aluminum	2.035
concrete	0.099
cement	0.719
stainless steel	3.325
silicon steel	1.563
copper	1.304
paper	1.685
PE	1.262
PVC	1.497
titanium steel	10.062
CFC	2.498
FRP	1.660
chrome steel	7.785
ammonia	1.107
silicone	86.241
glass	1.928
electricity[kWh]	0.438
coal	2.383
heavy oil	3.216

hydroelectric and PV, and 2%/year for OTEC, respectively, of the CO$_2$ emissions from their construction.   CO$_2$ emissions from fossil fuel-fired power plants during operation from sources other than fossil fuel combustion were not taken into account, as they are thought to be insignificant compared to that from fossil fuel combustion (Uchiyama *et al.*,1991).

Table 5. Power plant CO$_2$ emissions from construction and per kWh generated.

	CO$_2$ emissions [kg-CO$_2$/(kWh/year)]	CO$_2$ emissions per kWh [kg-CO$_2$/kWh]	[kg-C/kWh]
coal	0.0221	0.9159	0.2498
oil	0.0164	0.7557	0.2061
LNG	0.0160	0.5630	0.1536
hydro	0.3954	0.0171	0.0047
OTEC(2.5MW)	2.2305	0.1190	0.0324
OTEC(100MW)	0.2554	0.0136	0.0037
PV U	3.5400	0.1534	0.0418
PV I	3.4143	0.1480	0.0404
PV J	4.3070	0.1866	0.0509

CO$_2$ Payback Time

CO$_2$ payback time was calculated from estimates of CO$_2$ emissions from construction and during operation in a large scale of fossil fuel combustion during operation.

For example, the definition of CO$_2$ payback time, T for hydroelectric vs. coal-fired power plant is defined as follows:

$$T = (Chydro / Ehydro - Ccoal / Ecoal) / (Ocoal / Ecoal - Ohydro / Ehydro) \quad (1)$$

$$y = Ccoal + Ocoal \cdot T \quad (2)$$

$$Y = y / Ecoal = Ccoal / Ecoal + Ocoal / Ecoal \cdot T \quad (3)$$

The result of equations similar to Eq.2 for each combination of renewable energy and conventional power plants are represented by the intersection of the respective lines for CO_2 emissions shown in Fig. 1. CO_2 payback times obtained are shown in Table 6. CO_2 payback times of renewable energy electric power plants based on coal power plant were about 60% of

Table 6. CO_2 pay back time [year].

Items	coal	oil	LNG
hydro	0.41	0.50	0.68
OTEC(2.5MW)	2.54	3.12	4.28
OTEC(100MW)	0.26	0.32	0.43
PV U	4.00	4.90	6.69
PV I	3.85	4.71	6.43
PV J	4.91	6.03	8.26

those based on LNG power plant, which is mainly caused by the difference in carbon content / calorific value. CO_2 payback times of hydroelectric and OTEC (100MW) were very short, which was mainly smaller CO_2 emissions from construction compared to PV power plants. It was suggested that if the operating lifetime of the renewable energy power plant was longer than the CO_2 payback time, the plant was better with respect to the mitigation of global warming.

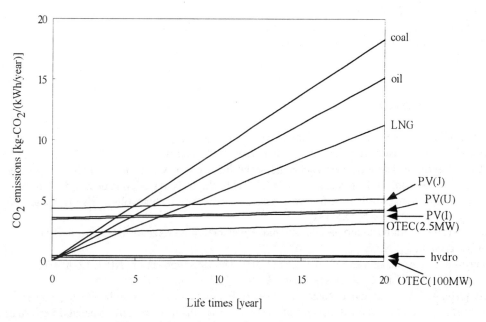

Fig. 1. CO_2 emissions of each power plant.

CO_2 Emissions per kWh

CO_2 emissions per kWh under the assumption of 30-year operating lifetime was calculated by Eq.(4).

$$Em = O \cdot Te + C / E \cdot Te \quad (4)$$

CO_2 emissions per kWh of each power plant are also shown in Table 5.

Japanese electricity supply is composed of nuclear (28.2%), hydroelectric (10.5%), coal (11.0%), oil (27.8%) and LNG (22.3%) CO_2 emissions from a nuclear power plant was assumed to be 0.00118 kg-CO2/kWh (BUWAL, 1991).   Hence, CO_2 emissions per kWh of Japanese electricity supplied was 0.439 kg-CO_2/kWh.

CONCLUSION

In this paper, CO_2 emissions from construction of each power plant was calculated by the LCA methodology using a bottom-up approach. CO_2 payback times for each renewable energy power plant were given by comparison with fossil fuel fired power plants. CO_2 pay back times of hydroelectric and OTEC (100MW) are very short, which is mainly caused by smaller CO_2 emissions during construction than that of PV power plants.   It is also suggested that CO_2 payback times of all renewable energy electric power plants evaluated in the present paper are much shorter than their typical operational lifetimes.

NOMENCLATURE

O = CO_2 emissions from generating plant [kg-CO_2/year]
E = Electricity generated annually [kWh/year]
C = CO_2 emissions from material product and construction [kg-CO_2]
T = CO_2 payback time [year]
y = Cumulative CO_2 emissions [kg-CO_2]
Y = Annual CO_2 emissions per electricity generated [kg-CO_2/(kWh/year)]
Em = CO_2 emission per kWh [kg-CO_2/kWh]
Te = Lifetime [year]

REFERENCES

Agency of Natural Resources and Energy (1995), Sougouenerugitoukei.

BUWAL (1991), Oekobilanz Von Packstoffen Stand 1990, Schriftenreihe
    Umwelt Nr.132, herausgegeben vom Bundessamt fur Umwelt, Wald und Landschaft, Bern, Februar .

Inaba, A., Y. Kondo, M. Kobayashi, K. Kita, N. Takahahi, S. Noda, S. Matsumoto, H. Morita
    and H. Komiyama (1995a), Life Cycle Assessment for Solar Photovoltaic Energy System. *Energy
    and resources*, 16, No.5, 525-531 .

Inaba, A (1995b) Raihusaikuruasesumento niokeru kisosozai no seizoudeta. *Environmental
    Management*, 31, No.6 .

Kobayashi, M., A. Inaba and T. Nakayama (1994), The Development of "NIRE-LCA"-The
    Software for the Life Cycle Assessment-. *J. JIE*, 73, (12), 1057-1079 .

Kobayashi, M., A. Inaba and T. Nakayama (1995), Evaluation of Effect of the Electric Power
    Generation Structure and Efficiencies of Thermal Power Plant on the Environmental Performanve of
    the Aluminum Production. *J .JIE*, 74, (1), 46-52 .

Resources Council, Science and Technology agency (1983), Natural energy and power generation
    technology, in Japanese.

Tahara, K., K. Horiuchi, T. Kojima and A. Inaba : "The role of ocean thermal energy conversion
    (OTEC) system in CO_2 problem. *MACRO REVIEW* 6, No.2, 35-43 (1993).

Uchiyama, Y. and H. Yamamoto (1991), Energy Analysis on Power Generation Plants,
    Economic Research Center Rep. No.Y90015 .

Uchiyama, Y. and H. Yamamoto (1992), Greenhouse Effect Analysis of Power Generation
    Plants. *Economic Research Center Rep.* No. Y91005.

Pergamon

Energy Convers. Mgmt Vol. 38, Suppl., pp. S621–S627, 1997
© 1997 Elsevier Science Ltd. All rights reserved
Printed in Great Britain
0196-8904/97 $17.00 + 0.00

PII: S0196-8904(97)00006-X

ECONOMIC ASSESSMENT OF CO_2 CAPTURE AND DISPOSAL

R.S. Eckaus, H.D. Jacoby, A.D. Ellerman, W-C. Leung and Z. Yang

Joint Program on the Science and Policy of Global Change,
Massachusetts Institute of Technology,
Cambridge, MA 02139-4307 USA

ABSTRACT

A multi-sector multi-region general equilibrium model of economic growth and emissions is used to explore the conditions that will determine the market penetration of CO2 capture and disposal technology. © 1997 Elsevier Science Ltd

KEYWORDS

Carbon dioxide; capture; disposal; economics

THE STUDY DESIGN

One method for controlling CO_2 emissions is to capture the gas at the point of combustion, and store it for the long term. The potential for this approach will depend not just on its own technical performance and cost, but also on patterns of economic growth and its energy intensity, on the structure of international agreements to control greenhouse gases, and on the availability of low-carbon alternatives to fossil fuels. Here we use the MIT Emissions Prediction and Policy Analysis (EPPA) model to evaluate the prospects for this method of emissions mitigation as applied to large-scale electric power generation.[1] Uncertainties in the costs of both the capture and the disposal stages of this process are very great, as are the future economic conditions that will influence its competitive position. Therefore the purpose of our inquiry is to illuminate the cost goals the technology will have to meet to achieve market penetration, not to predict its actual level of use.

Our analysis of the economics of market penetration divides the likely sources of future electric generation into three categories: conventional plants, non-carbon "backstop" technologies, and fossil-fired generation with CO_2 capture and disposal. In the future, these costs of fossil-fired generation will reflect any tax on CO_2 emissions or (equivalently) the price of any constraint on emissions which applies to the electric power industry. If the price is zero, as today, capture and disposal methods will never be used. A positive carbon price will tip the relative economics away from conventional fossil generation, however, and toward generation with capture and disposal. How either of these fossil-fired technologies fares will depend on competition with other forms of low- or no-carbon electricity generation, which are represented in the EPPA model by a non-carbon electric "backstop" technology. In our analysis, this backstop technology represents the possible expansion of solar power, nuclear power (through the development of new, advanced, and socially acceptable reactor designs), and biomass-based systems.

---

[1] The EPPA model has been developed with the support of a government-industry partnership including the U.S. Department of Energy (901214-HAR; DE-FG02-94ER61937; DE-FG02-93ER61713), U.S. Environmental Protection Agency (CR-820662-02), the Electric Power Research Institute (W02141-23), and a group of corporate sponsors from the United States, Europe and Japan.

Furthermore, the relevant input prices depend not only on evolving conditions within the electric sector, but on developments in each national economy as a whole and (because of the influence of international trade) on economic conditions in all other world regions. Thus the competition between electricity generation with capture and disposal and other technologies depends on many factors, all of which are changing over time. These include the technical development of all these technologies (in terms of their required inputs of capital, labor, fuel and other inputs per kWh of output), the relative prices of these inputs, and the stringency and structure of the CO$_2$ control policy.

To analyze the prospects for capture-and-disposal methods, we assume a policy environment which is as favorable to this technology as any now being discussed for adoption within the Framework Convention on Climate Change. Under it the OECD countries agree, by 2005, to limit their CO$_2$ emissions to 20% below their 1990 baselines. We explore two versions of such an agreement. In one case, each of the OECD regions agrees to meet this constraint on its own, and in the other a system of global trading in CO$_2$ emissions rights is agreed as well.

THE MIT EPPA MODEL

The MIT Emissions Prediction and Policy Analysis (EPPA) Model (Yang et al., 1996) is a recursive-dynamic computable general equilibrium (CGE) representation of multi-regional world economy, with supplies of input factors (e.g., labor, capital, land and other resources), consumer demand functions, and production technologies for each region. The world is divided into 12 regions, which are linked by multilateral trade. The economic structure of each region consists of nine fully-elaborated production sectors. Three of these are non-energy activities, and six are components of energy supply including fossil-fired electricity generation with capture and disposal of CO$_2$. There are four consumption sectors, one government sector, and one investment sector.

As applied in this study, the EPPA model incorporates two "backstop" energy supply technologies, represented as Leontief functions of labor and capital. These are a Carbon-Free Electric Backstop which can be used in any of the twelve EPPA regions, and a Carbon Liquids Backstop representing the industry producing liquid fuels from heavy oils, tar sands and shale, and which can be developed only in the three regions now known to have substantial amounts of the necessary resources. In the calculations below, we first consider the future for capture and disposal on the assumption that these alternatives do not materialize on large scale, and then discuss the difference in results if they are considered to be a real option.

The most important exogenous inputs to the EPPA model are population change, the rate of productivity growth (stated in terms of labor productivity), and a rate of autonomous energy efficiency improvement which reflects the effect of non-price-driven technical change on the energy intensity of economic activity. The rate of capital formation, which is another important influence on economic growth, is endogenous to the model. Finally, a key determinant of the carbon intensity of economic growth, which also has an important influence on the competitive position of the capture and disposal option, is the assumed costs relative to conventional sources of the backstop technologies. In the calculations reported below, the base-year (1985) cost of the carbon-free electric backstop is $0.15 per kWh, and the cost of the carbon fuels backstop is set at 1.4 times the 1985 cost of refined oil in each producing region. These costs change over time, relative to other energy sources, as input prices vary in response to changing economic conditions. More details of version of the model used here are provided in Eckaus et al. (1996).

REPRESENTATION OF CO$_2$ REMOVAL TECHNOLOGY

The Technical Options

Various methods have been developed for the capture of dilute CO$_2$ from gas streams, and many others are proposed (e.g., US DOE, 1993; IEA, 1992; Hendriks, 1994). Examples of the possibilities include chemical stripping, membrane systems, cryogenic separation, physical absorption, and physical absorption. Each of these technologies may be combined with various methods for fuel pre-processing and combustion. For example, the removal process may be applied directly to the flue gas from a conventional power plant; it may be combined with a coal gasifier as an add-on to an integrated coal gasifier combined cycle facility; or it may be integrated into a system based on fuel cells. The precise form of the separated CO_2 (purity; solid, liquid or gaseous state; and the temperature and/or pressure) is a function of the total system design, including the disposal method. The costs of each of these individual technical approaches are highly uncertain, particularly at the scale of a modern fossil power plant. Moreover, costs estimates may differ greatly depending on the type of fossil fuel used.

Uncertainty regarding the technical feasibility and cost of large-scale CO_2 disposal is even greater than that for the capture stage. A number of possible means and locations of disposal have been considered (US DOE, 1993; Ormerod, 1994) including discharge into depleted natural gas reservoirs or aquifers, discharge deep into the ocean, or conversion of the CO_2 back into useful organic compounds (a "re-use" option of which little is known except that it appears to be prohibitively expensive. The oil and gas industry has experience with gas storage in depleted natural gas fields, but the sites available for such use are limited. Little is known about the behavior of aquifers under CO_2 charge, or the aggregate capacity and cost. Thus although underground discharge may offer a way to test pilot installations, it may be that only the oceans can absorb the CO_2 quantities that substantial market penetration of capture and disposal methods would generate.

Because of these uncertainties and the widely differing disposal options faced by the various OECD countries, and the degree of regional and sectoral aggregation of the EPPA model, we do not conduct the analysis in terms of detailed designs of specific capture and disposal facilities. Rather, as shown in the next section, we formulate capture and disposal in terms of a single set of fuel, capital and M&O cost premia, over and above the cost of conventional generation. In this way we can explore the cost conditions that must be met by *any* combination of these technologies if it is to penetrate electricity markets.

Implementation in the EPPA Framework

As with the other CES production sectors in the EPPA model, conventional electricity is provided by a process that combines inputs of capital, labor, energy in the form of fossil fuels, intermediate goods other than fuel, and a fixed factor. The cost of electricity supply is a function of the quantities of the inputs used and their relative prices. These input prices evolve over time, as a result of a set of interdependent dynamic processes, including the growth of the twelve regional economies, technological change, depletion of oil and gas reserves, and the effects of CO_2 control policies. As a result, the relative price of conventional electricity also changes, as indeed it has done in most world regions over the *past* century, in a process driven by many of the same forces.

Applying the technology choices discussed above, the economic structure of capture and disposal is conceived as a two-stage process, described in more detail by Eckaus et al. (1996). The supply and generation stage has the same CES input structure as conventional generation, only it produces two products: electricity and the CO_2 that has been captured and prepared for pipeline transport. We assume the capture rate is 90%. Then, in a second stage, the captured CO_2 is transported to the location of the disposal facility, and placed in long-term storage.

Because of the power requirements of the CO_2 recovery equipment, and the process steps required to bring the captured gas to the level of purity and pressure required for shipment, the fuel efficiency of a facility with CO_2 capture will be lower than for a conventional power plant. If this efficiency loss is denoted *EL*, for example, then the fuel multiplier is $1.0/(1.0 - EL)$, so that a 20% degradation in fuel efficiency per kWh sent out requires 25% more fuel input. This fuel penalty varies according to the technology assumed. For example, the percentage loss is generally estimated to be less for an integrated coal gasifier combined cycle plant than for a conventional coal-fired facility. Given the historical data base of the EPPA model, the factor proportions (i.e., capital, labor, fuel, other intermediate inputs) which are implicit in its representation of the electric supply sector are closest to those of conventional fossil-fired generation. For this underlying generation technology, estimates of the loss in fuel efficiency for CO_2 capture range from around 20% (e.g., Hendriks, 1994) to as high as 35% (e.g., EPRI, 1991). In the calculations reported here we use multipliers of 1.25 and 1.33 (efficiency loses of 20% and 25%), because we are exploring the domain that these technologies must attain if they are to see use as part of a response to the climate change threat.

To meet the parasitic power demand of the capture equipment, the generation facility must be larger per kWh sent out (and thus more costly) than a conventional plant. Then, to the additional costs of the generation plant itself must be added the capital and M&O costs of the capture process facilities. For this analysis, all these additional costs (the generation plant and all ancillary facilities) are combined into a single premium applied to inputs of capital, labor and intermediate inputs, over and above the cost of an equivalent (in terms of kWh sent-out) conventional fossil-fired power plant.

As with the fuel multiplier, we explore that portion of the cost range where the capture and disposal option might possibly begin to take a significant role. Since it is unlikely that economies of scale, in cost per kW for the larger generation plant, are so strong as to overwhelm the additional costs of capture and disposal, the smallest capital and M&O premium we study is 25%, which is the lowest fuel (and therefore generation plant scale) multiplier. Then we also investigate generation premia of 33%, 50% and 58%.

PATTERNS OF MARKET PENETRATION

In the EPPA model, the OECD is represented by four countries or aggregated regions: the United States (USA), Japan (JPN), the European Community (EEC) and the Other OECD (OOE). We consider first a case where the backstop technologies are assumed not to be available. Also, no emissions trading is allowed, so each of the four OECD areas must meet the 20% emissions reduction on its own. The price of carbon in the resulting simulations is shown in Fig.1. For the United States, the European Community, and the Other OECD the carbon price rises to somewhere between $200 and $300 per ton (in $1990) by 2050. For Japan, the carbon price required to keep the economy on the reduced-emissions path is higher, around twice what it is for Europe. The difference results from the fact that, as modeled within EPPA, the Japanese economy operates at a higher level of energy and carbon efficiency from the start of the simulation period, and stringent internal measures are required to achieve and maintain the assumed emissions reduction. One way of thinking about the economics of CO₂ capture and disposal is in reference to this carbon price: in order to begin to take market share in a particular region, the technology must achieve emissions reductions at cost (per ton C) equal to or below the levels shown in the figure.

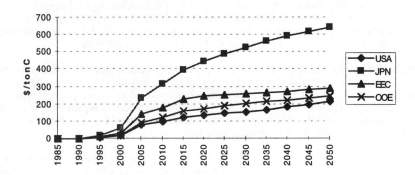

Fig.1. Carbon Price Under an AOSIS-Like Protocol, With No
Backstops and No Trade in Emissions Permits

Under these conditions, the market penetration of capture and disposal technology is as shown in Table 1. With a fuel multiplier of 1.33 (25% efficiency loss) or more, and a premium on capital and M&O of 50% or above, this technology does not enter at all between now and 2050 (nor in the period to 2100 either). With the smaller fuel multiplier, and a cost premium as low as 50%, the technology does take some role, achieving entry in the European Community in 2020. As can be seen with reference to Fig.1, under these cost conditions a carbon price of around $240 is required to bring this technology on line in the European Community. It does not penetrate first in Japan, despite the high carbon price there, because the cost of conventional electric power is high in Japan, relative to the other OECD regions, so a capture and disposal option which is some multiple of conventional plant cost has a hard time competing with other ways of reducing carbon emissions.

Table 1. Market Penetration in 2050 by Capture and Disposal Technology, Under CO_2 Reduction by the OECD to 20% Below 1990 Levels by 2010. No Backstops and No Emissions Trading. Year of first penetration shown in parentheses.

Fuel Multiplier per kWh Sent-Out	Premium on Generation Capital and M&O Cost			
	25%	33%	50%	58%
1.25	Moderate (2015)	Moderate (2015)	Small (2020)	Zero
1.33	N.A.	Small (2020)	Zero	Zero

Definitions:

Zero	=	No penetration anywhere in 2050.
Small	=	Penetration in only one region, at less than 15%.
Moderate	=	Penetration in 2 or more regions, none over 20%.
Large	=	Penetration in all regions at over 20%.
N.A.	=	Not applicable.

If the cost premium is lower, say 33% or 25%, the penetration is greater, first entering in the European Community in 2015, and in other regions some five to 15 years later. At the higher fuel penalty of 1.33, the prospects are poor: there is some introduction in the European Community in 2020 but, as noted earlier, a case where the premium on capital and M&O costs is no larger than the fuel multiplier is unlikely. The case where the cost penalty is even lower than the adjustment in plant size (required to compensate for the efficiency loss) is not applicable, for reasons discussed earlier.

Next, we contrast this "no backstops" case with one where both of the backstop technologies (non-carbon electricity and carbon-based fuels) are assumed to come into play if justified under the cost assumptions specified above. The availability of the carbon-free electric source will tend to lower overall emissions, and the new carbon-fuels source will tend to increase them. Under the particular cost assumptions imposed here, the carbon-free electric backstop has the greater influence, so that, in the case with no emissions restrictions, global CO_2 emissions in 2050 are 11% lower than in the no-backstop case. Further, because these baseline emissions are lower, the carbon prices needed to achieve the prescribed 20% reductions in OECD emissions also change. They are lower in three of the regions, but somewhat higher in the Other OECD.

The lower carbon prices would be expected to greatly diminish the prospects for the capture and disposal options, but there is a countervailing effect. The penetration of the backstop electric source tends to lower price pressure on the inputs to both conventional generation and generation with capture and disposal, and to enhance the attractiveness of the latter approach to carbon reduction. These factors combine to produce a picture similar to that without backstops in Table 1, only now it is the Other OECD that is the first place of entry. Indeed, only in the lowest-cost case studied (fuel multiplier of 1.25, cost premium of 25%) does the technology take market share in any other region. The timing of first entry is delayed by 15 to 20 years, as compared with the case with no backstops.

Finally, we consider the prospects for capture and disposal methods if the policy of 20% reduction in the OECD is supplemented by an agreement to full global trading in emissions rights. After 2010, each OECD region receives an annual allocation equal to its 1990 emissions *minus* 20%, and non-OECD regions receive an allocation equal to their emissions in the absence of policy, and trading in these rights is allowed. In this case the implicit carbon price rises only to about $50 per ton by 2050, because under the EPPA analysis there are so many low-cost ways to reduce carbon if the task is approached on a worldwide basis. The implied cost target is far below any level now being discussed for capture and disposal technology.

CONCLUSIONS

The use of a computable general equilibrium model like EPPA offers insights into the prospects for capture and disposal that would not appear in a project-level engineering analysis, or even in a partial equilibrium analysis of a country's electric power sector. First, it is reasonable to assume that nations will seek the lowest-cost ways to meet emissions targets, and therefore that capture and disposal methods will have to compete with other ways of reducing carbon in all different sectors of the economy. The intensity of this competition is summarized most simply in the carbon price, region by region, that is associated with possible future agreements to restrict CO$_2$ emissions. Because the capture and disposal of CO$_2$ from central-station electric power would in any case be only one component of regional carbon restriction, the calculated carbon price is not much influenced by different degrees of penetration by this technology. Therefore, a rough cost goal can be drawn from any study of the general-equilibrium response to policy, without the need for modeling the role of capture and disposal technology itself within the regional or global equilibrium.

Further, even with this single measure, one can see that the challenge faced by capture and disposal technology is dramatically increased if international agreements include cost-saving measures, such as emissions trading. In the calculations discussed here, it is assumed that the trading scheme is global, involving all regions. As shown by Jacoby et al. (1996) substantial cost reductions (and, therefore, reductions in the carbon price) are attainable from trade in emissions permits even if only a subset of nations participate. Thus, in the face of even limited emissions trading, the cost goals for generation with capture and disposal are very likely beyond reach, at least under policy conditions like those assumed here.

If a nation is under emissions restraint but does not have the opportunity to trade in emissions rights, then another important influence is the future performance of the so-called "backstop" technologies. If the backstops are available, the time of first penetration of the capture approach is delayed by 15 to 20 years, as compared to a world without them. Also, the regions where the new technology first becomes economic are different in the two cases. This latter result has implications for research on disposal methods. If different approaches to CO$_2$ disposal are likely to be preferred in different regions (e.g., some countries have extensive underground reservoirs, or some ocean regions are more stable than others), then the allocation of R&D effort should consider which world regions would be most likely to see first penetration if costs could be reduced, and attention should be focused on the approaches favored there.

Finally, the timing and degree of penetration of capture and disposal technology will be influenced by the general stringency of CO$_2$ emissions targets which nations accept. Advances in the scientific understanding of climate change could lead to stronger political commitments, and to more restrictive measures than those assume here. If there were no emissions trading, and commitments to speed development of the low-carbon backstop technologies were believed to be unaffected by an increasing stringency of emissions controls, then such developments would brighten the prospects for capture and disposal option. However, as made clear by this analysis, studies of the influence of policy stringency on the prospects for this technology need to take account of the effects of increased climate concern on the design of control policies (i.e., the inclusion of emissions trading), and of the pace of improvements in technologies competing with capture and disposal.

REFERENCES

Eckaus, R.S., H.D. Jacoby, A.D. Ellerman, W. Leung and Z. Yang (1996). Economic Assessment of CO$_2$ Capture and Disposal, MIT Joint Program on the Science and Policy of Global Change, Report No. 15, MIT, Cambridge, MA.

EPRI [Electric Power Research Institute] (1991). Engineering and Economic Evaluation of CO$_2$ Removal from Fossil-Fuel-Fired Power Plants, Volume 1: Pulverized-Coal-Fired Power Plants. EPRI IE-7365, Palo Alto, CA.

Hendriks, C. (1994). *Carbon Dioxide Removal from Coal-Fired Power Plants.* Utrecht University, the Netherlands.

IEA [International Energy Agency, Greenhouse Gas R &D Program] (1992). Carbon Dioxide Capture: An Examination of Potential Absorption Technologies for the Collection of CO$_2$ and Other Greenhouse Gases Arising from Power Generation Using Fossil Fuel, Study No. IEA/92/OE4.

Jacoby, H.D., R.S. Eckaus, A.D. Ellerman, R.G., Prinn, D.M. Reiner, and Z. Yang (1996). QELRO Impacts: Domestic Markets, Trade and Distribution of Burdens, and Climate Change, MIT Joint Program on the Science and Policy of Global Change, Report No. 9, MIT, Cambridge, MA.

Ormerod, B. (1994). The Disposal of Carbon Dioxide from Fossil Fuel Fired Power Stations, International Energy Agency, Greenhouse Gas R&D Program, Report IEAGHG/SR3, Cheltenham.

US DOE [U.S. Department of Energy] (1993). The Capture, Utilization and Disposal of Carbon Dioxide from Fossil Fuel-Fired Power Plants, DOE/NR Report No. 30194, Washington, D.C.

Yang, Z., R.S. Eckaus, A.D. Ellerman and H.D. Jacoby (1996). The MIT Emissions Prediction and Policy Assessment (EPPA) Model, MIT Joint Program on the Science and Policy of Global Change, Report No. 6, MIT, Cambridge, MA.

Pergamon

Energy Convers. Mgmt Vol. 38, Suppl., pp. S629–S634, 1997
© 1997 Elsevier Science Ltd. All rights reserved
Printed in Great Britain
0196-8904/97 $17.00 + 0.00

PII: S0196-8904(97)00007-1

SYNERGIES AND CONFLICTS OF SULFUR AND CARBON MITIGATION STRATEGIES

Sabine Messner

International Institute for Applied Systems Analysis
Schlossplatz 1, A-2361 Laxenburg, Austria

ABSTRACT

A technology-oriented optimization model of the world energy system is used to analyze sulfur and carbon mitigation strategies with respect to potential synergies and conflicts. A parametric approach with a stepwise reduction of maximum carbon dioxide (CO_2) concentrations down to 400 ppmv in 2100, combined with stringent limits on sulfur dioxide (SO_2) emissions, shows positive synergetic effects. Moreover, potential conflicts due to the negative contribution of sulfate aerosols to the climate change problem do not reach significant levels under the conditions investigated.

© 1997 Elsevier Science Ltd

KEYWORDS

Global change; world energy model; CO_2 mitigation; CO_2 removal; SO_2 emission reduction.

INTRODUCTION

A number of global scenarios for the next century were developed by the International Institute for Applied Systems Analysis (IIASA) and the World Energy Council (WEC) for the Tokyo conference of the WEC in 1995. These scenarios are grouped into three "families" (IIASA-WEC, 1995); all include energy efficiency improvements and some degree of decarbonization in the world. The scenarios are based on different economic and technological development trajectories, and their emissions range from very high levels to levels at which atmospheric carbon dioxide (CO_2) emissions are stabilized. The world energy model developed for this study, M11T, is based on the bottom-up energy optimization model MESSAGE (Messner and Strubegger, 1995). It consists of 11 regional models and interregional trade for the principal energy commodities, including methanol and hydrogen. The time horizon extends to the year 2100.

Since its use in the IIASA–WEC study, M11T has been extended to include linearized constraints for carbon concentrations in the atmosphere. These constraints utilize the energy-related CO_2 emissions in the model and assumptions concerning non-energy-related CO_2 emissions (due to land-use changes, cement production, etc.) from the IPCC IS92a scenario (IPCC, 1992). They have been calibrated using a simplified carbon cycle and climate change model, based on Wigley and Raper (1992). With respect to CO_2 mitigation measures, due to its nature as a technology-oriented optimization model, M11T includes all options related to the use of other fuels or energy sources. Additionally, CO_2 removal for coal- and gas-based power generation and CO_2 captured from fossil-based hydrogen production are included. These CO_2 removal options are modeled as add-on technologies for the corresponding basic technologies, including technical performance (efficiency of CO_2 removal, electricity use) and economic parameters (investment and operating costs), which are based on information available in the literature and accessible through an inventory of energy technologies relevant to climate change, CO2DB (Messner and Strubegger, 1991). Given the wide

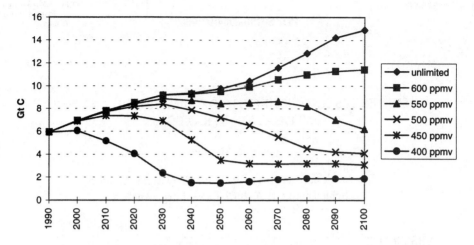

Fig. 1.   CO$_2$ emission trajectories for concentration limits from 600 to 400 ppmv and the unlimited case, with no limit on SO$_2$ emissions.

range of cost data for CO$_2$ disposal found in the literature (see Nakićenović and Messner, 1993 for a survey of this information) and the still rather uncertain quality of these data, an optimistic value in the range of existing estimates was used.

In the classification used in the second assessment report of the Intergovernmental Panel on Climate Change (IPCC, 1996), M11T includes the first class of mitigation options, all options eliminating or reducing CO$_2$ emissions. The other two classes, offsetting emissions and adaptation measures, are outside the scope of an energy model, as is the consideration of non-energy-related emissions of greenhouse gases. Concerning conventional pollutants, the latest version of M11T includes a representation of SO$_2$ and nitrogen oxide (NO$_x$) emissions and possible sulfur abatement strategies based on the technological options in the RAINS model (Alcamo *et al.*, 1990). These sulfur mitigation options include scrubbing at power plants, cleaning of fuels, and the import of low-sulfur coals, in addition to the options for fuel switching, which are inherent to the modeling approach.

LIMITING CO$_2$ CONCENTRATIONS

The new version of the world energy model M11T was applied to test the possible consequences of a series of concentration limits. Scenario definitions are taken from scenario A1 of the IIASA–WEC study (IIASA–WEC, 1995). This scenario, with high economic growth, strong technological development, and optimistic assumptions concerning the availability of fossil fuels, provides a good basis for investigating the flexibility of the energy system. In the unconstrained case based on scenario A1, CO$_2$ concentrations approach 620 ppmv by 2100. From this level, maximum allowed concentration levels are lowered in a step-wise manner to 600, 550, 500, 450, and 400 ppmv. CO$_2$ emissions are 15 Gt C by 2100 without a concentration limit; with declining concentrations, this figure decreases to reach approximately 2 Gt C in the 400 ppmv case (see Figure 1).

All these cases imply a certain amount of CO$_2$ removal. In the unlimited case, CO$_2$ is used up to an annual quantity of 1.5 Gt C for reinjection and enhanced oil recovery (EOR) measures. In the cases with limited concentration, the amount of CO$_2$ that is sequestered increases, reaching an annual removal rate of 4 Gt C by 2100 in the case with 400 ppmv. In this case, two-thirds of the CO$_2$ generated in the energy conversion processes is captured and stored; only one-third is emitted into the atmosphere. In an investigation of the time profile of CO$_2$ removal, steeper increases can be seen in the 450 ppmv case after 2030, whereas immediate employment of this technology is only observable in the rather extreme 400 ppmv case. Here CO$_2$ removal starts to grow before 2000,

Table 1. Marginal costs of the CO_2 concentration limit with no limit on
SO_2 emissions (in US\$/t C).

	Unlimited	600 ppmv	550 ppmv	500 ppmv	450 ppmv	400 ppmv
1990	0.0	0.1	0.5	1.1	7.5	123.4
2000	0.0	0.2	0.9	2.1	14.7	247.9
2010	0.0	0.3	1.5	3.6	25.3	440.0
2020	0.0	0.6	2.6	6.2	44.4	811.8
2030	0.0	1.0	4.5	10.7	78.7	1,489.2
2040	0.0	1.7	7.8	18.7	141.8	386.1
2050	0.0	2.9	13.6	33.2	261.7	116.5
2060	0.0	5.2	23.8	59.8	480.0	213.7
2070	0.0	9.2	42.2	110.3	71.1	116.5
2080	0.0	16.5	75.9	202.4	130.4	184.4
2090	0.0	30.4	140.1	183.0	214.0	202.4
2100	0.0	55.8	257.0	335.7	392.6	371.3

reaching a first maximum of 3.5 Gt in 2040 and the final maximum of 4 Gt in 2100. Marginal costs to reach these rather low concentrations are almost US\$1,500 per ton C for the year 2030, the highest single value observable (see Table 1). To reach a CO_2 concentration of 450 ppmv, the maximum marginal cost is US\$480 in the year 2060, and for 500 ppmv a local peak in marginal cost of US\$200 exists in 2080. Only in the cases with 600 and 550 ppmv is the exponentially growing slope of marginal costs to be expected due to the minimization of overall discounted cost. Only these cases manage to reach the concentration target in 2100 without additional costs in earlier periods.

Figure 2 shows SO_2 emissions for the same six cases included in the investigation of CO_2 concentrations. In the unlimited case, energy-related SO_2 emissions grow from the current level of 60 Mt S to 75 Mt S, an increase of 25%. In the long run, the increased use of natural gas and non-fossil sources of energy brings SO_2 emission levels down to 30 Mt S. Figure 2 also shows the effect that meeting the CO_2 concentration limits has on SO_2 emissions: they follow lower development paths, there is a positive effect of optimal CO_2 abatement on SO_2 emissions. However, all cases except the 400 and 450 ppmv cases have the same trajectories up to 2010 and start to diverge only thereafter. The final emission levels reached by 2100 are around 10 Mt S for the cases with 400–550 ppmv, a reduction of two-thirds and the 600 ppmv case has SO_2 emissions of 22 Mt S, 30% less than the 30 Mt S in the unlimited case.

From this exercise, it can be concluded that reducing CO_2 emissions improves the SO_2 problem. However, for higher concentration targets this improvement only applies in the longer run. SO_2 emissions are reduced in the nearer future only if concentration limits around 400–450 ppmv are applied.

LIMITING SO₂ EMISSIONS

To investigate the problem from the other side, an SO_2 reduction scenario has been designed with the help of the IIASA RAINS model of acidification (Alcamo et al., 1990). This scenario limits SO_2 emissions in Europe and Asia to levels that are environmentally acceptable. The assumptions for the rest of the world are based on the specific emission levels reached in European or Asian regions similar to the region considered. The SO_2 emission targets are rather stringent: by 2020 overall SO_2 emissions must be reduced to 25% of the 1990 value, and thereafter this level cannot be exceeded. Adapting M11T to include this SO_2 emission limit and providing the new options for SO_2 abatement described above, CO_2 emissions reach a level of 11.2 Gt C by 2100, 25% less

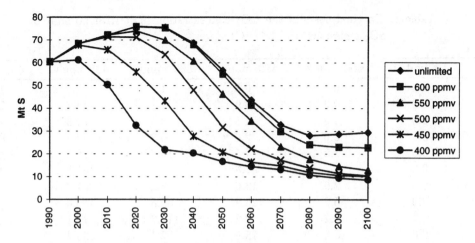

Fig. 2.   SO₂ emission trajectories for different CO₂ concentration limits
with no limit on SO₂ (in Mt S).

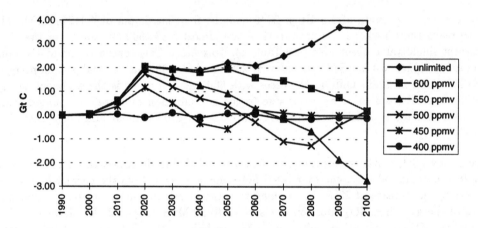

Fig. 3.   Difference in CO₂ emission trajectories between the cases with
and without SO₂ emission limit for concentration limits from 600
to 400 ppmv and the unlimited case.

than the original result of 15 Gt C. The CO₂ concentration in 2100 is 10% below the level of the unlimited case. These results show that a strategy of reducing SO₂ emissions, which includes fuel switching as an option in addition to scrubbing measures, can reduce CO₂ emissions. Figure 3 shows the difference in CO₂ emissions between the two cases (with and without SO₂ limits), and makes the same comparison for the runs with CO₂ concentration limits.

For the 600 ppmv case, the difference in annual CO₂ emissions is up to 2 Gt C. The trajectory is similar to that of the unlimited case up to 2050, but thereafter the reduction of CO₂ emissions due to the SO₂ limit is reduced and approaches zero in 2100. This result shows that the concentration limit of 600 ppmv is not binding if stringent SO₂ limits are applied: overall CO₂ emissions are different with and without limits on SO₂ emissions. For all other cases, overall CO₂ emissions and consequently CO₂ concentrations in 2100 are similar, only the paths of the trajectories differ. Larger reductions in the short run due to reductions in SO₂ emissions are offset by higher emissions in the longer run. The cases with 400 and 450 ppmv are somewhat different, because rather stable CO₂ emissions are reached around 2050. This achievement drastically reduces the degrees of freedom, and the CO₂ emission trajectories are rather similar with and without a limit on SO₂ emissions; in the case with 400 ppmv they are virtually identical.

Table 2. Marginal costs of the CO_2 concentration limit with limited SO_2 emissions (in US\$/t C).

	Unlimited	600 ppmv	550 ppmv	500 ppmv	450 ppmv	400 ppmv
1990	0.0	0.0	0.1	0.7	3.1	108.3
2000	0.0	0.0	0.1	1.3	6.1	217.5
2010	0.0	0.0	0.2	2.1	10.5	386.0
2020	0.0	0.0	0.3	3.7	18.4	712.2
2030	0.0	0.0	0.5	6.3	32.7	1,306.6
2040	0.0	0.0	0.9	10.8	58.9	418.1
2050	0.0	0.0	1.5	18.8	108.6	123.8
2060	0.0	0.0	2.6	33.0	199.3	227.1
2070	0.0	0.0	4.6	58.5	172.4	137.6
2080	0.0	0.0	8.2	105.3	106.3	153.9
2090	0.0	0.0	15.2	194.4	196.1	200.6
2100	0.0	0.0	27.8	356.5	359.7	367.9

Table 3. CO_2 concentration, SO_2 emissions, and resulting temperature change in 2100 for a parametric variation.

Case	CO_2 (ppmv)	SO_2 (Mt S)	Δt (°C)
1	619.0	52.4	2.31
2	558.0	35.4	2.17
3	619.0	35.4	2.44
4	558.0	52.4	2.04

An analysis of marginal costs of the CO_2 concentration constraint (see Table 2) shows a similar divergence: with a concentration limit of 400 ppmv, marginal costs of meeting the concentration limit are similar to the run without an SO_2 limit (Table 1). For 450 and 500 ppmv the final values are also similar, but the very high costs of US\$480 around 2060 for 450 ppmv are avoided here. Only above 550 ppmv is there a long-term difference in the marginal costs. The large difference in the 550 ppmv case is attributable to the fact that considerable CO_2 emission reductions occur due to the limit on SO_2 emissions before 2050, allowing higher emissions of nearly 3 Gt C in 2100.

CONSEQUENCES ON TEMPERATURE

In the following the effects of sulfate aerosols and CO_2 concentrations on temperature rise is investigated for the scenario used in this analysis. Table 3 shows a parametric variation based on the two model runs without a CO_2 concentration limit, one with and one without a limit on SO_2 emissions. A simplified carbon cycle and climate change model, based on Wigley and Raper (1992) was used to calculate global mean temperature change on the basis of CO_2 and SO_2 emission trajectories [other assumptions are taken from the IPCC IS92a scenario, (IPCC, 1992)].

Case 1 in this table refers to the basic run without any limitations; in case 2 the M11T model run with limited SO_2 emissions but without a CO_2 constraint is used. Cases 3 and 4 are combinations of these model runs: case 3 combines the CO_2 emissions of case 1 with the SO_2 emissions from case 2, and vice versa in case 4. A comparison of these results shows the impact SO_2 emissions have on temperature change: case 3, with 30% lower SO_2 emissions than case 1, results in a 0.13°C higher temperature increase. Reducing CO_2 concentrations by 10% (from case 1 to case 4) reduces the temperature effect by 0.27°C.

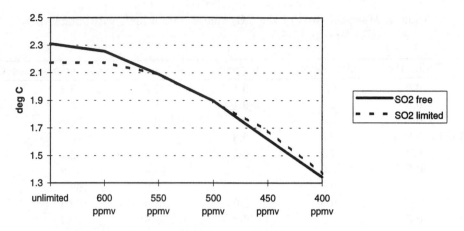

Fig. 4.   Temperature change (°C) for the different concentrations in 2100,
with and without a limit on SO_2 emissions.

Comparing cases 1 and 2, the original model runs with and without a limit on SO_2 emissions but with free CO_2 emissions, clearly shows that SO_2 mitigation has a positive effect on climate under these conditions. SO_2 emissions and the resulting negative forcing are reduced, but the implicit effect, reducing CO_2 emissions and consequently CO_2 concentrations by 10%, more than offsets this 0.14°C increase in radiative forcing. This effect can also seen from Figure 4, which compares temperature changes in 2100 for all concentration limits and the unlimited case for model runs with and without a limit on SO_2 emissions. Without an SO_2 limit, temperature increases by 1.3°C at 400 ppmv; this increase is 2.31°C in the completely unconstrainted case. If SO_2 emissions are constrained, the shape of the curve is slightly different: it starts at a lower temperature change in the unlimited case and reaches a slightly higher level at the CO_2 concentration level of 400 ppmv. Lower temperatures in the unconstrained and 600 ppmv cases are due to lower cumulative CO_2 emissions, whereas the higher temperatures in the case with SO_2 limits and lower CO_2 concentrations are due to the effect investigated above: the negative contribution of SO_2 to radiative forcing.

REFERENCES

Alcamo, J., R. Shaw and L. Hordijk, eds., (1990). *The RAINS Model of Acidification, Science and Strategies for Europe*. Kluwer, Dordrecht, Netherlands.

IIASA–WEC (International Institute for Applied Systems Analysis and World Energy Council), 1995. *Global Energy Perspectives to 2050 and Beyond*. WEC, London, UK.

IPCC (1992). *Climate Change 1992: The Supplementary Report to the IPCC Scientific Assessment*. J.T. Houghton, B.A. Callander, and S.K. Varney, eds., Cambridge University Press, Cambridge, UK.

IPCC (1996). *Climate Change 1995: Economic and Social Dimensions of Climate Change*. J.P. Bruce, H. Lee and E.F. Haites, eds., Cambridge University Press, Cambridge, UK.

Messner, S. and M. Strubegger (1991). User's Guide to CO2DB: The IIASA CO_2 Technology Data Bank – Version 1.0. WP-91-31a, International Institute for Applied Systems Analysis (IIASA), Laxenburg, Austria.

Messner, S. and M. Strubegger (1995). User's Guide for MESSAGE III. WP-95-69, International Institute for Applied Systems Analysis (IIASA), Laxenburg, Austria.

Nakićenović, N. and S. Messner (1993). Other Environmental Technologies. In: Long-Term Strategies for Mitigating Global Warming. *Energy – The International Journal* **18**(5):401–609.

Wigley, T.M.L, and S.C.B. Raper (1992). Implications for climate and sea level of revised IPCC emissions scenarios. *Nature*, 357:293–300.

Pergamon

Energy Convers. Mgmt Vol. 38, Suppl., pp. S635–S641, 1997
© 1997 Elsevier Science Ltd. All rights reserved
Printed in Great Britain
0196-8904/97 $17.00 + 0.00

PII: S0196-8904(97)00008-3

ASSESSMENT OF CO₂ REMOVAL UTILIZING THE CONCEPT OF SUSTAINABILITY LIMITATIONS

R. MATSUHASHI and H. ISHITANI

Department of Geosystem Engineering, Faculty of Engineering,
The University of Tokyo, 7-3-1 Hongo, Bunkyo-ku, Tokyo 113, Japan

ABSTRACT

This paper deals with sustainable energy technologies under restrictions on resources and environmental impact. We have developed a database to estimate life-cycle efficiencies and greenhouse-gas emissions. Next, we define sustainability limitations on resource depletion and emissions. Based on these concepts, the present world energy system is judged to be unsustainable. Mitigation measures such as disposal of CO_2 and solar power generation are evaluated utilizing the concept. We show how sustainability influences the cost-effectiveness of measures taken. At the same time CO_2 removal is shown to play a transitional but important role to realize sustainable energy scenario in ultra-long term.

© 1997 Elsevier Science Ltd

KEYWORDS

Life cycle assessments; sustainability limitations; global energy supply; CO_2 emissions; appropriate incentives.

SUSTAINABLE DEVELOPMENT

Integrated energy balances in the energy system

Energy supplies are obtained after mining, transportation, processing, and conversion for utilization. The net energy efficiency, μ is the product of component efficiencies. Each component efficiency is evaluated from an overall energy balance (Matsuhashi et al., 1995). Both energy use for fuel and energy required for construction must be considered. The integrated energy balance for a plant is given in eq. (1), where E_f is the input fuel energy, E_i the input energy required to construct the plant, and E_o the output energy from the plant; thus,

$$\mu_i = \frac{E_o}{(E_i + E_f)} \tag{1}$$

In Eq. (1), E_i is converted to the same unit as E_f. For example, if E_f is annual fuel consumption of a plant, then E_i can be evaluated by deviding the energy required to construct the plant by the lifetime in terms of years of the plant. Hereafter, we use IEB for the integrated energy balance μ_i. By introducing IEB, we can deal with renewable energy technologies in the same manner as for non-renewable energy technologies.

Quantitative definition of sustainability limitations

We derive the necessary conditions for sustainable limitations on renewable resources, non-renewable resources and environmental emissions by utilizing the concept of the life cycle IEB. The definition of sustainable consumption is obtained by investigating whether or not resource depletion and environmental crises can be avoided if the present rates of life-cycle IEBs and energy demands are continued.

R_0 : Reserves of the resources at initial time period.

r   : Rate of increase of R_0 by improvement of geophysical prospecting and mining.

s   : Rate of substitution by other resources.

μ_0 : Life cycle IEB of the resources at initial time period.

a   : Rate of increase of μ_0.

C_0 : Rate of recycle of the resources at initial time period. Although recycle is physically
              impossible in energy resources, it corresponds to the rate of cascading.

c   : Rate of increase of C_0.

P_0 : Production of the resources at initial time period.

D_0 : Demand of the resources at initial time period.

b   : Rate of increase of D_0.

As far as non-renewable resources are concerned, the sustainability condition is derived as follows. Suppose that grade of a resource is expressed in the function of $f(R,P)=R/P$, then the following equation is obtained by differentiating the function $f(R,P)$.

$$\frac{\partial f}{\partial a} = \lim_{\Delta t \to 0} \left[\frac{1}{\Delta t} \left[\frac{R_0 \exp(r\Delta t) - \dfrac{D_0\left\{1 - C_0 \exp(c\Delta t)\right\} \exp(b\Delta t)}{\mu_0 \exp(a\Delta t) \exp(s\Delta t)} \cdot \Delta t}{\dfrac{D_0\left\{1 - C_0 \exp(c\Delta t)\right\} \exp(b\Delta t)}{\mu_0 \exp(a\Delta t) \exp(s\Delta t)}} - \frac{R_0}{D_0(1 - C_0)} \Big/ \mu_0 \right] \right]$$

$$= \lim_{\Delta t \to 0} \left[\frac{1}{\Delta t} \left[\frac{\mu_0 R_0}{D_0} \times \frac{\exp\left\{(a + r + s - b)\Delta t\right\}}{\left\{1 - C_0 \exp(c\Delta t)\right\}} - \Delta t - \frac{\mu_0 R_0}{D_0(1 - C_0)} \right] \right] \qquad (2)$$

$$= \frac{\mu_0 R_0}{D_0} \left\{ \frac{a + r + s - b}{1 - C_0} + \frac{C_0 c}{(1 - C_0)^2} \right\} - 1$$

$\dfrac{\partial f}{\partial a} \geq 0$ Accordingly equation (3) is obtained.

$$a + r + s - b + \frac{C_0 c}{(1 - C_0)} \geq \frac{D_0}{\mu_0 R_0}(1 - C_0) = \frac{P_0}{R_0} \qquad (3)$$

Condition (3) indicates that depletion of a non-renewable resource can be avoided if the left hand side including the factors of technological improvement is larger than the reciprocal number of R_0/P_0. Therefore we define this as a sustainability condition of a non-renewable resource.

Renewable resources can also be dealt with as follows.   Stock type renewables are evaluated such as biomass resources, since flow-type renewables such as photo-voltaic and wind turbine systems do not deplete.   As conclusion, sustainability condition is the same as that of non-renewable resources except that r in (3) corresponds to a rate of growth.

Next we investigate environmental emissions such as anthropogenic CO_2 emissions. If we regard environmental emissions as negative resources, we are able to apply the same kind of condition as non-renewable resources. In evaluating CO_2 emissions, sustainability condition is the same as that of non-renewable resources except that C_0 in (3) corresponds to the rate of absorption by the environment and that both r and c in Equation (3) are zero. In particular, we should be aware of the fact that C_0, the rate of absorption by the environment is closely related with accumulation mechanism of CO2 emissions.

Thus the sustainability condition on renewables, non-renewables and environmental emissions are shown to be similar. Accordingly we can deal with various resources and emissions in the integrated framework. The sustainability conditions enable us to evaluate how the technologies of efficiency improvement, innovative mining or heat cascading contribute to the sustainability.

Fig.1 shows the sustainability conditions of various resources evaluated based on the condition (3) and the following assumptions.

(1) R/P of each resource is estimated based on proven reserves and production.
(2) Sustainability limitation of each resource is evaluated based on the above estimated R/P.
(3) The values in (3) are calculated as the average values between '70 and '90 for mineral resources and between '80 and '90 for energy resources.
(4) We can evaluate the distance between sustainable condition and actual situation of each resource as shown in Fig.1. This distance is defined as actual unsustainability.
(5) Reserves of those resources are supposed to increase as exploring and mining technologies   are improved. Therefore we evaluated the value of r in (3) assuming that the proven reserve of each resource will approach the ultimate reserve in fifty years.
(6) We can investigate the potential risk of depletion of each resource, which is defined as potential unsustainability.
(7) As far as CO_2 is concerned, sustainability limitations and present situation is assessed based on airborne fraction, which is the rate of CO2 accumulating in the atmosphere to anthropogenic CO2 emission and maximum permissible accumulation in the atmosphere. The value of airborne fraction is assumed to be 0.58 (Takahashi et al., 1987). Maximum permissible accumulation is assumed to be 560 ppm, twice of that in pre-industrial era.

The real line in Fig.1 expressed the sustainability limitation. Then a resource, of which the point is above the line, is judged to be sustainable.   For example, mercury is sustainable, although its R/P is only twenty five years. This is because toxicity of mercury is broadly recognized, and its production is rapidly decreasing. Toxicity of lead is also recognized globally, so that production of lead is stagnant. Therefore unsustainability of lead is very low. Copper is judged to be sustainable actually, since improvements in mining technologies increased the proven reserves. However, it is judged to be potentially unsustainable, since the ultimate reserve of copper is not so much. On the contrary, iron is judged to be potentially sustainable because of huge ultimate reserves, although it is actually unsustainable.
Whereas oil and natural gas is judged to be actually sustainable, all energy resources except for coal is potentially unsustainable. Uranium is evaluated based on once through utilization in light water reactors. If technical and social problems of fast breeder reactors are solved and it is penetrated globally, unsustainability of uranium becomes low. Unsustainability of CO_2 is lower than that of natural gas, and is comparable with that of oil and higher than that of coal.

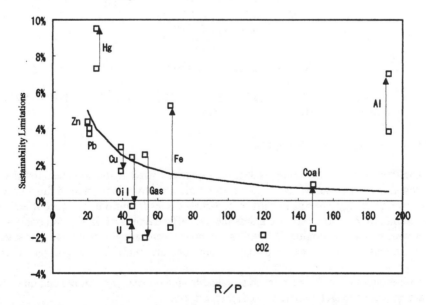

Mark indicates the direction from actual unsustainability to potential unsustainability.
Fig.1. Sustainability of the resources and the emissions

When a resource becomes depleted, the price of the resource generally goes up and it promotes the substitution to other resources. Sustainable use of resources are usually maintained by substitution of resources with high unsustainability to resources with low unsustainability.   If a resource begins to be depleted, the index of sustainability of the resource would deteriorate and be judged to be unsustainable. But if this resource could successfully be substituted by other resources, the index would be improved and judged to be sustainable. Therefore we have to be careful of resources that is judged to be unsustainable resulting from unsuccessful substitution, recycle or other reasons. More explicit consideration on substitution and resultant interaction among various resources is our future work. Fig.1 also indicates that energy resources and CO_2 emissions could threaten the sustainable development of humankind. Therefore we focus our analysis on energy resources and CO_2 in the next section.

<u>Measures to control energy system to sustainable area</u>

In order to investigate measures to control energy system to sustainable area, we developed a mathematical model and analyzed an optimal path to realize a sustainable energy system from the present unsustainable system (Matsuhashi <u>et al.</u>, 1996).   This model implied that we should impose economic incentives to reach sustainable area in long term before resource depletion or environmental catastrophe will eventually be caused and that these economic incentives should be increased at a certain rate. The values of economic incentives are determined so that non-renewable resources will not be depleted and CO_2 concentrations will not reach maximum permissible atmospheric concentration. Fig. 2 shows a schematic diagram of the path to sustainable area.

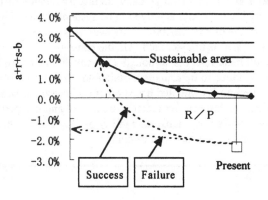

Fig. 2. Control of energy system to the sustainable area

ASSESSMENT OF ABATEMENT OPTIONS UTILIZING THE CONCEPT OF SUSTAINABILITY LIMITATIONS

<u>Assumptions on assessing abatement options</u>

The above analyses indicate that some economic incentives need to be imposed for the purpose of realizing sustainable energy use. If we impose economic incentives expressed in the above section, we could control overall energy systems to shift to sustainable ones by introducing cost-effective abatement options. According to this procedure, we assess the cost-effectiveness of mitigation options. Coal-fired power plants with systems removing CO_2 and solar power generating systems are investigated here as representative options.   The following assumptions are made for the purpose of assessing the systems.

● We impose economic incentives to reach sustainable area on energy resources and on CO_2 emissions, which are expressed as RTAX and CTAX, respectively.

● We compare the coal fired power stations with and without systems to remove CO_2.   Fuel input in power plants without CO_2 removal is normalized to be 1 toe/year.   In the case of power plants with CO_2 removal systems, fuel input has to increase due to efficiency reduction caused by the energy requirement of removing the CO_2.   This efficiency reduction is defined as the energy penalty (U.S. DOE., 1993).

- Carbon intensity of coal is assumed to be 1 t-C/toe.
- Several variables are defined as follows:
  EP the energy penalty, REC the rate of recovering CO_2, C_v the variable cost of coal fired power plant, ΔC_f the increase of fixed cost by adding equipment to recover CO_2

Assessment of abatement options utilizing the concept of sustainability limitations

Since CO_2 emissions can be reduced from 1 t-C/year to $(1-\mu)\,/\,(1-EP)$ t-C/year by installing systems removing CO_2, this reduction is expressed as follows.

$$\Delta(CO_2) = \frac{REC - EP}{1 - EP}(t - C\,/\,year) \tag{4}$$

Then incentives regarding CO_2 emissions are given as ΔCO_2 multiplied by CTAX. On the other hand, coal requirement increases from 1 toe/year to 1/(1-EP) toe/year. Therefore incentives regarding resource consumption have to be removed as fuel increase is multiplied by RTAX. Increase in the sum of capital and running cost is expressed as follows.

$$\Delta(Cost) = \frac{EP}{1 - EP}C_v + \Delta C_f \quad (yen\,/\,year) \tag{5}$$

Increase in the cost must be less than the above economic incentives, so that power plants with systems removing CO_2 are installed in the market:

$$\frac{EP}{1 - EP} \cdot C_v + \Delta C_f \;\leq\; \frac{REC - EP}{1 - EP} \cdot CTAX - \frac{EP}{1 - EP} \cdot RTAX \tag{6}$$

Eq. (6) indicates the following things. When depletion of resources are threatening sustainable development, an increase of RTAX makes CO_2 removal systems difficult to be installed in the market. However more than 4000 Gtoe of fossil fuels are thought to be ultimately recoverable. When the problem of CO_2 concentration is more serious than depletion of resources, CO_2 removal systems may be installed in the market.

Next we compare the cost-effectiveness of PV systems with that of conventional coal-fired power plants. $\Delta(CO_2)$ and $\Delta(Cost)$ are quantified as shown in eqs. (7) and (8). Then the necessary conditions can be derived as shown in eq. (9), in which PV systems replace conventional coal-fired power plants by the same procedures as discussed above.

EB : IEB of PV systems
CP : the rate of fixed cost of coal fired power plants to that of PV systems.

$$\Delta(CO_2) = 1 - \frac{1}{EB} = \frac{EB - 1}{EB} \quad (t - C\,/\,year) \tag{7}$$

$$\Delta(Cost) = (CP - 1)C_f - \frac{EB - 1}{EB}C_v \quad (yen\,/\,year) \tag{8}$$

$$(CP - 1)C_f - \frac{EB - 1}{EB}C_v \;\leq\; \frac{EB - 1}{EB}CTAX + \frac{EB - 1}{EB}RTAX \tag{9}$$

These evaluations have identified the region, in which each energy technology is the most economical. Fig. 3 shows the economical regions of the evaluated measures. The area A in the Fig. 3 expresses the economic region of conventional coal fired power stations. Regarding CO_2 sequestration systems, we assume IGCC with systems to remove and sequester CO_2 (Matsuhashi et al., 1992). The area B expresses its economic region. As far as PV systems are concerned, we quantitatively evaluate IEB and cost of roof-top PV systems by investigating manufacturing process of PV and its BOS (balance of the system) by bottom-up method (Matsuhashi et al., 1996). Recycling of steel frame and improvements in manufacturing process (Komiyama et. al., 1993) are taken into consideration to improve IEB of solar power generating system. Lifetime of photovoltaics is assumed to be 20 years and annual cost ratio is assumed to be 14.8%, including the cost of operation and maintenance.   The area C expresses its economic region.

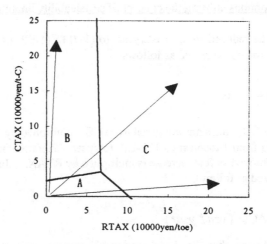

Fig. 3. Economical regions of the evaluated technologies

Fig. 3 indicates that CO_2 sequestration would not play a permanent but a transitional role.   It follows that renewable energy technologies such as PV systems can be dealt with as 'Back-stop technologies'. If we take account of the current exchange rate of 110 yen/US$, we can assess the levels of incentives, which make each energy technology competitive in the market.

CONCLUSIONS

In this paper, we first proposed the index of integrated energy balance (IEB).   Renewable energy systems can be dealt with as options to efficiently utilize non-renewable energies, utilizing this index.
Next   we quantitatively defined the concept of 'sustainability limitations' on resources and the environment.   This concept clarified the distances indicating how far the present energy system is from sustainability limitations. The distances are defined on the basis of   the   rate   of resource depletion and CO_2 concentration, respectively.
Then we evaluate the cost-effectiveness of mitigation options such as CO_2 disposal and PV systems for the case, where economic incentives are imposed to shift the present unsustainable energy system to a sustainable one.
As a result of evaluation, renewable energy systems such as photovoltaics were shown to play a 'Back-stop' role.   At the same time, CO_2 disposal was shown to play a transitional but significant role in the sustainable energy systems.

REFERENCES

Matsuhashi, R., K. Hikita, and H. Ishitani (1995).   Ultra-long term scenario for sustainable development considering CO_2 constraint. Proceedings of the 11th conference on energy system and economics. Japan Society of Energy and Resources, 217-222.
Takahashi, K. and K. Okamoto (1987). Global environment in 21st century. NHK books, 25-44.
US Department of Energy (1993).   A research needs assessment for the capture, utilization and disposal of carbon dioxide from fossil fuel-fired power plants. Vol. 1 Executive Summary. DOE/ER-30194, 29.
Komiyama, H. and K. Yamada (1993). Evaluation of solar power generation systems Ⅰ. Report of the first kind committee of the Japan Society for Chemical Engineering.
Maezawa, S., Y. Yanagisawa, Y. Nishigami and H. Sano (1996).   Analysis of chemical fixation of CO_2 utilizing solar energy. Proceedings of the 12th conference on energy system and economics. Japan

Society of Energy and Resources, 245-250.

Matsuhashi, R. and H.Ishitani (1992). Modification of energy system and CO_2 disposal for reducing CO_2 economically, Energy Convers. Mgmt, 33, 755-761.

Matsuhashi, R., K. Hikita and H. Ishitani (1996). An analysis of sustainable energy system taking resource depletion and environmental emissions into consideration. Proceedings of IEW/JSER International Conference, 263-268.

 **Pergamon**

Energy Convers. Mgmt Vol. 38, Suppl., pp. S643–S648, 1997
© 1997 Elsevier Science Ltd. All rights reserved
Printed in Great Britain
0196-8904/97 $17.00 + 0.00

PII: S0196-8904(97)00009-5

EXTERNALITY ESTIMATION OF GREENHOUSE WARMING IMPACTS

Bent Sørensen

Roskilde University, Institute 2, Energy & Environment Group,
DK-4000 Roskilde, Denmark

ABSTRACT

The impacts of greenhouse warming have been described by Working Group 2 of the 1995 IPCC assessment. They include the impacts of changing vegetation zones on agriculture and silviculture, changes in the occurrence of crop pests and disease-carrying insects, as well as estimates of the effects of increased frequencies of extreme events such as heat waves, dry spells and floods. The present study attempts to quantify and valuate each of these impacts, integrated over the 21st century, under the assumption of continued injection of greenhouse gases into the atmosphere, leading to a doubling of atmospheric content around the middle of the 21st century. The result is a greenhouse warming externality constituting a sizable fraction of the global gross national product, provided that impacts are valued in the way normally done in externality studies for the industrialized world. As the greenhouse gas emissions are chiefly from industrialized countries, while the largest impacts are in those regions of the world least likely to reach stages of high development during the 21st century, the results pose a difficult geo-political problem. The implications for estimating externalities of particular energy conversion activities are briefly discussed. © 1997 Elsevier Science Ltd

KEYWORDS

Greenhouse warming impacts, monetarization, environmental externalities.

MONETIZING GLOBAL WARMING IMPACTS

Table 1 gives a list of impacts identified by the IPCC (1996), in terms of the additional number of people exposed to a range of risks, due to the average warming, sea level rise and additional extreme events predicted by the climate studies reviewed by the IPCC. As many impacts involve human deaths, the most important monetizing assumption is the choice of a statistical value of life (SVL), that allows human death to be presented as an expense to society, without embarking into any ethical discussion of the value of life. In a recent externality study for industrialized countries (European Commission, 1996), this value is taken as 3.3 million US $ (2.6 million ECU), which coincides with the compensation given in some European countries if government employees die on the job. Clearly, not just a work days lost component is included, but also a welfare responsibility of society. In this study, a similar value of 3 M$ will be used, but seasoned with a discussion of the implications of selecting a much smaller value for developing countries.

The individual entries in Table 1 are based on the sources indicated, with monetization of impacts evaluated as follows:

Table 1. Estimated global warming impacts during 21st century, for IPCC reference case (CO_2 doubling) (US$). H: high SVL of 3 M$ used throughout. L: SVL put at zero for Third World.

Impact description:	Ref.	Valuation H (T$)	Valuation L (T$)
Additional heatwave deaths (doubling, additional 0.1M/y, valued at 3 M$ each)	a	30	20
Fires due to additional dryspells (doubling)	b	1	0.5
Loss of hardwood- and fuelwood-yields (20% relative to 1980 production)	cde	4	1
Increase in skin-cancer due to UV radiation	b	3	3
Additional asthma and allergy cases (increase in pollen and air pollution due to warming)	b	3	1.5
Financial impact of increase in extreme events (doubling)	b	3	0.8
Increase in insect attacks on livestock and humans	b	?	?
Food production (failure to adopt new crops, increased pests and insect attacks, production loss 10-30%), (population at risk increasing from present 640M to 700-1000M), additional deaths from starvation due to crop loss (100M additional deaths, chiefly in developing countries)	f g	300	0
Deaths connected with migration caused by additional droughts or floods (50M deaths, the affected population being over 300M)	b	150	0
Increased mortality and morbidity due to malaria (presently 2400M at risk and 400M affected, increase due to warming 100M cases, in tropical and partly in subtropical regions, a 7- fold increase potentially possible in temperate regions assumed to be curbed)	b h	300	75
Increased mortality and morbidity due to onchocerciasis (presently 120M at risk and 17.5M affected, increase and spread due to warming 10M additional cases, primarily in developing countries)	b b,i	30	5
Increased mortality and morbidity due to schistosomiasis (presently 600M at risk and 200M cases, increase due to warming 25%, in dev. countries)	b	150	20
Increased mortality and morbidity due to dengue (presently 1800M at risk and 20M cases, increase due to warming 25%, in developing countries)	b	15	0
Other effects of sanitation and freshwater problems connected with salt water intrusion, droughts, floods and migration (developing countries)	b	50	0
Loss of tourism, socioeconomic problems, loss of species, ecosystem damage, etc.	b	?	?
Total of valuated impacts (order of magnitude)		1000	100

Based on discussions in IPCC Working Group II, IPCC (1996) and on other sources as follows:
a. Kalkstein and Smoyer (1993)
b. McMichael et al., in IPCC (1996), chapter 18.
c. Zuidema et al. (1994)
d. Kirschbaum et al., in IPCC (1996), chapter 1.
e. Solomon et al., in IPCC (1996), chapter 15.

f. Reilly et al., in IPCC (1996), chapter 13 and Summary for Policymakers
g. Rosenzweig et al. (1993); Parry and Rosenzweig (1993)
h. Martens et al. (1994)
i. Walsh et al. (1981); Mills (1995)
The valuations involve further estimates and should be regarded as very uncertain.

The two valuation columns differ by the first using the high SVL globally, the second taking the SVL to be zero for developing countries, in order to display the geographical differences of impacts.

Heatwave deaths occur in major cities due to the heat island effect, possibly combined with urban pollution. The doubling estimated by Kalkstein and Smoyer (1993) is mostly due to increased occurrence at mid-latitudes (city temperature rise 2.5° assumed), and thus two-thirds are assumed to happen in industrialized countries. A case study in New York City (Kalkstein, 1993) finds an increased mortality of 4×10^{-5} over a 5-10 day period. The present annual New York rate of heat wave deaths is 320, and Kalkstein has collected similar numbers for a number of large cities around the world, experiencing days with temperatures above 33°C. The estimated doubling of incidence will thus imply an additional order of magnitude of 10^5 heatwave deaths, annually and globally, valued at 30 T$ over the 100 year period. Uncertainties come from possible acclimatization effects, and from increased populations in large cities expected over the 21st century.

The doubling of fires causes mainly economic loss, assumed to be evenly distributed between developed and developing countries, whereas the 20% losses of hardwood and fuelwood yields is predicted to follow a complex geographical pattern (Kirschbaum et al, IPCC, 1996, chapter 1), but with the highest losses occurring in tropical regions (Solomon et al., IPCC, 1996, chapter 15). It is here assumed that 75% of the economic losses pertains to developing countries. The increased number of skin cancer cases due to an assumed 10% loss of ozone is mainly occurring at higher latitudes (IPCC, 1996, chapter 18).

Additional allergy cases would be associated with increased levels of pollen and air pollution due to heating and would occur predominantly at lower latitudes, whereas asthma incidence is highest in humid climates, expected to be enhanced at higher latitudes (IPCC, 1996, chapter 18). The impacts are assumed to be equally divided between developed and developing regions. Due to expected shortfall of hardwood supply relative to demand during the 21st century, the actual economic value may be considerably higher than the estimate given in Table 1. The financial loss associated with a predicted doubling of extreme events (draughts and floods, IPCC, 1996, chapter 18) is assumed to occur 75% in developing countries. The predicted incidence of insect bites is not valued, but could have an economic impacts, e.g. on livestock production.

One major issue is the impact of climate change on agricultural production. Earlier evaluations (e.g. Hohmeyer, 1988) found food shortage to be the greenhouse impact of highest importance. However, the 1995 IPCC assessment suggests that in developed countries, farmers will be able to adapt crop choices to the slowly changing climate, such that the impacts will be entirely from Third World farmers lacking the skills to adapt. The estimated production loss amounts to 10-30% of current global production (Reilly et al., IPCC, 1996, chapter 13), increasing the number of people exposed to risk of hunger from the present 640 million to somewhere between 700 and 1000 million (Parry and Rosenzweig, 1993). There are also unexploited possibilities for increasing crop yields in the developing countries, so the outcome will depend on many factors, including speed of technology transfer and development. Assuming a lower estimate of 100 million additional starvation deaths during the 21st century, one arrives at the 300 T$ figure given in Table 1, all occurring in the Third World. This is also the case for deaths associated with migration induced by extreme events, estimated at 50 million.

The other major impact area is death from diseases transmitted by insect vectors influenced by climatic factors, such as mosquitoes. For Malaria, there are presently 2400 million people at risk (McMichael et al., IPCC, 1996, chapter 18), and an additional 400 million expected due to greenhouse warming and its implied expansion of the geographical area suited as habitat for the malaria-carrying mosquitoes (Martens et al, 1994). This will involve subtropical and even some temperate regions, but also in tropical regions, the incidence of malaria is predicted to increase. Assuming 100 million additional deaths from malaria during the 21st century, of which 75% in the tropics, one arrives at the figures given in Table 1. Large uncertainties are associated with the possibilities of mitigating actions in the subtropical and temperate regions. Also for onchocerciasis, vector populations are expected to both increase by 25% at current sites (Mills, 1995) and to spread to new sites (Walsh et al., 1981), leading to 10 million additional cases. Schistosomiasis may also spread in the subtropical regions, whereas dengue and yellow fever is expected to remain in the tropics. Table 1 reflects these expectations, through its distribution of impacts between developed and developing regions, and it also gives an estimate of deaths occurring due to sanitation problems connected with salt water intrusion into drinking water (due to sea-level rise) and larger migrating populations, mainly in the developing countries (assuming that immigration into industrialized countries continues to be controlled). The economic consequences of other identified impacts, such as loss of species, effect on tourism, etc., have not been estimated.

DISCUSSION

The overall impacts are of the order of magnitude 10^{15} US $, when using the SVL of industrialized countries, and one order of magnitude lower, if Third World impacts are valued at or near zero. This spells out the greenhouse impact dilemma, that 90% of the damage is in the Third World, much higher than their share in causing the problem. The IPCC Working Group 2 identification of a number of impacts specifically occurring in the low-latitude regions explains why the impact estimates are so much higher than those of early work based upon and extrapolated from industrialized regions (Cline, 1992; Frankhauser, 1990; Nordhaus, 1990). Other factors contributing to this result include the high SVL, and the assumptions of less ability to mitigate impacts in the Third World (by switching food crops, building dikes to avoid floods, and so on). If the impacts of the present study, evaluated with the uniform high SVL, are distributed equally over the 100-year period, the annual cost of greenhouse warming is found to be roughly 40% of the current global GNP.

Table 2. Impact from average passenger car (1990).

Enviromental impacts	type of impact: emissions (g per kWh of fuel)	monetised value (mUS$ per vehicle-km)	monetised value (mUS$ per kWh of fuel)	uncertainty & ranges
Environmental emissions:				
Car manufacture, decommissioning	average industry	18	28	H
Car maintenance	NQ			
Road construction and maintenance	NQ			
Operation:				
CO_2	277			L
NO_x (may form aerosols)	2.9			M
CO	17			M
HC	3.0			M
particles	0.06			M
Health effects from air pollutants		41	63	H,l,n
Greenhouse warming (cf. Table 1)	mainly from CO_2	72	109	H,g,m
Noise	av. increase 1.5 dB, variations are large	32	49	H,l,n
Environmental & visual degradation (from roads, signs, filling stations, etc.)		138	208	H,l,m
Health and injury				
Occupational (car/road construction and maintenance)	NQ			
Traffic accidents (incl. material damage, hospital and rescue costs):		107	161	M,l,n
Based on deaths (SVL=3 MUS$)	2.4×10^{-8} per kWh-fuel			
heavy injury	24×10^{-8} per kWh-fuel			
light injury (when reported)	16×10^{-8} per kWh-fuel			
Stress and inconvenience (e.g. to pedestrian passage)		33	50	H,l,n
Economic impacts				
Direct economy (cars, roads, gasoline, service and maintenance)	taxes excluded	276	420	L,l,n
Resource use	significant		NQ	
Labour requirements and import fraction (Denmark)	about 50% of direct costs are local		NQ	
Benefits (valued at cost of public transportation)		357	543	M,l,n
Time use (contingency valuation)		156	237	H,l,n
Other impacts	NA			

Sources: Danish Technology Council (1993); Danish Road Directorate (1981), Danish Statistical Office (1993), Danish Department of Public Works (1987), Christensen and Gudmundsen (1993), and own estimates. NA= not analysed, NQ= not quantified. Values are aggregated and rounded (to zero if below 0.1 mecu/kWh). (L,M,H): low, medium and high uncertainty. (l,r,g): local, regional and global impact. (n,m,d): near, medium and distant time frame.

APPLICATIONS OF GREENHOUSE WARMING EXTERNALITIES

The translation of the 10^{15} US$ impact from greenhouse emissions into externalities for specific energy activities may be done in the following way: It is assumed that about 75% of the forcing comes from energy-related emissions, and of these 17.5% are from natural gas, 40% from coal and 42.5% from oil. Modern power stations typically emit 0.27 kgC/kWh if coal-fired, 0.16 kgC/kWh if gas-fired and 0.21 kgC/kWh if oil-fired. The transportation sector mainly uses oil products and is responsible for 19.4% of the energy-related carbon emissions or 1.12 TkgC/y, which for gasoline driven automobiles corresponds to 659 gC/liter or about 49 gC per vehicle-km at an average of 13.5 km per litre of gasoline. The doubling of atmospheric greenhouse gases is now assumed to be due to 50 year's of emission at an average level 50% above the present one (corresponding to a doubling of fuel usage during the period). Using these assumptions to distribute the share of the total greenhouse externality on each activity, one obtains 0.40 US$/kWh for coal-fired power stations and 0.072 US$/vehicle-km for gasoline-driven cars.

The non-climate externalities associated with air pollution and mining accidents are about 0.06 US$/kWh for the fuel chain associated with a European state-of-the-art coal power plant (European Commission, 1996; Sørensen, 1996; these impacts would be larger for the current average power plant). In this case, the greenhouse warming impacts of 0.4 $/kWh are thus dominating.

Another example is that of externalities associated with ownership and use of motor vehicles, shown in Table 2. In this case, there are many externalities besides greenhouse warming, due to the impacts of traffic. Common assumptions for all the calculations are the use of an average-size car, driving 200000 km in 10 years with an average efficiency of 13.5 km per liter of gasoline (corresponding to mixed urban and highway driving). The greenhouse warming externality of 7.2 USc/km is evaluated above. The health effect caused by air pollution from car exhaust is taken as 4.1 USc/km, a number arrived at in several of the Scandinavian studies quoted. Also the accident statistics gives a firm basis for estimation (although the rate of accidents varies considerably between countries - the value used here corresponds to Denmark), but the value of police and rescue team efforts, hospital treatment, lost workdays and lives all have to be chosen. The SVL is again assumed to be 3 M$, and for time loss a figure of 9.4 $/h is used (based on an interview study on perceived values of waiting time; Danish Technology Council, 1993). This is a "recreational" value in the sense that it rather corresponds to unemployment compensation than to average salary. The "stress and inconvenience" line takes into account the barrier effect of roads with traffic, causing e.g. pedestrians to have to wait (e.g. at red lights) or to walk a larger distance to circumvent the road barrier. This may again be valued as time lost.

The noise impact is estimated at 3.2 USc/km based on hedonic pricing (i.e. the reduction in the value of property exposed to noise (e.g. houses along major highways as compared to those in secluded suburban locations) (Danish Transport Council, 1993). A similar approach is taken to estimate the visual degradation of the environment due to roads, signs, filling stations, parking lots and so on. Property values have been collected in 1996, for detached houses of similar standard, located at the same distances from the Copenhagen city centre (but outside the high rise area, at distances of 10 to 25 km from the centre), but exposed to different levels of visual and noise impact from traffic. The externality is then taken as the number of people exposed times the sum of property losses. The property loss found is 25-45%, and the total damage for 0.5 million people with 0.2 million houses and cars is 20.5 G$ or 54 USc/km, of which half is assumed to derive from visual impacts. A further reduction by a factor two is introduced by going from a suburban environment near Copenhagen to a country average. The value arrived at (for Denmark) is the 13.8 USc/km given in Table 2. The direct economic impacts include capital expenses and operation for cars and roads, as well as property value of parking space in garages, carports or open parking space, but omitting any taxes and duties, and the benefits from driving are taken at the value of public transport (considering that differences in convenience and inconveniences such as not being able to read when driving even out). Time use is as mentioned above derived from a contingency valuation (i.e. interviews). The "fairly calculated" cost of driving a passenger car, i.e. all direct costs and externality items except benefits and time use, add up to 0.65 US$/km, of which 0.07 are related to owning the car (purchase price without taxes plus environmental impacts of car manufacture), and the remaining 0.58 US$/km are related to driving the car. A fair tax level, reflecting external costs, would then be divided into a vehicle tax of around 4100 US$, and a kilometers-driven component, that levied onto the fuel would amount to about 5 US$ per litre.

REFERENCES

Christensen, L and Gudmundsen, H (1993). *Bilismens fremtid.* Danish Technology Council Report 93/3, Copenhagen

Cline, W. 1992. *Global Warming: the Economic Stakes.* Inst. Int. Economics. Washington DC.

Danish Department of Public Works (1987). *Trafikundersøgelse 1986.* Copenhagen

Danish Road Directorate (1981). *Trafikulykker 1980.* Copenhagen

Danish Statistical Office (1993). *Miljø.* Report 93/1, Copenhagen

Danish Technology Council (1993). *Trafikkens Pris.* Copenhagen.

Danish Transport Council (1993). *Externaliteter i transportsektoren.* Report 93-01, Copenhagen.

European Commission, 1996. DGXII, Science, Research and Development, *ExternE: Externalities of Energy,* Volumes 1-6. Report EUR 16520 EN, Brussels-Luxembourg.

Frankhauser, S. 1990. *Global warming damage costs - some monetary estimates.* CSERGE GEC Working paper 92/29. Univ. East Anglia

Hohmeyer, O. 1988. *Social costs of energy consumptions.* Springer, Berlin.

IPCC (1996). *Climate Change 1995. Impacts, adaptations amnd mitigation of climate change:Scientific-technmical analyses.* Contribution of Working Group II to the Second Assessment Report of the IPCC. Cambridge University Press.

Kalkstein, L (1993). Health and climate change: direct impacts in cities. *The Lancet,* **342,** 1397-1399.

Kalkstein, L and Smoyer, K (1993). The impact of climate change on human health: some international implications. *Experientia,* **49,** 969-979.

Martens, W, Rotmans, J, Niessen, L (1994). *Climate change and malaria risk.* GLOBO Report 3-461502003. Global Dynamics & Sustainable Development Programme, RIVM, Bilthoven, NL.

Mills, D (1995). A climate water budget approach to blackfly population dynamics. *Publications in Climatology,* **48,** 1-84

Nordhaus, W. 1990. *To slow or not to slow: the Economics of the Greenhouse Effect.* Yale Univ.

Parry, M and Rosenzweig, C (1993). Food supply and ths risk of hunger. *The Lancet,* **342,** 1345-7

Rosenzweig, C, Parry, M, Fischer, G, Frohberg, K (1993). *Climate change and world food supply.* Report 3, Environmental Change Unit, Oxford University, UK

Sørensen, B. 1996. Life-cycle approach to assessing environmental and social externality costs. pp. 297-331 in *Comparing energy technologies.* IEA/OECD Paris.

Walsh, J, Davis, J, Garms, R (1981). Further studies on the reinvasion of the onchocerciasis control programme by simulus damnosum s.l. *Tropical Medicine & Parasitology,* **32,** No. 4, 269-273.

Zuidema, G, v.d.Born, J, Alcoma, J, Kreileman, G (1994). Simulating changes in global land cover as affected by economic and climatic factors, *Water, Air and Soil Pollution,* **76,** 163-198.

Pergamon

Energy Convers. Mgmt Vol. 38, Suppl., pp. S649–S654, 1997
© 1997 Elsevier Science Ltd. All rights reserved
Printed in Great Britain
0196-8904/97 $17.00 + 0.00

PII: S0196-8904(97)00010-1

Japan's Cooperative Research of CO2 Mitigation with Developing Countries

S. Kusuda*, Y. Kurashige**, T. Higashikawa**, and I. Hayashi**

* Director for Global Environmental Technology, Agency of Industrial Science
and Technology, MITI, 1-3-1 Kasumigaseki, Chiyoda-ku, Tokyo, Japan
** Global Environmental Technology Department, New Energy and Industrial
Technology Development Organization, Sunshine 60, 3-1-1
Higashi-Ikebukuro, Toshima-ku, Tokyo, Japan

ABSTRACT

Asia has been the scene of rapid economic development. One result of this development has
been the mounting severity of environmental problems. The cooperation of advanced
industrial countries is indispensable for resolving the environmental problems in developing
countries. Japan is providing cooperation to Asia countries through the Green Aid Plan
conducted by NEDO and other groups, and promoted by MITI, Japan is also transferring
environmental and energy-saving technology that was developed to deal with its pollution
problems and the oil crisis. This report will describe the adoption of an energy-saving
system in a steel mill in Indonesia. © 1997 Elsevier Science Ltd

KEYWORDS

Green Aid Plan, international collaboration, NEDO, Japan, retrofit, energy-saving, SPH

JAPANESE COOPERATION IN THE ENVIRONMENTAL TECHNOLOGY

Environmental Problems in Developing Countries

Asia has a diversity of economies and social systems. In addition to a wealth of natural
resources, the region has achieved high economic growth. The rapid progression of
industrialization and urbanization has worsened industrial pollution in the developing
countries of Asia year by year, however. One challenge facing these countries in the future
is formulating a response to their sharp increases in energy consumption.

Developing countries lack the personnel and technology to solve these environmental
problems. They will require the assistance of the advanced industrialized countries. In the
past, Japan suffered from severe pollution problems and was harshly jolted by the oil crises
of the 70s. The country was able to overcome these problems by developing environmental
and energy-saving technology. Japan wants to put this experience and technology to use in
developing countries.

Overseas Cooperation Currently Provided by Japan in the Environmental Sector

Governmental assistance for development is the core of Japanese cooperative efforts. In the
UNCED at Brazil the Japanese government announced that they will provide from 900

billion to one trillion yen in official environmental assistance for the five-year period starting in 1992. This does not include the environmental assistance provided by Japan through a cooperative program with individual developing countries called the Green Aid Plan.

The Green Aid Plan is promoted by MITI, and conducted by NEDO, JETRO, and JICA. The objective of   the plan is to transfer environmental protection and energy-saving technology to developing countries.

Overseas Assistance Provide by NEDO

Established in 1980, NEDO is a government-affiliated body under the jurisdiction of MITI. They have been involved with developing and promoting the adoption of alternative energy sources for fossil fuels, including photovoltaic cells, fuel cells, and wind power. They also have been involved with research and development for industrial technology, including biotechnology and environmental technology.

The international cooperation they provide for environmental technology involves dispatching specialists, inviting researchers, sponsoring workshops and symposiums, and conducting model projects and research cooperation. Examples of this cooperative research include projects involving the technology for measuring auto exhaust in Thailand, waste water treatment for steel plants and paper manufacturing plants using straw pulp in China, and simple water purification systems for factories in Thailand, Indonesia, and Malaysia.

NEDO has been involved with 13 model energy-conservation projects. These include the use of waste heat from steel mills in China, energy conservation at cement plants in Indonesia, the use of residual heat from electric furnaces in Indonesia, and simple desulfurization devices in Chinese chemical plants.

This report will summarize the use of residual heat in electric furnace in Indonesia as a specific project in the Green Aid Plan.

THE SCRAP PREHEAT TECHNOLOGY FOR ELECTRIC ARC FURNACE

Technical Background

The steel industry in Japan has made great improvements in energy-saving technology since the energy crisis of the 70s. The scrap preheat (SPH) technology for waste gas from electric arc furnace (EAF) was developed to utilize a large amounts of waste heat. The quantity of heat from waste gas carry-over rises to 20 % of the total heat, in case of a typical 70 ton EAF. Japan's NKK Corp. transferred the technology and operating methods to PT. Budidharma in Jakarta (BDJ), Indonesia.

The SPH System

The process flow is shown in Fig. 1. The plant is equipped with two preheating chambers. The EAF capacity is 30 tons. Single-stage chamber heating with a simple bucket was chosen because it would fit in the existing plant facilities. The EAF waste gas causes the following chemical reaction in the combustion chamber : $CO + 1/2O_2 \rightarrow CO_2$. This is sent to the SPH to heat the scrap. In most SPH systems, the temperature range of the waste gas at the SPH inlet is usually lower than $500 \sim 600°C$. The maximum temperature in the demonstration test was $460°C$.

The gas is required to maintain the flow in the scrap inside the bucket to reduce pressure loss during preheating. A 160 kW fan was installed in the SPH line to measure the pressure loss at the main dust collector fan. This fan also controls the flow rate in synchronization with the control damper. The scrap bucket is transported by the transfer car. The temperature and flow rate of the waste gas at the preheating chamber inlet, in addition to the preheating time, are critical factors for efficient heat recovery. The SPH heating system is shown in Fig. 2. The waste gas is directed into the bucket from the top. The SPH preheating chamber is double-sealed, inside and outside to prevent the entry of air from outside and the leakage of waste gas. The joint between the scrap bucket and the preheating hood is sealed by thermal expansion (inside seal). The gap between the preheating chamber and the preheating hood is sealed by the water in the water sealing pit (outside seal).

The input energy to the SPH is controlled by opening the gas flow adjusting damper, positioning the sliding duct and controlling the rate of rotation of the preheating fan.

(1) The quantity and temperature of the gas are controlled by opening the gas flow adjusting damper and positioning the sliding duct.

(2) The gas feed to the SPH is controlled by the rate of rotation of the preheating fan based on the inlet gas temperature.

RESULTS OF THE OPERATION

Results of the Test Operation

The SPH system installation was completed in March 1995, and test operation was started in April. There were three charges for each production cycle for the standard furnace operation at this test plant. The bucket is preheated at each charge using the one-step preheating method. Each preheating chamber accommodates one bucket. The first bucket contains the largest amount of scrap. Fig. 3 shows that the first bucket is preheated when the waste gas temperature reaches the maximum in each cycle. It also shows the preheating pattern for test operations, 3 buckets - 1 preheating chamber operation. Fig. 4 shows that the temperatures inside the scrap bucket reach saturated levels after 30 minutes of preheating. Fig. 5 shows the thermocouples positions in the bucket. The amount of energy saved by SPH depends largely on the type of scrap, the SPH inlet temperatures, and the SPH preheating time. The power-saving effect of the SPH is increased with higher SPH inlet gas temperature, longer SPH preheating time, more suitable thickness of the preheating scrap and shorter post-charging time. Therefore, it is difficult to express the SPH effect quantitatively. For example, power savings can range from 15 to 50 kW/ton.

SPH Benefits

In March 1996, BDJ compared energy consumption with and without SPH after modifying the operation parameters for about a year. The benefits of the SPH system were as follows:
• The reduction in EAF power consumption was 38 kW / Tbt. (About 7 %)
• The reduction in tap to tap time was 6 minutes / heat. (About 6 %)
• Reduction in EAF electrode consumption was estimated at 0.2 Kg / Tbt. (About 5 %)
• Explosions during the charging process in the rainy season were avoided.
• The amount of carbon dioxide reduction per year is :
        4.5 Billion tons   (if electricity is generated from oil)
        5.8 Billion tons   (if electricity is generated from coal)

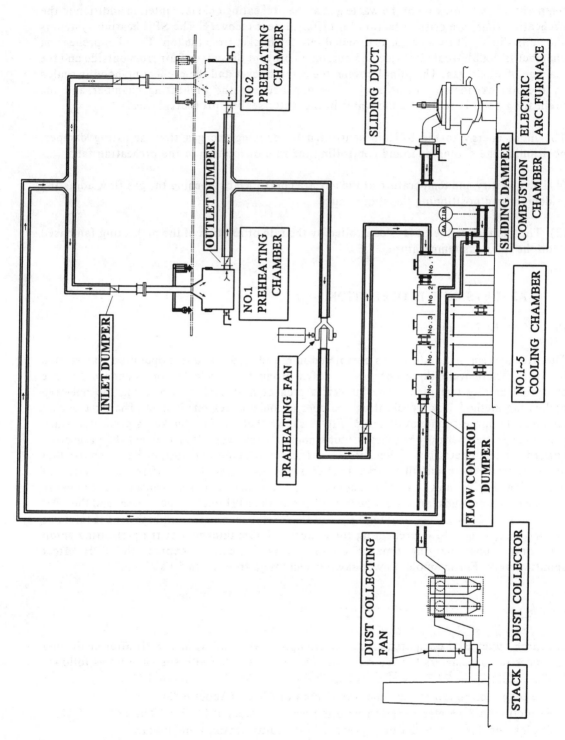

Fig. 1   **FLOW SHEET**

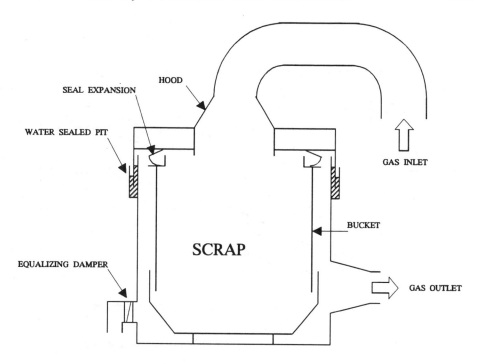

Fig. 2 PREHEATING CHAMBER

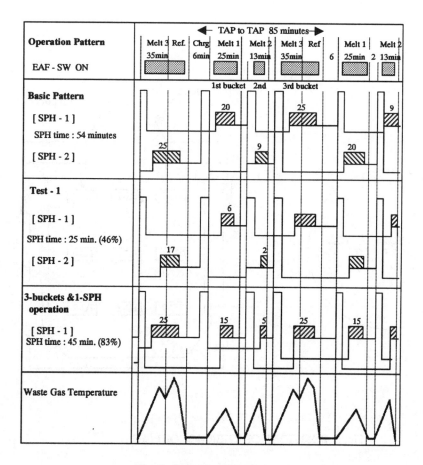

Fig. 3 Operation Pattern

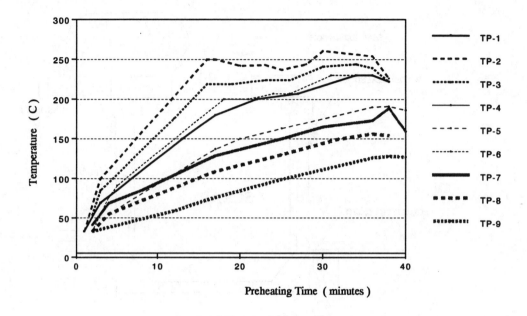

Fig. 4 SPH time v.s. Scrap temperature

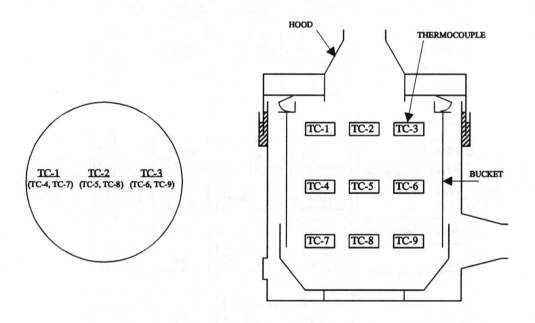

Fig. 5 POSITIONS OF THERMOCOUPLES

Pergamon

PII: S0196-8904(97)00011-3

Energy Convers. Mgmt Vol. 38, Suppl., pp. S655–S660, 1997
© 1997 Elsevier Science Ltd. All rights reserved
Printed in Great Britain
0196-8904/97 $17.00 + 0.00

CO₂ MITIGATION BY NEW ENERGY SYSTEMS

K. SAKAKI and K. YAMADA

Department of Chemical System Engineering, Faculty of Engineering
University of Tokyo, Hongo 7-3-1, Bunkyo-ku, Tokyo 113, Japan

ABSTRACT

Photovoltaic (PV) energy systems and power-generation systems using biomass are designed and evaluated in terms of energy, economics and CO_2 emissions. The calculated values of CO_2 emissions for dispersed and centralized PV systems are 5 and 15 g-C/kWh, respectively. Biomass gasification combined cycles (GCC) give lower CO_2 emissions of 3 g-C/kWh. These values are one or two orders smaller than those of existing thermal power plants. The evaluated electricity costs are $0.16 /kWh for dispersed PV systems, $0.29 /kWh for centralized PV systems, and $0.03-0.07 /kWh for GCC systems. Biomass can supply the most economical electricity, however, it requires 100 times more land than PV systems. © 1997 Elsevier Science Ltd

KEYWORDS

Energy Systems, CO_2 mitigation, Photovoltaic energy, Biomass, Gasification, EPT

INTRODUCTION

There are two technological methods to mitigate the increase in atmospheric CO_2 concentration. One is to reduce CO_2 emissions by energy-efficiency improvements (related to energy intensity) and/or fuel switching (related to carbon intensity), and the other is to collect and then immobilize atmospheric CO_2 in the ocean or on land. This paper focuses on the former method.

In Japan, energy efficiency has improved since the first oil crisis of 1973, especially in the industry sector. However, E/GNP (= primary energy consumed/GNP), which is the energy intensity index, has recently started to increase (Kaya, 1996). This suggests that it will be difficult in the future to reduce CO_2 emissions through only energy-efficiency improvements. Switching to non-fossil fuel will, thus, become more important. In particular, renewable sources of energy such as solar energy and biomass are the most promising non-fossil fuels, because they can meet a major part of the world's demand for energy in the long term.

Economic revenue over the short term is the most important factor in realizing renewable energy systems. However, energy consumption and environmental impact should also be taken into account for the long term planning of the system. In this paper, photovoltaic (PV) energy systems and power-generation systems by biomass are designed and evaluated in terms of energy, economics and CO_2 emissions.

PV SYSTEMS

In previous papers, we evaluated PV systems which were directly connected to the utility grid and included the balance of system (BOS) with supporting structure, inverter, and DC control device (Inaba *et al.*, 1993, Kato *et al.*, 1994, Yamada *et al.*, 1995). The results for the energy pay-back time

(EPT) and power-generation cost are shown in Table 1. The EPT and power-generation cost of the PV system using current technology are approximately 6 years and ¥600 /W (=¥90 /kWh), respectively. CO_2 emissions from the production of the PV system were also calculated at 60 g-C /kWh. These values appear to be too high. We, therefore, tried to reduce these values by improving the BOS.

Table 1.   EPT and power generation cost for PV systems using poly-Si solar cells.

Cell production scale (GW/y)	0.01		1		100	
Cell efficiency (%)	15		17		20	
	EPT (y)	Cost (¥/W)	EPT (y)	Cost (¥/W)	EPT (y)	Cost (¥/W)
Cell production	2.09	238	1.10	89	0.66	52
Module production	0.65	.106	0.40	69	0.35	58
BOS	2.93	203	2.56	156	2.24	124
Total	5.67	547	4.06	314	3.25	234

$1 ≈ ¥110

Application of PV Systems in Tokyo

PV systems with amorphous silicon (a-Si) cells may be used on roof tops or attached to walls. The total area of individual and apartment houses in Tokyo are 336 and 128 km², respectively. The roof area of individual and apartment houses and the wall area of apartment houses were calculated to be 42.8, 22.2 and 9.4 km², respectively. The EPTs and costs of the PV systems used on roofs and walls are shown in Table 2.

Table 2.   EPTs and costs for PV systems using a-Si solar cells in Tokyo.

	House roof		Apartment roof		Apartment wall	
Area (km²)	42.8		22.2		9.4	
Cell production scale (GW/y)	1	100	1	100	1	100
Maximum output (GW)	5.12	6.30	2.66	3.27	1.13	1.39
Electricity output (GWh/y)	5.12x10³	6.31x10³	2.66x10³	3.27x10³	6.44x10²	7.92x10²
Rate of demand of Tokyo	7.9%	9.7%	4.1%	5.0%	1.0%	1.2%
EPT (y)	0.64	0.49	0.75	0.56	1.19	0.89
Cost (¥/kWh)	26	18	29	21	45	32
CO₂ emissions (g-C/kWh)	5.9	4.5	6.9	5.2	8.1	5.9

Conversion efficiency of a-Si is 0.13 at the scale of 1 GW/y and 0.16 at 100 GW/y. $1 ≈ ¥110

The EPT, cost, and CO_2 emissions could be reduced by the use of light-weight BOS components and high-efficiency inverters. The CO_2 emissions of 5-8 g-C/kWh for PV systems is very low compared to the average value of 130 g-C/kWh produced by conventional systems in Japan. Roof-top PV systems could ultimately supply about 15% of the electricity required in Tokyo and reduce CO_2 emissions by $1.2x10^6$ t-C/y which corresponds to 12% of the total CO_2 emissions.

Application of PV systems in arid lands

A centralized PV system with poly-Si cells is constructed in a desert. Gibson desert (400 km x 840 km) in Australia where the insolation energy is 2,100 kWh/m²y, 1.8 times higher than in Tokyo, is selected for the power plant site. The electric power generated is transmitted 1,000 km to Perth. Batteries were installed for a stable power supply. Basic data used for the evaluation are shown in Table 3. We assumed that the modules produced in Japan were transported 7,000 km by a 50,000-t ship to Australia and then 1,000 km inland by 20-t trucks.

Table 3.   Calculation basis of PV systems in a desert.

Photovoltaic cell	Poly-Si (20 years Life)	Transmission efficiency	96 %
Efficiency of cells	20 %	Maximum power	145 GW
Cell production scale	100 GW/y	Efficiency of systems	90 %
Module surface area	4.72×10^8 m²	Area	860 km²
Module weight	2.86×10^6 t		

Energy input data for the production and construction of the BOS such as supporting racks, inverters, construction, electricity collectors, transmission towers, transmission wires and transformers were calculated according to the previous paper (Inaba *et al.*, 1993) and are shown in Table 4. The unit cost of inverter was assumed to be ¥500 /kg and that of the construction to be ¥1,000 /m² based on a labor cost of ¥10,000 /(man day). Inverters having a unit capacity of 18,000 kW were used and their unit cost was calculated to be ¥123,000,000 using a 0.8 power rule from the unit cost of a 3 kW inverter. Investment costs for transmission towers and transformers were calculated from reported values in Japan (MITI, 1992). A plantation was established around the power plant with a 100 m zone for a wind and sand break. Its cost was assumed to be ¥10,000 /ha. The calculated costs are shown in Table 4.

Table 4. Total annual energy input in terms of electricity (ITe), EPT and investment and electricity cost for PV systems using poly-Si solar cells. $1 ≈ ¥110

Maximum output	145 GW			Electricity output	1.10×10^{11} kWh/y	
	ITe (kWh/y)	EPT(y)	Rate	Investment (1000¥)	Cost (¥/kWh)	Rate
Supporting rack	2.61×10^{10}	0.24	14 %	3.27×10^9	4.4	4 %
Inverter	2.25×10^9	0.02	1 %	1.18×10^9	1.6	2 %
Construction	3.96×10^9	0.04	2 %			
Electricity collecting	1.01×10^8	0.00	0 %			
Transmission tower	1.41×10^9	0.01	1 %	5.00×10^8	0.7	1 %
Transmission	8.58×10^7	0.00	0 %			
First transformer	1.54×10^9	0.01	1 %	5.48×10^8	0.7	1 %
Final transformer	1.62×10^9	0.01	1 %	5.74×10^8	0.8	1 %
Module	6.95×10^{10}	0.63	37 %	8.75×10^9	11.8	12 %
Module transportation	7.57×10^8	0.01	0 %	6.33×10^6	0.0	0 %
Battery	7.80×10^{10}	0.71	42 %	5.91×10^{10}	80.5	80 %
Forestation				4.34×10^4	0.0	0 %
Total	1.85×10^{11}	1.7		7.39×10^{10}	101	

A power storage system is required to supply stable electric power from the centralized PV power plant . Here, a lead battery system was used. Electric power generated by the PV power plant was supplied through the battery system whose charge-discharge efficiency was assumed to be 0.87. The storage capacity of the system was assumed to be the power generated for two days considering weather conditions. The number of batteries needed was calculated based on 70% maximum capacity. The unit weight of a 2 V - 8,000 Ah battery without electrolyte was 498kg. The life time of the battery was assumed to be 10 years and its cost, ¥2,120,000 /unit (Komiyama *et al.*, 1995).

The EPT was calculated to be 1.7 years and the electricity cost to be ¥101 /kWh. As shown in Table 4, the transportation, transmission, civil construction and plantation did not strongly affect both the EPT and cost of the total system. The module, supporting rack and battery systems were the main cost factors. The battery system cost was as high as 80% of the total cost. In order to reduce costs, a sodium-sulfur (NaS) battery system instead of the lead battery system was investigated. The NaS battery system has a high energy utilization capacity of 100% of the maximum capacity and a high charge-discharge efficiency of 0.95. For the evaluation, a life time of 20 years and a cost of ¥200,000 /kW which is a target cost in Japan were used. The EPT and electricity cost of the PV system using NaS batteries were evaluated at 1.1 years and ¥32 /kWh, respectively.

Discussions

Roof-top PV systems are very effective for generating low CO_2 emission electricity and providing low electricity costs. However, the energy-conversion efficiency of a-Si should be increased from a present level of 8 to 16 % and the total system efficiency (mainly inverter) from 70 to 90%. Simultaneously, an annual production scale could be 1-100 GW which is 100-10000 times higher than the present level.

Centralized PV systems are also effective for electricity generation with low CO_2 emissions. However, the batteries required for energy storage raise the electricity cost remarkably and the result is that this type of electricity generation will not be economical even in the future. A partial solution would be to develop new energy-storage systems.

POWER-GENERATION SYSTEMS BY BIOMASS

The annual biomass consumption is estimated to be 880-1,080 MTOE (Million Tons Oil Equivalence, Yamaji et al, 1994). Biomass is mainly used as firewood, and the thermal efficiency is very low. So, the establishment of efficient biomass energy systems is very important for global sustainable development. Electric power generation by biomass is considered in this section.

Design of the Systems

The capacity of a power plant is assumed to be 100 MW. Fuel for the power plants is produced by the gasification of eucalyptus. Its heat of combustion is assumed to be 16.7 MJ/dry-kg. The annual growth rate of biomass in a plantation area is assumed to be 1.25 dry-kg/m^2. The harvesting cycle is 10 years. The area of the plantation site required for a 100 MW plant is calculated to be about 600 km^2 which depends on the type of fuel process.

A gasification process of biomass using a pilot scale fluidized-bed has been developed (Fujinami et al., 1991). We used their experimental data for our process design. The process flow diagram is shown in Fig. 1. Feed stock wood (25% Water) is fed to the drying step where water content is decreased to 10% using waste gas, and then the dried wood is forwarded to the gasification step operated at 850 °C. The heat of combustion of the product is 7.24 MJ/m^3. Steam turbine (GST) or combined cycle (GCC) was adopted as the electricity generation process.

Evaluation Results

The thermal efficiencies for electricity generation by each system were calculated based on the heat of combustion of the biomass (HHV) and are shown in Table 5. The GCC system shows a high efficiency of 30.5 %, and this value in very high compared with that (<20%) of current biomass systems due to an efficient heat recovery system.

The cost of biomass production was calculated using a rental cost basis for the plantation site in Thailand (Shinada et al., 1992). The rental cost of the land is assumed to be ¥24,060 /(ha y). The labor cost is assumed to be ¥1,500 /(day man). The investment cost of the processes and power plants are estimated based on Japanese costs. The investment cost of the processes are given in Table 5. The annual expenditure was assumed to be 12% of the total investment cost. The calculated electricity costs by each process are shown in Table 5 (¥10 /kWh in Japan). The result indicates that electricity generated by biomass energy systems is competitive to that of existing systems. The GST and GCC systems give a similar electricity cost.

Energy input and CO_2 emissions to construct the whole plant from the mining stage of the raw materials were calculated for steel and concrete using the weights of equipment. Unit energy consumption and CO_2 emissions are reported to be (1) 25.5 GJ/t, 0.59 t-C/t for steel, (2) 0.6 GJ/t, 0.05 t-C/t for concrete (AST, 1983). Herein, we used values three times higher than the above values considering the energy invested in the manufacturing of products from the raw materials.

Table 5. Thermal efficiencies, investment costs, EPTs and electricity costs of biomass energy systems of 100 MW scale. $1 ≈ ¥110

	GCC	GST		GCC	GST
Biomass conversion (%)	83.0	83.0	EIC (GJ)	2.7×10^5	9.2×10^5
Generation (%)	36.7	32.1	EOP (GJ/d)	200	200
Total (%)	30.5	26.7	CO₂ emissions		
Gasification process (¥10⁶)	480	480	for construction (t-C)	7,100	22,000
Boiler, turbine, others (¥10⁶)	5,639	4,037	for operation (t-C/d)	3	3
Total (¥10⁶)	6,119	4,517	Unit CO₂ emissions (g-C/kWh)	1.8	2.6
			EPT (d)	32	109
Electricity cost (¥/kWh)	3.4	3.3	Steel (t)	3,400	11,800
Electricity cost (¥/kWh)*	6.1	6.6	Concrete (t)	6,100	8,200

EIC: energy input for the construction, EOP: daily energy consumption for the operation.
* labor cost of ¥10,000 /(d man).

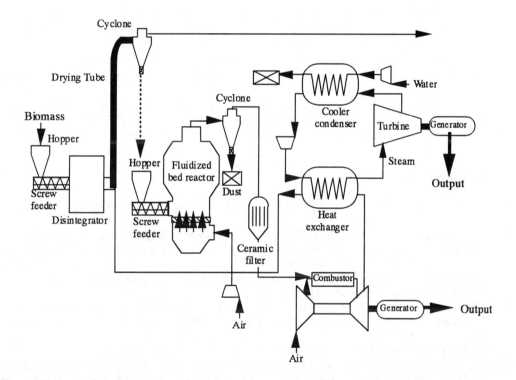

Fig. 1.   Process flow diagram of a biomass gasification-combined cycle (GCC). Composition of the product gas is N₂: 45 %, CO₂: 16 %, CO: 22 %, H₂: 8 %, CH₄: 6.5 %, and C₂H₄: 2.5 %.

Table 5 shows results on the weights of steel and concrete for power plants, CO₂ emissions and EPT. CO₂ emissions from biomass power plants are less than 3 g-C/kWh which are very low compared with those from a coal fired power plant (250 g-C/kWh) and somewhat lower than those from a photovoltaic (PV) system (10 g-C/kWh). EPT values are low, especially that of the GCC which is as low as 32 days based on the thermal energy. They are lower than that for a PV system (1 year), because the unit equipment sizes of the biomass systems are much larger than those of the PV system.

CONCLUSION

Table 6 summarizes the results obtained in this study. Biomass can supply the most economical electricity of the two types of systems examined. However, it requires 100 times more land than PV systems. If PV and/or biomass systems provided 20-30 % of the world energy supply, CO_2 emissions could be reduced by 5-8 % as shown in Table 6.

Table 6. Evaluation results on PV and biomass systems. $\$1 \approx \yen 110$

	PV		Biomass
	Dispersed (Roof-top type)	Centralized (with NaS-batteries)	GCC
Power generation per area (GWh/km² y)	147	128	1.3
EPT (d)	179	402	32
CO_2 emissions (g-C/kWh)	5	15	2
Electricity cost (¥/kWh)	18	32	3-7
Available capacity (TWh/y)	3.5x10³ [1]	8.4x10³ [2]	2.1-14x10³ [3]
Possible capacity (TWh/y)	2.4x10³ [4]	3.6x10³ [5]	3.6x10³ [5]
Possible CO_2 reduction (Mt-C/y) [6]	300	400	450
Rate of CO_2 reduction (%)	5.0	6.7	7.5

[1] based on the world roof area calculated using the population ratio of world and Tokyo
[2] based on the desert area, [3] based on possible biomass production
[4] 20% of the world electricity supply, [5] 30% of the world electricity supply
[6] based on the world mean value of CO_2 emissions (127 g-C/kWh) from existing power plant

Naturally, there exist large barriers towards realization of these new systems. Improvement of cell conversion efficiency and new energy storage system are necessary for the PV systems. A large scale sustainable plantation is necessary for the operation of biomass power plants.

REFERENCES

AST (Agency of Science and Technology) (1983). *Natural Energy and Generation Technology*. Taisei syuppansha, Tokyo.

Fujinami, S., S. Takashima, T. Nakamura, K. Nikawa, H. Hosoda, A. Deguchi and H. Takeuchi (1991). Fluidized-bed gasification of cellulosic wastes. *Ebara Jiho*, No. 151. 10-16.

Inaba, A., T. Shimatani, S. Tabata, S. Kawamura, H. Shibutani, K. Kato, T. Kakumoto, N. Kojima, K. Yamada and H. Komiyama (1993). Energy evaluation of solar photovoltaic energy systems. *Kagaku Kougaku Ronbunshu*, 19, 809-817.

Kato, K., K. Yamada, A. Inaba,T. Shimatani, S. Tabata, S. Kawamura, H. Shibuya, Y. Iwase, T. Kakumoto, N. Kojima and H. Komiyama (1994). Cost evaluation of photovoltaic energy systems. *ibid*, 20, 261-267.

Kaya, Y. (1996). Environmental management. *Hyojunka to Hinshitsukanri*, 49, 4-13.

Komiyama, H. and Research group on CO_2 and global environmental problems (1995). *Evaluation of Photovoltaic generation technology II*. The Society of Chemical Engineers, Japan, Tokyo.

MITI (Ministry of International Trade and Industry) (1992). *Outline of Power-Source Development*. MITI, Tokyo.

Shinada, Y., H. Matumura and I. Sakaguchi (1992). Reduction of atmospheric carbon dioxide by plants -forestation and marine plants and microorganisms-. *Denryoku Chuo Kenkyusyo Hokoku*, U92003.

Yamada, K., H. Komiyama, K. Kato and A. Inaba (1995). Evaluation of photovoltaic energy systems in terms of economics, energy and CO_2 emissions. *Energy Convers. Mgmt.*, 36, 819-822.

Yamaji, K., K. Okada, K. Nagano, E. Imamura, Y. Nagata, H. Yamamoto, T. Sugiyama and H. Honda (1994). World energy resources: endowments, supply/demand, economics, and related technology development. *Denryoku Chuo Kenkyusyo Hokoku*, Y94001, 73 pages.

Energy Convers. Mgmt Vol. 38, Suppl., pp. S661–S667, 1997
© 1997 Elsevier Science Ltd. All rights reserved
Printed in Great Britain
0196-8904/97 $17.00 + 0.00

Pergamon

PII: S0196-8904(97)00012-5

CO₂ MITIGATION THROUGH THE USE OF HYBRID SOLAR-COMBINED CYCLES

Y. Allani

COGENER, Parc Scientifique, EPFL
CH-1015 Lausanne, Switzerland

D. Favrat and M. R. von Spakovsky

Laboratoire d'Energétique Industrielle, Ecole Polytechnique Fédérale de Lausanne (EPFL)
CH-1015 Lausanne, Switzerland

ABSTRACT

The integration of a solar collector field generating steam into a conventional combined cycle in order to partially replace the fossil fuel required by the latter results in a substantial reduction in greenhouse gases, in an increase in the return on investments associated with the solar field and in an almost complete elimination of the need for solar energy storage. This paper discusses the design of such an integrated hybrid solar-fossil combined cycle with maximum daily and nightly power outputs of 88 MWe and 58 MWe, respectively. This cycle is currently being evaluated from a technical and economic risk feasibility standpoint for possible implementation as a pilot plant in Tunisia[1].

This paper outlines pertinent design considerations utilized in the thermoeconomic optimization approach employed for developing the hybrid combined cycle proposed here. The approach shows that there are several advantages to this type of design when compared with a purely solar steam cycle or any of the several other hybrid solar concepts which exist today. In addition to these advantages, the design presented revolves around the definition of a number of degrees of freedom which allow the solar energy part of the cycle to be highly integrated into the conventional part. A discussion of them is given.

Finally, from an environmental standpoint, the obvious advantage of this type of cycle is that due to the substitution of fossil fuel, there is a marked mitigation in CO_2 and NO_x emissions when compared to a conventional cycle and to other hybrid concepts. Pertinent results for these reductions are presented.

© 1997 Elsevier Science Ltd

KEYWORDS

integrated solar-fossil combined cycle; hybrid solar-fossil combined cycle; solar generated electricity; thermoeconomic optimization; CO_2 mitigation; pilot plant; Tunisia.

INTRODUCTION

It is obvious that increased reliance on renewable energies such as solar for the production of electricity has a high potential for CO_2 mitigation. However, the inherent variability and daily interruptibility of solar radiation are major hindrances to its broader economic use. Nevertheless, large scale thermal solar power plants have been shown to be the most economic of the solar technologies in use today (Cohen et al., 1995; Müller and Hennecke, 1993; Klaiss and Staiss, 1992).

The primary ways of compensating for these solar radiation drawbacks are either through the extensive use of thermal storage which turns out to be extremely expensive or through a coupling with fossil fuel equipment as proposed in this paper. This type of fuel though inherently the cheapest energy storage form known today

[1] This work is being done by COGENER and LENI-EPFL in collaboration with ELEKTROWATT (Zürich), KJC (Kramer Junction Corporation, Washington, D.C.) and STEG (Société Tunisienne de l'Électricité et du Gaz, Tunisia) and its industrial partners and is being financed by an agency of the Swiss government the DDC (Direction du Développement et de la Coopération, Berne).

must, nonetheless, be used with care primarily due to its associated environmental effects. When employed in connection with solar energy, fossil fuel must be utilized as effectively as possible in order to maximize the solar and fossil energy potential while minimizing the negatives associated with both energy forms. Hence the concept of the integrated hybrid solar-natural gas combined cycle under study by the authors (Allani, 1995; Allani, 1992; Allani and Favrat, 1991) and by others (e.g., Künstle (1994)) since the early 1990s. This concept guaranties a constant power availability and alleviates some of the perceived risks associated with the use of large solar concentrator fields. These characteristics provide the major arguments for convincing a large utility from one of the sun belt countries (Tunisia) to jointly investigate integration alternatives for a more *sustainable* power production through the increased use of indigenous energy sources and a progressive substitution of fossil fuels.

HYBRID COMBINED CYCLE CONCEPT

One of the integrated concepts proposed is shown in Fig 1. It consists of a field of cylindrical-parabolic concentrators feeding hot oil to a buffer tank and to an oil-steam boiler which is itself connected in parallel to the main steam boiler of the combined cycle. The combined cycle consists of two gas turbines (for reliability and for increased efficiency at partial load), one or two heat recovery steam generators and one steam turbine. A possible alternative design not yet mature consists of a partial generation of steam directly in the collector

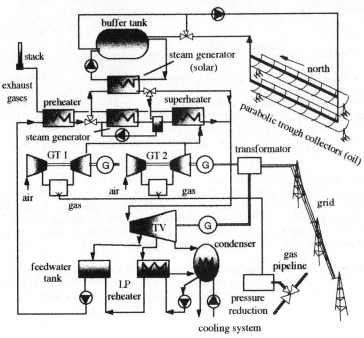

Fig. 1.   Proposed integrated hybrid solar-fossil combined cycle.

tubes (Lippke, 1994), thus, reducing the number of components and better exploiting the *exergy* potential of the working fluid. For larger plants, natural gas, oil or gas from coal gasification can be used as fossil fuel candidates. For smaller plants, biomass derived fuels comprise additional candidates with still further improvements in CO_2 mitigation. In these concepts, solar energy can only be considered as an additional energy source to the fossil fuel power plant. However, the concepts' advantage is that they provide a high solar conversion efficiency at a lower investment cost. Of course, care must be taken to not excessively penalize the fuel efficiency of the primary (or fossil) combined cycle.

The basic idea for the proposed integrated plant is to have a steam/turbine generator which is oversized when operated without solar input. As solar heating increases, the additional steam produced is fed to the turbine until its full capacity is reached. Fossil fuel consumption is then progressively reduced by decreasing the gas turbine load while keeping the steam turbine at full load.

It is clear from a thermodynamic standpoint that ideally one would use the solar energy directly in the toppinggas turbine cycle rather than in the bottoming steam cycle. However, present day technology for heating towers (judged not mature enough for heating compressed gases) as well as the inherent complexity of heater integration within standard gas turbines are major hindrances to the introduction of such advanced concepts.

THERMOECONOMIC DESIGN CONSIDERATIONS

The thermoeconomic design of such integrated hybrid plants is considerably more complex than for conventional plants since one must take into account variations in solar energy availability throughout the year as well as match it with the electricity rate structure. Solar energy availability for a typical site in Tunisia is shown in Fig. 2 above. It shows the cumulative duration expected for the various classes of solar radiation in a representative year and the corresponding usable energy per square meter. The choice of design radiation value is an important parameter in the thermoeconomic / environomic (von Spakovsky and Frangopoulos, 1996) assessment. If a design value too high is chosen, then the solar field will be small leading to correspondingly lower investments but an excessive reliance on fossil fuel throughout the year. Choosing a value too small results in excessive investments with the additional drawback of needing to defocalize the collector field to avoid overload on peak summer days.

Additional design parameters or degrees of freedom which are of importance are:

1. the number of gas turbines used, e.g., the use of two gas turbines provides a necessary flexibility in the operation of the cycle when solar energy is substituted for fossil;
2. a minimum partial load level for the gas turbines;
3. the level of operational losses associated with the combustion gases;
4. minimum and maximum non-nominal load levels for the steam turbines.

These design parameters are linked to two major operational strategies:

a) a maximum power strategy by which power is maximized in any given time interval without consideration for fossil fuel use or CO_2 emissions;
b) a maximum efficiency strategy whereby overall fuel efficiency is maximized in any given time interval.

The strategies used must be matched to national grid dispatching requirements, although, because of its inherently higher efficiency level, such a hybrid power plant is likely to benefit from a higher dispatching priority. With regards to CO_2 mitigation, another possibility for this type of plant is to stop operation during

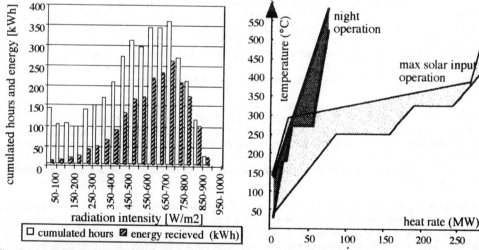

Fig. 2. Available solar radiation at a typical site in Gabes/Tunisia; duration and energy received are classed by radiation intensity.

Fig. 3. Pinch technology (temperature-heat rate) diagram for a dual pressure steam cycle design operating under the two extremes described in the text.

the lower rate hours of the night, resulting in an improved fuel efficiency at reduced cost penalties. The choice of any of the above strategies has a definite impact on the design values linked to the combined cycle itself. Figure 3 above illustrates, using a pinch technology (temperature-heat rate) diagram, the situation for a dual pressure steam cycle design operating under the following two extremes:

a) the composites for night operation with only the exhaust gases from the gas turbines generating steam;
b) the composites for a peak summer day with the combined heat of the exhaust gases and the solar heated oil generating steam.

Sliding steam pressure strategies are used to accommodate these various operating conditions at the steam turbine while verifying the law of cones for the steam turbine design (Kehlhofer, 1991).

PRELIMINARY RESULTS

Some optimal designs for several different plant configurations linked to the pilot plant project in Tunisia were determined using a quasi-stationary thermoeconomic approach which uses a hierarchical structure to determine system synthesis (configuration), component design (size and performance) and system operation in the following manner:

i) Two system configurations each with a steam cycle larger than that which would be found in a conventional combined cycle of the same size are evaluated. The amount of collector surface for the solar field which corresponds to each configuration is determined statistically based on the most often observed solar radiation events occurring during a typical year.

ii) For each configuration and size evaluated, two optimal command strategies are determined which maximize for each steady-state interval of operation (quasi-stationary approach) either the global electric power output or the global efficiency of the hybrid solar combined cycle.

iii) For each of the optimal command strategies, two operational policies are evaluated. They are based on an electricity rate structure which corresponds either to continuous (full time) operation or to operating only during the day and at peak nighttime hours, i.e. in Tunisia, the price of electricity is based on a daytime (middle), nighttime (low) and nighttime peak (high) rate structure.

Table 1 summarizes some preliminary results for two configurations with the same gas turbines but two different steam turbine sizes. Each column corresponds to the hierarchy outlined above: a particular configuration, optimal command strategy and rate-based operational policy. It is important to note that the numbers given result from an integration over time of steady-state operations for a typical year. For example, all plant simulations of the type shown in Fig. 4[2] are integrated over time and the results presented in Fig. 5 as monotonic curves. The integration is accomplished by determining levels of power production throughout the year along with the number of operating hours at each level and then in decreasing order determining the energy produced at each level in order to arrive at the curves shown in Fig. 5. The areas underneath the curves, thus, represent annual levels of electricity produced (solar and total). This type of approach is used, for example, by STEG not only for planning purposes but as a means for determining an operational policy for an existing plant. In this case, the monotonic curves are calculated on a specific basis.

Looking at Table 1 once more, the most economic plant corresponds to the smallest solar collector field and the longest operating time since the latter distributes the amortization over a larger energy output. However, the difference in ¢/kWh for the concepts examined and for the same number of operating hours is small. Also, note that the costs for the conventional part of the plant are based on Swiss prices and, therefore, somewhat elevated over the norm primarily due to civil engineering and a more restrictive regulatory environment. Thus, specific production costs could decrease by as much as 1 to 2 ¢/kWh in developing countries such as Tunisia.

In terms of the environment, CO_2 production for our hybrid concepts was compared to a conventional power plant with a 50% yearly mean conversion efficiency and the tons and percentage of CO_2 mitigated calculated. As shown in Table 1, the relative mitigation varies between 17% and 13% for continuous operation (left-hand-side of Fig. 4). These numbers can be pushed to 25% and more for a solar coverage of over 30% by stopping

---

[2] Note in this figure that during winter operations, the gas turbine operates at part load and for this reason can also act as an on-line backup system when insatbilities occur due to solar radiation fluctuations.

Table 1. Summary of preliminary results for the designs of the pilot plant project in Tunisia.

		67.5 MW steam cycle (overdim. by a factor of 4)				51.5 MW steam cycle (overdim. by a factor of 3)			
		max power strategy		max efficiency strategy		max power strategy		max efficiency strategy	
Operating time	[h/y]	7880	4930	7880	4930	7880	4930	7880	4930
ENERGY									
mean plant efficiency	[%]	60.2	67.1	60.3	68.9	57.	62.4	57.3	63.0
mean conventional efficiency	[%]	49.4	49.7	48.1	47.4	49.8	49.9	48.6	47.9
part of solar elec.on total production	[%]	18.0	26.0	20.3	31.2	13.4	20.0	15.1	24.0
solar field surface	$[10^3.m^2]$	344.2	344.2	344.2	344.2	244.4	244.4	244.4	244.4
specific gas consumption (day)	[-]	1.26	1.26	1.13	1.13	1.41	1.41	1.32	1.32
specific gas consumption (night)	[-]	2.04	2.04	2.04	2.04	2.02	2.02	2.02	2.02
specific gas consumption (mean)	[-]	1.66	1.49	1.66	1.45	1.74	1.60	1.75	1.59
annual solar electric output	[GWh]	98.5	98.5	98.1	98.1	70.1	70.1	69.8	69.8
annual conventional electric output	[GWh]	450.0	280.9	385.9	216.8	451.7	280.5	393.0	221.7
total annual electric output	[GWh]	548.5	379.4	484.0	314.9	521.8	350.6	462.8	291.6
CO$_2$ MITIGATION*									
CO$_2$ production by solar CC	$[10^3.t/y]$	192.3	119.3	169.4	96.4	191.5	118.6	170.6	97.6
CO$_2$ production by conventional plant (50% eff.)	$[10^3.t/y]$	231.5	160.1	204.3	132.9	220.3	148.0	195.4	123.1
annual CO$_2$ mitigation	$[10^3.t/y]$	39.2	40.8	34.9	36.4	28.7	29.4	24.8	25.4
annual CO$_2$ mitigation	[%]	17	25	17	27	13	20	13	21
CO$_2$ mitigation per installed collector surface	$[kg/m^2]$	114	118	101	106	117	120	101	104
ECONOMY**									
solar investment costs	[mio$]	105				74			
total investment costs	[mio$]	251				215			
invest./total installed capacity	[$/kW]	2303				2310			
annual invest. (20 years, 8%)	[mio$]	25.1	21.5						
operating and maintenance costs	[mio$]	5.70	4.01	5.70	4.01	5.37	3.66	5.37	3.66
gas consumption	[mio$]	7.84	4.87	6.90	3.93	7.81	4.83	6.95	3.98
specific electric. prod. cost	[¢/kWh]	7.04	8.96	7.79	10.49	6.64	8.55	7.30	9.99
electric cost with internalized CO$_2$ costs (2.42 ¢/kg CO$_2$)	[¢/kWh]	7.89	9.72	8.63	11.23	7.53	9.37	8.19	10.80
electric cost with internalized CO$_2$ costs (60 ¢/kg CO$_2$)	[¢/kWh]	28.08	27.83	28.67	28.87	28.67	28.85	29.42	30.01

(*)   The percentage in annual mitigation of CO$_2$ is calculated relative to a conventional combined cycle with the same total energy output as that of the hybrid solar combined cycle concept studied here.

(**)   The costs for the conventional part of the plant are based on Swiss prices and, therefore, somewhat elevated over the norm primarily due to civil engineering and a more restrictive regulatory environment

plant operation during nighttime hours (right-hand-side of Fig. 5). In Fig. 5, the solar electric power contribution is depicted by the dotted lines which monotonically correlate with the solar radiation availability.

It is interesting to note that even if the maximum efficiency strategy improves the yearly solar coverage figure, it does not significantly improve the yearly mean efficiency or the mitigation of CO$_2$. If the reduction in kilograms of CO$_2$ per square meter installed of collector surface is calculated, the maximum power strategy again beats the maximum efficiency strategy with 114 to 120 kg/m^2 for maximum power against 101 to 106 kg/m^2 for maximum efficiency.

Finally, the lowest specific CO$_2$ production corresponds to the highest electricity cost and even internalizing the CO$_2$ cost on the basis of 0.0242 $/kg CO$_2$ (Goswami, 1993) would not significantly change the comparative figures. It would require a considerably higher CO$_2$ tax to invert this trend. Note, however, that the cost for

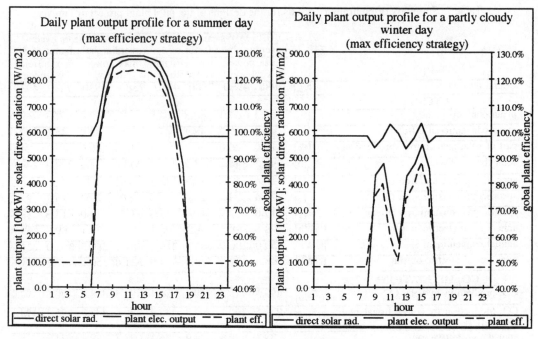

Fig. 4. Two examples for plant electric output over a single winter or summer day.

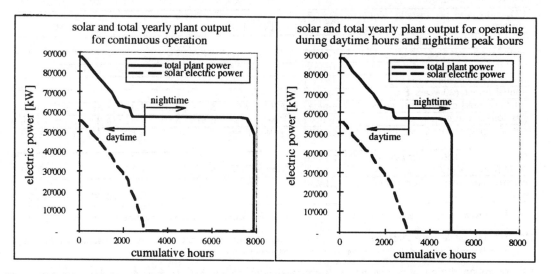

Fig. 5. Solar and total yearly electricity production at maximum efficiency and for two different rate-based operational strategies.

for CO_2 mitigation with hybrid solar-fossil combined cycle plants are only of the order of a few cents per kWh which is significantly cheaper than other CO_2 mitigation technologies based on renewable energies.

CONCLUSIONS

There are several advantages to the integrated hybrid solar-fossil combined cycle designs proposed in this paper when compared with a purely solar steam cycle or any of the several other hybrid solar concepts which exist today. The advantages are that

1. the conventional combined cycle part of our hybrid concepts operate at base load, guaranteeing a minimum of power. This leads to an increase in overall availability for the plant as compared to the other cycles mentioned above, to an increase in the return on investment for the solar part and to a better integration of the plant into the utility grid or network;

2. our hybrid cycle concepts have a higher dispatching priority from the standpoint of its annual average and peak efficiencies;

3. our concepts utilize the solar equipment better when compared to the other concepts mentioned above.

In addition, hybrid solar-fossil combined cycle plants have in general a high potential for CO_2 mitigation at reasonable costs. However, the strategies used in achieving reductions is important. For example, even if the maximum efficiency strategy improves the yearly solar coverage figure, it does not significantly improve the yearly mean efficiency or CO_2 mitigation. If the reduction in kilograms of CO_2 per square meter installed of collector surface is calculated, the maximum power strategy beats the maximum efficiency strategy with 114 to 120 kg/m2 for maximum power against 101 to 106 kg/m2 for maximum efficiency.

Finally, simple internalization of CO_2 production costs are unlikely to be sufficient for helping to identify the most *sustainable* path for achieving the major reduction targets which are likely to be needed in the future. More complex thermoeconomic / environomic optimization approaches (von Spakovsky and Frangopoulos, 1996) will have to be used.

REFERENCES

Allani, Y. (1995). PAESI Projet d'aménagement énergétique solaire intégré:Résultats majeures de l'étude de faisabilité technico-économique et environnemental d'une nouvelle centrale électro-thermo-solaire à cycle combiné bi-combustible, *Journées Internationales de Thermiques JITH '95*, Marrakech.

Allani, Y. (1992). A Global Concept of a New Type of Solar Combined Cycle Duel Fuel, *Proceedings of the 6th International Symposium on Solar Thermal Concentrating Technologies*, 9/28 - 10/2, v. II, p. 939-943.

Allani, Y. and Favrat, D. (1991). Concept Global d'une Nouvelle Centrale Solaire à Cycle Combiné Dual-Fuel, *Entropie*, 27, n° 164/165, pp121-122.

Cohen, G., Kearney, D.W. and Gable, R.G. (1995). Recent Improvements and Performance Experience at the Kramer Junction SEGS Plants, *Symposium: Solarthermische Kraftwerkw II (VDI)*, 9 - 12 Oct., Stuttgart.

Goswami, D.Y. (1993). Solar Energy and the Environment, *International Conference on Energy Systems and Ecology: ENSEC '93*, J. Szargut and G. Tsatsaronis editors, ASME, pp. 77-85, Poland.

Kehlhofer, R. (1991). *Combined-Cycle Gas & Steam Turbine Power Plants*, The Fairmount Press Inc.

Klaiss, H. and Staiss, F. (1992). *Solarthermische Kraftwerke für den Mittelmeerraum*, Bd. 2, DLR, Springer-Verlag.

Künstle, K., Lezuo, A. and Reiter, K. (1994). Solar Powered Combined Cycle Plant, *Power Gen Europ '94*, Cologne, 17-19 May.

Lippke, F. (1994). Numerische Simulation des Absorberdynamik von Parabolrinnen-Solar-Kraftwerken mit Direkter Dampferzeugung, *Energieerzeugung*, VDI-Verlag N°307, Reihe 6.

Müller, M. and Hennecke, K. (1993). Solare Farmkraftwerke und Direktverdampfung in Parabolinnen Kollektoren, *Forschungsverbund Sonnenergie*, pp 57-64, Themen 93/94.

von Spakovsky, M.R. and Frangopoulos, C.A. (1996). Environomic Modeling and Optimization of a Gas Turbine Cycle with Cogeneration, *Journal of Energy Resources Technology*, ASME, accepted for publication.

Pergamon

Energy Convers. Mgmt Vol. 38, Suppl., pp. S669–S672, 1997
© 1997 Elsevier Science Ltd. All rights reserved
Printed in Great Britain
0196-8904/97 $17.00 + 0.00

PII: S0196-8904(97)00013-7

MITIGATION IMPACT OF CO₂ IN DEVELOPING COUNTRIES.

K. BHAVAN-NA-RAYANA,

IAMMA, DAMAYANTHI CHAMBERS, Adarsh nagar, HYDERABAD-
pin. 500 063   INDIA.

ABSTRACT:

The developing countries facing different problem including increasing population and decrease of productivity. At this juncture, the increase of pupulation is one of the fundamental aspects of rising CO_2 emissions. At the outset to mitigate the impact of CO_2 in the developing countries are most problematic and mostly related to the public concern, which the public gives more importance to than the Government. Hence the policy to be adopted here to discuss in the event of involving the public body for crubing the issue. The Public Authority who should exercise the problematic things both from the Industrial and public side. Hence this aspect only sorted out by changing the policy by framing 5-10 years range by forcasting the upcomming issues in these countries. © 1997 Elsevier Science Ltd

The main constraints discussed here:

NO: I; The employment Vs. Industries with relation to the CO_2 utilisation.

NO: II; The constraint discussed here is the Population Vs. Food Vs. Mitigation of Co_2

Out of the above constraints the NO: I is concentrated in the Urban and semi-urban regions. The No: II constraint is mostly related to the Rural areas. However on studies as find that phytoremediation is best suited both constraints.

KEYWORDS:

Developing Countries, Population, Productivity, Public Authority.

METHODS:

Sampling was collected both in the urban and semi urban and also for the Rural areas of the India, Pakistan, China Phillipines etc., and studies carried therotically and in practically by approaching to the concerned authority and Public etc. The random of the influence as well as our study reveals for the policy on community basis and with Phytoremediation.

The Problem of population crubbed with Empployment-Industry-Food and CO_2 At this juncture the problems were analysed with Industry and also mitigation of the CO_2 . The more population which needs more employment which results increase of Industries in response to changing senerio of Economics, and Public needs. Most developing countries are thickly populated and have basic farming as the occupation for the population wherein the generation of income is below to $ 1000/- per year.

At his juncture, to survive the population usually migrates to the nearest urban or semi-urban areas. Further to this usually many agricultural labours who do not have land and smaller last holders will approach for the non-agriculture works due to growing expenditure both on farm as well as domestic.

Constraint I: Models collected from the different parts the countries found that more are less similar problems i.e, migrating the population in search of employment, in Industries. More increase of the industries leads 1. pollution, deforestration Food Secrecity, inflation, detring of the Standard of living. Out of all things again the increase of CO_2 and mitigation faces with socio-economic constraints.
Reasons: 1 Poor quality on the Enviornmental safety while building the industry, 2. population stays just behind to the Industrial belt, improper utilisation of land or Land grabbing for Industries, deforestration or spoiling of the Greenbelt.

Methods for remission with policy:

1. Proper utilization of Land.
2. Population to be educated properly on enviroment.
3. Specially referring to the Farmer must adequately advised from time to time about the Environment, and agriculture, Agro-forestry importance.
4. Train the farmer with moderated agronomic practices which is environments friendly to enable that the farmer to develop the farm to prevent the pollution in the urban and semi-urban areas besides the production of food.
5. Develop community basis approach and NGO for mitigation of CO_2 and similar gases.
6. Awareness of Nature importance on the daily life of Human being is one the main instruments.
7. Besides all the Government should allow Industry besdies having sufficient land and with full protection of the Environment free. Also the old industries should be shifted safe area by builting green belt.

8. A rule is necessary to plant a plant for each person.

9. Phytoremediation and green belt development is more important for every industry irrespective of size of it besides plantation in every house.

Constraint II : This is food Vs. population which mostly related to agrarian problem. Although it covers entire world, particularly to the Developing country relatively to be a major problems because of Average income of each family.

Population wherever thickly populated usually more of CO$_2$ because not only release by respiration, but also by consuming the fuels, running of vehicles, radiation exchanges Wastes - mostly household and substitution of envirofriendly.

To the result not only food problem but also leads for imbalance in weather conditions, results effect of Agrarian development. Besides this having increasing the population more houses required, for land grabbed from the Agricultural Land. Which results decrease in the Agriculture Land and imbalance weather conditions leads imbalance of food due low quality and quality / Food / grain production.

It is fact that although mostly the developing countries in tropical and substropical regions have ample scope of food production, but could not reach to minimum target and also poor quality results for Export Ban.

The policy to be adopted in this regard to be as follows:

1. Motivation of Rural population for decreasing or Zero population for certain period.
2. Community development with awareness
3. Participation of NGO's
4. Farmers to be trained with envirofriendly agronomic systems.
5. Rural based Agrobased industries to be developed
6. Frame a rule for planting a tree which adequate for Forestry and Food.
7. Stop deforestration
8. Ecofriendly Farming for Agricultural labours.
9. Awareness of standard of living without destruction of the Nature.

CONCLUSION:

Phytoremediation is best suited and recommended for this purpose. However the population control and Standard of living also to be improved by adopting the community development by framing a rule by the government.

Ref:

1.   Environfriendly Report - 1992 ., IAMMA -
     (Published by IAMMA)
2.   Narayana etal, Farm modles., IPI cooq. 1985 -
     page - 22 - 23.
3.   Vandana   biodiversity IET   1996 -
     page - 120 - 122.
4.   We Zheng   wind sand   science print Beijing.-
     page - 12 - 13.

Energy Convers. Mgmt Vol. 38, Suppl., pp. S673–S678, 1997
© 1997 Elsevier Science Ltd. All rights reserved
Printed in Great Britain
0196-8904/97 $17.00 + 0.00

Pergamon

PII: S0196-8904(97)00014-9

CO2 RELEASE OF MAIN INDUSTRIES IN CHINA: SITUATION AND OPTIONS

MA   JIANG

National Research Center for Science and Technology for Development
State Science and Technology Commission, China

ABSTRACT

This paper attempts to ascertain the overall situation of CO2 pollution of China's major industries by analyzing the status quo of CO2 and SO2 emission at major industries and several characteristics in China. The current approaches and options to reduce the emission volume of CO2 will be presented in the paper as well. © 1997 Elsevier Science Ltd

SITUATION

The Characteristics of CO2 Emission Situation at Major Industries of China:

The industrial CO2 emission volume centralizes in several major industries of China:

Table 1.   CO2 emission volume by regions in China (1990)

regions	sources of emission						order
	coal	oil	cement	nitrogen fertilizer	natural gas	total (Mt/a) (according to carbon)	
Liaoning	160.16	36.32	3.5926	0.5283	3.85	214.83	1
Shandong	140.83	31.11	6.3836	1.2081	2.72	204.50	2
Hebei	152.84	11.48	4.3103	1.0935	0.56	186.42	3
Sichuan	115.80	5.64	4.4478	1.2482	12.46	167.91	4

Heilongjiang	126.48	22.48	1.5553	0.3270	4.24	164.36	5
Shanxi	148.65	4.07	2.0150	0.5765	0.11	163.02	6
Henan	118.37	9.78	3.8786	1.1135	2.61	158.42	7
Jiangsu	108.43	21.55	5.0432	1.0642		153.82	8
Guangdong	52.12	32.39	6.8133	0.4206		108.37	9
Hunan	68.93	8.23	3.2958	0.8877		97.40	10
Jilin	77.92	10.30	1.2355	0.2844	0.18	97.06	11
Hubei	58.25	16.49	3.2472	0.8343	0.14	93.21	12
Inner Mongolia	76.72	2.89	0.7500	0.1277		86.16	13
Anhui	59.73	7.13	2.9137	0.8037		85.29	14
Shanghai	47.78	24.49	0.7577	0.2350	0.07	76.83	15
Zhejiang	43.32	9.82	4.4095	0.5608		69.03	16
Shannxi	52.95	4.03	1.7445	0.3547	0.01	67.78	17
Beijing	46.83	16.35	1.1152	0.035	0.08	67.31	18
Guizhou	47.20	2.56	0.9156	0.2966	0.05	59.58	19
Jiangxi	39.48	4.09	1.5434	0.1764		55.27	20
Yunnan	38.23	2.95	1.5487	0.4941		53.02	21
Xinjiang	35.61	10.15	0.9376	0.2673	0.94	51.92	22
Tianjing	34.70	13.05	0.4018	0.0609	0.63	51.16	23
Gansu	36.06	7.58	1.1791	0.1911	0.06	50.98	24
Guangxi	27.22	3.21	2.5178	0.2288		44.34	25
Fujian	22.77	3.75	1.7767	0.3362		36.59	26
Ningxia	17.18	1.36	0.3150	0.2438	0.01	20.34	27
Qinghai	9.14	1.52	0.1692	0.0025	0.10	12.00	28
Hainan	1.18	0.75	0.1408			3.81	29
Tibet			0.0435			0.62	30
total	1964.88	326.15	68.998	14.0509	28.81	2701.45	
amount to carbon	535.82	88.94	18.80	3.84	7.86	655.26	
(%)	81.80	13.6	2.8	0.6	1.2	100	

Source: "Research of Environmental Sciences" of China

From the Table 1, the top eight are the following provinces in China: Liaoning, Shandong, Hebei, Sichuan, Shanxi, Henan, Jiangsu, Guangdong;　Table 2 lists the sequence of products at different industries related to CO2 emission by regions in China in 1994.

Table 2　　　　Sequence of industrial products by regions

caustic sode (10000tn)	electricity (100000000 kwh)	iron & steel (1000tn)	cement (10000tn)	plateglass (10000 wht. cases)	sulfuric acid (10000tn)	soda ash (10000tn)	chemical fertilizers (10000tn)

Shandong	Shandong	Liaoning	Shandong	Henan	Sichuan	Shandong	Sichuan
51.31	678	2614	5068	1604	153.73	117.31	230.66
Jiangsu	Jiangsu	Shanghai	Guangdong	Hebei	Jiangsu	Liaoning	Anhui
42.99	632	2275	5019	1541	149.59	73.91	196.03
Shanghai	Hebei	Hebei	Jiangsu	Liaoning	Hubei	Hebei	Henan
38.18	550	1760	3087	1299	107.68	69.84	189.70
Liaoning	Liaoning	Shanxi	Henan	Shandong	Guangdong	Tianjin	Shandong
30.52	504	1674	3004	766	98.10	62.82	166.26
Tianjin	Henan	Hubei	Zhejiang	Guangdong	Liaoning	Jiangsu	Hebei
29.28	485	1346	2698	717	95.84	55.61	157.54
Sichuan	Shanxi	Sichuan	Hebei	Shanghai	Shandong	Sichuan	Jiangsu
28.36	457	1258	2687	665	89.69	49.51	153.09
Zhejiang	Shanghai		Sichuan	Heilongjiang	Hunan	Hubei	Hebei
22.9	399		2487	557	84.05	29.53	151.03
	Heilongjiang		Hunan	Beijing	Yunnan	Henan	Hunan
	382		1997	530	80.74	25.39	127.69

Source: China Statistical Yearbook 1995

It is obvious that, the CO2 emission by major industries in China is mainly concentrated in the industries of thermal electricity, iron & steel, chemicals and building materials. Based on the related statistics, it was estimated that the CO2 emission volume of these four industries occupied around 70% of that of all industries in China at present.

Mixed pollutants with high-positive coherency from the emission of CO2, industrial waste gas and SO2:

Table 3. Sequence of industrial emission by regions (1994)

CO2	Liaoning, Shandong, Hebei, Guangdong, Jiangsu, Henan, Sichuan, Shanxi
SO2	Shandong, Sichuan, Jiangsu, Hebei, Liaoning, Shanxi, Shannxi, Guizhou
industrial waste gas	Liaoning, Shandong, Hebei, Guangdong, Jiangsu, Henan, Sichuan, Shanxi

In Table 3, in the sequence of emission volume of CO2, SO2 and industrial waste gas, Liaoning, Shandong, Hebei, Sichuan and other provinces are in the front rank. This shows that the emission volume of CO2 and the emission volume of industrial waste gas and SO2 presents a direct ratio with high coherency.

Positive coherency of CO2 emission volume and coal consumption:

Table 4. Total energy consumption and coal consumption by main sectors

sector	total energy consumption (10000tn SCE)	coal consumption (10000tn)
Total Consumption	122736.85	128532.27
* industry	85853.40	107770.00
non-metal mineral products	12556.14	12218.90
smelting and pressing of ferrous metals	15338.62	11548.99
raw chemical materials and products	16196.26	9643.96
petroleum processing and coking products	3590.94	5477.61
electric power, gas and water production & supply	6598.24	40309.74
textile	3439.32	2532.54
smelting and pressing of non-ferrous metals	2555.12	1509.61
paper-making and paper products	1981.10	2013.77
others	23597.66	22514.88

Source: China Statistical Yearbook 1995

In Table 4, the top four industries of coal consumption in China are electric power, chemicals, building material and iron & steel, which shows the direct correlation between CO2 emission volume in China and coal consumption volume of the different industries. Based on analysis, the coal consumption volume of these four industries occupied 73.5% of that of all industries in China. This result is highly coincident with the estimation of CO2 emission volume of these four industries constituting around 70% of that of the whole, which presents a close connection.

CAUSALITY BETWEEN SITUATION AND OPTIONS:

Three characteristics of CO2 emission of major industries in China actually indicates the three key points for reducing CO2 emission in the country:
------Taking top four industries for CO2 emission as major industries to control its emission by adopting various methods;
------Considering and adopting comprehensive measures to reduce or control the industrial CO2, SO2 and waste gas, which are interrelated.
------Gradually changing the energy structure of China, decrease coal consumption volume and save coal energy for reducing CO2 emission in China.

CURRENT APPROACHES AND OPTIONS:

As a developing country, China has started industrialization and serious pollution in the country is an actual fact. To face this situation and control air pollution, China combines resources and efficient use of energy as the guiding principle. One fact is that it is not realistic to change coal as the main energy source within twenty years, which is decided by China's natural resources and economic conditions. So it becomes very important and more practical for China to conduct energy saving, efficient use of energy and new energy exploration, especially in the short term. Following are the main countermeasures for China:

Major Regulation:

------ "China's Act of Prevention for Air Pollution" and its concrete principle;
------ "Principle of Pollutant Prevention within certain period";
     For the serious pollutant sources --- the enterprises with bad pollution ---, the Environmental Protection Bureau at different levels has the power to conduct coercive measures, force these enterprises preventing and curing pollution before a deadline, close down large amounts of small-sized enterprises that are heavy polluters in the country;
------ Total amount control system of waste gas emission.
------ Emission standard of industrial dust;
------ Environmental quality standard of air;
There are 38 standards of air environmental protection in China.

Technological Approaches:

------ Central heat supply by adopting various sources including the after-heat from different industries;
------ Developing fuel gas in urban areas;
     use of natural gas;
     reclaim and use of combustible gas from enterprises especially iron & steel plants;
     - reclaim and use of gas from coal mine;
     use of liquefied petroleum gas;
     developing properly producing gas from heavy oil;
     developing actively producing gas from coal;
------ Clean production; etc.
     This is the most practical way for the industrial enterprises in China to control the pollution in the whole process of industrial production which emphasizes energy saving and consumption decrease by technology and macro-management. It could control the output of wastes and the emission of wastes to the minimization.

Economic Measures:

------ Polluter pays principle;

------ Raising emission charge for waste gas;

------ Emission charge of overproof;

------ Emission charge of SO2; etc.

INTERNATIONAL COOPERATION:

For global environmental issues, each country in the world should take a positive stance and be responsible for preserving the environment of the earth. Since developing countries are short of economic and technological strength, developed countries could help developing countries with many aspects to raise the capability of developing countries. Developed countries could assist:

------ To provide technologies in the fields of energy saving, use of new energy, and new technologies to reduce CO2 emission to developing countries;

------ To provide more financial support to conduct R & D on CO2 & SO2 removal, technical innovation in industrial enterprises, developing new energy;

------ To train professional personnel, etc.;

CONCLUSION:

For China, it is vital to control the emission volume of CO2 and air pollution by enhancing the technological innovation and management to the main industries in China. China is under its way to raise the effective use of energy and develop new energy which is more practical for China to reduce gradually the proportion of coal not only in main industries but also in other industries in the country. Various assistance from developed countries would help a great deal to developing countries, it is the only way to better the environment of the world by the common efforts from both developing and developed countries.

REFERENCES:

China Statistical Yearbook (1995).   National Statistical Bureau, China.

China Environmental Statistical Yearbook (1995). National Environmental Protection
    Agency of China

Li lei, Fang Zhi, Zhu Gang, Qiao Qi.   Development Strategy and Options for Controlling Air
    Pollution. China's Environmental Protection

Liu Yao Qi. Situation Analysis and Option Study on China's Industrial Pollution.
    China's Environmental Management

Industrial Development and Air Pollution in China.   National Research Center for Science
    and Technology for Development, China.

Energy Convers. Mgmt Vol. 38, Suppl., pp. S679–S684, 1997
© 1997 Elsevier Science Ltd. All rights reserved
Printed in Great Britain
0196-8904/97 $17.00 + 0.00

Pergamon

PII: S0196-8904(97)00015-0

TRENDS IN EVOLUTION OF CO_2 EMISSIONS IN ROMANIA AND PERSPECTIVES FOR DIMINISHING THEIR ENVIRONMENTAL IMPACT

L. Dragoş, N. Scarlat and C. Flueraru

ICPET CERCETARE

236 Vitan Street Sector 3, 74369 Bucharest, Romania

ABSTRACT

The paper presents the activities of Romanian specialists concerned with CO_2 emissions arising from human activities, primarily the main factors responsible for the rapid climate changes. The first section shows the CO_2 emissions' level for the last years, related to European emissions, concurrently with a comparative presentation of main sources and their contribution to the global CO_2 emissions in Romania. Furthermore the paper presents the prognosis for the CO_2 emissions' evolution in Romania up to 2020. Further, the research activity performed in Romania and the results obtained so far are presented for the development of a suitable CO_2 retention technology to reduce CO_2 emissions from flue gas generated by power plants. The paper also highlights the possibilities for further utilization in industry of the CO_2 retained through application of the technology developed by ICPET-CERCETARE for CO_2 retention.

© 1997 Elsevier Science Ltd

KEYWORDS

CO_2 emissions, CO_2 sources in Romania, chemical method.

INTRODUCTION

Global warming has become a problem that concerns the entire scientific world. If any important modification at greenhouse gas emissions will occur, it is foreseen to greatly increase the concentration of these gases in the atmosphere, with a large implication for the environment. Carbon dioxide is the most abundant and the most important of the greenhouse gases, with a very dynamic emissions growth in the last decades. The CO_2 concentration increased by about 25% in the last two centuries and by about 10% from 1958. As a result of the concerns in the world regarding climate changes, over than 150 nations signed the United Nations Framework Convention on Climate Change in Rio de Janeiro in 1992. This Convention, with the aim of stabilizing the concentration of greenhouse gases in the atmosphere at a level that prevent dangerous interference on the climate, has been ratified by Romania too. For Romania, as a country in transition to the market economy, the commitment for reducing the greenhouse gas emissions is quite flexible, regarding the reference level, 1989 being agreed by the Romanian Ministry for Environmental Protection as the reference year.

CO₂ EMISSIONS SOURCES

The development of the Romanian economy after The Second World War up to 1989, has lead to a continuous increase of the primary energy consumption and, as a consequence, to a continuous CO_2 emissions growth, that reached the peak in 1989 (fig. 1). After 1989, as a result of the serious reduction in the economic activity, a CO_2 emission' reduction, about 40%, took place up to 1994. The 1995 represents the beginning of an economic growth, passing over the minimum level from 1994 of the carbon dioxide emissions in the atmosphere.

An CO_2 emissions increase in Romania it is foreseen for the next years, as a result of the expanding economic activity in our country. So, it is estimated that up to 2000, the emissions level will reach about 200 mil t CO_2 / year, about 80% of the emissions level from 1989. The emissions level from 1989 will be reached approximately in 2015, and in the 2020 it will be about 260 -270 mil t CO_2 / year.

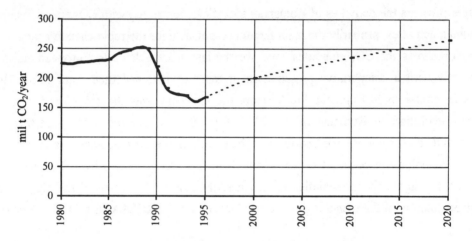

Figure. 1. Carbon Dioxide emission evolution in Romania

The primary energy sources consumed in Romania consist mainly on natural gas, (40%), oil (28%), coal (23%), hydraulic energy (6 %) and other sources. The CO_2 emissions in the atmosphere are coming especially from fuel combustion for electric and thermal power production and as a result of industry activity. Other important sources of the CO_2 emissions are represented by the transport sector and households (fig. 2).

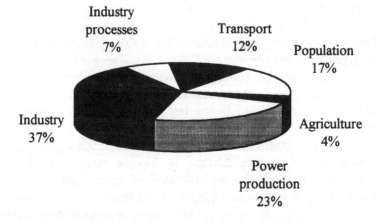

Fig. 2. CO₂ emission sources in Romania

From the overall power energy produced in Romania, about 39% comes from coal-firing power stations and 31% from hydrocarbons-fired power stations. An important portion of electric power (30%) is produced in Romania in hydroelectric stations, from the overall hydroelectric potential estimated to be 40 TWh, about 17 TWh representing the installed capacity of hydroelectric stations. At the same time, an important characteristic for Romania that must be taken into account is the development of district heating using hydrocarbons (66 %) and coals (34 %) in particular as fuels.

Among the most important sectors for CO_2 emissions generated in industry, there are the chemical industry with about 26 %, metallurgy (17 %), fuel processing (12 %), machine construction (10 %), ore extraction and preparation (8 %) and others (fig. 3).

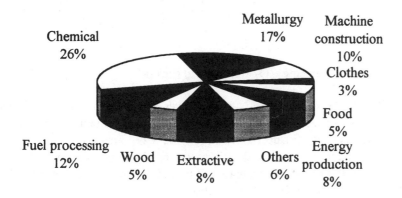

Fig. 3. Sources of CO_2 emissions from industry

Comparing CO_2 emissions levels, it could be noticed that Romania (fig. 4) has relatively high values compared with other European countries (Fontelle, 1995). Relating the specific quantity for annual CO_2 emissions per inhabitant (fig. 5), it could be seen that Romania is among the countries with the least CO_2 emissions. This is due especially to the reduction of economic activity in the last years and the high contribution of hydroelectric power in power production.

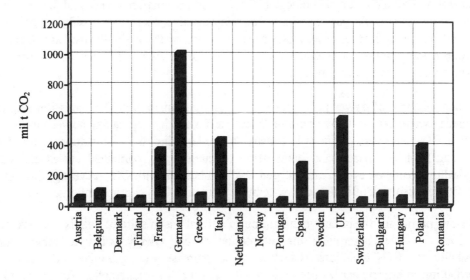

Figure 4. CO_2 emissions in Europe

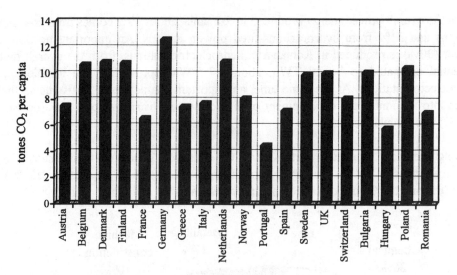

Figure 5. Specific CO_2 emissions in Europe

During 1996, the first unit of the 5×700 MWe Nuclear Power Station from Cernavodă, was brought into operation and it is foreseen that by the 2000 year the second unit also will be in operation. This will lead to the significant reduction of the CO_2 emissions, taking into consideration the high share of nuclear power in electric power production in Romania. Due to this fact, the CO_2 quantity released in the atmosphere will be reduced by about 4.5 mil t CO_2 per year up to 2000 and by about 9 mil t CO_2 after the 2000.

CO_2 EMISSION REDUCTION POSSIBILITIES

Among the main sources of carbon dioxide in Romania are the electric and thermal power plants. For this reason the main effort of the Romanian programme was focused on evaluating the possibilities for reducing the CO_2 emissions generated in these plants.

Over the world there are technologies for power production that could be important for reducing the greenhouse gas emissions. For this objective, there are considered the increase of plant efficiency, utilization of less CO_2 fuels, increase of the hydroelectric and nuclear power share in electric power production, the utilization of renewable sources and a more efficient utilization of energy by the consumers.

The improvement of power plant efficiency could determine high CO_2 emission reductions in Romania, where heat and power are produced in old and low efficiency power plants, resulting in high CO_2 emissions. This will consist in modernizing the old stations from Romania, switching from pulverized coal combustion to natural gas, combined cycles and integrated gasification combined cycles, and oxygen-coal combustion with CO_2 recycling, this will lead to high CO_2 concentrations in flue gas, with more attractive CO_2 capture options.

A serious barrier for the intensive utilization of natural gas in spite of coal is the lack of natural gas sources, which determines the orientation of our energy policy to other sources. A possibility for CO_2 emissions reduction is to increase the hydraulic and nuclear energy contribution to electric energy production. Utilization of hydroenergetic potential requires high hydrotechnical works however, and covers basically the peak energy load (Popescu, 1996).

CO$_2$ EMISSIONS REDUCTION BY APPLYING CAPTURE TECHNOLOGIES

Another possibility to reduce the CO$_2$ emissions is the CO$_2$ recovery from flue gas generated during fuel combustion by an adequate technology, and the subsequent utilization of CO$_2$ captured from technological processes in industry (petroleum, chemistry etc.), for biomass production or storage.

Considering the importance of lowering the greenhouse gas emissions, ICPET CERCETARE has begun in the last years an exploitative research programme for recovery the CO$_2$ generated by fuel combustion for thermal and power production. The research work included studies for elaboration of a CO$_2$ retention method and laboratory tests on a CO$_2$ removal laboratory scale facility, and it will continue on a large scale with experiments on a pilot facility. The research will be finished by identifying the appropriate possibilities for utilization and storage of the recovered CO$_2$. The aim of the program is to develop a suitable technical and economic method for CO$_2$ recovery from flue gas, to be implemented in the local industry.

From a number of carefully evaluated options, the chemical absorption appeared to be the most adequate and attractive for the local conditions from the economic and technical point of view for CO$_2$ recovery from the flue gas, in spite of large energy requirement for regeneration. The main efforts have been directed toward the evaluation of CO$_2$ chemical absorption efficiency in different aqueous solutions such as ethanolamine and hot potassium carbonate. The experimental works established the possibility for obtaining high CO$_2$ recovery using ethanolamine, more than 80 % at low liquid per gas ratios (1.5 - 2 l solution per STP m^3 gas), for CO$_2$ absorption by ethanolamine (Dragoş, 1992, 1995).

Experiments sought CO$_2$ removal only, not the CO$_2$ disposal and utilization to find a clean recovery process for an eventual further utilization in industry. The laboratory scale experiments aimed to select the most adequate absorbent to perform CO$_2$ recovery by chemical absorption, with high absorption efficiency at lower CO$_2$ partial pressure and high carrying capacity, to reduce the absorbent rate, easy regeneration and lower energy consumption for regeneration, non inflammable and non poisonous. On the pilot plant that will be in operation at the end of 1996 specific experiments on a larger scale will be performed to establish the main parameters and clarify some of the technical problems of this technology for CO$_2$ recovery from flue gas.

An important aspect of CO$_2$ capture is the energy required for CO$_2$ recovery, that reduces to a certain extent the energy generating efficiency, increasing the primary energy used. Thus, the CO$_2$ amount retained by different methods is diminished by using a higher amount of energy. At the same time, the long term consequences on the environment have to be carefully considered. Supplementary effort is required for identifying the possibilities for CO$_2$ capture and storage.

CO$_2$ UTILIZATION

Captured CO$_2$ could be used on a large scale in chemical industry. This option offers the advantage of using this gas and replacing other gases produced in different processes. CO$_2$ could be used in different ways in the food industry and petroleum industry, though in most cases it does not remain in products and it is released quickly in the atmosphere. CO$_2$ could be used for manufacturing chemicals, enhanced oil recovery, plants and algae culture.

Utilization of captured CO_2 for chemical production is attractive as long as the supplementary energy required is low. The technical and economical aspects of the process have to be evaluated also taking into account the impact on CO_2 emissions reduction. However, there is low interest now in using CO_2 by this way, because most of the CO_2 used on a commercial scale comes from natural sources and from different processes as by-product.

In Romania now, only a small amount of CO_2 could be used in the chemical industry. This small quantity when related to the overall emissions, could be increased by using in new developed chemical processes as a part of a strategy toward the goal of diminishing the CO_2 emissions.

CONCLUSIONS

The CO_2 emissions in Romania showed a continuous growth towards 1989 when it reached the maximum level. After 1989 and up to 1994 an important CO_2 emissions reduction took place due to decreasing economic activity. According to the predictions, over the next few years a growth in CO_2 emissions will be noticed, that will represent in the 2000 about 80 % of the 1989's emissions, the 1989 level being reached probably in 2015.

Implementing new and modern technologies for power production, the improving efficiency of existing plants, and moreover, using renewable sources, biomass, hydraulic and nuclear energy, solar and wind energy and other non CO_2 energy sources for power generation, could lead to clean and efficient production of energy with lower CO_2 emissions.

The Romanian research programme, developed for the aim of identification of methods for CO_2 emissions reduction, established the possibility to apply the chemical absorption technology using ethanolamine, that appears to be the most suitable for Romania. The experimental work already done allowed to obtain high CO_2 recovery efficiency, over 80 % at low liquid per gas ratios. The completion of the research Programme will allow to obtain partial reduction of CO_2 emissions from power plants by CO_2 retention for further utilization in industry.

CO_2 capture and disposal technology represent a method for reduction the CO_2 emissions from fuel combustion. There are more aspects to be clarified referring to the cost reduction for CO_2 capture. In the same time, those activity fields where the CO_2 could be used in large quantities have to be identified. Regarding the disposal of the captured CO_2 the impact and safety of disposal over the long term has to be estimated, beside the costs for implementation.

REFERENCES

U.S. Agency for International Development Washington, D.C. Global Climate Change: The USAID Response. *A Report to Congress*, June 1994.

Dragoș L. I., Nadă O., Flueraru C., Scarlat N. (1992, 1995). *Researches on CO₂ recovery from flue gas. Internal Reports*, Bucharest.

Fontelle J. P. (1995). Greenhouse gas emission inventories in Europe -CORINAIR 90 and 1994. *Proceedings of the International Energy Agency Greenhouse Gases: Mitigation Options Conference*, London, 783-788.

National Commission for Statistics (1995). *Romanian Statistical Yearbook*, Bucharest

Popescu A., Adler S., Ion C., Popovici D., Teodorescu S., Popescu M., Breazu F. (1996). Alternatives for minimizing the carbon dioxide emissions in the power and thermal sector.

Energy Convers. Mgmt Vol. 38, Suppl., pp. S685–S689, 1997
© 1997 Elsevier Science Ltd. All rights reserved
Printed in Great Britain
0196-8904/97 $17.00 + 0.00

Pergamon

PII: S0196-8904(97)00016-2

NON-CO2 EMITTING RENEWABLE ENERGY SOURCES IN NEPAL: PROBLEMS AND PROSPECTS

Jagan Nath Shrestha*
Toshinori Kojima**

* Department of Electronics Engineering, Institute of Engineering, Tribhuvan University, Pulchowk, Lalitpur, GPO Box 1175, Kathmandu, NEPAL.

** Department of Industrial Chemistry, Seikei University, Musashino, Tokyo 180, JAPAN

ABSTRACT

The renewable energy sources of Nepal is highlighted. The huge hydro-power potential of Nepal is explained. The problems associated with its wide exploitation are given. Nepal's contribution to the global greenhouse effect through the emission of CO_2, CH_4 and CFSs from industrial emissions(cement, brick, and other factories), automotive vehicles, indoor biomass combustion, outdoor biomass combustion, petroleum products consumption and fuel based thermal plant generating 85 GWh is estimated. Equivalent CO_2 saved by hydro and PV power is analyzed. Per capita consumption of CO_2 is mentioned. Early precaution needed to be taken in reducing CO_2 emissions from different sources is highlighted. © 1997 Elsevier Science Ltd

KEYWORDS

Industrial emissions, deforestation, hydro-power, renewable energy sources.

INTRODUCTION

Nepal is the second richest country possessing about 2.27% of the total world water resources after Brazil (Water and Energy Commission Secretariat, Ministry of Water Resources, Nepal, 1994). Despite this fact, Nepal's per capita energy consumption is estimated to be about 15 GJ in 1995, which is one of the five least energy consuming countries in the world. Fig 1 and Fig 2 indicate the consumption of energy scenario in the world and in Nepal respectively.

His Majesty's Government of Nepal is fully committed to develop energy sector essential for socio-economic development of the country. Utilization of conventional sources of energy with CO_2 emissions brings many

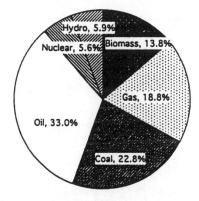

Fig.1 World Energy Consumption by Fuel Type
1992, Total Consumption; 398 EJ

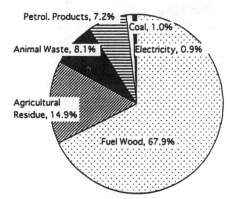

Fig. 2 Nepal: Energy Consumption by Fuel Type
1992/93, Total Consumption; 271 MGJ

problems related to climatic change, destruction of the ozone layer, soil, air and water pollution, desertification and deforestation, poverty and migration. Directly or indirectly all these environmental problems are related to energy problems in general. Realizing the fact that some of these problems could be tackled by replacing the conventional sources of energy with non-CO_2 emitting (renewable) sources of energy, Nepal in its Eighth Five Year Plan (1992-97) has provided some policy guidelines to develop alternate and decentralized energy resources (Water and Energy Commission Secretariat, Ministry of Water Resources, Nepal, 1994/95). The policy also points out to bring about the gradual replacement of imported fossil fuels with locally available renewable energy sources.

One important point about renewable energy sources is that it does not need big infrastructure   for example in case of photovoltaic technology. There is no need for transportation of energy generated   at a distant source if energy is generated locally. In this way, energy needed for   transporting the energy is   saved. When PV technology will be competitive with other forms of conventional sources of energy, which is very likely to happen in early 21 century, deforestation process will be retarded significantly and this will help in keeping environment clean. The other important point to be noted is that renewable energy sources could be used as the energy source needed for the capture, treatment and disposal of greenhouse gases such as CO_2. This fact should be considered also for developing countries where the energy demand is increasing at the rate of more than 10% annually.

In the Nepalese context, hydro-power, solar thermal, solar electricity, biomass and wind energy are considered to be non-CO_2 emitting sources of energy.

CO2 EMITTING SOURCES

World Resources Institute (National Planning Commission, Nepal, 1991) was the first international organization to estimate on Nepal's total contribution to the global greenhouse effect through the emission of CO_2, CH_4, and chlorofluorocarbons (CFSs). The per capita contribution of CO_2 in 1987 was quite low compared to the average Asian per capita contribution. But Nepal's per capita addition to the methane flux of 0.07 tonne was found to be more than the average Asian 0.04 tonne. The total CH_4 emission for Nepal included 38,000 tonnes from solid waste, 490,000 tonnes from livestock and 660,000 from wet rice in 1987. The extent to which Nepal is adding CFC to the atmospheric is virtually negligible and likely to remain so for a long time. But emissions of CH_4 from agriculture sector are significant. It is estimated that in 1995 the total CH_4 emission for Nepal is 1.5 million tonnes.

Industrial Emissions

In 1991/92 there were about 4270 industries employing about 213650 persons. Main CO_2 emitting industries are considered to be cement, leather/tanning, brick factories, pulp/paper, sugar mills/distilleries, textile mills and steel industries. The number of these industries are shown in Table 1 (Central Bureau of Statistics, National Planning Commission, Nepal, 1994) with rough estimate of CO_2 emissions per annum.

Table 1. CO2 emitting industries, 1991/92

Type of Industry	Number of Industries			Estimate of CO_2 emissions in kilo tonne (KT)
	Small Scale	Large Scale	Total	
Cement	134	3	137	35
Brick	560	4	564	28
Leather/tanning	16	2	18	2
Pulp and paper	42	2	44	2
Sugar mills/Distilleries	70	4	74	2
Textile	40	10	50	2
Steel/Metal	76	4	80	2

The main sources of CO_2 including other air pollutants from the industries mentioned in Table 1 are the combustion of coal and petroleum products for heating and power.

So far there has been no detail studies made to measure CO_2 emitted from different industries. The rough estimate of CO_2 (carbon content) shown in Table 1 is based on the assumption that the coal imported is used for these industries with 50% for the cement factories, 40% for brick factories and 2% each for the remaining industries.

Automobile Vehicles

In 1995 there were about 97,000 automotive vehicles on 341 km. roads of Kathmandu consuming 50% of total diesel and 80 % of total petrol imported (The Kathmandu Post, July 1, 1996). It is estimated that these vehicles besides emitting CO_2 produced the following pollutants in the atmosphere: CO- 71,000 tonnes/year; NO- 6,000 tonnes /year; Hydrocarbons - 1290 tonnes/year and SOx and other particulates-1090 tonnes/year.

The total energy consumption by fuel type (1992/93) is shown in Fig. 2. Based on these data, CO_2 is calculated and shown in Table 2 assuming that carbon content value for biomass is 10% of coal carbon content.

Table 2. CO_2 production in Nepal, 1992/93, by fuel type

Fuel Type	Energy consumed, Million GJ	CO_2 produced, kilo tonne (KT)
Fuelwood	184	1760
Agricultural Residue	40	366
Animal Waste	22	183
Petroleum Products	19.5	1334
Coal	2.7	260
Electricity (Thermal)	0.3	18
Total	268.5	3921
Per capita	14	205 Kg

Source: Water and Energy Commission Secretariat, Ministry of Water Resources, Nepal, 1994/95.

The sectoral energy consumption of Nepal in 1993 is shown in Table 3.

Table 3. Nepal energy consumption by sector, 1992/93, total consumption 271 Million GJ

Sector	Residential	Industrial	Transport	Commercial	Agriculture	Total
Use %	91.3	3.4	3.2	1.4	0.7	100

Source: Water and Energy Commission Secretariat, Ministry of Water Resources, Nepal, 1994/95.

NON-CO2 EMITTING SOURCES

Hydro-power

Nepal's theoretical hydro-power potential is estimated to be about 83,000 MW almost 300 times more than the present hydro-power generation. Economically feasible hydro-power potential is shown in table 4. The fact that there are about 6000 small and big rivers flowing from an elevation of more than 8000 meters in the northern Nepal to 150 meters in the south with a total length of about 45,000 km. indicates a vast potential of hydro- power generation.

Table 4: Economical hydro-power potential

River Basin	Number of Projects	Capacity, GW	Benefit/Cost Ratio, Range
Sapta Koshi	40	11	4.10-1.20
Sapta Gandaki	12	5	1.96-1.03
Karnali	7	24	up to 4
Mahakali	2	1	NA
Southern Rivers	5	1	up to 2
Total	66	42	

Source: Water and Energy Commission Secretariat, Ministry of Water Resources, Nepal, 1994/95.

If this vast potential could be tapped, generated and distributed among neighboring countries, an equivalent of 63 million metric tonnes of CO_2 could be saved from being exposed to the atmosphere assuming at 50% efficiency of hydro-power plants. But to generate this power, a sum of US$ 210 billion will be required at 1995 price. The country's economy does not provide enough basis for large scale investment for the exploitation of hydro-power potentials and laying transmission and distribution network in the immediate future.

At present, there are 924 micro-hydropower standalone plants generating about 3.3 MW of power reducing an equivalent of about 5 kilo tonnes of CO_2 from being emitted to the atmosphere per annum.

Forest

Nepal has an estimated area of 9.2 million hectares of potentially productive forest, shrubs and grassland, of which 3.4 million hectares are considered to be accessible for fuelwood collection in the year 1992/93. Sustainable yield from the accessible forest, farmland and non-cultivated inclusion is estimated to be about 7.5 million tonnes. The accurate rate of deforestation per annum is not known but it could be any number between 0.25% to 2.1% of the total forest area estimated at 37% of the country.

Solar Thermal

Nepal receives solar energy in the range of 3.6 to 6.2 kWh/m2/day. It has an average sunshine hour of 7 and sun shines for 300 days a year. This indicates that there is a very big potential for the utilization of solar energy. Among the solar thermal devices solar water heaters with 90 to 600 litres capacity are the most successful ones for domestic purposes in Nepal. They are being manufactured now by some 50 private mechanical workshops. Commercialization of solar dryers, cookers and stills have not yet taken place.

Solar Electricity

A total of about 800 kWp of photovoltaic power (about 0.15% of total electricity generated) is being used by different organizations such as Nepal Electricity Authority (NEA), Nepal telecommunications Corporation, Civil Aviation Authority , and individuals. There are some problems in the balance of systems in the NEA run centralized PV power system. If 0.01% of total available solar energy is converted into electricity using commercially available PV modules, 510 kilo tonnes of CO_2 could be reduced annually from being exposed to the atmosphere.

Biomass

Biogas, briquetting and improved cook stoves (ICS) are the three major biomass related energy technologies that are being widely developed and implemented in Nepal. Among these, biogas technologies are becoming very popular due to many reasons including significant government subsidy program. So far more than 12,000 biogas plants of various capacities ranging from 4 to 20 m3 have been installed in the hills and Tarai of 50 districts by Gobar Gas Company and other 14 private enterprises.

Wind Energy

Studies indicate large potential of wind energy in various parts of the country. A recent report estimated a potential of 200 MW between Kagbeni to Chusang in Western Nepal. If this wind power is fully utilized, 115 kilo tonnes of CO_2 emission can be reduced annually assuming that efficiency factor of wind generator is 10 %. So far wind mills have been used mainly for lift irrigation. Generation of electricity through wind turbines is not yet common. The major obstacle is that reliable data on availability of wind energy in the various parts of the country is still lacking.

Future Plan

Nepal's present Eighth Five Year Plan, which envisages government effort in developing alternative and decentralized energy resources, has provided detailed policy and programs including the provision for an institutional set up for the development of non-CO_2 emitting sources of energy. The budget layout for the non-CO_2 emitting sources of energy development in the Eighth Five Year Plan is shown in Table 5.

Table 5. Proposed plan for non-CO2 emitting sources of utilization in Nepal

Renewable Energy Category	Target	Investment Public Sector	in US $ Private Sector	total US$ Million
Microhydroelectricity	5 MW	1	5	6
Bio gas plant installation	30,000 nos.	5	16	21
Solar water heater	5000 nos.	0.1	2.8	2.9
Solar cooker	5000 nos.	0.05	0.5	0.55
solar dryer	2500 nos.	0.05	0.5	0.55
Wind Energy	preliminary study	0.1	0.5	0.6
Biomass energy(ICS)	250,000 nos.	0.4	1	1.4
Total		6.7	26.3	33

Source: Water and Energy Commission Secretariat, Ministry of Water Resources, Nepal, 1994.

Present Non-CO2 Emitting Sources
The present non-CO2 emitting sources of energy is shown in Table 6.

Table 6. Non-CO2 emitting sources of energy

Source Type	Organization	Capacity	Life Time (yr.)	Reduction of CO2 over life time, kilo tonne (KT)
Hydropower	Nepal Electricity Authority	836 GWh	30	8679
Solar PV power	Nepal Electricity Authority	0.78 MWh	25	1.9
Solar PV power	Nepal Telecom.	3 MWh	25	7.3
Solar PV power	Civil Aviation	0.084 MWh	25	0.2
Solar PV power	Water Supply	0.0324 MWh	25	0.08
Solar PV power	Private	0.606 MWh	25	1.47
Total				8689

CONCLUSIONS

Human being can have eternal peace with mother nature only if non-CO2 emitting sources of energy are properly planned, developed and implemented.
Nepal's vast hydropower potential should be tapped with a joint effort among neighboring countries and similarly share the resource and increase the physical quality of life index (PQLI) in the region.
Reliable data on emission of greenhouse gases should be collected in order to forecast the real danger to the environment and take early necessary action.
Establishment of a coordinating body vested with the authority and responsibility to disseminate non-CO2 emitting sources of energy is necessary for meeting the objective of reducing consumption of CO2 emitting sources of energy and to coordinate the activities of concerned private companies, NGOs, government institutions and research agencies.

REFERENCES

Central Bureau of Statistics, National Planning Commission, HMG of Nepal, 1994, *Statistical Pocket Book.*
National Planning Commission/IUCN, NCS Implementation Program, August 1991, *Environmental Pollution in Nepal*: A Review of Studies, pp. 34-42.
The Kathmandu Post, *local newspaper* dated July 1, 1996.
Water and Energy Commission Secretariat, Ministry of Water Resources, Nepal, 1994/95, Perspective Energy Plan, *Alternative Energy Technology: An Overview and Assessment*, Supporting Document No. 3, pp.   46-48.
Water and Energy Commission Secretariat, Ministry of Water Resources, Nepal, April 1994, Perspective Energy Plan, *Energy Sector Synopsis Report*, Supporting Document No. 1, pp. 12-13.

AUTHOR INDEX